Sourcebook of Bacterial Protein Toxins

Sourcebook of Bacterial Protein Toxins

edited by

J.E. Alouf
Bacterial Antigens Unit,
Institut Pasteur, Paris, France

and

J.H. Freer
Department of Microbiology,
University of Glasgow, Glasgow, UK

ACADEMIC PRESS

Harcourt Brace Jovanovich, Publishers

LONDON   SAN DIEGO   NEW YORK   BOSTON
SYDNEY   TOKYO   TORONTO

This book is printed on acid-free paper

ACADEMIC PRESS LIMITED
24–28 Oval Road
LONDON NW1 7DX

United States Edition published by
ACADEMIC PRESS INC.
San Diego, CA 92101

A catalogue record for this book is available from the British Library
ISBN 0–12–053078–3

Typeset by Photo·graphics, Honiton, Devon
Printed in Great Britain at The Bath Press, Avon

Contents

Contributors

David W.K. Acheson
Department of Medicine, Division of Geographic Medicine and Infectious Diseases, New England Medical Center and Tufts University School of Medicine, Boston, MA 02111, USA

Joseph E. Alouf
Bacterial Antigens Unit, CNRS-URA 557, Institut Pasteur, 75724 Paris Cédex 15, France

Jonas Ångström
Glycobiology, Department of Medical Biochemistry, University of Göteborg, PO Box 33031, S-400 33, Göteborg, Sweden

Alan W. Bernheimer
Department of Microbiology, New York University School of Medicine, 550 First Avenue, New York, NY 10016, USA

Patrice Boquet
Bacterial Antigens Unit CNRS-URA 557, Institut Pasteur, 75724 Paris Cédex 15, France

J. Brom
RUHR-Universität Bochum, Medizinische Mikrobiologie und Immunologie, Arbeitsgruppe für Infektabwehr, Universitätsstraße 150, Postfach 150, 4630 Bochum 1, Germany

John G. Coote
Department of Microbiology, The University of Glasgow, Glasgow G12 8QQ, UK

Arthur Donohue-Rolfe
Department of Medicine, Division of Geographic Medicine and Infectious Diseases, New England Medical Center and Tufts University School of Medicine, Boston, MA 02111, USA

Franz J. Fehrenbach
Department of Microbiology, Robert Koch-Institut, Nordufer 20, D-1000 Berlin 65, Germany

Timothy J. Foster
Microbiology Department, Moyne Institute, Trinity College, Dublin 2, Ireland

John H. Freer
Department of Microbiology, University of Glasgow, Glasgow G12 8QQ, UK

Christiane Geoffroy
Bacterial Antigens Unit, CNRS-URA 557, Institut Pasteur, 75724 Paris Cédex 15, France

D. Michael Gill†
Department of Molecular Biology and Microbiology, Tufts University, 136 Harrison Avenue, Boston, MA 21110, USA
†Deceased

Werner Goebel
Institute for Genetics and Microbiology, University of Würzburg, Würzburg, Germany

Emanuel Hanski
Department of Medical Microbiology, The Hebrew University-Hadassah Medical School, Jerusalem 91010, Israel

Timothy R. Hirst
The Biological Laboratory, University of Kent, Canterbury, Kent CT2 7NJ, UK

Dagmar Jürgens
Department of Microbiology, Robert Koch-Institut, Nordufer 20, D-1000 Berlin 65, Germany

Iwao Kato
The Second Department of Microbiology, Chiba University School of Medicine, Chiba 280, Japan

Karl-Anders Karlsson
Glycobiology, Department of Medical Biochemistry, University of Göteborg, PO Box 33031, S-400 33, Göteborg, Sweden

Gerald T. Keusch
Department of Medicine, Division of Geographic Medicine and Infectious Diseases, New England Medical Center and Tufts University School of Medicine, Boston, MA 02111, USA

Heide Knöll
Institute for Microbiology and Experimental Therapy, Jena, Germany

J. Knöller
RUHR-Universität Bochum, Medizinische Mikrobiologie und Immunologie, Arbeitsgruppe für Infektabwehr, Universitätsstraße 150, Postfach 150, 4630 Bochum 1, Germany

Werner Köhler
Institute for Microbiology and Experimental Therapy, Jena, Germany

W. König
RUHR-Universität Bochum, Medizinische Mikrobiologie und Immunologie, Arbeitsgruppe für Infektabwehr, Universitätsstraße 150, Postfach 150, 4630 Bochum 1, Germany

M. Köller
RUHR-Universität Bochum, Medizinische Mikrobiologie und Immunologie, Arbeitsgruppe für Infektabwehr, Universitätsstraße 150, Postfach 150, 4630 Bochum 1, Germany

Arnold S. Kreger
Department of Microbiology and Immunology, The Bowman Gray School of Medicine of Wake Forest University, Winston-Salem, North Carolina 27103, USA
Present address: The Musculoskeletal Sciences Research Institute, 2190 Fox Mill Road, Herndon, Virginia 22071, USA

Stephen H. Leppla
Laboratory of Microbial Ecology, National Institute of Dental Research, National Institutes of Health, Bethesda, Maryland 20892–0030, USA

Regina Linder
Hunter College School of Health Services, 425 East 25th Street, New York, NY 10010, USA

Albrecht Ludwig
Institute for Genetics and Microbiology, University of Würzburg, Würzburg, Germany

Gianfranco Menestrina
Dipartimento di Fisica, Università di Trento, I–38050 Povo, Trento, Italy

Cesare Montecucco
Centro CNR Biomembrane e Dipartimento di Scienze Biomediche, Università di Padova, Corso Trieste 75, Padova, Italy

John R. Murphy
Evans Department of Clinical Research and Department of Medicine, The University Hospital, Boston University Medical Center, Boston, MA 02118, USA

Heiner Niemann
Institute for Microbiology, Federal Research Center for Viral Diseases of Animals, PO Box 1149, D74 Tübingen, Germany

Masatoshi Noda
The Second Department of Microbiology, Chiba University School of Medicine, Chiba 280, Japan

Sjur Olsnes
Institute for Cancer Research at the Norwegian Radium Hospital, Montebello, Oslo, Norway

Emanuele Papini
Centro CNR Biomembrane e Dipartimento di Scienze Biomediche, Università di Padova, Corso Trieste 75, Padova, Italy

Mariagrazia Pizza
Sclavo Research Center, Via Fiorentina 1, 53100 Siena, Italy

Rino Rappuoli
Sclavo Research Center, Via Fiorentina 1, 53100 Siena, Italy

Kirsten Sandvig
Institute for Cancer Research at the Norwegian Radium Hospital, Montebello, Oslo, Norway

J. Scheffer
RUHR-Universität Bochum, Medizinische Mikrobiologie und Immunologie, Arbeitsgruppe für Infektabwehr, Universitätsstraße 150, Postfach 150, 4630 Bochum 1, Germany

W. Schönfeld
RUHR-Universität Bochum, Medizinische Mikrobiologie und Immunologie, Arbeitsgruppe für Infektabwehr, Universitätsstraße 150, Postfach 150, 4630 Bochum 1, Germany

Giampietro Schiavo
Centro CNR Biomembrane e Dipartimento di Scienze Biomediche, Università di Padova, Corso Trieste 75, Padova, Italy

Susann Teneberg
Glycobiology, Department of Medical Biochemistry, University of Göteborg, PO Box 33031, S-400 33, Göteborg, Sweden

Diane P. Willliams
Evans Department of Clinical Research and Department of Medicine, The University Hospital, Boston University Medical Center, Boston, MA 02118, USA

Preface

Great strides have been made in the depth of our understanding of the structure and mechanisms of action of bacterial toxins over the last decade. The current pace of this advance in knowledge is particularly impressive, and results largely from the power that gene manipulation techniques have offered in experimental biology.

Recent research achievements in the field of bacterial toxins, which consist of about 240 protein toxins as well as a relatively small number of non-protein toxins, reflect the extensive and productive blending of disciplines such as molecular genetics, protein chemistry and crystallography, immunology, neurobiology, pharmacology and biophysics. Furthermore, the exciting developments in many areas of cell biology, and particularly in membrane-associated mechanisms relating to signalling and communication, export and import of proteins and to cytoskeletal functions, have been facilitated because critical steps in these processes constitute the targets for bacterial toxins. Thus, we have toxins available which can be used to probe many fundamental aspects of eukaryotic cell biology.

Disruption of these same central cellular processes *in vivo* can also be the critical event in the pathogenesis of infectious diseases for man or domestic animals. Many such infectious diseases have major social or economic impacts on man, and such considerations have quickened the pace of the search for therapeutic agents. Currently, a number of physically inactivated bacterial or hybrid engineered toxoids are used as immunogens in vaccination programmes, and there is a major international effort to develop new and more effective vaccines based on our deeper understanding of the molecular events in pathogenesis and the host response to infection.

Since the publication of the excellent multi-volume treatise on *Bacterial Toxins* edited by S. Ajl, S. Kadis and T. Montie (Academic Press) in the early 1970s, most of the books published in the past twenty years have covered the subject by presenting individual toxins or groups of toxins in separate chapters. This is not the main approach followed in this book. Our aim is not to give an exhaustive review of the wide spectrum of the protein toxin repertoire but rather to give an 'in depth' critical review of the original and the newly expanding body of information accumulated during the past decade or so. The multifaceted aspects of toxin research and the multidisciplinary approaches adopted suggested to us that 'state of the art' toxin research might best be presented by putting together in several chapters the common structural and/or functional aspects of toxin 'families'. Other chapters highlight the various physiological or genetic mechanisms regulating toxin expression and the therapeutic or vaccine applications of genetically engineered toxins.

The 22 chapters of this book have been written by 44 internationally known specialists who have significantly contributed to the progress in the domains covered. It is hoped that this book will appeal to a wide readership, including microbiologists, biochemists, cell biologists and physicians. Also, we hope it will arouse the interest of students and scientists in other disciplines who see the power of these fascinating biological agents, either as exquisitely specific probes of cellular processes or as extremely potent agents of infectious disease.

Finally, we would like to thank all the authors for their contributions, and particularly to those who delivered their manuscripts by the first deadline. We also express our appreciation to the editorial staff at Academic Press for their help and patience throughout the preparation of this book.

J.E. Alouf and J.H. Freer

I

Structure and Evolutionary Aspects of ADP-Ribosylating Toxins

Rino Rappuoli and Mariagrazia Pizza

Sclavo Research Center, Via Fiorentina 1, 53100 Siena, Italy

Introduction

ADP-ribosylating bacterial toxins are proteins, produced by pathogenic bacteria, which are usually released into the extracellular medium and cause disease by killing or altering the metabolism of eukaryotic cells. The toxins are usually composed of two functionally distinct domains: a toxic moiety and a vector, which have been called domains A and B, respectively (Fig. 1). Within the holotoxin, the toxic domain (A) is in an inactive conformation and is carried by the vector (B) which binds the receptors on the surface of eukaryotic cells and delivers the toxic part (A) across the membrane of eukaryotic cells so that it can reach its target proteins (Middlebrook and Dorland, 1984). During this process A is released from the vector (B) and unfolds from an inactive to an active conformation.

This process may require the reduction of a disulphide bridge which often holds A in the inactive conformation (see Fig. 1). While the structure and the complexity of the vector (B) differ from toxin to toxin and often also within the same family of toxins, depending on the receptor they recognize on the surface of eukaryotic cells (see Fig. 2), all of the A domains have a common mechanism of action: they are enzymes which ADP-ribosylate eukaryotic target proteins which control crucial circuits of eukaryotic cells, such as protein synthesis, transmembrane signalling, oncogenesis and cytoskeleton structure. The target proteins also have a common feature and a

common structure: they are GTP-binding proteins. The only exception is actin, a protein which binds ATP instead of GTP. These concepts are illustrated in Fig. 2 and Table 1, which show the structural similarities and differences of the ADP-ribosylating toxins, and Fig. 4 which shows a scheme of the common properties of the target GTP-binding proteins and the amino acids which are ADP-ribosylated by each toxin.

The ADP-ribosylation reactions

Two types of ADP-ribosylation reactions are known to occur in nature: mono- and polyADP-ribosylation (Ueda and Hayaishi, 1985; Althaus and Richter, 1987). Mono-ADP-ribosylation, which is mediated by bacterial toxins, phage and *E. coli* proteins and many cytoplasmic and membrane-associated eukaryotic proteins, involves transfer of the ADP-ribose group to a nitrogen atom in the side chain of amino acids such as diphthamide, arginine, asparagine or cysteine, according to the reaction shown in Fig. 3.

In the case of poly-ADP-ribosylation, the enzymes are mostly found in the nucleus of eukaryotic cells where they modify histones and other nuclear target proteins by transferring first the ADP-ribose group to the carboxyl groups of glutamic or aspartic acids (or the carboxyl groups of C-terminal amino acids) and then elongate the ADP-ribose chain by adding further ADP-ribose groups. The eukaryotic GTP-binding proteins

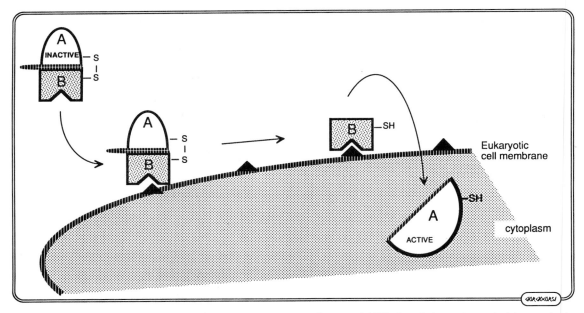

Figure I. Schematic representation of the basic A–B structure of bacterial ADP-ribosylating toxins and of the mechanism of entry of the toxic fragment A into the cell.

which are ADP-ribosylated by bacterial toxins are the following.

Elongation factor 2 (EF2)

Eukaryotic EF2 is a protein of 95 700 Da which is involved in protein synthesis (Kohno *et al.*, 1986). It contains a post-translationally modified histidine residue (diphthamide 715) (Ness *et al.*, 1980a, b), which is ADP-ribosylated by diphtheria and *Pseudomonas* exotoxin A (Honjo *et al.*, 1968; Gill *et al.*, 1969). Lysine 715 is not modified in the homologous bacterial EFG (see Fig. 4) (Kohno *et al.*, 1986), which is not a substrate for these toxins. ADP-ribosylation of EF2 causes inhibition of protein synthesis and cell death.

G proteins

G proteins are a family of membrane proteins composed of three subunits (α, β and γ), which bind GTP and regulate enzymes such as adenylate cyclase, phospholipase C and cyclic GMP-phosphodiesterase, which release secondary messengers into the cytoplasm as a response to external stimuli (Stryer and Bourne, 1986; Neer and Clapham, 1988). As these enzymes play a key role in cellular metabolism, they are regulated by a fine system.

Adenyl cyclase, for instance, is regulated by two GTP-binding proteins: G_s and G_i. G_s receives signals from stimulatory receptors located on the surface of eukaryotic cells and stimulates the activity of adenyl cyclase. G_i, on the other hand, receives signals from the inhibitory receptors and inhibits the adenyl cyclase activity (Gilman, 1984). Other known G proteins are G_o and transducin. Pertussis toxin ADP-ribosylates Cys^{352} of the α subunit of G_i and the equivalent cysteines in G_o and transducin (see Fig. 4) (Katada and Ui, 1982; Bokoch *et al.*, 1983; Sternweis and Robishaw, 1984; West *et al.*, 1985; Fong *et al.*, 1988). $G_s\alpha$, which has a tyrosine residue in this position, is not a target of pertussis toxin (Fong *et al.*, 1988). Until recently cholera toxin (and the related *E. coli* LT toxins) was believed to ADP-ribosylate only the Arg^{201} of G_s and transducin (Van Dop *et al.*, 1984). The reason for the selective ADP-ribosylation of these two proteins could not be understood, since this residue is also present in G_i and G_o (see Fig. 4). Recent observations, however, suggest that G_i and G_o can, in fact, be ADP-ribosylated by cholera toxin when they are coupled to their receptors (T. Mekada, pers. commun.).

ADP-ribosylation of G proteins by cholera and pertussis toxins causes a variety of effects in

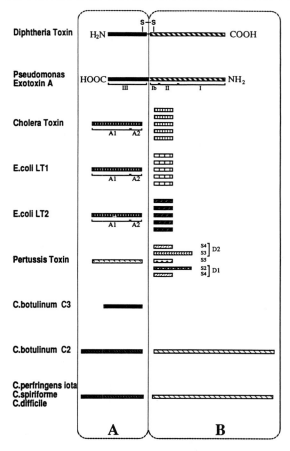

Figure 2. Schematic representation of the basic structure of the ADP-ribosylating toxins. The enzymatically active fragments A of each toxin are shown on the left. The B domains of each toxin are shown on the right. The size of each polypeptide is reported proportionally. When the B domains are oligomeric proteins composed of more than one polypeptide, we report those necessary to form one oligomer.

Diphtheria toxin

Diphtheria toxin (DT) is a protein encoded by a family of closely related bacteriophages which integrate into the chromosome of *Corynebacterium diphtheriae* and convert non-toxinogenic, non-virulent bacteria into toxinogenic, highly virulent species (Pappenheimer, 1977; Collier, 1982). Under iron-limiting conditions, lysogenic *C. diphtheriae* strains secrete into the culture medium large amounts of DT, a protein of 58 350 Da which is initially synthesized as a single polypeptide chain.

Following mild trypsin treatment and reduction of a disulphide bond, DT can be divided into two functionally different moieties: fragment A and fragment B of 21 150 and 37 200 Da, respectively. Fragment A ADP-ribosylates the post-translationally modified histidine residue (diphthamide) of elongation factor 2 (EF2), a GTP-binding protein involved in protein synthesis of eukaryotic cells (see Fig. 4) (Honjo *et al.*, 1968; Gill *et al.*, 1969).

The EF2–ADP-ribose complex is inactive and therefore diphtheria toxin causes inhibition of protein synthesis and cell death. It has been shown that *in vitro*, a single molecule of fragment A is enough to kill one eukaryotic cell (Yamaizumi *et al.*, 1978). *In vivo*, DT is one of the most potent bacterial toxins, with a minimal lethal dose of 0.1 μg/kg body weight (Pappenheimer, 1984). The entire lethal activity of DT resides in fragment A. Fragment B is required for recognition of specific receptors on the surface of eukaryotic cells and in the translocation of fragment A across the cell membrane.

Although the nature of the DT receptor has not yet been fully elucidated, recent studies suggest that it is a protein of 14 500 Da (Mekada *et al.*, 1988). Fragment B can also be divided into two functionally distinct domains: a C-terminal region involved in receptor binding and a hydrophobic N-terminal region required for membrane translocation of fragment A. During the cell intoxication process, DT binds the receptor through its C-terminal domain and is internalized into endosomes by receptor-mediated endocytosis. When the pH of the endosomes decreases below pH 5.5, fragment B undergoes a conformational change which allows the translocation of fragment A from the endosome into the cytosol (Olsnes and Sandvig, 1988; Sandvig

different tissues. In the case of adenyl cyclase, treatment with cholera toxin causes constitutive activation of the enzyme and accumulation of cAMP, while treatment with pertussis toxin uncouples G_i from its receptor so that it becomes unable to inactivate adenyl cyclase.

Ras and rho are two homologous low molecular weight GTP-binding proteins involved in oncogenesis: Ras is ADP-ribosylated by *Pseudomonas* exoenzyme S (Coburn *et al.*, 1989a; Pai *et al.*, 1989), while botulinum C3 toxin ADP-ribosylates the Asn^{41} of rho (Sekine *et al.*, 1989).

Table I. Main properties of bacterial ADP-ribosylating toxins

Bacterial toxin	Acceptor protein	Acceptor amino acid	Effect on eukaryotic cells	Primary structure	Eukaryotic cell receptor
Diphtheria toxin	EF2	Diphthamide 715	Inhibition of protein synthesis	Known	14.5-kDa protein?
Pseudomonas ETA	EF2	Diphthamide 715	Inhibition of protein synthesis	Known	Not known
Pertussis toxin	G_i, G_o, T	Cys^{352}	Alteration of transmembrane signal transduction	Known	160-kDa glycoprotein in CHO cells
Cholera toxin	G_s, T, (G_i and G_o)	Arg^{201}	Alteration of transmembrane signal transduction	Known	Ganglioside GMI > GDbI
E. coli LTI	G_s, T, (G_i and G_o)	Arg^{201}	Alteration of transmembrane signal transduction	Known	Ganglioside GMI > GDbI > GM2
E. coli LT2	G_s, T, (G_i and G_o)	Arg^{201}	Alteration of transmembrane signal transduction	Known	Ganglioside GDIa, GTIb
Botulinum C3 exoenzyme	Rho	Asn^{41}	No effect because does not enter in eukaryotic cells. When microinjected into cells changes the cell morphology	Known	Not known
Botulinum C2 toxin	Non-muscle actin	Arg^{177}	Inhibition of actin polymerization	Not known	Not known
Clostridium perfringens iota toxin	Monomeric skeletal and non-muscle actin	Arg^{177}	Inhibition of actin polymerization	Not known	Not known
Clostridium spiroforme toxin	Non-muscle actin	(Arg^{177})?	Inhibition of actin polymerization	Not known	Not known
Clostridium difficile transferase		(Arg^{177})?	Inhibition of actin polymerization	Not known	Not known
Pseudomonas exoenzyme S	Ras, vimentin	Arg residue	(Not known)	Not known	Not known

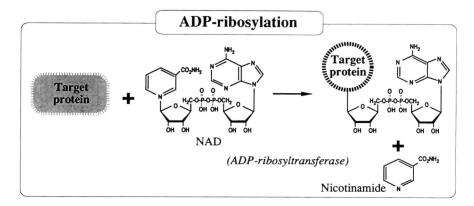

Figure 3. The ADP-ribosylation reaction.

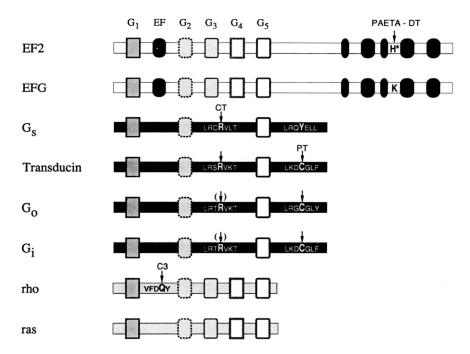

Figure 4. Schematic representation of the common properties of the GTP-binding proteins which are ADP-ribosylated by bacterial toxins. Three families of homologous protein are shown: the elongation factor II family (EF2 and EFG), the G proteins family (Gs, Go, Gi and transducin) and a family of Ras proteins (Rho and Ras). In addition to the strong homologies found within each family of GTP-binding protein, some regions are found to be homologous to most of the GTP-binding proteins (G1, G2, G3, G4 and G5) (Woolley and Clark, 1989). The amino acids which are ADP-ribosylated by the different toxins are indicated by an arrow. The modified histidine residue (diphthamide) ADP-ribosylated by DT and PAETA is indicated by an H*.

and Olsnes, 1981; Neville and Hudson, 1986; Papini *et al.*, 1987a, 1988).

DT was the first toxin for which structure–function relationships were elucidated, and it has been a model for all the other toxins. Most of the functional and structural properties of DT were initially deduced from the analysis of a number of non-toxic DT mutants (cross-reacting materials or CRMs), encoded by corynephages which had been mutagenized by nitrosoguanidine (Uchida *et al.*, 1971, 1973a,b,c). The sequence of the wild-type DT gene and of its mutants has then allowed the identification of the amino acids which are responsible for each phenotype. The most relevant DT mutants are shown in Fig. 5. CRM 197 contains a single amino acid mutation in fragment A (Gly^{52}→Glu) which completely abolishes the NAD$^+$ binding properties of fragment A and therefore its enzymatic activity (Giannini *et al.*, 1984). Being enzymatically inactive (and therefore non-toxic), but otherwise indistinguishable from diphtheria toxin, CRM 197 is an ideal candidate for the development of a new vaccine against diphtheria (Rappuoli, 1983). CRM 176 contains a Gly^{128}→Asp mutation in fragment A which reduces its enzymatic activity about ten-fold (Comanducci *et al.*, 1987). CRM 228 contains two mutations in fragment A and three in fragment B which result in inactive A and B fragments (Kaczorek *et al.*, 1983). CRM 1001 has an enzymatically active fragment A and a fragment B which binds the toxin receptor but is unable to translocate fragment

A across the eukaryotic cell membrane. This phenotype is due to a Cys^{471}→Tyr substitution which disrupts the Cys^{471}–Cys^{461} disulphide bridge, thus altering the conformation of this region and its ability to translocate fragment A (Dell'Arciprete *et al.*, 1988; Papini *et al.*, 1987e). CRM 103 and CRM 107 have an enzymatically active fragment A and a mutated fragment B which is still competent in translocating fragment A across the cell membrane. They are, however, non-toxic because they are unable to bind the toxin receptor. Ser^{508}→Phe in CRM 103 and Leu^{390}→-Phe, Ser^{525}→Phe in CRM 107 are the mutations responsible for this phenotype (Greenfield *et al.*, 1987). The other mutant which defines the C-terminal region as responsible for receptor binding is CRM 45, a prematurely terminated molecule with a stop codon in position 387; this truncated molecule is unable to bind to eukaryotic cells (Giannini *et al.*, 1984).

While the analysis of the above CRMs has allowed a rough demarcation of the three domains of DT involved in receptor binding, the membrane translocation and the EF2 ADP-ribosylation, a more precise definition of the amino acids involved in each of these functions is still awaited. So far, there are two amino acids for which a well-defined function has been established: Glu^{148} and His^{21}. Glu^{148}, initially identified as the only amino acid photolabelled by [carbonyl^{14}C]NAD (Carroll and Collier, 1984; Carroll *et al.*, 1985), has been subsequently shown to be necessary for catalysis because it cannot be substituted by any other amino acid (including the homologous aspartic acid) without losing enzymatic activity (Tweten *et al.*, 1985). His^{21} has been shown to be at the NAD$^+$-binding site because its chemical modification abolishes both NAD$^+$ binding and enzymatic activity (Papini *et al.*, 1989). The finding that His^{21}, Gly^{52} and Glu^{148} are essential for the enzymatic activity of fragment A is in agreement with the predicted involvement of these amino acids in the formation of the NAD$^+$-binding cavity and the catalytic site (see Fig. 8 for details). The domain with less information at the molecular level (the N-terminal region of fragment B, involved in translocation of fragment A) is now being investigated by two approaches, one involving *in vitro* translocation of DT deletion mutants containing variable parts of the DT molecule

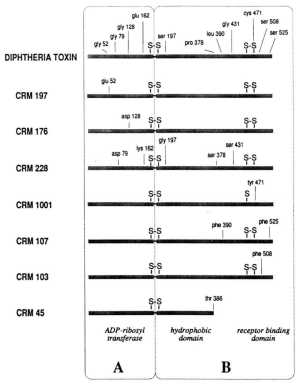

Figure 5. Schematic representation of diphtheria toxin and its cross-reacting (CRM) mutants. The amino acids which are changed in each of the CRM mutants are reported.

(McGill *et al.*, 1989; Stenmark *et al.*, 1989) and the other involving the screening of mutants which are unable to interact with cell membranes at acidic pH. Randomly generated mutants with such properties are easily identified because *E. coli* cells which express and secrete DT molecules into the periplasm, are killed when exposed at low pH if the DT molecule is competent in membrane interaction (O'Keefe and Collier, 1989).

The genetic organization of DT, with the region encoding for the receptor-binding domain at the 3'-end of the gene has allowed the development of a number of hybrid toxins in which the receptor-binding domain has been replaced by interleukin 2 or by the melanocyte hormone MSH (Murphy *et al.*, 1986; Williams *et al.*, 1987; Bacha *et al.*, 1988). These new molecules kill activated T cells and MSH receptor-bearing cells respectively. They might find therapeutic applications for the treatment of allograft rejection and melanomas.

Pseudomonas exotoxin A

Pseudomonas exotoxin A (PAETA) is secreted in the culture medium as a single polypeptide chain of 613 amino acids of which the sequence and the 3-dimensional structure at 3 Å of resolution is known (Gray *et al.*, 1984; Allured *et al.*, 1986). According to X-ray crystallography studies, the molecule can be divided into three domains (Hwang *et al.*, 1987; Siegall *et al.*, 1989; Pastan and FitzGerald, 1989). Domain I is composed of two non-contiguous regions: Ia comprising amino acids 1–252 and Ib composed of amino acids 365–404.

Domain II is composed of amino acids 253–364, while domain III comprises amino acids 405–613 (see Fig. 2). Four disulphide bridges are present, two located in domain Ia, one in domain Ib and one in domain II. Genetic studies, based mainly on the expression of mutated forms of the PAETA gene in *E. coli*, have shown that the deletion of domain Ia results in non-toxic, enzymatically active molecules which cannot bind the cells. A similar result can be obtained by mutating Lys^{57} to Glu (Jinno *et al.*, 1988). Deletions in domain II result in molecules which bind to the cells and which are enzymatically active but not toxic. A similar result can be obtained by mutagenizing Arg^{276}, Arg^{279} and Arg^{230} (Jinno *et al.*, 1989) or by

converting Cys^{265} and Cys^{268} to other amino acids, thus changing the structure of domain II (Siegall *et al.*, 1989). Deletions or mutations in domain III result in enzymatically inactive molecules (Siegall *et al.*, 1989).

Based on these observations it can be concluded that the three domains observed in the X-ray structure correspond exactly to three functional domains involved in cell recognition (domain I), membrane translocation (domain II) and ADP-ribosylation of elongation factor 2 (domain III). PAETA can be described as a typical bacterial toxin with an A–B structure, with a mechanism of action identical to diphtheria toxin. Although identical in function, DT and PAETA have an opposite structural organization: the enzymatically active domain is at the C-terminus of PAETA and at the N-terminus of DT. While the B domains of DT and PAETA do not have any structural similarity, the enzymes share a common structure of the catalytic site (see Fig. 8 for details) (Brandhuber *et al.*, 1988; Carroll and Collier, 1988). The structure of this site, well-described by X-ray crystallography, has been further corroborated by functional studies which have identified amino acids which play a key role in enzymatic activity. Glu^{553} was initially shown to be at the catalytic site because it was the only amino acid photoaffinity-labelled by NAD^{+} (Carroll and Collier, 1987). Later it was shown that substitution of Glu^{553} with any amino acid, including Asp, decreased the enzymatic activity by a factor of 1000 (Douglas and Collier, 1987) and that deletion of Glu^{553} completely abolished the toxicity of PAETA. Similarly, iodination of Tyr^{481} which is also at the catalytic site, was shown to abolish enzymatic activity (Brandhuber *et al.*, 1988). Other amino acids which, although not located at the catalytic site, have been shown to be essential for enzymatic activity are His^{426} and residues 405–408 (Galloway *et al.*, 1989). His^{426} has been proposed to be necessary for the interaction between PAETA and EF2.

The well-defined structural divisions in separated domains make PAETA an ideal candidate for the development of chimaeric toxins by replacing the gene parts encoding for domain I with others encoding for cell-binding domains with different specificities. So far, nucleotides encoding domain I have been replaced by sequences encoding interleukin 2, T cell growth factor, interleukin 6, interleukin 4 and the T cell antigen CD4. In all instances expression of these genes in *E. coli* has given new toxins which specifically kill the cells bearing the receptor recognized by the new domain I. Such molecules are promising candidates for the treatment of arthritis and allograft rejection (PAETA-IL2), AIDS (PAETA-CD4) and other diseases (Chaudhary *et al.*, 1987, 1988; Siegall *et al.*, 1988; Lorberboum-Galski *et al.*, 1988; Ogata *et al.*, 1989).

Pseudomonas exoenzyme S

Exoenzyme S is an ADP-ribosyltransferase produced by *Pseudomonas aeruginosa*. Although this enzyme has a demonstrated role in pathogenesis, its physiological role is not yet clear (Nicas and Iglewsky, 1985). It has recently been shown to be cytotoxic (Woods and Que, 1987) and to cause morphological changes in pulmonary tissue (Woods *et al.*, 1988).

It acts by ADP-ribosylating a small subset of cellular proteins in eukaryotic cell extracts, one of which is the p21 product of the proto-oncogene *c-H-ras* (Coburn *et al.*, 1989a). Recently it has been demonstrated that non-polymerized vimentin is one of the most abundant substrates (Coburn *et al.*, 1989b). Although in the cell most vimentin is filamentous and thus unavailable as a substrate, exoenzyme S is likely to ADP-ribosylate dissociated subunits, or the termini of the vimentin filament. The accumulation of non-polymerized vimentin could produce considerable effects on the cell architecture by a mechanism similar to that described for the *Clostridium botulinum* C2 toxin (see later and Fig. 7).

Exoenzyme S modifies all its substrates at an arginine residue (Coburn *et al.*, 1989a).

Purified exoenzyme S presents an amino acid composition different from that of the exoenzyme A. Its electrophoretic pattern shows different bands of different molecular weight, most of which are enzymatically active. The largest polypeptide capable of enzymatic activity has a molecular weight of 50 000 Da (Woods and Que, 1987). The estimated molecular weight of exoenzyme S obtained by gel filtration was 105 000. It has been proposed that the 105 000-Da measurement may

correspond to aggregates of the different species and that the exoenzyme S is composed of various subunits, as are many of the other bacterial toxins described in this chapter (Woods and Que, 1987).

Pertussis toxin

Pertussis toxin (PT) is a protein of 105 000 Da released into the extracellular medium by *Bordetella pertussis*, the aetiological agent of whooping cough. PT is a complex bacterial toxin composed of five different subunits which have been named S1 (21 220 Da), S2 (21 920 Da), S3 (21 860 Da), S4 (12 060 Da) and S5 (11 770 Da), according to their electrophoretic mobility (Tamura *et al.*, 1982; Sekura *et al.*, 1985). Exposure of PT to 2 M urea disassembles the PT into the monomer A (subunit S1), and the oligomer B (which comprises the subunits S2, S3, S4 and S5). Upon exposure to 5 M urea, the B oligomer can be dissociated into two dimers: dimer 1 (comprising S2 and S4) and dimer 2 (comprising S3 and S4), and the monomer S5 (see Fig. 2). With 8 M urea PT dissociates into five monomeric subunits (Tamura *et al.*, 1982).

As in the case of the other ADP-ribosylating toxins, the B oligomer of PT binds the receptors on the surface of eukaryotic cells and allows the toxic subunit S1 to reach its intracellular target proteins. In Chinese hamster ovary (CHO) cells, the PT receptor has been shown to be a 165-kDa glycoprotein which binds the PT B oligomer through a branched mannose structure containing sialic acid. Dimers S2–S4 and S3–S4 are also able to bind the same receptor (Witvliet *et al.*, 1989). S1 ADP-ribosylates the Cys^{352} of the α subunit of protein G_i, and the corresponding cysteine in G_o and transducin (see Fig. 4) (Katada *et al.*, 1983; West *et al.*, 1985). $G_s\alpha$, which contains a Tyr in place of the Cys residue, is not ADP-ribosylated by PT.

ADP-ribosylation of the above G proteins causes alteration in the response of eukaryotic cells to exogenous stimuli and results in a variety of phenotypes. *In vivo*, the most relevant consequences of PT intoxication are: leukocytosis, histamine sensitization, increased insulin production with consequent hypoglycaemia, and potentiation of anaphylaxis (Sekura *et al.*, 1985). *In vitro*, PT has a number of different activities, the most relevant of which is the change in cell morphology in CHO cells, a phenotype which is able to detect as little as 10 pg of active PT (Hewlett *et al.*, 1983). Further *in vitro* activities are haemagglutination, T-cell mitogenicity, inhibition of migration of peritoneal macrophages, enhancement of receptor-mediated accumulation of cAMP and many others (Sekura *et al.*, 1985). In contrast to the other bacterial toxins where all the activities are mediated by the enzymatically active subunit, in the case of PT, T-cell mitogenicity and haemagglutination are properties typical of the B oligomer alone (Tamura *et al.*, 1983).

Although PT has been the last toxin to be purified and characterized, today it is the one for which more information is available at the molecular level. The genes encoding for the five subunits of pertussis toxin are clustered in a fragment of DNA of 3200 base pairs, organized in an operon structure and in the following order: S1, S2, S5, S4 and S3 (Nicosia *et al.*, 1986; Locht and Keith, 1986). Each of the five subunits are co-translationally exported into the periplasmic space where the holotoxin is assembled and subsequently released into the extracellular medium. Analysis of the amino acid sequence of subunit S1 shows a significant homology with the A1 protomer of cholera and *E. coli* LT toxins (see Fig. 8) (Nicosia *et al.*, 1986; Locht and Keith, 1986). With the aim of developing enzymatically inactive PT molecules to be used for vaccination against whooping cough, an enzymatically active S1 subunit has been expressed in *E. coli*. By C- and N-terminal progressive deletion of the S1 subunit, it has been shown that amino acids 4–179 are necessary for enzymatic activity (Pizza *et al.*, 1988; Cieplak *et al.*, 1988; Barbieri and Cortina, 1988). Within this region, a substantial number of the amino acids have been changed by site-directed mutagenesis to identify those which were crucial for enzymatic activity (see Fig. 6) (Barbieri and Cortina, 1988; Pizza *et al.*, 1988; Burnette *et al.*, 1988a; Black *et al.*, 1988; Kaslow *et al.*, 1989; Locht *et al.*, 1989; Lobet *et al.*, 1989). Among more than 30 amino acids tested, Arg[9], Asp[11], Arg[13], Trp[26], His[35], Phe[50], Glu[129] and Tyr[130] were found to be essential for enzymatic activity. In fact their replacement with other amino acids reduced the activity of recombinant S1 molecules to levels equal or below 1%. Most of these amino acids

could be replaced without altering the overall structure of the recombinant S1 molecule which was still recognized by toxin-neutralizing monoclonal antibodies specific for a conformational epitope (see Fig. 6).

In addition to the amino acids described above, Cys[41] was shown to be close to the active site since its alkylation decreased enzymatic activity (Kaslow *et al.*, 1989). By photoaffinity labelling with NAD$^+$, Glu[129] was shown to be equivalent to Glu[148] of DT and Glu[553] of PAETA (Barbieri et al., 1989). When the above amino acid changes were introduced either alone or in combination into the PT operon in the *B. pertussis* chromosome, a number of mutant PT molecules were obtained; most of them had a polyacrylamide gel electrophoretic pattern indistinguishable from the wild-type PT and a reduced toxicity. In general, the molecules containing single amino acid mutations had a toxicity reduced from 4- to 1000-fold but none of them was completely non-toxic. The double mutants containing Glu[129] → Gly and another mutation in the N-terminal region (see Fig. 6) were free of any detectable toxicity (at least 10^6 times less than wild-type PT) (Pizza *et al.*, 1989). Unexpectedly, some mutants were unable to assemble the S1 subunit and released into the extracellular medium only the B oligomer. These mutants, labelled 'B' in Fig. 6, contain mutations in any of the two cysteines, in regions homologous to cholera toxin (8D/9G, 50E, 88E/89S), or contain three amino acid changes as in mutant 13L/26I/129G. The above mutations are believed to alter the structure of the S1 subunit and therefore prevent its assembly into the holotoxin (Pizza *et al.*, 1990). The non-toxic double mutants are now being tested in humans as new vaccines against whooping cough.

Cholera toxin and related proteins

Cholera toxin is the prototype of a number of functionally and serologically related toxins produced by a variety of bacteria including *Vibrio cholerae* (Van Heyningen, 1985; Betley *et al.*, 1986), *Vibrio mimicus*, enterotoxinogenic *E. coli*, *Salmonella*, *Pseudomonas*, *Campylobacter jejuni* and *Aeromonas hydrophyla* (Spira and Fedorka-Cray, 1984; Klipstein and Engert, 1985; Chopra *et al.*,

ADP/rib	Mab	TOX/holotox	Mutant	Sequence (residues 1–235)
100	+	100	S1 BP	DDPPATVYRYDSRPPEDVFQNGFTAWGNNDNVLDHLTGRSCQVGSSNSAFVSTSSSRRYTEVYLEHRMQEAVEAERAGRGTGHFIGYIYEVRADNNFYGAASSYFEYVDTYGDNAGRILAGALATYQSEYLAHRRIPPENIRRVTRVYHNGITGETTTTEYSNARYVSQQTRANPNPYTSRRSVASIVGTLVRMAPVIGACMARQAESSEAMAAWSERAGEAMVLVYYESIAYSF
100	+	n.d.	S1 BPP	E. I T PL T T T P T
100	+	n.d.	S1 BB	E. P T P
0.1	-	n.d.	8D	D.
100	+	n.d.	8F	F.
0.1	+	0.1	9K ←	K
0.2	+	n.d.	9H	H
0.2	+	n.d.	9A	A
0.2	-	n.d.	9(-)	(-)
0*	-	n.d.	8L/9E	LE
0*	-	B	8D/9G	DG
0.2	-	n.d.	8(-)/9H	-H
0*	-	n.d.	9Q/12G	Q ..G
50	+	n.d.	11E	E..
1*	+	n.d.	11S ←	S..
0*	-	n.d.	11P/14D	P..D
100	+	n.d.	12G	G.
1*	+	25	13L ←	L
30	+	n.d.	13K	K
1*	+	10	26I ←	I
3	+	n.d.	26L	L
0.4*	+	n.d.	26T	T
0.4*	+	n.d.	26(-)	(-)
100	+	n.d.	34Q	Q.
3	n.d.	n.d.	35N ←	N
100	n.d.	n.d.	39K	K
100	n.d.	n.d.	41A	A.
100	n.d.	n.d.	41S	S.
n.d.	n.d.	B	41Y	Y.
100	n.d.	n.d.	42E	E
<10	n.d.	n.d.	44E	E
0*	-	B	50E ←	E
40	+	n.d.	59K	K
<10	n.d.	n.d.	80E	E
n.d.	+	10	86E	E
n.d.	-	B	88E/89S	ES
100	+	n.d.	99E	E
100	+	n.d.	106D	D.
100	+	n.d.	106 (-)	(-)
1*	n.d.	1	107+VDGS	VDGS.
50	+	n.d.	109G	G.
150	+	n.d.	111G+DTGG	G+DTGG
50	+	n.d.	121E	E
50	+	n.d.	124D	D
1	+	5	129G ←	G
1	+	5	129Q	Q
n.d.	+	10	129 (-)	(-)
1	+	5	130G	
100	+	n.d.	135E	E
15	+	n.d.	139S	S
100	+	n.d.	159K	K.
70	+	n.d.	160G	G
100	n.d.	B	201S	S

Mutant	Toxicity on CHO cells (%)	ADP/ribosylation
PT 9K/129G	<0.0001	<0.01
PT 13L/129G	<0.0001	<0.01
PT 26I/129G	<0.0001	<0.01
PT 13L/26G	0.1	n.d.
PT 13L/26I/129G	<0.0001	B

1987a,b; Schultz and McCardell, 1988; Rose et al., 1989a,b). Although all of these toxins are supposed to play a role in bacterial virulence, so far, a clear relationship between toxin production and disease (diarrhoea) has been established only for *V. cholerae* and enterotoxinogenic *E. coli*. The molecular structure is known for cholera toxin (CT) (Mekalanos et al., 1983), *E. coli* enterotoxin I (LT1) (Dallas and Falkow, 1980), and *E. coli* enterotoxin II (LT2) (Pickett et al., 1987). Their genes, carried on a family of plasmids (LT1) or in the bacterial chromosome (CT and LT2), are organized in an operon which encodes for two proteins of 27 200 and 11 700 Da, called polypeptide A and B, respectively. Following co-translational export into the periplasm, five polypeptides B and one polypeptide A assemble into an oligomeric structure typical of the other bacterial toxins (Hardy et al., 1988). The assembled holotoxin is secreted into the extracellular medium by *V. cholerae*, but is mostly cell-associated in *E. coli* (Neill et al., 1983; Hirst et al., 1984).

Figure 6. Amino acid sequence of the S1 subunit of pertussis toxin and list of the amino acid changes which have been introduced by natural mutations or by site-directed mutagenesis. (S1BPP and S1BB derive from the chromosome of *B. parapertussis* and *B. bronchiseptica* respectively). From left to right: (1) ADP/rib = percentage of the enzymatic activity of each S1 mutant. (2) Mab = whether the mutant S1 is recognized by protective monoclonal antibodies recognizing a conformational protective epitope (Bartoloni et al., 1988). (3) TOX/holotox = the toxicity in CHO cells of the holotoxin which contains the mutation in the S1 subunit. (4) the name of each mutant. The mutants with enzymatic activity below 1% are indicated by an arrow. *An asterisk means that the ADP-ribosylating activity was below detectable levels but not necessarily 0% or 1% as originally reported. B indicates mutants where the amino acid changes in the S1 subunit prevented assembly of the S1 subunit into the holotoxin and therefore, only the B oligomer is secreted into the extracellular medium. The mutants are named by the number of the amino acid which has been mutagenized followed by a letter indicating the new amino acid, e.g. 9K = amino acid 9 has been changed to Lys. The mutants reported have been described in the following papers: 8D/9G, 50E, 99E, 109G, 111G+DTGG, 121E, 124D, 129G, 135E, 159K, 160G by Pizza et al. (1988); 44E, 80E, 88E/89G, 130G have not yet been described (Pizza and Rappuoli, unpublished data); 11S, 13L, 26I, 139S by Barbieri and Cortina (1988); 8F, 8L/9E, 9Q/12G, 11E, 11P/14D, 12G, 13K, 41S by Burnette et al. (1988a); 34Q, 35N, 39K, 42E, 201S by Kaslow et al. (1989); 26T, 26(−), 106D, 106(−) by Locht et al. (1989); 107+VDGS by Black et al. (1988).

Biochemical and electron microscopic studies have shown that the five B subunits are arranged in a ring around a central core which contains the A subunit (Ludwig et al., 1986; Ribi et al., 1988). As in the case of the other bacterial toxins, the B oligomer is involved in receptor binding and translocation of the A subunit across the cell membrane. The receptors bound by oligomers B of CT and LT1 are ganglioside GM1 and to a lesser extent ganglioside GDb1. LT2 oligomer B binds mostly gangliosides GD1 and GT1b.

Fragment A is proteolytically cleaved into two polypeptide chains (A1 and A2 of 22 and 5 kDa, respectively) which are held together by a disulphide bridge. After reduction A1 acquires enzymatic activity and becomes able to ADP-ribosylate $G_s\alpha$, thus inducing intracellular accumulation of cyclic AMP (cAMP). The accumulation of cAMP has been so far believed to be responsible for the water and electrolyte loss observed in the diarrhoea caused by cholera toxin. However, the recent finding that receptor-coupled $G_o\alpha$ and $G_i\alpha$ are also ADP-ribosylated by CT, suggests that this phenomenon might be more complex and could involve G proteins other than $G_s\alpha$. This is also supported by the observation that CT induces the release of prostaglandins (PGE) and that in the intestinal loop PGE but not cAMP can induce fluid accumulation (Peterson and Ochoa, 1989).

CT, LT1 and LT2, originally derived from the same ancestral gene, are the result of a long period of evolution under different selective pressure. While the enzymatically active subunit S1 is well-conserved in the three species (Pickett et al., 1987), the B oligomer of LT2 has evolved into a totally different structure with binding specificity different from CT and LT1 (Fukuta et al., 1988). At the nucleotide sequence level, fragments A1, A2 and B of CT and LT1 have a homology of approximately 78%, while the homology of LT2 for CT is 62% in A1, 36% in A2 and undetectable in B. Similar figures are also found at the amino acid level. The presence of toxins immunologically related to CT in *Salmonella*, *Pseudomonas*, *Campylobacter* and *Aeromonas* suggests that the same genes have followed different pathways of evolution in different species and that we should expect to find the genes homologous to CT A and CT B evolved and reassembled in a variety of manners.

In addition to the homologies described above

the A1 fragments share some homology (only at the amino acid level) with the subunit S1 of pertussis toxin (see Fig. 8). Although the crystallization of the CT B subunit has been reported (Maulik *et al.*, 1988) and the primary structures of CT, LT1 and LT2 have been known for a long time, very little is known about the role of key amino acids in the structure and function of these toxins. So far, one study has shown that simultaneous mutagenesis of two amino acids in the regions homologous to pertussis toxin (Ser^{61} → Phe and Gly^{79} → Lys) abolishes the enzymatic activity of LT1 (Harford *et al.*, 1989). An LT1 holotoxin containing these mutations is structurally and immunologically identical to the wild-type molecule, but is devoid of any detectable toxicity. Within the B oligomer it has been shown that the substitution of Ala^{64} with Val affects the ability of the B subunits to assemble into the oligomeric structure (Iida *et al.*, 1989).

Clostridial toxins

Clostridium botulinum produces two types of proteins which ADP-ribosylate eukaryotic proteins: exoenzyme C3 and the C2 toxin. Both enzymes are structurally and functionally distinct from botulinum neurotoxins. C3 is a 25-kDa ADP-ribosyltransferase which lacks the B moiety typical of the other toxins and therefore cannot enter eukaryotic cells and damage them (see Fig. 2) (Aktories *et al.*, 1987, 1988). Nevertheless, when microinjected into Swiss 3T3 cells, C3 induces rounding up of the cells, indicating that the ADP-ribosylation causes alteration of the cellular cytoskeleton (Aktories *et al.*, 1990a,b). Many studies have confirmed that the target protein of C3, initially identified in brain extracts, is the GTP-binding protein rho, a protein related to ras present in all tissues so far analysed (Chardin *et al.*, 1989; Kikuchi *et al.*, 1988; Narumiya *et al.*, 1988; Matsouka *et al.*, 1989; Aktories *et al.*, 1990a, b). C3 modifies the Asn^{41} of rho (see Table 1 and Fig. 4) (Sekine *et al.*, 1989). This modification does not alter either the GTPase or the GTP-binding activity of rho, but clearly alters its function. Similarly, amino acid substitutions in the homologous region of ras have been shown to reduce its biological effect, possibly by altering its

interaction with effector molecule, without altering the GTPase or GTP-binding activities.

Botulinum C2 toxin is the prototype of a number of botulinum toxins which ADP-ribosylate monomeric actin (Aktories *et al.*, 1986). Like most of the other toxins, they are composed of two components which are involved in binding (CII) and biological activity (CI), respectively. However, while the other toxins are composed of A and B moieties which are connected either by covalent bonds or by non-covalent interactions, botulinum C2 is composed of two separated proteins. In this regard it is more analogous to *Bacillus anthracis* protective and lethal factors than to the other toxins mentioned in this chapter. CII (100 kDa) is present in the culture supernatant in an inactive form which, following proteolytic cleavage, has been reported to assemble in an oligomer structure of high molecular weight possessing haemagglutinating and haemolytic activities. CI (50 kDa) ADP-ribosylates mainly non-muscle monomeric G-actin, but not polymerized F-actin, at Arg^{177} (see Table 1) (Aktories *et al.*, 1990a,b). The modified actin loses its ATPase activity and the ability to be polymerized. According to the model proposed by Aktories (Fig. 7) (Aktories *et al.*, 1990a,b), following ADP-ribosylation, the G-actin behaves as a capping protein, that is, it binds to the fast growing end of the F-actin and inhibits any further polymerization. As depolymerization at the other end of the actin increases the pool of G-actin, more actin is ADP-ribosylated until the microfilament network collapses. The result is a rounding up followed by lysis of the cells in tissue culture, and an increase in the vascular permeability and fluid accumulation in the mouse intestinal loop.

Clostridium perfringens iota toxin, *Clostridium spiroforme* and *Clostridium difficile* ADP-ribosylating toxins have a structure and a function very similar to C2 toxin of *C. botulinum*. However, they are usually classified in a separate family of iota-like toxins because they are immunologically related and do not show cross-reaction with the C2 toxin.

A common structure of the catalytic site

Comparison of the known amino acid sequences of ADP-ribosylating toxins reveals a significant

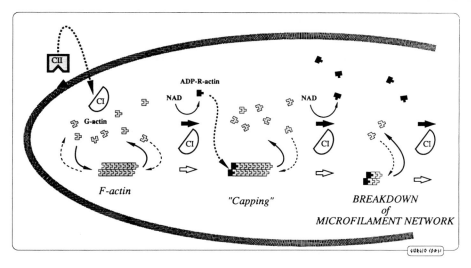

Figure 7. Model of the action of botulinum C2 toxin on the microfilament network of the cells. Component CII binds the receptor on the cell membrane and facilitates the entry of CI into the cells where it ADP-ribosylates monomeric G-actin. Following ADP-ribosylation, the G-actin is incorporated into the F-actin and prevents further polymerization. The depolymerization which occurs at the other end of the F-actin increases the pool of G-actin, which is the substrate of CI. Eventually, most of the F-actin is depolymerized and the microfilament network collapses.

homology between DT and PAETA, and between PT, CT and *E. coli* LT (Fig. 8). No obvious homology is detected between these two groups of enzymes. However, it should be noticed that the initial comparison of the amino acid sequences of DT and PAETA also did not show sequence homology until the functional identity of DT Glu[148] and PAETA Glu[553] was demonstrated by photoaffinity labelling. This initial observation gave a reference point for sequence comparison which allowed the detection of the strong homology shown in Fig. 8.

In the 3-dimensional structure, the most highly conserved residues of DT and PAETA have been found to form the main part of the active site cleft. This cleft is composed of amino acids deriving from three non-contiguous regions and comprises mainly: (I) His[440] in PAETA, homologous to His[21] in DT; (II) Trp[466], Phe[469], Tyr[470] and Tyr[481] in PAETA, homologous to Trp[50], Phe[53], Tyr[54] and Tyr[65] in DT, respectively; and finally, (III) Glu[553] in PAETA, homologous to Glu[148] in DT (Fig. 8B) (Carroll and Collier, 1987).

Although direct sequence comparison does not show any homologies of these regions in PT, CT and LT, functional studies show that region III is present at least in PT, and is probably present in all the enzymes. Purely on the basis of amino acid

similarity we propose here the presence of region II in PT, CT and LT. The rationale for this hypothesis is given below.

Region I: Histidine 21 of diphtheria toxin cannot tolerate any chemical modification without losing enzymatic activity (Papini *et al.*, 1989). Similarly, His[35] of PT S1 cannot be substituted with Asn without loss of enzymatic activity (Kaslow *et al.*, 1989). Remarkably, this histidine residue is conserved in all the enzymes (Fig. 8B).

Region II: The peculiarity of region II in DT and PAETA is the relative abundance of closely spaced aromatic amino acids which constitute the basic structure of the active site cleft (Fig. 8C). The only region in PT, CT and LT which is abundant in aromatic amino acids is the region comprising amino acids 82–98 of PT S1, which is well-conserved in all three enzymes. When this region is aligned with region II of DT and PAETA, there is a striking conservation of the position of the aromatic residues, suggesting that they might be functionally equivalent (Carroll and Collier, 1988).

Region III: Glu[129] of PT has been shown to be specifically photolabelled by NAD[+] and does not tolerate substitution, even with aspartic acid, without losing enzymatic activity (Barbieri *et al.*, 1989). These two properties, identical to those of

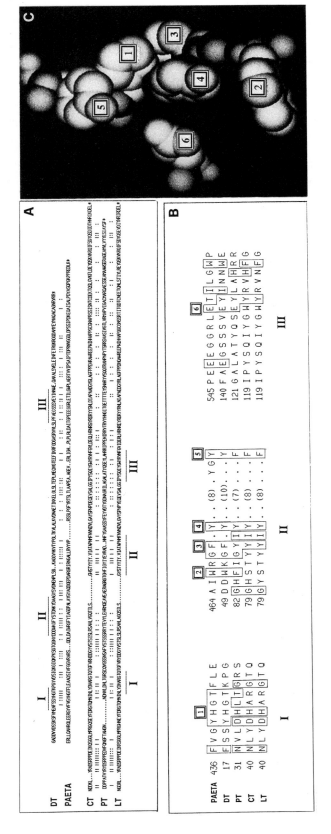

Figure 8. (A) Homologies between DT and PAETA (top) and between CT, PT and LTI (bottom). Identical amino acids are indicated by '|'; homologous amino acids are indicated by ':'. The following groups of homologous amino acids have been used: C; S, T, P, A, G; N, D, Q, E; M, I, L, V; F, Y, W; K, R, H. I, II and III are the regions which make the active site of PAETA and are conserved in all toxins, as shown in B. (B) Homologies between the three regions forming the active site of PAETA and corresponding regions in the other toxins. The conserved amino acids which form the active site cleft of PAETA shown in C are numbered from 1 to 6. (C) Structure of the active site cleft of PAETA, as deduced by the computer elaboration of X-ray crystallographic data. Fig. 8C has been kindly provided by J. Collier (Harvard Medical School, USA).

Glu[148] and Glu[553] of DT and PAETA, respectively, allow us to conclude unequivocally that these three glutamic acids are functionally equivalent (Fig. 8C). Region III in CT and LT has not yet been identified.

In conclusion, as initially happened for DT and PAETA, direct comparison of the amino acid sequences of the two groups of homologous enzymes does not show relevant similarities. However, as we learn more about the function of the key amino acids, more homologies are found and all enzymes seem to have a common structure at least in the catalytic site. The homologies that we have shown here are likely to be modified and improved as more functional and structural data are accumulated.

Endogenous mono-ADP-ribosylation of eukaryotic cells

Proteins showing the same enzymatic activity as bacterial toxins have been identified in several eukaryotic cells and some of them have also been purified. They can be classified into three main groups depending on whether they modify the diphthamide residue of EF2, the Arg[201] of $G_s\alpha$ protein or the Cys[352] of $G_i\alpha$ proteins. Another group which modifies the oncogene ras and ras homologous proteins has been described but not yet characterized.

To our knowledge, endogenous enzymes modifying Arg[177] of monomeric actin have not yet been identified. However, in the near future many other endogenous mono-ADP-ribosylating enzymes are likely to be discovered. Most of them will not have a corresponding bacterial toxin because they regulate cellular circuits which are not essential for cellular survival. One example which suggests the presence of such enzymes is a protein of 80 kDa which is heavily ADP-ribosylated in HeLa cells when they are starved of essential amino acids (Ledford and Jacobs, 1986). A summary of the known eukaryotic mono-ADP-ribosylases is reported in the following pages.

Endogenous modification of EF2

The presence of an enzymatic activity which modifies His[715] of EF2 into diphthamide, conserved

throughout eukaryotic cell evolution, suggests that this residue must be necessary for essential functions in eukaryotic cell metabolism and not only as a target of bacterial toxins which kill the eukaryotic cells. However, the search for an endogenous enzyme capable of ADP-ribosylating the diphthamide residues of EF2 has not been straightforward. When such an enzyme was finally described (Lee and Iglewski, 1984), most of the other groups were not able to reproduce the data. Recently, the same group has shown that polyoma virus-transformed baby hamster kidney cells grown in the presence of [α-^{32}P]orthophosphate accumulate ^{32}P in a protein of 97 kDa which is precipitated by antiserum against EF2 and that the peptide labelled in this protein is identical to that generated by trypsin digestion of diphtheria toxin-modified EF2 (Fendrick and Iglewski, 1989). The endogenous ADP-ribosylation was shown to be dependent on the culture conditions: while only a small fraction of EF2 was ADP-ribosylated when the cells were grown in 10% serum, up to 35% was ADP-ribosylated if the cells were grown in 2% serum.

ADP-ribosylation of the α chain of G_s and oncogenesis

While endogenous arginine-modifying mono-ADP-ribosylases were first described almost ten years ago, the endogenous enzyme equivalent to cholera toxin has not yet been purified. Recently, it has been shown that $G_s\alpha$ is ADP-ribosylated in platelets which had been electropermeabilized to introduce [α-^{32}P]NAD, and that this endogenous ADP-ribosylation is inhibited by nicotinamide and stimulated by the prostacyclin analogue, iloprost (Molina y Vedia et al., 1989). The stimulation of the endogenous ADP-ribosylation by iloprost induces a reduction of the $G_s\alpha$ available for cholera-mediated ADP-ribosylation, suggesting that $G_s\alpha$ is ADP-ribosylated at the Arg[201] by endogenous ADP-ribosylases.

The role of the ADP-ribosylation of the Arg[201] of $G_s\alpha$ has been recently elucidated by the discovery that three out of four growth hormone-secreting human pituitary tumours are caused by a change of Arg[201] of $G_s\alpha$ either to Cys or His (Landis et al., 1989). These mutations, like ADP-ribosylation of Arg[201], dramatically reduce the GTPase activity which turns off the protein's active conformation, and therefore cause a constituitive activation of the $G_s\alpha$ which in turn leads to the constituitive accumulation of cyclic AMP. The modified G_s proteins are therefore considered a new example of the family of the oncogenes and have been called GSP (G_s protein). Interestingly, the region corresponding to the Arg^{201} of G_s proteins is not present in the homologous GTP-binding proteins such as ras, rho and EF2. It is believed that while these last proteins need additional proteins to hydrolyse GTP (GAP or GTPase-activating proteins), G_s and G_i proteins have in the region of Arg^{201} a built-in device which activates the GTPase activity, thus turning off the active state of the molecule. In conclusion, ADP-ribosylation of $G_s\alpha$ by cholera toxin is equivalent to a mutation which transforms $G_s\alpha$ into an oncogene.

ADP-ribosylation of $G_i\alpha$ and related proteins

The ADP-ribosylation of the $G_i\alpha$ subunit of adenylate cyclase by an endogenous ADP-ribosylase is well-documented (Tanuma, 1990). An enzyme with a molecular weight of 28 kDa, present in all tissues tested, and purified from the cytosol of human erythrocytes was initially shown to ADP-ribosylate cysteine methyl ester and a protein of 41 kDa, identical in size to the substrate of pertussis toxin. This reaction could be inhibited specifically by previous incubation of the target membranes with pertussis toxin. The chemical stability of the ADP-ribose was also in agreement with a substrate modified at a Cys residue (Tanuma et al., 1987a,b; Tanuma and Endo, 1989).

As in the case of pertussis toxin, preincubation of human platelet membranes with the eukaryotic enzyme reduced the epinephrine-induced inhibition of adenylate cyclase, suggesting that pertussis toxin and the eukaryotic enzymes ADP-ribosylate the same target protein and that both of them attenuate receptor-mediated inhibition of adenylate cyclase.

It is not yet clear whether the eukaryotic enzyme purified by Tanuma and co-workers is specific for $G_i\alpha$ or whether it is able to ADP-ribosylate $G_o\alpha$ and the entire family of G proteins, as in the case of pertussis toxin. Therefore, while it is not clear whether we will find a different endogenous ADP-

ribosylase for each different G protein, it is already well-documented that there are eukaryotic ADP-ribosylases which by covalent modification of G proteins modulate the response of eukaryotic cells to external stimuli. As in the case of $G_s\alpha$ it is likely that mutations within the G_o and G_i proteins mimicking the effects of ADP-ribosylation will transform those proteins into oncogenes.

ADP-ribosylation of ras and other low molecular weight GTP-binding proteins

The purification from human platelets of an ADP-ribosyltransferase which modifies the oncogene ras, and other membrane proteins of 21 kDa and 26 kDa has been reported (Tanuma, 1990). Although it is not clear yet whether this enzyme modifies rho, as in the case of botulinum toxin CII, these findings confirm the expectation that for each bacterial toxin a corresponding eukaryotic ADP-ribosyltransferase will be found.

Conclusions

Mono-ADP-ribosylation is a post-translational modification widely used in nature to modify the structure and the function of proteins. Mono-ADP-ribosylating enzymes have been described in E. coli, bacteriophages, rhodospirillum (Ueda and Hayaishi, 1985; Fitzmaurice et al., 1989) and eukaryotic cells. While most of the ADP-ribosylases have their target protein within the cells in which they are produced, bacterial toxins have no function within the bacteria which produce them but have target proteins located inside eukaryotic cells. Remarkably, the same target proteins of bacterial toxins are also modified in an identical manner by ADP-ribosylases endogenous to the eukaryotic cells. This observation suggests a common origin of eukaryotic and bacterial mono-ADP-ribosylases. This could be explained either by the presence of a common ancestral gene or with a more likely common mechanism involving a recent transfer of eukaryotic genes to bacteria followed by a selective pressure which has favoured those bacteria which were able to express them. Cloning, sequencing and structural studies are necessary before we will know whether the eukaryotic mono-ADP-ribosyltransferases have an overall structure similar to that of their corresponding bacterial toxins.

Acknowledgements

We thank A. Prugnola, V. Scarlato, B. Arico', K. Aktories, C. Montecucco and R. Gross for critical reading of the manuscript, G. Corsi for graphic work.

References

Aktories, K., Bärmann, M., Ohishi, I., Tsuyama, S., Jakobs, K.H. and Habermann, E. (1986). Botulinum C2 toxin ADP-ribosylates actin. *Nature* **322**, 390–2.

Aktories, K., Weller, U. and Chhatwal, G.S. (1987). *Clostridium botulinum* produces a novel ADP-ribosyltransferase distinct from botulinum C2 toxin. *FEBS Lett.* **212**, 109–13.

Aktories, K., Rösener, S., Blaschke, U. and Chhatwal, G.S. (1988). Botulinum ADP-ribosyltransferase C3. *Eur. J. Biochem.* **172**, 445–50.

Aktories, K., Braun, U., Garrett, M., Habermann, B., Mohr, C., Rösener, S., Paterson, H., Steinbeißer, H. and Hall, A. (1990a). Botulinum ADP-ribosyltransferase C3 modifies the GTP-binding protein *rho*. In: *Bacterial Protein Toxins* (eds. R. Rappuoli and R. Gross), pp. 271–2. Gustav Fischer Verlag, Stuttgart.

Aktories, K., Geipel, U., Haas, T., Laux, M. and Just, I. (1990b). Clostridial actin-ADP-ribosylating toxins. In: *Bacterial Protein Toxins* (eds R. Rappuoli and R. Gross), pp. 219–26. Gustav Fischer Verlag, Stuttgart.

Allured, V.S., Collier, R.J., Carroll, S.F. and McKay, D.B. (1986).Structure of exotoxin A of *Pseudomonas aeruginosa* at 3.0-Angstrom resolution. *Proc. Natl. Acad. Sci. USA* **83**, 1320–4.

Althaus, F.R. and Richter, C. (1987). ADP-ribosylation of proteins. Enzymology and biological significance. *Molec. Biol. Biochem. Biophys.* **37**.

Bacha, P., Williams, D.P., Waters, C., Williams, J.M., Murphy, J.R. and Strom, T.B. (1988). Interleukin 2 receptor-targeted cytotoxicity: interleukin 2 receptor-mediated action of a diphtheria toxin-related interleukin 2 fusion protein. *J. Exp. Med.* **167**, 612–22.

Barbieri, J.T. and Cortina, G. (1988). ADP-ribosyltransferase mutations in the catalytic S-1 subunit of pertussis toxin. *Infect. Immun.* **56**, 1934–41.

Barbieri, J.T., Mende-Mueller, L.M., Rappuoli, R. and Collier, R.J. (1989). Photolabelling of Glu[129] of the S-1 subunit of pertussis toxin. *Infect. Immun.* **57**, 3549–54.

Bartoloni, A., Pizza, M., Bigio, M., Nucci, D., Ashworth, L.A., Irons, L.I., Robinson, A., Burns, D., Manclark, C., Sato, H. and Rappuoli, R. (1988). Mapping of a protective epitope of pertussis toxin by *in vitro* refolding of recombinant fragments. *Biotechnology* **6**, 709–12.

Betley, M.J., Miller, V.L. and Mekalanos, J.J. (1986). Genetics of bacterial enterotoxins. *Ann. Rev. Microbiol.* **40**, 577–605.

Black, W.J., Munoz, J.J., Peacock, M.G., Schad, P. A., Cowell, J.L., Burchall, J.J., Lim, M., Kent, A., Steinman, L. and Falkow, S. (1988). ADP-ribosyltransferase activity of pertussis toxin and immunomodulation by *Bordetella pertussis*. *Science* **240**, 656–9.

Bokoch, G.M., Katada, T., Northup, J.K., Hewlett, E.L. and Gilman, A.G. (1983). Identification of the predominant substrate for ADP-ribosylation by islet activating protein. *J. Biol. Chem.* **258**, 2072–5.

Brandhuber, B.J., Allured, V.S., Falbel, T.G. and McKay, D.B. (1988). Mapping the enzymatic active site of *Pseudomonas aeruginosa* exotoxin A. *Proteins* **3**, 146–54.

Burnette, W.N., Cieplak, W., Mar, V.L., Kaljot, K.T., Sato, H. and Keith, J.M. (1988a). Pertussis toxin S1 mutant with reduced enzyme activity and a conserved protective epitope. *Science* **242**, 72–4.

Burnette, W. N., Mar, V.L., Cieplak, W., Morris, C.F., Kaljot, K.T., Sato, H. and Keith, J.M. (1988b). Toward development of a recombinant pertussis vaccine. In: *Technological Advances in Vaccine Development*, pp. 75–85. Alan R. Liss, New York.

Carroll, S.F. and Collier, R.J. (1984). NAD binding site of diphtheria toxin: identification of a residue within the nicotinamide subsite by photochemical modification with NAD. *Proc. Natl. Acad. Sci. USA* **81**, 3307–11.

Carroll, S.F. and Collier, R.J. (1987). Active site of *Pseudomonas aeruginosa* exotoxin A. *J. Biol. Chem.* **262**, 8707–11.

Carroll, S.F. and Collier, R.J. (1988). Amino acid sequence homology between the enzymic domains of diphtheria toxin and *Pseudomonas aeruginosa* exotoxin A. *Molec. Microbiol.* **2**, 293–6.

Carroll, S.F., McCloskey, J.A., Crain, P.F., Oppenheimer, N.J., Marschner, T.M. and Collier, R.J. (1985). Photoaffinity labeling of diphtheria toxin fragment A with NAD: structure of the photoproduct at position 148. *Proc. Natl. Acad. Sci. USA* **82**, 7237–41.

Chardin, P., Bouquet, P., Madaule, P., Popoff, M.R., Rubin, E.J. and Gill, D.M. (1989). The mammalian G-protein zho is ADP-ribosylated by *Clostridium botulinum* exoenzyme C3 and affects active microfilaments *in vivo* cells. *EMBO J.* **8**, 1087–92.

Chaudhary, V.K., FitzGerald, D.J., Adhya, S. and Pastan, I. (1987). Activity of a recombinant fusion protein between transforming growth factor type α and *Pseudomonas* toxin. *Proc. Natl. Acad. Sci. USA* **84**, 4538–42.

Chaudhary, V.K., Mizukami, T., Fuerst, T.R., Fitz-Gerald, D.J., Moss, B., Pastan, I. and Berger, E.A. (1988). Selective killing of HIV-infected cells by recombinant human CD4-*Pseudomonas* exotoxin hybrid protein. *Nature* **335**, 369–72.

Chopra, A.K., Houston, C.W., Peterson, J.W. and Mekalanos, J.J. (1987a). Chromosomal DNA contains the gene coding for *Salmonella* enterotoxin. *FEBS Lett.* **43**, 345–9.

Chopra, A.K., Houston, C.W., Peterson, J.W., Prasad, R. and Mekalanos, J.J. (1987b). Cloning and expression of the *Salmonella* enterotoxin gene. *J. Bacteriol.* **169**, 5095–100.

Cieplak, W., Burnette, W.N., Mar, V.L., Kaljot, K.T., Morris, C.F., Chen, K.K., Sato, H. and Keith, J.M. (1988). Identification of a region in the S1 subunit of pertussis toxin that is required for enzymatic activity and that contributes to the formation of a neutralizing antigenic determinant. *Proc. Natl. Acad. Sci. USA* **35**, 4667–71.

Coburn, J., Wyatt, R.T., Iglewski, B.H. and Gill, D.M. (1989a). Several GTP-binding proteins, including p21$^{c\text{-}H\text{-}ras}$, are preferred substrates of *Pseudomonas aeruginosa* exoenzyme S. *J. Biol. Chem.* **264**, 9004–8.

Coburn, J., Dillon, S.T., Iglewski, B.H. and Gill D. M. (1989b). Exoenzyme S of *Pseudomonas aeruginosa* ADP-ribosylates the intermediate filament protein vimentin. *Infect. Immun.* **57**, 996–8.

Collier, R.J. (1982). Structure and activity of diphtheria toxin. In: *ADP-ribosylation Reactions* (eds D. Hayashi and K. Ueda), pp. 575–92. Academic Press, New York.

Comanducci, M., Ricci, S., Rappuoli, R. and Ratti, G. (1987). The nucleotide sequence of the gene coding for diphtheria toxoid CRM 176. *Nucleic Acids Res.* **15**, 5897.

Dallas, W.S. and Falkow, S. (1980). Amino acid homology between cholera toxin and *Escherichia coli* heat-labile toxin. *Nature* **288**, 499–501.

Dell'Arciprete, L., Colombatti, M., Rappuoli, R. and Tridente, G. (1988). A C terminus cysteine of diphtheria toxin B chain involved in immunotoxin cell penetration and cytotoxicity. *J. Immunol.* **140**, 2466–71.

Douglas, C.M. and Collier, R.J. (1987). Exotoxin A of *Pseudomonas aeruginosa*: substitution of glutamic acid 553 with aspartic acid drastically reduces toxicity and enzymatic activity. *J. Bacteriol.* **169**, 4967–71.

Fendrick, J.L. and Iglewski, W.J. (1989). Endogenous ADP-ribosylation of elongation factor 2 in polyoma virus-transformed baby hamster kidney cells. *Proc. Natl. Acad. Sci. USA* **86**, 554–7.

Fitzmaurice, W.P., Saari, L.L., Lowery, R.G., Ludden, P.W. and Roberts, G.P. (1989). Genes coding for the reversible ADP-ribosylation system of dinitrogenase reductase from *Rhodospirillum rubrum*. *Molec. Gen. Genet.* **218**, 340–7.

Fong, H.K.W., Yoshimoto, K.K., Eversole-Cire, P. and Simon M.I. (1988). Identification of a GTP-binding protein α subunit that lacks an apparent ADP-ribosylation site for pertussis toxin. *Proc. Natl. Acad. Sci. USA* **85**, 3066–70.

Fukuta, S., Magnani, J.L., Twiddy, E.M., Holmes, R.K. and Ginsburg, V. (1988). Comparison of the carbohydrate-binding specificities of cholera toxin

and *Escherichia coli* heat-labile enterotoxins LTh-I, LT-IIa, and LT-IIb. *Infect. Immun.* **56**, 1748–53.

Galloway, D.R., Hedstrom, R.C., McGowan, J.L., Kessler, S.P. and Wozniak, D.J. (1989). Biochemical analysis of CRM66. *J. Biol. Chem.* **264**, 14869–73.

Giannini, G., Rappuoli, R. and Ratti, G. (1984). The aminoacid sequence of two nontoxoid mutants of diphtheria toxin: CRM 45 and CRM 197. *Nucleic Acids Res.* **12**, 4063–9.

Gill, D.M., Pappenheimer, A.M. Jr., Brown, R. and Kurnick, J.J. (1969). Studies on the mode of action of *Diphtheria toxin*. VII. Toxin-stimulated hydrolysis of nicotinamide adenine dinucleotide in mammalian cell extracts. *J. Exp. Med.* **129**, 1–21.

Gilman, A.G. (1984). G proteins and dual control of adenylate cyclase. *Cell* **36**, 577–9.

Gray, G.L., Smith, D.H., Baldridge, J.S., Harkins, R.N., Vasil, M.L., Chen, E.Y. and Heyneker, H.L. (1984). Cloning, nucleotide sequence, and expression in *Escherichia coli* of the exotoxin A structural gene of *Pseudomonas aeruginosa*. *Proc. Natl. Acad. Sci. USA* **81**, 2645–9.

Greenfield, L., Johnson, V.G. and Youle, R.J. (1987). Mutations in diphtheria toxin separate binding from entry and amplify immunotoxin selectivity. *Science* **238**, 536–9.

Hardy, S.J.S., Holmgren, J., Johansson, S., Sanchez, J. and Hirst, T.R. (1988). Coordinated assembly of multisubunit proteins: oligomerization of bacterial enterotoxins *in vivo* and *in vitro*. *Proc. Natl. Acad. Sci. USA* **85**, 7109–13.

Harford, S., Dykes, C.W., Hobden, A.N., Read, M.J. and Halliday, I.J. (1989). Inactivation of the *Escherichia coli* heat-labile enterotoxin by *in vitro* mutagenesis of the A-subunit gene. *Eur. J. Biochem.* **183**, 311–16.

Hewlett, E.L., Sauer, K.T., Myers, G.A., Cowell, J.L. and Guerrant, R.L. (1983). Induction of a novel morphological response in Chinese hamster ovary cells by pertussis toxin. *Infect. Immun.* **40**, 1198–203.

Hirst, T.R., Sanchez, J., Kaper, J.B., Hardy, S.J.S. and Holmgren, J. (1984). Mechanism of toxin secretion by *Vibrio cholerae* investigated in strains harboring plasmids that encode heat-labile enterotoxins of *Escherichia coli*. *Proc. Natl. Acad. Sci. USA* **81**, 7752–6.

Honjo, T., Nishizuka, Y., Hayaishi, O. and Kato, I. (1968). Diphtheria toxin-dependent adenosine diphosphate ribosylation of aminoacyl transferase II and inhibition of protein synthesis. *J. Biol. Chem.* **243**, 3553–5.

Hwang, J., FitzGerald, D.J.P., Adhya, S. and Pastan, I. (1987). Functional domains of *Pseudomonas* exotoxin identified by deletion analysis of the gene expressed in *E. coli*. *Cell* **48**, 129–36.

Iida, T., Tsuji, T., Honda, T., Miwatani, T., Wakabayashi, S., Wada, K. and Matsubara, H. (1989). A single amino acid substitution in B subunit of *Escherichia coli* enterotoxin affects its oligomer formation. *J. Biol. Chem.* **264**, 14065–70.

Jinno, Y., Chaudhary, V.K., Kondo, T., Adhya, S., FitzGerald, D.J.P. and Pastan, I. (1988). Mutational analysis of domain I of *Pseudomonas* exotoxin. *J. Biol. Chem.* **263**, 13203–7.

Jinno, Y., Ogata, M., Chaudhary, V.K., Willingham, M.C., Adhya, S., FitzGerald, D. and Pastan, I. (1989). Domain II mutants of Pseudomonas exotoxin deficient in translocation. *J. Biol. Chem.* **264**, 15953–9.

Kaczorek, M., Delpeyroux, F., Chenciner, N., Streeck, R. E., Murphy, J. R., Boquet, P. and Tiollais, P. (1983). Nucleotide sequence and expression of the diphtheria *tox*[228] gene in *Escherichia coli*. *Science* **221**, 855–8.

Kaslow, H.R., Schlotterbeck, J.D., Mar, V.L. and Burnette, W.N. (1989). Alkylation of cysteine 41, but not cysteine 200, decreases the ADP-ribosyltransferase activity of the S1 subunit of pertussis toxin. *J. Biol. Chem.* **264**, 6386–90.

Katada, T. and Ui, M. (1982). ADP ribosylation of the specific membrane protein of C6 cells by islet-activating protein associated with modification of adenylate cyclase activity. *J. Biol. Chem.* **257**, 7210–16.

Katada, T., Tamura, M. and Ui, M. (1983). The A promoter of islet-activating protein, pertussis toxin, as an active peptide catalyzing ADP-ribosylation of a membrane protein. *Arch. Biochem. Biophys.* **224**, 290–8.

Kikuchi, A., Uamamoto, K., Fujita, T. and Takai, Y. (1988). ADP-ribosylation of the bovine brain *rho* protein by botulinum toxin type C1. *J. Biol. Chem.* **263**, 16303–8.

Klipstein, F.A. and Engert, R.F. (1985). Immunological relationship of the B subunits of *Campylobacter jejuni* and *Escherichia coli* heat-labile enterotoxins. *Infect. Immun.* **48**, 629–33.

Kohno, K., Uchida, T., Ohkubo, H., Nakanishi, S., Nakanishi, T., Fukui, T., Ohtsuka, E., Ikehara, M. and Okada, Y. (1986). Amino acid sequence of mammalian elongation factor 2 deduced from the cDNA sequence: homology with GTP-binding proteins. *Proc. Natl. Acad. Sci. USA* **83**, 4978–82.

Landis, C.A., Masters, S.B., Spada, A., Pace, A.M., Bourne, H.R. and Vallar, L. (1989). GTPase inhibiting mutations activate the α chain of G$_s$ and stimulate adenyl cyclase in human pituitary tumours. *Nature* **340**, 692–6.

Ledford, B.E. and Jacobs, D.F. (1986). Translations control of ADP-ribosylation in eucaryotic cells. *Eur. J. Biochem.* **161**, 661–7.

Lee, H. and Iglewski, W.J. (1984). Cellular ADP-ribosyltransferase with the same mechanism of action as diphtheria toxin and *Pseudomonas* toxin A. *Proc. Natl. Acad. Sci. USA* **81**, 2703–7.

Lobet, Y., Cieplak, W. Jr., Smith, S.G. and Keith, J.M. (1989). Effects of mutations in the S1 subunit of pertussis toxin on enzyme activity and immunoreactivity. *Infect. Immun.* **57**, 3660–2.

Locht, C. and Keith, J.M. (1986). Pertussis toxin gene: nucleotide sequence and genetic organization. *Science* **232**, 1258–64.

Locht, C., Capian, C. and Feron, L. (1989). Identification of amino acid residues essential for the enzymatic activities of pertussis toxin. *Proc. Natl. Acad. Sci. USA* **86**, 3075–9.

Lorberboum-Galski, H., FitzGerald, D.J.P., Chaudhary, V.K., Adhya, S. and Pastan, I. (1988). Cytotoxic activity of an interleukin 2-*Pseudomonas* exotoxin chimeric protein produced in *Escherichia coli*. *Proc. Natl. Acad. Sci. USA* **85**, 1922–6.

Ludwig, D.S., Ribi, H.O., Schoolnik, G.K. and Kornberg, R.D. (1986). Two-dimensional crystals of cholera toxin B subunit-receptor complexes: projected structure at 17-A° resolution. *Proc. Natl. Acad. Sci. USA* **83**, 8585–8.

Lukac, M. and Collier, R.J. (1988). *Pseudomonas aeruginosa* exotoxin A: effects of mutating tyrosine-470 and tyrosine-481 to phenylalanine. *Biochemistry* **27**, 7629–32.

McGill, S., Stenmark, H., Sandvig, K. and Olsnes, S. (1989). Membrane interactions of diphtheria toxin analyzed using *in vitro* synthesized mutants. *EMBO J.* **8**, 2843–8.

Matsuoka, I., Sakuma, H., Syuto, B., Moriishi, K., Kubo, S. and Kurihara, K. (1989). ADP-ribosylation of 24–26-kDa GTP-binding proteins localized in neuronal and non-neuronal cells by botulinum neurotoxin D. *J. Biol. Chem.* **264**, 706–12.

Maulik, P.R, Reed, R.A. and Shipley, G.G. (1988). Crystallization and preliminary X-ray diffraction study of cholera toxin B-subunit. *J. Biol. Chem.* **263**, 9499–501.

Mekada, E., Okada, Y. and Uchida, T. (1988). Identification of diphtheria toxin receptor and a nonproteinous diphtheria toxin-binding molecule in *vero* cell membrane. *J. Cell Biol.* **107**, 511–19.

Mekalanos, J.J., Swartz, D.J., Pearson, G.D.N., Harford, N., Groyne, F. and de Wilde, M. (1983). Cholera toxin genes: nucleotide sequence, deletion analysis and vaccine development. *Nature* **306**, 551–7.

Middlebrook, J.L. and Dorland, R.B. (1984). Bacterial toxins: cellular mechanisms of action. *Microbiol. Rev.* **48**, 199–221.

Molina y Vedia, L., Nolan, R.D. and Lapetina, E.G. (1989). The effect of iloprost on the ADP-ribosylation of G$_s$α (the α-subunit of G$_s$). *Biochem. J.* **261**, 841–5.

Murphy, J.R., Bishai, W., Borowski, M., Miyanohara, A., Boyd, J. and Nagle, S. (1986). Genetic construction, expression, and melanoma-selective cytotoxicity of a diphtheria toxin-related α-melanocyte stimulating hormone fusion protein. *Proc. Natl. Acad. Sci. USA* **83**, 8258–62.

Narumiya, S., Sekine, A. and Fukiwara, M. (1988). Substrate for botulinum ADP-ribosyltransferase, Gb, has an amino acid sequence homologous to a putative *rho* gene product. *J. Biol. Chem.* **263**, 17255–7.

Neer, E.J. and Clapham, D.E. (1988). Roles of G protein subunits in transmembrane signalling. *Nature* **133**, 129.

Neill, R.J., Ivins, B.E. and Holmes, R.K. (1983). Synthesis and secretion of the plasmid-coded heat-labile enterotoxin of *Escherichia coli* in *Vibrio cholerae*. *Science* **221**, 289–90.

Ness, B.G. van, Howard, J.B. and Bodley, J.W. (1980a). ADP-ribosylation of elongation factor 2 by diphtheria toxin. NMR spectra and proposed structures of ribosyl-diphthamide and its hydrolysis products. *J. Biol. Chem.* **255**, 10710–16.

Ness, B.G. van, Howard, J.B. and Bodley, J.W. (1980b). ADP-ribosylation of elongation factor 2 by diphtheria toxin. Isolation and properties of the novel ribosyl-amino acid and its hydrolysis products. *J. Biol. Chem.* **255**, 10717–20.

Neville, D.M. Jr. and Hudson, T.H. (1986). Transmembrane transport of diphtheria toxin, related toxins and colicins. *Ann. Rev. Biochem.* **55**, 195–224.

Nicas, T.I. and Iglewski, B.H. (1985). Contribution of exoenzyme S to the virulence of *Pseudomonas aeruginosa*. *Antibiot. Chemother.* **36**, 40–48.

Nicosia, A. and Rappuoli, R. (1987). Promoter of the pertussis toxin operon and production of pertussis toxin. *J. Bacteriol.* **169**, 2843–6.

Nicosia, A., Perugini, M., Franzini, C., Casagli, M.C., Borri, M.G., Antoni, G., Almoni, M., Neri, P., Ratti, G. and Rappuoli, R. (1986). Cloning and sequencing of the pertussis toxin genes: Operon structure and gene duplication. *Proc. Natl. Acad. Sci. USA* **83**, 4631–5.

Ogata, M., Chaudhary, V.K., FitzGerald, D.J. and Pastan, I. (1989). Cytotoxic activity of a recombinant fusion protein between interleukin 4 and *Pseudomonas* exotoxin. *Proc. Natl. Acad. Sci. USA* **86**, 4215–19.

O'Keefe, D.O. and Collier, R.J. (1989). Cloned diphtheria toxin within the periplasma of *Escherichia coli* causes lethal membrane damage at low pH. *Proc. Natl. Acad. Sci. USA* **86**, 343–6.

Olsnes, S. and Sandvig, K. (1988). How protein toxins enter and kill cells. In: *Immunotoxins* (ed. A.E. Frankel), p. 39. Kluwer Academic Publishers, New York.

Pai, E.F., Kabsch, W., Krengel, U., Holmes, K.C., John, J. and Wittinghofer, A. (1989). Structure of the guanine-nucleotide-binding domain of the Ha-*ras* oncogene product p21 in the triphosphate conformation. *Nature* **341**, 209–14.

Papini, E., Colonna, R., Cusinato, F., Montecucco, C., Tomasi, M. and Rappuoli, R. (1987a). Lipid interaction of diphtheria toxin and mutants with altered fragment B. Liposome aggregation and fusion. *Eur. J. Biochem.* **169**, 629–35.

Papini, E., Schiavo, G., Tomasi, M., Colombatti, M., Rappuoli, R. and Montecucco, C. (1987b). Lipid interaction of diphtheria toxin and mutants with altered fragment B. Hydrophobic photolabelling and cell intoxications. *Eur. J. Biochem.* **169**, 637–44.

Papini, E., Sandona, D., Rappuoli, R. and Montecucco, P. (1988). On the membrane translocation of diphtheria toxin: at low pH the toxin induces ion channels on cells. *EMBO J.* **7**, 3353–9.

Papini, E., Schiavo, G., Sandona, D., Rappuoli, R. and Montecucco, C. (1989). Histidine 21 is at the NAD$^+$ binding site of diphtheria toxin. *J. Biol. Chem.* **264**, 12385–8.

Pappenheimer, A.M. Jr. (1977). Diphtheria toxin. *Ann. Rev. Biochem.* **46**, 69–94.

Pappenheimer, A.M. Jr. (1984). Diphtheria. In: *Bacterial Vaccines* (ed. R. Germanier), pp. 1–36. Academic Press, New York.

Pastan, I. and FitzGerald, D. (1989). *Pseudomonas* exotoxin: chimeric toxins. *J. Biol. Chem.* **264**, 15157–60.

Peterson, J.W. and Ochoa, L.G. (1989). Role of prostaglandins and cAMP in the secretory effects of cholera toxin. *Science* **245**, 857–9.

Pickett, C.L., Weinstein, D.L. and Holmes, R.K. (1987). Genetics of type IIa heat-labile enterotoxin of *Escherichia coli*: operon fusions, nucleotide sequence, and hybridization studies. *J. Bacteriol.* **169**, 5180–7.

Pizza, M., Bartoloni, A., Prugnola, A., Silvestri, S. and Rappuoli, R. (1988). Subunit S1 of pertussis toxin: mapping of the regions essential for ADP-ribosyltransferase activity. *Proc. Natl. Acad. Sci. USA* **85**, 7521–5.

Pizza, M., Covacci, A., Bartoloni, A., Perugini, M., Nencioni, L., De Magistris, M.T., Villa, L., Nucci, D., Manetti, R., Bugnoli, M., Giovannoni, F., Olivieri, R., Barbieri, J.T., Sato, H. and Rappuoli, R. (1989). Mutants of pertussis toxin suitable for vaccine development. *Science* **246**, 497–500.

Pizza, M., Bugnoli, M., Manetti, R., Covacci, A. and Rappuoli, R. (1990). The subunit S1 is important for pertussis toxin secretion. *J. Biol. Chem.* **265**, 17759–63.

Rappuoli, R. (1983). Isolation and characterization of *Corynebacterium diphtheriae* non tandem double lysogens hyperproducing CRM 197. *Appl. Environ. Microbiol.* **45**, 560–4.

Ribi, H.D., Ludwig, D.S., Mercer, K.L., Schoolnick, G.K. and Koenberg, R.D. (1988). Three dimensional structure of cholera toxin penetrating a lipid membrane. *Science* **239**, 1272–6.

Rose, J.M., Houston, C.W., Coppenhaver, D.H., Dixon, J.D. and Kurosky, A. (1989a). Purification and chemical characterization of a cholera toxin-cross-reactive cytolytic enterotoxin produced by a human isolate of *Aeromonas hydrophila*. *Infect. Immun.* **57**, 1165–9.

Rose, J.M., Houston, C.W. and Kurosky, A. (1989b). Bioactivity and immunological characterization of a cholera toxin-cross-reactive cytolytic enterotoxin from *Aeromonas hydrophila*. *Infect. Immun.* **57**, 1170–6.

Sandvig, K. and Olsnes, S. (1981). Rapid entry of nicked diphtheria toxin into cells at low pH. Characterization of the entry process and effects of low pH on the toxin molecule. *J. Biol. Chem.* **256**, 9068–76.

Schultz, A.J. and McCardell, B.A. (1988). DNA homology and immunological cross-reactivity between *Aeromonas hydrophila* cytotonic toxin and cholera toxin. *J. Clin. Microbiol.* **26**, 57–61.

Sekine, A., Fujiwara, M. and Narumiya, S. (1989). Asparagine residue in the *rho* gene product is the modification site for botulinum ADP-ribosyltransferase. *J. Biol. Chem.* **264**, 8602–5.

Sekura, R.D., Moss, J. and Vaughan, M. (1985). In: *Pertussis Toxin*. Academic Press, New York.

Siegall, C.B., Chaudhary, V.K., Fitzgerald, D.J. and Pastan, I. (1988). Cytotoxic activity of an interleukin 6-*Pseudomonas* exotoxin fusion protein on human myeloma cells. *Proc. Natl. Acad. Sci. USA* **85**, 9738–42.

Siegall, C.B., Chaudhary, V.K., FitzGerald, D.J. and Pastan, I. (1989). Functional analysis of domains II, Ib, and III of *Pseudomonas* exotoxin. *J. Biol. Chem.* **264**, 14256–61.

Spira, W.M. and Fedorka-Cray, P.J. (1984). Purification of enterotoxins from *Vibrio mimicus* that appear to be identical to cholera toxin. *Infect. Immun.* **45**, 679–84.

Stenmark, H., McGill, S., Olsnes, S. and Sandvig, K. (1989). Permeabilization of the plasma membrane by deletion mutants of diphtheria toxin. *EMBO J.* **8**, 2849–53.

Sternweis, P.C. and Robishaw, J.D. (1984). Isolation of two proteins with high affinity for guanine nucleotides from membranes of bovine brain. *J. Biol. Chem.* **259**, 13806–13.

Stryer, L. and Bourne, H.R. (1986). G. proteins: a family of signal transducers. *Ann. Rev. Cell Biol.* **2**, 391–419.

Tamura, M., Nogimori, K., Murai, S., Yajima, M., Ito, K., Katada, T., Ui, M. and Ishii, S. (1982). Subunit structure of islet-activating protein, pertussis toxin, in conformity with the A-B model. *Biochemistry* **21**, 5516–20.

Tamura, M., Nogimori, K., Yajima, M., Ase, K. and Ui, M. (1983). The role of the B-oligomer moiety of islet-activating protein, pertussis toxin, in development of the biological effects on intact cells. *J. Biol. Chem.* **258**, 6756–61.

Tanuma, S. (1990). Physiological role of endogenous mono(ADP-ribosyl)ation of G proteins. In: *Bacterial Protein Toxins* (eds. R. Rappuoli and R. Gross), pp. 211–18. Gustav Fischer Verlag, Stuttgart.

Tanuma, S. and Endo, H. (1989). Mono (ADP-ribosyl)ation of Gi by eukaryotic cysteine specific mono (ADP-ribosyl)transferase attenuates inhibition of adenylate cyclase by epinephrine. *Biochim. Biophys. Acta* **1010**, 246–9.

Tanuma, S., Kawashima, and Endo, H. (1987a). An NAD: cysteine ADP-ribosyltransferase is present in human erythrocytes. *J. Biochem.* **101**, 819–22.

Tanuma, S., Kawashima, K. and Endo, H. (1987b). Eukaryotic mono(ADP-ribosyl)transferase that ADP-ribosylates GTP-binding regulatory Gi protein. *J. Biol. Chem.* **263**, 5485–9.

Tweten, R.K., Barbieri, J.T. and Collier, R.J. (1985). Diphtheria toxin. Effect of substituting aspartic acid for glutamic acid 148 on ADP-ribosyltransferase activity. *J. Biol. Chem.* **260**, 10392–4.

Uchida, T., Gill, D.M. and Pappenheimer, A.M. Jr. (1971). Mutation in the structural gene for diphtheria toxin carried by temperate phage β. *Nature New Biol.* **233**, 8–11.

Uchida, T., Pappenheimer, A.M. Jr. and Greany, R. (1973a). Diphtheria toxin and related proteins: Isolation and properties of mutant proteins serologically related to diphtheria toxin. *J. Biol. Chem.* **248**, 3838–44.

Uchida, T., Pappenheimer, A.M. Jr. and Harper, A.A. (1973b). Kinetic studies on intoxication of HeLa cells by diphtheria toxin and related proteins. *J. Biol. Chem.* **248**, 3845–50.

Uchida, T., Pappenheimer, A.M. Jr. and Harper, A.A. (1973c). Reconstitution of hybrid 'diphtheria toxin' from nontoxic mutant proteins. *J. Biol. Chem.* **248**, 3851–4.

Ueda, K. and Hayaishi, O. (1985). ADP-ribosylation. *Ann. Rev. Biochem.* **54**, 73–100.

Van Dop, C., Tsubokawa, M., Bournet, H.R. and Ramachandran, J. (1984). Amino acid sequence of retinal transducin at the site ADP-ribosylated by cholera toxin. *J. Biol. Chem.* **259**, 696–8.

Van Heyningen, S. (1985). Cholera and related toxins. In: *Molecular Medicine* (ed. A.D.B. Malcolm), pp. 1–16, IRL Press, Oxford.

West, R.E., Jr., Moss, J., Vaughan, M., Liu, T. and Liu, T.-Yung. (1985). Pertussis toxin-catalyzed ADP-ribosylation of transducin. *J. Biol. Chem.* **260**, 14428–30.

Williams, D.P., Parker, K., Bacha, P. Bishai, W., Borowski, M., Genbauffe, F., Strom, T.B. and Murphy, J.R. (1987). Diphtheria toxin receptor binding domain substitution with interleukin-2: genetic construction and properties of a diphtheria toxin-related interleukin-2 fusion protein. *Protein Engng* **1**, 493–8.

Witvliet, M.H., Burns, D.L., Brennan, M.J., Poolman, J.T. and Manclark, C.R. (1989). Binding of pertussis toxin to eucaryotic cells and glycoproteins. *Infect. Immun.* **57**, 3324–30.

Woods, D.E. and Que, J.U. (1987). Purification of *Pseudomonas aeruginosa* exoenzyme S. Alteration of pulmonary structure by *Pseudomonas aeruginosa* exoenzyme S. *Infect. Immun.* **55**, 579–86.

Woods, D.E., Hwang, W.S., Shahzabadi, M.S. and Que, J.V. (1988). Alteration of pulmonary structure by *Pseudomonas aeruginosa* exoenzyme S. *J. Med. Microbiol.* **26**, 133–41.

Woolley, P. and Clark, B.F.C. (1989). Homologies in the structures of G-binding proteins: an analysis based on elongation factor EF-TU. *Biotechnology* **7**, 913–20.

Yamaizumi, M., Mekada, E., Uchida, T. and Okada, Y. (1978). One molecule of diphtheria toxin fragment A introduced into a cell can kill the cell. *Cell* **15**, 245–50.

2

Modulation of Cell Functions by ADP-Ribosylating Bacterial Toxins

Patrice Boquet[1] and D. Michael Gill[2]†

[1] Bacterial Antigens Unit (URA CNRS 0557), Institut Pasteur, 75724 Paris Cédex 15, France
[2] Department of Molecular Biology and Microbiology, Tufts University, 136 Harrison Avenue, Boston, MA 21110, USA
† Deceased

Introduction

Bacterial protein toxins can be classified into three types according to their mode of action on animal cells. Type I act by binding to a cell surface protein and induce on this molecule a transmembrane signal somewhat analogous to growth factors (Yarden and Ullrich, 1988). *E. coli* thermostable toxin (ST) (Greenberg and Guerrant, 1986) or toxic shock syndrome toxin 1 from *Staphylococcus aureus* (see Chapter 17) act by this kind of mechanism. Type II toxins act directly on the cell membrane either by channel formation or by disruption of the lipid bilayer (see Menestrina, Chapter 10). Toxins of type III act by translocating an enzymatic component inside the cytosol which modifies an 'acceptor' molecule by a post-translational reaction. ADP-ribosylation is such a post-translational mechanism and was discovered some 20 years ago for diphtheria toxin (Honjo *et al.*, 1968; Gill *et al.*, 1969). A second toxin-induced modification (*N*-glycosidase activity) was shown for either the plant toxin ricin (Endo *et al.*, 1987) or for shiga and shiga-like toxins (Endo *et al.*, 1988).

The purpose of this chapter is to review the effects induced in cell regulation by ADP-ribosylating toxins, with special emphasis on newly discovered bacterial molecules which carry this enzymatic activity, such as *Clostridium botulinum* C2 and C2-like toxins and *Clostridium botulinum* exoenzyme C3.

The characteristic of a bacterial toxin is its extraordinary potency. Only one molecule of diphtheria toxin fragment A inside a cell can kill it (Yamaizumi *et al.*, 1978). The enzymatic essence of diphtheria toxin was first suspected by Roux and Yersin in 1888, who postulated that diphtheria toxin was a 'diastase' (the French name at that time for enzyme). In the 1960s NAD was found to be an essential cofactor for diphtheria toxin to inhibit incorporation of amino acids into trichloroacetic acid-precipitable proteins in mammalian cell extracts (Collier and Pappenheimer, 1964). Then the inhibitory activity of diphtheria toxin on protein synthesis was found to be due to specific inactivation of elongation factor 2 (EF2) (Collier, 1968). Finally, it was concluded that diphtheria toxin in the presence of NAD catalysed the conversion of EF2 into its inactive ADP-ribose derivative with release of free nicotinamide (Honjo *et al.*, 1968; Gill *et al.*, 1969).

$$EF_2 + NAD \underset{\underset{\substack{diphtheria \\ toxin}}{2}}{\overset{1}{\rightleftharpoons}} EF2.ADPR + N^+ + H^+$$

A few years later, with the introduction of polyacrylamide gel electrophoresis (PAGE), the structure of diphtheria toxin was examined and was shown to consist of a unique polypeptide chain of about 60 kDa which was cleaved (nicked) by a serine protease into two subunits still held by a disulphide bridge (Gill and Pappenheimer, 1971; Gill and Dinius, 1971; Drazin et al., 1971). The 20-kDa N-terminus chain (fragment A) was shown to contain all the enzymatic activity (Gill and Pappenheimer, 1971; Drazin et al., 1971).

In 1990 ten bacterial toxins or bacterial exoenzymes were found to carry the enzymatic activity of ADP-ribosylation, and probably other microbial molecules will be found in the future to have the same action. One very puzzling observation is the apparent fit between ADP-ribosylating toxins (or exoenzymes) and GTP-binding proteins (also called G proteins) which are involved in a variety of diverse metabolic processes of central importance for eukaryotic organisms. These G proteins serve as initiation and elongation factors in protein synthesis, they comprise a large family of regulatory proteins that mediate transmembrane signalling and presumably also are involved in regulating growth, cytoskeleton assembly and intracellular transport. Indeed all the ADP-ribosylating toxins known to date except one (the C. botulinum C2 and C2-like toxins family) act on G proteins that have their function altered on ADP-ribosylation.

The activity of ADP-ribosylating toxins provides a useful tool to investigate the metabolic pathway controlled by certain G proteins. We have therefore constructed this review not on the description of each ADP-ribosylating toxin but instead according to the size of the G proteins which are their targets. The G protein family can be divided into three groups according to their molecular functions (Fig. 3A). The first one is represented by the ribosome interacting molecules; the second encompasses the 39–46 kDa molecules of regulatory elements (heteromeric) which modulate the transmembrane signalling; and the third group is the ras superfamily of 21–25 kDa proteins (Chardin, 1988). In a separate section we have examined the activity of the C2 toxin family which ADP-ribosylates G-actin (Ohishi and Tsuyama 1986; Aktories et al., 1986), an ATP-binding molecule, and which represents an 'odd' case. However, we must keep in mind that the protein which is the target for the C2

toxin family is nonetheless a nucleotide-binding molecule. Finally, we have discussed several hypotheses which could explain why G proteins are the preferred substrates for ADP-ribosylating toxins.

ADP-ribosylation of a large GTP-binding protein

Two bacterial toxins are known to modify by ADP-ribosylation a protein of 96 kDa which binds and hydrolyses GTP. This protein is involved in the mechanism of protein synthesis and is referred to as EF2 (elongation factor 2) (Weissbach and Ochoa, 1976). The two toxins which act on EF2 are diphtheria toxin (Honjo et al., 1968; Gill et al., 1969) and Pseudomonas aeruginosa exotoxin A (Iglewski and Kabat, 1975). They both block protein synthesis of their target cells.

The reactions involved in the translation of mRNA into proteins at the ribosomal level are divided into two steps. Initiation, which results in the formation of 40S and 80S ribosomal complexes, is followed by elongation. The first reaction of elongation is the EF1-dependent binding of aminoacyl tRNA to the ribosomal A site (Arlinghaus et al., 1963). EFtu is the prokaryotic analogue of EF1. For several years and until very recently, EFtu was the only GTP-binding protein whose GTP-binding structure had been elucidated at high resolution (Lacour et al., 1985; Jurnak, 1985). A great interest in EFtu has arisen since it has been found that proteins controlling signal transmission in such different functions as cell proliferation, hormone response, neurotransmission and protein synthesis share common features both structurally and functionally (Bourne, 1986; Gilman, 1987).

No bacterial toxin up to now is known to ADP-ribosylate EF1.

The second reaction of elongation is translocation, which involves the motion of the messenger RNA of one codon and transfer of the aminoacyl tRNA from the ribosomal (A) site of the peptidyl (P) site. The mechanism of translocation is poorly understood and probably very complex. A soluble factor EF2 and GTP are required for translocation. EF2-GTP binds on the same ribosome site as EF1, the binding of EF1 and EF2 being mutually exclusive. EFG is the factor corresponding to the eukaryotic EF2 in prokaryotes. This molecule catalyses the transfer of peptidyl tRNA from the ribosomal A site to the P site. EF2 association with the ribosome is tightly coupled to GTP binding. The release of EF2 from the ribosome is due to GTP hydrolysis with the loss of the deacylated tRNA from the P site.

EF2 of rat liver has a molecular weight of 96 500 (Raeburn et al., 1971). The most sensitive assay for EF2 is based on its specific ADP-ribosylation by diphtheria toxin (Gill and Dinius, 1973). ADP-ribosylation of EF2 by diphtheria or *Pseudomonas aeruginosa* toxins occurs at a unique amino acid. This amino acid in EF2 is a post-translationally modified histidine 2-(3-carboxyamino-3-(trimethyl-ammonio)propyl histidine) which has been given the name of diphthamide (Van Ness et al., 1980a).

EFG has no diphthamide and therefore is not a substrate for diphtheria or *Pseudomonas aeruginosa* A toxins. However, diphthamide is present in archebacteria elongation factor (Kessel and Klink, 1980; Pappenheimer et al., 1983). The post-translational modification of one EF2 histidine residue is a complex reaction which involves several separate steps. Using yeast grown on potential labelled amino acid precursors it has been shown that the four carbon sidechains attached to the imidazole ring of histidine derive from methionine (Dunlop and Bodley, 1983). Two types of CHO-K1 cells (Chinese hamster ovary cells) isolated after mutagenesis were found resistant to both diphtheria and *Pseudomonas aeruginosa* A toxins. Both lack the side carbon chain attached to histidine. By somatic hybridization these two Mod-CHO-K1 cells could complement each other (Moehring et al., 1984) suggesting that at least two enzymes are involved in the reaction catalysing addition of the methionine backbone to the histidine imidazole ring. The three methyl groups of the trimethylammonio group are then added from 5-adenosyl methionine (Dunlop and Bodley, 1983) to form the deaminated form of diphthamine which has been called diphthine (Dunlop and Bodley, 1983). It is not clear if one or three different methyltransferases are involved in the process of trimethylation of the diphthine precursor (Moehring and Moehring, 1988). The final step of conversion of diphthine to diphthamide is the

addition of amide to the carboxyamide group which renders EF2 ADP-ribosylatable by diphtheria or *Pseudomonas aeruginosa* A toxins (Moehring et al., 1984). Therefore, if three methyltransferases are required for the trimethylation step, at least six distinct enzymes are necessary for the post-translational modification of EF2. This complex post-translational modification seems only to occur in EF2 (Dunlop and Bodley, 1983).

Hamster and rat EF2 genes have been cloned and sequenced (Kohno et al., 1986). A peptide of 15 amino acids produced by tryptic digestion of ADP-ribosylated rat and bovine EF2, for which the sequence was known (Robinson et al., 1974; Brown and Bodley, 1979), allowed the localization of histidine 715 as the amino acid modified to diphthamide (Kohno et al., 1986). Histidine 715 is far from the GTP-binding domain of EF2 which is localized at the N-terminal end of EF2 (Kohno et al., 1986). The sole presence of the modified histidine to diphthamine does not allow diphtheria or *Pseudomonas aeruginosa* toxins to ADP-ribosylate this residue. Indeed a tryptic peptide derived from EF2 which contains the diphthamine was not ADP-ribosylated by diphtheria toxin (Van Ness et al., 1980b). Furthermore, the amino acids in the near vicinity of the histidine 715 of EF2 are important for the ADP-ribosylation reaction.

A variety of cells resistant to diphtheria or *Pseudomonas aeruginosa* A toxins, due to a direct mutation of EF2, have been isolated in several laboratories and these cells are named Tox R (the Mod−cell mutant appelations are for mutants of the histidine 715 post-translational pathway leading to diphthamide) (Gupta and Siminovitch, 1978, 1980; Moehring and Moehring, 1979; Moehring et al., 1979, 1980; Kohno et al., 1985). EF2 from Tox R mutants have been sequenced (Kohno and Uchida, 1987) and it has been found that most often a change of the arginine 717 to glycine by a G to A transition in the first nucleotide of this codon made EF2 resistant to ADP-ribosylation. Amino acid 713 with conversion to proline was also found (Kohno et al., 1987). No point mutation on codon 715 could be found (Kohno and Uchida 1987). By site-directed mutagenesis of codon 715 of EF2 it has been established that His[715] is essential for the biological activity of EF2 (Omura et al., 1989). The role of His[715] in the functionality of EF2 was shown by the inability of L cells transfected with Arg[715] EF2 DNA to be resistant to *Pseudomonas aeruginosa* exotoxin A (Omura et al., 1989). On the other hand, L cells transfected with Gly[717] EF2 DNA or Pro[713] EF2 DNA did render them resistant to *Pseudomonas aeruginosa* exotoxin A (Omura et al., 1989). These results indicate that Arg[715] EF2 was inactive whereas the toxin-resistant variant (Gly[717] or Pro[713]) could replace the wild type of EF2.

ADP-ribosylated EF2 is still able to bind GTP and to form a complex with the ribosome although with 50% less efficiency than the non-modified factor (Nygard and Nilsson, 1985). On the other hand, Arg[715] EF2 seems to bind to the ribosome with the same efficiency as EF2 diphthamide 715 but does not allow translocation (Omura et al., 1989). Therefore, ADP-ribosylation of the diphthamide residue 715 in EF2 provokes a more important modification than transition of His[715] to Arg[715] since both binding to ribosome and translocation ability of EF2 are altered.

The ADP-ribosylation of EF2 does not modify the binding of GTP to this factor (Nilsson and Nygard, 1985) but affects its GTPase activity which is linked to the translocase activity (Bermek, 1976). ADP-ribosylation has no direct effect on the GTP hydrolysis domain of EF2 but rather blocks EF2 function, making the ribosome unable to induce EF2 GTPase activity. The block in protein synthesis induced by ADP-ribosylation of EF2 by *Pseudomonas aeruginosa* or diphtheria toxin is amplified by the fact that the number of EF2 molecules present in the cytosol is roughly equal to the amount of ribosomes (1.1–1.5 molecules of EF2 per ribosome) (Gill and Dinius, 1973). Therefore, ADP-ribosylation of EF2 will reduce first the amount of factor for translocation, leading to a reduced number of ribosomes in activity, and when the pool of ADPR-EF2 is large enough to compete efficiently with native EF2, it definitively halts protein synthesis.

Two important applications of the studies on the mode of action of both diphtheria and *Pseudomonas aeruginosa* A toxins are (i) the generation of chimeric toxins in which the binding subunits of these toxins have been modified to recognize specific membrane receptors and thus can kill certain types of cells such as melanoma (Murphy et al., 1986), cells harbouring high numbers of IL2 receptor (Waltz et al., 1989) or cells expressing the gp120 surface protein of the AIDS virus (Chaudhary et al., 1988); and (ii) the use of the ADP-ribosylating fragment of diphtheria toxin in developmental biology studies. In a very elegant piece of work the diphtheria toxin A gene was put under the control of a tissue-specific promoter (elastase promoter) and injected into fertilized egg pronuclei in order to make transgenic mice. It was shown that transgenic mouse embryos displayed selective destruction of the pancreas when the elastase gene was turned on during development (Palmiter et al., 1987). Fragment A thus can be used to selectively destroy certain cell populations during embryogenesis when controlled by a tissue-specific promoter and can be utilized to study their influence on the development of neighbouring tissues.

ADP-ribosylation of heterotrimeric GTP-binding proteins

This family of proteins, which bind and hydrolyse GTP, serve as intermediaries in transmembrane signalling pathways. They consist of three proteins: receptors, G protein and effector (Gilman, 1987). Many membrane receptors are supposed to mediate their effect, when a ligand is bound, via a large G protein. The best characterized of these receptors are those for the β-adrenergic agonists and light-activated rhodopsin. The effectors known up to now, which are modulated by large G proteins, are adenylate cyclase, phosphodiesterase, potassium channel in heart, phospholipase C and A2 in certain cells and calcium channels. These G proteins are heterotrimers associated with the inner face of plasma membrane. The subunits have been named α, β and γ. How do the large G proteins work? Using the non-hydrolysable GTP analogue GTPγs it has been shown that GTPγs promotes G protein subunit dissociation by binding to the α molecule (Sternweiss et al., 1981; Hanski and Gilman, 1982; Bokoch et al., 1983; Codina et al., 1983; Northup et al., 1983).

$$G\alpha\beta\gamma + GTP\gamma s \rightleftharpoons G\alpha GTP\gamma s + \beta\gamma$$

The βγ complex increases the dissociation of GTPγs to the Gα subunit (Higashijima et al., 1987).

The natural ligand of the Gα subunit is of course

GTP. $G\alpha\beta\gamma$ has a low GTPase activity since it has a low affinity for GTP and exists mainly as GDP-bound form. Interaction of the receptor (R) with its ligand (L) drives the dissociation of GDP from G-GDP and allows the binding of GTP to the $G\alpha$ subunit (Brandt and Ross, 1986), inducing therefore the GTPase activity. One receptor–ligand complex (RL) can interact within a few seconds with a dozen molecules of $G\alpha\beta\gamma$ (Pedersen and Ross, 1982; Hekman et al., 1984). The $\beta\gamma$ subunits associated with the $G\alpha$ subunit are necessary for exchange of the GDP bound to GTP (Asano and Ross, 1984; Brandt and Ross, 1985).

Four different α subunits have been characterized, cloned and sequenced and are involved in distinct regulatory pathways: α_s activates adenylate cyclase and calcium channels, it has an M_r of 46 000 (Robishaw et al., 1986). Recently, an α_s (.olf.) specific for regulation of the adenylate cyclase involved in the olfaction process has been described (Reed, 1990). The α_i, which was first thought to exclusively inhibit adenylate cyclase, now appears very ubiquitous (α_{i1}, α_{i2}, α_{i3}) and also activates K^+ channels in heart, and is involved in certain cases in phospholipase C and A2 regulation (Neer and Clapham, 1988). It has an M_r of 40 000 (Nukada et al., 1986). Transducin (α_t) activates the photosensitive cyclic GMP-specific phosphodiesterase of retinal rod cells (α_{t1}) or of retinal cone cells (α_{t2}). Transducin α is a molecule of M_r 40 000 (Tanabe et al., 1985; Yatsunami and Khorana, 1985; Medynski et al., 1985). Finally, α_o was found in brain where it possibly activates phospholipase C and inactivates calcium channels (Neer and Clapham, 1988). This molecule has an M_r of 39 000 (Sternweiss and Robishaw, 1984) (brain cells contain up to 5% of α_o).

The β subunits (M_r 36 000) of G_i, G_s, G_o and transducin seem similar (Gilman, 1987). The γ subunit (M_r 8 400) of G_i, G_s, G_o and G_t could be heterogeneous (Gilman, 1987). α_s can be ADP-ribosylated by cholera toxin (Gill and Meren, 1978; Cassel and Pfeuffer, 1978) whereas α_i is ADP-ribosylated by pertussis toxin (Katada and Ui, 1982). α_o is also a substrate for pertussis toxin (Sternweiss and Robishaw, 1984). α_t can be ADP-ribosylated both by cholera and pertussis toxins at different residues (Van Dop et al., 1984). Once the α is associated with GTP, the complex $G\alpha\beta\gamma$ is broken and, as shown in the equation above,

αGTP is released free of $G\beta\gamma$. Now α, in its active conformation, can interact with the effector (adenylate cyclase, phosphodiesterase, etc.). The GTPase activity of α hydrolyses GTP into GDP. When α has hydrolysed the bound GTP it returns in the GDP form and reassociates with $G\beta\gamma$ and can interact back with the receptor–ligand complex (RL).

Cholera toxin ADP-ribosylates α_s or α_t at arginine 201 (Yatsunami and Khorana, 1985) and blocks the α_s or α_t GTPase activity (Cassel and Selinger, 1977; Abood et al., 1982). ADP-ribosylation of α_s decreases, therefore, the affinity of α_s for $\beta\gamma$ and α_s remains thus associated with GTP in interaction with the effector, activating it permanently (leading to elevation of cyclic AMP). However, cholera toxin cannot act directly on the $G_s\alpha\beta\gamma$ complex to ADP-ribosylate α_s. Indeed, another factor found either in solution in the cytosol (CF for cytosolic factor) (Enomoto and Gill, 1980) or membrane bound (ARF for ADP-ribosylating factor) (Schleifer et al., 1982) is required for transfer of the ADPR moiety of NAD to the Arg^{201} of α_s by cholera toxin. ARF is a small GTP-binding protein (20 kDa) (Kahn and Gilman, 1984a) distinct from ras. ARF must be liganded with GTP to interact with cholera toxin and to allow ADP-ribosylation of α_s. ARF does not interact with α_s but forms a complex with the cholera toxin enzymic subunit and α_s (Gill and Coburn, 1987). ARF is most abundant in neural tissue (1–2% of cellular protein) and it acts catalytically in the ADP-ribosylation process. As for the small G protein, ARF has very low GTPase activity.

What is the role of ARF beside being necessary for cholera toxin to ADP-ribosylate α_s? Recent observations suggest that a membrane is the site of action of ARF (Kahn et al., 1988; Botstein et al., 1988). The gene of ARF has been cloned in a mammalian cell line (Sewell and Kahn, 1988) and yeast (Botstein et al., 1988). Disruption of ARF genes in yeast is lethal (Botstein et al., 1988). Yeast ARF protein has 59% homology with the yeast CIN4 protein (another small G protein) which is involved in microtubule function during chromosome segregation in mitosis (Botstein et al., 1988). Recently, ARF has been localized in the Golgi apparatus (Stearns et al., 1990).

Pertussis toxin ADP-ribosylates α_i, α_o and α_t

without requiring an additional ARF (Gilman, 1987). The site of ADP-ribosylation is a cysteine residue located four amino acids before the end of the α_i, α_o or α_t (Hurley *et al.*, 1984a) (see Fig. 1). The C-terminus of α_i and α_t interacts with the receptor (R) molecule and ADP-ribosylation by pertussis toxin blocks the binding of G_i or G_t with

R (Van Dop *et al.*, 1984) leading to the loss of inhibitory control of adenylate cyclase or blocking the photon-induced signal transmitted from rhodopsin to cGMP phosphodiesterase. ADP-ribosylation of isolated α_i, α_o or α_t by pertussis toxin is not possible (Neer *et al.*, 1984; Tsai *et al.*, 1984; Watkins *et al.*, 1985; Huff and Neer, 1986). The

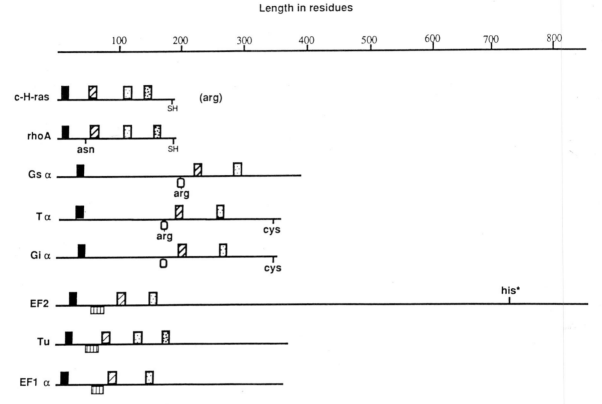

Figure I. Features of GTP-binding proteins: residues that become ADP-ribosylated bear no evident relationship to the sequences involved in GTP-binding or to other regions of homology. Residues that are ADP-ribosylated are marked: exoenzyme S ADP-ribosylates an unidentified arginine in ras proteins and other small G proteins (Coburn *et al.*, 1989b). C3 labels Asn[41] in rho. Cholera toxin, and the heat-labile enterotoxins (LT) of *E. coli* and other enterics, label Arg[201] of G_s and the equivalent arginine of transducin. Pertussis toxin ADP-ribosylates a cysteine near the C-terminus of several G proteins, including α_{i1}, α_{i2}, α_{i3}, α_o, and transducin α. The target for diphtheria toxin and *Pseudomonas* exotoxin A is a modified His[715], diphthamide. Tu and yeast elongation factor I α subunit (Nagata *et al.*, 1984), which are not targets for any known toxins, are included for comparison. Thus the top two sequences are from the p21-ras-related small G protein family, the next three from the 40-kDa heterotrimeric G protein group, and the bottom three are involved in ribosomal protein synthesis.

The shadings denote blocks of similar sequence. The solid-filled blocks represent a region involved in binding the phosphates of GTP: GXXXXGKS (in ras: GAGGVGKS, G protein α subunit GAGESGKS, ribosomal factors consensus GHVDHGKT). Diagonal shadings indicate regions involved in binding the α phosphate of GTP and possibly in a conformational switch between on and off states (Holbrook and Kim, 1989). Ras: DTAGQE, G protein α subunits DVGGQR, ribosomal factors consensus: DXPGHXD. Since the latter similarity is weak, it is shown with a lighter shading. Residues in the stippled regions bind the guanine ring of GTP. Light stippling: typically NKXD. Heavy stipping: ETSAKT--GV---F in ras and rho, and the weakly similar -TCA-D--NV---F in G proteins. Corresponding regions of the ribosomal factors show even fainter similarity. Vertically-hatched boxes represent an additional region of similarity among the ribosomal elongation factors, consensus EXERGITI (Kohno *et al.*, 1986).

αβγ association is required for ADP-ribosylation of α and proteolysis of the first N-terminal residues of α_i prevents its association with βγ and therefore inhibits ADP-ribosylation of α_i by pertussis toxin (Neer et al., 1988).

To stress the importance of cholera and pertussis toxins as tools to study cell biology, ADP-ribosylation by these molecules is one of the criteria (among five others) which has been proposed to define the involvement of a heterotrimeric Gα protein in the control of a regulatory pathway (Stryer and Bourne, 1986).

One final important question is the ability of G proteins to hydrolyse rapidly GTP to GDP. Indeed in EF2 the GTPase activity is induced by ribosome (Weissbach and Ochoa, 1976). The small G proteins, such as a p21ras, have a very low GTPase activity. A special protein called GTPase-activating protein (GAP) augments the GTPase activity of ras by 100 times (Trahey and McCormick, 1987). ADP-ribosylation of Arg^{201} in α_s blocks the GTPase activity of this molecule, leaving it 'active' (GTP-bound form). The effect of cholera toxin on α_s is thus reminiscent of the events that convert *ras* genes to oncogenes. Indeed mutations of *p21ras* which 'activate' it and render it transforming for NIH/3T3 cells are found in the protein domain required for the activity of the GAP molecule (effector domain) (Adari et al., 1988). The idea that a GAP-like domain could be inserted in α_s proteins has been studied (Landis et al., 1989). This hypothesis was supported by comparison of the α_s amino acid sequence around arginine 201 and that of the C-terminal part of p21ras GAP (which contains the GTPase activity) (Landis et al., 1989). A certain homology was observed, and to stress the analogy with p21ras it was found that mutation of α_s arginine 201 into cysteine was the cause of the transformation of secreting human pituitary cells (Landis et al., 1989).

ADP-ribosylation of small G proteins

Besides the heterotrimeric G proteins that we have described in the preceding section, a family of smaller molecules which have the ability to bind GTP is known. The prototype of these small G proteins is p21ras (Barbacid, 1987). As for the EF2 molecule or the heterotrimeric G proteins, small G proteins have four domains involved in the binding of GTP, making them about 30% homologous to each other. The small G proteins have a very low GTPase activity. The ras protein was discovered by its ability to transform NIH/3T3 cells (Ellis et al., 1981). The transforming potential of p21ras is due to point mutations on amino acids 12–13 or 61 which make them permanently GTP-bound (Barbacid, 1987). The biochemical role of p21ras is still unknown but the localization of this protein appears to be on the cytoplasmic face of the plasma membrane anchored to the lipid bilayer by palmitylation of its C-terminus (Hancock et al., 1989).

Recently the ras family has been expanded, first by the discovery of two ras-related proteins: the YPT gene product in yeast (Gallwitz et al., 1983) and the rho gene product in the marine snail *Aplysia* (Madaule and Axel, 1985). Secondly, by noticing that p21ras, YPT and rho had a conserved sequence (Asp–Thr–Ala–Gly–Gln–Gly) an oligonucleotide strategy was designed to screen various cDNA libraries and this work led to the discovery of many other members of the p21ras family (Chardin and Tavitian, 1986; Chardin et al., 1988). The ras superfamily of p21–p24 G proteins can be subdivided into three groups (Chardin, 1988), namely ras, rho and rab (Fig. 3B). As we already pointed out, the exact role of ras is unknown in higher eukaryotic cells. In yeast (*Saccharomyces cerevisiae*) p21ras can be substituted for the p35 yeast protein RAS1 which is involved in the control of adenylate cyclase activity (Toda et al., 1985). The YPT protein seems to be involved in the control of cell secretion. Indeed the YPT gene product has been shown to be localized in the Golgi apparatus (Segev et al., 1988) and one other member of the YPT protein family, the yeast SEC4 gene product, has been shown to play a crucial role in post-Golgi vesicle fusion with the plasma membrane (Salminen and Novick, 1987; Goud et al., 1988).

One characteristic of these small G proteins is their C-terminus, which binds lipid membrane in different ways (Hancock et al., 1989). Attachment to a membrane seems to be necessary for the small G proteins to express their activity. A model to explain the mode of action of the small G proteins has been proposed (Bourne, 1988). Basically, the small G protein is associated with GDP. When it binds to a lipid membrane it can associate (or be in contact) with a protein whose role is to exchange the bound GDP for GTP (exchange protein). When the GTP is bound a change of conformation of the small G protein structure is induced (Pai *et al.*, 1989) and renders it 'active'. The small G proteins in active conformation either 'turn on' the activity of a membrane protein or possibly a cytosolic one (effector).

The effector(s) of the activated small G protein is 'turned on' only when the complex G protein–GTP–effector exists. Since the small G molecules have a very low GTPase activity they must encounter a GTPase-activating protein (GAP) to hydrolyse the bound GTP and become changed to bound GDP state. In the bound GDP form the complex G protein–GDP–effector is thought to fall apart and no longer give a signal. Two GAP proteins have been isolated, one for ras (Trahey and McCormick, 1987) and one for rho (Garrett *et al.*, 1989).

It was reported first that *C. botulinum* D neurotoxin could ADP-ribosylate a 21 000-Da membrane protein of bovine adrenal gland tissue (Ohashi and Narumiya, 1987). The neurotoxin preparation used in this work was obtained from a Japanese Chemical Company (Wako) and was described as homogeneous by PAGE analysis (Ohashi and Narumiya, 1987). A large amount of *C. botulinum* D neurotoxin was, however, required for ADP-ribosylation (20 µg in a volume of 200 µl) (Ohashi and Narumiya, 1987). It was shown subsequently that in fact the ADP-ribosyltransferase activity on the 21 000-Da cellular protein was not the neurotoxin but an exoenzyme of molecular weight 26 000 called exoenzyme C3 present in *C. botulinum* C and D serotypes (Rösener *et al.*, 1987; Aktories *et al.*, 1987, 1988a; Rubin *et al.*, 1988). The exoenzyme C3 ADP-ribosyltransferase was unrelated to either D or C1 neurotoxin by the fact that (i) highly purified neurotoxin did not exhibit the enzymatic activity, (ii) antibodies raised against exoenzyme C3 did not react by Western blot with the neurotoxins and did not neutralize their activity in mice (Aktories *et al.*, 1987; Rubin *et al.*, 1988). Recently the gene of exoenzyme C3 has been shown to be localized on the same phage as the neurotoxin coding DNA. The C3 gene was subsequently sequenced (Popoff *et al.*, 1990) and

it was shown that the C1 and D neurotoxins do not share homology with the C3 structural gene (Hauser *et al.*, 1990; Binz *et al.*, 1990). However, the present literature dealing with the ADP-ribosyltransferase activity found in *C. botulinum* C1 or D neurotoxin preparations is still very confused and some authors have even shown an immunological cross-reactivity between exoenzyme C3 and *C. botulinum* neurotoxins C1 or D (Toratani *et al.*, 1989).

It is clear that introducing exoenzyme C3 and therefore provoking the ADP-ribosylation of the 21 000-Da protein in PC12 cells did not block the release of norepinephrine induced by high concentration of K^+ (Rubin *et al.*, 1989). Therefore exoenzyme C3 does not act in the same way as *C. botulinum* C1 or D neurotoxins.

The target protein for exoenzyme C3 was found to be the rho protein, a small G protein (Kikuchi *et al.*, 1988; Narumiya *et al.*, 1988; Aktories *et al.*, 1989; Chardin *et al.*, 1989). Recently, possible second substrates for exoenzyme C3 have been found; they are small GTP-binding proteins that have been named rac (Didsbury *et al.*, 1989). The rac proteins (rac1 and rac2) are highly homologous with rho and are therefore members of the rho family.

The rho G proteins encompass three different molecules in large eukaryotic cells (rho A, B and C) (Chardin *et al.*, 1988; Yeramian *et al.*, 1987). In yeast two *RHO* genes have been isolated, *RHO*1 and *RHO*2 (Madaule *et al.*, 1987). A gene called *CDC42* is also a member of the yeast *RHO* family (Johnson and Pringle, 1990). The amino acid acceptor of ADP-ribose in rho induced by the action of exoenzyme C3 was shown not to be arginine, cysteine or a modified histidine (Aktories *et al.*, 1988b). An asparagine residue at position 41 of rhoA was found to be the amino acid acceptor of ADP-ribosylation (Sekine *et al.*, 1989). When rho is ADP-ribosylated it can still bind and hydrolyse GTP and its GAP protein can interact with ADPR–rhoA (Paterson *et al.*, 1990). Asparagine 41 in rho is in close proximity to the so-called effector domain which is involved in the recognition of effector molecules by small G proteins (the effector domain in ras is localized between amino acids 29 and 39). Therefore, a working hypothesis would be that upon ADP-ribosylation, the rho protein could not interact with its effector.

Exoenzyme C3 does not enter the cell with high efficiency since no B chain (such as component II of *C. botulinum* C2) was found which allows binding to the cell surface and transport across the membrane (Rubin *et al.*, 1988). Nonetheless by the technique of osmotic shock, C3 can be introduced into cultured cells (Rubin *et al.*, 1988); by this technique NIH/3T3 cells were found more sensitive to C3 than PC12, CV1, HEp2 or HeLa cells (Rubin *et al.*, 1988). When C3 was introduced into the cells about 90% of rho was ADP-ribosylated (Rubin *et al.*, 1988). NIH/3T3 cells treated with C3 rounded profoundly and became refractile with long processes attached to the plate (Rubin *et al.*, 1988). The first changes were visible after 1 h (Rubin *et al.*, 1988). Binucleate cells accumulated, indicating that cytokinesis was blocked (Rubin *et al.*, 1988).

Direct microinjection of C3 into *Xenopus laevis* oocytes did not provoke germinal vesicle break-down (GVBD) but induced germinal migration to the animal pole of the nucleus and potentiated progesterone activity on GVBD (Rubin *et al.*, 1988). In Vero cells, after treatment with C3 (in which it was shown that only rho was ADP-ribosylated and no ADP-ribosylation of G-actin by a possible C2 toxin contamination could be found) staining by the fluorescent dye NDB-phalloidin, which recognizes F-actin exclusively, showed a disappearance of stress fibre or bundle actin filament (Chardin *et al.*, 1989). These data indicate that rho is probably involved in the regulation of actin microfilament organization. Similar effects on cell cytoskeleton have been observed by direct microinjection of ADP-ribosylated rho into cells (Paterson *et al.*, 1990).

Yeast RHO1 protein can be ADP-ribosylated by exoenzyme C3 (Rubin *et al.*, 1988). Using this property the possible localization of the RHO1 protein in a cell compartment has been undertaken (McCaffrey *et al.*, unpublished results).

The particulate form of RHO1 in yeast is not associated with the plasma membrane, SEC4-bearing secretory vesicles, mitochondrion, endoplasmic reticulum or the vacuole, but with late Golgi elements and early post-Golgi vesicles (McCaffrey *et al.*, unpublished results). This is an unexpected result considering the effect of rho on the cytoskeleton. How can the Golgi apparatus and post-Golgi vesicles be involved in the control of microfilament assembly/disassembly? One hypothesis would be that RHO1-GDP in the cytosol has to move to the Golgi membrane to become associated with a protein which exchanges GDP for GTP, making the RHO 'activated'. When bound with GTP it could leave the Golgi apparatus with early post-Golgi vesicles and stimulate an actin-binding protein which regulates the assembly of the microfilament (a severin-bundling protein such as gelsodin – see next section). Rho-GTP could block the severing activity of this particular actin-binding protein, leading to stabilization of the microfilaments. After ADP-ribosylation, the rho protein could not interact with its effector, leaving the actin-binding protein with full severing activity and able to disassemble the microfilaments. Although this is still a working hypothesis several experiments using mutants of the *rho* gene suggest that this small G protein could control actin assembly/disassembly by such a mechanism (Paterson *et al.*, 1990).

Exoenzyme S of *Pseudomonas aeruginosa* is an ADP-ribosyltransferase distinct from *Pseudomonas aeruginosa* toxin A. Exoenzyme S is able to ADP-ribosylate a large variety of proteins (Iglewski *et al.*, 1978). However, at low concentrations, exoenzyme S has some preferred substrates such as vimentin (an intermediary filament) (Coburn *et al.*, 1989a) and especially membrane-bound proteins of molecular weight 21000–25 000 which appear to be GTP-binding proteins (Coburn *et al.*, 1989). The small G protein p21ras is one of the preferred substrates for exoenzyme S (Coburn *et al.*, 1989a). Curiously, it has been shown that the recombinant p21ras produced in bacteria was unable to be ADP-ribosylated by Exo-S (Coburn and Gill, unpublished results). However, by adding a small amount of cytosol from eukaryotic cells the enzyme recovered its activity (Coburn and Gill, unpublished results). This result indicates that an ADP-ribosylating factor such as ARF for cholera toxin is required for exoenzyme S. Apparently this ADP-ribosylating factor is not a G protein (Coburn and Gill, unpublished results).

ADP-ribosylation of actin, 'the odd case'

Actin, a molecule of molecular weight 43 000, is the major protein component of eukaryotic cells

(5% of the total cell protein). Actin is involved in muscular power, cell form and certain other functions such as anchoring cells on extracellular matrices (focal adhesion). Actin molecules have the property of self-assembly, forming various structures (bundle, stress fibres, etc.). The assembly of actin monomers (called G-actin) into filaments (called F-actin) is controlled by proteins called 'acting-binding proteins' (Stossel et al., 1985). Actin is divided into two classes: skeletal actin and non-muscular actin. Skeletal actin is found in muscles and is involved in contraction. Non-muscular actin is found in all eukaryotic cells and is mainly localized in the peripheral cytoplasm (Ishikawa et al., 1969). Two important roles for non-muscular actin are formation of focal adhesion plaques by which a cell can bind to extracellular matrix (such as fibronectin) and cytokinesis. Assembly of acting monomers to form linear polymers (F-actin) which can be several micrometres in length is a reversible process. Actin monomers bind ATP and actin polymerization is associated with the conversion of actin-bound ATP to ADP–P. The rate and extent of actin polymerization depend greatly on whether ATP or ADP–P is bound to individual G-actin monomers (Carlier et al., 1984; Pollard, 1984).

Actin filaments are bipolar and have a 'barbed' (+) and a 'pointed' end (−) (Huxley, 1963). Actin monomers liganded with ATP are added preferentially at the barbed ends of the filament. ATP monomers at the barbed ends stabilize the growing part of the polymer (Carlier et al., 1984). Dissociation of the F-actin filament takes place when ATP is hydrolysed into ADP with formation of unstable G-actin-ADP. G-actin-ADP is mostly localized at the pointed end of the filament. Actin-ATP assembly at the barbed end, and disassembly at the pointed end of the microfilament is called 'end to tail polymerization' or 'treadmilling' (Neuhaus et al., 1983).

Actin-binding proteins are essential in in vivo modulation of the polymerization of the G-actin monomer to F-actin (Stossel et al., 1985). Actin-binding proteins (ABP) are of two types: (i) ABPs which sequester actin monomers do not associate with F-actin; their major effect is to delay the incorporation of monomeric G-actin into F-actin (prolifin is the major ABP-sequestering acid monomer); (ii) ABPs which block the ends of

acting filaments are called capping proteins. Most of them block the + end of the filaments. Gelsolin, villin and fragmin are the most well-known capping proteins (Stossel et al., 1985). The capping proteins have two activities: they promote nucleation of filaments and the breakage of long F-actin polymers in smaller ones (severing activity).

Clostridium botulinum C2 toxin was found to ADP-ribosylate arginine residues when polyarginine was used as an acceptor of ADP-ribosylation (Simpson, 1984). This toxin is composed of two non-covalently associated subunits: light chain or component I (M_r 45 000) and heavy chain or component II (M_r 88 000) (Ohishi et al., 1980). It is able to induce a particular cytopathic effect on cultured cells which consists in the profound rounding with progressive detachment of the plates (Ohishi et al., 1984). The light chain of C2 toxin was shown to contain all the enzymatic activity (Simpson, 1984). Actin is the target protein of C. botulinum C2 toxin ADP-ribosyltransferase and monomeric G-actin is the substrate not F-actin (Ohishi and Tsuyama, 1986; Aktories et al., 1986). Curiously it was found that skeletal actin was a poor substrate compared to the non-muscular form (Aktories et al., 1986). Arginine 177 is the site for ADP-ribosylation of G-actin induced by C2 toxin (Vandekerckhove et al., 1988). The poor activity of C2 toxin on skeletal actin is not caused by a change of the amino acid acceptor since there is an arginine at position 177 in skeletal actin (Vandekerckhove and Weber, 1979). Upon ADP-ribosylation by C2 toxin the F-actin content of cells disappeared, as evidenced by immunofluorescence studies using the specific F-acting fluorescent dye NBD phallicidin (Reuner et al., 1987). Direct measurements done on actin polymerization in vitro have shown that ADP-ribosylated G-actin behaves as a capping ABP (Wegner and Aktories, 1988).

Indeed ADP-ribosylated actin on arginine 177 is still able to be liganded with ATP and bind to the barbed end of F-actin (Wegner and Aktories, 1988), but has no nucleating activity such as other actin capping proteins (gelsolin or villin). ADP-ribosylation of actin by C2 toxin has a reduced ATPase activity (Geipel et al., 1989). ADPR-actin, which caps the barbed end of F-actin, blocks the polymerization while depolymerization can still happen at the pointed end. The result is the progressive disappearance of the microfilament network of the cell, causing rounding up since focal adhesion plaques are ruptured. When a high amount of G-actin is ADP-ribosylated, actin-binding proteins can no longer induce microfilament formation.

Besides the C2 toxin found in *C. botulinum* C and D strains other *Clostridium* species are able to produce actin ADP-ribosyltransferases. *C. perfringens* iota toxin has the same structure and ativity as C2 (Simpson *et al.*, 1987). However, iota toxin is able to ADP-ribosylate skeletal actin in addition to non-muscular actin (Schering *et al.*, 1988). Iota toxin does not cross-react immunologically with C2 toxin (both components I and II) and there is no possible complementation between isolated chains of C2 and iota (Popoff and Boquet, 1988; Popoff *et al.*, 1988). *C. spiroforme* toxin is also an ADP-ribosyltransferase directed toward G-actin (Popoff and Boquet, 1988; Simpson *et al.*, 1989; Popoff *et al.*, 1989). *C. spiroforme* is an iota-like toxin (Popoff and Boquet, 1988). Component I of *C. spiroforme* toxin can complement component II of iota toxin and vice versa (Popoff and Boquet, 1988). An isolated iota-like component has been found in one strain of *C. difficile* (Cd 196); no component II was found to be produced by this strain (Popoff *et al.*, 1988). This finding could indicate that the DNA coding for components I and II could be borne by non-contiguous genes.

Recently, ADP-ribosyltransferases able to modify G-actin have been found in one strain of *C. botulinum* A and in two strains of *C. botulinum* B (Popoff and Boquet, unpublished results). These new ADP ribosyltransferases did not cross-react immunologically with C2 component I, iota, spiroforme or the ADP-ribosyltransferase found in Cd 196, indicating that these enzymes probably belong to a third family (Popoff and Boquet, unpublished results).

The fact that actin ADP-ribosyltransferases are found in human *C. botulinum* pathogenic strains such as A and B raises the possibility that C2-like toxin could play a role in certain botulinum diseases. Finally, a new actin ADP-ribosyltransferase of smaller molecular weight (M_r 26 000) was found in a *C. limosum* strain (Popoff and Boquet, unpublished results).

Why are ADP-ribosylatable proteins related to each other?

In this final section we ask why so many of the toxin targets are of one type. This question is illustrated by Fig. 1: except for actin and some less-efficient substrates of exoenzyme S and cholera toxin, *all* of the ADPR substrates are GTP-binding proteins and have common amino acid sequences which imply a common ancestry. They are represented in all three major groups of GTP-binding proteins: protein synthesis factors, heterotrimeric ($\alpha\beta\gamma$) G proteins of around 40 kDA, and the ras family. The current list of substrates may be biased by active searching among the GTP-binding proteins but even so the tendency is beyond coincidence. It is not difficult to understand that GTP-binding proteins, operational at pressure points of cell physiology, make suitable targets for bacterial toxins, but many other proteins could be inactivated with equally devastating results so there must be an additional reason for the high frequency. One possibility is that the present pairs of toxins with GTP-binding proteins evolved from one ancestral pair, the partners of which radiated in parallel. An alternative possibility, convergence, would be reasonable as long as there existed some special attribute of the GTP-binding protein's structure that toxins could select. The former view seems more compatible with the existence of faint sequence similarities of one small region (Fig. 2) but in either case we want to identify the constant features of GTP-binding proteins which the toxins recognize.

The only common motif obvious in the linear sequences constitutes blocks of residues which make up the GTP-binding site itself (Fig. 1). The folding to make up the GTP site is probably similar in all cases (Holbrook and Kim, 1989). Is the common 3-dimensional GTP-binding site what the toxins see? If so, what nucleotide should it contain?

Distinction between target sequences and sequences that define GTP-binding proteins

ADP-ribosylation affects a variety of amino acids (Arg, Asn, Cys, His) located at quite different positions in the linear sequences (Fig. 1). That

```
Cholera A1      26:    G Q S E Y F D - - R G T Q M N I N L Y D H A R G T - - - - - - - Q T G F V R H D D G
Coli LT-I       26:    G H N E Y F D - - R G T Q M N I N L Y D H A R G T - - - - - - - Q T G F V R Y D D G
Coli LT-IIA     26:    G Q Q E A Y E - - R G T P I N I N L Y D H A R G T - - - - - - - A T G N T R Y N D G
Coli LT-IIB     26:    G Q D E A Y E - - R G T P I N I N L Y E H A R G T - - - - - - - V T G N T R Y N D G
Pertussis S1    27:    G N N D - - - - - - - - - - - N V L D H L T G - - - - - - R S C Q V G S S - - N S A

Diphtheria A     1:    G A D D V V D S S K S F V M E N F S S Y H - - G T K P G Y V D S I Q K G I Q K P K S G
Ps. aeruginosa A 425:  A H R Q L - E - E R G Y V - - - F V G Y H - - G T F L E A A Q S I V F G G V R A R S Q

Rh. rubrum draT 63:    G E A F Y K Y M I A M F G L D P E N N D H R P G - - - - - - - - - E G G A V R R F H A
```

Figure 2. Regions of the enzymically active portions of bacterial ADP-ribosyltransferases with similar sequences. Cholera toxin A1, the closely related heat-labile enterotoxins LT and LT-II, and pertussis toxin, ADP-ribosylate similar substrates and have substantially similar sequences in the region shown. Likewise, diphtheria toxin and *Pseudomonas aeruginosa* exotoxin A both ADP-ribosylate EF2. Their active domains are obviously related. Over the region of the active domains illustrated above, but nowhere else in these molecules, there are also certain positions where the same amino acid appears in one or more members of the cholera toxin group and one or both of the EF2 group. Such positions are indicated by bold print. This region is probably involved in NAD binding. The alignment is updated from an earlier version (Gill, 1988): the sequence of *E. coli* heat-labile enterotoxin (LT) has been revised (Yamamoto *et al.*, 1987) and the sequences of LT-II (Pickett *et al.*, 1989) added. The sequence of an ADP-ribosyltransferase encoded by the *draT* gene of *Rhodospirillum rubrum* is shown for comparison (Fitzmaurice *et al.*, 1989). It modifies dinitrogenase reductase.

would not be the case if the toxins modify residues that are involved in GTP binding. Indeed, if something about the GTP site is recognized, it must be recognized allosterically. We would have to suppose that a part of the toxin recognized the GTP site while the active site of the toxin ADP-ribosylated some residue elsewhere in the target molecule. It could be that the folded substrates all have their target residues in some special position *vis-à-vis* the GTP site which is not evident from the linear sequence, but we do not yet have enough crystal structures of the substrates to check this idea. Only ras has been solved (Tong *et al.*, 1989).

Several observations indicate that the toxins recognize more than the simple sequence around their ADPR sites. Denatured substrates do not work and even rather subtle conformational shifts of the G proteins seem to affect ADP-ribosylation dramatically. The abilities of G_s or G_i to serve as substrates for cholera or pertussis toxins are readily destroyed by chemical modifications which have no apparent effects on less stringent substrates (Gill and Woolkalis, 1985). $G_i\alpha$ has an apparent cholera toxin target sequence (RT*R*VKTT) yet G_i is not a substrate. Ligands such as $\beta\gamma$ and hormone receptors (or photolysed rhodopsin) affect ADP-ribosylation rates of the G proteins, probably by causing conformational shifts (see below).

GTP is not needed

It is not generally necessary to add GTP for ADP-ribosylation. It does not seem that the toxin needs to recognize a bound nucleotide directly or a binding site occupied by GTP.

This statement seems to contradict some published data, but we argue that the apparent benefits of GTP can be ascribed to stabilization of the substrates rather than to direct effects on ADP-ribosylation. There is also the special case of cholera toxin, which certainly does require GTP for its enzyme activity, but this GTP is needed to bind to the cofactor protein ARF, not to the substrates.

Thus, it seems that the substrate may have an empty GTP-binding site and still react. It is more difficult to tell what may be contained in the nucleotide site. To answer this we will examine how GTP, GDP and the hydrolysis-resistant GTP analogues GppNHp and GTPγs affect ADP-ribosylation rates.

Ras and ras-related proteins

Pseudomonas aeruginosa exoenzyme S ADP-ribosylates H-ras and other small G proteins equally well with no added nucleotide or with 1 mM GTP or

(A)

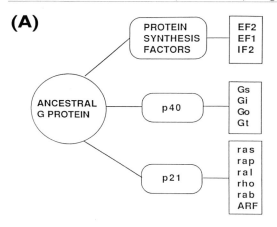

BACTERIAL G PROTEINS:

Iep, EFTu, EFG, Era, obg

(B)

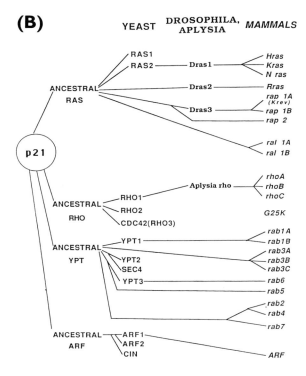

Figure 3. (A)The GTP-binding proteins family. (B) The p21ras proteins superfamily.

GTPγS. This remains true when the H-ras protein is purified by immunoprecipitation and any bound GTP is stable (not rapidly hydrolysed to GDP). Mutant forms of ras which are expected to be bound to GTP or GDP or no nucleotide or which are defective in GTP hydrolysis are all ADP-ribosylated well (J. Coburn and L. Feig, unpublished data).

The C3-catalysed labelling of rho proteins in tissue homogenates is little affected by adding extranucleotides. However, the extent of labelling of partially or wholly purified rho protein may be increased by GTP. For example, membrane-bound rho acquires a requirement for GTP when endogenous nucleotides are adsorbed out with charcoal (Ohasi and Narumiya, 1987). We have known for some time that guanylyl nucleotides stabilize rho proteins against heat and chemical denaturation (Rubin *et al.*, 1988), and, we suspect, against proteolysis, all of which destroy its substrate activity. Recent experiments of ours indicate that rho is quite unstable even at 30°C unless it is bound to GTP. If this stabilization is controlled for, it is clear that GTP has no additional effect on ADP-ribosylation *per se*.

EDTA (not EGTA) inhibits labelling of rho protein, and magnesium ions reverse this. Furthermore, rho is permanently inactivated as a C3 substrate more rapidly in the presence of EDTA than in the mere absence of GTP. Note that a magnesium ion is involved in the GTP binding (at least to related small G proteins), and so the addition of EDTA implies the loss of bound GTP. A simple interpretation is that bound magnesium, as well as bound guanylyl nucleotide, restrict rho denaturation.

Rho in PC12 membrances (4 mM Mg^{2+}) (Matsuoka *et al.*, 1987) and platelet membranes (Aktories and Frevert, 1987) is labelled less well when 10 or 20 μM GTPγS is present. Other nucleotides have less or no effect. It is not clear why such an effect of GTPγS has been demonstrated on soluble rho protein.

G protein substrates for pertussis toxin: G_i, G_o and transducin

Nucleotide effects on pertussis toxin catalytic rates must be interpreted with care because adenyl nucleotides (Lim *et al.*, 1985; Watkins *et al.*, 1985) and to some extent guanylyl nucleotides, directly bind and activate pertussis toxin by dissociating the S1 subunit from its complex. This effect is controlled for in the experiments discussed below.

Guanylyl nucleotides sometimes increase the

ADP-ribosylation of G_i. This was shown for GTP, GDP and GDPγS by Tsai *et al.* (1984) and Wong *et al.* (1985). However, it was recently shown that these nucleotides act by preventing denaturation of G_i (George Holz, unpublished), as already discussed for rho proteins.

Isolated α_i subunits will not react unless bound to $\beta\gamma$ subunits. This has been shown for (detergent-solubilized) G_i (Neer *et al.*, 1984; Tsai *et al.*, 1984; Mattera *et al.*, 1987), G_o (Huff and Neer, 1986) and soluble transducin (which is soluble without detergent) (Watkins *et al.*, 1985). Thus the $\alpha\beta\gamma$ forms are the actual substrates.

If we start with the $\alpha\beta\gamma$ complex the addition of a relatively high concentration of GTPγS reduces the ADP-ribosylation by pertussis toxin. This was shown for α_i in membranes (Wong *et al.*, 1985) and in solution (Huff and Neer, 1986). GTPγS is not tolerated in the case of transducin (Van Dop *et al.*, 1984) unless there is no activated rhodopsin present (Watkins *et al.*, 1985). Rhodopsin facilitates the entry of GTPγS into its binding site. The converse observation is that photolysed rhodopsin inhibits ADP-ribosylation of membrane transducin (Van Dop *et al.*, 1984), and even of G_i (Tsai *et al.*, 1984). With resolved components, the inhibition by photolysed rhodopsin requires GppNHp or GTPγS (Watkins *et al.*, 1985). The interpretation first suggested that GTPγS blocked ADP-ribosylation by causing the dissociation of α (GTPγS) from $\beta\gamma$ subunits necessary for labelling. The dissociation would certainly prevent ADP-ribosylation but Mattera *et al.* (1987) showed that GTPγS-liganded pure G_i could retain its heterotrimeric size even after it had lost its ability to serve as a substrate. Also fluoride (as AlF^{4-}) activates G proteins without destroying their substrate activity. Thus neither subunit dissociation nor activation *per se* seem to be responsible for the GTPγS effect. Mattera *et al.* (1987) propose that the GTPγS induces a shift in the conformation of α_i to a form that is unavailable to the toxin, active, and yet still binds $\beta\gamma$. It is indeed likely that G_i can adopt a variety of conformations, but the data do not demand such an explanation. It might simply be that the very presence of GTPγS in the nucleotide site prevents ADP-ribosylation.

G_s and transducin as cholera toxin substrates

Cholera toxin ADP-ribosylates arginine residues on a great variety of proteins, possibly most. What distinguishes G_s from the others is the relative ease with which ADP-ribosylation of G_s can be taken to completion. There are dozens of secondary substrates ADP-ribosylated at comparatively slow rates. Some GTP-binding proteins such as tubulin can be shown to be ADP-ribosylated by cholera toxin, especially when they are provided in high concentration, but their reaction is not obviously better than that of any random protein.

Guanylyl nucleotides are absolutely required for ADP-ribosylation by cholera toxin but the major effect is indirect. Cholera toxin subunit A, to be enzymically active, must interact with a GTP-liganded ras-like protein that is now generally called ARF (for ADP-ribosylation factor; Kahn and Gilman, 1984a). Brain membranes have much endogenous ARF and can be ADP-ribosylated directly, but most membranes have little ARF and are only labelled well in the presence of cytosol. The cytosolic factor, which we called CF (Enomoto and Gill, 1980; Woolkalis *et al.*, 1988) is a soluble form of ARF. In the presence of membranes and GTPγS or GppNHp, an activated species is formed on the membrane which we now interpret to be a membrane-bound ARF bound to a guanylyl nucleotide. Thus ARF(GTP) in nature, or ARF(GppNHp) experimentally, is the species that activates cholera toxin. If dissolved off membranes with detergent, ARF(GppNHp) still activates cholera toxin (Gill and Coburn, 1987).

We once believed that ARF interacted with the cyclase system (Gill and Meren, 1983) but we saw that this could not be so when we found that ARF increases the activity of cholera toxin towards all of its substrates, all in parallel (Gill and Coburn, 1987; cf. Kahn and Gilman, 1984a). Clearly, activated ARF interacts with cholera toxin, not with the cholera toxin substrates.

In a crude membrane system the guanylyl nucleotide that binds ARF and a guanylyl nucleotide that binds α_s can be distinguished on the basis of sensitivities to divalent cations, temperature, response to hormones, and by a different order of effectiveness of GTP and analogues (Gill and

Coburn, 1987). In particular, GppNHp is good for ARF, not so good for G_s; GTPγS is better for G_s. Also, although magnesium ions are needed for the binding of guanylyl nucleotides to $α_s$, the magnesium requirement for ARF is less demanding. Consequently, magnesium does not have to be supplied for ADP-ribosylation (e.g. Neer et al., 1987). However, sufficient EDTA can block ADP-ribosylation (Gill and Meren, 1983). This again shows that ADP-ribosylation of G_s does not require endogenous GTP.

When the effect of G nucleotides on ARF is discounted, it is possible to recognize a second effect of G nucleotides on the ADP-ribosylation rate. Non-hydrolysable analogues, especially GTPγS, inhibit labelling, particularly in the presence of a hormone to accelerate GTP for GDP exchange (J. Coburn and D.M. Gill, unpublished observations). Our interpretation is that, for G_s as for G_i, the insertion of a GTP analogue into the GTP-binding site prevents the $α_s$ from responding to cholera toxin. As expected, GTPγS plus isoproterenol inhibits the labelling of $α_s$ but not any secondary substrate of cholera toxin. Likewise, the ADP-ribosylation of purified G_s in detergent solution is completely inhibited by GTPγS alone. Fluoride does not block ADP-ribosylation although it activates G_s (Kahn and Gilman, 1984b). Thus again it is not evident that the effects of GTPγS are to be attributed to a conformation change in $α_s$ or to mere occupancy of the GTP site by the triphosphate analogue.

Transducin is another cholera toxin substrate (Fig. 1). Again, the cholera toxin reaction needs ARF and a supporting G nucleotide (Navon and Fung, 1984). Transducin labelling in rod outer segment is stimulated by light (Abood et al., 1982; Navon and Fung, 1984; Watkins et al., 1984; Vandenberg and Montal, 1984). Possible explanations are a conformation change in transducin when it binds photolysed rhodopsin, or the consequent loss of bound GDP. GTPγS reduces the labelling of illuminated membranes, recalling the inhibition effect of G_s labelling by GTPγS plus isoproterenol (Abood et al., 1982; Navon and Fung, 1984; Vandenberg and Montal, 1984). This effect applied to transducin but not to secondary toxin substrates in the same membranes.

It is clear that βγ subunits are needed for the ADP-ribosylation of transducin (Navon and Fung, 1984). It has been claimed that βγ are not needed for the ADP-ribosylation of G_s (Kahn and Gilman, 1984b), but ADP-ribosylation was not measured directly. Rather the conclusion rests on the change in adenyl cyclase activity consequent on ADP-ribosylation, which is subject to a variety of downstream influences.

Elongation factor 2

When GTP binds to EF2 a conformational change in the protein allows the complex to bind to a specific site on the ribosome. The GTP is split in a ribosome-dependent way, coupled to polypeptidyl translocation, presumably during or after the translocation event. EF2-GDP is released.

Some early experiments indicate that the binding of GTP in the absence of ribosomes reduces the ability of diphtheria toxin to ADP-ribosylate EF2 (Raeburn et al., 1968; Sperti et al., 1971). (Note the trivial exception: EF2 can bind non-specifically to ribosomes or to RNA where it cannot be ADP-ribosylated. GTP prevents this binding and thus may appear to increase ADP-ribosylation.) These experiments need to be repeated with pure components, with GTP analogues, and with Pseudomonas exotoxin A in parallel.

Van Ness et al. (1980b) found that the yeast tryptic fragment of EF2 containing diphthamide was not a substrate. Moreover, the fragment formed from previously ADP-ribosylated EF2 could not participate in the reverse reaction. Nilsson and Nygard (1985) have identified the tryptic fragment as the 34-kDa C-terminal piece, which evidently lacks the GTP-binding domain (see Fig. 1). This is certainly what we would expect if toxin binding to the GTP site were required for enzymic activity, but of course other explanations can be found.

Summary and conclusion

Many of the relevant experiments were performed before confounding influences, such as the nucleotide effects of the toxins, and on ARF, were known and understood. Consequently the original interpretations were often wrong. In most cases it seems clear that substrate proteins can be ADP-ribosylated with empty nucleotide sites. GDP is always tolerated (fortunately since this is the ligand usually bound to these proteins inside the cell) but GTP analogues or even GTP seem to be detrimental at least for the 40-kDa G proteins and probably also for EF2. The possibility exists that guanylyl nucleotides induce a conformation unfavourable to ADP-ribosylation, but the fact that the fluoride does not block ADP-ribosylation of G_s or G_i indicates that an activated G protein can be a substrate. On the other hand, $\beta\gamma$ subunits are needed for the ADP-ribosylation of most, if not all, $G\gamma$ subunits. This can be shown independently of G nucleotide effects and so may involve another conformational shift.

The hypothesis that toxins may recognize GTP-binding sites seems worthy of further experimentation. Although it seems that certain ligands are tolerated in the GTP site, we do not really know that they stay there when the toxin binds. It is possible that the toxins create empty sites by displacing GDP, for example. Perhaps GTPγS, which has the highest avidity of all the GTP analogues for GTP-binding proteins, sometimes simply cannot be displaced and this is why it blocks ADP-ribosylation.

We wish to point out another possible explanation for the inhibition by GTPγS. Suppose that the toxins bind empty guanylyl nucleotide binding sites and only efficiently ADP-ribosylate when they are so bound. Suppose further that the toxins displace without difficulty GDP in residence, but have difficulty in displacing non-hydrolysable analogues, particularly GTPγS. This would be apparent as an inhibition by GTPγS and seems to fit the accumulated evidence.

In this case the toxins would recognize GTP-binding sites from the 'inside'. Such recognition could serve to direct the toxins to their target sequence and, for example, distinguish G_s from other (secondary) cholera toxin substrates which react more slowly.

References

Abood, M.D., Hurley, J.B., Pappone, MC., Bourne, H.R. and Stryer, L. (1982). Functional homology between signal-coupling proteins. Cholera toxin inac-

tivates the GTPase activity of transducin. *J. Biol. Chem.* **257**, 10540–3.

Adari, H., Lowy, D.R., Willumsen, B.M., Der, C.J. and McCormick, R. (1988). Guanosine triphosphatase activating protein (GAP) interacts with the p21 effector binding domain. *Science* **240**, 518–21.

Aktories, K. and Frevert, J. (1987). ADP-ribosylation of a 21–24 kDa eukaryotic protein(s) by C3, a novel botulinum ADP-ribosyltransferase is regulated by guanine nucleotide. *Biochem. J.* **247**, 363–8.

Aktories, K., Barman, M., Ohishi, I., Tsuyama, S., Jakobs, K.H. and Habermann, E. (1986). Botulinum C2 toxin ADP-ribosylates actin. *Nature* **322**, 390–2.

Aktories, K., Weller, U. and Chhatwal, G.S. (1987). Clostridium botulinum type C produces a novel ADP-ribosyltransferase distinct from botulinum C2 toxin. *FEBS Lett.* **212**, 109–13.

Aktories, K., Rosener, S., Blaschke, U. and Chhatwal, G.S. (1988a). Botulinum ADP-ribosyl transferase C3. Purification of the enzyme and characterization of the ADP-ribosylation reaction in platelet membranes. *Eur. J. Biochem.* **172**, 445–50.

Aktories, K., Just, I. and Rosenthal, W. (1988b). Different types of ADP-ribose protein bonds form by botulinum C2 toxin, botulinum ADP-ribosyltransferase C3 and pertussis toxin. *Biochem. Biophys. Res. Commun.* **156**, 361–7.

Aktories, K., Braun, U., Rosener, S., Just, I. and Hall, A. (1989). The rho gene product expressed in *E. coli* is a substrate of botulinum ADP-ribosyltransferase C3. *Biochem. Biophys. Res. Commun.* **158**, 209–13.

Arlinghaus, R., Favelukes, G. and Schweet, R. (1963). A ribose bound intermediate in polypeptide synthesis. *Biochem. Biophys. Res. Commun.* **11**, 92–96.

Asano, T. and Ross, E.M. (1984). Catecholamine-stimulated guanosine 5′ O-(3 triphosphate) binding to the stimulatory GTP-binding protein of adenylate cyclase: Kinetic analysis in reconstituted phospholipid vesicles. *Biochemistry* **23**, 5467–71.

Barbacid, M. (1987). Ras genes. *Ann. Rev. Biochem.* **56**, 779–827.

Bermek, E. (1976). Interaction of adenosine diphosphate-ribosylated elongation factor 2 with ribosomes. *J. Biol. Chem.* **251**, 6544–9.

Binz, T., Kurazono, H., Popoff, M.R., Eklund, M.W., Sakaguchi, G., Kosaki, S., Krieglstein, K., Henschen, A., Gill, D.M. and Niemann, H. (1990). Nucleotide sequence of the gene encoding *Clostridium botulinum* neurotoxin type D. *Nucleic Acids Res.* **18**, 5556.

Bokoch, G.M., Katada, T., Northup, J., Hewlett, E.L. and Gilman, A.G. (1983). Identification of the predominant substrate for ADP-ribosylation by islet activating protein. *J. Biol. Chem.* **258**, 2072–9.

Botstein, D., Segev, N., Stearns, T., Hoyt, M.A. and Kahn, R.A. (1988). Diverse biological functions of small GTP-binding proteins in Yeast. *Cold Spring Harbour Symp. Quant. Biol.* **LIII**, 629–36.

Bourne, H.R. (1986). One molecular machine can

transduce diverse signals. *Nature* **321**, 814–16.

Bourne, H.R. (1988). Do GTPases direct membrane traffic in secretion? *Cell* **53**, 669–71.

Brandt, D.R. and Ross, E.M. (1985). GTPase activity of the stimulatory GTP-binding regulatory protein of adenylate cyclase Gs. *J. Biol. Chem.* **260**, 266–72.

Brandt, D.R. and Ross, E.M. (1986). Catecholamine-stimulated GTPase cycle; multiple sites of regulation by beta adrenergic receptor and Mg^{2+} studied in reconstituted vesicles. *J. Biol. Chem.* **261**, 1656–64.

Brown, B.A. and Bodley, J.W. (1979). Primary structure at the site in beef and wheat elongation factor 2 of ADP-ribosylation by diphtheria toxin. *FEBS Lett.* **103**, 253–5.

Carlier, M.F., Pantaloni, D. and Korn, E.D. (1984). Steady state length distribution of F-actin under controlled fragmentation and mechanism of length redistribution following fragmentation. *J. Biol. Chem.* **259**, 9987–91.

Cassel, D. and Pfeuffer, T. (1978). Mechanism of cholera toxin action: covalent modification of the guanyl nucleotide binding protein of the adenylate cyclase system. *Proc. Natl. Acad. Sci. USA* **75**, 2669–73.

Cassel, D. and Selinger, Z. (1977). Mechanism of adenylate cyclase activation by cholera toxin: inhibition of GTP hydrolysis at the regulatory site. *Proc. Natl. Acad. Sci. USA* **74**, 3307–11.

Chardin, P. (1988). The ras superfamily proteins. *Biochimie* **70**, 865–8.

Chardin, P. and Tavitian, A. (1986). The ral gene: a new ras related gene isolated by the use of a synthetic probe. *EMBO J.* **5**, 2203–8.

Chardin, P., Madaule, P. and Tavitian, A. (1988). Coding sequence of human rho cDNAs clone 6 and clone 9. *Nucleic Acids Res.* **16**, 2717.

Chardin, P., Boquet, P., Madaule, P., Popoff, M.R., Rubin, E.J. and Gill, D.M. (1989). The mammalian protein rhoC is ADP-ribosylated by *Clostridium botulinum* exoenzyme C3 and affects actin microfilaments in Vero cells. *EMBO J.* **8**, 1087–92.

Chaudhary, V.K., Mizukami, T., Fuerst, T., R. Fitzgerald, D.J., Moss, B., Pastan, I. and Berger, E.A. (1988). Selective killing of HIV-infected cells by recombinant CD4-*Pseudomonas* exotoxin hybrid protein. *Nature* **335**, 369–72.

Chuang, D.M. and Weissbach, H. (1972). Studies on elongation factor 11 from calf brain. *Arch. Biochem. Biophys.* **152**, 114–24.

Coburn, J., Dillon, S.T., Iglewski, B.H. and Gill, D.M. (1989a). Exoenzyme S of *Pseudomonas aeruginosa* specifically ADP-ribosylates the intermediate filament protein vimentin. *Infect. Immun.* **57**, 996–8.

Coburn, J. Wyatt, R.T., Iglewski, B.H. and Gill, D.M. (1989b). Several GTP-binding proteins, including p21 c-H-ras, are preferred substrates of *Pseudomonas aeruginosa* exoenzyme S. *J. Biol. Chem.* **264**, 9004–8.

Codina, J., Hildebrandt, J., Iyengaar, R., Birnbaumer, L., Sekura, R.D. and Manclark, C.R. (1983). Pertussis toxin substrate, the putative Ni component

of adenylate cyclase is an alpha beta heterodimer regulated by guanine nucleotide and magnesium. *Proc. Nat. Aad. Sci. USA* **80**, 4276–80.

Collier, R.J. (1968) Effect of diphtheria toxin on protein synthesis: Inactivation of one of the transfer factors. *J. Mol. Biol.*, **25**, 83–98.

Collier, R.J. and Pappenheimer, A.M. Jr. (1964). Studies on the mode of action of diphtheria toxin: II Effect of toxin on amino acid incorporation in cell-free systems. *J. Exp. Med.* **120**, 1019–39.

Didsbury, J.R., Weber, R.F., Bokoch, G.M., Evans, T. and Snyderman, R. (1989). Rac, a novel ras related family of proteins that are botulinum toxin substrates. *J. Biol. Chem.* **264**, 16378–82.

Drazin, R., Kandel, J. and Collier, R.J. (1971). Structure–activity of diphtheria toxin: II Attack by trypsin at a specific site within the intact molecule. *J. Biol. Chem.* **246**, 1504–10.

Dunlop, P.C. and Bodley, J.W. (1983). Biosynthetic labeling of diphtamide in *Saccharomyces cerevisiae*. *J. Biol. Chem.* **258**, 4754–8.

Ellis, R.W., DeFeo, D., Shih, T.Y., Gonda, H.A., Young, N., Tsuchida, N., Lowy, D.R. and Scolnick, E.M. (1981). The p21 src genes of Harvey and Kirsten sarcoma viruses originate from divergent members of a family of normal vertebrate genes. *Nature* **292**, 506–12.

Endo, Y., Mitsui, K., Motizuki, M. and Tsurugi, K. (1987). The mechanism of action of ricin and related toxic lectin on eukaryotic ribosomes. *J. Biol. Chem.* **262**, 5908–12.

Endo, Y., Tsurugi, K., Yutsudo, T., Takeda, Y., Ogasawara, I. and Igarashi, K. (1988). The site of action of a Verotoxin (VT2) from *Escherichia coli* 0157:H7 and of Shiga toxin on eukaryotic ribosomes: RNA N-glycosidase activity of the toxins. *Eur. J. Biochem.* **171**, 45–60.

Enomoto, K. and Gill, D.M. (1980). Cholera toxin activation of adenylate cyclase: roles of nucleosides triphosphates and a macromolecular factor in the ADP-ribosylation of the GTP-dependent regulatory component. *J. Biol. Chem.* **255**, 1252–8.

Fitzmaurice, W.P., Saari, L.L., Lowery, R.G., Ludden, P.W. and Roberts, G.P. (1989). Genes coding for the reversible ADP-ribosylation system of dinitrogenase reductase from *Rhodospirillum rubrum*. *Molec Gen. Genet.* **218**, 340–7.

Gallwitz, D., Donath, C. and Sander, C. (1983). A yeast gene encoding a protein homologous to the human C-has-bas protooncogene product. *Nature* **306**, 704–7.

Garrett, M.D., Self, A.J., van Oers, C. and Hall, A. (1989) Identification of distinct cytoplasmic targets for ras/R-ras and rho regulatory proteins. *J. Biol. Chem.* **264**, 10–13.

Geipel, U. Just, I., Schering, B., Haas, D. and Aktories, K. (1989) ADP-ribosylation of actin causes increase in the rate of ATP exchange and inhibition of ATP hydrolysis. *Eur. J. Biochem.* **179**, 229–32.

Gill, D.M. (1988). Sequence homologies among the enzymically active portions of ADP-ribosylating toxins. In: *Bacterial Protein Toxins* (eds F. Fehrenbach, J.E. Alouf, P. Falagne, W. Goebel, J. Jeljaszewicz, D. Jurgen and R. Rappuoli), Zbl. Bakt. Suppl. 17, pp. 315–23. Gustav Fischer Verlag, Stuttgart and New York.

Gill, D.M. and Coburn, J. (1987). ADP-ribosylation by cholera toxin: Functional analysis of a cellular system that stimulates the enzymatic activity of cholera toxin fragment A1. *Biochemistry* **26**, 6364–71.

Gill, D.M. and Dinius, L.L. (1971). Observation on the structure of diphtheria toxin. *J. Biol. Chem.* **246**, 1485–91.

Gill, D.M. and Dinius, L.L. (1973). The elongation factor 2 content in mammalian cells. *J. Biol. Chem.* **248**, 654–8.

Gill, D.M. and Meren, R. (1978). ADP-ribosylation of membrane proteins catalyzed by cholera toxin: basis of the activation of adenylate cyclase. *Proc. Natl. Acad. Sci. USA* **75**, 3050–4.

Gill, D.M. and Meren, R. (1983). A second guanyl nucleotide-binding site associated with adenylate cyclase. Distinct nucleotides activate adenylate cyclase and permit ADP-ribosylation by cholera toxin. *J. Biol. Chem.* **259**, 11908–14.

Gill, D.M. and Pappenheimer, A.M. Jr. (1971). Structure–activity relationship in diphtheria toxin. *J. Biol. Chem.* **246**, 1492–5.

Gill, D.M. and Woolkalis, M. (1985). Toxins which activate adenylate cyclase. *CIBA Foundation Symposium* **112**, 57–73.

Gill, D.M., Pappenheimer, A.M. Jr., Brown, R. and Kurnick, J.T. (1969). Studies on the mode of action of diphtheria toxin VII: Toxin-stimulated hydrolysis of nicotinamide adenine dinucleotide in mammalian cell extracts. *J. Exp. Med.* **129**, 1–21.

Gilman, A. (1987). G-Proteins: Transducers of receptor-generated signals. *Ann. Rev. Biochem.* **56**, 615–49.

Goud, B., Salminen, A., Walworth, N.C. and Novick, P.J. (1988). A GTP-binding protein required for secretion rapidly associates with secretory vesicles and the plasma membrane in yeast. *Cell* **57**, 1167–77.

Greenberg, R.N. and Guerrant, R.L. (1986). In: *Heat stable enterotoxin in pharmacology of bacterial toxins* (eds, F. Dorner and J. Drews) Pergamon Press, Oxford.

Gupta, R.S. and Siminovitch, L. (1978). Diphtheria-toxin-resistant mutants of CHO cells affected in protein synthesis: a novel phenotype. *Somatic Cell Genet.* **6**, 553–71.

Gupta, R.S. and Siminovitch, L. (1980). Diphtheria toxin resistance in chinese hamster cells: biochemical characterizatics of the mutants affected in protein synthesis. *Somatic Cell Genet.* **6**, 361–79.

Hancock, J.F., Magee, A.I., Childs, J.E. and Marshall, C.J. (1989). All ras proteins are polyisoprenylated but only some are palmitoylated. *Cell* **57**, 1167–77.

Hanski, E. and Gilman, A.G. (1982). The guanine nucleotide-binding regulatory component of adenylate cyclase in human erythrocytes. *J. Cyclic Nucleotides Res.* **8**, 323–36.

Hauser, D., Eklund, M.W., Kuruzono, H., Binz, T., Niemann, H., Gill, D.M., Boquet, P. and Popoff, M.R. (1990). Nucleotide sequence of *Clostridium botulinum* C1 neurotoxin. *Nucleic Acids Res.* **18**, 4924.

Hekman, M., Feder, D., Keenan, A., Gal, A. and Klein, H.W. (1984). Reconstitution of beta-adrenergic receptor with component of the adenylate cyclase. *EMBO J.* **3**, 3339–45.

Higashijima, T., Ferguson, K.M., Sternweiss, P.C., Smigel, N.D. and Gilman, A.G. (1987). Effects of Mg^{2+} and the beta gamma-subunit complex on the interaction of guanine nucleotides with G proteins. *J. Biol. Chem.* **262**, 762–6.

Holbrook, S.R. and Kim, S.H. (1989). Molecular model of the G-protein alpha subunit based on the crystal structure of the H-ras protein. *Proc. Natl. Acad. Sci. USA* **86**, 1751–5.

Honjo, T., Nishizuka, Y., Hayaishi, O., and Kato, I. (1968). Diphtheria-toxin dependent adenosine diphosphate ribosylation of amino-acyl transferase II and inhibition of protein synthesis. *J. Biol. Chem.* **243**, 3553–5.

Huff, R.M. and Neer, E.J. (1986). Subunit interactions of native and ADP-ribosylated alpha 39 and alpha 41, two guanine nucleotide-binding proteins from bovine cerebral cortex. *J. Biol. Chem.* **261**, 1105–10.

Hurley, J.B., Simon, M.I., Teplow, D.B., Robishaw, J.D. and Gilman, A.G. (1984a). Homologies between signal transducing G protein and ras gene product. *Science* **226**, 860–2.

Hurley, J.B., Fong, H.K.W., Teplow, D.B., Dreyer, W.J. and Simon, M.I. (1984b). Isolation and characterization of a cDNA clone for the gamma subunit of bovine retinal transducin. *Proc. Natl. Acad. Sci. USA* **81**, 6948–52.

Huxley, H.E. (1963). Electron microscope studies on the structure of natural and synthetic protein filaments from striate muscle. *J. Molec. Biol.* **7**, 281–308.

Ishikawa, H., Bischoff, R. and Holtzer, H. (1969). Formation of arrowhead complexes with heavy meromyosin in a variety of cell types. *J. Cell Biol.* **43**, 312–28.

Iglewski, B.H. and Kabat, D. (1975). NAD-dependent inhibition of protein synthesis by *Pseudomonas aeruginosa* toxin. *Proc. Natl. Acad. Sci. USA* **72**, 2284–8.

Iglewski, B.H., Sadoff, J., Bjorn, M.J. and Maxwell, E.S. (1978). *Pseudomonas aeruginosa* exoenzyme S: An adenosine diphosphate ribosyl transferase distinct from toxin A. *Proc. Natl. Acad. Sci. USA* **75**, 3211–15.

Johnson, D.I. and Pringle, J.R. (1990). Molecular characterization of CDC42 a *Saccharomyces cerevisiae* gene involved in the development of cell polarity. *J. Cell Biol.* **4**, 1440–8.

Jurnak, F. (1985). Structure of the GDP domain of EFtu and localisation of the amino-acid homologous to ras oncogene proteins. *Science* **230**, 32–6.

Kahn, R.A. and Gilman, A.G. (1984a). Purification of a protein cofactor required for ADP-ribosylation of the stimulatory regulatory component of adenylate cyclase by cholera toxin. *J. Biol. Chem.* **259**, 6228–34.

Kahn, R.A. and Gilman, A.G. (1984b). ADP-ribosylation of Gs promotes the dissociation of its alpha and beta subunits. *J. Biol.Chem.* **259**, 6235–40.

Kahn, R.A., Goddard, C. and Newkirk, M. (1988). Chemical and immunological characterization of the 21 kDa ADP-ribosylation factor (ARF) of adenylate cyclase. *J. Biol. Chem.* **263**, 8282–7.

Kaziro, Y. (1978). The role of guanosine-5'-triphosphate in polypeptide chain elongation. *Biochim. Biophys. Acta* **505**, 95–127.

Katada, T. and Ui, M. (1982). ADP-ribosylation of the specific membrane protein of C6 cells by Islet activating protein associated with modification of adenylate cyclase activity. *J. Biol. Chem.* **257**, 7210–16.

Kessel, M. and Klink, F. (1980). Archebacterial elongation factor is ADP-ribosylated by diphtheria toxin. *Nature* **287**, 250–1.

Kohno, K. and Uchida, T. (1987). High frequency single amino acid substitition in mammalian elongation factor 2 (EF2) results in expression of resistance to EF2-ADP-ribosylation. *J. Biol. Chem.* **262**, 12298–305.

Kohno, K., Uchida, T., Mekada, E. and Okada, Y. (1985). Characterization of diphtheria-toxin resistant mutants lacking receptor function or containing nonribosylatable elongation factor 2. *Somatic Cell Genet.* **11**, 421–31.

Kohno, K., Uchida, T., Ohkubo, H., Nakanishi, S., Nakanishi, T., Fukui, T., Ohtsuka, E., Ikehara, M. and Okada, Y. (1986). Amino acid sequence of mammalian elongation factor 2 deduced from the cDNA sequence: homology with GTP-binding proteins. *Proc. Natl. Acad. Sci USA* **83**, 4978–82.

Kikuchi, A., Yamamoto, K., Fujita, T. and Takai, Y. (1988). ADP-ribosylation of the bovine brain rho protein by botulinum toxin type C1. *J. Biol. Chem.* **263**, 16303–8.

Lacour, T.F.N., Nyborg, J., Thirup, S. and Clark, B.F.C. (1985). Structural details of the binding of guanosine diphosphate to elongation factor tu from *E. coli* as studied by X-ray crystallography. *EMBO J.* **4**, 2385–8.

Landis, C.A., Masters, S.B., Spada, A., Pace, A.M., Bourne, H.R. and Vallar, L. (1989). GTPase inhibiting mutations activate the alpha chain of Gs and stimulate adenylyl cyclase in human pituitary tumours. *Nature* **340**, 692–6.

Lim, K.L., Sekura, R.D. and Kaslow, H.R. (1985). Adenine nucleotide directly stimulates pertussis toxin. *J. Biol. Chem.* **260**, 2585–8.

Madaule, P. and Axel, R. (1985). A novel ras-related gene family. *Cell* **41**, 31–40.

Madaule, P., Axel, R. and Myers, A.M. (1987). Characterization of two members of the rho gene family from the yeast *Saccharomyces cerevisiae*. *Proc. Natl. Acad. Sci. USA* **84**, 779–83.

Matsuoka, I., Syuto, B., Kurihara, K. and Kubo, S. (1987). ADP-ribosylation of specific membrane protein in pheochromocytoma and primary-cultured brain cells by botulinum neurotoxin type C and D. *FEBS Lett.* **216**, 295–9.

Mattera, R., Codina, J., Sekura, R.D. and Birnbaumer, L. (1986). The interaction of nucleotides with pertussis toxin. Direct evidence for a nucleotide binding site on the toxin regulating the rate of ADP-ribosylation of Ni, the inhibitory regulatory component of adenylate cyclase. *J. Biol. Chem.* **261**, 11173–9.

Mattera, R., Codina, J., Sekura, R.D. and Birnbaumer, L. (1987). GTP reduces ADP-ribosylation of the inhibitory guanine nucleotide binding regulatory protein adenylyl cyclase (Ni) by pertussis toxin without causing dissociation of the subunits. *J. Biol. Chem.* **262**, 11247–51.

Meddynski, D.C., Sullivan, K., Smith, D., Van Dop, C. and Chang, F.M. (1985). Amino acid sequence of the alpha subunit of transducin deduced from cDNA sequence. *Proc. Natl. Acad. Sci. USA* **82**, 4311–15.

Moehring, J.M. and Moehring, T.J. (1977). Selection and characterization of cells resistant to diphtheria toxin and *Pseudomonas* exotoxin A: presumptive translational mutants. *Cell* **11**, 447–54.

Moehring, J.M. and Moehring, T.J. (1979). Characterization of the diphtheria toxin resistance system in chinese hamster ovary cells. *Somatic Cell Genet.* **5**, 453–68.

Moehring, J.M. and Moehring, T.J. (1988). The post-translational trimethylation of diphthamide studied in vitro. *J. Biol. Chem.* **263**, 3840–4.

Moehring, T.J., Danley, D.E. and Moehring, J.M. (1979). Codominant translational mutants of chinese ovary cells selected with diphtheria toxin. *Somatic Cell Genet.* **5**, 469–80.

Moehring, J.M., Moehring, T.J. and Danley, D.E. (1980). Post-translational modificational of elongation factor 2 in diphtheria toxin mutants of CHO-K1 cells. *Proc. Natl. Acad. Sci. USA* **77**, 1010–14.

Moehring, J.M., Danley, D.E. and Moehring, J.M. (1984). In vitro biosynthesis of diphthamide: Studied with mutant chinese hamster ovary cells resistant to diphtheria toxin. *Molec. Cell. Biol.* **4**, 642–50.

Montanaro, L., Sperti, S and Mattioli, A. (1971). Interaction of ADP-ribosylated amino-acyl transferase II with GTP and with ribosomes. *Biochim. Biophys. Acta* **238**, 493–7.

Murphy, J.R., Bishai, W., Borowski, M., Miyanohaia, A., Boyd, J. and Nagle, S. (1986). Genetic construction, expression and melanoma-selective cytotoxicity of a diphtheria toxin related alpha melanocyte stimulating hormone fusion protein. *Proc. Natl. Acad. Sci. USA* **83**, 8258–62.

Nagata, S., Nagashima, K., Tsunetsugu-Yokata, Y., Fujimura, K., Miyazaki, M. and Kaziro, Y. (1984). Polypeptide chain elongation factor 1 alpha (EF-1 alpha) from yeast: nucleotide sequence of one of the two gene for EF-1 alpha from yeast. *EMBO J.* **3**, 1825–30.

Narumiya, S., Sekine, A. and Fujiwara, M. (1988). Substrate for botulinum ADP-ribosyltransferase Gb has an amino acid sequence homologous to a putative rho gene product. *J. Biol. Chem.* **263**, 17255–7.

Navon, S.E. and Fung, B.K.K. (1984). Characterization of transducin from bovine retinal rod outer segments: Mechanism and effects of cholera toxin-catalysed ADP-ribosylation. *J. Biol. Chem.* **259**, 6686–93.

Navon, S.E. and Fung, B.K.K. (1987). Characterization of transducin from bovine retinal rod outer segments. Participation of the amino terminal region of T alpha in subunit interactions. *J. Biol. Chem.* **262**, 15746–51.

Neer, E.J. and Clapham, D.E. (1988). Roles of G proteins subunits in transmembrane signalling. *Nature* **333**, 129–34.

Neer, E.J., Lok, J.M. and Wolf, L.G. (1984). Purification and properties of the inhibitory guanine regulatory unit of brain adenylate cyclase. *J. Biol. Chem.* **259**, 1422–9.

Neer, E.J., Wolf, L.G. and Gill, D.M. (1987). The stimulatory guanine-nucleotide regulatory unit of adenylate cyclase from bovine cerebral cortex. ADP-ribosylation and purification. *Biochem. J.* **241**, 325–36.

Neer, E.J., Pulsifer, L. and Wolf, L.G. (1988). The amino terminus of G protein alpha subunits is required for interaction with beta gamma. *J. Biol. Chem.* **263**, 8996–9000.

Neuhaus, J.M., Wanger, M., Keisler, T. and Wegener, A. (1983). Treadmilling of actin. *J. Muscle Res. Cell Motil.* **4**, 507–27.

Nilsson, L. and Nygard, O. (1985). Localization of the sites of ADP-ribosylation and GTP binding in the eukaryotic elongation factor EF2. *Eur. J. Biochem.* **148**, 299–304.

Nilsson, L. and Nygard, O. (1988). Structural and functional studies of the interaction of the eukaryotic elongation factor EF2 with GTP and ribosomes. *Eur. J. Biochem.* **171**, 293–9.

Northrup, J.K., Smigel, M.D., Sternweiss, P.C. and Gilman, A.G. (1983). The subunits of the stimulatory regulatory component adenylate cyclase: resolution of the activated 45 000 dalton (α) subunit. *J. Biol. Chem.* **258**, 11369–76.

Nukada, T., Tanabe, T., Takahashi, N., Noda, M. and Hirose, T. (1986). Primary structure of the alpha subunit of bovine adenylate cyclase-inhibiting G-protein deduced from the cDNA sequence. *FEBS Lett.* **197**, 305–10.

Nygard, O. and Nilsson, L. (1985). Reduced ribosomal binding of eukaryotic elongation factor 2 following ADP-ribosylation. Difference in binding selectivity between polyribosomes and reconstituted monoribosomes. *Biochim. Biophys. Acta* **824**, 152–62.

Ohashi, Y. and Narumiya, S. (1987). ADP-ribosylation of a Mr 21 000 membrane protein by type D botulinum toxin. *J. Biol. Chem.* **262**, 1430–3.

Ohashi, Y., Kamiya, T., Fujiwara, M. and Narumiya, S. (1987). ADP-ribosylation by type C1 and D botulinum neurotoxins: stimulation by guanine

nucleotides and inhibition by guanidino-containing compounds. *Biochem. Biophys. Res. Commun.* **142**, 1032–8.

Ohishi, I. and Tsuyama, S. (1986). ADP-ribosylation of non-muscle actin with component I of C2 toxin. *Biochem. Biophys. Res. Commun.* **136**, 802–6.

Ohishi, I., Iwasaki, M. and Sakaguchi, G. (1980). Purification and characterization of two components of botulinum C2 toxin. *Infect. Immun.* **30**, 668–73.

Ohishi, I., Miyake, M., Ogura, K. and Nakamura, S. (1984). Cytopathic effect of botulinum C2 toxin on tissue culture cell lines. *FEMS Microbiol. Lett.* **23**, 281–4.

Omura, F., Kohno, K. and Uchida, T. (1989). The histidine residue of codon 715 is essential for the function of elongation factor 2. *Eur. J. Biochem.* **180**, 1–8.

Palmiter, R.D., Behringer, R.R., Quaife, J., Maxwell, F., Maxwell, I.A. and Brinster, R.L. (1987). Cell lineage ablation in transgenic mice by cell-specific expression of a toxin gene. *Cell* **50**, 435–43.

Pappenheimer, A.M. Jr., Dunlop, P.C., Adolph, K.W. and Bodley, J.W. (1983). Occurrence of diphthamide in archebacteria. *J. Bacteriol.* **153**, 1342–7.

Pai, E.F., Kabsch, W., Krengel, U., Holmes, K.C., John, J. and Wittinghoffer, A. (1989). Structure of the guanine-nucleotide binding domain of the Ha-ras oncogene product p21 in the triphosphate conformation. *Nature* **341**, 209–14.

Paterson, H.F., Self, A.J., Garrett, M.D., Just, I. Aktories, K. and Hall, A. (1990). Microinjection of recombinant p21 rho induces rapid changes in cell morphology. *J. Cell Biol.* **111**, 1001–7.

Pedersen, S.E. and Ross, E.M. (1982). Functional reconstitution of beta adrenergic receptors and the stimulatory GTP-binding protein of adenylate cyclase. *Proc. Natl. Acad. Sci. USA* **79**, 7228–32.

Pickett, C.L., Weinstein, E.L. and Holmes, R.K. (1987). Genetics of type IIa heat-labile enterotoxin of *Escherichia coli*: operon fusion, nucleotide sequence, and hybridization studies. *J. Bacteriol.* **169**, 5180–7.

Pickett, C.L., Twiddy, E.M., Coker, C. and Holmes, R.K. (1989). Cloning, nucleotide sequence, and hybridization studies of the type IIb Heat-labile enterotoxin gene of *Escherichia coli*. *J. Bacteriol.* **171**, 4945–52.

Pollard, T.D. (1984). Polymerization of ADP-actin. *J. Cell Biol.* **99**, 1970–80.

Popoff, M.R. and Boquet, P. (1988). *Clostridium spiroforme* toxin is a binary toxin which ADP-ribosylates cellular actin. *Biochem. Biophys. Res. Commun.* **152**, 1361–8.

Popoff, M.R., Rubin, E.J., Gill, D.M. and Boquet, P. (1988). Actin-specific ADP-ribosyl transferase produced by a *Clostridium difficile* strain. *Infect. Immun.* **56**, 2299–306.

Popoff, M.R., Milward, F.W., Bancillon, B. and Boquet, P. (1989). Purification of the *Clostridium spiroforme* binary toxin and activity of the toxin on HEp2 cells. *Infect. Immun.* **57**, 2462–9.

Popoff, M.R., Boquet, P., Gill, D.M. and Eklund, M.W. (1990). DNA sequence of exoenzyme C3, an ADP-ribosyltransferase encoded by *Clostridium botulinum* C and D phages. *Nucleic Acids Res.* **18**, 1291.

Reed, R.R. (1990). How does the nose know? *Cell* **60**, 1–2.

Raeburn, S., Collins, J.F., Moon, H.M. and Maxwell, E.S. (1971). Aminoacyltransferase II from rat liver. Purification and enzymatic activity. *J. Biol. Chem.* **246**, 1041–8.

Raeburn, S., Goor, R. Schneider, J.A. and Maxwell, E.S. (1986). Interaction of aminoacyl tranferase II and guanosine triphosphate: inhibition by diphtheria toxin and nicotinamide adenine dinucleotide. *Proc. Natl. Acad. Sci. USA* **61**, 1428–34.

Reuner, K.H., Presek, P., Boschek, C.B. and Aktories, K. (1987). Botulinum C2 toxin ADP-ribosylates actin and disorganizes the microfilament network in intact cells. *Eur. J. Cell Biol.* **43**, 134–40.

Robinson, E.A., Enriksen, O. and Maxwell, E.S. (1974). Elongation factor 2: aminoacid sequence at the site of adenosine diphosphate ribosylation. *J. Biol. Chem.* **249**, 5088–93.

Robishaw, J.D., Smigel, M.D. and Gilman, A.G. (1986). Molecular basis for two forms of the Gα protein that stimulates adenylate cyclase. *J. Biol. Chem.* **261**, 9587–50.

Rösener, S., Chhatwal, G.S. and Aktories, K. (1987). Botulinum ADP-ribosyl transferase C3 but not botulinum neurotoxins C1 and D ADP-ribosylates a low molecular mass GTP-binding proteins. *FEBS Lett.* **224**, 38–42.

Roux, E. and Yersin, A. (1888). Contribution a l'étude de la diphterie. *Ann. Inst. Pasteur* **2**, 629–61.

Rubin, E.J., Gill, D.M., Boquet, P. and Popoff, M.R. (1988). Functional modification of a p21 ras-like protein when ADP-ribosylated by exoenzyme C3 of *Clostridium botulinum*. *Molec. Cell. Biol.* **8**, 418–26.

Rubin, E.J., Boquet, P., Popoff, M.R. and Gill, D.M. (1989). ADP-ribose transfer reactions: mechanism and biological significance. In: (eds M.K. Jacobson and E.L. Jacobson), pp. 422–9. Springer-Verlag, New York.

Salminen, A. and Novick, P.J. (1987). A ras-like protein is required for a post Golgi event in yeast secretion. *Cell* **49**, 527–38.

Schering, B.M., Barman, G.S., Chhatwal, U., Geipel, U. and Aktories, K. (1988). ADP-ribosylation of skeletal muscle and non-muscle actin by *Clostridium perfringens* iota toxin. *Eur. J. Biochem.* **171**, 225–9.

Schleifer, L.S., Kahn, R.A., Hanski, E., Northup, J.K., Sternweis, P.C. and Gilman, A.G. (1982). Requirements for cholera toxin dependent ADP-ribosylation of the purified regulatory component of adenylate cyclase. *J. Biol. Chem.* **257**, 20–3.

Segev, N., Mulholland, J. and Botstein, D. (1988). The yeast GTP-binding YPT1 protein and a mammalian counterpart are associated with the secretion machinery. *Cell* **52**, 915–24.

Sekine, A., Fujiwara, M. and Narumiya, S. (1989). Asparagine residue in the rho gene product is the modification site for botulinum ADP-ribosyltransferase. *J. Biol. Chem.* **264**, 8602–5.

Sewell, J.L. and Khan, R.A. (1988). Sequences of the bovine and yeast ADP-ribosylation factor and comparison to other GTP-binding proteins. *Proc. Natl. Acad. Sci. USA* **85**, 4620–4.

Simpson, L.L. (1984). Molecular basis for the pharmacological actions of *Clostridium botulinum* type C2 toxin. *J. Pharmacol. Exp. Ther.* **223**, 695–701.

Simpson, L.L., Stiles, B.G., Zapeda, H.H. and Wilkins, T.D. (1987). Molecular basis for the pathological actions of *Clostridium perfringens* iota toxin. *Infect. Immun.* **55**, 118–22.

Simpson, L.L., Stiles, B.G., Zapeda, H.H. and Wilkins, T.D. (1989). Production by *Clostridium spiroforme* of an iotalike toxin that possesses mono(ADP-ribosyl)transferase activity: identification of a novel class of ADP-ribosyltransferases. *Infect. Immun.* **57**, 255–61.

Sperti, S., Montanaro, L. and Mattioli, A. (1971). Studies on diphtheria toxin. The effect of GTP on the toxin-dependent adenosine diphosphate ribosylation of rat-liver aminoacyl transferase II. *Chem. Biol. Interaction* **3**, 141–8.

Stearns, T., Willingham, M.C., Botstein, D. and Khan, R.A. (1990). ADP-ribosylation factor is functionally and physically associated with the Golgi complex. *Proc. Natl. Acad. Sci. USA* **87**, 1238–42.

Sternweiss, P.C. and Robishaw, J.D. (1984). Isolation of two proteins with high affinity for guanine nucleotides from membranes of bovine brain. *J. Biol. Chem.* **259**, 13806–13.

Sternweiss, P.C., Northup, J.K., Smigel, M.D. and Gilman, A.G. (1981). The regulatory component of adenylate cyclase. Purification and properties. *J. Biol. Chem.* **256**, 11517–26.

Stossel, T.P., Chaponnier, C., Ezzel, R.M., Hartwig, J.H., Janmey, P.A., Kwiatkowski, D.J., Lind, S.E., Smith, D.B., Southwick, F.S., Yin, H.L. and Zaner, K.S. (1985). Nonmuscle actin-binding proteins. *Ann. Rev. Cell Biol.* **1**, 353–402.

Stryer, L. and Bourne, H.R. (1986). G proteins: A family of signal transducers. *Ann. Rev. Cell Biol.* **2**, 391–419.

Tanabe, T., Nukada, T., Nishikawa, Y., Sugimoto, K., Suzuki, M., Takahashi, H., Noda, M., Haga, T., Ichiyama, A., Kangawa, K., Minamino, N., Matsuo, M. and Numa, S. (1985). Primary structure of the alpha subunit of transducin and its relationship to ras protein. *Nature* **315**, 242–3.

Toda, T., Uno, I., Ishikawa, S., Powers, S., Kataoka, T., Broek, D., Cameron, S., Broach, J., Matsumoto, K. and Wigler, M. (1985). In yeast RAS proteins are controlling elements of adenylate cyclase. *Cell* **40**, 27–36.

Tong, L., Milburn, M.V., de Vos, A.M. and Kim,

S.H. (1989). Structure of ras protein. *Science* **245**, 244.

Toratani, S., Yokosawa, N., Yokosawa, H., Ishii, S. and Oguma, K. (1989). Immuno-crossreactivity between botulinum neurotoxin type C1 or D and exoenzyme C3. *FEBS Lett.* **252**, 83–7.

Trahey, M. and McCormick, F. (1987). A cytoplasmic protein stimulates normal N-ras p21 GTPase, but does not affect oncogenic mutants. *Science* **242**, 542–5.

Tsai, S.C., Adamik, R., Kanaho, Y., Hewlett, E.L. and Moss, J. (1984). Effects of guanyl nucleotides and rhodopsin on ADP-ribosylation of the inhibitory GTP-binding component of adenylate cyclase by pertussis toxin. *J. Biol. Chem.* **259**, 15320–3.

Vandekerckhove, J.B. and Weber, K. (1979). The complete aminoacid sequence of actins from bovine aorta, bovine heart, bovine fast skeletal muscle and rabbit slow skeletal muscle. *Differentiation* **14**, 123–33.

Vandekerckhove, J.B., Schering, J.B., Barman, M. and Aktories, K. (1988). Botulinum C2 toxin ADP-ribosylates cytoplamic β/γ actin in arginine 177. *J. Biol. Chem.* **263**, 696–700.

Vandenberg, C.A. and Montal, M. (1984). Light regulated biochemical events in invertabrate photoreceptor. I Light-activated GTPase, guanine nucleotide binding and cholera toxin-catalyzed labeling of squid photoreceptor membranes. *Biochemistry* **23**, 2339–47.

Van Dop, C., Yamanaka, G., Steinberg, F., Sekura, R.D., Manclark, C.R., Stryer, L. and Bourne, H.R. (1984). ADP-ribosylation of transducin by pertussis toxin blocks the light-stimulated hydrolysis of GTP and cGMP in retinal photoreceptors. *J. Biol. Chem.* **259**, 23–6.

Van Ness, B.G., Howard, J.B. and Bodley, J.W. (1980a). ADP-ribosylation of elongation factor 2 by diphtheria toxin. Isolation and properties of the novel ribosyl-amino acid and hydrolysis products. *J. Biol. Chem.* **255**, 10717–20.

Van Ness, B.G., Borrowclough, B. and Bodley, J.W. (1980b). Recognition of elongation factor 2 by diphtheria toxin is not solely defined by the presence of diphthamide. *FEBS Lett.* **120**, 4–6.

Waltz, G., Zanken, B., Brand, K., Waters, C., Gen-bauffe, F., Zeldis, J.B., Murphy, J.R. and Strom, T.B. (1989). Sequential effects of interleukin 2 diphtheria toxin fusion protein on T-cell activation. *Proc. Natl. Acad. Sci. USA* **86**, 9485–8.

Watkins, P.A., Moss, J., Burns, D.L., Hewlett, E.L. and Vaughan, M. (1984). Inhibition of bovine rod outer segment GTPase by *Bordetella pertussis* toxin. *J. Biol. Chem.* **260**, 13478–82.

Watkins, P.A., Burns, D.L., Kanaho, Y., Liu, T.Y., Hewlett, E.L. and Moss, J. (1985) ADP-ribosylation of transducin by pertussis toxin. *J. Biol.Chem.* **260**, 13478–82.

Wegner, A. and Aktories, K. (1988). ADP-ribosylated actin caps the barbed ends of actin filaments. *J. Biol. Chem.* **263**, 13739–42.

Weissbach, H. and Ochoa, S. (1976). Soluble factors required for eukaryotic protein synthesis. *Ann. Rev. Biochem.* **45**, 191–216.

Wong, S.K.F., Martin, R. and Tolkovsky, A.M. (1985). Pertussis toxin substrate is a guanosine 5′(thio) diphosphate-N-ethyl maleinemide, Mg^{2+} and temperature-sensitive GTP-binding protein. *Biochem. J.* **232**, 191–7.

Woolkalis, M., Gill, D.M. and Coburn, J. (1988). Assay and purification of cytosolic factor required for cholera toxin activity. *Methods Enzymol.* **165**, 246–9.

Yamamoto, T., Gojoboi, T. and Yokota, T. (1987). Evolutionary origin of pathogenic determinants in enteric *Escherichia coli* and *Vibrio cholerae* O1. *J. Bacteriol* **169**, 1352–7.

Yamaizumi, M., Mekada, E., Uchida, T. and Okada, Y. (1978). One molecule of diphtheria toxin fragment A introduced into a cell can kill the cell. *Cell* **15**, 245–50.

Yarden, Y. and Ullrich, A. (1988). Molecular analysis of signal transduction by growth factors. *Biochemistry* **27**, 311–18.

Yatsunami, K. and Khorana, H.G. (1985). GTPase of bovine rod outer segments. The amino acid sequence of the alpha subunit as derived from cDNA sequence. *Proc. Natl. Acad. Sci. USA* **82**, 4316–20.

Yeramian, P., Chardin, P., Madaule, P. and Tavitian, A. (1987). Nucleotide sequence of human rho cDNA clone 12. *Nucleic Acids Res.* **15**, 1869.

3

Molecular Models of Toxin Membrane Translocation

Cesare Montecucco, Emanuele Papini and Giampietro Schiavo

Centro CNR Biomembrane e Dipartimento di Scienze Biomediche, Università di Padova, Corso Trieste 75, Padova, Italy

Introduction

The mechanism of cell intoxication by bacterial protein toxins with intracellular targets can be divided into three steps: (i) cell binding, (ii) membrane penetration and, in some cases, translocation, (iii) target modification. The second step is the least understood and the most remarkable from the point of view of protein chemistry and membrane topology. In fact these toxins are water-soluble and hence presumably endowed with a surface fully complemented with hydrophilic, even if not charged, residues. Yet, these proteins are somehow able to insert into the hydrophobic milieu of biological membranes and, in some cases, also to translocate across the membrane into the cytoplasm. This property is shared by other pathogenic agents such as certain viruses and by protein complexes involved in the immune response such as complement and perforin proteins.

An intense effort has been made in the last ten years to understand this property at the molecular level. Both cellular and membrane model systems have been advantageously used: this chapter will focus on the picture of toxin membrane translocation emerging from biochemical and biophysical studies with model membranes.

Toxins acting inside cells can be divided into three groups: (a) toxins with a target located on the cytosolic side of the plasma membrane, such as cholera and pertussis toxins; (b) toxins acting on a cytoplasmic structure, such as diphtheria toxin and shiga toxins; and (c) toxins whose target is still unknown, such as tetanus and botulinum neurotoxins. We will follow this order and discuss first the membrane interaction of toxins of group a.

Membrane penetration by bacterial protein toxins with target located on the cytoplasmic face of the membrane

Cholera (CLT) and pertussis toxins (PTx) are the two most studied and employed of all protein toxins. This is because of the central role in cell function of their target: the G proteins. A large group of G proteins are localized on the plasma membrane and expose part of their surface to the cytoplasm. The active subunits of CLT, and of the closely related *E. coli* heat-labile enterotoxin (termed A1) and of PTx (termed S1) catalyse the transfer of ADP-ribose from their cytoplasmic substrate NAD to the α subunit of G proteins.

Cholera toxin

CLT and *E. coli* heat-labile enterotoxin differ in only a few residues and are formed by one enzymic subunit A1 joined by a single disulphide bond to A2. These two subunits are non-covalently bound to five B subunits, each one possessing a single binding site for the oligosaccharide portion of ganglioside GM1, the specific cellular receptor of CLT (van Heyningen, 1977).

Since binding occurs on the external surface of the membrane and toxin activity is exerted on the cytoplasmic face of the membrane with the use of cytoplasmic substrates, it necessarily follows that at least their enzymatic portion must insert into the membrane and expose both its NAD and G protein-binding sites to the cytoplasm.

All mechanisms proposed to account for the cell penetration of CLT are based on the concept that the toxin, upon binding to GM1, undergoes a gross conformational change resulting in the membrane insertion of the active subunit A1 (Brady and Fishman, 1979) and, in other models, also of the B subunits (Gill, 1976). When oligosaccharide GM1 binding to CLT was studied with differential UV, circular dichroism and fluorescence spectroscopies (Fishman et al., 1978; Tomasi et al., 1984) no evidence for a large conformational rearrangement of the toxin structure was found.

Membrane photolabelling with photoreactive phospholipids (Tomasi and Montecucco, 1981; Tomasi et al., 1986) and, more recently, 3-dimensional image reconstruction of the structure of membrane-bound CLT (Ribi et al., 1988) have shown that reduction of the interchain disulphide bridge between A1 and A2 is sufficient to cause the insertion of A1 into the lipid bilayer.

Membrane photolabelling is a very useful method for probing lipid–protein interactions including those resulting from the insertion of a protein into the lipid bilayer (Bayley, 1983; Bisson and Montecucco, 1985; Montecucco, 1988). It is based on the use of radioactive lipid molecules, carrying a photoactivatable group stable under dark conditions, interdispersed among the other lipids of the membrane under study. On illumination a highly reactive intermediate is formed that is able to form a covalent, stable bond with neighbour molecules such as lipids or proteins. If the photoactive group is placed at different levels of a phospholipid molecule, different depths of penetration of a protein in the lipid bilayer can be monitored (Bisson and Montecucco, 1981; Montecucco, 1988).

With CLT it was found that, after binding to GM1-containing membranes, only B subunits were photolabelled and only with a shallow probe that reacts with those protein regions interacting with phospholipids at the membrane surface level. No labelling was found with a deep probe, whose photoactive group is linked to the fatty acid methyl terminus and hence probes the hydrophobic core of the lipid bilayer. This result suggested that the toxin, after binding to the oligosaccharide portion of GM1, remains at the membrane surface with its B subunits interacting with the polar head groups of phospholipids. On the other hand, the pattern of photolabelling of CLT changed completely after treatment with glutathion that reduces the A1–A2 disulphide bridge and causes the activation of the toxin: while B subunits remained labelled only at the membrane surface level, A1 was labelled with both the surface and the deep probes. This result led to the suggestion that only A1 penetrates into the lipid bilayer and that the trigger of A1 membrane insertion is the reduction of the interchain disulphide bond (Tomasi and Montecucco, 1981). The apparent disagreement with another membrane photolabelling experiment, where no need for reduction in the entry of A1 into viral membranes was found (Wisnieski and Bramhall, 1981), is explained by the intrinsic reducing activity of the viral membrane itself.

Very recently the structure at 2.3 Å resolution of the E. coli heat-labile enterotoxin (LT) has been described and this has provided a long awaited molecular basis for the understanding of membrane binding, insertion and immunological properties of the LT and CLT toxins (Sixma et al., 1991). Figure 1 shows the two alternative modes of toxin membrane binding compatible with the structure: with A1 inserted in the hydrophobic core of the lipid bilayer or with A1 facing the aqueous solvent. In both cases the B protomer remains at the membrane surface, as suggested by all available data, but it interacts with lipids with two different regions in the two cases.

Presently available experimental evidence does not allow discrimination between the two possibilities. Some results (Gill and King, 1975; Wisnieski and Bramhall, 1981; Dwyer and Bloomfield, 1982; Ribi et al., 1988) support the former mode (Figure 1A) while others (Gill, 1976; Tomasi and Montecucco, 1981; Tomasi et al., 1986) support the latter view (Figure 1B) and another study can be interpreted in both ways (Goins and Freire, 1985). At this stage one should also consider the possibility suggested by the structure of LT itself that the toxin binds in

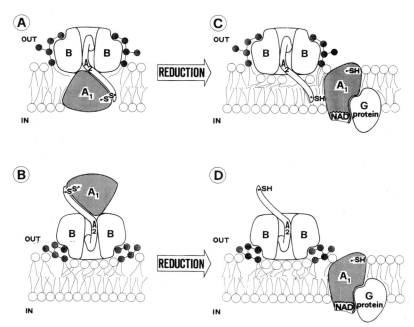

Figure 1. Possible mechanisms of membrane binding and translocation of cholera and cholera-related toxins. On the basis of the 3-D structure of LT (Sixma *et al.*, 1991) and on the flexibility of the oligosaccharide portion of the ganglioside G_{M1} the toxin can bind in two opposite ways. (A) with A1–A2 penetrating and spanning the lipid bilayer exposing part of A1 to the cytoplasm (panel A) or (B) with A1–A2 facing the water solvent on the cell exterior. Reduction of the single interchain disulphide bridge joining A1 to A2 activates the toxin and triggers a structural change of A1 that acquires an hydrophobic character and it is released from A2. In (A) reduction can be performed by cytoplasmic reducing agents and it leads to the detachment and lateral movement of A1, which is now a integral component of the membrane, to reach it target G protein (C). In (B) reduction can be performed by chemical or enzymic reducing agents present in the extracellular medium and A1 may reach the hydrophobic part of the lipid bilayer either by rolling on B away from the water solvent or with the help of a protein (not depicted in panel D).

different modes to different membranes depending on variable physico-chemical conditions. For example, rigid phospholipid bilayers, high lateral surface pressures or low temperatures are expected to not favour the membrane insertion of A1. Additional important roles can be predicted for negatively charged lipids and bivalent cations. Clearly more experiments are needed to define the mode of membrane binding of the CLT and LT toxins.

One point where there is substantial agreement is that the trigger for the establishment of A1 as a integral component of the membrane is the reduction of the single interchain disulphide joining A1 to A2 (Tomasi and Montecucco, 1981; Tomasi *et al.*, 1986; Ribi *et al.*, 1988). While A1 does not have large hydrophobic surfaces when disulphide connected to A2, it appears that this reduction causes a major structural rearrangement of A1 which becomes hydrophobic. This acquired hydrophobicity leads to its aggregation if reduction is performed in

solution or to its insertion into the lipid bilayer if reduction is performed on membrane bound toxin.

No information is available on the nature of the reducing agent acting in the intestinal lumen nor on the accessibility of the interchain disulphide bond. In principle either a protein or a chemical reductant could perform this step and release A1 (Moss *et al.*, 1980) and it is possible that the time required for reduction accounts for at least part of the time interval between toxin binding and adenylate cyclase activation (van Heyningen, 1983). The two above-described modes of CLT and LT binding to membranes differ substantially with respect to the nature of the possible reductant. In the mode with A1 facing the aqueous solvent at the cell exterior both chemical and enzymatic agents present in the intestinal lumen will be effective. In the alternative mode with A1–A2 inserted in the membrane their dimensions are such that the disulphide bridge can reach the cytoplasm and be reduced by cytoplasmic reducing agents both of a chemical or enzymatic nature, as depicted in Figure 1A.

Another open question is how does A1 reach the NAD, accessory factors and the G proteins on the other side of the membrane, once the disulphide bond is reduced. It is thought to involve a simple detachment and lateral displacement within the membrane plane to reach the target G protein shown in the model in Figure 1B. The simplicity of such a step is the major appeal of this model which adds to the available experimental evidence discussed previously. In the model of binding with A1 facing the intestinal lumen one may consider two possibilities. Firstly the reduced hydrophobic A1 rolls over the B surface away from the water solvent to partition into the hydrophobic core of the lipid bilayer. A second possibility envisages the involvement of a protein of the microvillar membrane, possibly the same causing toxin reduction. What appears to be unlikely, on the basis of the structure of LT (Sixma et al., 1991) is a protomer B tunnel mechanism (Gill, 1976) with the penetration of the unfolded A1 into the central pore of the protomer B ring because of its insufficient size.

Many hydrophilic and charged residues are present on A1. If these residues are shielded during their passage across the hydrophobic barrier of the membrane via formation of ionic couples with phospholipid head groups or other ions, the energetic cost of the process will be greatly lowered. After membrane insertion, A1 can be considered as an integral protein of the plasma membrane with its substrate binding sites exposed to the cytoplasm and a membrane-embedded hydrophobic sector.

It is difficult at this stage to describe the kind of forces involved in the cell penetration of CLT. One difficulty for the direct insertion of A1–A2 upon binding (Figure 1A) derives from the observation that they lack large hydrophobic patches, rather the two subunits have an overall hydrophilic surface (Sixma et al., 1991). The exact position of the interchain disulphide of LT could not be determined, but it appears to have a superficial location and its hydrophobicity could play a substantial role in favour of a partition into the lipid bilayer. It could well be the first part of the toxin that reaches the cytoplasm, as suggested for diphtheria toxin (see below). A1–B interaction is rather weak and the A1 detachment does not appear to require a substantial energy input. It is likely that binding to G protein and to NAD

is involved in driving the correct membrane location of A1 on the cytosolic side.

Another as yet unanswered aspect of the mechanism of cell intoxication by CLT is related to the plasma membrane polarization of the epithelial cell of the intestine. These cells present ganglioside GM1 on the external leaflet of the microvillar plasma membrane exposed to the intestine lumen, while their adenylate cyclase, the ultimate target of the action of the toxin, is on the basolateral plasma membrane (Dominguez et al., 1985). CLT could either be taken up in an endosomal vesicle that will then fuse with the basolateral membrane or the toxin ADP-ribosylated G protein could travel via the same way to the basolateral membrane. For the present, it is sufficient to note that in any case A1 will preserve the same membrane topology with its NAD and G protein-binding sites exposed to the cytoplasm.

Pertussis toxin

The difficulty in obtaining large amounts of purified PTx and of its receptor have so far prevented comparable experiments with model membranes. However, there are two major similarities between CLT and PTx: reduction is required to release their enzymic activity (Moss et al., 1983) and ADP-ribosylation of G proteins takes place on the cytoplasmic face of the membrane. In the case of PTx, ATP may play an important role because it efficiently promotes the dissociation of the active subunit from the binding part (Burns and Manclark, 1986).

To understand the mechanism of membrane insertion of PTx it will be important to determine the exact mode of cell binding, particularly with respect to lipids. In a model system it was found that the two larger subunits of the binding part are able to interact with lipids and may be involved in assisting the membrane penetration of the active subunit (Montecucco et al., 1986a).

Membrane penetration and translocation of toxins with targets located in the cytoplasm

Another group of bacterial protein toxins, including diphtheria toxin (DT), exotoxin A from *Pseudomonas aeruginosa* (Exo-A) and the shiga and shiga-like toxins, kill cells by blocking protein synthesis. DT and Exo-A ADP-ribosylate specifically elongation factor 2 thereby blocking its activity (Honjo et al., 1968; Uchida, 1982). Shiga toxins remove a single adenine residue (A-4324) from the 28S ribosomal RNA, thus inhibiting polypeptide chain elongation because of a block of EF1-dependent aminoacyl-tRNA binding to ribosomes (Endo and Tsurugi, 1987; Endo et al., 1988).

From the cytoplasmic location of the target sites of all these toxins it follows that at least their catalytic subunit must leave the plasma membrane, thus showing only a transient membrane interaction.

Diphtheria toxin

After the discovery that DT can be induced to enter into cells from the plasma membrane by lowering the pH of the external medium (Draper and Simon, 1980; Sandvig and Olsnes, 1980, 1981), different techniques have been employed in the study of the membrane insertion and translocation of DT.

By applying DT to planar lipid bilayers at different pHs it was found that (i) at low pH DT, its truncated mutant CRM 45 (lacking a 16.5 kDa C-terminal fragment) and its B fragment form ion channels of low conductance, (ii) more than one toxin molecule is involved in the formation of a channel, and (iii) negatively charged phospholipids are required (Donovan et al., 1981, 1982; Kagan et al., 1981; Hoch et al., 1985). Analysis of the change of permeability induced by DT on potassium-loaded liposomes led essentially to the same conclusions (Boquet and Duflot, 1982; Shiver and Donovan, 1987). These findings were taken as evidence for the tunnel model for the membrane translocation of diphtheria toxin proposed by Boquet et al. (1976). Fragment B was suggested to insert into the lipid bilayer at low pH and to form a transmembrane hydrophilic pore large enough to allow the passage of the enzymic A chain in an unfolded form (Kagan et al., 1981; Hoch et al., 1985).

Subsequently it was found that DT at acidic pHs forms ion channels also across the plasma membrane of living cells; these channels show a specificity for monovalent cations, allow the uptake only of very small molecules and remain open as long as there is a pH gradient across the membrane (Sandvig and Olsnes, 1988; Papini et al., 1988). These results, obtained with cells in vitro, cannot

simply be taken as evidence for a tunnel model because no close correlation was found between the two cellular effects of DT, namely block of protein synthesis and channel formation, as one would have expected if formation of a transmembrane hydrophilic tunnel is a prerequisite for the translocation of chain A into the cytoplasm (Papini et al., 1988; Alder et al., 1990).

Several studies have demonstrated and characterized the low-pH-induced structural change of DT in relation to model membranes: liposome binding of DT (Alving et al., 1980), membrane photolabelling with photoreactive lipids (Zalman and Wisnieski, 1984; Hu and Holmes, 1984) or with photoreactive phospholipids (Montecucco et al., 1985; Papini et al., 1987a, c) or with a small lipophilic probe (Dumont and Richards, 1988), DT fluorescence (Blewitt et al., 1985; Zhao and London, 1988; Chung and London, 1988) and circular dichroism (Hu and Holmes, 1984; Blewitt et al., 1985), DT-induced liposome fusion assayed by light scattering (Cabiaux et al., 1984; Papini et al., 1987b) or by fluorescence energy transfer (Papini et al., 1987b), DT-induced liposome leakage measured by ESR (Lai et al., 1984) or membrane potential (Shiver and Donovan, 1987) or calcein fluorescence recovery (Defrise-Quertain et al., 1989; Jang et al., 1989a, b), scanning calorimetry (Ramsay et al., 1989), polarized infrared spectroscopy (Cabiaux et al., 1989) and lipid monolayer (Demel et al., 1991). Given the wide variety of experimental approaches listed above, it is astonishing to note their agreement on the main findings and it is also reassuring that, whenever a comparison was possible, a close correlation was found between results obtained with model systems and with cells in culture, as summarized in Fig. 2.

Biologically relevant results derived from the above-quoted papers are summarized here:

(1) At neutral pH DT interacts with negatively charged liposomes; this interaction involves both A and B chains and it is limited to the head group level of phospholipids; as a consequence there is a 10% increase of α-helix content.

As discussed elsewhere, this membrane surface interaction of DT may be relevant to its cell surface binding (Papini et al., 1987a). It was proposed that DT first lands on negative lipidic domains of the cell plasma membrane, where it binds with a low affinity constant K_{lipids}, and then moves laterally

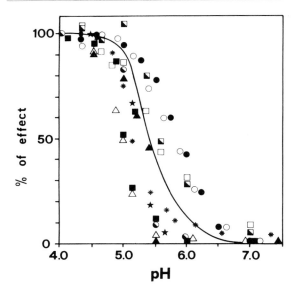

Figure 2. pH dependence of the membrane interaction of diphtheria toxin. The solid curve represents the pH dependence of DT cell intoxication, measured as inhibition of protein synthesis or alteration of monovalent cation content of cells exposed to DT in media of different pH values (Sandvig and Olsnes, 1986; Papini et al., 1987c, 1988). Points refer to the interaction of DT with model membranes assayed at different pH values with the following techniques: light scattering (○), liposome fusion measured by fluorescence energy transfer (●), membrane photolabelling with a surface (□) and a deep probe (◪), fluorescence assay of membrane potential (★), differential scanning calorimetry (◓), protein fluorescence (■), micelle protein fluorescence quenching (△), calcein release (▲) and lipid monolayers (∗).

to hit and bind to its protein receptor (Cieplak et al., 1987; Mekada et al., 1988), with a higher affinity constant $K_{protein}$. At this stage DT is bound to the cell both via a lipid interaction and a protein interaction and its cell association constant will be:

$$K_{cell} = K_{lipids} \times K_{protein}$$

Such a model explains the protective effect against DT cell intoxication of phospholipase C that, by removing the negatively charged phosphate group of phospholipids, prevents the lipid interaction of the toxin (Moehring and Crispell, 1974; Olsnes et al., 1985). Such a multiple mode of cell binding, with single low-affinity interactions that result in a strong overall binding, appears to be a general binding strategy of several pathogenic agents including enterotoxins, clostridial neurotoxins and several viruses such as influenza and polio viruses (Montecucco, 1986).

(2) The lowering of pH induces a true structural change in DT from a neutral conformation to an acid conformation. While the neutral form is water-solvated and stable for prolonged periods in water solution, the acidic form is characterized by the exposure of previously hidden hydrophobic surfaces, that are responsible for its aggregation in water solutions. The neutral form of DT is 30% β-sheet, around 5% low-frequency β-sheet and around 25% α-helix, that increases to over 35% in the presence of negatively charged liposomes. The acidic form of DT has a lower overall α-helix content and an increased amount of β-sheet structure with strong hydrogen bonds; its hydrophobicity makes it able to penetrate into the hydrophobic core of lipid micelles or bilayers with a strong preference for negatively charged lipids.

(3) Both the A and B fragments of DT are involved in the low pH-induced interaction of the toxin with the fatty acid portion of phospholipids; such interaction extends all along the hydrocarbon chains. In this process α-helices of the inner part of the B chain orient parallel to the lipid acyl chains, while β-sheets of the N-terminal part of the B chain orient parallel to the membrane surface.

(4) The neutral–acid pH transition of DT is characterized by the exposure of the interchain disulphide bridge, that becomes accessible to reduction by thioreductase (Moskaug et al., 1987).

(5) Figure 2 is a collection of data obtained with different techniques and it shows that the transition between the neutral and acidic forms of DT occurs between pH 6 and 5 in the presence of negatively charged liposomes; the fitting of these data with the pH dependence of DT cell intoxication (solid line) is remarkable, particularly if one considers that they have been collected in different laboratories with different toxin batches under a variety of experimental set-ups. It should be noted that the acid transition of DT is virtually complete before pH 5, the lowest pH value that has been detected inside endosomes in living cells.

(6) Negatively charged lipids may play a multiple role in this process: (a) to lower the pH near the membrane surface, though the relevance of this local pH effect is difficult to estimate because it decreases sharply with distance and the arrangement with respect to the membrane of DT residues involved in the pH-induced transition is not known (Chung and London, 1988); (b) to concentrate

DT onto the membrane surface via electrostatic interactions; and (c) to shield some toxin charges during its insertion into the hydrophobic core of the lipid bilayer (Montecucco *et al.*, 1985).

(7) The low pH-induced conformational change of chain A of DT is fully reversible, while that of chain B is partial; this result provides experimental evidence for the escape of chain A from the membrane when pH is returned to neutrality, while chain B remains membrane embedded.

(8) Chain B appears to be the first part of DT that inserts into the membrane, as suggested by the different kinetic and pH dependence of the lipid interaction of the two chains; the membrane insertion of chain B is not the rate-limiting step of DT membrane translocation.

(9) It is likely that the membrane translocation competent form of DT is actually a multimer-aggregated toxin, even though there are no precise data on the number of molecules involved.

To account for all these findings a 'cleft' model for the membrane translocation of DT has been proposed (Bisson and Montecucco, 1987). The various steps envisaged by such a model are depicted in Fig. 3. Cell binding is the result of the interactions of DT with both a protein receptor and negatively charged lipids. The C-terminal region of protomer B is mainly responsible for the interaction with the protein receptor, while the interchain disulphide bond is hidden in the protein interior. The receptor-bound DT is internalized via coated pits, that eventually merge into an endosome. Following the acidification of the endosomal lumen, DT changes conformation and detaches from its protein receptor because of a reduced affinity of the acid form of DT.

Fragment B is the first part of DT that changes structure and becomes hydrophobic, thereby inserting into the lipid bilayer with its newly exposed hydrophobic surfaces interacting with lipids. DT may do so as a single molecule or in a multimeric form: Fig. 3 shows a dimer though there are no precise data on this point.

Given that the apparent pK_a of B chain membrane insertion is about 6, the protonation of histidine residues (15 out of the 16 such residues present in DT are clustered in chain B) is expected to play a major role in the conformational transition of chain B. Part of the 16-kDa C-terminal fragment remains outside the membrane, exposed to the endosomal lumen.

In such a process fragment(s) B forms a hydrophilic cleft(s) that drives the insertion of chain(s) A with their hydrophobic segments exposed to lipids and their hydrophilic regions interacting with corresponding regions of the hydrophilic cleft(s) of fragment(s) B. Such a matching of hydrophobic and hydrophilic protein–lipid and protein–protein interactions is required in order to minimize the energetic cost of the translocation process.

This model can accommodate the possibility that the DT receptor participates in the membrane translocation of protomer A (Stenmark *et al.*, 1988) by assuming that it cooperates with fragment B in forming the hydrophilic cleft that nests chain A during its passage through the lipid bilayer; it is possible that the efficiency of the chain-A–membrane-translocating unit formed by chain B plus DT receptor in cells is higher than that formed by fragment B alone in model membranes.

The C-terminal part of protomer A is the first part to reach the cytoplasm driven by its disulphide binding with the N-terminal portion of protomer B. In such a process the interchain disulphide bond, previously hidden in the hydrophobic protein interior, becomes exposed to the water solvent of the cytoplasm and hence accessible to both chemical and proteic cytosolic reducing agents. The membrane insertion of the enzymic fragment A has an apparent pK_a near to 5 and hence the protonation of carboxyl residues is expected to play an important role in rendering chain A able to penetrate into the membrane.

It is likely that the membrane-penetrating form of DT has a net positive charge and that some Lys, Arg or His residues are not in the position to form salt bridges with corresponding negatively charged Asp and Glu residues. Due to the high energetic cost of driving charged groups across the hydrophobic core of the lipid bilayer (particularly the positive charges of Lys and Arg), membrane insertion and translocation of DT would be made less energetically unfavourable if these charges are neutralized and shielded by formation of ionic couples with phospholipid head groups and small anions such as chloride, bromide or thiocyanate. This effect would help to explain the finding that anions are involved in the penetration of DT into cells and that the more hydrophobic membrane-permeant anions are the more efficient (Sandvig and Olsnes, 1986; Moskaug

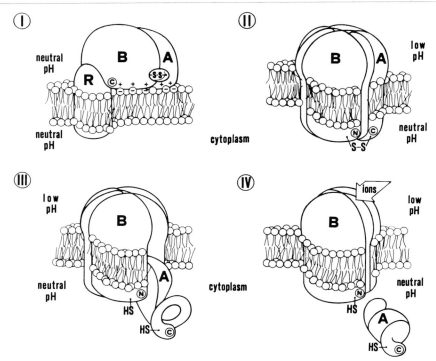

Figure 3. Membrane binding and translocation of diphtheria toxin. (I) DT binds to the cell surface of DT-sensitive cells via a protein–protein interaction with its proteic receptor (R) and via electrostatic interactions with negatively charged lipids. The C-terminal region of protomer B of DT is involved in binding R, while its N-terminal part is kept near the C-terminus of protomer A by the interchain disulphide bond, which is hidden in the hydrophobic protein interior. (II) At low pH DT changes conformation and detaches from R. Protomer B becomes hydrophobic before A and inserts into the lipid bilayer. In a monomeric or multimeric form, alone or together with R, chain B forms a hydrophilic cleft that wraps the hydrophilic surfaces of protomer A during its membrane penetration, while the hydrophobic surfaces of chain A are exposed lipids. It is proposed that some toxin charges are neutralized during membrane translocation by counterions of phospholipid head groups or small anions. A matching of hydrophilic (protein–protein and protein–water) and hydrophobic (protein–protein and protein–lipid) interactions is required to render the process energetically feasible. The first part of chain A that crosses the membrane is its C-terminus linked to the N-terminus of chain B with the exposure of the interchain disulphide to the cytosol. (III) Reduction by glutathione or by other cytoplasmic reducing agents, neutral cytosolic pH and possibly the transmembrane potential and the presence of NAD induce the refolding of protomer A to its neutral and catalytically active form. (IV) The now water-soluble chain A leaves the hydrophilic wrapping cleft, made of B or B plus receptor, the margins of the cleft are expected to tighten up in order to minimize the energetically unfavourable interaction with lipids. Even so this will leave a transmembrane alignment of hydrophilic residues that will constitute a transmembrane ion channel. Due to its reduced size, this channel will accommodate only small ions and it is expected to have a low conductance. A transmembrane proton gradient is required for the opening of this channel.

et al., 1989). In other words, it is proposed that the membrane-translocating form of DT is actually a transient anion–phospholipid–multimeric toxin complex that assembles on the low-pH lumen side of the endosomal membrane and lasts until it reaches the cytoplasmic face of the membrane, where the complex disassembles because its ion couples are released by water solvation. Such a proposal emphasizes the role of non-bilayer lipidic configurations in DT membrane translocation, though

direct evidence for such structures is lacking. However, this suggestion is in keeping with the well-documented ability of DT to induce fusion of lipid vesicles at low pH that is believed to occur via zones of DT-induced lipid destabilization.

The finding that chain A can escape from the liposomes once the pH is returned to neutrality suggests that, after reduction of the interchain disulphide bond (Moskaug *et al.*, 1987), fragment A can reacquire its neutral water-soluble form again when it faces the neutral cytoplasmic pH. In contrast, fragment B will remain embedded in the membrane because its membrane insertion is irreversible.

After chain A has reached the cytoplasm, the margins of the hydrophilic cleft(s) embedded in the lipid bilayer, formed by promoter(s) B (or promoter B plus DT receptor), are expected to tighten up in order to reduce to a minimum the amount of hydrophilic protein surface exposed to lipids. This will leave a transmembrane alignment of hydrophilic residues that would constitute a peculiar flat-shaped ion channel with more or less rigid protein walls and a very flexible oily side made of lipids. This channel is expected to have a low conductance because of its reduced size and at the same time to be able to accommodate large charged objects provided that they have a hydrophobic flat portion such as an aromatic ring. Such features would account for the finding that the DT channel has a low permeability to choline or glucosamine and a higher permeability to tryptophan methyl ester or to TEMPO-choline (Lai *et al.*, 1984; Sandvig and Olsnes, 1988; our unpublished data). It appears that the DT ion channel does not contribute significantly to the killing of cells by DT (Papini *et al.*, 1988; Alder *et al.*, 1990).

In this simplified model the major driving force for the membrane insertion and translocation of DT is the transmembrane proton gradient because it is the low pH that causes the conformational transition on one side of the membrane and it is the neutral pH that induces the refolding of chain A on the other side of the membrane. However, a contribution of the plasma membrane potential is suggested by cellular studies (Marnell *et al.*, 1984; Neville and Hudson, 1986; Hudson *et al.*, 1988) and the involvement of a protein electrophoretic effect in the membrane translocation of fragment A should be tested in future studies. Also, the role of disulphide reduction and of NAD binding in the refolding of protomer A in the cytoplasm remains to be investigated. Since NAD is a substrate of fragment A, it is very likely that it accelerates the reacquisition of the active water-soluble and NAD-binding conformation of fragment A, as suggested by a differential scanning calorimetry study of the thermal denaturation of DT (Kyger and Wright, 1984). In such a way also the transmembrane NAD gradient could contribute to the translocation of fragment A into the cytoplasm.

Pseudomonas aeruginosa exotoxin A

Much less is known about the mechanism of cell entry of Exo-A. The sequence homology between the two toxins is limited to the segments forming the NAD-binding site with little or no similarity elsewhere. Fluorescence and membrane photolabelling studies indicate that this toxin also undergoes an acid-driven structural transition with exposure of hydrophobic surfaces that enables it to interact with the hydrocarbon chains of lipids (Farahbakhsh *et al.*, 1987; Farahbakhsh and Wisnieski, 1989).

Shiga toxins

Cell binding of shiga toxins resembles that of CLT because it is mediated by a multi-subunit binding part with specific sites for a Gal–Gal sugar moiety and because reduction appears to expose hydrophobic surfaces (Sandvig and Brown, 1987). However, no model studies devoted to unravel its mechanism of membrane translocation have been performed so far.

Membrane interaction of toxins with unknown intracellular targets

There is a large group of bacterial protein toxins, very relevant for their pathogenic effects, whose targets inside the cells are unknown. This group includes the *Bacillus anthracis* lethal toxin and the clostridial neurotoxins (tetanus neurotoxin and seven different botulinum neurotoxins) that are known to block neurotransmitter release. As discussed by Simpson (1986), there is evidence that these neurotoxins also enter into cells via receptor-mediated endocytosis.

Experiments with model membrane systems suggest that the clostridial neurotoxins at neutral pH interact with the surface of liposomes via: (i) an unspecific interaction with negatively charged lipids and (ii) a more specific interaction with polysialogangliosides (Montecucco *et al.*, 1988). At low pH the clostridial neurotoxins become more hydrophobic and able to penetrate into lipid bilayers with both their heavy and light chains interacting with lipids. In this process a low conductance channel is formed with the involvement of a 50-kDa N-terminal fragment of the heavy subunit; this channel is voltage-dependent and permeable to molecules up to 700 Da (Boquet and Duflot, 1982; Boquet *et al.*, 1984; Roa and Boquet, 1985; Hoch *et al.*, 1985; Montecucco *et al.*, 1986b, 1989; Donovan and Middlebrook, 1986; Blaustein *et al.*, 1987; Shone *et al.*, 1987; Gambale and Montal, 1988; Menestrina *et al.*, 1989). Taken together, these data, though less extensive, support a mechanism of cell entry similar to that of DT.

Acknowledgements

We thank Drs R. Bisson and G. Menestrina for critical reading of the manuscript and we acknowledge the support of grants from the CNR Target Project on Biotechnology and Bioinstrumentation.

References

Alder, G.M., Bashford, C.L. and Pasternak, C.A. (1990). Action of diphtheria toxin does not depend on the induction of large, stable pores across biological membranes. *J. Membr. Biol.* **113**, 67–74.

Alving, C.E., Iglewski, B.H., Urban, K.A., Moss, J., Richards, R.L. and Sadoff, J.C. (1980). Binding of diphtheria toxin to phospholipids in liposomes. *Proc. Natl. Acad. Sci. USA* **77**, 1986–90.

Bayley, H. (1983). *Photogenerated Reagents in Biochemistry and Molecular Biology*. Elsevier Biomedical Press, Amsterdam.

Bisson, R. and Montecucco, C. (1981). Photolabelling of membrane proteins with photoactive phospholipids. *Biochem. J.* **193**, 757–93.

Bisson, R. and Montecucco, C. (1985). Use of photoreactive phospholipids for the study of lipid–protein interactions. In: *Progress in Protein–Lipid Interactions* (eds A. Watts and J.J.H.H.M. DePont), pp. 259–87. Elsevier, Amsterdam.

Bisson, R. and Montecucco, C. (1987). Diphtheria toxin membrane translocation: an open question. *Trends Biochem. Sci.* **12**, 181–2.

Blaustein, R.O., Germann, W.J., Finkelstein, A. and DasGupta, B.R. (1987). The N-terminal half of the heavy chain of botulinum type A neurotoxin forms channels in planar phospholipid bilayer. *FEBS Lett.* **226**, 115–20.

Blewitt, M.G., Chung, L.A. and London, E. (1985). Effect of pH on the conformation of diphtheria toxin and its implications for membrane penetration. *Biochemistry* **24**, 5458–64.

Boquet, P. and Duflot, E. (1982). Tetanus toxin fragment forms channels in lipid vesicles at low pH. *Proc. Natl. Acad. Sci. USA* **79**, 7614–18.

Boquet, P., Silverman, M.S., Pappenheimer, A.M. and Vernon, W.B. (1976). Binding of Triton X-100 to diphtheria toxin, crossreacting material 45 and their fragments. *Proc. Natl. Acad. Sci. USA* **73**, 4449–53.

Boquet, P., Duflot, E. and Hauttecoeur, B. (1984). Low pH induces hydrophobic domain in the tetanus toxin molecule. *Eur. J. Biochem.* **144**, 339–44.

Brady, R.O. and Fishman, P.H. (1979). Biotransducers of membrane-mediated information. *Adv. Enzymol.* **50**, 303–21.

Burns, D. and Manclark, C.R. (1986). Adenine nucleotides promote dissociation of pertussis toxin subunits. *J. Biol. Chem.* **261**, 4324–7.

Cabiaux, V., Vandenbranden, M., Falmagne, P. and Ruysschaert, J.M. (1984). Diphtheria toxin induces fusion of small unilamellar vesicles at low pH. *Biochim. Biophys. Acta* **775**, 31–6.

Cabiaux, V., Brasseur, R., Wattiez, R., Falmagne, P., Ruysschaert, J.M. and Goormaghtigh (1989). Secondary structure of diphtheria toxin and its fragments interacting with acidic liposomes studied by polarized infrared spectroscopy. *J. Biol. Chem.* **264**, 4928–38.

Chung, L.A. and London, E. (1988). Interaction of diphtheria toxin with model membranes. *Biochemistry* **27**, 1245–53.

Cieplak, W., Gaudin, H.M. and Eidels, L. (1987). Diphtheria toxin receptor. *J. Biol. Chem.* **262**, 13246–53.

Defrise-Quertain, F., Cabiaux, V., Vandenbranden, M., Wattiez, R., Falmagne, P. and Ruysschaert, J.M. (1989). pH-dependent bilayer destabilization and fusion of phospholipidic large unilamellar vesicles induced by diphtheria toxin and its fragments A and B. *Biochemistry* **28**, 3406–13.

Demel, R., Schiavo, G., de Kruijff, B. and Montecucco, C. (1991). Lipid interaction of diphtheria toxin and mutants: a study with phospholipid and protein monolayers. *Eur. J. Biochem.* **197**, 481–6.

Dominguez, P., Barros, F. and Lazo, P.S. (1985). The activation of adenylate cyclase from small intestinal epithelium. *Eur. J. Biochem.* **146**, 533–8.

Donovan, J.J. and Middlebrook, J.L. (1986). Ion-conducting channels produced by botulinum toxin in planar lipid membranes. *Biochemistry* **25**, 2872–6.

Donovan, J.J., Simon, M.I., Draper, R.K. and Montal, M. (1981). Diphtheria toxin forms transmembrane channels in planar lipid bilayers. *Proc. Natl. Acad. Sci. USA* **78**, 172–6.

Donovan, J.J., Simon, M. and Montal, M. (1982). Insertion of diphtheria toxin into and across membranes: role of phosphoinositide asymmetry. *Nature* **298**, 669–72.

Draper, R.K. and Simon, M.I. (1980). The entry of diphtheria toxin into the mammalian cell cytoplasm: evidence for lysosomal involvement. *J. Cell Biol.* **87**, 849–54.

Dumont, M. and Richards, F.M. (1988). The pH-dependent conformational change of diphtheria toxin. *J. Biol. Chem.* **263**, 2087–97.

Dwyer, J.D. and Bloomfield, V.A. (1982). Subunit arrangement of cholera toxin in solution and bound to receptor-containing model membranes. *Biochemistry* **21**, 3227–31.

Endo, Y. and Tsurugi, K. (1987). RNA N-glycosidase activity of ricin A chain. *J. Biol. Chem.* **262**, 8128–30.

Endo, Y., Tsurugi, K., Yutsudo, T., Takeda, Y., Ogasawara, T. and Igarashi, K. (1988). Site of action of a Vero toxin from *Escherichia coli* 0157:H7 and of Shiga toxin on eukaryotic ribosomes. *Eur. J. Biochem.* **171**, 45–50.

Farahbakhsh, Z.T. and Wisnieski, B.J. (1989). The acid-triggered pathway of *Pseudomonas* exotoxin A. *Biochemistry* **28**, 580–5.

Farahbakhsh, Z.T., Baldwin, R.L. and Wisnieski, B.J. (1987). Effect of low pH on the conformation of *Pseudomonas* exotoxin A. *J. Biol. Chem.* **262**, 2256–61.

Fishman, P., Moss, J. and Osborne, J.C. (1978). Interaction of choleragen with the oligosaccharide of ganglioside G_{M1}: evidence for multiple oligosaccharide binding sites. *Biochemistry* **17**, 711–16.

Gambale, F. and Montal, M. (1988). Characterization of the channel properties of tetanus toxin in planar lipid bilayers. *Biophys. J.* **53**, 771–83.

Gill, D.M. (1976) The arrangements of subunits in cholera toxin. *Biochemistry* **15**, 1242–8.

Gill, D.M. and King, C.A. (1975). The mechanism of action of cholera toxin in pigeon erythrocyte lysates. *J. Biol. Chem.* **250**, 6424–32.

Groins, B. and Freire, E. (1985). Lipid phase separations induced by the association of cholera toxin to phospholipid membranes containing ganglioside GM1. *Biochemistry* **24**, 1791–7.

Hoch, D.H., Romero-Mira, M., Ehrlich, B.E., Finkelstein, A., DasGupta, B.R. and Simpson, L.L. (1985). Channels formed by botulinum, tetanus, and diphtheria toxins in planar lipid bilayers: relevance to translocation of proteins across membranes. *Proc. Natl. Acad. Sci. USA* **82**, 1692–6.

Honjo, T., Nishizuka, Y., Hayaishi, O. and Kato, I. (1968). Diphtheria toxin-dependent adenosine diphosphate ribosylation of aminoacyl transferase II and inhibition of protein synthesis. *J. Biol. Chem.* **243**, 3553–5.

Hu, V.W. and Holmes, R.K. (1984). Evidence for direct insertion of fragment A and B of diphtheria toxin into model membranes. *J. Biol. Chem.* **259**, 12226–33.

Hudson, T.H., Scharff, J., Kimak, A.G. and Neville, D.M. (1988). Energy requirements for diphtheria toxin translocation are coupled to the maintenance of a plasma membrane potential and a proton gradient. *J. Biol. Chem.* **263**, 4773–81.

Jiang, G., Solow, R. and Hu, V.W. (1989a). Characterization of diphtheria toxin-induced lesions in liposomal membranes. *J. Biol. Chem.* **264**, 13424–9.

Jiang, G., Solow, R. and Hu, V.W. (1989b). Fragment A of diphtheria toxin causes pH-dependent lesions in model membranes. *J. Biol. Chem.* **264**, 17170–3.

Kagan, B.L., Finkelstein, A. and Colombini, M. (1981). Diphtheria toxin fragment forms large pores in phospholipid bilayer membranes. *Proc. Natl. Acad. Sci. USA* **78**, 4950–4.

Kyger, E. and Wright, H.T. (1984). Thermal stability of different forms of diphtheria toxin. *Arch. Biochem. Biophys.* **228**, 569–76.

Lai, C.S., Kushnaryov, V., Panz, T. and Basosi, R. (1984). Diphtheria toxin induces leakage of acidic liposomes. *Arch. Biochem. Biophys.* **234**, 1–6.

Ludwig, D.S., Ribi, H.O., Schoolnik, G.K. and Kornberg, R.D. (1986). Two-dimensional crystals of cholera toxin B-subunit-receptor complexes: projected structure at 17-Å resolution. *Proc. Natl. Acad. Sci. USA* **83**, 8585–8.

Marnell, M.H., Shia, S.P., Stookey, M. and Draper, R.K. (1984). Evidence for penetration of diphtheria toxin to the cytosol through a prelysosomal membrane. *Infect. Immun.* **44**, 145–50.

Mekada, E., Okada, Y. and Uchida, T. (1988). Identification of diphtheria toxin receptor and a nonproteinous diphtheria toxin-binding molecules in Vero cell membrane. *J. Cell Biol.* **107**, 511–19.

Menestrina, G., Forti, S. and Gambale, F. (1989). Interaction of tetanus toxin with lipid vesicles: effect of pH, surface charge and transmembrane potential on the kinetics of channel formation. *Biophys. J.* **55**, 393–405.

Moehring, T.J. and Crispell, J.P. (1974). Enzyme treatment of KB cells: the altered effect of diphtheria toxin. *Biochem. Biophys. Res. Commun.* **60**, 1446–52.

Montecucco, C. (1986). How do tetanus and botulinum toxins bind to neuronal membranes? *Trends Biochem. Sci.* **11**, 314–17.

Montecucco, C. (1988). Photoreactive lipids for the study of membrane-penetrating toxins. *Methods Enzymol.* **165**, 347–57.

Montecucco, C., Schiavo, G. and Tomasi, M. (1985). pH-dependence of the phospholipid interaction of diphtheria-toxin fragments. *Biochem. J.* **231**, 123–8.

Montecucco, C., Schiavo, G., Brunner, J., Duflot, E., Boquet, P. and Roa, M. (1986a). Tetanus toxin is labeled with photoactivatable phospholipids at low pH. *Biochemistry* **25**, 919–24.

Montecucco, C., Tomasi, M., Schiavo, G. and Rappuoli, R. (1986b). Hydrophobic photolabelling of pertussis toxin subunits interacting with lipids. *FEBS Lett.* **194**, 301–4.

Montecucco, C., Schiavo, G., Gao, Z., Bauerlein, E., Boquet, P. and DasGupta, B.R. (1988). Interaction of botulinum and tetanus toxins with the lipid bilayer surface. *Biochem. J.* **251**, 379–83.

Montecucco, C., Schiavo, G. and DasGupta, B.R. (1989). Effect of pH on the interaction of botulinum neurotoxins A, B and E with liposomes. *Biochem. J.* **259**, 47–53.

Moskaug, J.O., Sandvig, K. and Olsnes, S. (1987). Cell-mediated reduction of the interfragment disulfide in nicked diphtheria toxin. *J. Biol. Chem.* **262**, 10339–45.

Moskaug, J.O., Sandvig, K. and Olsnes, S. (1989). Role of anions in low-pH-induced translocation of diphtheria toxin. *J. Biol. Chem.* **264**, 11367–72.

Moss, J., Stanley, S.J., Morin, J.E. and Dixon, J.E. (1980).

Activation of choleragen by thiol:protein disulfide oxidoreductase. *J. Biol. Chem.* **255**, 11085–7.

Moss, J., Stanley, S.J., Burns, D.L., Hsia, J.A., Yost, D.A., Myers, G.A. and Hewlett, E.L. (1983). Activation by thiol of the latent NAD glycohydrolase and ADP ribosyltransferase activities of *Bordetella pertussis* toxin. *J. Biol. Chem.* **258**, 11879–82.

Neville, D.M. and Hudson, T.H. (1986). Transmembrane transport of diphtheria toxin, related toxins and colicins. *Ann. Rev. Biochem.* **55**, 195–224.

Olsnes, S., Carvajal, E., Sundan, A. and Sandvig, K. (1985). Evidence that membrane phospholipids and protein are required for binding of diphtheria toxin in Vero cells. *Biochim. Biophys. Acta* **846**, 334–41.

Papini, E., Colonna, R., Schiavo, G., Cusinato, F., Tomasi, M., Rappuoli, R. and Montecucco, C. (1987a). Diphtheria toxin and its mutant crm 197 differ in their interaction with lipids. *FEBS Lett.* **215**, 73–8.

Papini, E., Colonna, R., Cusinato, F., Montecucco, C., Tomasi, M. and Rappuoli, R. (1987b). Lipid interaction of diphtheria toxin and mutants with altered fragment B: liposome aggregation and fusion. *Eur. J. Biochem.* **169**, 629–35.

Papini, E., Schiavo, G., Tomasi, M., Colombatti, M., Rappuoli, R. and Montecucco, C. (1987c). Lipid interaction of diphtheria toxin and mutants with altered fragment B: hydrophobic photolabelling and cell intoxication. *Eur. J. Biochem.* **169**, 637–44.

Papini, E., Sandonà, D., Rappuoli, R. and Montecucco, C. (1988). On the membrane translocation of diphtheria toxin: at low pH the toxin induces ion channels on cells. *EMBO J.* **7**, 3353–9.

Ramsay, G., Montgomery, D., Berger, D. and Freire, E. (1989). Energetics of diphtheria toxin membrane interaction and translocation: calorimetric characterization of the acid pH induced transition. *Biochemistry* **28**, 529–33.

Reed, R.A., Mattai, J. and Shipley, G.G. (1987). Interaction of cholera toxin with ganglioside G_{M1} receptors in supported lipid monolayers. *Biochemistry* **26**, 824–32.

Ribi, H.O., Ludwig, D.S., Mercer, K.L., Schoolnik, G.K. and Kornberg, R.D. (1988). Three-dimensional structure of cholera toxin penetrating a lipid membrane. *Science* **239**, 1272–6.

Roa, M. and Boquet, P. (1985). Interaction of tetanus toxin with lipid vesicles at low pH. *J. Biol. Chem.* **260**, 6827–35.

Sandvig, K. and Brown, J.E. (1987). Ionic requirements for entry of Shiga toxin from *Shigella dysenteriae* 1 into cells. *Infect. Immun.* **55**, 298–303.

Sandvig, K. and Olsnes, S. (1980). Diphtheria toxin entry into cells is facilitated by low pH. *J. Cell Biol.* **87**, 828–32.

Sandvig, K. and Olsnes, S. (1981). Rapid entry of nicked diphtheria toxin into cells at low pH. *J. Biol. Chem.* **256**, 9068–76.

Sandvig, K. and Olsnes, S. (1986). Interactions between diphtheria toxin entry and anion transport in Vero cells: evidence that entry of diphtheria toxin is dependent on efficient anion transport. *J. Biol. Chem.* **261**, 1570–5.

Sandvig, K. and Olsnes, S. (1988). Diphtheria toxin-induced channels in Vero cells selective for monovalent cations. *J. Biol. Chem.* **263**, 12352–9.

Shiver, J.W. and Donovan, J.J. (1987). Interactions of diphtheria toxin with lipid vesicles: determinants of ion channel formation. *Biochim. Biophys. Acta* **903**, 48–55.

Shone, C.C., Hambleton, P. and Melling, J. (1987). A 50-kDa fragment from the NH_2-terminus of the heavy subunit of *Clostridium botulinum* type A neurotoxin forms channels in lipid vesicles. *Eur. J. Biochem.* **167**, 175–80.

Simpson, L.L. (1986). Molecular pharmacology of botulinum toxin and tetanus toxin. *Ann. Rev. Pharmacol. Toxicol.* **26**, 427–53.

Sixma, T.K., Pronk, S.E., Kalk, K.H., Wartna, E.S., van Zanten, B.A.M., Witholt, B. and Hol, W.G.J. (1991). Crystal structure of a cholera toxin-related heat-labile enterotoxin from *E. coli*. *Nature* **351**, 371–7.

Stenmark, H., Olsnes, S. and Sandvig, K. (1988). Requirement of specific receptors for efficient translocation of diphtheria toxin A fragment across the plasma membrane. *J. Biol. Chem.* **263**, 13449–55.

Tomasi, M. and Montecucco, C. (1981). Lipid insertion of cholera toxin after binding to G_{M1}-containing liposomes. *J. Biol. Chem.* **256**, 11177–81.

Tomasi, M., Battistini, A., Cardelli, M., Sonnino, S. and D'Agnolo, G. (1984). Interaction of cholera toxin with gangliosides: differential effects of the oligosaccharide of ganglioside G_{M1} and of micellar gangliosides. *Biochemistry* **23**, 2520–6.

Tomasi, M., Gallina, A., D'Agnolo, G. and Montecucco, C. (1986). Interaction of cholera toxin with lipid model membrane. In: *Bacterial Protein Toxins* (eds P. Falmagne, J.E. Alouf, F.J. Feherenbach, J. Jeljaszewicz and M. Thelestam), pp. 19–25. Gustav Fischer Verlag, Stuttgart.

Uchida, T. (1982). Diphtheria toxin: biological activity. In: *Molecular Action of Toxins and Viruses* (eds P. Cohen and S. Van Heyningen), pp. 1–31. Elsevier Biomedical Press, Amsterdam.

van Heyningen, S. (1977). Cholera toxin. *Biol. Rev.* **52**, 509–49.

van Heyningen, S. (1983). A conjugate of the A1 peptide of cholera toxin and the lectin of *Wisteria floribunda* that activates the adenylate cyclase of intact cells. *FEBS Lett.* **164**, 132–4.

Wisnieski, B.J. and Bramhall, J.S. (1981). Photolabelling of cholera toxin subunits during membrane penetration. *Nature* **289**, 319–21.

Zalman, L.S. and Wisnieski, B.J. (1984). Mechanism of insertion of diphtheria toxin: peptide entry and pore size determinations. *Proc. Natl. Acad. Sci. USA* **81**, 3341–5.

Zhao, J.M. and London, E. (1988). Conformation and model membrane interactions of diphtheria toxin fragment A. *J. Biol. Chem.* **263**, 15369–77.

4

Membrane Translocation of Diphtheria Toxin

Kirsten Sandvig and Sjur Olsnes

Institute for Cancer Research at the Norwegian Radium Hospital, Montebello, Oslo, Norway

Introduction

Diphtheria toxin is secreted by strains of *Corynebacterium diphtheriae* carrying the toxin gene on a
lysogenic phage (Pappenheimer and Gill, 1973;
Pappenheimer, 1977). In the disease the toxin is
produced by bacteria in the throat where large
membranes (Greek: *diphtheria*) consisting of leukocytes, fibrin and bacteria are formed and may
choke the patient. The toxin is taken up into the
general circulation and may damage a number of
organs. Most important is damage to cardiac cells
that often leads to heart failure and death. Routine
vaccination of children has made diphtheria a rare
disease in developed countries.

Diphtheria toxin is produced as a single polypeptide chain of 535 amino acid residues. Its molecular
weight is 58 342 (Greenfield *et al.*, 1983). The
toxin is easily cleaved into two polypeptides
by proteolytic enzymes present in cultures of
Corynebacterium diphtheriae and by trypsin. Such
cleavage is required for intoxication of cells (Collier
and Kandel, 1971). The two-chain structure is
held together by an interchain disulphide bond
between cysteine 186 and cysteine 201.

The two fragments formed are denoted A and
B (Fig. 1) and play different roles in the intoxication
process. The B fragment binds the toxin to cell
surface receptors (Everse *et al.*, 1977) and facilitates
the entry of fragment A into the cytosol where
this polypeptide rapidly inhibits protein synthesis
by inactivation of elongation factor 2
(Pappenheimer, 1977). Fragment A is an enzyme
that transfers ADP-ribose from NAD to an unusual

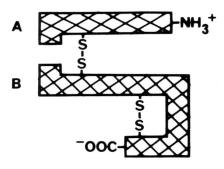

Figure I. Schematic structure of diphtheria toxin.

amino acid, diphthamide, present in the elongation
factor (Oppenheimer and Bodley, 1981; Bodley *et
al.*, 1984). Diphthamide is a post-translationally
modified histidine residue which is found in
elongation factor 2 of all eukaryotes and in
archaebacteria, but not in eubacteria (Van Ness
et al., 1980; Kessel and Klink, 1980; Moehring
et al., 1984; Moehring and Moehring, 1988; Chen
and Bodley, 1988). ADP-ribosylated diphthamide
has a strongly reduced affinity for ribosomes
(Sitikov *et al.*, 1984a; Nygård and Nilsson, 1985).
The A fragment of diphtheria toxin is so efficient
in inactivating EF2 that one single A fragment in
the cytosol is sufficient to kill the cell (Yamaizumi
et al., 1978). During early stages of cell intoxication
a peculiar pattern of synthesized protein is observed
(Battistini *et al.*, 1988). Fragment A is easily
renatured after treatment with various denaturing
agents including incubation at 100°C (Gill and
Dinius, 1971).

The ADP-ribosyltransferase activity of the A

fragment seems to involve glutamic acid 148. Photoaffinity labelling experiments first suggested that glutamic acid 148 is a constituent of the NAD-binding site (Carroll and Collier, 1984; Carroll et al., 1985). When this amino acid was substituted with aspartic acid, the toxin molecule still bound NAD, but the enzyme activity was reduced by a factor of at least 160 (Tweten et al., 1985).

The role of diphthamide in the cell is unknown. Evidence of enzymes able to ADP-ribosylate EF2 in rat and beef liver, in reticulocytes and in polyoma virus transformed baby hamster kidney cells has been presented (Iglewski et al., 1984; Sitikov et al., 1984b; Sayhan et al., 1986; Fendrick and Iglewski, 1989) suggesting a normal regulatory function for this process.

The NAD-binding site in the molecule seems to be located close to a cationic phosphate-binding region (the P site) on the B fragment (Proia et al., 1980; Collins and Collier, 1984). Polyphosphates binding to this region not only block the binding of diphtheria toxin to its receptor, but also inhibit the binding of the dinucleotide ApUp which in itself is able to inhibit binding of NAD (Collins and Collier, 1984). Also, binding of NAD site-ligands and P site-ligands is competitive (Lory et al., 1980).

After binding of diphtheria toxin to the cell surface receptors the toxin is endocytosed and exposed to low endosomal pH. The low pH induces a conformational change in the toxin molecule and translocation of fragment A into the cytosol. Since the naturally occurring receptors seem to be required for efficient toxin translocation, we will first discuss in detail what is known about the diphtheria toxin receptor and its interaction with the toxin. We next discuss the endocytic uptake of the toxin, and we then go through the requirements for membrane translocation as studied in a model system that involves induction of diphtheria toxin entry directly from the cell surface by exposure to low external pH. Finally, we discuss diphtheria toxin-induced channel formation.

The diphtheria toxin receptor

Nature of the receptor

Diphtheria toxin binds specifically to a number of different cell lines. However, there is a large variation in the number of receptors per cell as well as in the sensitivity of the cells to the toxin. Mouse cells have few if any receptors and are resistant to the toxin (Boquet and Pappenheimer, 1976; Proia et al., 1979b; Heagy and Neville, 1981; Keen et al., 1982; Didsbury et al., 1983; Chang and Neville, 1978). A factor rendering mouse cells sensitive to the toxin is located on human chromosome 5 (Athwal et al., 1985). A small fraction of erythrocytes were reported to bind diphtheria toxin specifically (Kushnaryov et al., 1984). Also, HeLa cells bind the toxin (Ittelson and Gill, 1973) and they contain about 4000 receptors per cell (Boquet and Pappenheimer, 1976). Vero cells, which are derived from African green monkey kidney, have $\sim10^5$ binding sites per cell (Middlebrook et al., 1978). The number of diphtheria toxin-binding sites both on Vero cells and on the less-sensitive CHO-K1 cells was recently shown by Schaefer et al. (1988) to vary with the cell density. These authors found that, depending on the cell density, the number of binding sites on Vero cells could vary from 50 000 to 370 000 per cell, being highest on cells grown at low density. Due to the high number of receptors and the sensitivity of these cells to the toxin, Vero cells are often used in studies of diphtheria toxin entry. Also, some other cell types seem to have a relatively high number of binding sites. The sensitivity of both A431 cells and MDCK cells to diphtheria toxin is high, and measurements with ^{125}I-labelled diphtheria toxin suggest that A431 cells have about 50 000 receptors per cell (unpublished results).

Treatment of Vero cells with proteases was shown to inhibit the binding of toxin (Moehring and Crispell, 1974; Olsnes et al., 1985), suggesting that the binding site involves a membrane protein. Also, treatment of cells with the compounds DIDS and SITS, which react with amino groups at the cell surface, strongly reduces the binding (Sandvig and Olsnes, 1984). Glycoproteins with molecular weights of $\sim150 000$ in hamster thymocytes and guinea-pig lymphocytes and glycoproteins of $\sim140 000$ and $\sim70 000$ in Vero cells have been shown to bind diphtheria toxin (Proia et al., 1979b, 1980, 1981; Eidels and Hart, 1982; Eidels et al., 1982, 1983; Hranitzky et al., 1985). However, those studies involved binding of diphtheria toxin to components in detergent extracts of cells and it is not clear whether these molecules are also involved in binding of the toxin to intact cells. Recently, Cieplak et al. (1987) showed that when ^{125}I-labelled diphtheria toxin was first bound to Vero and BS-C-1 cells which were then solubilized, and the extracts were then treated with anti-diphtheria toxin serum, toxin-binding proteins with molecular weights of 10 000–20 000 were precipitated. Furthermore, when cells with radio-iodinated surface-bound diphtheria toxin were treated with cross-linking reagents, a predominant band migrating with a molecular weight of 80 000 appeared. Unlabelled toxin inhibited the formation of this band, supporting the view that a protein of molecular weight ~10 000–20 000 is involved in the binding of diphtheria toxin. Also, Rolf et al. (1989) found that anti-idiotypic antibodies against the combining site of specific anti-diphtheria toxoid antibodies protected cells against diphtheria toxin and immunoprecipitated a cell surface protein from Vero cells with molecular weight ~15 000. However, the antiserum did not inhibit the binding of diphtheria toxin to the cells and it would be interesting to know whether this antiserum could coprecipitate from detergent-lysed cells diphtheria toxin bound to the receptor.

It is possible that the binding site for the toxin on the cell surface consists of more than one molecule. Recently, Mekada et al. (1988) showed that a membrane fraction from Vero cells obtained by an alkali extraction method contained two substances able to bind diphtheria toxin. One of these substances was sensitive to RNAase treatment and could also be isolated from mouse cells, which are known to contain few if any diphtheria toxin receptors. Furthermore, although the substance inhibited the binding of diphtheria toxin, it had no effect on binding of CRM 197 which also binds to the cell surface receptors. CRM 197 is the product of a mutated diphtheria toxin gene (Uchida et al., 1973). The authors suggest that this factor, which seems to contain RNA, could function as a second receptor on the cell surface. It could be involved in stabilization of the binding, or it could induce a conformational change in the toxin molecule. The second substance isolated could bind both diphtheria toxin and CRM 197, and such binding activity could not be observed in membrane preparations from mouse cells. Furthermore, this second substance is sensitive to proteases and the data suggest that this protein,

which has molecular weight of ~14 500, represents the diphtheria toxin receptor. The finding that neuraminidase treatment increases the sensitivity of the cells to toxin (Sandvig et al., 1978; Mekada et al., 1979) whereas tunicamycin treatment was reported to reduce the binding affinity (Hranitzky et al., 1985), suggests that a glycoprotein is involved in the binding. Consistent with this idea is the finding that both concanavalin A and wheatgerm agglutinin protected Chinese hamster V79 cells against diphtheria toxin (Draper et al., 1978). However, concanavalin A had no effect on diph-theria toxin binding to Vero cells, and the protec-tion against the toxin could be due to inhibition of endocytic uptake of the toxin (Middlebrook et al., 1979).

Also, treatment of cells with the enzymes phospholipase C and D strongly reduces the binding of diphtheria toxin to cells (Olsnes et al., 1985). It is, however, not clear if lipids are indeed involved in the binding or if the enzyme treatment leads to modification of the protein receptor. Interestingly, Friedman et al. (1982) reported that antibodies against phosphatidylinositol phosphate inhibited intoxication of CHO cells with diphtheria toxin, and Papini et al. (1987a,b,c) proposed that binding of diphtheria toxin to cells occurs both to a protein receptor and to phospholipid head groups. Important in this context is the finding that deletion mutants of diphtheria toxin lacking most of the A fragment or the whole A fragment bind 5–10 times more strongly to cells than intact diphtheria toxin (McGill et al., 1989). This could be due to better interaction of the lipid-associating domain of the N-terminal part of the B fragment with cell surface lipids after removal of the A fragment.

Interactions between diphtheria toxin and the receptor

Diphtheria toxin binds specifically to cell surface receptors with an association constant of $K_a = 9 \times 10^8$ M^{-1} at 4°C (Middlebrook et al., 1978). The binding is strongly dependent on the composition of the medium surrounding the cells. The binding is stronger in the presence of di- and trivalent cations (Sandvig and Olsnes, 1982), and it has been reported by Basosi et al. (1986) that diphtheria toxin binds divalent cations. Diphtheria toxin binds best to cells in the presence of permeant anions. In the absence of anions, in the presence of impermeant anions or in the presence of anion transport inhibitors, the binding is strongly reduced (Sandvig and Olsnes, 1984, 1986). Also, polyphosphates inhibit the binding of diphtheria toxin to its receptor (Middlebrook et al., 1978; Chang and Neville, 1978). Polymers of L-lysine likewise blocked binding of diphtheria toxin to Chinese hamster ovary cells, possibly by an interaction with negatively charged residues on the cell surface (Eidels and Hart, 1982).

When diphtheria toxin is produced by the bacteria, part of the toxin molecules bind the dinucleotide ApUp (Lory and Collier, 1980; Barbieri et al., 1981). This dinucleotide is strongly bound at 5°C, $K_d = 9$ pM (Collins et al., 1984). As mentioned in the introduction, ApUp seems to bind both to the NAD-binding site of the A fragment as well as to the P site of the molecule. Barbieri et al. (1986) found that ApUp is able to bind to toxin molecules that are already bound to the receptor, supporting the view that the P site is separate from the binding domain. However, Proia et al. (1981) found a much lower toxic effect on cells of toxin with bound ApUp than of toxin without this compound. Since ApUp dissociated at low pH the nucleotide should not interfere with translocation of A fragment from the endosome to the cytosol. The possibility exists that prebound ApUp inhibits binding of diphtheria toxin to its receptor by inducing a conformational change in the molecule. Proia et al. (1981) found that ApUp inhibited binding of diphtheria toxin to a soluble diphtheria toxin-binding molecule. Some preparations of diphtheria toxin contain dimeric and multimeric forms of the toxin. The dimer is unable to bind to the toxin receptors, but in neutral solutions it slowly dissociates to form monomers (Carroll et al., 1986).

Diphtheria-sensitive cells can be subjected to a number of treatments that lead to inhibition of toxin binding. These treatments include K+-depletion of cells and treatment of the cells with the tumour promoter TPA, with vanadate, fluoride, salicylate and menadione (vitamin K$_3$) (Middlebrook, 1981; Sandvig and Olsnes, 1981a; Sandvig et al., 1985; Olsnes et al., 1986). Some of these treatments lead to elevation of the level of phosphorylation in the cells, and the reduced binding could be due to phosphorylation of a

protein involved in the binding of diphtheria toxin. In fact, the affinity of cell surface receptors for epidermal growth factor is decreased by phosphorylation (Shoyab et al., 1979). Also, treatment with the energy inhibitors azide and deoxyglucose resulted in apparent loss of receptors for the toxin (Middlebrook, 1981).

It has been known for many years that the C-terminal 17 000-Da part of the B fragment of diphtheria toxin is involved in receptor binding (Uchida et al., 1973). However, even though the B fragment alone binds to diphtheria toxin receptors in a specific manner, fragment A also plays a role for the binding of whole toxin (Mekada and Uchida, 1985). CRM 197, which contains a single mutation in the A fragment (glycine 52 has been substituted with glutamic acid (Giannini et al., 1984)), binds 50 times more strongly to the receptors on sensitive cells than wild-type diphtheria toxin. Furthermore, as will be discussed below, removal of most of the A fragment also increases the affinity of the toxin for the receptor. The binding site may be within the last 50 amino acids (Murphy et al., 1986), and small deletions at the C-terminus suggest that removal of as few as the terminal 12 amino acids of the toxin will abolish its ability to bind to cells (our unpublished data).

After removal of 54 amino acids from the C-terminal end the protein toxin still binds ATP, supporting the view that the cationic site (the P site) which binds nucleotides is different from the receptor-binding domain. It has been suggested that the P site is involved in low-affinity interactions with membrane phospholipids (Alving et al., 1980), and it is possible that this site somehow facilitates intoxication of cells. Myers and Villemez (1988) found a higher toxic effect of hybrid toxins consisting of diphtheria toxin and concanavalin A when the hybrid contained the P site than when this site was deleted.

Wright et al. (1984) suggested that the internal disulphide bond in the B fragment plays a role in the binding of the toxin since reduction led to reduced binding. On the other hand, the mutant toxin CRM 1001 – where the last cysteine is substituted with tyrosine – was found to bind normally to cells (Papini et al., 1988).

Endocytosis of diphtheria toxin

Ultrastructural studies of receptor-mediated entry of diphtheria toxin into Vero cells were carried out by Morris and co-workers (1985). They bound biotinylated diphtheria toxin to cells and subsequently visualized the conjugate with avidin gold. They found that the toxin is slowly internalized by the coated pit/coated vesicle pathway. Methylamine prevented the movement of the toxin to coated pits, but the toxin was still internalized to about the same extent as in the absence of methylamine, presumably from uncoated areas of the cell membrane. In mouse cells diphtheria toxin appeared to be internalized mainly from non-clathrin-coated regions of the membrane (Morris and Saelinger, 1983). Only rarely was diphtheria toxin observed in coated pits in these cells. However, since ligands entering cells by different endocytic mechanisms seem to end up in the same endosomes (see below), different pathways of endocytosis in sensitive and resistant cells is probably not the reason for the different sensitivities. Keen et al. (1982) demonstrated endocytic uptake of diphtheria toxin by using fluorescently labelled toxin. These authors found colocalization of toxin and α_2-macroglobulin after 15–30 min incubation in both a human cell line and in mouse 3T3 fibroblasts. However, since it was recently shown by Tran et al. (1987) that ligands entering cells from uncoated areas of the cell membrane can after some time be found in the same vesicles as ligands entering the cell from coated pits, no conclusion as to the endocytic entry mechanism can be drawn from the study by Keen et al. (1982).

Experiments performed by Marsh (1988) suggest that conjugates containing diphtheria toxin can also enter by endocytosis both from uncoated areas of the membrane and from coated pits and in both cases intoxicate cells. Both the Thy antigen, which is excluded from coated pits, and transferrin, which is taken up by the coated pit/coated vesicle pathway, effectively route the toxin to an intracellular site from which the toxin fragment A can be translocated to the cytosol.

Transport of endocytosed diphtheria toxin fragment A into the cytosol most likely occurs from acidic endosomes. As discussed below, the low endosomal pH is required for intoxication, and the time from addition of toxin until inhibition of protein synthesis is observed is short (Sandvig and Olsnes, 1980, 1981b). Furthermore, no additional processing of nicked toxin appears to be required since translocation of toxin across the plasma membrane is induced when cells with surface-bound toxin are exposed to low pH. Also, when the temperature is lowered to inhibit fusion of endocytic vesicles with the Golgi apparatus and with lysosomes, there is still efficient entry of diphtheria toxin (Sandvig et al., 1984; Marnell et al., 1984b). Studies performed by Moynihan and Pappenheimer (1981) indicate that only a minor fraction of diphtheria toxin molecules bound to the cell surface reach the cytosol. This is in agreement with later studies on the efficiency of the membrane translocation. Even when direct transport of surface-bound toxin into the cytosol is induced by low external pH, only about 5% of the bound toxin enters the cell (Moskaug et al., 1987, 1988, 1989a,b).

Requirements for membrane penetration

Requirement for low pH and for a pH gradient

The first indication that low intracellular pH might be required for intoxication of cells by diphtheria toxin was reported by Kim and Groman (1965) who showed that addition of NH_4Cl to cells protected them against the toxin. This compound increases the pH in intracellular acidic compartments. Later, Dorland et al. (1981) found that addition of ammonium chloride to Vero cells did not inhibit the endocytic uptake of diphtheria toxin. Leppla et al. (1980) obtained similar results with chloroquine. That low pH can induce transport of receptor-bound toxin across the cell surface was first shown by Draper and Simon (1980), and by Sandvig and Olsnes (1980) who demonstrated that cells with surface-bound nicked toxin are rapidly intoxicated when exposed to pH 5.3. This low pH-induced translocation was demonstrated more directly in recent experiments performed by Moskaug et al. (1987, 1988, 1989a,b), who showed that when cells with surface-bound ^{125}I-labelled diphtheria toxin are exposed to low pH, two polypeptides with molecular weights of 25 kDa and 20 kDa become protected against externally added protease (Fig. 2). A number of compounds and conditions which protect cells against diphtheria toxin prevented the protection against pronase. The 20-kDa polypeptide appeared to be the A fragment since it was able to ADP-ribosylate elongation factor 2. When the cells were permeabilized with saponin, the 20-kDa fragment was released into the medium, suggesting that the translocated A fragment is free in the cytosol. The 25-kDa fragment, which (because of its size) must be derived from the B fragment, remained associated with the permeabilized cells, suggesting that it is inserted into the cell membrane.

Results from studies of mutant cell lines resistant to diphtheria toxin and defective in endosomal acidification support the view that acidification is crucial for intoxication. The finding that the resistance to diphtheria toxin was overcome when cells with surface-bound toxin were exposed to low pH suggests that the lack of endosomal acidification was the reason for the resistance in these cases (Robbins et al., 1983, 1984; Merion et al., 1983; Marnell et al., 1984a; Roff et al., 1986).

Several studies have shown that low pH induces conformational changes in the toxin molecule so that otherwise hidden hydrophobic areas become exposed (Sandvig and Olsnes, 1981b; Sandvig and Moskaug, 1987; Hu and Holmes, 1984; Blewitt

et al., 1984, 1985a,b; London, 1986; Zhao and London, 1986; Dumont and Richards, 1988). Thus, at low pH the toxin is able to bind Triton X-100 and Triton X-114 (Sandvig and Olsnes, 1981; Sandvig and Moskaug, 1987), and the peptide comprising amino acid residues 193–229, which Lambotte et al. (1980) showed has sequence similarity to human apolipoprotein A1, becomes susceptible to proteolytic cleavage (Dumont and Richards, 1988). Thermodynamic characterization of the pH stability of diphtheria toxin suggested that a massive unfolding of the toxin molecule takes place at low pH (Ramsay et al., 1989), and it has been suggested that the conformational change is dependent on isomerization of prolines located close to the C-terminus of the B fragment (Deleers et al., 1983; Brasseur et al., 1986). An infrared spectroscopy study of the interaction of diphtheria toxin with asolectin vesicles at pH 4 suggested that the secondary structure of diphtheria toxin is characterized by the appearance of β-sheet structure (Cabiaux et al., 1989a,b).

Photolabelling studies show that at low pH the toxin penetrates into lipid bilayers, and it has been suggested that both the A and the B fragments come into contact with the lipids (Hu and Holmes, 1984; Zalman and Wisnieski, 1984; Montecucco et al., 1985; González and Wisnieski, 1988). However, according to Dumont and Richards (1988) it is difficult to exclude that some of the

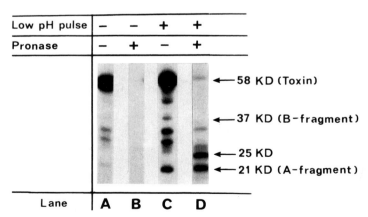

Figure 2. Translocation of nicked diphtheria toxin across the cell membrane. Lane A: ^{125}I-labelled nicked toxin was bound to cells, which were then dissolved under non-reducing conditions and analysed by polyacrylamide gel electrophoresis in the presence of sodium dodecyl sulphate. Lane B: As lane A, but the cells were treated with pronase before they were lysed. Lane C: As lane A, but the cells were exposed to pH 4.5 for 2 min before lysis. Lane D: As lane C, but after being exposed to low pH, the cells were treated with pronase and then lysed. (Data from Moskaug et al., 1987.)

labelling occurs outside the membrane. Gonzalez and Wisnieski (1988) found that the peptide region between the A and B fragment in intact diphtheria toxin is translocated across the vesicle bilayer at low pH. In contrast, when cells with surface-bound unnicked toxin are exposed to low pH, insertion of the toxin into the membrane is inhibited. The possibility exists, however, that the insertion process studied in liposomes containing no specific receptor differs from the receptor-mediated entry of diphtheria toxin into cells.

Studies on liposomes indicate that the conformational change that increases the hydrophobicity of the toxin depends on the lipid charge, and that the interaction of toxin with lipid vesicles at low pH has both hydrophobic and electrostatic components (Chung and London, 1988). The low pH-dependent conformational change of diphtheria toxin is most likely also responsible for the ability of the toxin to induce fusion of vesicles at low pH (Cabiaux et al., 1984; Papini et al., 1987a).

A single mutation in the A fragment of diphtheria toxin, such as that found in CRM 197, seems to facilitate the insertion of this molecule into lipid bilayer (Hu and Holmes, 1987). Analysis of such mutants with monoclonal antibodies suggested that single amino acid substitutions in the A fragment also affect the conformation of the B fragment (Bigio et al., 1987).

At low pH diphtheria toxin forms channels in lipid bilayers (Donovan et al., 1981; Kagan et al., 1981), in liposomes (Kagan et al., 1981; Zalman and Wisnieski, 1984) and in cells (Papini et al., 1988; Sandvig and Olsnes, 1988). Furthermore, Escherichia coli strains that secrete into the periplasm certain cloned diphtheria toxin-related proteins are rapidly killed when exposed to pH 5, probably by a related mechanism (O'Keefe and Collier, 1989). The pore-forming ability of the toxin is described in more detail below.

When translocation of diphtheria toxin across the cell membrane is induced by low pH, there is an inward-directed pH gradient. This pH gradient seems to be required for the translocation of the A fragment to the cytosol. When the pH gradient was reduced by acidification of the cytosol, the cells were protected against intoxication (Sandvig et al., 1986) and translocation of ^{125}I-labelled A fragment was inhibited as well (Moskaug et al., 1987). Experiments where the cytosolic pH was reduced to different extents suggested that an inward-directed gradient of about 1.5 pH units is required for efficient translocation of the A fragment to occur. Interestingly, experiments with ^{125}I-labelled toxin showed that the B fragment was inserted into the membrane even in the absence of a pH gradient. Thus, insertion of the B fragment can occur without concomitant transport of fragment A.

The possibility exists that the electrical membrane potential (negative inside) might be of importance in pulling the positively charged A fragment into the cytosol, and Hudson et al. (1988) found that cell depolarization slows down the rate of diphtheria toxin translocation. However, in experiments where we dissipated the potential by increasing the extracellular potassium concentration, low pH-induced translocation of fragment A occurred as efficiently as in the presence of the membrane potential (Sandvig et al., 1986; Moskaug et al., 1987). Since a strong reduction in cellular ATP also had no effect on the translocation process, the pH gradient could be the only driving force for diphtheria toxin entry.

Role of the diphtheria toxin receptor

Studies on the sensitivity of different cell lines to diphtheria toxin and their ability to bind the toxin suggested a direct correlation between binding and intoxication. However, the naturally occurring receptors for diphtheria toxin seem to play a role not only for the binding as such, but also for efficient toxin translocation at low pH. When the receptor-binding step was circumvented by binding of biotinyl-diphtheria toxin to avidin-treated cells or when the toxin was added at low pH to induce non-specific binding, there was no increase in the toxic effect, and there was also no increase in the amount of A fragment translocated to the cytosol (Stenmark et al., 1988). Also, although unspecific binding was found under such conditions, no toxin was protected against protease when the specific binding was abolished by treatment of the cells with the tumour promoter TPA, with the anion transport inhibitor DIDS, or by pretreatment of the cells with trypsin (Stenmark et al., 1988). Similarly, exposure to low pH of L cells with non-specifically bound diphtheria toxin did not lead to insertion of the B fragment, translocation of the

A fragment or to pore formation in these cells. Although toxic effect on the cells was seen at high toxin concentrations, this is most likely due to toxin that entered by a different and less efficient mechanism than that observed in Vero cells.

L cells were also intoxicated when they were incubated with a diphtheria toxin–concanavalin A conjugate (Guillemot et al., 1985). However, even though the cells could be protected by addition of ammonium chloride and monensin, suggesting a low pH-dependent mechanism of translocation, exposure of cells with surface-bound conjugate to low pH did not induce translocation across the plasma membrane. This strongly suggests that the entry mechanism of the conjugate in mouse cells is different from that mediated by the diphtheria toxin receptor in Vero cells.

A number of toxic conjugates consisting of surface binding ligands such as interleukin 2, transferrin, human chorionic gonadotropin, epidermal growth factor, melanocyte-stimulating hormone, and monoclonal antibodies specific for cell surface receptors on the one side, and whole diphtheria toxin, parts of the toxin, or mutated toxin molecules with strongly reduced binding affinity on the other have been prepared (Chang and Neville, 1977; Chang et al., 1977; Uchida et al., 1978; Masuho et al., 1979; Gilliland and Collier, 1980; Shimizu et al., 1980; Bacha et al., 1983, 1988; Villemez and Carlo, 1984; Oeltman, 1985; Columbatti et al., 1986; Murphy et al., 1986; Pastan et al., 1986; Bishai et al., 1987; Williams et al., 1987; Kelley et al., 1988; Perentesis et al., 1988). In some cases conjugates consisting of a mutated diphtheria toxin which in itself is unable to bind to cells is as toxic or even more toxic than the same conjugate containing whole diphtheria toxin (Greenfield et al., 1987; Gray Johnson et al., 1988). However, compared to the few minutes it takes from addition of diphtheria toxin until protein synthesis starts to decline in Vero cells, the lag time observed with the conjugates is rather long, suggesting that the entry mechanism is different from that of native diphtheria toxin. It would be interesting to study whether receptor-bound hybrid toxins are inserted into the membrane and whether there is a direct translocation of fragment A through the membrane upon exposure to low pH.

Mouse cells which are normally very resistant to diphtheria toxin are intoxicated not only when

concanavalin A is used as a binding moiety, but also when the toxin or a mutated toxin unable to bind to receptors is bound by means of transferrin or monoclonal antibodies to determinants at the cell surface (O'Keefe and Draper, 1985; Gray Johnson et al., 1988; Marsh, 1988). However, the entry mechanism of the conjugates is different from that of native diphtheria toxin in sensitive cells. When diphtheria toxin was coupled to anti-murine Thy 1 antibodies, the conjugate was highly efficient in killing mouse cells but, unlike intoxication of Vero cells with diphtheria toxin, the intoxication of the mouse cells was highly temperature-dependent (Marsh, 1988). A strong reduction in toxicity was observed when the cells were incubated below 20° C, suggesting that fusion of endosomes with another cellular compartment is required for intoxication by the conjugate. Furthermore, the intoxication occurred after a rather long lag time, and the temperature-dependent step preceded an ammonium-sensitive step. As Marsh pointed out, a temperature-sensitive step followed by vesicular acidification occurs both in the process of delivery of endosomal vesicles to the lysosomes and to the trans-Golgi compartment. Acidification did not induce rapid entry of cell surface-bound conjugate. It is not known whether these toxin conjugates require processing for entry, or if they require for translocation some cellular element present only in certain intracellular compartments.

Role of anions in the translocation process

The presence in the medium of permeant anions is not only required for the binding step, but also for translocation of the A fragment to the cytosol (Sandvig and Olsnes, 1984, 1986). When chloride is removed from the medium and sulphate is added instead, the cells are protected against the toxin, and the A fragment is no longer translocated across the membrane when the pH in the medium is lowered. Also, inhibitors of anion transport strongly inhibit the translocation of the A fragment. Since translocation of the toxin was also found to affect anion transport (Olsnes and Sandvig, 1986), we suggested that the anion exchanger could constitute part of the diphtheria toxin receptor. The relatively small membrane molecules which become cross-linked to the toxin are unlikely to be candidates for such an exchanger. However, the possibility still exists that a complex of molecules are involved in the binding and translocation processes.

When Vero cells are depleted for chloride, the toxin can enter even in the absence of external chloride, suggesting that the chloride gradient across the cell membrane is important for translocation (Moskaug et al., 1989b). However, inhibitors of chloride transport like SITS and DIDS inhibited entry even in chloride-depleted cells.

The role of the site of cleavage between the A and the B fragments

In the loop region between fragments A and B there are three possible cleavage sites for trypsin: the molecule can be cleaved next to Arg^{190}, Arg^{192} or Arg^{193}. This should give rise to A fragments with three different pI values. When trypsin-treated toxin is run on an isoelectric focusing gel there are three main bands as expected, and a minor fourth band of unknown nature (Moskaug et al., 1989a). When translocated A fragments were analysed, only the two most acidic species of A fragments were found to enter, suggesting that both Arg^{192} and Arg^{193} must be removed to allow translocation of the A fragment to occur. Carboxypeptidase B treatment of nicked toxin to remove C-terminal arginines increased the fraction of A fragments lacking the two last arginines, and this treatment also increased the amount of A fragment translocated to the cytosol and the toxic effect on cells, as well as the ability of the toxin to form cation-selective channels in the membrane.

Translocation of mutant molecules

Even small changes in the diphtheria toxin molecule are sufficient to affect translocation of fragment A to the cytosol. In the mutant protein CRM 1001 cysteine 471 is substituted with tyrosine, and this single change is sufficient to give a strong reduction in the intoxication of cells. The ability of this molecule to bind to cell surface receptors and to ADP-ribosylate EF2 is unchanged, and experiments performed by Dell'Arciprete et al. (1988) suggested that the low toxicity is due to impaired penetration of the cell membrane. When cells with surface-bound CRM 1001 were exposed to low

pH, the intoxication was 5000 times less efficient than with natural diphtheria toxin.

In order to study which parts of the diphtheria toxin molecule are essential for insertion of the B fragment into the membrane and therefore probably also necessary for translocation of the A fragment across the membrane, we have recently studied the membrane interactions of in vitro synthesized mutants of diphtheria toxin (McGill et al., 1989). Both the B fragment alone and incomplete B fragment (deleted from the N-terminus) were able to bind to cells and become inserted into the membrane when exposed to low pH. The pore-forming abilities of these molecules are discussed in the next section.

Diphtheria toxin-induced channel formation in membranes

When cells with surface-bound diphtheria toxin are exposed to low pH the cell membrane becomes permeable to monovalent cations (Fig. 3), but not to anions and uncharged molecules such as sucrose and glucose (Sandvig and Olsnes, 1988; Papini et al., 1988). Channel formation seems to occur only under conditions which permit A fragment translocation and intoxication of the cells. Thus, when the pH gradient is abolished to inhibit A fragment translocation, there is no channel formation although the B fragment is inserted into the membrane under these conditions (Moskaug et al., 1987).

Not only the formation of channels, but also the maintenance of open channels seems to be dependent on a pH gradient. The channels close when the pH gradient is diminished due to influx of protons, and they close even faster when the gradient is rapidly reduced when the cytosol is acidified by addition of acetic acid (Sandvig and Olsnes, 1988). Also, other conditions which inhibit A fragment translocation, such as addition of Cd^{2+} or addition of anion transport inhibitors inhibit channel formation (Sandvig and Olsnes, 1988). Since the channel formation in cells seems to be dependent on naturally occurring diphtheria toxin receptors, the channel formation obtained in artificial lipid membranes must be of a different type.

The ability of deletion mutants of diphtheria toxin to form pores in cells has recently been

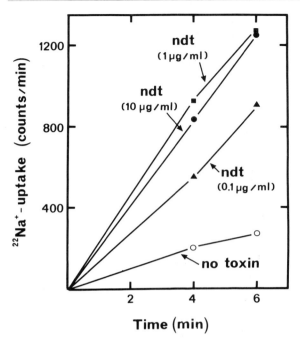

Figure 3. Effect of diphtheria toxin on the permeability of Vero cells to sodium. Vero cells were incubated with the indicated concentrations of nicked diphtheria toxin for 15 min at 37°C. The medium was then removed, and the cells were exposed to buffer containing 1 mM amiloride, 50 μM ouabain, 0.14 M NaCl, 20 mM MES/Tris, 5 mM gluconic acid, pH 4.8, and [^{22}Na]Cl. The cell-associated radioactivity was measured after increasing periods of time.

studied. One protein comprising most of the B fragment (the nine N-terminal residues of the B fragment have been replaced with the tripeptide Met–Ala–Leu as well as two proteins consisting of the whole B fragment and small parts – the last 16 and 57 amino acids – of the A fragment) were more efficient in permeabilizing the plasma membrane to monovalent cations than intact diphtheria toxin (Stenmark et al., 1989).

In spite of the lack of the whole A fragment or most of the A fragment, the requirements for pore formation were similar to those described for intact toxin. Thus, the channels were specific for cations, and a pH gradient and specific receptors were required for channel formation. Furthermore, channel formation by the mutants could be inhibited by inhibitors of anion transport as well as Cd^{2+}, as in channel formation by whole toxin. It is not clear why these mutant molecules are more efficient in channel formation than intact

diphtheria toxin. The high efficiency cannot only be due to the somewhat higher binding (the affinity for the receptor is 5–10 times higher than with whole toxin) or to an increased fraction of inserted molecules. It therefore appears that intact fragment A somehow reduces the efficiency of the channels.

The cation specificity of the channel-forming molecules described here seems to be associated with the N-terminal end of the B fragment. Thus, when truncated B fragments lacking 31 and 97 amino acids from the N-terminal end were added to cells, they formed pores with completely different properties from those obtained with diphtheria toxin or with mutants containing most of the B fragment. These molecules formed pores that were permeable to cations and anions, as well as to uncharged molecules, and they were formed at all pH values tested between pH 4.7 and 8.0. Furthermore, unlike diphtheria toxin these molecules permeabilize even mouse cells. Both mutants showed some specific binding to diphtheria toxin receptors, but in addition, they also bound unspecifically to cells. The formation of this type of pore is unlikely to be associated with A fragment translocation since the requirements for the formation and the properties of the pores are different from those described with intact toxin.

The channels formed in cells by diphtheria toxin and some of the toxin-related molecules have some resemblance to channels obtained in artificial lipid membranes. Thus, both Kagan et al. (1981) and Donovan et al. (1981) found that diphtheria toxin and CRM 45 formed channels in lipid membranes at low pH. Kagan et al. (1981) estimated the pore diameter to be about 18Å, sufficiently large for an extended A fragment to pass through the channel. The formation of channels in lipid bilayers and in liposomes has since been studied in considerable detail (Misler, 1983, 1984; Zalman and Wisnieski, 1984; Shiver and Donovan, 1987). Donovan et al. (1985) and Kagan et al. (1984) found that inositol hexaphosphate added on the trans side of membranes greatly stimulated channel formation obtained by diphtheria toxin, suggesting translocation of the P site of the B fragment to the inner surface of the membrane. Also, the studies by Kagan et al. (1981) suggested that CRM 45 spanned the membrane since protease added to the trans side abolished pore activity. Donovan et al. (1985) claimed that diphtheria toxin was able to ADP-ribosylate elongation factor 2 which was entrapped in lipid vesicles.

It is possible that both the putative amphipathic region at the N-terminus of the B fragment and a hydrophobic membrane-spanning region in the middle of the B fragment (Lambotte *et al.*, 1980; Greenfield *et al.*, 1983; Eisenberg *et al.*, 1984) play a role in the pore formation observed with intact diphtheria toxin as well as with the mutants containing most of the B fragment. A 12-kDa cyanogen bromide fragment derived from the middle region of the B fragment containing part of the membrane-spanning region has been shown to induce conductance changes in planar lipid membranes (Kayser *et al.*, 1981). Interestingly, this increased conductance occurred at neutral pH.

Even pure fragment B was found to increase the ionic conductance of planar lipid bilayer both at pH 7.2 and at pH 4.2. The results obtained with this membrane system therefore differ from studies with cells where the B fragment only forms channels at low pH. A mutant molecule consisting of only the 16 000 Da C-terminal piece of the B fragment does not contain this hydrophobic area and does not induce pore formation. The results obtained so far suggest that the receptor-binding domain of diphtheria toxin, cell surface receptors, as well as the N-terminal piece of the B fragment are required to confer specificity on the toxin-formed channels seen in intact Vero cells.

To what extent is diphtheria toxin a model system for toxin translocation in general?

A growing number of bacterial and plant toxins exert enzymatic action in the cytosol implying that an enzymatically active polypeptide is translocated into the cells. These toxins comprise *Pseudomonas aeruginosa* exotoxin A, shigella toxin, cholera toxin, pertussis toxin, abrin, ricin, modeccin and a number of others (van Heyningen, 1977; Boquet and Duflot, 1982; Critchley *et al.*, 1985; Hoch *et al.*, 1985; Middlebrook, 1986; Montecucco 1986; Olsnes and Sandvig, 1988; Poulain *et al.*, 1988; Janicot *et al.*, 1988). As far as is known, binding to surface receptors and endocytic uptake is required in all cases. Acidification of intracellular compartments is required only in some cases, and even in these cases the role of the low pH is not clear. Thus, unlike diphtheria toxin, it has not been possible to induce translocation of any of the other toxins from the cell surface by exposing the cells to low pH. Furthermore, in most cases the time required from addition of toxin until the toxic response is developed is longer than with diphtheria toxin. Also, temperatures below 20°C that inhibit intracellular vesicular fusion, have a much stronger inhibitory effect on other toxins than on diphtheria toxin. This indicates that while diphtheria toxin is translocated from endosomes, the other toxins must be transported to more distal intracellular compartments before translocation can occur. It may, therefore, be dangerous to extrapolate from the mechanism of diphtheria toxin translocation to that of other toxins.

References

Alving, C.R., Iglewski, B.H., Urban, K.A., Moss, J., Richards, R.L. and Sadoff, J.C. (1980). Binding of diphtheria toxin to phospholipids in liposomes. *Proc. Natl. Acad. Sci. USA* 77, 1986–90.

Athwal, R.S., Searle, B.M. and Jansons, V.K. (1985). Diphtheria toxin sensitivity in a monochromosomal hybrid containing human chromosome 5. *J. Heredity* 76, 329–34.

Bacha, P., Murphy, J.R. and Reichlin, S. (1983). Thyrotropin-releasing hormone-diphtheria toxin-related polypeptide conjugates: Potential role of the hydrophobic domain in toxin entry. *J. Biol. Chem.* 258, 1565–70.

Bacha, P., Williams, D.P., Waters, C., Williams, J.M., Murphy, J.R. and Strom, T.B. (1988). Interleukin 2 receptor-targeted cytotoxicity: Interleukin 2 receptor-mediated action of a diphtheria toxin-related interleukin 2 fusion protein. *J. Exp. Med.* 167, 612–22.

Barbieri, J.T., Carroll, S.F. and Collier, R.J. (1981). An endogenous dinucleotide bound to diphtheria toxin: Adenylyl-(3′, 5′)-uridine 3′-monophosphate. *J. Biol. Chem.* 256, 12247–51.

Barbieri, J.T., Collins, C.M. and Collier, R.J. (1986). Diphtheria toxin can simultaneously bind to its receptor and adenylyl-(3′, 5′)-uridine 3′-monophosphate. *Biochemistry* 25, 6608–11.

Basosi, R., Kushnaryov, V., Panz, T. and Lai, C-S. (1986). Evidence that divalent cations bind to diphtheria toxin: An ESR approach. *Biochem. Biophys. Res. Commun.* 139, 991–5.

Battistini, A., Curatola, A.M., Gallinari, P. and Rossi, G.B. (1988). Inhibition of protein synthesis by diphtheria toxin induces a peculiar pattern of synthesized protein species. *Exp. Cell. Res.* 176, 174–9.

Bigio, M., Rossi, R., Nucci, D., Antoni, G., Rappuoli,

R. and Ratti, G. (1987). Conformational changes in diphtheria toxoids: Analysis with monoclonal antibodies. *FEBS Lett.* **218**, 271–6.

Bishai, W.R., Myianohara, A. and Murphy, J.R. (1987). Cloning and expression in *Escherichia coli* of three fragments of diphtheria toxin truncated within fragment B. *J. Bacteriol.* **169**, 1554–63.

Blewitt, M.G., Zhao, J-M., McKeever, B., Sarma, R. and London, E. (1984). Fluorescence characterization of the low pH-induced change in diphtheria toxin conformation: Effect of salt. *Biochem. Biophys. Res. Commun.* **120**, 286–90.

Blewitt, M.G., Chung, L.A. and London, E. (1985). Effect of pH on the conformation of diphtheria toxin and its implications for membrane penetration. *Biochemistry* **24**, 5458-64.

Bodley, J.W., Upham, R., Crow, F.W., Tomer, K.B. and Gross, M.L. (1984). Ribosyl-diphthamide: Confirmation of structure by fast atom bombardment mass spectrometry. *Arch. Biochem. Biophys.* **230**, 590–3.

Boquet, P. and Duflot, E. (1982). Tetanus toxin fragment forms channels in lipid vesicles at low pH. *Proc. Natl. Acad. Sci. USA* **79**, 7614–18.

Boquet, P. and Pappenheimer, A.M. Jr. (1976). Interaction of diphtheria toxin with mammalian cell membranes. *J. Biol. Chem.* **251**, 5770–8.

Brasseur, R., Cabiaux, V., Falmagne, P. and Ruysschaert, J-M. (1986). Dependent insertion of a diphtheria toxin B fragment peptide into the lipid membrane: A conformational analysis. *Biochem. Biophys. Res. Commun.* **136**, 160–8.

Cabiaux, V., Vandenbranden, M., Falmagne, P. and Ruysschaert, J-M. (1984). Diphtheria toxin induces fusion of small unilammelar vesicles at low pH. *Biochim. Biophys. Acta* **775**, 31–6.

Cabiaux, V., Goormaghtigh, E., Wattiez, R., Falmagne, P. and Ruysschaert, J-M. (1989a). Secondary structure changes of diphtheria toxin interacting with asolectin lipsomes: an infrared spectroscopy study. *Biochimie* **71**, 153–8.

Cabiaux, V., Brasseur, R., Wattiez, R., Falmagne, P., Ruysschaert, J-M. and Goormaghtigh, E. (1989b). Secondary structure of diphtheria toxin and its fragments interacting with acidic liposomes studied by polarized infrared spectroscopy. *J. Biol. Chem.* **264**, 4928–38.

Carroll, S.F. and Collier, R.J. (1984). NAD binding site of diphtheria toxin: Identification of a residue within the nicotinamide subsite by photochemical modification with NAD. *Proc. Natl. Acad. Sci. USA* **81**, 3307–11.

Carroll, S.F., McCloskey, J.A., Crain, P.F., Oppenheimer, N.J., Marschner, T.M. and Collier, R.J. (1985). Photoaffinity labeling of diphtheria toxin fragment A with NAD: Structure of the photoproduct at position 148. *Proc. Natl. Acad. Sci. USA* **82**, 7237–41.

Carroll, S.F., Barbieri, J.T. and Collier, R.J. (1986).

Dimeric form of diphtheria purification and characterization. *Biochemistry* **25**, 2425–30.

Chang, T-M. and Neville, D.M. (1977). Artificial hybrid protein containing a toxic protein fragment and a cell membrane receptor-binding moiety in a disulfide conjugate: I. Synthesis of diphtheria toxin fragment A-S-S-human placental lactogen with methyl-5-bromovalerimidate. *J. Biol. Chem.* **252**, 1505–14.

Chang, T. and Neville, Jr., D.M. (1978). Demonstration of diphtheria toxin receptors on surface membranes from both toxin-sensitive and toxin-resistant species. *J. Biol. Chem.* **253**, 6866–71.

Chang, T-M., Dazord, A. and Neville, Jr., D.M. (1977). Artificial hybrid protein containing a toxic protein fragment and a cell membrane receptor-binding moiety in a disulfide conjugate: II. Biochemical and biologic properties of diphtheria toxin fragment A-S-S-human placental lactogen. *J. Biol. Chem.* **252**, 1515–22.

Chen, J-Y.C. and Bodley, J.W. (1988). Biosynthesis of diphthamide in *Saccharomyces cerevisiae*: Partial purification and characterization of a specific S-adenosylmethionine: elongation factor 2 methyltransferase. *J. Biol. Chem.* **263**, 11692–6.

Chung, L.A. and London, E. (1988). Interaction of diphtheria toxin with model membranes. *Biochemistry* **27**, 1245–53.

Cieplak, W., Gaudin, H.M. and Eidels, L. (1987). Diphtheria toxin receptor. *J. Biol. Chem.* **262**, 13246–53.

Collier, R.J. and Kandel, J. (1971). Structure and activity of diphtheria toxin: I. Thiol-dependent dissociation of a fraction of toxin into enzymically active and inactive fragments. *J. Biol. Chem.* **246**, 1496–503.

Collins, C.M. and Collier, R.J. (1984). Interaction of diphtheria with adenylyl-(3′, 5′)-uridine 3′-monophosphate: II. The NAD-binding site and determinants of dinucleotide affinity. *J. Biol. Chem.* **259**, 15159–62.

Collins, C.M., Barbieri, J.T. and Collier, R.J. (1984). Interaction of diphtheria toxin with adenylyl-(3′, 5′)-uridine 3′-monophosphate: I. Equilibrium and kinetic measurements. *J. Biol. Chem.* **259**, 15154–8.

Colombatti, M., Greenfield, L. and Youle, R.J. (1986). Cloned fragment of diphtheria toxin linked to T Cell-specific antibody identifies regions of B chain active in cell entry. *J. Biol. Chem.* **261**, 3030–5.

Critchley, D.R., Nelson, P.G., Habig, W.H. and Fishman, P.H. (1985). Fate of tetanus toxin bound to the cell surface of primary neurons in culture: Evidence for rapid internalization. *J. Cell Biol.* **100**, 1499–507.

Deleers, M., Beugnier, N., Falmagne, P., Cabiaux, V. and Ruysschaert, J-M. (1983). Localization in diphtheria toxin fragment B of a region that induces pore formation in planar lipid bilayers at low pH. *Arch. Int. Physiol. Biochim* **91**, BP6–BP7.

Dell'Arciprete, L., Colombatti, M., Rappuoli, R. and Tridente, G. (1988). A C terminus cysteine of diphtheria toxin B chain involved in immunotoxin cell penetration and cytotoxicity. *J. Immunol.* **140**, 2466–71.

Didsbury, J.R., Moehring, J.M. and Moehring, T.J. (1983). Binding and uptake of diphtheria toxin by toxin-resistant Chinese hamster ovary and mouse cells. *Molec. Cell. Biol.* **3**, 1283–94.

Donovan, J.J., Simon, M.I., Draper, R.K. and Montal, M. (1981). Diphtheria toxin forms transmembrane channels in planar lipid bilayers. *Proc. Natl. Acad. Sci. USA* **78**, 172–6.

Donovan, J.J., Simon, M.I. and Montal, M. (1985). Requirements for the translocation of diphtheria toxin fragment A across lipid membranes. *J. Biol. Chem.* **260**, 8817–23.

Dorland, R.B., Middlebrook, J.I. and Leppla, S.H. (1981). Effect of ammonium chloride on receptor-mediated uptake of diphtheria toxin by vero cells. *Expl Cell Res.* **134**, 319–27.

Draper, R.K. and Simon, M.I. (1980). The entry of diphtheria toxin into the mammalian cell cytoplasm: Evidence for lysosomal involvement. *J. Cell Biol.* **87**, 849–54.

Draper, R.K., Chin, D. and Simon, M.I. (1978). Diphtheria toxin has the properties of a lectin. *Proc. Natl. Acad. Sci. USA* **75**, 261–5.

Dumont, M.E. and Richards, F.M. (1988). The pH-dependent conformational change of diphtheria toxin. *J. Biol. Chem.* **263**, 2087–97.

Eidels, L. and Hart, D.A. (1982). Effect of polymers of L-lysine on the cytotoxic action of diphtheria toxin. *Infect. Immun.* **37**, 1054–8.

Eidels, L., Ross, L.L. and Hart, D.A. (1982). Diphtheria toxin–receptor interaction: A polyphosphate-insensitive diphtheria toxin-binding domain. *Biochem. Biophys. Res. Commun.* **109**, 493–9.

Eidels, L., Proia, R.L. and Hart, D.A. (1983). Membrane receptors for bacterial toxins. *Microbiol. Rev.* **47**, 596–620.

Eisenberg, D., Schwartz, E., Komaromy, M. and Wall, R. (1984). Analysis of membrane and surface protein sequences with the hydrophobic moment plot. *J. Molec. Biol.* **179**, 125–42.

Everse, J., Lappi, D.A., Beglau, J.M., Lee, C-L. and Kaplan, N.O. (1977). Investigations into the relationship between structure and function of diphtheria toxin. *Proc. Natl. Acad. Sci. USA* **74**, 472–6.

Fendrick, J.L. and Iglewski, W.J. (1989). Endogenous ADP-ribosylation of elongation factor 2 in polyma virus-transformed baby hamster kidney cells. *Proc. Natl. Acad. Sci. USA* **86**, 554–7.

Friedman, R.L., Iglewski, B.H., Roerdink, F. and Alving, C.R. (1982). Suppression of cytotoxicity of diphtheria toxin by monoclonal antibodies against phosphatidylinositol phosphate. *Biophys. J.* **37**, 23–4.

Giannini, G., Rappuoli, R. and Ratti, G. (1984). The amino-acid sequence of two non-toxic mutants of diphtheria toxin: CRM45 and CRM197. *Nucleic Acids Res.* **12**, 4063–9.

Gill, D.M. and Dinius, L.L. (1971). Observations on the structure of diphtheria toxin. *J. Biol. Chem.* **246**, 1485–91.

Gilliland, D.G. and Collier, R.J. (1980). A model system involving anti-concanavalin A for antibody targeting of diphtheria toxin fragment A. *Cancer Res.* **40**, 3564–9.

González, J.E. and Wisnieski, B.J. (1988). An endosomal model for acid triggering of diphtheria toxin translocation. *J. Biol. Chem.* **263**, 15257–9.

Gray Johnsen, V., Wilson, D., Greenfield, L. and Youlet, R.J. (1988). The role of the diphtheria toxin receptor in cytosol translocation. *J. Biol. Chem.* **263**, 1295–300.

Greenfield, L., Bjorn, M.J., Horn, G., Fong, D., Buck, G.A., Collier, R.J. and Kaplan, D.A. (1983). Nucleotide sequence of the structural gene for diphtheria toxin carried by corynebacteriophage. *Proc. Natl. Acad. Sci. USA* **80**, 6853–7.

Greenfield, L., Gray Johnson, V. and Youle, R.J. (1987). Mutations in diphtheria toxin separate binding from entry and amplify immunotoxin selectivity. *Science* **238**, 536–9.

Guillemot, J.C., Sundan, A., Olsnes, S. and Sandvig, A. (1985). Entry of diphtheria toxin linked to concanavalin A into primate and murine cells. *J. Cell. Physiol.* **122**, 193–9.

Heagy, W.E. and Neville, D.M. Jr. (1981). Kinetics of protein synthesis inactivation by diphtheria toxin in toxin-resistant L cells. *J. Biol. Chem.* **256**, 12788–92.

Hoch, D.H., Romero-Mira, M., Ehrich, B.E., Finkelstein, A., DasGupta, B.R. and Simpson, L.L. (1985). Channels formed by botulinum, tetanus, and diphtheria toxins in planar lipid bilayers: Relevance to translocation of proteins across membranes. *Proc. Natl. Acad. Sci. USA.* **82**, 1692–6.

Hranitzky, K.W., Durham, D.L., Hart, D.A. and Eidels, L. (1985). Role of glycosylation in expression of functional diphtheria toxin receptors. *Infect. Immun.* **49**, 336–43.

Hu, V.W. and Holmes, R.K. (1984). Evidence of direct insertion of fragments A and B of diphtheria toxin into model membranes. *J. Biol. Chem.* **259**, 12226–33.

Hu, V.W. and Holmes, R.K. (1987). Single mutation in the A domain of diphtheria toxin results in a protein with altered membrane insertion behaviour. *Biochim. Biophys. Acta* **902**, 24–30.

Hudson, T.H., Scharff, J., Kimak, M.A.G. and Neville, Jr., D.M. (1988). Energy requirements for diphtheria toxin translocation are coupled to the maintenance of a plasma membrane potential and a proton gradient. *J. Biol. Chem.* **263**, 4773–81.

Iglewski, W.J., Lee, H. and Muller, P. (1984). ADP-ribosyltransferase from beef liver with ADP-ribosylates elongation factor-2. *FEBS Lett.* **173**, 113–18.

Ittelson, T.R. and Gill, D.M. (1973). Diphtheria toxin: Specific competition for cell receptors. *Nature* **242**, 330–1.

Janicot, M., Clot, J.-P. and Desbuquois, B. (1988). Interactions of cholera toxin with isolated hepatocytes. *Biochem. J.* **253**, 735–43.

Kagan, B.L., Finkelstein, A. and Colombini, M. (1981). Diphtheria toxin fragment forms large pores in phospholipid bilayer membranes. *Proc. Natl. Acad. Sci. USA* **78**, 4950–4.

Kagan, B.L., Reich, K.A. and Collier, R.J. (1984). Orientation of the diphtheria toxin channel in lipid bilayers. *Biophys. J.* **45**, 102–4.

Kayser, G., Lambotte, P., Falmagne, P., Capiau, C., Zanen, J. and Ruysschaert, J-M. (1981). A CNBR peptide located in the middle region of diphtheria toxin fragment B induces conductance change in lipid bilayers. *Biochem. Biophys. Res. Commun.* **99**, 358–63.

Keen, J.H., Maxfield, F.R., Hardegree, M.C. and Habig, W.H. (1982). Receptor-mediated endocytosis of diphtheria toxin by cells in culture. *Proc. Natl. Acad. Sci. USA,* **79**, 2912–16.

Kelley, V.E., Bacha, P., Pankewycz, O., Nichols, J.C., Murphy, J.R. and Strom, T.B. (1988). Interleucin 2-diphtheria toxin fusion protein can abolish cell-mediated immunity in vivo. *Proc. Natl. Acad. Sci. USA* **85**, 3980–4.

Kessel, M. and Klink, F. (1980). Archaebacterial elongation factor is ADP-ribosylated by diphtheria toxin. *Nature* **287**, 250–1.

Kim, K. and Groman, N.B. (1965). In vitro inhibition of diphtheria toxin action by ammonium salts and amines. *J. Bacteriol.* **90**, 1552–6.

Kushnaryov, V.M., MacDonald, H.S., Sedmak, J.J. and Grossberg, S.E. (1984). Diphtheria toxin receptor sites on membranes of cultured cells and erythrocytes demonstrated by fluorescence and electron microscopy. *Cytobios* **41**, 7–22.

Lambotte, P., Falmagne, P., Capiau, C., Zanen, J., Ruysschaert, J.-M. and Dirkx, J. (1980). Primary structure of diphtheria toxin fragment B: Structural similarities with lipid-binding domains. *J. Cell Biol.* **87**, 837–40.

Leppla, S.H., Dorland, R.B. and Middlebrook, J.L. (1980). Inhibition of diphtheria toxin degradation and cytotoxic action by chloriquine. *J. Biol. Chem.* **255**, 2247–50.

London, E. (1986). A fluorescence-based detergent binding assay for protein hydrophobicity. *Analyt. Biochem.* **154**, 57–63.

Lory, S. and Collier, R.J. (1980). Diphtheria toxin: Nucleotide binding and toxin heterogeneity. *Proc. Natl. Acad. Sci. USA* **77**, 267–71.

Lory, S., Carrolls, S.F. and Collier, R.J. (1980). Ligand interactions of diphtheria toxin: II. Relationships between the NAD site and the P site. *J. Biol. Chem.* **255**, 12016–19.

McGill, S., Stenmark, H., Sandvig, H., Collier, R.J. and Olsnes, S. (1989). Membrane interactions of diphtheria toxin analysed using *in vitro* synthesised mutants. *EMBO J.* **8**, 2843–48.

Marnell, M.H., Mathis, L.S., Stookey, M., Shia, S-P., Stone, D.K. and Draper, R.K. (1984a). A Chinese hamster ovary cell mutant with a heat-sensitive conditional-lethal defect in vacuolar function. *J. Cell. Biol.* **99**, 1907–16.

Marnell, M.H., Shia, S-P., Stookey, M. and Draper, R.K. (1984b). Evidence for penetration of diphtheria toxin to the cytosol through a prelysosomal membrane. *Infect. Immun.* **44**, 145–50.

Marsh, J.W. (1988). Antibody-mediated routing of diphtheria toxin in murine cells results in a highly efficacious immunotoxin. *J. Biol. Chem.* **263**, 15993–9.

Masuho, Y., Hara, T. and Noguchi, T. (1979). Preparation of a hybrid of fragment Fab' of antibody and fragment A of diphtheria toxin and its cytotoxicity. *Biochem. Biophys. Res. Commun.* **90**, 320–6.

Mekada, E. and Uchida, T. (1985). Binding properties of diphtheria toxin to cells are altered by mutation in the fragment A domain. *J. Biol. Chem.* **260**, 12148–53.

Mekada, E., Uchida, T. and Okada, Y. (1979). Modification of the cell surface with neuraminidase increases the sensitivities of cells to diphtheria toxin and pseudomonas aeruginosa exotoxin. *Expl Cell. Res.* **123**, 137–46.

Mekada, E., Okada, Y. and Uchida, T. (1988). Identification of diphtheria toxin receptor and a nonproteinous diphtheria toxin-binding molecule in vero cell membrane. *J. Cell Biol.* **107**, 511–19.

Merion, M., Schlesinger, P., Brooks, R.M., Moehring, J.M., Moehring, T.J. and Sly, W.S. (1983). Defective acidification of endosomes in Chinese hamster ovary cell mutants "cross-resistant" to toxins and viruses. *Proc. Natl. Acad. Sci. USA* **80**, 5315–19.

Middlebrook, J.L. (1981). Effect of energy inhibitors on cell surface diphtheria toxin receptor numbers. *J. Biol. Chem.* **256**, 7898–904.

Middlebrook, J.L. (1986). Cellular mechanisms of action of botulinum neurotoxin. *J. Toxicol.-Toxin Rev.* **5**, 177–90.

Middlebrook, J.L., Dorland, R.B. and Leppla, S.H. (1978). Association of diphtheria toxin with vero cells: Demonstration of a receptor. *J. Biol. Chem.* **253**, 7325–30.

Middlebrook, J.L., Dorland, R.B. and Leppla, S.H. (1979). Effects of lecithins on the interaction of diphtheria toxin with mammalian cells. *Expl Cell Res.* **121**, 95–101.

Misler, S. (1983). Gating of ion channels made by diphtheria toxin fragment in phospholipid bilayer membranes. *Proc. Natl. Acad. Sci. USA* **80**, 4320–4.

Misler, S. (1984). Diphtheria toxin fragment channels in lipid bilayer membranes: Selective sieves or discarded wrappers? *Biophys. J.* **45**, 107–9.

Moehring, T.J. and Crispell, J.B. (1974). Enzyme treatment of KB cells: The altered effect of diphtheria toxin. *Biochem. Biophys. Res. Commun.* **60**, 1446–52.

Moehring, J.M. and Moehring, T.J. (1988). The post-translation trimethylation of diphthamide studied in vitro. *J. Biol. Chem.* **263**(8), 3840–4.

Moehring, T.J., Danley, D.E. and Moehring, J.M. (1984). In vitro biosynthesis of dipthamide, studied with mutant Chinese hamster ovary cells resistant to diphtheria toxin. *Molec. Cell. Biol.* **4**, 642–50.

Montecucco, C. (1986). How do tetanus and botulinum toxins bind to neuronal membranes? *Trends Biochem. Sci.* **11**, 314–17.

Montecucco, C., Schiavo, G. and Tomasi, M. (1985). pH-dependence of the phospholipid interaction of diphtheria-toxin fragments. *Biochem. J.* **231**, 123–8.

Morris, R.E. and Saelinger, C.B. (1983). Diphtheria toxin does not enter cells by receptor-mediated endocytosis. *Infect. Immun.* **42**, 812–17.

Morris, R.E., Gerstein, A.S., Bonventre, P.F. and Saelinger, C.B. (1985). Receptor-mediated entry of diphtheria toxin into monkey kidney (vero) cells: Electron microscopic evaluation. *Infect. Immun.* **50**, 721–7.

Moskaug, J.Ø., Sandvig, K. and Olsnes, S. (1987). Cell-mediated reduction of the interfragment disulfide in nicked diphtheria toxin: A new system to study toxin entry at low pH. *J. Biol. Chem.* **262**, 10339–45.

Moskaug, J.Ø., Sandvig, K. and Olsnes, S. (1988). Low pH-induced release of diphtheria toxin A-fragment in vero cells. Biochemical evidence for transfer to the cytosol. *J. Biol. Chem.* **263**, 2518–25.

Moskaug, J.Ø., Sletten, K., Sandvig, K. and Olsnes, S. (1989a). Translocation of diphtheria toxin A-fragment to the cytosol. Role of site of interfragment cleavage. *J. Biol. Chem.* **264**, 15709–13.

Moskaug, J.Ø., Sandvig, K. and Olsnes, S. (1989b). Role of anions in low pH-induced translocation of diphtheria toxin. *J. Biol. Chem.* **264**, 11367–72.

Moynihan, M.R. and Pappenheimer, A.M., Jr. (1981). Kinetics of adenosinediphosphoribosylation of elongation factor 2 in cells exposed to diphtheria toxin. *Infect. Immun.* **32**, 575–82.

Murphy, J.R., Bishai, W., Borowski, M., Miyanohara, A., Boyd, D. and Nagle, S. (1986). Genetic construction, expression, and melanoma-selective cytotoxicity of a diphtheria toxin-related a-melanocyte-stimulating hormone fusion protein. *Proc. Natl. Acad. Sci. USA* **83**, 8258–62.

Myers, D.A. and Villemez, C.L. (1988). Specific chemical cleavage of diphtheria toxin with hydroxylamine: Purification and characterization of the modified proteins. *J. Biol. Chem.* **263**, 17122–7.

Nygård, O. and Nilsson, L. (1985). Reduced ribosomal binding of eukaryotic elongation factor 2 following ADP-ribosylation. Difference in binding selectivity between polyribosomes and reconstituted monoribosomes. *Biochim. Biophys. Acta* **824**, 152–62.

Oeltman, T.N. (1985). Synthesis and in vitro activity of a hormone-diphtheria toxin fragment a hybrid. *Biochem. Biophys. Res. Commun.* **133**, 430–5.

O'Keefe, D.O. and Collier, R.J. (1989). Cloned diphtheria toxin within the periplasm of *Escherichia coli* causes lethal membrane damage at low pH. *Proc. Natl. Acad. Sci. USA* **86**, 343–6.

O'Keefe, D.O. and Draper, R.K. (1985). Characterization of a transferrin-diphtheria toxin conjugate. *J. Biol. Chem.* **260**, 932–7.

Olsnes, S. and Sandvig, K. (1986). Interactions between diphtheria toxin entry and anion transport in vero cells: II. Inhibition of anion antiport by diphtheria toxin. *J. Biol. Chem.* **261**, 1553–61.

Olsnes, S. and Sandvig, K. (1988). How protein toxins enter and kill cells. In: *Immunotoxins* (ed. A.E. Frankel), pp. 39–73. Kluwer Academic Publishers, Boston.

Olsnes, S., Carvajal, E., Sundan, A. and Sandvig, K. (1985). Evidence that membrane phospholipids and protein are required for binding of diphtheria toxin in vero cells. *Biochim. Biophys. Acta* **846**, 334–41.

Olsnes, S., Carvajal, E. and Sandvig, K. (1986). Interactions between diphtheria toxin entry and anion transport in vero cells: III. Effect on toxin binding and anion transport of tumor-promoting phorbol esters, vandate, fluoride, and salicylate. *J. Biol. Chem.* **261**, 1562–9.

Oppenheimer, N.J. and Bodley, J.W. (1981). Diphtheria toxin: Site and configuration of ADP-ribosylation of diphthamide in elongation factor 2. *J. Biol. Chem.* **256**, 8579–81.

Papini, E., Colonna, R., Cusinato, F., Montecucco, C., Tomasi, M. and Rappuoli, R. (1987a). Lipid interaction of diphtheria toxin and mutants with altered fragment B: 1. Liposome aggregation and fusion. *Eur. J. Biochem.* **169**, 629–35.

Papini, E., Schiavo, G., Tomasi, M., Colombatti, M., Rappuoli, R. and Montecucco, C. (1987b). 2. Hydrophobic photolabelling and cell intoxication. Lipid interaction of diphtheria toxin and mutants with altered fragment B. *Eur. J. Biochem.* **169**, 637–44.

Papini, E., Colonna, R., Schiavo, G., Cusinato, F., Tomasi, M., Rappuoli, R. and Montecucco, C. (1987c). Diphtheria toxin and its mutant crm 197 differ in their interaction with lipids. *FEBS Lett.* **215**, 73–8.

Papini, E., Sandoná, D., Rappuoli, R. and Montecucco, C. (1988). On the membrane translocation of diphtheria toxin: at low pH the toxin induces ion channels on cells. *EMBO J.* **7**, 3353–9.

Pappenheimer, A.M. Jr. (1977). Diphtheria toxin. *Ann. Rev. Biochem.* **46**, 69–94.

Pappenheimer, A.M. Jr. and Gill, D.M. (1973). Diphtheria: recent studies have clarified the molecular mechanisms involved in its pathogenesis. *Science* **182**, 353–8.

Pastan, I., Willingham, M.C. and FitzGerald, D.J.P. (1986). Immunotoxins. *Cell* **47**, 641–8.

Perentesis, J.P., Genbauffe, F.S., Veldman, S.A., Galeotti, C.L., Livingston, D.M., Bodley, J.W. and Murphy, J.R. (1988). Expression of diphtheria toxin fragment A and hormone-toxin fusion proteins in toxin-resistant yeast mutants. *Proc. Natl. Acad. Sci. USA* **85**, 8386–90.

Poulain, B., Tauc, L., Maisey, E.A., Wadsworth, J.D.F., Mohan, P.M. and Dolly, J.O. (1988). Neurotransmitter release is blocked intracellularly by botulinum neurotoxin, and this requires uptake of both toxin polypeptides by a process mediated by the larger chain. *Proc. Natl. Acad. Sci. USA* **85**, 4090–4.

Proia, R.L., Hart, D.A. and Eidels, L. (1979a). Interaction of diphtheria toxin with phosphorylated molecules. *Infect. Immun.* **26**, 942–8.

Proia, R.L., Hart, D.A., Holmes, R.K., Holmes, K.V. and Eidels, L. (1979b). Immunoprecipitation and partial characterization of diphtheria toxin-binding glycoproteins from surface of guinea pig cells. *Proc. Natl. Acad. Sci. USA* **76**, 685–9.

Proia, R.L., Wray, S.K., Hart, D.A. and Eidels, L. (1980). Characterization and affinity labeling of the cationic phosphate-binding (nucleotide-binding) peptide located in the receptor-binding region of the B-fragment of diphtheria toxin. *J. Biol. Chem.* **255**, 12025–33.

Proia, R.L. Eidels, L. and Hart, D.A. (1981). Diphtheria toxin: receptor interaction. Characterization of the receptor interaction with the nucleotide-free toxin, the nucleotide-bound toxin, and the B-fragment of the toxin. *J. Biol. Chem.* **256**, 4991–7.

Ramsay, G., Montgomery, D., Berger, D. and Freire, E. (1989). Energetics of diphtheria toxin membrane insertion and translocation: Calorimetric characterization of the acid pH induced transition. *Biochemistry* **28**, 529–33.

Robbins, A.R., Peng, S.S. and Marshall, J.L. (1983). Mutant Chinese hamster ovary cells pleiotropically defective in receptor-mediated endocytosis. *J. Cell Biol.* **96**, 1064–71.

Robbins, A.R., Oliver, C., Bateman, J.L., Krag, S.S., Galloway, C.J. and Mellman, I. (1984). A single mutation in Chinese hamster ovary cells impairs both golgi and endosomal functions. *J. Cell Biol.* **99**, 1296–308.

Roff, C.F., Fuchs, R., Mellman, I. and Robbins, A.R. (1986). Chinese hamster ovary cell mutants with temperature-sensitive defects in endocytosis. I. Loss of function on shifting to the nonpermissive temperature. *J. Cell Biol.* **103**, 2283–97.

Rolf, J.M., Gaudin, H.M., Tirrell, S.M., MacDonald, A.B. and Eidels, L. (1989). Anti-idiotypic antibodies that protect cells against the action of diphtheria toxin. *Proc. Natl. Acad. Sci. USA* **86**, 2036–9.

Sandvig, K. and Moskaug, J.Ø. (1987). Pseudomonas toxin binds Triton X-114 at low pH. *Biochem. J.* **245**, 899–901.

Sandvig, K. and Olsnes, S. (1980). Diphtheria toxin entry into cells is facilitated by low pH. *J. Cell Biol.* **87**, 828–32.

Sandvig, K. and Olsnes, S. (1981a). Effects of retinoids and phorbols esters on the sensitivity of different cell lines to the polypeptide toxins modeccin, abrin, ricin and diphtheria toxin. *Biochem. J.* **194**, 821–7.

Sandvig, K. and Olsnes, S. (1981b). Rapid entry of nicked diphtheria toxin into cells at low pH: Characterization of the entry process and effects of low pH on the toxin molecule. *J. Biol. Chem.* **256**, 9068–76.

Sandvig, K. and Olsnes, S. (1982). Entry of the toxic proteins abrin, moideccin, ricin and diphtheria toxin into cells: I. Requirement for calcium. *J. Biol. Chem.* **257**, 7495–503.

Sandvig, K. and Olsnes, S. (1984). Anion requirement and effect of anion transport inhibitors on the response of vero cells to diphtheria toxin and modeccin. *J. Cell. Physiol.* **119**, 7–14.

Sandvig, K. and Olsnes, S. (1986). Interactions between diphtheria toxin entry and anion transport in vero cells: IV. Evidence that entry of diphtheria toxin is dependent on efficient anion transport. *J. Biol. Chem.* **261**, 1570–5.

Sandvig, K. and Olsnes, S. (1988). Diphtheria toxin-induced channels in vero cells selective for monovalent cations. *J. Biol. Chem.* **263**, 12352–9.

Sandvig, K., Olsnes, S. and Pihl, A. (1978). Binding, uptake and degradation of the toxic proteins abrin and ricin by toxin-resistant cells. *Eur. J. Biochem.* **82**, 13–23.

Sandvig, K., Sundan, A. and Olsnes, S. (1984). Evidence that diphtheria toxin and modeccin enter the cytosol from different vesicular compartments. *J. Cell Biol.* **98**, 963–70.

Sandvig, K., Sundan, A. and Olsnes, S. (1985). Effect of potassium depletion of cells on their sensitivity to diphtheria toxin and pseudomonas toxin. *J. Cell. Physiol.* **124**, 54–60.

Sandvig, K., Tønnessen, T.I., Sand, O. and Olsnes, S. (1986). Requirement of a transmembrane pH gradient for the entry of diphtheria toxin into cells at low pH. *J. Biol. Chem.* **261**, 11639–44.

Sayhan, O., Özdemirli, Nurten, R. and Bermek, E. (1986). On the nature of cellular ADP-ribosyltransferase from rat liver specific for elongation factor 2. *Biochem. Biophys. Res. Commun.* **139**, 1210–14.

Schaeffer, E.M., Moehring, J.M. and Moehring, T.J. (1988). Binding of diphtheria toxin to CHO-K1 and vero cells is dependent on cell density. *J. Cell. Physiol.* **135**, 407–15.

Shimizu, N., Miskimins, W.K. and Shimizu, Y. (1980). A cytotoxic epidermal growth factor cross-linked to diphtheria toxin A-fragment. *FEBS Lett.* **118**, 274–8.

Shiver, J.W. and Donovan, J.J. (1987). Interaction of diphtheria toxin with lipid vesicles: determinant of ion channel formation. *Biochim. Biophys. Acta* **903**, 48–55.

Shoyab, M., DeLarao, J.E. and Todaro, G.T. (1979). Biologically active phorbol esters specifically alter affinity of epidermal growth factor membrane receptors. *Nature* **279**, 387–91.

Sitikov, A.S., Davydova, E.K., Bezlepkina, T.A., Ovchinnikov, L.P. and Spirin, A.S. (1984a). Eukaryotic elongation factor 2 loses its non-specific affinity for RNA and leaved polyribosomes as a result of ADP-ribosylation. *FEBS Lett.* **76**, 406–10.

Sitikov, A.S., Davydova, E.K. and Ovchinnikov, L.P. (1984b). Endogenous ADP-ribosylation of elongation factor 2 in polyribosome fraction of rabbit reticulocytes. *FEBS Lett.* **176**, 261–3.

Stenmark, H., Olsnes, S. and Sandvig, K. (1988).

Requirement of specific receptors for efficient translocation of diphtheria toxin A fragment across the plasma membrane. *J. Biol. Chem.* **263**, 13449–55.

Stenmark, H., McGill, S., Sandvig, K., Collier, R.J. and Olsnes, S. (1989). Permeabilisation of the plasma membrane by deletion mutants of diphtheria toxin. *EMBO J.* **8**, 2849–53.

Tran, D., Carpentier, J.-L., Sawano, F., Gorden, P. and Orci, L. (1987). Ligands internalized through coated or noncoated invaginations follow a common intracellular pathway. *Proc. Natl. Acad. Sci. USA* **84**, 7957–61.

Tweten, R.K., Barbieri, J.T. and Collier, R.J. (1985). Diphtheria toxin: Effect of substituting aspartic acid for glumatic acid 148 on ADP-ribosyltransferase activity. *J. Biol. Chem.* **260**, 10392–4.

Uchida, T., Pappenheimer, A.M. Jr. and Greany, R. (1973). Diphtheria toxin and related proteins. I. Isolation and some properties of mutant proteins serologically related to diphtheria toxin. *J. Biol. Chem.* **248**, 3838–44.

Uchida, T., Yamaizumi, M., Mekada, E. and Okada, Y. (1978). Reconstitution of hybrid toxin from fragment A of diphtheria toxins as a subunit of *Wistaria floribunda* lectin. *J. Biol. Chem.* **253**, 6307–10.

van Heyningen, S. (1977). Cholera toxin. *Biol. Rev.* **52**, 509–49.

Van Ness, B.G., Howard, J.B. and Bodley, J.W. (1980). ADP-ribosylation of elongation factor 2 by diphtheria toxin: NMR spectra and proposed structures of ribosyl-diphthamide and its hydrolysis products. *J. Biol. Chem.* **255**, 10710–16.

Villemez, C.L. and Carlo, P.L. (1984). Preparation of an immunotoxin for *Acanthamoeba castellani*. *Biochem. Biophys. Res. Commun.* **125**, 25–9.

Williams, D.P., Parker, K., Bacha, P., Bishai, W., Borowski, M., Genbauffe, F., Strom, T.B. and Murphy, J.R. (1987). Diphtheria toxin receptor binding domain substitution with interleukin-2: genetic construction and properties of a diphthheria toxin-related interleukin-2 fusion protein. *Protein Engng* **1**, 493–8.

Wright, H.T., Marston, A.W. and Goldstein, D.J. (1984). A functional role for cysteine disulfides in the transmembrane transport of diphtheria toxin. *J. Biol. Chem.* **259**, 1649–54.

Yamaizumi, M., Mekada, E., Uchida, T. and Okada, Y. (1978). One molecule of diphtheria toxin fragment A introduced into a cell can kill the cell. *Cell* **15**, 245–50.

Zalman, L.S. and Wisnieski, B.J. (1984). Mechanism of insertion of diphtheria toxin: Peptide entry and pore size determinations. *Proc. Natl. Acad. Sci. USA* **81**, 3341–5.

Zhao, J-M. and London, E. (1986). Similarity of the conformation of diphtheria toxin at high temperature to that in the membrane-penetrating low-pH state. *Proc. Natl. Acad. Sci. USA* **83**, 2002–61.

5

Assembly and Secretion of Oligomeric Toxins by Gram-Negative Bacteria

Timothy R. Hirst

The Biological Laboratory, University of Kent, Canterbury, Kent CT2 7NJ, UK

Introduction

Oligomeric toxins are a major group of biological macromolecules that count amongst their number, cholera toxin, shiga toxin and pertussis toxin. All of them consist of two or more different types of subunits which assemble together into non-covalently linked multimeric complexes comprised of six polypeptide chains. Each toxin consists of an A subunit (termed S1 for pertussis toxin) which exhibits enzymatic activity, and a B oligomer that is responsible for binding the toxin to eukaryotic cell surface receptors (for reviews on the structure and biological activity of these toxins, see Chapters 1, 2 and 18 of this volume).

Production of oligomeric toxins occurs exclusively in Gram-negative bacterial species (see next section). This may be attributable to the evolutionary origin of the various toxin families or to a requirement for the particular biosynthetic and assembly processes found in these organisms.

The genes which encode the toxin subunits are organized into polycistronic operons located either in the bacterial chromosome (e.g. for cholera toxin), in plasmids (e.g. for *Escherichia coli* heat-labile enterotoxin) or on temperate bacteriophages (e.g. for *E. coli* shiga-like toxin). The subunits are expressed as precursor polypeptides and it is presumed that each subunit is exported independently, followed by processing, subunit folding and toxin assembly. Some oligomeric toxins remain within the envelope of the producing organism (e.g. *E. coli* heat-labile enterotoxin and shiga toxin) whereas others are secreted across the bacterial outer membrane into the extracellular milieu (e.g. cholera toxin and pertussis toxin). Subunit translocation across the cytoplasmic membrane involves the participation of a 'general export pathway' that is utilized by all proteins exported to the envelope compartments of Gram-negative bacteria.

Secretion of proteins across the outer membrane into the external milieu is a much rarer occurrence which appears to be governed by a different mechanism for each type of secreted protein. Studies on the export, assembly and secretion of oligomeric toxins will not only provide an insight into the intriguing events associated with these biological phenomena, but they will also open up the prospect of engineering novel oligomeric toxoids for use in new vaccines.

Production of oligomeric toxins

The limited number of categories of oligomeric toxins, exemplified by cholera toxin, shiga toxin and pertussis toxin, belies the fact that numerous Gram-negative bacteria produce such toxins. Cholera toxin, for example, is the prototype of a large family of related enterotoxins produced by organisms responsible for causing secretory diarrhoea, including *Vibrio cholerae*, *V. mimicus*, certain strains of *E. coli*, *Salmonella*, *Campylobacter*, and *Aeromonas* (Finkelstein and LoSpalluto, 1969;

SOURCEBOOK OF BACTERIAL PROTEIN TOXINS
ISBN 0-12-053078-3

Holmgren, 1981; Ruiz-Palacious *et al.*, 1983; Spira and Fedorka-Cray, 1984; Klipstein and Engert, 1985; Betley *et al.*, 1986; Chopra *et al.*, 1987; Yamamoto *et al.*, 1987; Fernandez *et al.*, 1988; Schultz and McCardell, 1988; Rose *et al.*, 1989a, b). Shiga toxin and shiga-like toxins (SLTs; also called Vero toxins) are another major group of oligomeric toxins, that have a cytotoxic activity on eukaryotic cells, and are produced by *Shigella dysenteriae* and other *Shigella* spp., as well as certain *E. coli*, *Salmonella typhimurium*, *Vibrio cholerae* and *Campylobacter jejuni* (O'Brien and Holmes, 1987; Jackson *et al.*, 1987; Strockbine *et al.*, 1988; Weinstein *et al.*, 1988).

Cholera toxin and related enterotoxins are comprised of two different types of subunits; a single A subunit (M_r approx. 28 000) which has ADP-ribosyltransferase activity, and five B subunits (M_r appprox. 12 000) which bind to ganglioside or glycoprotein receptors on eukaryotic cell surfaces (Holmgren, 1973; Gill, 1976; Moss and Richardson, 1978; Gill *et al.*, 1981; Holmgren *et al.*, 1985; Fukuta *et al.*, 1988). The genes for *Escherichia coli* heat-labile enterotoxin type I (LT-I, with subtypes LTh-I and LTp-I produced by human and porcine isolates of *E. coli* respectively) and cholera toxin have been cloned and sequenced, and shown to be organized as contiguous operons which specify a single polycistronic mRNA encoding the A and B subunits (Dallas and Falkow, 1980; Spicer and Noble, 1982; Yamamoto and Yokota, 1983; Lockman and Kaper, 1983; Mekalanos *et al.*, 1983; Yamamoto *et al.*, 1984, 1987). The A subunits are initially synthesized as 258 amino acid precursor polypeptides containing at their N-termini an 18-residue signal sequence (Spicer and Noble, 1982; Mekalanos *et al.*, 1983). The B subunits are expressed as 124 amino acid precursors with a 21-residue signal sequence (Dallas and Falkow, 1980; Yamamoto and Yokoto, 1983; Lockman and Kaper, 1983; Mekalanos *et al.*, 1983). The A and B subunits of cholera toxin and LT-I are approximately 80% homologous.

A second group of heat-labile enterotoxins, type II (LT-II, subtypes LT-IIa and LT-IIb) have been identified by Holmes and co-workers, which have ADP-ribosyltransferase activities, but differ in their ganglioside-binding specificities from cholera toxin and LT-I (Holmes *et al.*, 1986; Fukuta *et al.*, 1988). Nucleotide sequence analysis of the genes encoding LT-IIa revealed that they were organized in a similar manner to cholera toxin and LT-I, except that the A and B genes of LT-IIa overlapped by 11 base pairs instead of four (Pickett *et al.*, 1987).

The precursor A subunit of LT-IIa contains an 18-residue signal sequence, whilst the mature protein is longer by an additional amino acid compared to the A subunit of LT-I. The A gene shows an overall 57% homology with the gene of LT-I and cholera toxin. The B gene encodes a polypeptide with no significant primary structural homology to the B subunit of LT-I, and this may account for its altered ganglioside-binding specificity. The B subunit of LT-IIa is synthesized as a precursor polypeptide with a 19-residue signal sequence, and with a mature sequence of 104 amino acids (Pickett *et al.*, 1987).

The structure of the assembled toxins (1A:5B subunits) implies that there may be stoichiometric control of subunit synthesis. In the case of cholera toxin this appears to be governed by the relative efficiency of the Shine–Dalgarno (SD) sequences * just upstream of the start codons of the A and B genes. The A gene has two overlapping SD sequences, whilst the B gene has a theoretically perfect SD sequence (TAAGGA) which lies within the C-terminal coding sequence of the A cistron. Deletion of the majority of the A gene to give an in-frame fusion between amino acid 17(−1) of the A subunit signal sequence and amino acid 19(−3) of the B signal sequence, maintained the normal cleavage site of the B subunit but made B subunit expression dependent on the efficiency of translational initiation sequences of the A cistron (Mekalanos *et al.*, 1983). Under these circumstances, only 1/9th of the B subunit was produced compared with the expression in the wild-type operon, suggesting that the SD sequence controlling expression of the B cistron is up to ninefold more efficient (Mekalanos *et al.*, 1983). This may account for the expected fivefold difference in A and B subunit synthesis which gives rise to the

---

* The Shine–Delgarno sequence or ribosome-binding site is complementary to the 3′-end of the 16S ribosomal RNA and is thought to be directly involved in the reaction sequence which leads to translational initiation at the start codon within the mRNA.

1A:5B stoichiometry in the assembled holotoxin molecule.

Shiga toxin and related cytotoxins are also composed of two different types of subunits, in a ratio of 1A:5B subunits; the A subunit (M_r approx. 32 000) inhibits protein synthesis in eukaryotic cells by cleaving the N-glycosidic bond of adenine 4324 in the 28S rRNA, and the B subunits (M_r approx. 7 700) bind to galactose-α1-4-galactose containing Gb_3 receptors on eukaryotic cell surfaces (Jackson et al., 1987; Lindberg et al., 1987; Endo et al., 1988; Strockbine et al., 1988). Genes encoding the A and B subunits are organized into polycistronic operons, but unlike the cholera toxin operon the cistrons do not overlap. The shiga toxin operon has been cloned from the chromosome of Shigella dysenteriae type 1, and several shiga-like toxin operons have been cloned from toxin-converting phages of E. coli (Newland et al., 1985; Willshaw et al., 1985; Huang et al., 1986; Calderwood et al., 1987; Newland et al., 1987; Strockbine et al., 1988). Shiga toxin and shiga-like toxin type I (SLT-I) are essentially identical with only a single amino acid difference at position 45 of the A subunit (Strockbine et al., 1988). The A subunits of shiga toxin and SLT-I are synthesized with a 22-residue signal sequence which, after cleavage, gives a 293 amino acid polypeptide. The mature B subunits consist of 69 amino acids and the initial translation product carries a 20-residue signal sequence. An antigenic variant of SLT-I, designated SLT-II or Vero toxin type 2, shares an overall 55% homology with the A and B subunits of SLT-I (Jackson et al., 1987). The B subunits of SLT-II are 70 amino acids in length and are synthesized with an additional 19-residue signal sequence. More recently, a variant of SLT-II has been identified (SLT-IIv) in E. coli which causes swine oedema disease; unlike SLT-II it is unable to bind to Gb_3 receptors (Weinstein et al., 1988). The A subunit genes of SLT-II and SLT-IIv are highly homologous (94%) whereas the B subunits are only 79% homologous. Of particular interest has been the finding that SLT-IIv is not exclusively cell-associated like other members of the shiga toxin family, but instead a significant fraction of the toxin appears to be released into the extracellular milieu when expressed in E. coli (Weinstein et al., 1989). O'Brien and co-workers have used this observation to analyse the role of the A and B subunits in determining cellular location and release, by expressing heterologous mixtures of the subunits of shiga toxin and the SLTs.

Pertussis toxin is a protein of 105 kDa produced by Bordetella pertussis. It is the most structurally complex of the oligomeric toxins, and is comprised of five different subunits (S1, S2, S3, S4 and S5) (Tamura et al., 1982; Nicosia et al., 1986; Locht and Keith, 1986). Like other oligomeric toxins it has a single enzymatically active subunit (S1) which in this case catalyses ADP-ribosylation of the G_i protein of adenylate cyclase, and a B oligomer consisting of S2, S3, S4 and S5 subunits which bind the toxin to eukaryotic cell surfaces (see Chapter 1 of this volume). In vitro dissociation and reconstitution experiments have revealed that the B oligomer consists of two dimers (S2, S4) and (S3, S4), which are non-covalently associated with a monomer of S5 (Tamura et al., 1982). The genes encoding the five subunits have been cloned and sequenced, and shown to be organized into a polycistronic operon with a gene order of S1, S2, S4, S5 and S3 (Nicosia et al., 1986; Locht and Keith, 1986). The molecular weights of the subunits are: S1, 26 220; S2, 21 950; S3, 21 860; S4, 12 060; and S5, 10 940. The amino acid sequences of S2 and S3 are 75% homologous (Nicosia et al., 1986). Each subunit is synthesized as a precursor containing a signal sequence; these exhibit the structural characteristics found in the majority of signal sequences, with the exception of the S4 protein which has an unusually long (42-residue) signal sequence with seven positively charged arginine residues in the n-region (Nicosia et al., 1986). The significance of this highly charged N-terminus for export across the B. pertussis cytoplasmic membrane is unknown.

Expression of many of the oligomeric toxins is transcriptionally controlled by positive regulatory factors.

Protein export and secretion in Gram-negative bacteria

The Gram-negative bacterial envelope

The envelopes of Gram-negative bacteria have a remarkably complex structure and an appreciation of this is necessary to understand the sequence of events which leads to the secretion of oligomeric toxins. The distinction between Gram-positive and Gram-negative bacteria is based on a staining technique developed by Christian Gram in 1884, in which bacteria differentially retain crystal violet/iodine precipitates when rinsed in alcohol. This difference is due to the structure of their envelopes. Gram-positive bacteria have a thick peptidoglycan wall which retains the crystal violet/iodine precipitates, and which surrounds a single (cytoplasmic) membrane. Gram-negative bacteria have a much thinner peptidoglycan layer from which the precipitates easily leach away. This layer, however, is bounded by two membranes – the inner (cytoplasmic) membrane and an outer membrane; between them can be found a periplasmic compartment that contains soluble proteins, membrane-derived oligosaccharides, as well as the peptidoglycan (Fig. 1; Inouye, 1979; Hobot et al., 1984; Nakae, 1986; Nikaido and Vaara, 1987; Kennedy, 1987; Oliver, 1987; Park, 1987).

The various layers of the Gram-negative envelope serve different functions. The cytoplasmic membrane is the true osmotic barrier of the cell and consists of a lipid bilayer that is rich in biosynthetic and transport proteins. It is through this bilayer

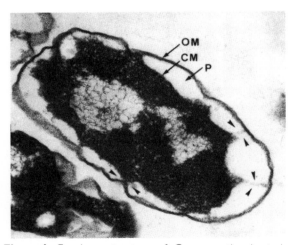

Figure 1. Envelope structure of Gram-negative bacteria. Micrograph of *Escherichia coli* B, plasmolysed in sucrose and fixed with formaldehyde. The outer membrane (OM), cytoplasmic membrane (CM) and periplasm (P) are indicated. Zones of adhesion between the cytoplasmic and outer membranes are indicated by arrowheads. (Used with permission, Hirst and Welch, 1988; copyright Elsevier Trends Journals, Cambridge, UK.)

that all exported proteins must pass if they are to be incorporated into the other layers of the envelope or be secreted into the external medium. The outer membrane is a protective permeability barrier, permitting the diffusion of small solutes (<1000 Da) whilst shielding the cell from attack by external degradative enzymes and antibiotics (for reviews see Nakae, 1986; Nikaido and Vaara, 1987). It consists of an asymmetric bilayer of lipids, lipopolysaccharides, proteins and lipoproteins. The periplasm contains a host of enzymes and transport-binding proteins involved in the degradation and uptake of small solutes such as sugars and amino acids (Oliver, 1987).

Electron microscopic studies of Gram-negative envelopes have revealed that the cytoplasmic and outer membranes are in intimate contact at sites which have been termed zones of adhesion or Bayers junctions (Fig. 1; Bayer, 1968, 1979). The existence and function of adhesion zones have been the subject of some controversy (Bayer, 1979; Hobot et al., 1984) although they are generally accepted to play a role in the export of envelope components such as lipopolysaccharides and carbohydrate capsules (Muhlradt et al., 1973; Bayer and Thurow, 1977). Adhesion zones have also been implicated in the export and secretion of proteins by Gram-negative bacteria since outer membrane porins emerge on the bacterial cell surface at distinct sites which correspond to adhesion zones (Smit and Nikaido, 1978; Bayer, 1979; Randall et al., 1987; Hirst and Welch, 1988). In addition, certain proteins appear to be preferentially localized within adhesion zones, e.g. thioredoxin, which may play a role in catalysing the formation of disulphide bonds in exported proteins (Bayer et al., 1987).

The export and secretion of 'oligomeric' proteins is a phenomenon that occurs in Gram-negative bacteria, but not in Gram-positive organisms. All layers of the Gram-negative envelope have oligomeric proteins, including the periplasm which contains enzymes such as the dimeric alkaline phosphatase (PhoA), and in certain instances, oligomeric toxins, such as the heat-labile enterotoxin (LT) produced by certain *Escherichia coli* strains (Hirst et al., 1984b; Hofstra and Witholt, 1984). Outer membranes contain trimeric porins, and the surfaces of outer membranes are the sites of attachment of multimeric adhesins.

The explanation why only Gram-negative bacteria are able to produce oligomeric proteins is not obvious. One possibility is that the structure of the envelope provides an environment which favours subunit–subunit interactions. Alternatively, novel factors found only in Gram-negative bacteria may be involved in facilitating the assembly of multimeric proteins. What these factors might be, and how the structure of the envelope might influence subunit interactions is discussed in relation to the production of oligomeric toxins in the next section.

In order to understand the export of toxin subunits across the cytoplasmic membranes of Gram-negative bacteria, it is important to realize that virtually all proteins translocated across this membrane utilize a general export pathway and machinery (for reviews, see Randall *et al.*, 1987; Randall and Hardy, 1989; for the few exceptions, see Holland *et al.*, 1990). During the past 15 years considerable progress has been made in examining the export of proteins in prokaryotic systems, and one of the most important and striking features to have emerged is that protein translocation across the cytoplasmic membranes of bacteria shares many fundamental features with protein translocation across the endoplasmic reticulum of eukaryotes and with import of proteins into organelles, such as chloroplasts and mitochondria (Randall *et al.*, 1987; Singer *et al.*, 1987; Verner and Schatz, 1988; Ellis and Hemmingsen, 1989; Keegstra, 1989; Pelham, 1989; Randall and Hardy, 1989). Since these mechanisms are fundamentally similar, I shall assume that polypeptide transfer in all Gram-negative bacterial species is identical to the processes which have been characterized in *Escherichia coli*. This is particularly important for any discussion of toxin export and secretion, because protein export in most toxinogenic organisms has not been investigated.

The general export pathway

Protein export is governed by the properties and characteristics of the exported polypeptide itself and by the cellular machinery required to accomplish the export event.

Virtually all exported and secreted proteins, including the subunits of oligomeric toxins, are synthesized as precursor proteins with N-terminal extensions called leader or signal sequences which are usually 15–30 amino acids★ in length (von Heijne, 1985; Randall and Hardy, 1989; Gierasch, 1989). The signal sequences are proteolytically removed during the translocation of the polypeptide across the membrane to yield the so-called 'mature' protein. This processing step is achieved by one of two signal peptidases found in the membrane; signal peptidase I (leader peptidase) cleaves most precursor proteins whilst signal peptidase II removes signal sequences from lipoproteins (Wolfe *et al.*, 1983; Tokunaga *et al.*, 1984). Von Heinje (1985) has analysed the residue variability of signal sequences and identified three domains which have a consensus of characteristics. Amino acids adjacent to the mature polypeptide, i.e. the cleavage site (the c-region) comprise 5–7 residues and exhibit a '−3, −1 rule', in which these positions are occupied by residues of high polarity, usually alanine. Adjacent to this region are 7–13 hydrophobic amino acids (the h-region), and at the N-terminus (the n-region) are found a sequence of amino acids of variable length and composition which always include one or more positively charged residues. The signal sequences in the precursors of oligomeric toxins obey these general rules, with the exception of the S4 subunit of pertussis toxin which has an unusually long (42-residue) signal sequence (Nicosia *et al.*, 1986).

The role of signal sequences in protein export has been the source of considerable conjecture and discussion (Wickner, 1979; Ferenci and Silhavy, 1987; Randall and Hardy, 1989; Gierasch, 1989). They are essential for the correct and efficient targeting of exported proteins to the membrane, but how they achieve this remains to be resolved. In addition to the signal sequence, the mature portion of the protein almost certainly plays a role in the export process. This may involve passive properties such as the absence of amino acid sequences of exceptional hydrophobicity, hydrophilicity or domains which might fold rapidly, thereby inhibiting or aborting polypeptide transfer across the membrane. A more active role would be in the interaction of the mature portion with the signal sequence or with cytosolic export factors.

★ Each residue in the signal sequence is designated a negative number starting from the cleavage site (−1), and proceeding through to the N-terminus.

Randall and co-workers have shown that the cytosolic export factor, SecB (see below), interacts with regions of the mature portion of maltose-binding protein, and this is thought to be important in preserving the nascent precursor in a conformation that can successfully engage the export machinery (Lui *et al.*, 1989).

Researchers have identified numerous cytoplasmic and membrane-associated factors which are thought to constitute the cellular machinery involved in polypeptide transfer across membranes (Randall *et al.*, 1987; Crooke and Wickner, 1987; Fandl and Tai, 1987; Bieker and Silhavy, 1989; Bertstein *et al.*, 1989; Cunningham *et al.*, 1989; Ellis and Hemmingsen, 1989; Kumamoto, 1989; Kusukawa *et al.*, 1989; Watanabe and Blobel, 1989). In *E. coli*, several cytoplasmic export factors have been characterized, some of which are clearly essential proteins, whilst others appear to influence the efficiency of the export process. One of the most well-characterized is SecA, a protein of 102 kDa, found both in the cytoplasm and in association with the cytoplasmic membrane, which is essential for protein export (Oliver and Beckwith, 1982; Schmidt *et al.*, 1988; Cunningham *et al.*, 1989). It has been postulated that the molecule is involved in interacting with nascent precursor proteins and in targeting the ribosome–precursor complex to the membrane. Other cytoplasmic factors, including SecB, trigger factor and GroES/EL affect the efficient export of certain polypeptides in *E. coli* (Kumamoto and Beckwith, 1983; Crooke and Wickner, 1987; Kusukawa *et al.*, 1989). These proteins appear to interact with the mature portion of precursor proteins to modulate folding and preserve them in conformations which can successfully enter the export apparatus (Crooke and Wickner, 1987; Kumamoto, 1989; Lui *et al.*, 1989).

The only unequivocally identified component of the export machinery within the membrane of *E. coli* is SecY (Ito *et al.*, 1983; Shiba *et al.*, 1984; Fandl and Tai, 1987), with the exception of the signal peptidases which are required for precursor cleavage. This is a 49-kDa hydrophobic polypeptide found in the cytoplasmic membrane. SecY mutants accumulate precursor forms of exported proteins, and allelic mutations of *secY* (called *prlA*) can suppress the export deficiency of precursor proteins which have defective signal sequences (Emr *et al.*,

1981). Several other gene products may also be required for protein export in *E. coli*. These include SecD (Gardel *et al.*, 1987), SecE (Riggs *et al.*, 1988), PrlC (Emr *et al.*, 1981) and PrlD (Bankaitis and Bassford, 1985). Further evidence confirming the role of these proteins and their function is needed (see the Note added in proof, page 100).

Protein export not only involves the functioning of specialized proteins but also requires energy in the form of proton-motive force and ATP (Date *et al.*, 1980; Daniels *et al.*, 1981; Enequist *et al.*, 1981; Chen and Tai, 1985; Geller *et al.*, 1986). The dissipation of proton-motive force by uncouplers causes the accumulation of precursor forms of exported proteins within the cell. Chen and Tai (1985) demonstrated that ATP was obligatory for import of periplasmic and outer membrane precursors into inverted inner membrane vesicles, and more recently Wickner and co-workers have demonstrated that both proton-motive force and ATP are involved in the efficient import of proOmpA into such vesicles (Geller *et al.*, 1986). The precise role of ATP and proton-motive force in the translocation process remains to be determined.

The subsequent sorting of exported proteins into the periplasm and outer membrane seems to be dependent on information contained within the mature polypeptide (Benson *et al.*, 1984; Nikaido and Wu, 1984). This may determine the pathway intermediates take during their export through the envelope, as well as key folding and release steps. It has been postulated that both maltose-binding protein and outer membrane porin are translocated via membrane sites which are of a higher density than bulk cytoplasmic membrane, and which may correspond to adhesion zones (Thom and Randall, 1988). For outer membrane proteins, the subsequent insertion and assembly into the outer membrane may be governed by lateral diffusion from the adhesion sites, whereas for periplasmic proteins, release from the membrane is required to yield a soluble protein. Some outer membrane proteins (e.g. TonA) may be released into the periplasm as membrane-free intermediate forms, followed by their partitioning into the outer membrane (Jackson *et al.*, 1986). Understanding the molecular events and sorting signals which lead to the correct incorporation of exported proteins into their appropriate membrane compartments remains one of the intriguing and unresolved aspects of the general export pathway.

Multiple pathways for secretion of extracellular proteins

Secretion of polypeptides into the surrounding medium by Gram-negative bacteria was thought to be a rare phenomenon, mainly because of the virtual absence of such secretion events in *E. coli*. However, it is now clear that this is not the case, and that many Gram-negative bacteria specifically secrete proteins to the medium, including certain pathogenic strains of *E. coli* (for reviews, see Pugsley and Schwartz, 1985; Hirst and Welch, 1988; Holland *et al.*, 1990). Moreover, some bacterial species, especially those belonging to the Pseudomonads and Vibrionaceae are prolific secretors of extracellular proteins. In the majority of cases, these proteins are synthesized with typical N-terminal signal sequences, and are presumed to be exported across the cytoplasmic membrane by the general export pathway. However, they often require the activity of additional gene products which encode accessory secretion factors in order for the extracellular protein to traverse the outer membrane into the external medium.

This raises two important issues: what are the identities and functions of these accessory factors and what pathways and mechanisms are involved?

One of the most well-studied secretion systems is the production of a starch-debranching lipoprotein, called pullulanase by *Klebsiella pneumoniae*. Pullulanase (PulA) is synthesized as a precursor with a typical signal sequence which is cleaved off by signal peptidase II and then a proportion of the mature protein is acylated (Pugsley *et al.*, 1986). Studies on the expression of the cloned *pulA* gene in *E. coli* revealed that pullulanase was exported to the outer membrane but was not secreted into the medium. Further investigation revealed the critical importance of DNA which flanks the *pulA* gene, in accomplishing pullulanase secretion (d'Enfert and Pugsley, 1989; Kornacher *et al.*, 1989). This identified at least two loci which encode five or more polypeptides which participate in translocating pullulanase across the outer membrane and in releasing it into the medium (see the Note added in proof, page 100). The mechanism

by which these accessory secretion proteins achieve efficient secretion of pullulanase is not yet known.

In contrast to the large number of accessory secretion proteins required for pullulanase secretion, secretion of α-haemolysin by *E. coli* requires two gene products (HlyB and HlyD) (see the Note added in proof), whilst IgA protease production by *Neisseria gonorrhoeae* and *Haemophilus influenzae* appears not to require any accessory secretion factors or additional 'help', other than the structural and functional characteristics found in the proteases themselves (Pohlner *et al.*, 1987; Holland *et al.*, 1990). IgA proteases are produced as approximately 150-kDa polypeptides with N-terminal signal sequences. Translocation into the medium has been shown to be dependent on the presence of the carboxyl (49 kDa) domain, since deletions in this region result in the accumulation of protease within the periplasm (Pohlner *et al.*, 1987). Expression of the wild-type neisserial IgA protease gene in *E. coli* was sufficient for protease secretion. These observations have led to a model for protease secretion in which the precursor is translocated across the cytoplasmic membrane by the general export pathway and the mature polypeptide released into the periplasm. The C-terminal 49-kDa domain has then been postulated to partition into the outer membrane and cause the translocation of the N-terminal (105-kDa protease) portion to the external surface of the cell, where autoproteolytic cleavage results in the liberation of the active protease (Pohlner *et al.*, 1987).

In the case of pullulanase and haemolysin secretion from *K. pneumoniae* and *E. coli* respectively, there is no evidence for the polypeptides transiently entering the periplasm during their translocation across the envelope. However, for many other proteins, both indirect and direct evidence has been obtained which indicates that secretion is a two-step phenomenon involving export across the cytoplasmic membrane to the periplasm followed by a subsequent secretory event across the outer membrane. Analyses of mutant strains which are defective in the secretion of extracellular proteins often results in the accumulation of the polypeptides in the periplasm (Howard and Buckley, 1983; Andro *et al.*, 1984; Hirst and Holmgren, 1987b). Moreover, when cloned genes specifying extracellular proteins from a variety of

Gram-negative bacteria are expressed in *E. coli*, most of the polypeptides are only exported as far as the periplasm (Pearson and Mekalanos, 1982; Bricker *et al.*, 1983; Douglas *et al.*, 1987; Gobius and Pemberton, 1988). This confirms a general rule, that for most extracellular proteins, with the exception of certain proteases, additional gene products produced by the parent bacteria are required for secretion. The structural variation of these additional factors implies that they do not function in a pleiotropic fashion but are restricted to secreting a single type or class of polypeptides through the outer membrane.

Biochemical studies on the secretion of oligomeric toxins from *V. cholerae* has provided definitive evidence of polypeptide entry into the periplasm during secretion across the envelope (Hirst and Holmgren, 1987b). The flux of newly synthesized toxin subunits through the envelope was investigated by a kinetic pulse-labelling and cell-fractionation experiment. This involved radioactively labelling polypeptides by adding [^{35}S]methionine to the cells for a pulse period of 30 s, followed by addition of a chase of non-radioactive methionine which quenches further incorporation of the radiolabel. Medium was separated from the cells at different times after addition of the chase, the bacteria fractionated to isolate the periplasm and the labelled polypeptides analysed by SDS–polyacrylamide gel electrophoresis and autoradiography (Fig. 2). This demonstrated that newly synthesized toxin subunits were rapidly released into the periplasmic compartment and that this was followed by their slow efflux (half-time, 13 min) from the periplasm into the medium.

Assembly and secretion of oligomeric toxins

Subunit translocation across the cytoplasmic membrane

Since all of the subunits of oligomeric toxins are synthesized as longer precursors with N-terminal signal sequences it is reasonable to assume that they utilize a general export pathway to cross the cytoplasmic membrane. It is nevertheless important to realize that the definitive experiments have yet to be undertaken which show, for example, that

SecA or SecY are obligatory for export of *E. coli* enterotoxins and cytotoxins. Furthermore, the identity of the export machinery of Gram-negative organisms other than *E. coli* and *Salmonella typhimurium* has yet to be investigated. It would be quite conceivable that additional cytosolic export factors will be identified, especially in organisms like *B. pertussis*, which produce the S4 subunit with such an unusual signal peptide.

Another comparison that is worth noting, is that most of the exported proteins which have been extensively investigated in *E. coli* have molecular weights in excess of 25 000. In contrast, some of the subunits of the oligomeric toxins are extremely small (<8 000 for shiga and shiga-like toxin B subunits; 12 000 for enterotoxin B subunits). These polypeptides will scarcely have emerged from the ribosome before elongation has been completed. It is therefore quite likely that these molecules are exported post-translationally. Their export may be independent of such cytosolic factors as SecB, or they may require as yet unidentified factors.

The rate of subunit translocation across the membrane, and release into the periplasm has been studied by kinetic pulse-labelling techniques (Hofstra and Witholt, 1984). This showed that the mature B subunits of LTp-I appeared in the periplasm 4–6 s after polypeptide chain termination. The A subunits required considerably longer, with processing having a half-time of approximately 30 s and release not being complete even after an extensive chase period of 3 min (Hofstra and Witholt, 1984). The reason for this delay in release is not known.

Processing of the A and B subunits of LT-I and of the S1 subunit of pertussis toxin is inhibited by membrane perturbing or uncoupling reagents (Palva *et al.*, 1981; Hofstra and Witholt, 1984; Perera and Freer, 1987). This indicates that energy, presumably in the form of proton motive force, is required for subunit translocation to the site of processing by signal peptidase I. However, the precise role of energy in subunit translocation, as for other exported proteins, remains to be elucidated.

Subunit release from the cytoplasmic membrane

Polypeptides which are translocated across the cytoplasmic membrane emerge on the trans or

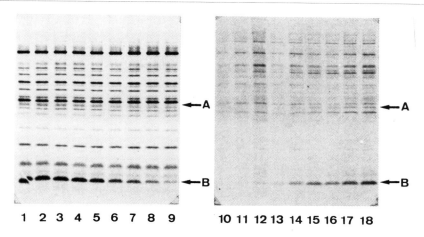

Figure 2. *V. cholerae* secrete enterotoxin subunits via the periplasm. *V. cholerae* strain TRH7000 ($\triangle$CT) harbouring a plasmid encoding LTh-I was labelled for 0.5 min with [^{35}S]methionine, followed by addition of a chase of I mM L-methionine to quench radiolabel uptake. The periplasm (lanes 1–9) and medium (lanes 10–18) were obtained from cells chased for the following lengths of time: 0.25 min (lanes 1 and 10); 0.5 min (lanes 2 and 11), 1 min (lanes 3 and 12); 2 min (lanes 4 and 13); 4.5 min (lanes 5 and 14); 9.5 min (lanes 6 and 15); 14.5 min (lanes 7 and 16); 19.5 min (lanes 8 and 17) and 29.5 min (lanes 9 and 18). Equivalent amounts of each fraction were analysed by SDS–polyacrylamide gel electrophoresis and an autoradiograph was obtained of the dried gel. The migration positions of the LTA and B subunits are indicated. (From Hirst and Holmgren, 1987; ASM Publications.)

periplasmic face of the membrane, and must then fold up and be released. The events of folding and release are important in the translocation of all proteins, yet these processes have been largely overlooked and as such remain poorly understood. Polypeptide release has nevertheless been recognized as a distinct step in the export process, through studies at lower temperatures in which release is slowed down, and by analysis of truncated and mutant proteins which fail to complete the export process (Ito and Beckwith, 1981; Koshland and Botstein, 1982; Hengge and Boos, 1985; Minsky *et al.*, 1986; Fitts *et al.*, 1987; Sandkvist *et al.*, 1987, 1990).

To further understand polypeptide folding and release it will be necessary to know considerably more about how proteins fold *in vivo*, particularly within the context of environment of the membrane. For some proteins, such as β-lactamase and maltose-binding protein, it has been suggested that various mutants are defective in protein folding, thus causing the formation of partially folded intermediates or aberrantly folded aggregates, which remain associated with the trans-side of the cytoplasmic membrane (Koshland and Botstein, 1982; Hengge and Boos, 1985; Fitts *et al.*, 1987; Duplay and Hofnung, 1988). A similar phenotype

has recently been observed with mutant B subunits of heat-labile enterotoxin (LTh-I) from *Escherichia coli* (Sandkvist *et al.*, 1987, 1990). Genetic deletion of the last two amino acid residues from the C-terminus of LTB was found to cause an approximately 50% reduction in the release of B subunits from the cytoplasmic membrane, whilst deletion of the last three amino acids inhibited B subunit release altogether. Since the mutant B subunits had translocated across the membrane and appeared to have folded sufficiently to allow intra-chain disulphide formation, it is not obvious why minimal alterations at the C-terminus of LTB should profoundly influence subunit release from the membrane. It may be that loss or alteration of C-terminal amino acids results in a more hydrophobic molecule, or that there is exposure of a hydrophobic pocket that is normally masked by the last few amino acids; thus causing the molecule to remain associated with the membrane. Alternatively the last few amino acids may be important in stabilizing subunit–subunit interactions, which may be an important prerequisite for release from the membrane. A further possibility is that the mutant B subunits may not be able to interact with additional 'release' or 'assembly factors' which may be important in the export and assembly of multimeric proteins. Whilst this latter possibility is highly speculative, it has recently emerged that the assembly of multimeric proteins in the endoplasmic reticulum of eukaryotic cells and in mitochondria requires a class of protein chaperonins, which participate at some stage of the folding, release or assembly of exported proteins (Gething *et al.*, 1986; Hurtley *et al.*, 1989; Cheng *et al.*, 1989; Ellis and Hemmingsen, 1989). No such 'release' or 'assembly factor' has so far been identified on the trans-side of the cytoplasmic membrane or in the periplasm of bacteria. This clearly represents a potentially important area for future research in toxin translocation and assembly.

An intriguing homology exists between cholera and related enterotoxins, and linear sequences found in other proteins; that is a four-residue sequence, Lys–Asp–Glu–Leu (KDEL)* present in the C-terminus of cholera toxin A subunit, which is identical to the retention signal found on ER proteins, such as BiP (immunoglobulin-binding protein) and protein disulphide isomerase (Pelham, 1989). This sequence is thought to be involved in causing the selective retrieval of proteins back into the ER. What then can be the significance of an ER retention signal in the A subunits of cholera and related toxins? One possible explanation is that the A subunits penetrate into the ER during their intoxication of eukaryotic cells. However, it is generally thought that only the N-terminal fragment (A_1-peptide, M_r 22 000) crosses the membrane of eukaryotic cells, with the C-terminal domain remaining in association with B subunits bound to GM1 (Ribi *et al.*, 1988). Another explanation would be that the sequence imposes a transient retention of the A polypeptide at the cytoplasmic membrane of the bacterial cell and that this plays an important role in the assembly process. Such transient retention would account for the relatively slow appearance of newly synthesized A subunit in the periplasm of *E. coli* (Hofstra and Witholt, 1984). Thus, the sustained presence of A subunits on the membrane may ensure its interaction with newly emerging and folding B subunits. Whether or not these interactions occur

before the B subunits have fully folded remains to be resolved. It is conceivable that the A subunit exhibits properties that are chaperonin-like, influencing the efficacy and rate of B subunit–B subunit interactions.

Assembly of oligomeric toxins

The requirement for subunit–subunit interactions to occur at some stage in the biosynthesis of oligomeric toxins distinguishes them from most of the other exported and secreted proteins which have so far been investigated. Not only does this add another level of complexity to toxin production, it provides a convenient structural state than can be readily monitored throughout synthesis and secretion and thereby give an insight into the conformation of the molecule before and after each event. To understand the assembly process requires an appreciation of the cellular location in which it occurs, the folding events and amino acid residues involved in stabilizing subunit interactions, the sequential pathway of interactions that lead from monomer to dimer to oligomer, and the identity of any cellular factors that may influence the efficiency of each step of the assembly process.

Like many other multimeric proteins, such as the well-characterized tetrameric dehydrogenases of eukaryotes, the purified oligomeric toxins can be dissociated into their constituent subunits and reassembled again *in vitro* (Finkelstein *et al.*, 1974; Takeda *et al.*, 1981; Tamura *et al.*, 1982; Hardy *et al.*, 1988; Ito *et al.*, 1988). Work with other systems has shown that such denaturation and refolding/reassociation experiments can generate oligomers that are structurally and functionally indistinguishable from the original native protein (for a review, see Jaenicke, 1987). This has led to a generally accepted paradigm that acquisition of 3-dimensional structure is determined by the primary sequence of amino acids in a protein, solvent conditions and subunit concentration. However, an increasing number of oligomeric proteins have been found to be assisted in their assembly *in vivo* by what may be termed assembly factors or chaperonins (Gething *et al.*, 1986; Cheng *et al.*, 1989; Ellis and Hemmingsen, 1989; Pelham, 1989). It is therefore important that *in vitro* experiments are viewed only as indicative of possible events which may occur *in vivo*, and that they do not

---

* The first residue of retention signals appears to be variable, being HDEL in ER proteins from *Saccharomyces cerevisiae* (Pelham, 1989). The corresponding sequence in the last four residues of the A subunits of *E. coli* LT-I is RDEL.

exclude an important role for cellular processes, e.g. polypeptide elongation or the participation of membrane surfaces or assembly factors in achieving assembly *in vivo*.

In the bacterial cell, toxin assembly will occur after the constituent polypeptide chains have folded into secondary structures and tertiary domains, exposing complementary interfaces that permit stable subunit–subunit interactions. Many of the enterotoxin and cytotoxin subunits contain a single intra-chain disulphide bond and this is likely to have a significant stabilizing effect on their tertiary structures. The reduction of the disulphide bond in the B subunits of cholera toxin or LTh-I inhibited their reassociation *in vitro* (Hardy et al., 1988). Site-specific mutagenesis of either or both of the cysteine residues in the B subunit of LTh-I prevented the formation of assembled toxin (M.G. Jobling, and R.K. Holmes, Abstr. ASM Meeting, abstract no. B-185, 1989). Thus the formation of the disulphide bond in each B subunit is probably an essential step in the folding pathway leading to subunits which can stably interact.

The location within the cell in which folding and subunit interactions occur is somewhat equivocal. This is because folding will occur immediately after the polypeptide emerges on the trans-side of the cytoplasmic membrane, whereas it has been proposed that assembly takes place in the periplasm (Hirst et al., 1984a; Hirst and Holmgren, 1987a; Nicosia and Rappuoli, 1987; Hardy et al., 1988). However, recent observations on the assembly of mutant B subunits of LTh-I which have short carboxyl deletions, indicate that assembly also occurs on the trans-side of the cytoplasmic membrane (M. Sandkvist, M. Bagdasarian and T.R. Hirst, unpublished results). Mutant B subunits that were defective in release from the membrane, but which had folded sufficiently to form intra-chain disulphides, were released from the membrane when the cells were induced to express the A subunit polypeptide. This implies that the A and B subunits are interacting at the membrane surface.

The precise pathway of subunit interactions which lead to the formation of holotoxin complexes is not known. The B subunits of cholera toxin and LT-I, as well as the B oligomer of pertussis toxin assemble *in vivo* in the absence of concomitant A subunit or S1 synthesis. This was taken to imply

that assembly proceeded via sequential B subunit interactions to yield the B pentamer, followed by association with the A subunit (Hirst et al., 1984a). However, pulse-labelling experiments, which monitored the rate of appearance of B pentamers of LTh-I in the periplasm of *E. coli*, revealed that co-expression of the A subunit caused an approximate fourfold acceleration in the rate of B subunit pentamerization (Fig. 3; Hardy et al., 1988). This implies that the A subunit stabilizes an intermediate in B subunit assembly. Moreover, *in vitro* experiments showed that A subunit interaction with assembled B subunits is precluded (Hardy et al., 1988). In the case of pertussis toxin formation, the presence of mutated S1 subunits unable to fold correctly reduced the yield of B oligomer, again suggesting that S1 stabilizes a

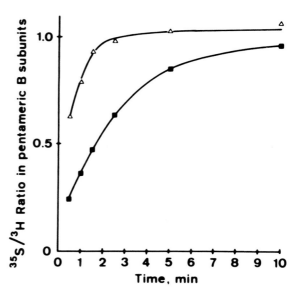

Figure 3. The A subunit of LTh-I accelerates B pentamer assembly in vivo. *E. coli* harbouring plasmid pWD600 (encoding A and B subunits) or pWD615 (encoding B subunits) (Dallas, 1983) were prelabelled with [³H] methionine before pulse-labelling with [³⁵S]methionine. Periplasmic samples obtained from cells chased for different lengths of time were subjected to SDS–polyacrylamide gel electrophoresis and the quantity of ³⁵S-labelled B pentamers determined. The radioactivity in the B pentamers is given as a ³⁵S/³H ratio, since the amount of ³H-labelled pentamers remains constant during the pulse-chase and thereby serves as an internal control for sample loss. The half-time of B subunit assembly in the strain with pWD600 (△) was approximately 30 s, whilst the half-time for B subunit assembly in the *E. coli* containing pWD615 (■) was approximately 2 min. (From Hardy et al., 1988; National Academy of Sciences, USA.)

B oligomer intermediate (R. Rappuoli, personal communication). For cholera toxin and related enterotoxins the A subunit may initially interact with a single B subunit, thereby causing a conformational alteration in the B subunit interface which favours further interaction with other B subunits. In so doing the A subunit would accelerate the assembly process and ensure its own inclusion within the toxin molecule.

Precise molecular information on the residues involved in subunit–subunit interactions is limited. Clearly, a most significant advance in this area will come when high-resolution 3-dimensional crystal structures become available (see the Note added in proof). Comparisons of the amino acid sequences of LT-I and LT-IIa reveal that the A_1-peptides (the N-terminal 22-kDa domain of the A subunits) show 64% homology, whereas the C-terminal domains (7 kDa) are only 20% homologous (Pickett *et al.*, 1987). Given the almost complete lack of homology between the B subunits of LT-I and LT-IIa, it would appear that B subunits are most likely to interact with the carboxyl domain of the A polypeptides as suggested by structural studies (Ribi *et al.*, 1988). Several researchers have been able to assemble hybrid oligomeric toxins *in vitro*, which show a toxicity characteristic of the A component, and receptor binding characteristics of the B oligomer. This has been achieved both for the A and B subunits of cholera toxin and LT-I (Takeda *et al.*, 1981) and for the A and B subunits of SLT-I and SLT-II (Ito *et al.*, 1988). Shiga toxin and related cytotoxin hybrids have been obtained *in vivo* by mixing A and B subunit expression (Weinstein *et al.*, 1989). This implies that the amino acid changes that have occurred in the subunits of the respective toxin families are not in sites critical for A–B subunit interaction. C-terminal extensions of seven amino acids on the B subunit of LTh-I were found to abolish A–B subunit interaction but not to affect the formation of B oligomers (Sandkvist *et al.*, 1987). Mixed assembly *in vitro* of purified B subunits of cholera toxin and LTh-I occurs with a similar efficiency to that of the homologous B subunit, indicating that the 19 amino acid differences between these polypeptides are again not at sites essential for complementary interaction of the subunits (T.R. Hirst, unpublished results). A single amino acid in the B subunit of LTp-I, alanine 64, has been identified as an important residue in the formation of B oligomers (Tetsuya *et al.*, 1989). Substitution of this residue with valine destabilizes B subunit–B subunit interaction, and no holotoxin forms. This residue is fully conserved in other members of the enterotoxin family.

Structural studies of cholera and related enterotoxins are well advanced and a high-resolution structure should be available soon (see the Note added in proof). Low-resolution, electron microscopic image analyses of cholera toxin and its B oligomer have shown a ring of five subunits (Fig. 4) with approximately one-third of the A subunit (possibly corresponding to the carboxyl domain) occupying the centre of the ring (Ribi *et al.*, 1988). Diffraction patterns of various crystal forms of LTp-I have revealed a five-fold axis of symmetry (Witholt *et al.*, 1988). However, solving the crystal structures of cholera toxin and LT appears to have been hindered by microheterogeneity in the purified toxins and by difficulties in obtaining

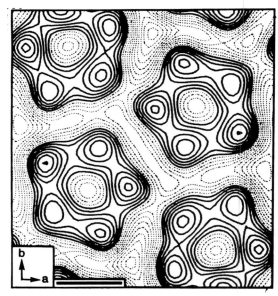

Figure 4. Low-resolution 3-dimensional structure of cholera toxin B oligomers. This was determined using electron microscopy and image processing of 2-dimensional crystals of B subunit bound to GM1 receptors in a lipid monolayer. The projected structure for the B oligomer is represented as solid and dashed contour lines of increasing levels of protein density and stain accumulation. The unit cell contains four pentagonally-shaped regions of protein density. Scale bar corresponds to 30 Å. (Used with permission from Ribi *et al.*, 1988; copyright 1988 by the AAAS.)

stable heavy-metal atom derivatives. However, significant progress has recently been made with two derivatives of LTp-I and this has resulted in a low-resolution structure determination which confirms the pentameric organization of the B oligomer (T. Sixma, S. Pronk and W. Hol, pers. comm.). When high-resolution structures become available it should be possible to identify and evaluate the role of each residue in stabilizing subunit–subunit interactions.

For several of the toxins, once assembly has occurred and the toxin released into the periplasm, the biosynthetic pathway is over. This is the case for shiga toxin in *S. dysenteriae*, the SLTs in *E. coli* (with the exception of SLT-IIv; see next section) and LT-I and LT-II in *E. coli* since these oligomeric proteins remain within the periplasm (Hirst *et al.*, 1984b; Hofstra and Witholt, 1984; Weinstein *et al.*, 1989). In contrast *V. cholerae* efficiently secretes cholera toxin across its outer membrane (Hirst and Holmgren, 1987b), and *B. pertussis* secretes pertussis toxin (Nicosia and Rappuoli, 1987). This additional step represents a remarkable phenomenon, for which there are no precedents. The possible mechanisms and role of the individual subunits are discussed in the following sections.

The role of the A and B subunits in toxin secretion across the outer membrane

The role of the various subunits in toxin secretion has been investigated by studying engineered bacterial strains which produce all the subunits, only one or other of the subunits or mixtures of subunits from different toxins (Yamamoto and Yokota, 1982; Mekalanos *et al.*, 1983; Hirst *et al.*, 1984a; Nicosia and Rappuoli, 1987; Weinstein *et al.*, 1989). This has led to the conclusion that the B subunit or B oligomer complex determines the extracellular location of oligomeric toxins.

In studies of toxin secretion by *V. cholerae* it was found that expression of only the B subunit of either cholera toxin or heat-labile enterotoxin (LT) resulted in their efficient secretion into the extracellular medium (Table 1; Mekalanos *et al.*, 1983; Hirst *et al.*, 1984a). In contrast, expression of only the A subunit of heat-labile enterotoxin in *V. cholerae* resulted in the A subunit remaining associated with the cells (Table 1). Therefore, B

subunits must contain the necessary structural information for translocation across the outer membrane of *V. cholerae*, and association between the A and B subunits cannot be a prerequisite for B subunit entry into the secretory step. The mechanism and possible involvement of accessory factors in translocation across the outer membrane are discussed in the next section.

Release of pertussis toxin B oligomer (containing S2, S3, S4 and S5 subunits) from *B. pertussis* occurred in the absence of assembly with subunit S1, if S1 could not fold correctly (R. Rappuoli, pers. comm.; see Chapter 1 of this volume). This demonstrated that the B oligomer contains the requisite information for translocation across the outer membrane of *B. pertussis*. However, given the complexity of the B oligomer of pertussis toxin, the role of each of the individual subunits remains to be fully evaluated. Evidence has, however, been obtained that the S3 subunit may play an important role in pertussis toxin secretion (Nicosia and Rappuoli, 1987), since a mutant of *B. pertussis* (BP356) which carries a Tn5-insertion in the gene encoding S3 (Weiss *et al.*, 1983) failed to secrete pertussis toxin[*] (Fig.5). As S3 is the last of the pertussis toxin genes and it has been shown that BP356 produced cell-associated S1, S2, S4 and S5 subunits that were active in Chinese hamster ovary cell assays (Nicosia and Rappuoli, 1987), this suggested that S3 was either directly involved in the secretory step, or that its assembly into the toxin complex gave the necessary structural configuration for pertussis toxin to cross the outer membrane.

Heterologous expression of shiga toxin and shiga-like toxins in *E. coli* has recently been reported (Table 2; Weinstein *et al.*, 1989). Shiga toxin and shiga-like toxins, SLT-I and SLT-II, are normally cell-associated in *E. coli*, whereas the variant SLT-IIv toxin was released into the culture medium (Table 2). Expression of hybrid toxins revealed that it was the origin of the B subunit component which determined the cellular or extracellular location of the hybrid molecule. The possibility that the B subunit of SLT-IIv causes perturbation of the *E. coli* outer membrane and a non-specific leakage of periplasmic contents was not excluded

---

[*] The Tn5-transposon insertion interrupts the gene encoding S3 such that only the first 136 amino acids of the subunit are produced (Nicosia and Rappuoli, 1987).

Table I. Cellular location of LT and its subunits in *Vibrio cholerae* harbouring various plasmids

Toxin (subunits expressed)	Plasmid	Toxin (subunit) concentration (ng/ml)[a]	
		Extracellular	Cell-associated
A subunit and B subunit	pWD600	1110	< 67
A subunit	pWD605	0	1200
B subunit	pWD615	1380	170

[a]*V. cholerae* strain TRH7000 harbouring various plasmids (Dallas, 1983) encoding the above subunits were cultured in syncase broth (Finkelstein *et al.*, 1966) and the concentration of subunits in the extracellular medium, or remaining cell-associated was determined for the B subunit by GM1-ELISA (Svennerholm and Holmgren, 1978) and for the A subunit by a pigeon erythrocyte lysate assay (Kunkel and Robertson, 1979). Data from Hirst *et al.* (1984a).

in this study. This information is clearly important for assessing the possible mechanism of SLT-IIv secretion. It will be interesting to evaluate which of the amino acid differences in the B subunits of SLT-II and SLT-IIv are responsible for this phenomenon.

Mechanism of secretion across the outer membrane

The efflux of enterotoxin subunits from the periplasmic compartment of *V. cholerae* into the

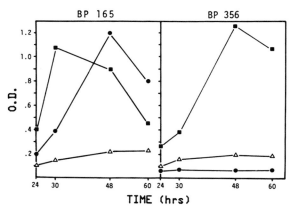

Figure 5. Distribution of pertussis toxin in a wild-type strain of *B. pertussis* (BP165) and a mutant (BP356) carrying a Tn-5 insertion in the S3 gene. Concentration of toxin (expressed as arbitrary optical density units) was determined by an enzyme-linked immunosorbant assay for samples taken at various times. Toxin concentration in the medium (●), in a Tris-extract of the cells (control)(△) and a polymyxin B extract (which releases the periplasmic proteins)(■) is shown. (Used with permission from Nicosia and Rappuoli, 1987; ASM Publications.)

medium was found to be highly specific, and was not accompanied by the release or leakage of periplasmic proteins, such as β-lactamase (Hirst and Holmgren, 1987b). Expression of cholera toxin or LT in *E. coli* resulted in the assembled holotoxin remaining within the periplasm (Pearson and Mekalanos, 1982; Hirst *et al.*, 1984a; Hofstra and Witholt, 1984). These observations suggest that the secretory event observed in *V. cholerae* is unlikely to be due solely to the inherent structural properties of the toxin, but suggests a requirement for a specific secretory machinery that is present in the *V. cholerae* outer membrane, but absent from *E. coli*. Genetic evidence for the presence of a secretory apparatus was provided by Holmes *et al.* (1975) when a mutant of *V. cholerae* was obtained which was defective in cholera toxin secretion and which accumulated the holotoxin within the periplasm (Hirst and Holmgren, 1987b). The gene(s) which is responsible for this secretory defect has reportedly been cloned (H. Marcus and R.K. Holmes, ASM Meeting Abstr. abstract no. B-190, 1989). Analysis of this gene(s) should provide a considerable insight into cholera toxin secretion, although the mechanism will require an understanding of the dynamics of interaction of the toxin with this gene product, as well as the molecular basis for the translocation event. The finding that B subunits of LT or cholera toxin are efficiently secreted from *V. cholerae* would imply that amino acid residues within these subunits engage the secretory apparatus. Interestingly, the secretion of the B subunits of LTh-I is more efficient than those of LTp-I which differ only by

Table 2. Cellular distribution of hybrid shiga and shiga-like toxins in *E. coli* HB101

Toxin		% Extracellular toxin[a]
A subunit	B subunit	
Shiga	Shiga	2
I	I	I
II	II	14
IIv	IIv	62
Shiga	II	ND[b]
Shiga	IIv	ND
I	II	ND
I	IIv	ND
II	Shiga	2
II	I	<1
II	IIv	62
IIv	Shiga	2
IIv	I	<1
IIv	II	14

[a] Toxin distribution in cells and medium determined by a cytotoxicity assay on Vero cells (Marques *et al.*, 1986). % Extracellular (extracellular/cell-associated + extracellular) × 100.
[b] ND, No cytotoxicity detected in cells or medium above background.
Adapted from Weinstein *et al.* (1989).

four amino acid residues (T.R. Hirst, unpublished results). This may reflect a reduced 'fit' by LTp-I B subunits in the secretory apparatus, although other differences may account for the behaviour, such as the relative extent of concentration-dependent aggregation of B subunits within the periplasm (Hirst *et al.*, 1984b).

Translocation, or entry into the secretory apparatus, was inhibited by low temperatures, e.g. when going from 37°C to 15°C (T.R. Hirst, unpublished results). The reason for this inhibition is unknown. Attempts have been made to evaluate the effect of energy poisons and sulphydryl-modifying reagents on the secretion of LTB subunits from *V. cholerae*. These proved unsuccessful because the reagents perturbed the integrity of the outer membrane and caused the concomitant non-specific release of periplasmic proteins (Hirst and Holmgren, 1987b). However, a requirement

for either energy or an outer membrane ion gradient cannot be excluded, especially in the light of recent findings by Wong and Buckley (1989), showing that proaerolysin secretion across the outer membrane of *Aeromonas salmonicida* was inhibited by the proton ionophore, carbonyl cyanide *m*-chlorophenyl hydrazone (CCCP).

Mechanistically the process of enterotoxin secretion from *V. cholerae* should prove to be a most remarkable phenomenon since it is likely to involve translocation of the fully assembled quaternary complex (of 90 kDa) through either the asymmetric membrane itself or through the postulated secretory apparatus. One possible mechanism would be for the toxin molecule to undergo facilitated partitioning into and across the membrane. This could be a reversible phenomenon involving interaction with the membrane, followed either by release back into the periplasm or translocation through the membrane. This model would predict that the interaction at the membrane was transiently stabilized by components of the secretory apparatus. Such an interaction and partitioning ought to be favoured by the high concentration of toxin subunits found in the periplasm (Hirst and Holmgren, 1987a,b). An alternative model would be to postulate a gated pore, generated by the components of the secretory apparatus, which would open upon interaction with B subunits and facilitate passage across the membrane. The conformational state of the pore could be determined by the energized state of the membrane, and energy could be expended in returning the pore to a conformation able to interact with further toxin molecules from the periplasm. If such a gated pore were to exist, its dimensions would have to accommodate not only the toxin oligomers, but also extensions onto the subunits, since these do not appear to inhibit secretion across the outer membrane. A third possible model would be one involving dissociation and unfolding of the subunits, with a translocation mechanism analogous to the general export pathway, followed by reassembly on the other side of the outer membrane and release into the medium. However, this would seem to be most unlikely, given the stability of the periplasmic quarternary complex of assembled pentamers or holotoxin, the absence of a signal sequence, and the unavailability of ATP.

In *E. coli* and *B. pertussis* it has been proposed

that vesicles formed from the outer membrane play an important role in secretory process (Gankema et al., 1980; Perera and Freer, 1987). E. coli K-12 mutants with a leaky phenotype produce copious quantities of outer membrane blebs (Lazzaroni and Portalier, 1981; J.-C. Lazzaroni, pers. comm.). However, this type of 'secretion' is pleiotropic and has neither the efficiency nor specificity which characterize the secretion of cholera toxin from V. cholerae. Freer and co-workers had examined the release of outer membrane vesicular material from B. pertussis and concluded that pertussis toxin is not released in association with such vesicles (Perera and Freer, 1987). It would seem likely that B. pertussis also has a secretory apparatus in its outer membrane that facilitates toxin translocation into the medium. The generation of mutants of B. pertussis which are defective in toxin secretion and the identity of any genes required are clearly major goals for future studies.

Assembly and secretion of recombinant oligomeric toxoids which have biotechnological and vaccine potential

The potent immunogenicity of cholera toxin B subunits and related enterotoxoids has provoked considerable interest in their use as 'carriers' for delivery of other antigens or epitopes to the immune system (Klipstein et al., 1983; Klipstein 1986; Sanchez et al., 1988; Guzman-Verduzco and Kupersztoch, 1987; Czerkinsky et al., 1989; Dertzbaugh and Macrina, 1989; Schödel and Will, 1989; Verheul et al., 1989; Dertzbaugh et al., 1990; Fergusson et al., 1990; Hirst et al., 1991a, b). Two alternative approaches have been used for the attachment of antigens and epitopes onto the toxoids: (1) by using chemical coupling techniques and (2) by genetic manipulation of the genes which encode the B subunit polypeptide (for examples see Czerkinsky et al., 1989; Schödel and Will, 1989). The latter approach has the merit of generating structurally defined vaccine antigens which can be used in a purified form or be expressed in live attenuated strains, but also has the potential drawback of adversely affecting the export, folding and assembly properties of the B subunit portion of the fusion molecule. Nonethe-

less, promising results have been obtained with this approach (Table 3).

This has resulted in the development of a number of plasmid vectors for insertion of heterologous antigens and epitopes at the 5'- or 3'-ends of the genes that encode the B subunits of either cholera toxin or E. coli heat-labile enterotoxin LTh-I (Table 4). Each vector or family of vectors provides convenient restriction endonuclease sites for insertion of additional DNA sequences. In the case of cholera toxin B subunits, the engineered modifications have been introduced adjacent to the region encoding the cleavage site between the signal sequence and the mature protein, permitting insertion of antigens or epitopes onto the N-terminus of the mature B subunit molecule (Table 4). In the vector developed by Sanchez and Holmgren (1989) a signal sequence precedes the site for insertion of additional DNA, thus permitting the hybrid gene product to be exported across the cytoplasmic membrane of the producing organism. In contrast, the vectors of Dertzbaugh and Macrina (1989) lack an N-terminal signal sequence, thus giving the potential to produce hybrid antigens which are located in the cytoplasm. This is likely to prevent oxidation of the disulphide bond found between Cys^9 and Cys^{86} of the B subunit, and may preclude the B subunit domain from exhibiting its characteristic receptor-binding and oligomeric properties. This can, however, be circumvented by inserting DNA that encodes antigens which carry their own N-terminal signal sequence or by engineering a heterologous signal sequence onto the inserted DNA (Dertzbaugh et al., 1990).

The vectors which utilize the E. coli heat-labile enterotoxin B subunit, permit insertion of DNA at the 3'-end of the gene, adjacent to the C-terminus of the B subunit polypeptide (Table 4; Sandkvist et al., 1987; Schodel and Will, 1989; Fergusson et al., 1990; Hirst et al., 1991a,b). This means that the N-terminal signal sequence of the B subunit remains intact and thereby ensures that the hybrid antigens are translocated across the cytoplasmic membrane of the producing organism.

One of the striking features that has emerged from studies on these toxoid-hybrid antigens, is that the extremities of the B subunits can successfully accommodate extensions which do not perturb the export, oligomerization and GM1-binding proper-

Table 3. Epitopes and antigens fused to CTB or LTB

Epitope/antigen – Toxoid	Vector used	Hybrid properties			Reference[e]
		GM1-binding	Anti-B	Anti-epitope	
Heat-stable enterotoxin (decapeptide derivative) – CTB	pJS162	+	+	+	1
Glycosyltransferase B (Ser[345]–Leu[359]) epitope – CTB	pVA1542	+	+	+	2
Heat-stable enterotoxin – LTB	pGK26	+	+	–	3
Heat-stable enterotoxin – LTB	pMMB68[b]	–	–	–	4
LTB–hepatitis B virus pre S2 surface antigen (1311–1367)[a]	pFS2.2	NT[d]	+	+	5
LTB–hepatitis B virus core antigen (84–522)[a]	pFS2.2	NT	+	+	5
LTB–woodchuck hepatitis B virus core antigen (2029–2533)[a]	pFS2.2	NT	–	–	5
LTB–heat-stable enterotoxin (Pro[19]–Tyr[72])	pMMB138	+	+	+	6
LTB–heat-stable enterotoxin (Gly[49]–Tyr[72])	pMMB138	+	+	+	6
LTB–circumsporozoite coat protein (Asn–Ala–Asn–Pro)$_3$ epitope	pTRH102	+	+	+	7
LTB–*B. pertussis* OMP P.69 epitope	pBRD026[c]	+	+	+	8
LTB–*S. mansoni* Smp20 Ca-binding protein	pMMB138	–	NT	–	9

[a] Numbers in parentheses refer to the nucleotide number in the gene (see Schödel and Will, 1989).
[b] Vector from Sandkvist et al. (1987).
[c] Vector from Maskell et al. (1987).
[d] Not tested.
[e] (1) Sanchez et al., 1988; (2) Dertzbaugh et al., 1990; (3) Guzman-Verduzco and Kupersztoch, 1987; (4) T.R. Hirst and B.-E. Uhlin, unpublished observations; (5) Schödel and Will, 1989; (6) Hirst et al., 1991a; (7) Fergusson et al., 1990; T.R. Hirst, L. Fergusson and G. del Guidice, unpublished observations; (8) M. Lipscombe, pers. commun.; (9) K. Johnson and C.E. Hormache, pers. commun.

Table 4. Plasmid vectors for fusion of antigens and epitopes to CTB or LTB

Plasmid	Nucleotides adjacent to the site(s) for insertion of DNA	Toxoid	Vector origin[a]	Reference
pJS152	... ... −1 +1 Gly Ala Pro Gly GGA GCT CCC GGG ... SacI SmaI −4 −3 −2 −1 +1 +2 Tyr Ala His Gly Thr Pro TAT GCA CAT GGA ACA CCT ...	CTB	RSF1010 (1)	Sanchez and Holmgren (1989)

pVA1542	Glu Phe Gly Thr Asp Pro Ser
	... GAA TTC GGT ACC GAT CCT TCA
	   EcoRI KpnI
	
	−4 −3 −2 −1 +1 +2
	Tyr Ala His Gly Thr Pro
	TAT GCA CAT GGA ACA CCT ...
	  NdeI
		CTB	pBR322 (2)	Dertzbaugh and Macrina (1989)

pVA1543	Asn Ser Asp Pro Ser Arg Ala
	.G AAT TCG GAT CCT TCT AGA GCA
	  EcoRI XbaI
	
	−4 −3 −2 −1 +1 +2
	Tyr Ala His Gly Thr Pro
	TAT GCA CAT GGA ACA CCT ...
	  NdeI
		CTB	pBR322 (2)	Dertzbaugh and Macrina (1989)

pVA1544	Ile Leu Asp Pro Arg Ser
	.GA ATT CTG GAT CCT AGG TCA
	   EcoRI AcrII
	
	−4 −3 −2 −1 +1 +2
	Tyr Ala His Gly Thr Pro
	TAT GCA CAT GGA ACA CCT ...
	  NdeI
		CTB	pBR322 (2)	Dertzbaugh and Macrina (1989)

pMMB138	+102 Glu Lys Leu Ala Pro Gln Lys Arg Trp ... GAA AAG CTT GCC CCC CAG AAA CGC TGG TGA ... HindIII	LTB	RSF1010 (1)	Sandkvist et al. (1987)

pMMB141

+102
Glu Ser Leu Ala Val Leu Ala Asp Glu Arg Arg Phe Ser Ala
... GAA AGC TTG GCT GTT TTG GCG GAT GAG AGA AGA TTT TCA GCC TGA
　　　　HindIII

LTB　　RSF1010 (1)　　Sandkvist et al. (1987)

pFS2.2

+102 +103
Glu Asn Tyr Pro Gln Asp Pro Ile Ser Asn
... GAA AAC TAC CCT CAG GAT CCG ATA TCT AAT TAA TAG ... ...
　　　　　　SauI　BamHI　　　EcoRI

LTB　　pUC19 (3)　　Schödel and Will (1989)

pTRHI00

+102
Glu Lys Leu Gly Pro Gln Ala Gly Asp
... GAA AAG CTG GGT CCG CAG GCC GGC GAC TAG TGG ATC CTC TAG AAG
　　　　　　　　NaeI　　SpeI　　BamHI　　Xbal HindIII

LTB　　RSF1010 (1)　　Hirst et al. (1991)

pTRHI01

+102
Glu Lys Leu Gly Pro Gln Arg His
... GAA AAG CTG GGT CCG CAG GGC CGG CAC TAG TGG ATC CTC TAG AAG
　　　　　　　　NaeI　　SpeI　　BamHI　　Xbal HindIII

LTB　　RSF1010 (1)　　Hirst et al. (1991)

pTRHI02

+102
Glu Lys Leu Gly Pro Gln Pro Ala Asp
... GAA AAG CTG GGT CCG CAG CCG GCT GAC TAG TGG ATC CTC TAG AAG
　　　　　　　　NaeI　　SpeI　　BamHI　　Xbal HindIII

LTB　　RSF1010 (1)　　Hirst et al. (1991)

The modified nucleotide sequences adjacent to the 5'-end of CTB and the 3'-end of LTB are shown. Engineered restriction endonuclease sites enable insertion of additional DNA sequences, which may encode antigens or epitopes. The amino acids (shown in bold) are from the native CTB or LTB polypeptide, and the numbers above them correspond to the residue number in either the signal sequence (− numbers) or mature sequence (+ numbers) of the protein.

[a]References for the origin of the plasmid vector used: (1) Furste et al., 1986; Sandkvist et al., 1987; (2) Bolivar et al., 1977; Kaper et al., 1984; (3) Yanisch-Peron et al., 1985; Schodel and Will, 1989.

ties of the B subunit portion of the molecule (Table 3). Therefore, the N- and C-termini of the B subunits must be exposed in the folded subunit, and neither can participate in B subunit assembly or GM1 binding.

Several of the hybrid antigens, including CTB/heat-stable enterotoxin derivatives and LTB/malarial repetitive epitope-(Asn–Ala–Asn–Pro)₃ fusions have been expressed in *Vibrio cholerae*, and have been shown to be efficiently secreted into the culture medium (Fig. 6; Sanchez and Holmgren, 1989; Hirst, unpublished results). This means that the N- and C-termini of CTB and LTB are not functionally involved in enterotoxin translocation across the outer membrane, and that

the translocating mechanism can accommodate a significant expansion in the size of the toxoid which crosses the membrane. These observations herald the prospect of delivering hybrid antigens *in situ* to mucosal surfaces by their expression in attenuated strains of *V. cholerae* or other secreting organisms.

Concluding remarks

Appearance of toxin molecules in the culture medium is the culmination of a complex biosynthetic pathway that includes protein translation, export, processing, folding, assembly and secretion. The precise location of toxin assembly, the pathway of subunit interactions and the possible involvement of assembly factors remains to be determined. Nevertheless, compelling evidence has been obtained demonstrating that subunit assembly occurs before translocation across the outer membrane. This makes the secretory step a most intriguing biological phenomenon. Understanding both assembly and outer membrane translocation processes will require a thorough knowledge of all of the accessory factors and folding events involved if an explanation of how a large hydrophilic toxin composed of six polypeptides can possibly cross an outer membrane bilayer. This major enigma remains a continuing challenge for the 1990s.

Acknowledgements

I wish to thank R. Aitken for helpful discussion and for critically reading the manuscript, and J. Whiteman and J. Warriner for preparing the typescript. T.R.H. is supported by the Wellcome Trust.

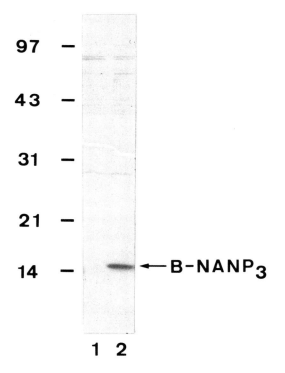

Figure 6. Secretion of recombinant LTB-(Asn–Ala–Asn–Pro)₃ antigen by *V. cholerae*. 50 μl of media from *V. cholerae* strain TRH7000 harbouring a plasmid for controlled expression of the fusion protein, were obtained after growth for 5½ h in the absence of inducer (IPTG)(lane 1) or in the presence of the inducer (lane 2). The samples were analysed by SDS–polyacrylamide gel electrophoresis, and the migration position of the fusion protein (B-NANP₃) is indicated. Molecular weight marker positions (×10⁻³Da) are shown on the left-hand-side.

References

Andro, T., Chambost, J.-P., Kotoujansky, A., Cattaneo, J., Bertheau, Y., Barras, F., Van Gijsejem, F. and Coleno, A. (1984). Mutants of *Erwinia chrysanthemi* defective in secretion of pectinase and cellulase. *J. Bacteriol.* **160**, 1199–203.

Bankaitis, V.A. and Bassford, P.J. (1985). Proper interaction between at least two components is required for efficient export of proteins to the

Escherichia coli cell envelope. *J. Bacteriol.* **161**, 169–78.

Bayer, M.E. (1968). Areas of adhesion between wall and membrane of *Escherichia coli J. Gen. Microbiol.* **53**, 395–404.

Bayer, M.E. (1979). The fusion sites between the outer membrane and cytoplasmic membrane of bacteria: their role in membrane assembly and virus infection. In: *Bacterial Outer Membranes* (ed. M. Inouye), pp. 167–202. John Wiley & Sons, New York.

Bayer, M.E. and Thurow, H. (1977). Polysaccharide capsule of *Escherichia coli*: microscopic study of its size, structure and sites of synthesis. *J. Bacteriol.* **130**, 911–36.

Bayer, M.E., Bayer, M.H., Lunn, C.A. and Pigiet, V. (1987). Association of thioredoxin with the inner membrane and adhesion sites in *Escherichia coli. J. Bacteriol.* **169**, 2659–66.

Benson, S.A., Bremer, E. and Silhavy, T.J. (1984). Intragenic regions required for LamB export. *Proc. Natl. Acad. Sci. USA* **81**, 3830–4.

Betley, M.J., Miller, V.L. and Mekalanos, J.J. (1986). Genetics of bacterial enterotoxins. *Ann. Rev. Microbiol.* **40**, 577–605.

Bieker, K.L. and Silhavy, T.J. (1989). PrlA is important for translocation of exported proteins across the cytoplasmic membrane of *Escherichia coli. Proc. Natl. Acad. Sci. USA* **86**, 968–72.

Bertstein, H.D., Rapoport, T.A. and Walter, P. (1989). Cytosolic protein translocation factors – is SRP still unique? *Cell* **58**, 1017–19.

Bolivar, F., Rodriguez, R.L., Greene, P.J., Betlach, M.C., Heyneker, H.L., Boyer, H.W., Crosa, J.H. and Falkow, S. (1977). Construction and characterization of new cloning vehicles: II. A Multipurpose cloning system. *Gene* **2**, 95–113.

Bricker, J., Mulks, M.H., Plaut, A.G., Moxon, E.R. and Wright, A. (1983). IgA1 proteases of *Haemophilus influenzae*: cloning and characterization in *Escherichia coli* K-12. *Proc. Natl. Acad. Sci. USA* **80**, 2681–5.

Calderwood, S.B., Auclair, F., Donohue-Rolfe, A., Keusch, G.T. and Mekalanos, J.J. (1987). Nucleotide sequence of the shiga-like toxin genes of *Escherichia coli. Proc. Natl. Acad. Sci. USA* **84**, 4304–68.

Chen, L. and Tai, P.C. (1985). ATP is essential for protein translocation into *Escherichia coli* membrane vesicles. *Proc. Natl. Acad. Sci. USA* **82**, 4384–8.

Cheng, M.Y., Hartl, F.-U., Martin, J., Pollock, R.A., Kalousek, F., Neupert, W., Hallberg, E.M., Hallberg, R.L. and Horwich, A.L. (1989). Mitochondrial heat-shock protein hsp60 is essential for assembly of proteins imported into yeast mitochondria. *Nature* **337**, 620–5.

Chopra, A.K., Houston, C.W., Peterson, J.W., Prasad, R. and Mekalanos, J.J. (1987). Cloning and expression of the *Salmonella* enterotoxin gene. *J. Bacteriol.* **169**, 5095–100.

Crooke, E. and Wickner, W. (1987). Trigger factor: a soluble protein that folds pro-OmpA into a membrane-assembly-competant form. *Proc. Natl. Acad. Sci USA* **84**, 5216–20.

Cunningham, K., Lill, R., Crooke, E., Rice, M., Moore, K., Wickner, W. and Oliver, D. (1989). SecA protein, a peripheral protein of the *Escherichia coli* plasma membrane, is essential for the functional binding and translocation of ProOmpA. *EMBO J.* **8**, 955–9.

Czerkinsky, C., Russel, M.W., Lycke, N., Lindblad, M. and Holmgren, J. (1989). Oral administration of a streptococcal antigen coupled to cholera B subunit evokes strong antibody responses in salivary glands and extramucosal tissues. *Infect. Immun.* **57**, 1072–7.

Dallas, W.S. (1983). Conformity between heat-labile toxin genes from human and porcine enterotoxigenic *Escherichia coli. Infect. Immun.* **40**, 647–52.

Dallas, W.S. and Falkow, S. (1980). Amino acid sequence homology between cholera toxin and *Escherichia coli* heat-labile toxin. *Nature (Lond.)* **288**, 499–501.

Daniels, C.J., Bole, D.G., Quay, S.C. and Oxender, D.L. (1981). Role for membrane potential in the secretion of protein into the periplasm of *Escherichia coli. Proc. Natl. Acad. Sci. USA* **78**, 5396–400.

Date, T., Zwizinski, C., Ludmerer, S. and Wickner, W. (1980). Mechanisms of membrane assembly: effects of energy poisons on the conversion of soluble M13 coliphage procoat to membrane-bound coat protein. *Proc. Natl. Acad. Sci. USA* **77**, 827–31.

d'Enfert, C. and Pugsley, A.P. (1989). *Klebsiella pneumoniae pulS* gene encodes an outer membrane lipoprotein required for pullalanase secretion. *J. Bacteriol.* **171**, 3673–9.

Dertzbaugh, M.T. and Macrina, F.L. (1989). Plasmid vectors for constructing translational fusions to the B subunit of cholera toxin. *Gene* **82**, 335–42.

Dertzbaugh, M.T., Peterson, D.L. and Macrina, F.L. (1990). Cholera toxin B-subunit gene fusion: structural and functional analysis of the chimeric protein. *Infect. Immun.* **58**, 70–9.

Douglas, C.M., Guidi-Rontani, C. and Collier, R.J. (1987). Exotoxin A of *Pseudomonas aeruginosa*: active, cloned toxin is secreted into the periplasmic space of *Escherichia coli. J. Bacteriol.* **169**, 4962–6.

Duplay, P. and Hofnung, M. (1988). Two regions of mature periplasmic maltose-binding protein of *Escherichia coli* involved in secretion. *J. Bacteriol.* **170**, 4445–50.

Ellis, R.J. and Hemmingsen, S.M. (1989). Molecular chaperones: proteins essential for the biogenesis of some macromolecular structures. *Trends Biol. Sci.* **14**, 339–42.

Emr, S.D., Hanley-Way, S. and Silhavy, T.J. (1981). Suppressor mutations that restore export of a protein with a defective signal sequence. *Cell* **23**, 79–88.

Endo, Y., Tsurugi, K., Yutsudo, T., Takada, Y., Ogasawara, T. and Igarashi, K. (1988). The site of action of a verotoxin (VT2) from *Escherichia coli* 0157:H7 and of shiga toxin on eukaryotic ribosomes:

RNA N-glycosidase activity of the toxins. *Eur. J. Biochem.* **171**, 45–50.

Enequist, H.G., Hirst, T.R., Harayama, S., Hardy, S.J.S. and Randall, L.L. (1981). Energy is required for maturation of exported proteins in *Escherichia coli. Eur. J. Biochem.* **116**, 227–33.

Fandl, J.P. and Tai, P.C. (1987). Biochemical evidence for secY24 defect in *Escherichia coli* protein translocation and its suppression by soluble cytoplasmic factors. *Proc. Natl. Acad. Sci. USA* **84**, 7448–52.

Ferenci, T. and Silhavy, T.J. (1987). Sequence information required for protein translocation from the cytoplasm. *J. Bacteriol.* **169**, 5339–42.

Fergusson, L.F., McDiarmid, N. and Hirst, T.R. (1990). A multivalent carrier for delivery of epitopes and antigens based upon the B subunit enterotoxoid of *Escherichia coli.* In: *Bacterial Protein Toxins* (eds R. Rappuoli and R. Gross), Zbl. Bakt. Suppl., **19**, 519–520, Gustav Fischer, Stuttgart.

Fernandez, M., Sierra-Madero, J., de la Vega, H., Vazquez, M., Lopez-Vidal, Y., Ruiz-Palacios, G.M. and Calva, E. (1988). Molecular cloning of *Salmonella typhi* LT-like enterotoxin gene. *Molec. Microbiol.* **2**, 821–5.

Finkelstein, R.A. and LoSpalluto, J.J. (1969). Pathogenesis of experimental cholera. Preparation and isolation of choleragen and choleragenoid. *J. Infect. Dis.* **130**, 145–50.

Finkelstein, R.A., Atthasampunu, P., Chulasmya, M. and Charunmethee, P. (1966). Pathogenesis of experimental cholera: biologic activities of purified procholeragen. *J. Immunol.* **96**, 440–9.

Finkelstein, R.A., Boseman, M., Neoh, S.H., LaRue, M.K. and Delaney, R. (1974). Dissociation and recombination of the subunits of the cholera enterotoxin (choleragen). *J. Immunol.* **113**, 145–50.

Fitts, R., Reuveny, Z., Van Amsterdam, J., Mulholland, J. and Botstein, D. (1987). Substitution of tyrosine for either cysteine in β-lactamase prevents release from the membrane during secretion. *Proc. Natl. Acad. Sci. USA* **84**, 8540–3.

Fukuta, S., Magnani, J.L., Twiddy, E.M., Holmes, R.K. and Ginsberg, V. (1988). Comparison of the carbohydrate-binding specificities of cholera toxin and *Escherichia coli* heat-labile enterotoxins LTh-I, LT-IIa, and LT-IIb. *Infect. Immun.* **56**, 1748–53.

Furste, J.P., Pansegrau, W., Frank, R., Blocker, H., Scholz, P., Bagdasarian, M. and Lanka, E. (1986). Molecular cloning of the plasmid RP4 primase region in a multi-host-range *tac*P expression vector. *Gene,* **48**, 119–31.

Gankema, H., Wensink, J., Guinee, P.A.M., Jansen, W.H. and Witholt, B. (1980). Some characteristics of outer membrane material released by growing enterotoxinogenic *Escherichia coli. Infect Immun.,* **29**, 704–13.

Gardel, G., Benson, S., Hunt, J., Michaelis, S. and Beckwith, J. (1987). *SecD*, a new gene involved in protein export in *Escherichia coli. J. Bacteriol.* **169**, 1286–90.

Geller, B.L., Movva, N.R. and Wickner, W. (1986). Both ATP and electrochemical potential are required for optimal assembly of pro-OmpA into *Escherichia coli* inner membrane vesicles. *Proc. Natl. Acad. Sci. USA* **83**, 4219–22.

Gething, M.-J., McGammon, K. and Sambrook, J. (1986). Expression of wild-type and mutant forms of influenzae haemagglutinin: the role of folding in intracellular transport. *Cell* **46**, 939–50.

Gierasch, L.M. (1989). Signal sequences. *Biochemistry* **28**, 923–30.

Gill, D.M. (1976). The arrangement of subunits in cholera toxin. *Biochemistry* **15**, 1242–8.

Gill, D.M., Clements, J.D., Robertson, D.C. and Finkelstein, R.A. (1981). Subunit number and arrangement in *Escherichia coli* heat-labile enterotoxin. *Infect. Immun.* **33**, 677–82.

Gobius, K.S. and Pemberton, J.M. (1988). Molecular cloning, characterization, and nucleotide sequence of an extracellular amylase gene from *Aeromonas hydrophilia. J. Bacteriol.* **170**, 1325–32.

Guzman-Verduzco, L.-M. and Kupersztoch, V.M. (1987). Fusion of *Escherichia coli* heat-stable enterotoxin and heat-labile enterotoxin B subunit. *J. Bacteriol.* **169**, 5201–8.

Hardy, S.J.S., Holmgren, J., Johansson, S., Sanchez, J. and Hirst, T.R. (1988). Coordinated assembly of multisubunit proteins: Oligomerization of bacterial enterotoxins *in vivo* and *in vitro. Proc. Natl. Acad. Sci. USA* **85**, 7109–13.

Hengge, R. and Boos, W. (1985). Defective secretion of maltose- and ribose-binding proteins caused by a truncated periplasmic protein in *Escherichia coli. J. Bacteriol.* **162**, 972–8.

Hirst, T.R. and Holmgren, J. (1987a). Conformation of protein secreted across bacterial outer membranes: a study of enterotoxin translocation from *Vibrio cholerae. Proc. Natl. Acad. Sci. USA* **84**, 7418–22.

Hirst, T.R. and Holmgren, J. (1987b). Transient entry of enterotoxin subunits into the periplasm occurs during their secretion from *Vibrio cholera. J. Bacteriol.* **169**, 1037–45.

Hirst, T.R. and Welch, R.A. (1988). Mechanisms for secretion of extracellular proteins by Gram-negative bacteria. *Trends Biochem. Sci.* **13**, 265–9.

Hirst, T.R., Sanchez, J., Kaper, J., Hardy, S.J.S. and Holmgren, J. (1984a). Mechanism of toxin secretion by *Vibrio cholerae* investigated in strains harboring plasmids that encode heat-labile enterotoxins of *Escherichia coli. Proc. Natl. Acad. Sci. USA* **81**, 7752–6.

Hirst, T.R., Randall, L.L. and Hardy, S.J.S. (1984b). Cellular location of heat-labile enterotoxin in *Escherichia coli. J. Bacteriol.* **157**, 637–42.

Hirst, T.R., Sandkvist, M., Aitken, R. and Bagdasarian, M. (1991a). Assembly and secretion of oligomeric toxins. In: *Microbial Surface Components and Toxins* (ed. E. Ron). Plenum Press, New York (in press).

Hirst, T.R., Fergusson, L.F., Aitken, R. and DelGuidice, G. (1991b). Plasmid vectors for expression and secretion of antigens and epitopes fused to the B subunit of heat-labile enterotoxin (in preparation).

Hobot, J.A., Carlemalm, E., Villiger, W. and Kellenberger, E. (1984). Periplasmic gel: new concept resulting from the reinvestigation of bacterial cell envelope ultrastructure by new methods. *J. Bacteriol.* **160**, 143–52.

Hofstra, H. and Witholt, B. (1984). Kinetics of synthesis, processing and membrane transport of heat-labile enterotoxin, a periplasmic protein in *Escherichia coli.* *J. Biol. Chem.* **259**, 15182–7.

Holland, I.B., Blight, M.A. and Kenny, B. (1990). The mechanism of secretion of hemolysin and other polypeptides from Gram negative bacteria. *J. Bioenerget. Biomem.* **22**, 473–91.

Holmes, R.K., Vasil, M.L. and Finkelstein, R.A. (1975). Studies on toxigenesis in *Vibrio cholerae* III. Characterization of nontoxinogenic mutants *in vitro* and in experimental animals. *J. Clin. Invest.* **55**, 551–60.

Holmes, R.K., Twiddy, E.M. and Pickett, C.L. (1986). Purification and characterization of type II heat-labile enterotoxin of *Escherichia coli. Infect. Immun.* **53**, 464–73.

Holmgren, J. (1973). Comparison of the tissue receptors for *Vibrio cholerae* and *Escherichia coli* enterotoxins by means of gangliosides and natural cholera toxoid. *Infect. Immun.* **8**, 851–9.

Holmgren, J. (1981). Actions of cholera toxin and the prevention and treatment of cholera. *Nature (Lond.)* **292**, 413–17.

Holmgren, J., Lindblad, M., Fredman, P., Svennerholm, L. and Myrvold, H. (1985). Comparison of receptors for cholera and *Escherichia coli* enterotoxins in human intestine. *Gastroenterology* **89**, 27–35.

Howard, S.P. and Buckley, J.T. (1983). Intracellular accumulation of extracellular proteins by pleiotropic export mutants of *Aeromonas hydrophila. J. Bacteriol.* **154**, 413–8.

Huang, A., DeGrandis, S., Friesen, J., Karmali, M., Petric, M., Congi, R. and Brunton, J.L. (1986). Cloning and expression of the genes specifying shiga-like toxin production in *Escherichia coli* H19. *J. Bacteriol.* **166**, 375–9.

Hurtley, S.M., Bole, D.G., Hoover-Litty, H., Helenius, A. and Copeland, S. (1989). Interactions of malfolded influenza virus hemagglutinin with binding protein (Bip). *J. Cell. Biol.* **108**, 2117–26.

Inouye, M. (1979). *Bacterial Outer Membranes*. John Wiley & Sons, New York.

Ito, K. and Beckwith, J.R. (1981). Role of the mature protein sequence of maltose binding protein in its secretion across the *E. coli* cytoplasmic membrane. *Cell* **25**, 143–50.

Ito, K., Wittekind, M., Nomura, M., Shiba, K., Yura, T., Miura, A. and Nashimoto, H. (1983). A temperature-sensitive mutant of *E. coli* exhibiting slow processing of exported proteins. *Cell* **32**, 789–97.

Ito, H., Yutsudo, T., Hirayama, T. and Takeda, Y. (1988). Isolation and some properties of A and B

subunits of Vero toxin 1 and Vero toxin 2 from *Escherichia coli* 0157:H7. *Microb. Pathogen.* **5**, 189–95.

Jackson, M.E., Pratt, J.M. and Holland, I.B. (1986). Intermediates in the assembly of the TonA polypeptide into the outer membrane of *Escherichia coli* K12. *J. Molec. Biol.* **189**, 477–86.

Jackson, M.P., Neill, R.J., O'Brien, A.D., Holmes, R.K. and Newland, J.W. (1987). Nucleotide sequence analysis and comparison of the structural genes for shiga-like toxin I and shiga-like toxin II encoded by bacteriophages from *Escherichia coli* 933. *FEMS Lett.* **44**, 109–14.

Jaenicke, R., (1987). Folding and association of proteins. *Prog. Biophys. Molec. Biol.* **49**, 117–237.

Kaper, J., Lockman, H., Baldini, M. and Levine, M. (1984). Recombinant nontoxinogenic *Vibrio cholerae* strains as attenuated cholera vaccine candidates. *Nature (Lond.)* **308**, 655–8.

Keegstra, K. (1989). Transport and routing of proteins in chloroplasts. *Cell* **56**, 247–53.

Kennedy, E.P. (1987). Membrane-derived oligosaccharides. In: *Escherichia coli* and *Salmonella typhimurium*, Vol. I. (eds. F.C. Niedhardt, J.L. Ingraham, K.B. Low, B. Magasanik, M. Schaechter and H.E. Umbarger), pp. 672–9. American Society for Microbiology, Washington, DC.

Klipstein, F.A. (1986). Development of *Escherichia coli* vaccines against diarrhoeal disease in humans. In: *Development of Vaccines and Drugs against Diarrhoea*. 11th Nobel Conf. (eds J. Holmgren, A. Lindberg and R. Mollby), pp.62–7. Studentliltteratur, Lund, Sweden.

Klipstein, F.A. and Engert, R.F. (1985). Immunological relationship of the B-subunits of *Campylobacter jejuni* and *Escherichia coli* heat-labile enterotoxins. *Infect. Immun.* **48**, 629–33.

Klipstein, F.F., Engert, R.F., Clements, J.D. and Houghten, R.A. (1983). Vaccine for enterotoxigenic *Escherichia coli* based on synthetic heat-stable toxin cross-linked to the B subunit of heat-labile toxin. *J. Infect. Dis.* **147**, 318–26.

Kornacher, M.G., Boyd, A., Pugsley, A.P. and Pastow, G.S. (1989). *Klebsiella pneumoniae* strain K21: evidence for the rapid secretion of an unacylated form of pullulanase. *Molec. Microbiol.* **3**, 497–503.

Koshland, D. and Botstein, D. (1982). Evidence for post-translational translocation of β-lactamase across the bacterial inner membrane. *Cell* **30**, 893–902.

Kumamoto, C.A. (1989). *Escherichia coli* SecB protein associates with exported protein precursors *in vivo*. *Proc. Natl. Acad. Sci. USA* **86**, 5320–4.

Kumamoto, C. and Beckwith, J. (1983). Mutations in a new gene, *secB*, cause defective protein localization in *Escherichia coli. J. Bacteriol.* **154**, 253–60.

Kunkel, S.L. and Robertson, D.C. (1979). Purification and chemical characterization of the heat-labile enterotoxin produced by enterotoxigenic *Escherichia coli. Infect. Immun.* **25**, 586–96.

Kusukawa, N., Yura, T., Ueguchi, C., Akiyama, Y. and Ito, K. (1989). Effects of mutation in heat-shock genes *groES* and *groEL* on protein export in *E. coli*. *EMBO J.* **8**, 3517–21.

Lazzaroni, J.-C. and Portalier, R.C. (1981). Genetic and biochemical characterization of periplasmic-leaky mutants of *Escherichia coli* K-12. *J. Bacteriol.* **145**, 1351–8.

Lindberg, A.A., Brown, J.E., Stromberg, N., Westling-Ryd, M., Schultz, J.E. and Karlsson, K.-A. (1987). Identification of the carbohydrate receptor for shiga toxin produced by *Shigella dysenteriae* type 1. *J. Biol. Chem.* **262**, 1779–85.

Locht, C. and Keith, J.M. (1986). Pertussis toxin gene: nucleotide sequence and genetic organization. *Science* **232**, 1258–64.

Lockman, H. and Kaper, J.B. (1983). Nucleotide sequence analysis of the A2 and B subunits of *Vibrio cholerae* enterotoxin. *J. Biol. Chem.* **258**, 13722–6.

Lui, G., Topping, T.B. and Randall, L.L. (1989). Physiological role during export for the retardation of folding by the leader peptide of maltose-binding protein. *Proc. Natl. Acad. Sci. USA* **86**, 9213–17.

Marques, L.R.M., Moore, M.A., Wells, J.G., Wachsmuth, I.K. and O'Brien, A.D. (1986). Production of shiga-like toxin by *Escherichia coli*. *J. Infect. Dis.* **154**, 338–41.

Maskell, D.J., Sweeney, K.J., O'Callaghan, D., Hormaeche, C.E., Liew, F.V. and Dougan, G. (1987). *Salmonella typhimurium aroA* mutants as carriers of the *Escherichia coli* heat-labile enterotoxin B subunit to the murine secretory and systemic immune systems. *Microb. Pathogen.* **2**, 211–21.

Mekalanos, J.J., Swartz, D.J., Pearson, G.D.N., Harford, N., Groyne, F. and deWilde, M. (1983). Cholera toxin genes: nucleotide sequence, deletion analysis and vaccine development. *Nature (Lond.)* **306**, 551–7.

Minsky, A., Summers, R.G. and Knowles, J.R. (1986). Secretion of β-lactamase into the periplasm of *Escherichia coli*: Evidence for a distinct release step associated with a conformational change. *Proc. Natl. Acad. Sci. USA* **83**, 4180–4.

Moss, J. and Richardson, S.H. (1978). Activation of adenylate cyclase by *Escherichia coli* enterotoxin. Evidence for ADP-ribosyltransference activity similar to that of choleragen. *J. Clin. Invest.* **62**, 281–5.

Muhlradt, P.F., Menzel, J., Golecki, J.R. and Speth, V. (1973). Outer membrane of *Salmonella*. Sites of export of newly synthesized lipopolysaccharide on the bacterial surface. *Eur. J. Biochem.* **35**, 471–81.

Nakae, T. (1986). Outer-membrane permeability of bacteria. *CRC Crit. Rev. Microbiol.* **13**, 1–62.

Newland, J.W., Strockbine, N.A., Miller, S.F. and O'Brien, A.D. (1985). Cloning of shiga-like toxin structural genes from a toxin-converting phage of *Esherichia coli*. *Science* **230**, 179–81.

Newland, J.W., Strockbine, N.A. and Neill, R.J. (1987). Cloning of genes for production of *Escherichia coli* shiga-like toxin type II. *Infect. Immun.* **55**, 2675–80.

Nicosia, A. and Rappuoli, R. (1987). Promoter of the pertussis toxin operon and production of pertussis toxin. *J. Bacteriol.* **169**, 2843–6.

Nicosia, A., Perugini, M., Franzini, C., Casagli, M.C., Borri, M.G., Antoni, G., Almoni, M., Neri, P., Ratti, G. and Rappuoli, R. (1986). Cloning and sequencing of the pertussis toxin genes; operon structure and gene duplication. *Proc. Natl. Acad. Sci. USA* **83**, 4631–5.

Nikaido, H. and Vaara, M. (1987). Outer membrane. In: *Escherichia coli* and *Salmonella typhimurium*, Vol. I (eds F.C. Niedhardt, J.L. Ingraham, K.B. Low, B. Magasanik, M. Schaechter and H.E. Umbarger), pp.7–22. American Society for Microbiology, Washington, DC.

Nikaido, H. and Wu, H.C.P. (1984). Amino acid sequence homology among the major outer membrane proteins of *Escherichia coli*. *Proc. Natl. Acad. Sci. USA* **81**, 1048–52.

O'Brien, A.D. and Holmes, R.K. (1987). Shiga and shiga-like toxins. *Microbiol. Rev.* **51**, 206–20.

Oliver, D.B. (1987). Periplasm and protein secretion. In: *Escherichia coli* and *Salmonella typhimurium*, Vol. I (eds F.C. Niedhardt, J.L. Ingraham, K.B. Low, B. Magasanik, M. Schaechter and H.E. Umbarger), pp.56–69. American Society for Microbiology, Washington, DC.

Oliver, D. and Beckwith, J. (1982). Identification of a new gene (secA) and gene product involved in the secretion of envelope proteins. *J. Bacteriol.* **150**, 686–91.

Palva, E.T., Hirst, T.R., Hardy, S.J.S., Holmgren, J. and Randall, L.L. (1981). Synthesis of a precursor to the B subunit of heat-labile enterotoxin in *Escherichia coli*. *J. Bacteriol.* **146**, 325–30.

Park, J.T. (1987). The murein sacculus. In: *Escherichia coli* and *Salmonella typhimurium*, Vol. I (eds F.C. Niedhardt, J.L. Ingraham, K.B. Low, B. Magasanik, M. Schaechter and H.E. Umbarger), pp.23–30. American Society for Microbiology, Washington, DC.

Pearson, G.D.N. and Mekalanos, J.J. (1982). Molecular cloning of *Vibrio cholerae* enterotoxin genes in *Escherichia coli* K-12. *Proc. Natl. Acad. Sci. USA* **79**, 2976–80.

Pelham, H.R.B. (1989). Heat shock and sorting of luminal ER proteins. *EMBO J.* **8**, 3171–6.

Perera, V.Y. and Freer, J.H. (1987). Accumulation of precursor of subunit S1 of pertussis toxin in cell envelopes of *Bordetella pertussis* in response to the membrane perturbant phenethyl alcohol. *J. Med. Microbiol.* **23**, 269–74.

Pickett, C.L., Weinstein, D.L. and Holmes, R.K. (1987). Genetics of type IIa heat-labile enterotoxin of *Escherichia coli*: operon fusions, nucleotide sequence, and hybridisation studies. *J. Bacteriol.* **169**, 5180–7.

Pohlner, J., Halter, R., Beyreuther, K. and Meyer, T.F. (1987). Gene structure and extracellular secretion of *Neisseria gonorrhoeae* IgA protease. *Nature (Lond.)* **325**, 458–62.

Pugsley, A.P. and Schwartz, M. (1985). Export and

secretion of proteins by bacteria. *FEMS Microbiol. Rev.* **32**, 3–38.

Pugsley, A.P., Chapron, C. and Schwartz, M. (1986). Extracellular pullulanase of *Klebsiella pneumoniae* is a lipoprotein. *J. Bacteriol.* **166**, 1083–8.

Randall, L.L. and Hardy, S.J.S. (1989). Unity in function in the absence of consensus in sequence: role of the leader peptides in export. *Science* **243**, 1156–9.

Randall, L.L., Hardy, S.J.S. and Thorn, J.R. (1987). Export of protein: a biochemical view. *Ann. Rev. Microbiol.* **41**, 507–41.

Ribi, H.O., Ludwig, D.S., Mercer, K.L., Schoolnik, G.K. and Kornberg, R.D. (1988). Three-dimensional structure of cholera toxin penetrating a lipid membrane. *Science* **239**, 1272–6.

Riggs, P.D., Darman, A.I. and Beckwith, J. (1988). A mutation affecting the regulation of a *secA-lacZ* fusion defines a new *sec* gene. *Genetics* **118**, 571–9.

Rose, J.M., Houston, C.W., Coppenhaver, D.H., Dixon, J.D. and Kurosky, A. (1989a). Purification and chemical characterization of a cholera toxin-cross-reactive cytolytic enterotoxin produced by a human isolate of *Aeromonas hydrophila*. *Infect. Immun.* **57**, 1165–9.

Rose, J.M., Houston, C.W. and Kurosky, A. (1989b). Bioactivity and immunological characterization of a cholera-toxin-cross-reactive cytolytic enterotoxin from *Aeromonas hydrophila*. *Infect. Immun.* **57**, 1170–6.

Ruiz-Palacious, G.M., Torres, J., Torres, N.I., Escamilla, E., Ruiz-Palacious, B.R. and Tamayo, J. (1983). Cholera-like enterotoxin produced by *Campylobacter jejuni*. *Lancet* **ii**, 250–2.

Sanchez, J. and Holmgren, J. (1989). Recombinant system for overexpression of cholera toxin B subunit in *Vibrio cholerae* as a basis for vaccine development. *Proc. Natl. Acad. Sci. USA* **86**, 481–5.

Sanchez, J., Svennerholm, A.-M. and Holmgren, J. (1988). Genetic fusion of a non-toxic heat-stable enterotoxin-related decapeptide antigen to cholera toxin B subunit. *FEBS Lett.* **241**, 110–14.

Sandkvist, M., Hirst, T.R. and Bagdasarian, M. (1987). Alterations at the carboxyl terminus change assembly and secretion properties of the B subunit of *Escherichia coli* heat-labile enterotoxin. *J. Bacteriol.* **169**, 4570–6.

Sandkvist, M., Hirst, T.R. and Bagdasarian, M. (1990). Minimal deletion of amino acids from the carboxyl terminus of the B subunit of heat-labile enterotoxin causes defects in its assembly and release from the cytoplasmic membrane of *Escherichia coli*. *J. Biol. Chem.* **265**, 15239–44.

Schmidt, M.G., Rollo, E.E., Grodberg, J. and Oliver, D.B. (1988). Nucleotide sequence of the *secA* gene and *secA*(ts) mutations preventing protein export in *Escherichia coli*. *J. Bacteriol.* **170**, 3404–14.

Schodel, F. and Will, H. (1989). Construction of a plasmid for expression of foreign epitopes as fusion proteins with subunit B of *Escherichia coli* heat-labile enterotoxin. *Infect. Immun.* **57**, 1347–50.

Schultz, A.J. and McCardell, B.A. (1988). DNA homology and immunological cross-reactivity between *Aeromonas hydrophila* cytotonic toxin and cholera toxin. *J. Clin. Microbiol.* **26**, 57–61.

Singer, S.J., Maher, P.A. and Yaffe, M.P. (1987). On the translocation of proteins across membranes. *Proc. Natl. Acad. Sci. USA* **84**, 1015–19.

Shiba, K., Ito, K., Yura, T. and Cerreti, D. (1984). A defined mutation in the protein export gene within the spc ribosomal protein operon of *Escherichia coli*: isolation and characterization of a new temperature-sensitive sec Y mutant. *EMBO J.* **3**, 631–5.

Smit, J. and Nikaido, H. (1978). Outer membrane of gram negative bacteria, XVIII. Electron microscopic studies on porin insertion sites and growth of cell surface of *Salmonella typhimurium*. *J. Bacteriol.* **135**, 687–702.

Spicer, E.K. and Noble, J.A. (1982). *Escherichia coli* heat-labile enterotoxin. Nucleotide sequence of the A subunit gene. *J. Biol. Chem.* **257**, 5716–21.

Spira, W.M. and Fedorka-Cray, P.J. (1984). Purification of enterotoxins from *Vibrio mimicus* that appear to be identical to cholera toxin. *Infect. Immun.* **45**, 679–84.

Strockbine, N.A., Jackson, M.P., Sung, L.M., Holmes, R.K. and O'Brien, A.D. (1988). Cloning and sequencing of the genes for shiga toxin from *Shigella dysenteriae* type 1. *J. Bacteriol.* **170**, 1116–22.

Svennerholm, A.-M. and Holmgren, J. (1978). Identification of *Escherichia coli* heat-labile enterotoxin by means of ganglioside immunosorbent assay (GM1-ELISA) procedure. *Curr. Microbiol.* **1**, 19–27.

Takeda, Y., Honda, T., Taga, S. and Miwatani, T. (1981). In vitro formation of hybrid toxins between subunits of *Escherichia coli* heat-labile enterotoxin and those of cholera enterotoxin. *Infect. Immun.* **34**, 341–6.

Tamura, M., Nogimori, K., Murai, S., Yajima, M., Ito, K., Katada, T., Ui, M. and Ishii, S. (1982). Subunit structure of islet-activating protein, pertussis toxin, in conformity with the A-B model. *Biochemistry* **21**, 5516–20.

Tetsuya, I., Tsuji, T., Honda, T., Miwatani, T., Wakabayashi, S., Wada, K. and Matsubura, H. (1989). A single amino acid substitution in B subunit of *Escherichia coli* enterotoxin affects its oligomer formation. *J. Biol. Chem.* **264**, 14065–70.

Thom, J.R. and Randall, L.L. (1988). Role of the leader peptide of maltose-binding protein in two steps of the export process. *J. Bacteriol.* **170**, 5654–61.

Tokunaga, M., Loranger, J. and Wu, H. (1984). Prolipoprotein modification and processing enzymes in *Escherichia coli*. *J. Biol. Chem.* **259**, 3825–30.

Verheul, A.F.M., Versteeg, A.A., De Reuver, M.J., Jansze, M. and Snippe, H. (1989). Modulation of the immune response to pneumococcal type 14 capsular polysaccharide protein conjugates by the adjuvant quilA depends on the properties of the conjugates. *Infect. Immun.* **57**, 1078–83.

Verner, K., and Schatz, G. (1988). Protein translocation across membranes. *Science* **241**, 1307–13.

von Heijne, G. (1985). Signal sequence: limits of variation. *J. Molec. Biol.* **184**, 99–105.

Watanabe, M. and Blobel, G. (1989). Binding of a soluble factor of *Escherichia coli* to preproteins does not require ATP and appears to be the first step in protein export. *Proc. Natl. Acad. Sci. USA* **86**, 2248–52.

Weinstein, D.L., Jackson, M.P., Samuel, J.E., Holmes, R.K. and O'Brien, A.D. (1988). Cloning and sequencing of a shiga-like type II variant from an *Escherichia coli* strain responsible for edema disease of swine. *J. Bacteriol.* **170**, 4223–30.

Weinstein, D.L., Jackson, H.P., Perera, L.P., Holmes, R.K. and O'Brien, A.D. (1989). *In vivo* formation of hybrid toxins comprising shiga toxin and shiga-like toxins and role of the B subunit in localization and cytotoxic activity. *Infect. Immun.* **57**, 3743–50.

Weiss, A.A., Hewlett, E.L., Myers, G.A. and Falkow, S. (1983). Tn5-induced mutation affecting virulence factors of *Bordetella pertussis*. *Infect. Immun.* **42**, 33–41.

Wickner, W. (1979). The assembly of proteins into biological membranes: the membrane trigger hypothesis. *Ann. Rev. Biochem.* **48**, 23–45.

Willshaw, G.A., Smith, H.R., Scotland, S.M. and Rowe, B. (1985). Cloning of genes determining the production of vero cytotoxin by *Escherichia coli*. *J. Gen. Microbiol.* **131**, 3047–53.

Witholt, B., Hofstra, H., Kingma, J., Proak, S.E., Hol., W.G.J. and Drenth, J. (1988). Studies on the synthesis, assembly and structure of the heat-labile enterotoxin (LT) of *Escherichia coli*. In: *Bacterial Protein Toxins* (eds F. Fehrenbach, J.E. Alouf, P. Falmagne, W. Goebel, J. Jeljaszewicz, D. Jurgen and R. Rappuoli), Zbl. Bakt. Suppl. **17**, pp.3–12. Gustav Fischer, Stuttgart.

Wolfe, P.B., Wickner, W. and Goodman, J.M. (1983). Sequence of the leader peptidase gene of *Escherichia coli* and the orientation of leader peptidase in the bacterial envelope. *J. Biol. Chem.* **258**, 12073–80.

Wong, K.R. and Buckley, J.T. (1989). Proton motive force involved in protein transport across the outer membrane of *Aeromonas salmonicida*. *Science* **246**, 654–6.

Yamamoto, T. and Yokota, T. (1982). Release of heat-labile enterotoxin subunits by *Escherichia coli*. *J. Bacteriol.* **150**, 1482–4.

Yamamoto, T. and Yokota, T. (1983). Sequence of heat-labile enterotoxin of *Escherichia coli* pathogenic for humans. *J. Bacteriol.* **155**, 728–33.

Yamamoto, T., Tamura, T. and Yokota, T. (1984). Primary structure of heat-labile enterotoxin produced by *Escherichia coli* pathogenic for humans. *J. Biol. Chem.* **259**, 5037–44.

Yamamoto, T., Gojobori, T. and Yokota, T. (1987). Evolutionary origin of pathogenic determinants in enterotoxinogenic *Escherichia coli* and *Vibrio cholerae* 01. *J. Bacteriol.* **169**, 1352–7.

Yanisch-Peron, C., Vieira, J. and Messing, J. (1985). Improved M13 phage cloning vectors and host strains: nucleotide sequences of the M13mp18 and pUC19 vectors. *Gene* **33**, 103–19.

Note added in proof

In the past two years considerable advances have been made in the molecular characterization of the protein components in the cytoplasmic membrane which are involved in the general protein export pathway of bacteria (for a review, see; Genetic analysis of protein export in *Escherichia coli* by Schatz, P.J. & Beckwith, J. (1990). *Annu. Rev. Genet.* **24**, 215–48). In addition, significant progress has been made in the analysis of pullanase secretion by *K. oxytoca* (formally *K. pneumoniae*). This has revealed that the secretion of PulA to the extracellular medium requires the components of the general export pathway and an enormous array of additional gene products encoded by DNA which flanks the *pulA* gene. This region has been found to encode 15 different polypeptides, 13 of which have been definitively demonstrated to be required for accomplishing pullanase secretion (for a review, see: Genetics of extracellular protein secretion by Gram-negative bacteria by Pugsley, A.P. *et al.* (1990). *Annu. Rev. Genet* **24**, 67–90). W. Hol and coworkers have recently reported the spectacular three-dimensional structure of LT-I from *E. coli* which causes diarrhoea in pigs (Sixma, T.K., Pronk, S.E., Kalk, K.H., Wartna, E.S., van Zantan, B.A.M., Witholt, B. and Hol, W.G.J. (1991). Crystal structure of a cholera toxin-related heat-labile enterotoxin from *E. coli*. *Nature*, **351**, 371–77).

6

Comparative Toxinology of Bacterial and Invertebrate Cytolysins

Regina Linder[1] and Alan W. Bernheimer[2]

[1]Hunter College School of Health Sciences, 425 East 25th Street, New York, NY 10010, USA and [2]Department of Microbiology, New York University School of Medicine, 550 First Avenue, New York, NY 10016, USA

Introduction

The mechanisms by which the cytolytic products of some bacteria damage mammalian membranes have been the subject of much investigation in the past two decades. This interest is related in part to the ability of many cytolytic toxin-producing organisms to cause human disease. The results of studies in this area have been summarized in numerous review articles describing the nature and actions of bacterial cytolysins (Freer and Arbuthnott, 1976; Bernheimer and Rudy, 1986; Bhakdi and Tranum-Jensen, 1987, 1988; Harshman, 1988). Similarly much attention has been focused on the membrane-damaging polypeptides of higher forms of life. These agents contribute to tissue damage in humans and animals, most notably following bites of venomous snakes as well as stings of insects and marine invertebrates (Tu, 1977; Bernheimer and Rudy, 1986). Examination of the overall findings of these two lines of inquiry reveals surprising relationships. One can identify mechanistic 'themes' by which individual cytotoxic agents or groups of agents strongly resemble members of phylogenetically distant organisms in the details of their activity. An interesting example is the observation that antibody produced in response to melittin, the well-characterized lysin of the honey bee, inhibits haemolysis by complement (Laine et al., 1988).

It is our intention to consider the characterized mechanisms of cytolytic products of taxonomically diverse organisms. A brief general description of the well-defined classes of mechanisms which have been identified will be followed by a discussion of selected individual cytolytic agents in the broad taxonomic groups: bacteria, insects (Hymenoptera) and marine invertebrates (Cnidaria). The most thoroughly characterized agents will be treated in some detail, while less studied toxins will be introduced together with agents to which they appear similar in activity. We will not attempt to review exhaustively, but rather to discuss recent data which elucidate the mechanistic relationships among agents. Table 1 summarizes the data by assigning agents to broad mechanistic categories. It should be noted that in some cases, assignment is somewhat arbitrary.

Mechanisms of membrane damage by cytolytic polypeptides

Channel formation

The ability of a small polypeptide which is amphiphilic, i.e. containing regions of relatively hydrophilic character and relatively hydrophobic character, to traverse the bilayer and cause changes in cation conductance has been clearly associated with disruption of mammalian membranes. Many

Table I. Membrane disruption by cytolytic agents

Class of membrane disruption	Toxin	Molecular size	Source	Special properties
Oligomeric pores (well defined)	Gramicidin A	15 amino acids	*Bacillus brevis*	Non-haemolytic, ion-conducting pores
	Alamethicin	20 residues	*Trichoderma viride*	Cyclic
	Gramicidin S	10 amino acids	*Bacillus brevis*	Cyclic and haemolytic
	Peptide lysin	15 amino acids	*Streptoverticillium griseoverticillatum*	Cyclic and haemolytic
	Delta-haemolysin	26 amino acids	*Staphylococcus aureus*	α-Helical oligomeric channels
	Sphingomyelin-inhibitable lysins	15–21 kDa	*Stichodactyla, Condylactis* and related cnidarians	Oligomeric channels
Probable transmembrane pores	Thiol-activated agents	40–80 kDa	Gram-positive bacteria, *Klebsiella, Metridium*	Bind membrane cholesterol, visible aggregates
	Alpha-haemolysin	34 kDa	*Staphylococcus aureus*	Hexameric aggregates
	E. coli haemolysin and related lysins	110 kDa	Enterobacteriaceae, *Bordetella*	Cation-conducting pores

Category	Name	Size	Organism	Mechanism
Subtle alterations in phospholipid bilayer	Melittin	26 amino acids	*Apis mellifera*	Activates PLA_2, tetrameric aggregates
	Mastoparan	14 amino acids	*Vespula lewsii*	Activates PLA_2
	Bombolitins	17 amino acids	*Megabombus pennsylvanicus*	Activates PLA_2
	Barbatolysin	34 amino acids	*Pogonomyrmex barbatus*	Activates PLA_2
	Toxin A-III	95 amino acids	*Cerebratulus lacteus*	Tetramer induced by PLA_2
Enzymatic modification of membrane components	Beta-haemolysin	39 kDa	*Staphylococcus aureus*	Sphingomyelinase C, hot-cold haemolysin
	Sphingomyelinase D	30 kDa	*Corynebacterium, Loxosceles*	Cooperative haemolysis with cholesterol oxidase
	Cholesterol oxidase	35 kDa	*Rhodococcus* and related genera	Cooperative haemolysis with phospholipases
	Venom	Mixture	*Aiptasia pallida*	Haemolytic mixture of PLAs and other proteins
Mechanism of lysis unknown	Gamma-haemolysin	32 kDa and 36 kDa	*Staphylococcus aureus*	Cooperative haemolysis by two polypeptides
	CAMP protein	23 kDa	*Streptococcus agalactiae*	Cooperative haemolysis with β-haemolysin
	Portuguese man-of-war and box jellyfish lysins	100 kDa	*Physalia* and *Chironex*	Glycosylated, subunit, rod-like particles

of the better understood examples of cytolytic agents can broadly be defined as producing aqueous channels in target membranes, often in the form of oligomeric complexes. Evidence in such extensively studied systems as Gramicidin A (Wallace and Ravikumar, 1988) demonstrates that channels composed of amphiphilic molecules arrayed as oligomers display a hydrophilic interior lining, with hydrophobic residues oriented to the remainder of the membrane.

Probable pore formation

Recent evidence is strongly suggestive of membrane pore formation by a diverse group of full-sized cytotoxic proteins from bacteria and other sources. The much studied SH-activated group of cytotoxins, as well as the haemolytic toxins of Enterobacteriaceae and other Gram-negative bacteria, are thought to produce membrane lesions by insertion. Recombinant DNA technology has been used to demonstrate regions of amino acid homology and amphiphilicity in these large proteins (Felmlee and Welch, 1988), providing evidence for pore-type mechanisms of membrane disruption in contrast to the induction of lipid phase transitions. It will be evident, however, that the characterization of a 'pore' or lesion usually lacks the clear physical definition which has been obtained for the smaller peptide amphiphiles.

Detergent-like mechanisms

Early observations of surface activity in such lysins as staphylococcal delta-toxin (Bernheimer, 1974) and melittin (Sessa et al., 1969) suggested that their activities resembled that of detergents. Accumulated evidence, however, does not in our opinion support micelle formation (detergent action) as the mechanism of membrane damage. The lysins in question seem rather to be channel or pore formers, or to produce more subtle membrane interactions.

Enzymatic mechanisms

This category refers to those enzymatically active products of microorganisms and higher forms which catalyse modifications in the lipid components of target membranes. In general, rather

than producing cell damage independently, these agents (usually phospholipases) have been observed to accentuate (and sometimes prevent) the action of non-enzymatic cytotoxins. An example is the phospholipase A_2 of bee venom which is activated by melittin and in turn enhances cell destruction by the venom (Yunes et al., 1977). Bacterial cholesterol oxidase has similarly been shown to participate in cell damage in concert with phospholipases of other bacteria (Linder, 1984).

Less well-characterized agents

For those cytolytic products which cannot yet be classified in the categories described, discussion will be aimed at a core of questions which speak to the underlying nature of the activity. Examples of these questions are: Are lesions on target cells visible by electron microscopy? Is lysis direct or of the colloid osmotic type?, i.e. does the leakage of macromolecules occur only after the transmembrane migration of small ions and water? Is lysis inhibitable by a particular membrane component, thereby implicating a specific receptor? What is known of the primary, secondary and tertiary structure of the protein? Such techniques as genetic cloning and site-specific mutagenesis have been especially valuable in revealing similarities between well-characterized and less well-understood agents. Regions of amino acid homology among proteins, as well as the demonstration of amphiphilic domains are important for our understanding of the relationship of structure to function of cytolysins.

Cytolytic products of bacteria

Gramicidin and related peptides

Gramicidin A, the product of *Bacillus brevis*, is a cation-selective hydrophobic peptide of 15 amino acids which have alternating L and D chirality. The structure of the peptide is as follows:

Formyl-L-Val–Gly–L-Ala–D-Leu–L-Ala–D-Val–
Val–D-Val–L-Trp–D-Leu–L-Trp—D-Leu–L-Trp-
ethanolamine

(Sarges and Witkop, 1965; Anderson, 1984). While gramicidin A is antibacterial (especially to Gram-positive bacteria, Hotchkiss, 1944) it is not haemo-lytic. It has, however, been much studied as a model of channel formation, and stands as our best understood example of transmembrane conductance.

Physical studies have demonstrated that ion conductance by the molecule is related to its ability to dimerize in a head-to-head fashion to form a transmembrane pore (Urry, 1971; Veatch and Stryer, 1977). Veatch et al. (1974) proposed the functional form of the dimer to be an antiparallel double helix, with hydrophobic residues oriented to the membrane exterior of the pore, and carbonyl groups lining the hydrophilic shaft.

Two recent X-ray crystallographic studies, one of free gramicidin A (Langs, 1988), and the other of a caesium complex in which the ion acts as a conducted ligand (Wallace and Ravikumar, 1988), have confirmed the double-stranded beta-helical structure, stabilized by double bonds between the adjacent edges of the coil. Free vs. Cs-complexed gramicidin molecules differ in helical dimensions. Wallace and Ravikumar propose that the transmembrane ion-conducting channel is composed of two such dimers stacked one above the other to span the bilayer, as is illustrated in Fig. 1.

Alamethicin, a similar peptide produced by the fungus *Trichoderma viride*, is also thought to form oligomers lining hydrophilic cation-conducting pores in the manner of the staves of a barrel (Latorre and Alvarez, 1981; Hall et al., 1984). A cyclic 15 amino acid peptide of *Streptoverticillium griseoverticillatum*, a streptomycete, contains several unusual hydrophobic amino acids including internal cross-linkages (Choung et al., 1988). It is disruptive to red cells of a variety of species and to liposomes containing phosphatidylethanolamine (Suzuki et al., 1988), and is therefore similar to gramicidin S the cyclic peptide product of *B. brevis*.

Thiol-activated agents

The family of cytotoxic agents of which streptolysin (SLO) is the prototype represents the products of more than 15 bacterial species. The bacteria producing them are predominantly, but not exclusively Gram-positive and they vary considerably in morphology, and in pathogenic potential (Albesa et al., 1985). An extensive literature of original

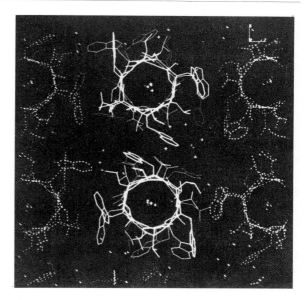

Figure 1. The packing of gramicidin pores in the gramicidin–caesium crystal. In this view, along the crystallographic *b* axis, the two independent dimers in the asymmetric unit and their symmetry-related molecules are shown. The dimers form end-to-end stacks of tubes perpendicular to the plane of this figure. (From Wallace and Ravikumar, 1988).

and review articles describes these, and indeed a chapter in the present volume is devoted to them (see Chapter 8). They will be discussed only as they relate to key themes of this contribution.

Although non-identical in amino acid composition, the lysins are thought to share a mechanistic mode. Similarities in their properties include sensitivity to mild oxidation, a common membrane binding site (cholesterol), the production of characteristic visible membrane lesions, and cross-neutralization by hyperimmune sera. The genetic determinant for one of these proteins, alveolysin has been cloned and expressed in *E. coli* (Geoffroy and Alouf, 1988). The product demonstrated significant amino acid homology with several other SH-activated lysins, especially near the C-terminus.

Several lines of evidence indicate that aggregation of toxin monomers leads to membrane disruption by SH-activated lysins. The long observed 'rings' and 'arcs' in electron micrographs of treated membranes or cells are composed of oligomers varying between 25 and 100 molecules of toxin (Bhakdi et al., 1984; Niedermeyer, 1985). They are thought to comprise toxin-lined aqueous pores which constitute the lesions of membrane damage.

Hugo *et al.* (1986) used a monoclonal antibody against SLO which prevented both oligomerization and lysis, but not binding of toxin monomer to membranes. A protease-nicked fragment of perfringiolysin was shown to bind to target membranes, but it brought about neither lysis nor the development of visible oligomeric complexes (Ohno-Iwashita *et al.*, 1988). In another study, detailed isoelectric focusing analysis of SLO revealed substantial heterogeneity which was correlated with differing efficiency of lysis (Suzuki *et al.*, 1988). Differences were interpreted as a reflection of an unequal ability to aggregate for molecules carrying varying charges.

An interesting example of taxonomically distant agents which are mechanistically similar is the product of the common large sea anemone, *Metridium senile*. Medtridiolysin is an SH-activated protein which is inhibited by cholesterol and very similar in its biological activity to the bacterial lysins of this group (Bernheimer and Avigad, 1978; Bernheimer *et al.*, 1979). It does not, however, demonstrate antigenic relatedness to the bacterial agents.

Cytotoxins of *Staphylococcus aureus*

Alpha-toxin

As a lethal, cytotoxic secreted product of an important bacterial pathogen, α-toxin has been the subject of much experimental observation (Bernheimer, 1974; Harshman, 1979; Rogolsky, 1979). The polypeptide molecule of 34 kDa (Gray and Kehoe, 1984) has long been considered a contributor to the virulence of the bacterium. In a previous investigation related to fulminant staphylococcal endocarditis, a site-specific α-toxin mutant was observed to be much reduced in ability to damage cultured endothelial cells (Vann and Proctor, 1988).

The mode of membrane disruption by α-toxin has proven elusive. It was early recognized that toxin binding preceded haemolysis (Cooper *et al.*, 1964; Cassidy and Harshman, 1976), and that ion transport brought about a colloid-osmotic rupture (Wilbrandt, 1941). Another crucial observation was the ability of the toxin in solution to aggregate to a hexamer form which appeared identical to structures observed on lysed target membranes (Freer *et al.*, 1968; Arbuthnott *et al.*, 1973). The

relationship of these hexameric complexes to ion-conducting pores has been addressed by a number of recent studies. Measurements of fluorescence energy transfer in liposomes disrupted by α-haemolysin detected a two-step process consisting of binding and hexamer formation (Ikigai and Nakae, 1987a,b). Conductivity measurements in planar lipid membranes by Belmonte *et al.* (1987) were consistent with the hypothesis that reversible toxin dimerization in the membrane is followed by dimer aggregation to form transmembrane channels and the typical hexameric array.

Beta-haemolysin

The gene specifying the sphingomyelin-specific phospholipase C (Bernheimer *et al.*, 1974) has been recently cloned and sequenced (Projan *et al.*, 1989). A 39-kDa protein, it is typical of lipid-hydrolysing cytotoxic agents in that its enzymatic action alone on susceptible erythrocytes does not bring about haemolysis. A second condition, such as chilling or the action of an additional agent is necessary for lysis (Smyth *et al.*, 1975; Linder, 1984). While lysis follows a molecular disorientation of the lipid bilayer, the exact nature of the change is not clear. There is no evidence to date for aggregation or defined channel formation. An interesting exception to the non-haemolytic character of the enzymatic agents is the sphingomyelinase of *Bacillus cereus*, a haemolytic protein which is differentially stimulated by calcium and magnesium ion (Tomita *et al.*, 1982). The recent cloning and sequencing of the *B. cereus* enzyme (Yamada *et al.*, 1988) should aid our understanding of membrane alteration through hydrolytic change.

Gamma-haemolysin

The least well-characterized lysin of *S. aureus* is also difficult to clearly associate with bacterial virulence. It is thought to function in the pathogenesis of staphylococcal osteomyelitis (Taylor and Plommet, 1973), and is produced by most strains associated with toxic shock syndrome (Clyne *et al.*, 1987). The early observation that two polypeptides are required for haemolytic activity (Guyonnet and Plommet, 1970; Taylor and Bernheimer, 1974) has recently been confirmed by physical mapping of the cloned genetic determinant (Cooney *et al.*, 1988). Hopefully, further attention will clarify the means of membrane disruption brought about by this unusual bacterial product.

Delta-haemolysin

The product of many strains of pathogenic *S. aureus*, δ-haemolysin is a 26 amino acid amphiphilic peptide which is non-selectively membrane disruptive. Its properties have been reviewed and a mechanism of membrane insertion of α-helical molecules proposed (Freer *et al.*, 1984). The length of the helices allows a model in which molecules span the membrane arranged as the staves of a barrel, similar to the hypothesized insertion of the fungal product alamethicin (Menestrina *et al.*, 1986; Lee *et al.*, 1987). A recent study of the properties of channel formation in planar lipid membranes supports such a model, and proposes that six α-helices form a channel (Mellor *et al.*, 1988).

Like alamethicin, δ-lysin also shows activity against bacterial protoplasts, and therefore it is interesting to speculate on its role in the physiology of the organism which produces it. In an investigation of penicillinase-independent resistance to β-lactam antibiotics, clinical isolates of *S. aureus* demonstrating an intermediate or 'heteroresistant' response were less likely to be δ-lysin producers than either fully sensitive or resistant strains (Linder and Salton, 1986). The results suggest a possible role for this agent in permeability to antibiotics.

Membrane-active agents of corynebacteria and related Gram-positive rods

This category is intended to illustrate the many examples of membrane disruption brought about by the combined activity of the products of two bacteria (Fraser, 1964; Linder, 1984). The best understood example of cooperative haemolysis is the CAMP phenomenon which has long served in the presumptive identification of group B streptococci (Christie *et al.*, 1944). The β-haemolysin of *S. aureus*, not haemolytic by itself, renders sheep erythrocytes susceptible to the lytic action of the CAMP protein of the streptococcus by hydrolysis of membrane sphingomyelin. While it has been shown to have a specific affinity for ceramide (the sphingomyelinase product) in liposomes (Bernheimer *et al.*, 1979b), the mechanism by which CAMP disorders the hydrolysed leaflet has yet to be clarified.

By analogy with the CAMP system, a number of Gram-positive bacteria of the genus *Corynebacterium*, *Rhodococcus* and related genera secrete agents which participate in cooperative lysis or indeed prevent lysis by the products of other bacteria (Linder, 1984). A discussion of several examples will attempt to illustrate the effects of enzyme combinations on susceptible membranes.

The phospholipases D of *C. pseudotuberculosis* and *C. ulcerans* protect erythrocytes from the action of phospholipases C and heliantholysin (a lytic product of a marine invertebrate; Bernheimer and Avigad, 1976). A very similar PLD from the brown recluse spider, *Loxosceles reclusa* can substitute for the bacterial agent in the protection assay (Bernheimer *et al.*, 1985), and like them is thought to participate in the tissue damage produced by the organisms which elaborate them. Steric blockage of receptors and/or their modification has been considered a likely explanation for the inhibition (Bernheimer and Avigad, 1976).

Rhodococcus equi, an animal pathogen increasingly responsible for opportunistic infections in humans (Fierer *et al.*, 1987) secretes cholesterol oxidase in the course of its growth. The combined action of this enzyme with bacterial phospholipases D produces haemolysis of sheep cells (Linder and Bernheimer, 1982). The requirement for prior treatment with detergent, phospholipase action or cholesterol enhancement in order for cholesterol oxidase to reach its substrate has been demonstrated in human red cells, cultured fibroblasts and an enveloped virus (Gottlieb, 1977; Moore *et al.*, 1977; Patzer *et al.*, 1978; Linder and Bernheimer, 1984). Presumably it is the inaccessibility of cholesterol in native membranes which underlies the cooperative reaction involving this enzyme.

Escherichia coli haemolysin and related proteins

Strains of *E. coli* responsible for extra-intestinal infections, often of the urinary tract in adults, and septicaemias in newborns, secrete a haemolysin termed α-haemolysin (Cavalieri *et al.*, 1984). Screening for antibodies in healthy humans has shown that this agent is a common immunogen

(Hugo *et al.*, 1987). Recent data have resulted in our emerging understanding of the structure, genetics, activity and secretion not only of this lysin, but of the closely related agents produced by a variety of pathogenic bacteria.

Four genes constitute the genetic determinant for haemolysin activity in *E. coli*, and they are termed *hlyA, B, D* and *C*. Nucleotide sequences have been determined and haemolytic activity found to reside in the 110-kDa product of *hlyA* (Wagner *et al.*, 1983; Felmlee *et al.*, 1985). Site-specific mutagenesis of the polypeptide revealed an internal hydrophobic region of 21 amino acids which is required for haemolytic activity (Goebel *et al.*, 1988). The C-terminus of the molecule, 37 residues of which can be lost without loss of activity, is necessary for export of the molecule (Ludwig *et al.*, 1987). *HlyA* contains an eight-residue sequence which is repeated 13 times, and is conserved in the haemolysins of *Pasteurella hemolytica*, *Bordetella pertussis*, *Proteus mirabilis*, *Proteus vulgaris* and *Morganella morganii* (Koronakis *et al.*, 1987; Ludwig *et al.*, 1987; Welch, 1987; Felmlee and Welch, 1988). The repeat sequence is thought to function in the binding of Ca^{2+}, which is considered a factor in efficient red cell disruption (Goebel *et al.*, 1988). *HlyB* and *D* function in secretion, and the gene product of *hlyC* participates in a post-translational modification of the haemolysin, the nature of which is not entirely clear (Noegel *et al.*, 1981).

Results of experiments with *E. coli* haemolysin on artificial membranes were consistent with the production of cation-selective functional pores, and displayed a voltage dependence similar to colicin action (Menestrina *et al.*, 1987; Menestrina, 1988). No evidence for aggregation, or the requirement for a specific membrane receptor was obtained. On erythrocyte membranes, bound toxin could not be removed by treatments which remove all but integral membrane proteins, and detergent solubilization resulted only in monomer recovery (Bhakdi *et al.*, 1986). The authors propose the integration of haemolysin to form a monomeric pore, the opening of which leads to Ca^{2+} uptake and ultimately, colloid osmotic lysis. While more work remains, the absence of electron microscopic lesions, oligomerization of toxin and specific receptors point to a distinct mechanism of action among these bacterial pathogens. Of particular note among

the lysins displaying genetic homologies, are the bifunctional protein of *B. pertussis* which possesses both haemolytic and adenyl cyclase activity (Glaser *et al.*, 1988), and the leukotoxin of *P. hemolytica* (Strathdee and Lo, 1987).

Cytolytic products of arthropods

Melittin

The cytolytic component of the venom of the honey bee, *Apis mellifera* stands together with gramicidin A as one of the two best characterized peptide lytic agents. In part because of its central contribution to the tissue damage which follows the sting of a bee (Haberman, 1972), and because melittin has served as a model for the interaction of peptides with membranes, a substantial literature is available for study (Banks *et al.*, 1981; Levin, 1984; Shipolini, 1984; Bernheimer and Rudy, 1986). Rather than attempting to do justice to this literature, we will try to summarize those findings which most directly impact on our understanding of the cytolytic mechanism.

A 26 amino acid peptide, melittin is an amphiphile, largely hydrophobic, with a cluster of positive charges at its C-terminus (Fitton *et al.*, 1980). Circular dichroism (Knoppel *et al.*, 1979; Shipolini, 1984) and X-ray diffraction (Terwilliger and Eisenberg, 1982a,b) of the tetramer which spontaneously forms in high ionic strength and in crystals, reveals a perpendicular packing pattern illustrated in Fig. 2. Chains are largely in an α-helical conformation, and the tetramer itself derives the properties of an amphiphile from the segregation of hydrophobic and hydrophilic sidechains on opposite sides of the structure.

The relationship of the crystalline form to the mechanism of membrane disruption remains controversial. While proposals of tetramers or dimers in the membrane have often been put forward (DeGrado *et al.*, 1982; Georghiou *et al.*, 1982; Hider *et al.*, 1983), direct lytic activity of the monomer has not been excluded. Spectroscopic studies, however, concur that the backbone of the molecule is an α-helix (Dawson *et al.*, 1978; Lavialle *et al.*, 1982), that the tryptophan at residue 19 penetrates only slightly into the membrane, and that the positively charged C-terminus of the

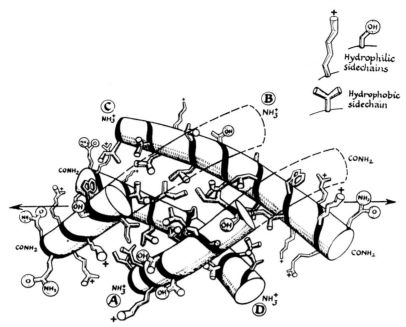

Figure 2. Cut-away drawing showing the packing of melittin chains in the tetramer. (From Terwilliger and Eisenberg, 1982a).

molecule (residues 20–26) is exposed to the aqueous environment (DeBony *et al.*, 1979; Georghiou *et al.*, 1982; Dufourq *et al.*, 1984). The latter sequence has been shown to be required for haemolysis (Schroder *et al.*, 1971).

The orientation of the remaining 18 residues with respect to the plane of the membrane is yet to be established with certainty. A channel-type of mechanism involving oligomeric spanning of the membrane is advocated by some investigators (DeGrado *et al.*, 1982; Kempf *et al.*, 1982). Clearly such models have much in common with the channel-forming peptides gramicidin and alamethicin in which polar sidechains define aqueous pores spanning the bilayer. Alternatively, melittin has been proposed to lie parallel to the plane of the membrane, or to penetrate only slightly into its surface (DeGrado *et al.*, 1982; Terwilliger *et al.*, 1982; Vogel *et al.*, 1983). Lysis according to this type of hypothesis involves the disruption of phospholipid structure that has been termed a 'leaky patch' mechanism (Laine *et al.*, 1988).

Cell disruption due to small areas of membrane distortion is in fact proposed by some investigators to explain cytolysis by a diverse group of agents including haemolytic viruses, complement, melittin

and a number of bacterial toxins (Bashford *et al.*, 1986). That lysis can generally be prevented by divalent cations is particularly persuasive to those who view the generation of aqueous pores as an oversimplified explanation of lysis. The observation of synergistic kinetics when pairs of agents were applied together likewise may not be expected if each agent produced its distinct membrane channel.

In a study using antibody raised in response to melittin, antigenic cross-reactivity was detected between this peptide and the C9 component of complement (Laine *et al.*, 1988). Anti-melittin protected erythrocytes from complement lysis, and the cross-reacting epitope was identified as a nine-residue shared sequence. The authors favour a mechanism of membrane lysis in which strong associations of protein agents with lipid components of the bilayer occur at or near the external surface of the membrane. The resulting structural distortion is envisioned to produce changes in cation transport and ultimately, colloid-osmotic cytolysis. The surprising finding of homology between two such overtly different agents suggests that, in those cases where channels have not been directly observed, a 'leaky patch' model may require fewer assumptions.

Other amphiphilic arthropod lysins

A variety of hymenopteran insects other than the honey bee produce venom constituents showing striking similarities with melittin. The wasp *Vespula lewsii* produces mastoparan, a 14 amino acid basic polypeptide cytolysin possessing the following sequence (Hirai *et al.*, 1979):

Ile–Asn–Leu–Lys–Ala–Leu–Ala–Ala–Leu–Ala–
Lys–Lys–Ile–Leu–NH$_2$

The molecule was observed to undergo conformational change upon interaction with a phospholipid membrane (Higashijima *et al.*, 1983; Wakamatsu *et al.*, 1983), resulting in a highly structured α-helix. Exposed to the aqueous medium in this form are four positive charges, leading the authors to propose a structure with two basic intracellular loops close to the surface of the bilayer (Higashijima *et al.*, 1988). It should be noted that this model is reminiscent of the 'leaky patch' view of cytolysis due to melittin.

Also sharing properties with melittin are the bombolitins, the closely related group of 17-residue lytic peptides of the bumblebee, *Megabombus pennsylvanicus* (Argiolas and Pisano, 1985). Barbatolysin, a constituent of the venom of the red harvester ant, *Pogonomyrmex barbatus* is substantially larger than the other hymenopteran lysins described, having 34 amino acid residues (Bernheimer *et al.*, 1980). It is a basic molecule, and although lacking a number of common amino acids, is amphiphilic and thought to act in a fashion similar to melittin and its congeners (Bernheimer and Rudy, 1986).

In addition to non-specific membrane disruption, melittin and the other hymenopteran agents are known to stimulate the action of phospholipase A$_2$, a component of bee venom and other biological sources (Yunes *et al.*, 1977; Shier, 1979). The activation is thought to involve increased access of the phospholipase to its membrane-bound substrate, due to the action of the lysin (Mollay and Kriel, 1974; Yunes *et al.*, 1977). Presumably cytolysis is in turn enhanced by the catalytic modification of bilayer phospholipids.

Cytolysins of marine invertebrates

Sea anemones

Among marine invertebrates, cnidarians (coelenterates) commonly produce cytolytic polypeptides among their venom constituents. The lysins of sea anemones in particular have been examined systematically in the past decade, and a recent review summarizes the results (Bernheimer, 1990). What follows is a brief overview of these agents with particular emphasis on comparison with non-marine lysins.

Sphingomyelin-inhibitable agents

Derived from 16 species of sea anemones, these lysins are grouped for their common property of lysis inhibition by small quantities of sphingomyelin (Bernheimer, 1990). The molecules are isoelectric at pH 9 or above, and while there is significant variation in amino acid sequence, there is overall similarity. Molecular size varies between 15 and 21-kDa. Immunologic cross-neutralization was observed between two members of the group (Bernheimer and Lai, 1985).

The best characterized of the group, *Stichodactyla (Stoichactis) helianthus* lysin is comprised of four isotoxins differing in their N-termini (Kem and Dunn, 1988). Membrane sphingomyelin appears to be the favoured receptor for heliantholysin judging from interaction with liposomes, as well as from experiments in which ruminant erythrocytes (rich in this phospholipid) were modified with phospholipase (Bernheimer and Avigad, 1976; Linder and Bernheimer, 1978). Membrane disruption is thought to involve channel formation following toxin aggregation (Michaels, 1979; Shin *et al.*, 1979; Varanda and Finkelstein, 1980).

Aiptasia pallida lysin

The venom from the nematocysts of *Aiptasia pallida* contains a haemolytic agent in addition to phospholipase A, as well as two activating proteins (Hessinger and Lenhoff, 1976). While the detailed mechanism of lysis remains to be clarified, the component activities are reminiscent of arthropod and snake lysins in which synergy with phospholipase participates in membrane damage.

Metridiolysin

Produced by the common large anemone *Metridium senile*, this agent was mentioned earlier because of similarities with the SH-activated bacterial lysins. It is similar to them also in its lethality for mice (Bernheimer and Avigad, 1978) and the formation of 33 nm rings on treated membranes (Bernheimer *et al.*, 1979a). In spite, therefore, of its somewhat larger molecular size (80 kDa), the data suggest a similar mode of action to the streptolysin O prototype agents.

Other invertebrate cytolysins

Physalia and Chironex toxins

The marine invertebrates, *Physalia physalix* (Portuguese man-of-war) and *Chironex fleckeri* ('box jellyfish') are notorious for the hazard they present to swimmers. The venom of each contains a toxic protein which is both cytolytic and lethal. The *Physalia* agent is a large (240-kDa), rod-like, glycosylated, subunit protein which is inactivated by concanavalin A (Tamkun and Hessinger, 1981). The smaller *Chironex* lysin has been observed to be inhibited by gangliosides. A recently isolated lysin from the jellyfish *Rhizostoma pulmo* is similar in molecular size (260 kDa) to that of *Physalia*, and in addition possesses subunits and a rod-shaped appearance (Cariello *et al.*, 1988).

Cerebratulus lacteus cytolysin

Toxin A-III is a component of the mucus of the skin of the heteronemertine marine invertebrate *C. lacteus*. Analysis reveals the lysin to be a 95-residue polypeptide (Blumenthal, 1980; Blumenthal and Kem, 1980), possessing distinct amphiphilicity. The C-terminus is relatively hydrophobic and the N-terminal approximately two-thirds of the molecule are quite polar and positively charged (Blumenthal, 1982; Dumont and Blumenthal, 1985).

Similar to lysins of other sources, a synergistic relationship exists with phospholipase A_2, which has been explored in some detail (Liu and Blumenthal, 1988). Using a fluorescent probe in erythrocyte membranes, kinetic analysis implied that the binding of lysin A-III activates membrane-bound phospholipase. The resulting free fatty acid in turn brings about the aggregation of toxin monomers probably to a tetramer. The latter has substantial α-helix content and is thought to induce lysis by a channel type of mechanism. This mechanism contrasts with the commonly accepted idea that in melittin synergism, lysis and the resulting phospholipid disorganization allows increased access of phospholipase to its substrate (Mollay and Kriel, 1974; Yunes *et al.*, 1977; Shier, 1979). The finding that fatty acid is able to induce conformational change in a lysin (Liu and Blumenthal, 1988) is clearly of significance for this and other agents.

Acknowledgements

We are grateful to Mila Dela Torre and Leah Trachten for assistance with the preparation of the manuscript.

References

Albesa, I., Barberis, L.I., Pajaro, M.C., Farnochi, M.C. and Eraso, A.J. (1985). A thiol-activated hemolysin in Gram-negative bacteria. *Can. J. Microbiol.* **31**, 297–300.

Anderson, O.S. (1984). Gramicidin channels. *Ann. Rev. Physiol.* **46**, 531–48.

Arbuthnott, J.P., Freer, J.H. and Bernheimer, A.W. (1973). Physical state of staphylococcal α-toxin. *J. Bacteriol.* **94**, 1170–7.

Argiolas, A. and Pisano, J.J. (1985). Bombolitins, a new class of mast cell degranulating peptides from the venom of the Bumblebee, *Megabombus pennsylvanicus. J. Biol. Chem.* **260**, 1437–44.

Banks, B.E.C., Dempsey, C., Pearce, F.L., Vernon, C.A. and Wholly, T.E. (1981). New method of isolating bee venom peptides. *Analyt. Biochem.* **116**, 48–52.

Bashford, C.L., Alder, G.M., Menestrina, G., Micklem, K.J., Murphy, J.J. and Pasternak, C.A. (1986). Membrane damage by hemolytic viruses, toxins, complement, and other cytotoxic agents. *J. Biol. Chem.* **261**, 9300–8.

Belmonte, G., Cescatti, L., Ferrari, B., Nicolussi, T., Ropele, M. and Menestrina, G. (1987). Pore formation by *Staphylococcus aureus* alpha-toxin in lipid bilayers. *Eur. J. Biophys.* **14**, 349–58.

Bernheimer, A.W. (1974). Interactions between membranes and cytolytic bacterial toxins. *Biochim. Biophys. Acta.* **344**, 27–50.

Bernheimer, A.W. (1990). Cytolytic peptides of sea anemones. In: *Marine Toxins, Origins, Structure and*

Molecular Pharmacology (eds S. Hall and G. Strichartz) pp. 304–311. Amer. Chem. Soc., Washington, DC.

Bernheimer, A.W. and Avigad, L.S. (1976). Properties of a toxin from the sea anemone *Stoichactis helianthus*, including specific binding to sphingomyelin. *Proc. Natl. Acad. Sci. USA* **73**, 467–71.

Bernheimer, A.W. and Avigad, L.S. (1978). A cholesterol-inhibitable cytolytic protein from the sea anemone *Metridium senile*. *Biochim. Biophys. Acta* **541**, 96–106.

Bernheimer, A.W. and Lai, C.Y. (1985). Properties of a cytolytic protein from a sea anemone, *Stoichactis kenti*. *Toxicon* **23**, 791–9.

Bernheimer, A.W. and Rudy, B. (1986). Interactions between membranes and cytolytic peptides. *Biochim. Biophys. Acta* **864**, 123–41.

Bernheimer, A.W., Avigad, L.S. and Kim, K.S. (1974). Staphylococcal sphingomyelinase (β-hemolysin). *Ann. N.Y. Acad. Sci.* **236**, 292–306.

Bernheimer, A.W., Avigad, L.S. and Kim, K.S. (1979a). Comparison of metridiolysin from the sea anemone with thiol-activated cytolysins from bacteria. *Toxicon* **17**, 69–75.

Bernheimer, A.W., Linder, R. and Avigad, L.S. (1979b). Nature and mechanism of action of the CAMP protein of group B streptococci. *Infect. Immun.* **23**, 838–44.

Bernheimer, A.W., Avigad, L.S. and Schmidt, J.O. (1980). A hemolytic polypeptide from the venom of the red harvester ant, *Pogonomyrmex barbatus*. *Toxicon*. **18**, 271–8.

Bernheimer, A.W., Campbell, B.J. and Forrester, L.J. (1985). Comparative toxinology of *Loxosceles reclusa* and *Corynebacterium pseudotuberculosis*. *Science* **228**, 590–1.

Bhakdi, S. and Tranum-Jensen, J. (1987). Damage to mammalian cells by proteins that form transmembrane pores. *Rev. Physiol. Biochem. Pharmacol.* **107**, 147–223.

Bhakdi, S. and Tranum-Jensen, J. (1988). Damage to cell membranes by pore-forming bacterial cytolysins. *Prog. Allergy* **40**, 1–43.

Bhakdi, S., Roth, M., Sziegoleit, A. and Tranum-Jensen, J. (1984). Isolation of two hemolytic forms of streptolysin-O. *Infect. Immun.* **46**, 394–400.

Bhakdi, S., Mackman, N., Nicaud, J.-M. and Holland, I.B. (1986). *Escherichia coli* hemolysin may damage target cell membranes by generating transmembrane pores. *Infect. Immun.* **52**, 63–9.

Blumenthal, K.M. (1980). Structure and action of heteronumertine polypeptide toxins. Disulfide bonds of *Cerebratulus lacteus* toxin A-III. *J. Biol. Chem.* **255**, 8273–4.

Blumenthal, K.M. (1982). Structure and action of heteronumertine polypeptide toxins. Membrane penetration by *Cerebratulus lacteus* toxin A-III. *Biochemistry* **21**, 4229–33.

Blumenthal, K.M. and Kem, W.R. (1980). Structure and action of heteronumertine polypeptide toxins.

Primary structure of *Cerebratulus lacteus* toxin A-III. *J. Biol. Chem.* **255**, 8266–72.

Cariello, L., Romano, G., Spagnuolo, A. and Zanetti, L. (1988). Isolation and partial characterization of rhizolysin, a high molecular weight protein with hemolytic activity, from the jellyfish *Rhizostoma pulmo*. *Toxicon* **11**, 1057–65.

Cassidy, P. and Harshman, S. (1976). Studies on the binding of staphylococcal ^{125}I-labeled α-toxin to rabbit erythrocytes. *Biochemistry* **15**, 2348–55.

Cavalieri, S.J., Bohach, G.A. and Snyder, I.S. (1984). *Escherichia coli* α-hemolysin: characteristics and probable role in pathogenicity. *Microbiol. Rev.* **48**, 326–43.

Christie, R., Atkins, N.E. and Munch-Petersen, E. (1944). A note on a lytic phenomenon shown by group B streptococci. *Aust. J. Exp. Biol.* **22**, 197–200.

Choung, S.Y., Kobayashi, T., Inoue, J., Takemoto, K., Ishitsuka, H. and Inoue, K. (1988). Hemolytic activity of a cyclic peptide Ro09-0198 isolated from *Streptoverticillium*. *Biochim. Biophys. Acta*. **940**, 171–9.

Clyne, M., De Azevedo, J. and Arbuthnott, J.P. (1987). Production of a gamma haemolysin by toxic shock syndrome associated strains of *Staphylococcus aureus*. *J. Clin. Biol.* **26**, 535–9.

Cooney, J., Mulvey, M., Arbuthnott, J.P. and Foster, T.J. (1988). Molecular cloning and genetic analysis of the determinant for gamma-lysin, a two-component toxin of *Staphylococcus aureus*. *J. Gen. Microbiol.* **134**, 2179–88.

Cooper, L.Z., Madoff, M.A. and Weinstein, L. (1964). Hemolysis of rabbit erythrocytes by purified staphylococcal alpha-toxin. I. Kinetics of the lytic reaction. *J. Bacteriol.* **87**, 127–35.

Dawson, C.R., Drake, A.F., Heliwell, J. and Hider, R.C. (1978). The interaction of bee-melittin with lipid bilayer membrane. *Biochim. Biophys. Acta* **510**, 75–86.

DeBony, J., Dufourq, J. and Clin, B. (1979). Lipid–protein interactions: nmr study of melittin and its binding to lysophosphatidylcholine. *Biochim. Biophys. Acta*. **552**, 531–4.

DeGrado, W.F., Musso, G.F., Lieber, M., Kaiser, E.T. and Kezdy, F.J. (1982). Kinetics and mechanism of hemolysis induced by melittin and a synthetic melittin analogue. *Biophys. J.* **37**, 329–38.

Dufourq, J., Dasseux, J.L. and Faucon, J.F. (1984). An illustrative model for lipid-protein interactions in membranes. In: *Bacterial Protein Toxins* (eds J.E. Alouf, F.J. Fehrenbach, J.H. Freer and J. Jeljaszewicz), pp. 127–38. Academic Press, London.

Dumont, J.A. and Blumenthal, K.M. (1985). Structure and action of heteronumertine polypeptide toxins: Importance of amphipathic helix for activity of *Cerebratulus lacteus* toxin A-III. *Arch. Biochem. Biophys.* **236**, 167–75.

Felmlee, T. and Welch, R.A. (1988). Alterations of amino acid repeats in the *Escherichia coli* hemolysin affect cytolytic activity and secretion. *Proc. Natl. Acad. Sci. USA* **85**, 5269–73.

Felmlee, T., Pellett, S. and Welch, R.A. (1985). Nucleotide sequence of an *Escherichia coli* chromosomal hemolysin. *J. Bacteriol.* **163**, 94–105.

Fierer, J., Wolf, P., Seed, L., Gay, T., Noonan, K. and Haghighi, P. (1987). Non-pulmonary *Rhodococcus equi* infections in patients with AIDS. *J. Clin. Pathol.* **40**, 556–8.

Fitton, J.E., Dell, A. and Shaw, W.V. (1980). The amino acid sequence of the delta haemolysin of *Staphylococcus aureus*. *FEBS Lett.* **115**, 209–12.

Fraser, G. (1964). The effect on animal erythrocytes of combinations of diffusible substances produced by bacteria. *J. Pathol. Bacteriol.* **88**, 43–53.

Freer, J.H. and Arbuthnott, J.P. (1976). Biochemical and morphologic alterations of membranes by bacterial toxins. In: *Mechanisms in Bacterial Toxinology* (ed. A.W. Bernheimer), pp. 169–93. John Wiley & Sons, New York.

Freer, J.H., Arbuthnott, J.P. and Bernheimer, A.W. (1968). Interaction of staphylococcal α-toxin with artificial and natural membranes. *J. Bacteriol.* **95**, 1153–68.

Freer, J.H., Birkbeck, T.H. and Bhakoo, M. (1984). Interaction of staphylococcal δ-lysin with phospholipid monolayers and bilayers. In: *Bacterial Protein Toxins* (eds J.E. Alouf, F.J. Fehrenbach, J.H. Freer and J. Jeljaszewicz), pp. 181–9. Academic Press, London.

Geoffroy, C. and Alouf, J.E. (1988). Molecular cloning and characterization of *Bacillus alvei* thiol-dependent cytolytic toxin expressed in *Escherichia coli*. *J. Gen. Microbiol.* **134**, 1961–70.

Georghiou, S., Thompson, M. and Mukhopadhyay, A.K. (1982). Melittin–phospholipid interaction studied by employing the single tryptophan residue as an intrinsic fluorescent probe. *Biochim. Biophys. Acta.* **668**, 441–52.

Glaser, P., Sakamoto, H., Bellalou, J., Ullman, A. and Danchin, A. (1988). Secretion of cyclolysin, the calmodulin-sensitive adenylate cyclase-haemolysin bifunctional protein of *Bordetella pertussis*. *EMBO J.* **7**, 3997–4004.

Goebel, W., Chakraborty, T. and Kreft, J. (1988). Bacterial hemolysins as virulence factors. *Ant. van. Leeuw.* **54**, 453–63.

Gottlieb, M.H. (1977). The activity of human erythrocyte membrane cholesterol with a cholesterol oxidase. *Biochim. Biophys. Acta.* **466**, 422–8.

Gray, G.S. and Kehoe, M. (1984). Primary sequence of the δ-toxin gene from *Staphylococcus aureus* Wood 46. *Infect. Immun.* **46**, 615–8.

Guyonnet, F. and Plommet, M. (1970). Hemolysine gamma de *Staphylococcus aureus*: purification et proprietes. *Ann. Inst. Pasteur* **118**, 19–33.

Haberman, E. (1972). Bee and wasp venoms. *Science* **117**, 314–22.

Hall, J.E., Vodyanoy, I., Balasuvramanian, T.M. and Marshall, G. (1984). Alamethicin. A rich model for channel behavior. *Biophys. J.* **45**, 233–47.

Harshman, A. (ed.) (1988). *Microbial Toxins: Tools in Enzymology. Methods in Enzymology*, Vol. 165. Academic Press, San Diego.

Harshman, S. (1979). Action of staphylococcal α-toxin on membranes: some recent advances. *Molec. Cell. Biochem.* **23**, 143–52.

Hessinger, D.A. and Lenhoff, H.M. (1976). Mechanism of hemolysis induced by nematocyst venom: roles of phospholipase A and direct lytic factor. *Arch. Biochem. Biophys.* **173**, 603–13.

Hider, R.C., Khader, F. and Tatham, A.S. (1983). Lytic activity of monomeric and oligomeric melittin. *Biochim. Biophys. Acta.* **728**, 206–14.

Higashijima, T., Wakamatsu, K., Takemitsu, M., Fujino, M., Nakajima, T. and Miyazawa, T. (1983). Conformational change of mastoparan from wasp venom on binding with phospholipid membrane. *FEBS Lett.* **152**, 227–30.

Higashijima, T., Uzu, S., Nakajima, T. and Ross, E.M. (1988). Mastoparan, a peptide toxin from wasp venom, mimics receptors by activating GTP-binding regulatory proteins (G proteins). *J. Biol. Chem.* **263**, 6491–4.

Hirai, Y., Yashuhara, T., Yoshida, H., Nakajima, T., Fujino, M. and Kitada, C. (1979). A new mast cell degranulating peptide "mastoparan" in the venom of *Vespula lewsii*. *Chem. Pharmac. Bull.* **27**, 1942–4.

Hotchkiss, R.D. (1944). Gramicidin, tyrocidine and thyrothricin. *Adv. Enzymol.* **4**, 153–99.

Hugo, F., Reichwein, J., Arvand, M., Kramer, S. and Bhakdi, S. (1986). Mode of transmembrane pore formation by streptolysin-O analysed with a monoclonal antibody. *Infect. Immun.* **54**, 641–5.

Hugo, F., Arvand, M., Reichwein, J., Mackman, N., Holland, I.B. and Bhakdi, S. (1987). Identification of hemolysin produced by clinical isolates of *E. coli* with monoclonal antibodies. *J. Clin. Microbiol.* **25**, 26–30.

Ikigai, H. and Nakae, T. (1987a). Interaction of the α-toxin of *Staphylococcus aureus* with the liposome membrane. *J. Biol. Chem.* **262**, 2150–5.

Ikigai, H. and Nakae, T. (1987b). Assembly of the α-toxin hexamer of *Staphylococcus aureus* with the liposome membrane. *J. Biol. Chem.* **262**, 2156–60.

Kem, W.R. and Dunn, B.M. (1988). Separation and characterization of four different amino acid sequence variants of sea anemone (*Stichodactyla helianthus*) protein cytolysin. *Toxicon* **26**, 997–1008.

Kempf, C., Klausner, R.D., Weinstein, J.N., Renswoude, J.V., Pincus, M. and Blumental, R. (1982). Voltage-dependent *trans*-bilayer orientation of melittin. *J. Biol. Chem.* **257**, 2469–76.

Knoppel, E., Eisenberg, D. and Wickner, W. (1979). Interactions of melittin, a preprotein model, with detergents. *Biochemistry* **18**, 4177–81.

Koronakis, V., Cross, M., Senor, B., Koronakis, E. and Hughes, C. (1987). The secreted hemolysins of *Proteus mirabilis*, *Proteus vulgaris* and *Morganella morganii* are genetically related to each other and to the alpha-hemolysin of *E. coli*. *J. Bacteriol.* **169**, 1509–15.

Laine, R.O., Morgan, B.P. and Esser, A.F. (1988). Comparison between complement and melittin hemolysis: anti-melittin antibodies inhibit complement lysis. *Biochemistry* **27**, 5308–14.

Langs, D.A. (1988). Three-dimensional structure at 0.86A of the uncomplexed form of the transmembrane ion channel peptide Gramicidin A. *Science* **241**, 188–91.

Latorre, R. and Alvarez, O. (1981). Voltage-dependent channels in planar bilayer membranes. *Physiol. Rev.* **61**, 77–150.

Lavialle, F., Adams, R.G. and Levin, I.W. (1982). Infrared spectroscopic study of the secondary structure of melittin in water, 2-chloroethanol and phospholipid bilayer dispersions. *Biochemistry* **21**, 2305–12.

Lee, K.H., Fitton, J.E. and Wuthrich, K. (1987). Nuclear magnetic resonance investigation of the conformation of δ-hemolysin bound to dodecylphosphocholine micelles. *Biochim. Biophys. Acta* **911**, 144–53.

Levin, I.W. (1984). Vibrational studies of model membrane-melittin interactions. In: *Handbook of Natural Toxins* Vol. 2 (ed. A.T. Tu), pp. 87–108. Marcel Dekker, New York.

Linder, R. (1984). Alteration of mammalian membranes by the cooperative and antagonistic actions of bacterial proteins. *Biochim. Biophys. Acta* **779**, 423–35.

Linder, R. and Bernheimer, A.W. (1978). Effect on sphingomyelin-containing liposomes of phospholipase D from *Corynebacterium ovis* and cytolysin from *Stoichactis helianthus*. *Biochim. Biophys. Acta* **530**, 236–46.

Linder, R. and Bernheimer, A.W. (1982). Enzymatic oxidation of membrane cholesterol in relation to lysis of sheep erythrocytes by corynebacterial enzymes. *Arch. Biochem. Biophys.* **213**, 395–404.

Linder, R. and Bernheimer, A.W. (1984). Action of bacterial cytotoxins on normal mammalian cells and cells with altered membrane lipid composition. *Toxicon* **22**, 641–51.

Linder, R. and Salton, M.R.J. (1986). Production of hemolytic toxins by *Staphylococcus aureus* in relation to susceptibility to B-lactam antibiotics. Abstract B-47 Ann. Mtg, Amer. Soc. Microbiol.

Liu, J. and Blumenthal, K.M. (1988). Functional interaction between *Cerebratulus lacteus* cytolysin A-III and phospholipase A₂. *J. Biol. Chem.* **263**, 6619–24.

Ludwig, A., Vogel, M. and Goebel, W. (1987). Mutations affecting activity and transport of haemolysin in *Escherichia coli*. *Molec. Gen. Genet.* **206**, 238–45.

Mellor, I.R., Thomas, D.H. and Sansom, M.S.P. (1988). Properties of ion channels formed by *Staphylococcus aureus* δ-toxin. *Biochim. Biophys. Acta* **942**, 280–94.

Menestrina, G. (1988). *E. coli* hemolysin permeabilizes small unilameller vesicles loaded with calcein by a single-hit mechanism. *FEBS Lett.* **232**, 217–20.

Menestrina, G., Voges, K.P., Jung, G. and Boheim, G. (1986). Voltage-dependent channel formation by rods of helical polypeptides. *J. Membr. Biol.* **93**, 111–32.

Menestrina, G., Mackman, N., Holland, I.B. and Bhakdi, S. (1987). *Escherichia coli* haemolysin forms voltage-dependent ion channels in lipid membranes. *Biochim. Biophys. Acta.* **905**, 109–17.

Michaels, D.W. (1979). Membrane damage by a toxin from the sea anemone *Stoichactis helianthus*. I. Formation of transmembrane channels in lipid bilayers. *Biochim. Biophys. Acta* **555**, 67–78.

Mollay, C. and Kriel, G. (1974). Enhancement of bee venom phospholipase A₂ activity by melittin, direct lytic factor from cobra venom and polymixin B. *FEBS Lett.* **46**, 141–4.

Moore, N.F., Patzer, E.J., Barenholz, Y. and Wagner, R.R. (1977). Effect of phospholipase C and cholesterol oxidase on membrane integrity, microviscosity and infectivity of vesicular stomatitis virus. *Biochemistry* **16**, 4708–15.

Niedermeyer, W. (1985). Interaction of streptolysin-O with biomembranes: kinetic and morphological studies on erythrocyte membranes. *Toxicon* **23**, 425–39.

Noegel, A., Rdest, U. and Goebel, W. (1981). Determination of the function of hemolytic plasmid pHly152 of *Escherichia coli*. *J. Bacteriol.* **145**, 233–47.

Ohno-Iwashita, Y., Iwamoto, M., Mitsui, K., Ando, S. and Nagai, Y. (1988). Protease-nicked theta-toxin of *Clostridium perfringens*, a new membrane probe with no cytolytic effect, reveals two classes of cholesterol as toxin-binding sites on sheep erythrocytes. *Eur. J. Biochem.* **176**, 95–101.

Patzer, E.J., Wagner, R.R. and Barenholz, Y. (1978). Cholesterol oxidase as a probe for studying membrane organisation. *Nature* **274**, 394–5.

Projan, S.J., Kornblum, J., Kreiswirth, B., Moghazeh, S.L., Eisner, W. and Novick, R.P. (1989). Nucleotide sequence: the beta-hemolysin gene of *Staphylococcus aureus*. *Nucleic Acids Res.* **17**, 3305.

Rogolsky, M. (1979). Nonenteric toxins of *Staphylococcus aureus*. *Microbiol. Rev.* **43**, 320–60.

Sarges, R. and Witkop, B. (1965). Gramicidin A. V. The structure of valine and isoleucine-gramicidin A. *J. Am. Chem. Soc.* **87**, 2011–27.

Schroder, E., Lubke, K., Lehmann, M. and Beetz, I. (1971). Haemolytic activity and action on the surface tension of aqueous solutions of synthetic melittins and their derivatives. *Experientia* **27**, 764–5.

Sessa, G., Freer, J.H., Colacicco, G. and Weissman, G. (1969). Interaction of a lytic polypeptide, mellitin, with lipid membrane systems. *J. Biol. Chem.* **244**, 3575–82.

Shier, W.T. (1979). Activation of high levels of endogenous phopholipase A₂ in cultured cells. *Proc. Natl. Acad. Sci. USA* **76**, 195–9.

Shin, M.L., Michaels, D.W. and Mayer, M.M. (1979). Membrane damage by a toxin from the sea anemone, *Stoichactis helianthus*. *Biochim. Biophys. Acta* **555**, 79–88.

Shipolini, R.A. (1984). Biochemistry of bee venom. In:

Handbook of Natural Toxins Vol. 2 (ed. A.T. Tu), pp. 49–86. Marcel Dekker, New York.

Smyth, C.J., Mollby, R. and Wadstrom, T. (1975). Phenomenon of hot-cold hemolysis: chelator-induced lysis of sphingomyelinase-treated erythrocytes. *Infect. Immun.* **12**, 1104–11.

Strathdee, C.A. and Lo, R.Y.C. (1987). Extensive homology between the leukotoxin of *Pasteurella haemolytica* A_1 and the alpha-hemolysin of *E. coli*. *Infect. Immun.* **55**, 3233–6.

Suzuki, J., Kobayashi, S., Kagaya, K. and Fukazawa, Y. (1988). Heterogeneity of hemolytic efficiency and isoelectric point of streptolysin O. *Infect. Immun.* **56**, 2474–8.

Tamkun, M.M. and Hessinger, D.A. (1981). Isolation and partial characterization of a hemolytic and toxic protein from the nematocyst venom of the Portuguese Man-of-War, *Physalia physalis*. *Biochim. Biophys. Acta* **667**, 87–98.

Taylor, A.G. and Bernheimer, A.W. (1974). Further characterization of staphylococcal gamma-hemolysin. *Infect. Immun.* **10**, 54–9.

Taylor, A.G. and Plommet, M. (1973). Anti-gamma haemolysin as a diagnostic test in staphylococcal osteomyelitis. *J. Clin. Pathol.* **26**, 409–12.

Terwilliger, T.C. and Eisenberg, D. (1982a). The structure of melittin I. Structure determination and partial refinement. *J. Biol. Chem.* **257**, 6010–15.

Terwilliger, T.C. and Eisenberg, D. (1982b). II. Interpretation of the structure. *J. Biol. Chem.* **257**, 6016–22.

Terwilliger, T.C., Weissman, L. and Eisenberg, D. (1982). The structure of melittin in the form I crystals and its implication for melittin's lytic and surface activities. *Biophys. J.* **37**, 353–61.

Tomita, M., Taguchi, R. and Ikezawa, H. (1982). Molecular properties and kinetic studies on sphingomyelinase of *Bacillus cereus*. *Biochim. Biophys. Acta* **704**, 90–9.

Tu, A.T. (ed.) (1977). *Venoms: Chemistry and Molecular Biology*. John Wiley & Sons, New York.

Urry, D.W. (1971). The Gramicidin A transmembrane channel: A proposed $\pi_{(L,D)}$ helix. *Proc. Natl. Acad. Sci. USA* **68**, 672–6.

Vann, J.M. and Proctor, R.A. (1988). Cytotoxic effects of ingested *Staphylococcus aureus* on bovine endothelial cells: role of *S. aureus* α-hemolysin. *Microb. Pathogen.* **4**, 443–53.

Varanda, W. and Finkelstein, A. (1980). Ion and nonelectrolyte permeability properties of channels formed in planar lipid bilayer membranes by the cytolytic toxin from sea anemone, *Stoichactis helianthus*. *J. Membr. Biol.* **55**, 203–11.

Veatch, W.R. and Stryer, L. (1977). The dimeric nature of the gramicidin A transmembrane channel: conductance and fluorescence energy transfer studies of hybrid channels. *J. Molec. Biol.* **113**, 89–102.

Veatch, W.R., Fossel, E.T. and Blout, E.R. (1974). The conformation of gramicidin A. *Biochemistry* **13**, 5249–56.

Vogel, H., Jahnig, F., Hoffman, V. and Stumpel, J. (1983). The orientation of melittin in lipid membranes. A polarized infrared spectroscopy study. *Biochim. Biophys. Acta* **733**, 201–9.

Wagner, W., Vogel, M. and Goebbel, W. (1983). Transport of hemolysin across the outer membrane of *Escherichia coli* requires two functions. *J. Bacteriol.* **154**, 200–10.

Wakamatsu, K., Higashijima, T., Fujino, M., Nakajima, T. and Miyazawa, T. (1983). Transferred NOE analysis of conformations of peptides as bound to membrane bilayer of phospholipid; mastoparan-X. *FEBS Lett.* **162**, 123–6.

Wallace, B.A. and Ravikumar, K. (1988). The gramicidin pore: crystal structure of a cesium complex. *Science* **241**, 182–7.

Welch, R.A. (1987). Identification of two different hemolysin determinants in uropathogenic *Proteus* isolates. *Infect. Immun.* **55**, 2183–90.

Wilbrandt, W. (1941). Osmotische Natur sogennanter nicht-osmotischer Haemolysin (kolloidosmotische Haemolyse). *Arch. Physiol. Pfluger* **245**, 22–52.

Yamada, A., Tsukagoshi, N., Udaka, S., Sasaki, T., Makino, S., Nakamura, S., Little, C. and Ikezawa H. (1988). Nucleotide sequence and expression in *E. coli* of the gene coding for sphingomyelinase of *Bacillus cereus*. *Eur. J. Biochem.* **175**, 213–20.

Yunes, R., Goldhammer, A.R., Garner, W.K. and Cordes, E.H. (1977). Melittin facilitation of bee venom phospholipase A_2 catalyzed hydrolysis of unsonicated liposomes. *Arch. Biochem. Biophys.* **183**, 105–12.

7

Genetic Determinants of Cytolytic Toxins from Gram-Negative Bacteria

Albrecht Ludwig and Werner Goebel

Institute for Genetics and Microbiology, University of Würzburg, Würzburg, Germany

Introduction

Haemolysins represent a group of bacterial toxins which, unlike many other toxins, are not internalized by the mammalian target cells but act as membrane active proteins forming transmembrane pores into susceptible cell membranes. The most common cell used for assaying the pore-forming activity of these proteins is the red blood cell. The osmotic lysis of the erythrocytes ('haemolysis') which follows the pore formation in hypotonic media has led to the designation 'haemolysins' for these toxins. The pore formation by haemolysins is in most cases not restricted to erythrocytes but may also occur with other haemopoietic cells (lymphocytes and leukocytes) and even with fibroblasts and epithelial cells.

The production of these extracellular toxin proteins is observed in many Gram-negative (and Gram-positive) bacteria. The frequent association of the haemolytic phenotype with pathogenic isolates has led to the assumption that haemolysins may represent virulence factors. The biochemical properties of these extracellular proteins and their mode of action have been extensively studied over the past decades. The introduction of modern genetic techniques into this research field has significantly increased the knowledge on these bacterial toxins and has provided convincing evidence for the involvement of haemolysins in the pathogenesis of infections. The actual contribution of haemolysins to the pathogenesis of infection remains, however, still poorly defined in most cases.

There is growing evidence from the molecular analyses of the haemolysin genes that some of the determinants for these cytolytic toxins have a common origin. These genes have probably been spread among different bacteria by horizontal transfer. Other haemolysin determinants from Gram-negative bacteria are apparently unique. This suggests that the evolution of polypeptides able to form transmembrane pores in a eukaryotic cell membrane has not been a single genetic event. In fact the present data rather indicate that different amino acid sequences may form structures that are active in pore formation. Since 3-dimensional structural data are lacking for haemolysins one cannot predict whether a specific structural motive is common to the pore-forming domains of haemolysins.

This chapter will describe the recent advances in the genetic analysis of haemolysin determinants from various Gram-negative bacteria. The conclusions which can be drawn from these results concerning the biochemical and functional properties of haemolysins will also be discussed.

Haemolysin (HlyA) of *Escherichia coli* and HlyA-related bacterial toxins ('Repeat toxins', RTX)

The genetics and biochemistry of *E. coli* haemolysin have been extensively studied (for recent reviews see also Cavalieri *et al.*, 1984; Hacker and Hughes,

1985; Mackman *et al.*, 1986; Welch, 1991). *E. coli* haemolysin is now considered to be the prototype of a family of structurally and functionally related Ca^{2+}-dependent bacterial toxins. All of these toxins possess a repeat region which consists of several highly conserved basic repeat units, containing nine amino acids. The number of the basic repeat units varies between six (leukotoxin of *Pasteurella haemolytica*) and 47 (adenylate cyclase-haemolysin of *Bordetella pertussis*). Based on this conserved structural feature the designation 'RTX-cytotoxins' (for 'repeats in the structural *toxin*') has been proposed for these proteins. The RTX-cytotoxins are transported across the double membrane of the Gram-negative bacterial cell by a unique and highly conserved transport machinery in a signal peptide-independent fashion.

The members of the RTX family include haemolysins from *Escherichia coli*, *Proteus vulgaris*, *Morganella morganii*, *Actinobacillus pleuropneumoniae* and *Actinobacillus suis*, the leukotoxins from *Pasteurella haemolytica* and *Actinobacillus actinomycetemcomitans* and the bifunctional adenylate cyclase-haemolysin from *Bordetella pertussis*. Interestingly, other extracellular bacterial proteins like the metalloproteases B and C from *Erwinia chrysanthemi* or the protease SM from *Serratia marcescens* are partially related to the pore-forming RTX-cytotoxins and the transport system allowing the secretion of these proteins is homologous to that of the repeat toxins.

Haemolysin of *Escherichia coli*

E. coli *haemolysin as a virulence factor*

Escherichia coli strains producing an extracellular haemolysin which forms clear lysis zones on blood agar were first reported in 1903 by Kayser. Subsequently it has been shown that the haemolytic phenotype is associated predominantly with *E. coli* isolates from extra-intestinal infections, especially urinary tract infections (UTI), peritonitis, meningitis and septicaemia. The haemolytic phenotype occurs less frequently in *E. coli* strains isolated from patients with acute enteritis (enterotoxigenic, enteropathogenic and enteroinvasive *E. coli* strains) or in those from the normal faecal flora (Dudgeon *et al.*, 1921; Minshew *et al.*, 1978; DeBoy *et al.*, 1980; Green and Thomas, 1981; Hughes *et al.*, 1982a,b, 1983; Hacker *et al.*, 1983b). This obser-

vation has led to the conclusion that haemolysin is a virulence factor in the pathogenesis of these extra-intestinal infections.

It has been shown that supernatant from haemolytic *E. coli* grown in broth is lethal to mice when injected intravenously (Smith, 1963; Smith and Huggins, 1985). In addition, the virulence of haemolytic wild-type strains is considerably higher than that of non-haemolytic mutants in several animal model systems, further supporting the assumption of haemolysin being a virulence factor (Fried *et al.*, 1971; Minshew *et al.*, 1978; Van den Bosch *et al.*, 1981, 1982a,b; Hull *et al.*, 1982; Waalwijk *et al.*, 1982; Smith and Huggins, 1985). Evidence for the involvement of haemolysin in the pathogenesis of extra-intestinal *E. coli* infections has been demonstrated by infecting mice and rats with isogenic *E. coli* strains with and without cloned haemolysin genes (Welch *et al.*, 1981; Hacker *et al.*, 1983a; Welch and Falkow, 1984; Marre *et al.*, 1986). These studies have also shown that cloned plasmid- and chromosome-inherited haemolysin genes (see below) contribute to the virulence of *E. coli* strains to a different extent. These differences are due to the level of expression of the *hly* genes (Welch and Falkow, 1984) and to often subtle differences in the amino acid sequence of the haemolysin proteins (Hacker *et al.*, 1983a; Ludwig *et al.*, 1987; Hacker *et al.*, submitted).

The precise function of *E. coli* haemolysin in the pathogenesis of a urinary tract infection is, however, still unclear. It has been shown not only that haemolysin lyses erythrocytes but also that it is cytotoxic and cytolytic for human peripheral leukocytes (Cavalieri and Snyder, 1982a,b), especially monocytes and granulocytes (Gadeberg *et al.*, 1983; Gadeberg and Orskov, 1984), for epithelial cells of the urinary tract and parenchymal kidney cells (De Pauw *et al.*, 1971; Fry *et al.*, 1975) and for mouse fibroblasts (Cavalieri and Snyder, 1982c). The application of isogenic *E. coli* strains with and without cloned haemolysin genes has further shown that haemolysin causes the release of histamine from rat mast cells and the induction and release of specific leukotrienes from polymorphic nuclear granulocytes (Scheffer *et al.*, 1985). Haemolysin may cause tissue lesions and may facilitate thereby the penetration of the bacteria into deeper tissue layers in pyelonephritis, the most serious form of an UTI. In addition, haemolysin may be responsible for the symptoms of inflammation commonly observed in an acute pyelonephritis. It has been further proposed that the lysis of erythrocytes may provide iron to the infecting bacteria (Linggood and Ingram, 1982; Waalwijk et al., 1983; Grünig and Lebek, 1988). Haemolysin synthesis has been shown to be regulated by iron in some haemolytic E. coli strains (Lebek and Grünig, 1985; Grünig et al., 1987; Grünig and Lebek, 1988).

The genes determining haemolysin production

Smith and Halls (1967) first reported that the genes encoding haemolysin production in E. coli may be extrachromosomal. In fact, plasmids containing haemolysin genes (Hly plasmids) have been isolated (Goebel and Schrempf, 1971; Goebel et al., 1974; Le Minor and Le Coueffic, 1975; Monti-Bragadin et al., 1975; Royer-Pokora and Goebel, 1976; De la Cruz et al., 1979). Most of the Hly plasmids are self-transmissible and vary in size between 50 kb and 160 kb. These plasmids belong predominantly to the incompatibility groups incI2, incFIII, incFIV and incFVI (Monti-Bragadin et al., 1975; De la Cruz et al., 1979, 1980). Haemolysin plasmids are readily isolated from haemolytic E. coli strains of animals. In contrast, most haemolytic E. coli isolates from UTI or other extra-intestinal infections in humans and also from the normal human faecal flora carry the haemolysin genes on the chromosome (Welch et al., 1981, 1983; Berger et al., 1982; Hull et al., 1982; De la Cruz et al., 1983; Müller et al., 1983; Hacker and Hughes, 1985) and haemolysin plasmids occur only rarely in these E. coli isolates (Höhne, 1973; Tschäpe and Rische, 1974; De la Cruz et al., 1980; Waalwijk et al., 1982; Hacker and Hughes, 1985; Grünig et al., 1987; Grünig and Lebek, 1988).

The genetic determinant for haemolysin synthesis and export has been first cloned from a plasmid (Goebel and Hedgpeth, 1982). The construction of mutant derivatives of this clone and complementation analysis has led to the conclusion that the haemolysin determinant of E. coli consists of four structural genes arranged in the order hlyC, hlyA, hlyB and hlyD (Noegel et al., 1979, 1981; Goebel and Hedgpeth, 1982; Wagner et al., 1983). These studies have further shown that the genes hlyC and hlyA are necessary for the synthesis of active haemolysin and that hlyA is the structural gene encoding the haemolysin protein (HlyA). The genes hlyB and hlyD encode functions required for the transport of haemolysin across the two membranes of E. coli (Noegel et al., 1979, 1981; Wagner et al., 1983). Hybridization of other haemolytic plasmids and chromosomal DNAs from uropathogenic E. coli strains with the cloned haemolysin (hly) genes has revealed extensive homology between all four structural genes of these hly determinants (De la Cruz et al., 1980, 1983; Müller et al., 1983). This similarity in the four structural hly genes has been further confirmed by the cloning and mapping of a variety of plasmid- and chromosome-inherited hly determinants (Berger et al., 1982; Stark and Shuster, 1982, 1983; Welch et al., 1983; Mackman and Holland, 1984b; Waalwijk et al., 1984; Mackman et al., 1985a).

The complete nucleotide sequence of a plasmid and a chromosomal hly determinant (Felmlee et al., 1985b; Hess et al., 1986) have confirmed the earlier genetic data for the existence of the four structural hly genes and the extensive sequence homology between different E. coli hly determinants within the structural genes. The G+C content of the hly genes has been shown to be unusually low (about 40%) for E. coli genes (Felmlee et al., 1985b; Hess et al., 1986) arguing for the possibility that the hly genes have been transferred from another organism into E. coli.

From the sequence data, hlyC consists of 510 bp, hlyA of 3072 bp, hlyB of 2121 bp and hlyD of 1434 bp (Hess et al., 1986). These genes could encode proteins of 20 kDa (HlyC), 110 kDa (HlyA), 80 kDa (HlyB) and 54.6 kDa (HlyD). Whereas the sequence homology is high (between 95% (hlyC) and 98.5% (hlyB)) within the structural hly genes between plasmid and chromosomal hly determinants, little homology is found in the 5′- and 3′-noncoding regions flanking the structural genes. Even these flanking regions are remarkably conserved in many haemolysin plasmids and include an IS2 element in the 5′-noncoding region (Knapp et al., 1985) and IS91-like elements at both ends of the hly determinant (Zabala et al., 1982, 1984). The discovery of these mobile genetic elements at both ends of many plasmid hly determinants has led to the assumption that the hly determinant itself may be a mobile genetic structure (Zabala et al., 1984; Knapp et al.,

1985). In some uropathogenic *E. coli* isolates the chromosomal *hly* genes are localized to large specific inserts in the chromosome (Knapp *et al.*, 1984, 1986). More than one such *hly* insert may occur within the chromosome (Knapp *et al.*, 1986). These inserts may carry at both termini short, directly repeated sequences which by homologous recombination cause the deletion of these *hly* inserts at a high rate under normal culture conditions. The loss of the *hly* insert(s) may not only eliminate the *hly* genes but may also inhibit the expression of other virulence factors such as adhesins and serum resistance in these uropathogenic *E. coli* strains (Knapp *et al.*, 1986).

Expression of the haemolysin genes

Insertions of Tn3, Tn5 and Mud*lacZ* into the plasmid-encoded *hly* determinants have demonstrated that *hlyC*, *A*, and *B* are organized in an operon which is transcribed from a promoter upstream of *hlyC* (Noegel *et al.*, 1979, 1981; Wagner *et al.*, 1983, Juarez *et al.*, 1984; Mackman *et al.*, 1985a,b). The gene *hlyD* appears to be transcribed from an independent promoter in the same direction as the *hlyC,A,B* operon. Three functional transcriptional start sites located 69 bp, 166 bp and 256 bp upstream of the ATG start codon of *hlyC* have been mapped in the cloned plasmid-encoded *hly* determinant (Koronakis and Hughes, 1988), but only two such sites at positions -264 bp and -266 bp have been found by another group (Welch and Pellett, 1988). Most transcripts stop at a *rho*-independent termination signal downstream of *hlyA* (Koronakis *et al.*, 1988a; Welch and Pellett, 1988).

A significant increase (50–100 fold) in the amount of extracellular haemolysin (HlyA) is observed with clones carrying a larger region (about 4.5 kb) of the original Hly plasmid (pHly152) upstream of the *hlyC* gene (Gonzalez-Carrero *et al.*, 1985; Ludwig *et al.*, 1987). This region includes a sequence, *hlyR*, of about 650 bp, which functions as a long-distance activator causing antitermination of the transcription at the *rho*-independent termination signal downstream of *hlyA* (Vogel *et al.*, 1988; Koronakis *et al.*, 1989a). This allows the efficient transcription of *hlyB* to occur. *HlyR* acts in cis but only in one orientation. The *hlyR* sequence is separated from the first structural *hly* gene (*hlyC*) by 1.8 kb (Vogel *et al.*, 1988). This

spacer region includes a complete IS2 sequence identified at this position in many Hly plasmids (see above) and the promoters for the transcription of the *hlyC,A,B* operon. Based on the nucleotide sequence *hlyR* may contain binding site(s) for one or more host regulator proteins (Vogel *et al.*, 1988). Transcription of *hlyD* in this plasmid-encoded *hly* determinant proceeds in the presence or absence of *hlyR* from a single promoter which is located in *hlyB* (M. Vogel, pers. commun.).

The regulation of transcription of chromosomal *hly* genes seems to be different from that of the above described plasmid-encoded *hly* genes. The sequences in the non-coding region 5′-upstream of *hlyC* between these two *hly* determinants are different and may contain different regulatory sites. Transcription of the *hly* genes of a chromosomal *hly* determinant has been shown to start 462 bp and 464 bp upstream of *hlyC* (Welch and Pellett, 1988). Transcription appears to terminate downstream of *hlyA* and after *hlyD* generating two transcripts of 4.0 kb (*hlyC,A* transcript) and 8.0 kb (*hlyC,A,B,D* transcript). An activator sequence similar to *hlyR* has not been detected in chromosomal *hly* determinants.

However, a region of 1.1 kb flanking a chromosomal *hly* determinant seems to activate, both in cis and in trans, the specific HlyB/HlyD-dependent secretion of haemolysin. This sequence is highly conserved among chromosomal *hly* determinants from *E. coli* strains of different serotypes (Cross *et al.*, 1990).

Figure 2 summarizes the data on the transcription of plasmid and chromosomal *hly* determinants.

The haemolysin gene products

Previous biochemical studies on the molecular nature of *E. coli* haemolysin have suggested that this haemolysin is an extracellular labile protein which may be complexed with lipids (for a review see Cavalieri *et al.*, 1984). This view of complex formation has been confirmed by demonstrating that the haemolytic activity is sensitive to phospholipase C and ultrasonication (Wagner *et al.*, 1988). Both conditions destroy haemolytic activity but do not alter the size of the denatured HlyA protein.

As indicated above, the synthesis of active haemolysin requires the functions of the genes *hlyC* and *hlyA* (Goebel and Hedgpeth, 1982; Welch *et al.*, 1983). The gene product of *hlyA* has been identified as a 107 kDa protein (Goebel and Hedgpeth, 1982; Welch *et al.*, 1983) and the haemolytic activity is associated with an extracellular protein of similar size (Mackman and Holland, 1984a,b; Felmlee *et al.*, 1985a,b; Gonzalez-Carrero *et al.*, 1985; Nicaud *et al.*, 1985b; Wagner *et al.*, 1988). Since HlyA by itself is haemolytically inactive one has to assume that HlyA must be activated in the cytoplasm by HlyC prior to export (Goebel and Hedgpeth, 1982; Welch *et al.*, 1983; Nicaud *et al.*, 1985a). HlyC has been identified as a cytoplasmic protein of 18–20 kDa (Noegel *et al.*, 1979; Härtlein *et al.*, 1983; Felmlee *et al.*, 1985b; Nicaud *et al.*, 1985a). The biochemical function of HlyC is poorly understood. Recent data indicate, however, that the HlyC-mediated activation of HlyA consists of a covalent modification which is accompanied by the loss of a negative charge in the haemolysin protein (Garcia *et al.*, in press). In addition, results obtained by Issartel *et al.* (1991) suggest that a fatty acyl group is transferred from acyl carrier protein to HlyA during the modification reaction. The site of the HlyC-mediated modification was narrowed down to a region located immediately proximal to the repeat domain of HlyA (Pellett *et al.*, 1990; Garcia *et al.*, in press). Interestingly, this region contains a putative α-helix–loop–α-helix structure. The modification of HlyA by HlyC is necessary to increase the general binding affinity of haemolysin to the target membrane (Garcia *et al.*, in press).

The activation of HlyA is not required for the secretion of haemolysin from the *E. coli* cell but export of modified (haemolytically active) and unmodified (haemolytically inactive) HlyA depends on HlyB and HlyD (Noegel *et al.*, 1979, 1981; Härtlein *et al.*, 1983; Wagner *et al.*, 1983; Welch *et al.*, 1983; Mackman *et al.*, 1985b). Both gene products have been identified as membrane proteins in mini and maxi cells of *E. coli* of 80 kDa (HlyB) and 54.6 kDa (HlyD), respectively. In addition to the 80 kDa HlyB protein, cells carrying the *hlyB* gene express a 46 kDa protein (Härtlein *et al.*, 1983; Felmlee *et al.*, 1985b; Mackman *et al.*, 1985b). It is, however, unknown whether this 'smaller HlyB protein' also has a transport function or represents merely a degradation product of HlyB.

The organization of the HlyB/HlyD transport machinery is not completely understood. Structural considerations suggest a transmembrane organization for HlyB, with at least three pairs of transmembrane segments spanning the cytoplasmic membrane, thereby generating three periplasmic loops. This membrane-integrated portion of HlyB is represented by the N-terminal half of the protein. The C-terminal half of HlyB probably faces the cytoplasm (Oropeza-Wekerle *et al.*, 1990). This cytoplasmic region of HlyB contains a sequence which resembles the conserved ATP-binding site of prokaryotic and eukaryotic protein kinases (Koronakis *et al.*, 1988b). Non-conservative amino acid exchanges in this sequence motive (Gly–X–Gly–X–X–Gly–X_{14-24}–Lys) eliminate the transport function of HlyB (Koronakis *et al.*, 1988b). In addition, this C-terminal part of HlyB carries putative nucleotide (presumably ATP)-binding sites which exhibit extensive sequence homology to that of several bacterial proteins involved in transport (Gerlach *et al.*, 1986; Higgins *et al.*, 1986). Interestingly, HlyB shows significant homology to the P-glycoprotein of mammalian cells (Gerlach *et al.*, 1986). The 170 kDa P-glycoprotein is localized to the membrane and causes, when amplified, resistance against a variety of anticarcinogenic drugs in tumour cells. Based on the predicted structure and the similarity to the other transmembrane proteins it has been postulated that HlyB may form a transmembrane channel through which HlyA is translocated. In fact, recent results indicate that HlyB is able to recognize the C-terminal secretion signal of HlyA (see below) and to initiate its translocation across the inner and outer membrane of the *E. coli* cell (Oropeza-Wekerle *et al.*, 1990). HlyD does not directly interact with HlyA but it seems to be required for pulling the N-terminal part of HlyA through the envelope and for releasing haemolysin from the cell surface (Wagner *et al.*, 1983; Oropeza-Wekerle *et al.*, 1990). A single putative transmembrane region has been found in HlyD which may anchor this protein in the cytoplasmic membrane. The isolation of a chromosomal mutation which specifically blocks the transport of HlyA in the presence of functional HlyB and HlyD (Juarez and Goebel, 1984) suggested that in addition to these transport proteins other cellular components may be necessary for haemolysin secretion. This presumption was confirmed recently. TolC, a minor *E. coli* outer membrane protein which is encoded by a gene not located in the *hly* cluster, has been found to be specifically required for the secretion of haemolysin (Wandersman and Delepelaire, 1990). This protein might permit a specific interaction between the inner and outer membrane of *E. coli*.

HlyA does not contain a conventional signal sequence for transport at the N-terminus as expected for an exported protein (Härtlein *et al.*, 1983; Felmlee *et al.*, 1985a; Hess *et al.*, 1986). No proteolytic processing of HlyA occurs during transport (Felmlee *et al.*, 1985a). A sequence of about 60 amino acids at the C-terminal end of HlyA appears to be the only element of HlyA required for the HlyB/HlyD/TolC transport machinery. This has been shown by deletions of C-terminal sequences of HlyA (Gray *et al.*, 1986; Ludwig *et al.*, 1987) and by generation of C-terminal peptides from HlyA (T. Jarchau, pers. commun.). The construction of fusion proteins consisting of non-transportable proteins and C-terminal sequences of HlyA consisting of 60–100 amino acids further supports the view that the last 60 amino acids of HlyA carry the complete information required for HlyB/HlyD/TolC-mediated transport (Mackman *et al.*, 1987; J. Hess, pers. commun.). The comparison of the C-terminal amino acid sequence of HlyA with the corresponding sequences of other HlyA-related toxins (see below) indicates the presence of common structural domains within all C-terminal sequences, although there are significant differences in the primary amino acid sequence (Koronakis *et al.*, 1989b; J. Hess, pers. commun.; Kenny and Holland, pers. commun.).

The N-terminal sequence of HlyA is not required for HlyB/HlyD/TolC-dependent transport. This sequence has putative amphiphilic α-helical properties and resembles the transit peptides of proteins imported into mitochondria and chloroplasts (Colman and Robinson, 1986; Von Heijne, 1986). When N-terminal sequences of HlyA are fused to alkaline phosphatase of *E. coli*, which lacks its own N-terminal transport signal, the fusion proteins are translocated into and in part across the cytoplasmic membrane of *E. coli*, indicating that the N-terminal region of HlyA is membrane active (Erb *et al.*, 1987). The biological function of this portion of HlyA seems to be the regulation of the stability of the haemolysin pore generated in the target membrane. The deletion of the N-terminal amphi-

philic region results in the formation of pores which have a higher conductance and a much longer lifetime than that generated by wild-type haemolysin suggesting that the pore-forming structure of the mutant haemolysin is more stably inserted into the target membrane (Ludwig *et al.*, 1991). We assume that the N-terminal, amphiphilic, α-helical region of wild-type haemolysin competes with its pore-forming structure for insertion into the target membrane, resulting in the formation of typical transient cation-selective pores with a lifetime of about 2 seconds (Benz *et al.*, 1989).

Binding of haemolysin (activated HlyA) to erythrocytes is dependent on the intact repeat region of HlyA (Ludwig *et al.*, 1988; Boehm *et al.*, 1990a,b). This domain is located in the C-terminal half of the haemolysin protein and contains twelve short repeat units each consisting of nine amino acids with the consensus sequence X–Leu–X–Gly–Gly/X–X–Gly–Asn/Asp–Asp (Felmlee and Welch, 1988; Ludwig *et al.*, 1988). As discussed below, this repeat sequence is conserved in all HlyA-related bacterial toxins. It has been suggested that the repeat region of HlyA is involved in export of haemolysin (Felmlee and Welch, 1988). Site-specific deletions of single repeat units and of combinations of several repeat units in HlyA indicate, however, that the repeat domain is required for the Ca^{2+}-dependent binding of haemolysin to erythrocytes. The pore-forming property and the HlyB/HlyD/TolC-dependent transport of the mutant HlyA proteins are not impaired (Ludwig *et al.*, 1988).

HlyA is a negatively charged, hydrophilic protein. There are, however, three pronounced hydrophobic domains in the N-terminal half of HlyA. Evidence has been provided that these three hydrophobic regions participate in the haemolysin-induced pore formation. Deletion of the first hydrophobic domain leads to a significant decrease in haemolytic activity and deletions of the second and third hydrophobic domain destroy the haemolytic and pore-forming activity entirely (Ludwig *et al.*, 1987; Ludwig *et al.*, 1991). Computer-assisted analyses predict the existence of four hydrophobic membrane-spanning α-helices of 21 amino acids each within the three hydrophobic domains, one in domain I, another one in domain II and two in domain III. Interestingly, the amino acid sequences upstream of and between these hydrophobic α-helices have amphipathic properties and may form four additional α-helices. Thus, we propose a model in which the pore-forming structure of *E. coli* haemolysin consists of four hydrophobic and four amphipathic α-helices (Ludwig *et al.*, 1991). The polar sides of the amphipathic α-helices may provide the hydrophilic coat of the water-filled haemolysin pore. The functional diameter of the *E. coli* haemolysin pore is about 1–2 nm and the conductance of the pore is 500–600 pS (Bhakdi *et al.*, 1986; Benz *et al.*, 1989).

The arrangement of functional regions in the HlyA protein of *E. coli* is summarized in Fig. 1.

The Ca²⁺-dependent haemolysins from *Proteus vulgaris* and *Morganella morganii*

The *Proteus* species are, next to *E. coli*, the most frequent aetiological agents of urinary tract infections in humans (Stamm *et al.*, 1977). *Morganella morganii* (previously called *Proteus morganii*) causes similar urinary tract infections, albeit less frequently. This pathogen, like *P. vulgaris* and *P. mirabilis*, is also frequently isolated from infected wounds and may cause septicaemia (Stamm *et al.*, 1977; Parker, 1984). Most clinical isolates belonging to these three bacterial species are haemolytic (Koronakis *et al.*, 1987). These haemolysins have therefore been considered as virulence factors in *Proteus* together with urease and specific adhesins (Koronakis *et al.*, 1987; Welch, 1987).

Two distinct extracellular haemolysins are produced by *proteus* spp., a Ca²⁺-independent haemolysin and (less frequently) a Ca²⁺-dependent haemolysin (Welch, 1987; Koronakis *et al.*, 1987; Uphoff and Welch, 1990; Swihart and Welch, 1990a). The latter type of haemolysin is also synthesized by *Morganella morganii*. Hybridization of genomic DNA from *P. vulgaris* and *M. morganii* isolates secreting the Ca²⁺-dependent haemolysin with specific *hly* gene probes of *E. coli* indicates extensive sequence homology with all four structural *hly* genes (Koronakis *et al.*, 1987; Welch, 1987). The extracellular haemolysin of these strains has a similar size as HlyA of *E. coli* (110 kDa) and cross-reacts with anti-HlyA antibodies (Welch, 1987). In addition, the cloned *hly* determinants have the same arrangement of four structural *hly* genes as the *E. coli hly* determinant (Koronakis *et al.*, 1987) (Fig. 2). The DNA sequence of the *hlyB* genes from *P. vulgaris* and *M. morganii* (Koronakis *et al.*, 1988b) indicates high sequence homology with the *hlyB* gene of *E. coli*. The typical structural features of the deduced HlyB proteins (transmembrane domains, ATP-binding site) are highly conserved (Koronakis *et al.*, 1988b). Not surprisingly, the non-coding regions 5'-upstream of *hlyC* are quite different as compared with the corresponding sequences in the *E. coli hly* determinants. The transcription of the four structural *hly* genes in *P. vulgaris* and *M. morganii* may also be differently regulated (Koronakis and Hughes, 1988).

The cytolytic toxis from *Pasteurellaceae*

Leukotoxin (LktA) from Pasteurella haemolytica

Pasteurella haemolytica biotype A, serotype 1 is the aetiological agent of bovine pneumonic pasteurellosis. In addition, this organism may cause pneumonia and septicaemia in sheep and goats. *P. haemolytica* produces an extracellular, heat-labile cytotoxin which specifically reacts with leukocytes of ruminants (Kaehler *et al.*, 1980b; Shewen and Wilkie, 1982). The cytotoxic effect of this leukotoxin has been extensively studied using bovine alveolar macrophages, peripheral blood monocytes, neutrophiles and peripheral blood

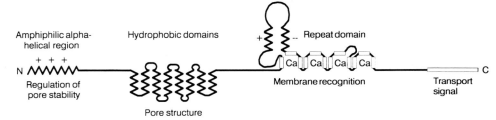

Figure 1. Location of functional regions in the haemolysin protein (HlyA) of *Escherichia coli*.

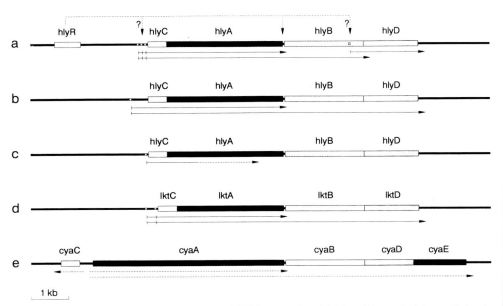

Figure 2. Comparison of the genetic determinants of RTX-cytotoxins. (a) Plasmid-encoded haemolysin determinant of *E. coli* PM152 (pHly152). (b) Chromosomally encoded haemolysin determinant of *E. coli* J96. (c) Haemolysin determinant (*hly*) of *Proteus vulgaris*. (d) Leukotoxin determinant of *Pasteurella haemolytica*. (e) Adenylate cyclase–haemolysin determinant of *Bordetella pertussis*. The transcriptional organization of the determinants is indicated by arrows. Functional promoters are represented by small boxes.

lymphocytes (Benson *et al.*, 1978; Kaehler *et al.*, 1980a,b; Markham and Wilkie, 1980; Baluyut *et al.*, 1981; Shewen and Wilkie, 1982). The toxin causes efflux of K⁺ ions from bovine lymphoma cells and extensive membrane damage (Clinkenbeard *et al.*, 1989). Protection studies using carbohydrates with increasing sizes suggest that this leukotoxin forms small transmembrane pores with a diameter of 0.9 nm upon contact with target cells (Clinkenbeard *et al.*, 1989).

The genetic determinant for leukotoxin has been recently cloned in *E. coli* (Lo *et al.*, 1985; Chang *et al.*, 1987, 1989a; Strathdee and Lo, 1989a). The clones synthesize active leukotoxin that is exported to the culture medium. The leukotoxin activity is found associated with a protein of 100–105 kDa. Nucleotide sequence analysis of the cloned DNA determining this activity indicates four open reading frames (ORFs) encoding putative proteins of 166 amino acids (19.8 kDa), 953 amino acids (101.9 kDa), 708 amino acids (79.7 kDa) and 478 amino acids (54.7 kDa), respectively (Lo *et al.*, 1987; Strathdee and Lo, 1989a). The first two genes are required for the synthesis of active leukotoxin. Sequence data have indicated a signifi-

cant homology between these two genes of *P. haemolytica* and the *hly* genes, *hlyC* and *hlyA* of *E. coli* (Lo *et al.*, 1987; Strathdee and Lo, 1987). The *Pasteurella* genes therefore have been named *lktC* and *lktA*. The LktC protein shows almost 70% homology (conserved amino acid substitutions included) with HlyC. Interestingly, LktC can be replaced by HlyC from *E. coli* in the activation of LktA without altering the activity or target cell specificity of leukotoxin (Forestier and Welch, 1990). LktC, on the other hand, is not able to activate HlyA from *E. coli*. LktA exhibits more than 60% homology (conserved amino acid substitutions included) with HlyA. The hydrophobicity profiles of the corresponding proteins are identical and the specific functional domains recognized in HlyA (see above) are conserved in LktA (Strathdee and Lo, 1987). This similarity includes the three hydrophobic regions, necessary for the generation of pores, and the repeat region involved in Ca²⁺-dependent binding to erythrocytes. This latter region consists in LktA of only six repeat units instead of the twelve present in HlyA of *E. coli*. The consensus sequence of the repeat unit is identical in both proteins (Strathdee and Lo, 1987).

Since the leukotoxin activity is Ca^{2+}-dependent like the *E. coli* haemolysin, the repeat region in LktA may likewise represent the Ca^{2+}-binding domain of this toxin. The two ORFs downstream of *lkt*A exhibit high homology to *hly*B and *hly*D of *E. coli* (Strathdee and Lo, 1989a). These genes, named *lkt*B and *lkt*D, are required for export of leukotoxin (LktA) in *P. haemolytica* and *E. coli*. Both genes are expressed in *E. coli* to yield an 80 kDa (LktB) and a 54 kDa (LktD) protein, respectively. The amino acid sequences of LktB and LktD are highly homologous to those of HlyB and HlyD. Functionally, HlyB and HlyD from *E. coli* can replace LktB and LktD in the export of leukotoxin (Chang *et al.*, 1989a), in spite of considerable differences in the amino acid sequence of the C-termini of HlyA and LktA which represent the recognition signal for the HlyB/HlyD/TolC transport system (see above). The expression of the four *lkt* genes seems to be regulated in a similar way to that of the chromosomal *hly* genes of *E. coli* (Lo *et al.*, 1987; Strathdee and Lo, 1989b) (Fig. 2). The sequence upstream of *lkt*C is, however, different from the comparable region of sequenced *hly* determinants (Strathdee and Lo, 1989a).

Cytolytic toxins from Actinobacillus species

Actinobacillus actinomycetemcomitans is a Gram-negative, facultative anaerobe which is commonly isolated from locale juvenile periodontitis (Tanner *et al.*, 1979; Slots *et al.*, 1980; Mandell and Socransky, 1981). This pathogen is also frequently associated with periodontitis in adults, but occurs rarely in the normal oral flora (Slots *et al.*, 1980; Zambon *et al.*, 1983). The occasional occurrence of *A. actinomycetemcomitans* in other infectious diseases has also been reported (Heinrich and Pulverer, 1959; Vandepitte *et al.*, 1977).

A. *actinomycetemcomitans* produces a variety of putative virulence factors including a leukotoxin with cytotoxic activity against human polymorphic nuclear leukocytes and monocytes. This leukotoxin is inactive on human lymphocytes, erythrocytes, thrombocytes and fibroblasts and also inactive on leukocytes from rodents (Tsai *et al.*, 1979, 1984; Taichman *et al.*, 1980; Taichman and Wilton, 1981; Zambon *et al.*, 1983; Slots and Genco, 1984). The leukotoxin is heat-labile and has a relative molecular mass of 115 kDa (Tsai *et al.*, 1984). It

is able to generate pores with a diameter of about 0.96 nm in target cell membranes (Iwase *et al.*, 1990). The leukotoxin determinant has been isolated from a gene bank by its homology with the *lkt*A gene of *Pasteurella haemolytica* (Kolodrubetz *et al.*, 1989; Lally *et al.*, 1989). DNA sequence analysis of the cloned fragment confirmed that the leukotoxin of *A. actinomycetemcomitans* is another member of the RTX family (Lally *et al.*, 1989; Kraig *et al.*, 1990). Two genes, termed *lkt*C and *lkt*A, were identified in the 5'-terminal half of the leukotoxin determinant. These genes encode proteins of 168 and 1055 amino acids, respectively, which are homologous to the C and A proteins encoded by the RTX determinants from other Gram-negative bacteria.

Interestingly, LktA (= AaLtA) from *Actinobacillus actinomycetemcomitans* is more related to HlyA from *E. coli* (51% identity) than to LktA from *P. haemolytica* (43% identity). The same is true for LktC (Kraig *et al.*, 1990). Partial DNA sequence analysis of the region downstream of *lkt*A suggests the existence of two genes (*lkt*B and *lkt*D) in the *A. actinomycetemcomitans lkt* determinant which are homologous to *hly*B and *hly*D from *E. coli* (Kraig *et al.*, 1990). However, in contrast to the other RTX-cytotoxins the leukotoxin of *A. actinomycetemcomitans* seems to remain cell-associated (membrane-bound) and not to be secreted into the culture medium (Tsai *et al.*, 1984; Lally *et al.*, 1989). The expression of the *lkt* genes of *A. actinomycetemcomitans* is associated with the generation of 4.3 kb *lkt*C,A transcripts and 9.3 kb *lkt*C,A,B,D transcripts (Spitznagel *et al.*, 1991).

Actinobacillus pleuropneumoniae, the aetiological agent of porcine pleuropneumonia (Shope, 1964) produces several different extracellular haemolysins. The best characterized toxin is a 105 kDa protein originally identified in serotype 1 strains, which has both haemolytic and cytotoxic activities (Frey and Nicolet, 1988a,b; Chang *et al.*, 1989b). This haemolysin (HlyIA, HppA) seems to be produced by all serotypes of *A. pleuropneumoniae* (Devenish *et al.*, 1989). HppA has been purified (Frey and Nicolet, 1988a) and the gene encoding this protein (*hpp*A) has been cloned and sequenced (Chang *et al.*, 1989b). These analyses demonstrated that the 105 kDa HppA protein of *A. pleuropneumoniae* is related to the RTX-toxins. The highest homology exists between HppA and LktA from

P. haemolytica (>65% identity) (Chang *et al.*, 1989b). When HppA is expressed in *E. coli*, it can be activated by HlyC and secreted by the *E. coli* haemolysin transport system (HlyB/HlyD/TolC) (Gygi *et al.*, 1990). The 105 kDa haemolysin from *A. pleuropneumoniae* is able to generate discrete pores in phospholipid membranes which have a conductance of 350–400 pS (Lalonde *et al.*, 1989).

Recent studies indicate that an extracellular haemolysin belonging to the RTX family is also produced by *Actinobacillus suis* (Devenish *et al.*, 1989).

The adenylate cyclase–haemolysin of *Bordetella pertussis*

Bordetella pertussis, the causative pathogen of whooping cough, produces several well-characterized extracellular proteins that have been shown or considered to be virulence factors (Weiss and Hewlett, 1986). Among those is an extracellular adenylate cyclase which is internalized by human macrophages, neutrophils, lymphocytes and other cell types. It is activated intracellularly by calmodulin and causes an increase of the intracellular level of cAMP (Hewlett *et al.*, 1976; Wolff *et al.*, 1980; Confer and Eaton, 1982; Shattuck and Storm, 1985; Friedman *et al.*, 1987; Hanski, 1989). Mutants of *B. pertussis* which lack adenylate cyclase are avirulent (Weiss *et al.*, 1984). This toxin is closely linked to a haemolysin as first shown by transposon mutagenesis (Weiss *et al.*, 1983). Most non-haemolytic mutants also lack adenylate cyclase activity or produce significantly reduced amounts of this enzyme (Weiss *et al.*, 1983).

The cloning and sequencing of the adenylate cyclase–haemolysin locus showed that both activities are encoded by a single gene (*cyaA*) (Glaser *et al.*, 1988a,b). The gene product is a large protein of 177.3 kDa (1706 amino acids) which has been named 'cyclolysin' (Glaser *et al.*, 1988b). The adenylate cyclase activity resides within the N-terminal 450 amino acids of CyaA (Glaser *et al.*, 1988a). Results obtained by Masure and Storm (1989) suggested that this part of CyaA is cleaved from the polypeptide posttranslationally at the cell surface of *Bordetella pertussis* and released into the culture medium. In agreement to that the relative molecular mass of extracellular adenylate cyclase was reported to be 45 kDa (Ladant *et al.*, 1986;

Masure and Storm, 1989). Recent investigations demonstrated, however, that CyaA is not processed and that the toxic form of adenylate cyclase which is able to penetrate eukaryotic cells corresponds to the full-length protein (Rogel *et al.*, 1989; Hewlett *et al.*, 1989; Bellalou *et al.*, 1990a, Leusch *et al.*, 1990). The N-terminal 45 kDa fragment of CyaA probably originates from proteolytic degradation of CyaA. It has adenylate cyclase activity but it is neither haemolytic nor invasive (Bellalou *et al.*, 1990b). The C-terminal 1250 amino acids of the adenylate cyclase holotoxin (CyaA) exhibit striking sequence similarity to HlyA of *E. coli* and LktA of *P. haemolytica* which led to the conclusion that this part of CyaA mediates haemolysin activity (Glaser *et al.*, 1988b). The haemolytic moiety of CyaA seems to be necessary for the Ca^{2+}-dependent penetration of the host cell plasma membrane (E. Hanski, personal communication). The homology between the putative haemolysin region of CyaA and HlyA and LktA, respectively, is pronounced in the hydrophobic domains and the repeat domain (see above), whereas less homology is found in the adjacent sequences of CyaA. The repeat domain of CyaA consists of 47 repeat units, each having the same consensus amino acid sequence X–Leu–X–Gly–Gly/X–X–Gly–Asn/Asp–Asp as previously identified in HlyA and LktA. This extended repeat domain of HlyA makes up more than half of the putative haemolysin moiety of CyaA (Glaser *et al.*, 1988a,b). The functional significance of the conserved hydrophobic region in CyaA for the haemolytic activity has been shown by a site-specific deletion covering this part of CyaA (Glaser *et al.*, 1988b).

The adenylate cyclase toxin activity as well as the haemolytic activity of CyaA require a second protein, CyaC, for activation. The gene *cyaC* encoding this protein was recently identified upstream of *cyaA* and shown to be homologous to *hlyC* from *E. coli* (Barry *et al.*, 1991). In contrast to *hlyC*, however, *cyaC* is oriented oppositely from the adenylate cyclase toxin structural gene. In addition, it is separated from *cyaA* by a larger spacer region (Barry *et al.*, 1991).

For the export of CyaA a DNA region of 5.5 kb located 3′-downstream of *cyaA* has been identified (Glaser *et al.*, 1988b). This chromosomal segment consists of three genes, *cyaB*, *cyaD* and *cyaE* (Fig. 2). The CyaB protein (712 amino acids) shares more than 50% sequence homology with HlyB of *E. coli* and CyaD (440 amino acids) exhibits striking structural similarities with HlyD of *E. coli*, although the overall sequence homology between CyaD and HlyD is relatively low (32%). CyaE (474 amino acids) is also essential for the export of CyaA. This protein might have a function analogous to that of TolC in the secretion of *E. coli* haemolysin although there is only very little homology between these two proteins (Glaser *et al.*, 1988b; Wandersman and Delepelaire, 1990). In contrast to HlyB/CyaB and HlyD/CyaD, TolC and CyaE are synthesized with typical N-terminal signal sequences.

It is interesting to note that the C-terminal sequence, which functions as transport signal in HlyA, seems to be absent in CyaA. Nevertheless, when CyaA is expressed in *E. coli*, it can be secreted from the cell by the *E. coli* haemolysin transport machinery and the signal allowing this export has been demonstrated to reside at the C-terminal end of CyaA (Masure *et al.*, 1990).

Other proteins related to *E. coli* haemolysin

Recent investigations demonstrated the existence of several extracellular proteins in Gram-negative bacteria, which, although they do not act as pore-forming toxins, are partially related to *E. coli* haemolysin and secreted by a similar mechanism as HlyA.

The secretion of the extracellular metalloproteases B (53 kDa) and C (55 kDa) of *Erwinia chrysanthemi*, for example, depends on three closely linked genes, *prt*D, *prt*E and *prt*F (Letoffe *et al.*, 1990). These genes share significant homology with the *hly*B, *hly*D and *tol*C genes required for haemolysin secretion in *E. coli*. Interestingly, PrtF is not only related to TolC, but it has homologies with both TolC and CyaE (about 20% identity in each case) (Letoffe *et al.*, 1990). The secretion functions necessary for the export of the 50 kDa metalloprotease SM from *Serratia marcescens* must be analogous to the PrtD, PrtE and PrtF proteins. When the *prt*SM gene encoding this protease was cloned and expressed in *E. coli*, the secretion of protease SM from the *E. coli* cell could be accomplished by the *Erwinia chrysanthemi* metalloprotease B secretion apparatus (Letoffe *et al.*,

1991). The proteases B and SM share 60% identity and as in the case of protease C and *E. coli* haemolysin the secretion signal is located at the C-terminal end of the proteins (Delepelaire and Wandersman, 1990; Letoffe *et al.*, 1991). In addition, the homologous metalloproteases B and C from *Erwinia chrysanthemi* and the protease SM from *Serratia marcescens* share partial amino acid homology with HlyA from *E. coli* in the region containing a tandem array of a nine amino acid repeat with the consensus sequence X–L–X–G–G/X–X–G–N/D–D (Nakahama *et al.*, 1986; Delepelaire and Wandersman, 1989, 1990). A secretion system similar to that of *E. coli* haemolysin also seems to be involved in the export of other bacterial proteins. These proteins include, for example, the *Pseudomonas aeruginosa* alkaline protease (Guzzo *et al.*, 1991) and the *Rhizobium leguminosarum* NodO protein (Economou *et al.*, 1990).

Haemolysins unrelated to *E. coli* haemolysin

Haemolysins are described in this section which are apparently unrelated to the *E. coli* haemolysin. The genetic determinants for these haemolysins fail to exhibit sequence homology to the *hly* genes of *E. coli*. Furthermore, the haemolysin proteins show no immunological cross-reactivity with the *E. coli* haemolysin. There are, however, some striking functional similarities between the *E. coli* haemolysin system and some of the haemolysins described here. For example, they may require additional genes for activation of a non-haemolytic precursor protein and/or additional transport functions.

The Ca^{2+}-independent haemolysins from *Serratia marcescens* and *Proteus* spp.

Serratia marcescens may cause urinary tract infections, bacteraemia, keratitis, arthritis and meningitis in immunocompromised people or patients suffering from other diseases like diabetes mellitus, heart or kidney damage (Maki *et al.*, 1973; König *et al.*, 1987). Virtually all clinical isolates of *S. marcescens* tested produce a haemolysin (Roland, 1977; Braun *et al.*, 1985; Ruan and Braun, 1990). Although earlier investigations suggested that the haemolytic activity of *S. marcescens* is cell-associated and requires the direct contact between the bacteria and the red blood cell (Braun *et al.*, 1985) it is now clearly established that the haemolysin is really secreted into the culture medium (Schiebel *et al.*, 1989). Ca^{2+} or other divalent cations are not necessary for the activity of the *S. marcescens* haemolysin (Braun *et al.*, 1985) but like the *E. coli* haemolysin it induces the synthesis and release of leukotriene C_4 (and to a lesser extent B_4) in polymorphic nuclear leukocytes and the release of histamine from rat mast cells (König *et al.*, 1987). Interestingly, haemolysin synthesis by *S. marcescens* appears to be regulated by iron (Braun *et al.*, 1985; Poole and Braun, 1988). Starvation of the bacteria for iron causes a 3–4 fold increase in haemolytic activity.

The genetic determinant for the production of *S. marcescens* haemolysin was cloned in *E. coli* on an 8 kb fragment (Braun *et al.*, 1987). This DNA fragment appears to carry the entire genetic information for synthesis and secretion of active haemolysin since the haemolytic activity expressed in *E. coli* is also found in the culture medium and independent of Ca^{2+} ions (Braun *et al.*, 1987; Schiebel *et al.*, 1989). DNA sequence analysis of the *Serratia marcescens* haemolysin determinant demonstrated the existence of two ORFs of 1671 bp (*shlB*) and 4824 bp (*shlA*) coding for proteins of 62 kDa (ShlB) and 165 kDa (ShlA), respectively (Poole *et al.*, 1988). Both polypeptides are expressed in *E. coli* minicells (Braun *et al.*, 1987) and represent precursors containing typical N-terminal signal sequences which are processed to the mature forms during transport across the cytoplasmic membrane (Poole *et al.*, 1988). Putative ribosome binding sites and promoters are identified upstream of both *shl* genes but only one putative transcription termination signal (*rho*-independent) appears to be present downstream of *shlA*. Deletion analyses further suggest that the promoter 5′-upstream of *shlB* is predominantly used for the expression of both genes (Poole and Braun, 1988) (Fig. 3). At least in *E. coli*, transcription from this promoter seems to be regulated by iron and the Fur protein, which is supported by the identification of a sequence overlapping the −35 region of the promoter, resembling strikingly the *E. coli* consensus sequence for binding of the Fur protein (Poole and Braun, 1988).

The product of the *shlA* gene (ShlA) corresponds to the *S. marcescens* haemolysin protein. It is secreted to the culture supernatant in an active form when synthesized together with ShlB (Poole *et al.*, 1988). In the absence of ShlB, ShlA stays in the periplasm and shows no haemolytic activity (Schiebel *et al.*, 1989). Consequently, ShlB is required for the activation of ShlA and for the transport of the haemolysin protein across the outer membrane. ShlB is a typical integral outer membrane protein. Computer-assisted analyses of the secondary structure of ShlB predict an amphipathic β-pleated sheet structure for the central and carboxyterminal parts (Schiebel *et al.*, 1989) which is also the predominant conformation of the integral major outer membrane proteins OmpA, OmpC and OmpF. The activation of ShlA by ShlB probably takes place during the export across the outer membrane or at the cell surface (Poole *et al.*, 1988). The yet unknown mechanism of this activation step – formally similar to the activation of *E. coli* HlyA by HlyC (see above) – involves a conformational change of ShlA since activated ShlA is more resistant to tryptic digestion than non-activated ShlA (Schiebel and Braun, 1989). Only activated ShlA binds to erythrocytes and causes haemolysis. Proteolytic processing is not involved in the activation of ShlA.

The analysis of several deletion mutants of ShlA allowed the localization of regions required for export and haemolytic activity of this toxin. The information for the extracellular secretion resides in the N-terminal region of ShlA. The domain required for the binding of ShlA to erythrocytes was also found in the N-terminal half of the protein. The C-terminal half of ShlA, on the other hand, seems to contribute to pore formation (Schiebel and Braun, 1989; Schiebel *et al.*, 1989).

Recently, a haemolysin determinant highly homologous to the *shl* determinant from *S. marcescens* was identified in a clinical isolate of *Proteus mirabilis* (Welch, 1987; Uphoff and Welch, 1990). This determinant was cloned and sequenced (Uphoff and Welch, 1990). It contains two genes, *hpmB* and *hpmA* (in that transcriptional order), which encode proteins with predicted molecular masses of 63 kDa (HpmB) and 166 kDa (HpmA), respectively. The amino acid sequence identity between HpmB and ShlB is 55.4% and that between HpmA and ShlA is 46.7%. As in the case of ShlB and

Figure 3. Physical map of the haemolysin determinant from *Serratia marcescens*. The regulation of transcription of *shlB* and *shlA* is delineated by arrows. The location of functional promoters is shown by small boxes.

ShlA, HpmB and HpmA have typical N-terminal leader peptides. HpmA is the structural haemolysin protein and its activity is independent of Ca^{2+} or other cofactors. HpmB is probably located in the outer membrane, like ShlB, and required for the activation and extracellular secretion of HpmA. The expression of HpmB and HpmA seems to be regulated in a similar way as that of ShlB and ShlA (Uphoff and Welch, 1990). HmpA is produced by all tested isolates of *Proteus mirabilis* and most isolates of *P. vulgaris* (Swihart and Welch, 1990a) and it has both haemolytic and cytotoxic activity (Swihart and Welch, 1990b).

In conclusion, the *shl* and *hpm* determinants constitute a new haemolysin gene family among Gram-negative opportunistic pathogens. Interestingly, a sequence of 115 amino acids in the aminoterminal portion of the filamentous haemagglutinin protein (FHA) from *Bordetella pertussis* shows striking homology to the N-terminal region of ShlA and HpmA (Delisse-Gathoye *et al.*, 1990), suggesting that ShlA, HpmA and FHA may be members of a larger group of virulence-associated proteins.

Vibrio cholerae haemolysin

Vibrio cholerae serotype O1 is the most common cause of cholera in humans. The O1 serotype is subdivided into the 'classic' biotype and the 'El Tor' biotype. One test to distinguish between the two biotypes is based on the ability or inability of the O1 *V. cholerae* strains to produce an extracellular haemolysin. Strains belonging to the classic biotype are considered non-haemolytic (Gallut, 1974), whereas strains belonging to the El Tor biotype are haemolytic (Honda and Finkelstein, 1979). In addition, El Tor strains are generally more resistant to polymyxin B and produce a cell-associated haemagglutinin. Recent genetic studies

also indicate differences between the two biotypes with respect to the cholera toxin genes and the restriction/modification systems. The distinction between the classic and the El Tor biotypes based on the haemolytic phenotype is, however, rather ambiguous since haemolytic classic strains as well as non-haemolytic El Tor strains have been isolated (Liu, 1959; De Moor, 1963; Gallut, 1971; Kaper *et al.*, 1982; Goldberg and Murphy, 1984; Richardson *et al.*, 1986). Haemolytic El Tor strains can give rise to non-haemolytic variants and vice versa (Roy and Mukerjee, 1962; Barrett and Blake, 1981; Goldberg and Murphy, 1984).

The haemolysin determinant of a *V. cholerae* El Tor strain (RV79) has been first cloned by Goldberg and Murphy (1984) on a 3.6 kb chromosomal fragment. Manning *et al.* (1984) have identified the haemolysin determinant from a different El Tor strain (strain 017) on a 6.2 kb chromosomal fragment. Both clones produce cell-associated haemolysin in *E. coli*, whereas the original *V. cholerae* strains export virtually the entire haemolysin (Goldberg and Murphy, 1984; Manning *et al.*, 1984). In *E. coli* the localization of the *V. cholerae* haemolysin is most likely to the periplasm (Mercurio and Manning, 1985), indicating that the genetic information required for transport of this haemolysin across the outer membrane is absent in *E. coli*. Both recombinant plasmids express in minicells or maxicells of *E. coli* a protein of approximately 80 kDa (Manning *et al.*, 1984; Goldberg and Murphy, 1985) which is similar in size to haemolysin immunoprecipitated from the supernatant of a haemolytic El Tor strain (Goldberg and Murphy, 1985). Two other proteins of 71 kDa and 22 kDa are expressed by the 6.2 kb fragment (Manning *et al.*, 1984). The genes for these proteins have been termed *hlyA* (encoding the 80 kDa protein), *hlyB* (encoding the 71 kDa protein) and *hlyC* (encoding the 22 kDa protein). The three genes form an operon (in the order *hlyC, A, B*) which seems to be transcribed in the direction *hlyC* to *hlyB* (Fig. 4). The functions of *hlyB* and *hlyC* are still unknown. Transposon insertions in *hlyB* lead to a reduction of the haemolytic activity and insertions in *hlyC* abolish the haemolytic activity entirely, as do insertions in *hlyA* (Manning *et al.*, 1984). This observation suggests that HlyA from *V. cholerae* may undergo a similar activation as HlyA from *E. coli* or ShlA from *S. marcescens* (see above) for haemolytic activity.

Hybridization with *hlyA*-specific gene probes shows homology in chromosomal fragments of similar size in all tested *V. cholerae* O1 strains, regardless of whether they belong to the classic or the El Tor biotype and independent of their haemolytic phenotype (Goldberg and Murphy, 1984; Brown and Manning, 1985). This indicates that all O1 strains carry a *hlyA* or *hlyA*-related gene which maps on the chromosome of *V. cholerae* between *arg* and *ilv* (Goldberg and Murphy, 1984).

The cloned *hlyA* genes from a non-haemolytic El Tor strain and a non-haemolytic classic strain express in *E. coli* proteins of 64 kDa and 27 kDa, respectively (Goldberg and Murphy, 1985; Alm *et al.*, 1988). Both proteins are non-haemolytic. Interestingly the nucleotide sequence of the *hlyA* gene from the non-haemolytic El Tor strain is identical to that of the haemolytic counterpart (Rader and Murphy, 1988). The occurrence of the 64 kDa protein instead of the normal 80 kDa HlyA protein remains unclear. One may speculate that the non-haemolytic El Tor strain expresses a protease which degrades HlyA to the 64 kDa protein (Rader and Murphy, 1988) or alternatively that the putative activation of HlyA (by HlyC, see above) does not properly function in the non-haemolytic El Tor strain and the non-activated HlyA is more sensitive to protease(s) which degrade it to the 64 kDa protein.

The *hlyA* gene from the non-haemolytic classic strain carries an 11 bp deletion in the 5′-terminal part generating a truncated protein of 27 kDa (Alm *et al.*, 1988; Rader and Murphy, 1988). This is in agreement with the size of the protein expressed by this *hlyA* gene in *E. coli*.

Recent investigations suggest that the 80 kDa haemolysin of *V. cholerae* El Tor is processed posttranslationally in the culture medium at its amino-terminal end to a final protein with a molecular weight of 65.3 kDa (Alm *et al.*, 1988; Hall and Dvasar, 1990).

The regions located upstream and downstream of the *hlyA* gene from the wild-type El Tor strain, the non-haemolytic El Tor variant and the non-haemolytic classic strain are highly conserved (Alm *et al.*, 1988; Rader and Murphy, 1988). They contain a putative promoter upstream of *hlyA* and a potential *rho*-independent termination signal immediately downstream of *hlyA*. In addition, the upstream region carries three copies of the sequence motive TATTTTA and it has been suggested that this sequence may serve a similar function as the repeat sequence TTTTGAT which has been identified upstream of the *ctxAB* genes as binding site for the positive regulator ToxR in *V. cholerae* (Miller *et al.*, 1987). A gene, termed *hlyR*, located close to the *toxR* gene of *V. cholerae* has been described (Von Mechow *et al.*, 1985). This regulatory gene seems to encode a trans acting protein which activates the transcription of *hlyA*. The HlyR protein may be identical to the ToxS regulatory protein which is necessary for the expression of cholera toxin by activating ToxR (Alm *et al.*, 1988).

As stated above, haemolytic O1 strains of *V. cholerae* belonging to the classic biotype can be isolated (Liu, 1959; Richardson *et al.*, 1986). Their haemolytic activity is, however, lower than that of the El Tor strains. Nevertheless, an *hlyA* gene with extensive sequence homology to the El Tor *hlyA* has been identified in a haemolytic classic

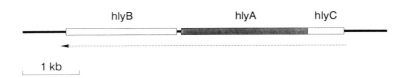

Figure 4. Arrangement of genes in the haemolysin determinant of *Vibrio cholerae* El Tor (strain O17). The putative orientation of transcription is shown below the operon consisting of *hlyC, hlyA* and *hlyB*.

strain in addition to another haemolysin determinant which is apparently unrelated to the *hlyA* gene (Richardson *et al.*, 1986).

Many other *V. cholerae* strains not belonging to the O1 serotype are also haemolytic (Sakazaki *et al.*, 1967) and export a haemolysin which is indistinguishable from the El Tor haemolysin (Yamamoto *et al.*, 1986). These strains also possess an *hlyA*-related haemolysin gene since an El Tor *hly*-specific probe hybridizes with similar chromosomal fragments (Brown and Manning, 1985).

Haemolysins from other *Vibrio* species

Other frequently isolated virulent non-cholera *Vibrio* species are also haemolytic. Most of these *Vibrio* species are marine bacteria and are therefore frequently encountered as contaminants in seafood. The spectrum of diseases can range from gastro-Menteritis to wound infections, septicaemia, pneumonia and meningitis (Miwatani and Takeda, 1976; Farmer, 1979; Blake *et al.*, 1980; Love *et al.*, 1981; Morris *et al.*, 1982; Tacket *et al.*, 1982; Wickboldt and Sanders, 1983; Tison and Kelly, 1984; Morris and Black, 1985). The genetics of the haemolysins from these *Vibrio* species have been quite extensively studied in the halophilic species *V. parahaemolyticus* and *V. vulnificus*. Little is known on the genetics of the haemolysins synthesized by the halophilic *V. damsela*, *V. hollisae*, *V. mimicus*, *V. metschnikovii* and *V. fluvialis*. It is assumed that these haemolysins are virulence factors functional in the development of the diseases caused by the non-cholera *Vibrio* species.

The best studied haemolysin determinant of the non-cholera *Vibrio* species is that encoding the extracellular 'thermostable direct haemolysin' (TDH) of *Vibrio parahaemolyticus*, also known as 'Kanagawa phenomenon (KP)-associated haemolysin'. This haemolysin is produced by 88–96% of all clinical *V. parahaemolyticus* isolates, whereas only 1–2% of the non-clinical environmental strains are haemolytic. TDH is therefore considered as one of the major virulence factors of *V. parahaemolyticus* (Sakazaki *et al.*, 1968; Miyamoto *et al.*, 1969). The TDH determinant was recently cloned in *E. coli* (Kaper *et al.*, 1984; Nishibuchi and Kaper, 1985; Taniguchi *et al.*, 1985) and

sequenced (Nishibuchi and Kaper, 1985). The only open reading frame detected in this sequence codes for a protein of 189 amino acids with a relative molecular mass of 21.1 kDa (Nishibuchi and Kaper, 1985). The N-terminal 24 amino acids represent a transport signal which is apparently cleaved off during transport across the cytoplasmic membrane (Tsunasawa *et al.*, 1983; Nishibuchi and Kaper, 1985). TDH expressed in *E. coli* seems to accumulate in the cytoplasm (Nishibuchi and Kaper, 1985). A possible explanation for this is the occurrence of three lysine residues in the middle of the signal sequence which may not be recognized by the *E. coli* protein transport machinery. The biochemical characterization of purified TDH suggests its composition of two identical subunits (18.5 kDa), each of which represents the mature product of the *tdh* gene (Takeda *et al.*, 1978; Nishibuchi and Kaper, 1985). In this context it is interesting to note that all KP-positive strains of *V. parahaemolyticus* carry two similar but not identical *tdh* gene copies named *tdh*1 and *tdh*2 (Nishibuchi and Kaper, submitted).

Sequences homologous to the *tdh* gene were also detected in *V. hollisae*, *V. mimicus* and in some non-O1 *V. cholerae* strains, but not in a variety of other *Vibrio* species (Nishibuchi *et al.*, 1985, 1986, 1988, 1990; Honda *et al.*, 1986). The recent isolation of a plasmid from a non-O1 *V. cholerae* strain carrying *tdh*-homologous sequences (Honda *et al.*, 1986) might explain the wide distribution of *tdh*-related genes among *Vibrio* species. The *tdh* plasmid gene expresses in *E. coli* a haemolysin (NAG-rTDH) which has a size similar to that of TDH of *V. parahaemolyticus* and is biochemically and immunologically highly related to this TDH (Honda *et al.*, 1986; Yoh *et al.*, 1986).

Several clinical isolates of *V. parahaemolyticus* produce a TDH-related haemolysin (TRH) which is not associated with the Kanagawa phenomenon (i.e. it does not cause haemolysis on Wagatsuma agar) (Honda *et al.*, 1988; Shirai *et al.*, 1990). TRH is immunologically cross-reacting with TDH but in contrast to TDH, it has been shown to be heat-labile (Honda *et al.*, 1988). The gene encoding the TRH protein hybridizes with the *tdh* gene only under reduced stringencies. Cloning and sequence analysis of the *trh* gene demonstrated 68% nucleotide sequence homology with the *tdh* gene indicating that both genes were probably derived from a common ancestor (Nishibuchi *et al.*, 1989). The TRH protein has the same length as the *tdh* gene product and it also seems to form dimers (Honda *et al.*, 1988; Nishibuchi *et al.*, 1989). DNA sequences homologous to the *trh* gene are only rarely found in environmental strains of *V. parahaemolyticus* and they do not exist at all in most other *Vibrio* species (Shirai *et al.*, 1990).

In addition to TDH and TRH, *V. parahaemolyticus* strains produce at least one other heat-labile haemolysin. Taniguchi *et al.* (1985) cloned the genetic determinant for a heat-labile haemolysin of *V. parahaemolyticus* in *E. coli* and showed that this determinant is present in all tested KP$^+$ and KP$^-$ *V. parahaemolyticus* strains. No homology is observed between this haemolysin determinant and the *tdh* gene.

Vibrio vulnificus, a marine bacterium that can cause gastroenteritis and other diseases in humans, produces another type of heat-labile haemolysin which is cytotoxic (demonstrated on CHO cells) and haemolytic on a large number of mammalian erythrocytes (Gray and Kreger, 1985). The genetic determinant encoding this haemolysin has been identified in a *V. vulnificus* DNA library constructed in *E. coli* (Wright *et al.*, 1985). Using the cloned haemolysin determinant as gene probe it has been shown that all clinical and environmental strains of *V. vulnificus* carry homologous sequences (Wright *et al.*, 1985; Morris *et al.*, 1987).

The DNA fragment encoding the haemolysin of *V. vulnificus* was sequenced and found to contain two genes, *vvh*A and *vvh*B (Yamamoto *et al.*, 1990). Gene *vvh*A is the structural gene of the haemolysin itself and carries the information for a mature protein of 51 kDa, which is synthesized as a precursor with an N-terminal signal peptide of 20 amino acids. Surprisingly, regions of *vvh*A show homology to the structural gene of the *Vibrio cholerae* El Tor haemolysin in both sequence and organization, suggesting that the two genes have a common evolutionary origin. The second gene, *vvh*B, is located immediately upstream of *vvh*A and encodes a protein of unknown function with a deduced molecular mass of 18 kDa (Yamamoto *et al.*, 1990). The genes *vvh*A and *vvh*B seem to be organized in an operon structure. Recent results obtained by Wright and Morris (1991) suggeest that the 51 kDa haemolysin possibly is not a major virulence factor in *V. vulnificus* infections, because

the inactivation of this toxin did not affect the pathogenicity of *V. vulnificus* in several animal models.

The other reported haemolysins from *Vibrio* species, e.g. 'damselysin' from *Vibrio damsela*, a phospholipase D (Kreger, 1984; Kothary and Kreger, 1985; Kreger *et al.*, 1987) and the haemolysins from *Vibrio metschnikovii* (Miyake *et al.*, 1988, 1989) and *Vibrio fluvialis* (Wall *et al.*, 1984) seem to be biochemically and immunologically unrelated to the above discussed *Vibrio* haemolysins. Their genetic determinants remain to be identified.

Haemolysins from *Aeromonas hydrophila* and *Aeromonas sobria*

Bacteria belonging to the genus *Aeromonas* are frequently found in rivers and other aquatic environments throughout the world. The two species, *A. hydrophila* and *A. sobria*, in particular, may cause severe infections in neonates, immunoM-compromised and elderly people. The most common symptoms are severe diarrhoea, septicaemia and meningitis (Daily *et al.*, 1981; Champsaur *et al.*, 1982; Gracey *et al.*, 1982; Janda *et al.*, 1983; Freij, 1984; George *et al.*, 1985). Clinical and environmental isolates of these two *Aeromonas* species are often haemolytic due to the production of one or more haemolysins (Buckley *et al.*, 1981; Ljungh and Wadström, 1983; Asao *et al.*, 1984, 1986). Among these the cytotoxin–haemolysin, known as aerolysin, is the best studied. Using a genetic approach it has been shown that aerolysin is a major virulence factor in *Aeromonas sobria* and most likely also in *Aeromonas hydrophila* (Chakraborty *et al.*, 1987).

Aerolysin is an extracellular protein which is synthesized in both *Aeromonas* species as a preprotoxin of 54 kDa. During transport across the cytoplasmic membrane the N-terminal 23 amino acid signal peptide is removed (Howard and Buckley, 1985a,b, 1986; Husslein *et al.*, 1988) and the protoxin is further transported across the outer membrane. Proaerolysin is haemolytically inactive. Activation requires the removal of the last 25 amino acids from the C-terminus, presumably by an extracellular protease (Howard and Buckley, 1985b). The mature aerolysin (48 kDa) binds to a specific glycoprotein receptor (glycophorin) on the surface of mammalian cells (e.g. erythrocytes) and forms, in contact with the membrane, protein aggregates which are inserted into the membrane (Garland and Buckley, 1988). This results in the formation of stable anion-selective transmembrane pores with a diameter of at least 0.7 nm (Howard and Buckley, 1982; Chakraborty et al., 1990). The C-terminal 25 amino acids seem to be necessary to prevent aggregation of the proaerolysin molecules during transport across the bacterial membrane (T. Chakraborty, pers. commun.).

Proaerolysin is able to bind to erythrocyte membranes, but it does not aggregate and it cannot form transmembrane pores indicating that aggregation of the toxin is a necessary step in pore formation (Garland and Buckley, 1988). Three histidine residues in the aerolysin protein (AerA) have been shown to be involved in receptor binding and aggregation, respectively (Green and Buckley, 1990).

The genetic determinants for aerolysin production in A. hydrophila and A. sobria strains have been extensively investigated (Chakraborty et al., 1986; Howard and Buckley, 1986; Howard et al., 1987; Husslein et al., 1988). The structural genes (aerA) for the two aerolysins encode proteins of 486 and 492 amino acids, respectively (Howard et al., 1987; Husslein et al., 1988), consistent with the size of preproaerolysin. The predicted amino acid sequence of the mature aerolysin protein (AerA) is extremely hydrophilic and devoid of α-helical structures. Consequently, the pore structure of aerolysin is probably formed by β-pleated sheets, as in the case of the bacterial outer membrane porins. In fact, the transmembrane channels generated by aerolysin and the porins have very similar properties (Chakraborty et al., 1990).

The aerA genes from A. hydrophila and A. sobria show substantial homology (77%) and the homology on the protein level is even higher (Howard et al., 1987; Husslein et al., 1988). The regulatory sequences immediately upstream and downstream of aerA are also conserved between the two determinants. A putative promoter (P_A) upstream of the start of aerA (Husslein et al., 1988) and a potential rho-independent termination signal downstream of aerA (Howard et al., 1987; Husslein et al., 1988) are common to both genes. Transposon mutagenesis has identified regions further upstream and downstream of aerA in the aerolysin determi-

nant of A. sobria which influence the expression and activity of aerolysin. The regions have been termed aerC and aerB, respectively (Chakraborty et al., 1986). Sequence analysis of these regions has shown that aerC contains, in addition to the promoter P_A, an eight-fold repeated sequence motive (aATAAAa) and an ORF for a putative protein of 61 amino acids. There is little homology (46%) in the aerC region between the two cloned aerolysin determinants (Husslein et al., 1988) and the corresponding region of the A. hydrophila determinant lacks the repeat motive and the ORF. A physical map of the aerolysin determinant of A. sobria summarizing the described features is shown in Fig. 5.

The aerolysin determinant is widely distributed among Aeromonas isolates (clinical and environmental) as determined by hybridization analysis using an aerA-specific gene probe (Chakraborty et al., 1988). However, whereas all clinical isolates exhibited strong hybridization with this probe, a few environmental isolates did not hybridize even though these strains showed haemolytic and cytotoxic activity (Chakraborty et al., 1988). This is consistent with the isolation of haemolysins from Aeromonas spp. which differ from the above described aerolysin (Asao et al., 1984, 1986; Notermans et al., 1986; Chakraborty et al., 1988; Nomura et al., 1988).

Haemolysin (phospholipase C) of Pseudomonas aeruginosa

Pseudomonas aeruginosa is a ubiquitously occurring opportunistic pathogen which can cause severe infections of burns, but can also infect immunocompromised people and neonates (Parker, 1984).

Among the extracellular toxins which are produced by P. aeruginosa (Liu, 1979; Young, 1980) two may be considered as haemolysins. One is a heat-labile, haemolytically active phospholipase C (Liu, 1974, 1979; Ostroff et al., 1989) and the other, the heat-stable haemolysin, consists of the two glycolipids rhamnose–rhamnose–β–hydroxydecanoic acid–β–hydroxydecanoic acid and rhamnose–β–hydroxydecanoic acid–β–hydroxydecanoic acid (Jarvis and Johnson, 1949; Liu, 1979; Johnson and Boese-Marrazzo, 1980; Fujita et al., 1988). Although neither of the two haemolysins belongs to the non-enzymatic proteinaceous,

Figure 5. Structure of the aerolysin determinant of *Aeromonas sobria*. Putative promoters are represented by small boxes. The direction of transcription is depicted by arrows.

pore-forming cytolysins, the phospholipase C (PLC-H) is discussed in this context. The genetics of this virulence factor (Liu, 1974, 1979; Berka *et al.*, 1981; Berk *et al.*, 1987; Coutinho *et al.*, 1988; Ostroff *et al.*, 1989) has been extensively investigated in the past (Vasil *et al.*, 1982; Coleman *et al.*, 1983; Lory and Tai, 1983; Ding *et al.*, 1985). The synthesis of PLC-H (and of the heat-stable haemolysin) is regulated by phosphate together with at least two other extracellular proteins, namely a non-haemolytic phospholipase C (PLC-N) and an alkaline phosphatase (Liu, 1979; Stinson and Hayden, 1979; Johnson and Boese-Marrazzo, 1980; Gray *et al.*, 1981; Ostroff and Vasil, 1987). Low phosphate concentration leads to the induction of these proteins, whereas high phosphate concentration represses its synthesis. These phosphate-regulated products may be part of a system to provide inorganic phosphate from phospholipids to the cell (Liu, 1979). PLC-H and PLC-N may generate phosphorylcholine from glycolipid-solubilized phospholipids (e.g. phosphatidylcholine) and the alkaline phosphatase may then cleave the phosphorylcholine into inorganic phosphate and choline. This reaction may be important in *P. aeruginosa* pathogenesis.

The haemolytic phospholipase C (PLC-H) has an apparent molecular mass of 78 kDa (Berka and Vasil, 1982). The gene for PLC-H cloned in *E. coli* also expresses a protein of 78–80 kDa (Vasil *et al.*, 1982; Coleman *et al.*, 1983; Lory and Tai, 1983). The 78 kDa protein from *E. coli* has haemolytic and phospholipase C activity. PLC-H is found extracellularly in *P. aeruginosa* (Stinson and Hayden, 1979; Vasil *et al.*, 1982) but it remains cell-bound and probably localized to the outer membrane in *E. coli* (Vasil *et al.*, 1982; Coleman *et al.*, 1983; Lory and Tai, 1983).

Recent studies show that the gene for PLC-H is part of the *plc* operon that is phosphate-regulated (Shen *et al.*, 1987). The nucleotide sequence of

the *plc* operon and the expression of this operon in *E. coli* indicate the presence of three genes (Pritchard and Vasil, 1986; Shen *et al.*, 1987): *plcS* is the structural gene for phospholipase C (PLC-H or PlcS), *plcR1* is a gene encoding a 23 kDa periplasmic or extracellular protein, and *plcR2* encodes a 19 kDa cytoplasmic protein (Fig. 6). The sequence data further indicate the existence of an N-terminal signal sequence of 38 amino acids for transport of PLC-H (Pritchard and Vasil, 1986). The start site of the ORF for PlcR1 is 21 bp downstream from the stop codon of *plcS* and there is overlapping of the *plcR2* gene with *plcR1*. Both genes, *plcR1* and *plcR2*, use the same reading frame and the same stop codon (Shen *et al.*, 1987). The ATG start codon of *plcR2* is located 168 bp downstream of the start of *plcR1*. The PlcR1 protein (207 amino acids) carries a putative N-terminal transport signal sequence of 20 amino acids but PlcR2 (151 amino acids) lacks a transport signal. These analyses may explain the localization of PlcR1 to the periplasm and PlcR2 in the cytoplasm, when expression occurs in *E. coli* (Shen *et al.*, 1987). The transcriptional start of the *plc* operon maps 26 bp upstream of the beginning of *plcS*. The putative promoter does not exhibit sequence similarities to the canonical −10 or −35 regions of *E. coli* promoters (Pritchard and Vasil, 1986). A putative transcription terminator was identified downstream of the common stop codon of the *plcR1* and *plcR2* genes (Shen *et al.*, 1987). The function of PlcR1 and PlcR2 is poorly understood. Preliminary data suggest their involvement in the regulation and/or activation of PlcS (PLC-H) (Ostroff *et al.*, 1989, 1990). One or both of the products of the *plcR* genes seems to modify PlcS posttranslationally. Modified PlcS is more haemolytic than the unmodified version. The nature of the modification is, however, not known (Ostroff *et al.*, 1990).

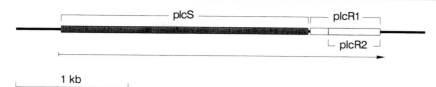

Figure 6. Physical map of the *plc* operon from *Pseudomonas aeruginosa*. The orientation of the transcription of *plcS*, *plcR1* and *plcR2* is indicated.

'Haemolysins' from other Gram-negative bacteria

It is remarkable that all haemolysins described above are synthesized and secreted by Gram-negative bacteria which normally adhere and colonize the surface of cells and tissues but are usually non-invasive. The haemolytic phenotype of these bacteria is readily recognized on blood agar and substantial amounts of these haemolysins are frequently secreted into the culture media. Antibodies against these haemolysins can be found in the sera of patients suffering from infections caused by these bacteria.

On the other hand, facultatively intracellular (invasive) Gram-negative bacteria like *Shigella*, *Salmonella*, *Yersinia*, *Brucella* or *Legionella* are non-haemolytic or only weakly haemolytic when grown on blood agar. Recent work from Sansonetti's group suggests the existence of a 'contact haemolysin' which presumably lyses the host membrane enclosing the intracellular bacterium. This type of haemolysin may be only required for intracellular growth (Sansonetti *et al.*, 1986; Kuhn *et al.*, 1988). Such a role has been also suggested for listeriolysin, the haemolysin produced by the Gram-positive bacterium *Listeria monocytogenes* (Gaillard *et al.*, 1987; Kuhn *et al.*, 1988). Most clinical isolates of *L. monocytogenes* are only weakly haemolytic but haemolysin production can be induced under stress conditions (Sokolović and Goebel, 1989; Sokolović and Goebel, unpublished results), which the bacteria may also encounter during intracellular growth. It is intriguing to speculate that invasive bacteria may preferentially produce cytolytic products as a response to the intracellular environment. The recent cloning of a haemolysin–cytotoxin gene from *Salmonella typhimurium* in *E. coli* (Libby and Heffron, pers. commun.) appears to be in line with this assumption. This gene is highly conserved in all *Salmonella typhimurium* isolates tested so far

but is permanently turned off under extracellular growth conditions. In *E. coli* expression of this haemolysin gene is observed only under certain conditions.

Conclusions and perspectives

During the past decade our knowledge on the genetics of bacterial haemolysins has been considerably widened. The tools of modern molecular genetics have allowed the detailed analysis of even those haemolysins which could hardly be investigated by the conventional methods of protein biochemistry. In addition, these studies have revealed the existence of accessory proteins required for the activation of inactive precursors and/or the export of the haemolysins. The relationship between several haemolysins and leukotoxins from Gram-negative bacteria has been clearly established. Homology between the transport genes of this class of toxins (RTX) and other prokaryotic (and even eukaryotic) genes whose products are involved in other transport processes has been demonstrated. This suggests that some haemolysin determinants may have evolved by fusion of genes which may have served different functions in their original host. More work will be needed to further clarify this interesting genetic aspect.

The analysis of the functional domains of haemolysins involved in specific binding to target cells and in pore formation has just begun mainly by the techniques of site-specific mutagenesis. This genetic approach has proven to be successful and further interesting results can be expected. The final answer to these basic questions of haemolysin–cell interactions will have to await the 3-dimensional structure of haemolysins and membrane–haemolysin complexes. The mechanism(s) by which some haemolysin precursor proteins are activated by additional proteins remains to be unravelled. It is interesting to note that this requirement for activation is apparently a specific feature of haemolysins from Gram-negative bacteria which has never been described for haemolysins from Gram-positive bacteria.

Although considerable genetic information has been accumulated concerning the specific transport systems for the RTX-class of haemolysins and leukotoxins the biochemistry of this alternative protein transport machinery is still poorly understood.

It has been shown that haemolysins represent important virulence factors. Much work needs still to be done to fully understand their function(s) in the pathogenesis of infections. The construction of genetically defined mutant strains may help to solve this important problem.

As pointed out, haemolysins from invasive Gram-negative bacteria have not been detected until now. Yet lysis of mammalian cell membranes is a requirement for the intracellular lifecycle. Invasive bacteria have to destroy the membrane of the phagosome (endosome) in order to gain access to the cytoplasm. Even those invasive Gram-negative bacteria which remain enclosed in the vacuole for most of their intracellular lifespan and replicate in this compartment, like *Salmonella*, have to ultimately lyse the cytoplasmic membrane to evade from the host cell. Membrane-active proteins such as cytolysins are good candidates for performing these steps. Such proteins will probably also lyse erythrocyte membranes and hence will be haemolysins by definition. The fact that invasive Gram-negative bacteria are non-haemolytic when grown extracellularly does not rule out the existence of haemolysin-encoding genes in these bacteria. The expression of such genes may, however, be only induced under specific intracellular conditions.

References

Alm, R.A., Stroeher, U.H. and Manning, P.A. (1988). Extracellular proteins of *Vibrio cholerae*: nucleotide sequence of the structural gene (*hly*A) for the haemolysin of the haemolytic El Tor strain 017 and characterization of the *hly*A mutation in the non-haemolytic classical strain 569B. *Molec. Microbiol.* **2**, 481–8.

Asao, T., Kinoshita, Y., Kozaki, S., Uemura, T. and Sakaguchi, G. (1984). Purification and some

properties of *Aeromonas hydrophila* hemolysin. *Infect. Immun.* **46**, 122–7.

Asao, T., Kozaki, S., Kato, K., Kinoshita, Y., Otsu, K., Uemura, T. and Sakaguchi, G. (1986). Purification and characterization of an *Aeromonas hydrophila* hemolysin. *J. Clin. Microbiol.* **24**, 228–32.

Baluyut, C.S., Simonson, R.R., Bemrick, W.J. and Maheswaran, S.K. (1981). Interaction of *Pasteurella haemolytica* with bovine neutrophils: identification and partial characterization of a cytotoxin. *Am. J. Vet. Res.* **42**, 1920–6.

Barrett, T.J. and Blake, P.A. (1981). Epidemiological usefulness of changes in hemolytic activity of *Vibrio cholerae* biotype El Tor during the seventh pandemic. *J. Clin. Microbiol.* **13**, 126–9.

Barry, E.M., Weiss, A.A., Ehrmann, I.E., Gray, M.C., Hewlett, E.L. and Goodwin, M. St. M. (1991). *Bordetella pertussis* adenylate cyclase toxin and hemolytic activities require a second gene, *cya*C, for activation. *J. Bacteriol.* **173**, 720–6.

Bellalou, J., Ladant, D., and Sakamoto, H. (1990a). Synthesis and secretion of *Bordetella pertussis* adenylate cyclase as a 200-kDa protein. *Infect. Immun.* **58**, 1195–1200.

Bellalou, J., Sakamoto, H., Ladant, D., Geoffroy, C. and Ullmann, A. (1990b). Deletions affecting hemolytic and toxin activities of *Bordetella pertussis* adenylate cyclase. *Infect. Immun.* **58**, 3242–7.

Benson, M.L., Thompson, R.C. and Valli, V.E.O. (1978). The bovine alveolar macrophage. II. In vitro studies with *Pasteurella haemolytica*. *Can. J. Comp. Med.* **42**, 368–9.

Benz, R., Schmid, A., Wagner, W. and Goebel, W. (1989). Pore formation by the *Escherichia coli* hemolysin: evidence for an association-dissociation equilibrium of the pore-forming aggregates. *Infect. Immun.* **57**, 887–95.

Berger, H., Hacker, J., Juarez, A., Hughes, C. and Goebel, W. (1982). Cloning of the chromosomal determinants encoding hemolysin production and mannose-resistant hemagglutination in *Escherichia coli*. *J. Bacteriol.* **152**, 1241–7.

Berk, R.S., Brown, D., Coutinho, I. and Meyers, D. (1987). In vivo studies with two phospholipase C fractions from *Pseudomonas aeruginosa*. *Infect. Immun.* **55**, 1728–30.

Berka, R.M. and Vasil, M.L. (1982). Phospholipase C (heat-labile hemolysin) of *Pseudomonas aeruginosa*: purification and preliminary characterization. *J. Bacteriol.* **152**, 239–45.

Berka, R.M., Gray, G.L. and Vasil, M.L. (1981). Studies of phospholipase C (heat-labile hemolysin) in *Pseudomonas aeruginosa*. *Infect. Immun.* **34**, 1071–4.

Bhakdi, S., Mackman, N., Nicaud, J.-M. and Holland, I.B. (1986). *Escherichia coli* hemolysin may damage target cell membranes by generating transmembrane pores. *Infect. Immun.* **52**, 63–9.

Blake, P.A., Weaver, R.E. and Hollis, D.G. (1980). Diseases of humans (other than cholera) caused by vibrios. *Ann. Rev. Microbiol.* **34**, 341–67.

Boehm, D.F., Welch, R.A. and Snyder, I.S. (1990a). Calcium is required for binding of *Escherichia coli* hemolysin (HlyA) to erythrocyte membranes. *Infect. Immun.* **58**, 1951–8.

Boehm, D.F., Welch, R.A. and Snyder, I.S. (1990b). Domains of *Escherichia coli* hemolysin (HlyA) involved in binding of calcium and erythrocyte membranes. *Infect. Immun.* **58**, 1959–64.

Braun, V., Günther, H., Neuss, B. and Tautz, C. (1985). Hemolytic activity of *Serratia marcescens*. *Arch. Microbiol.* **141**, 371–6.

Braun, V., Neuss, B., Ruan, Y., Schiebel, E., Schöffler, H. and Jander, G. (1987). Identification of the *Serratia marcescens* hemolysin determinant by cloning into *Escherichia coli*. *J. Bacteriol.* **169**, 2113–20.

Brown, M.H. and Manning, P.A. (1985). Haemolysin genes of *Vibrio cholerae*: presence of homologous DNA in non-haemolytic O1 and haemolytic non-O1 strains. *FEMS Microbiol. Lett.* **30**, 197–201.

Buckley, J.T., Halasa, L.N., Lund, K.D. and MacIntyre, S. (1981). Purification and some properties of the hemolytic toxin aerolysin. *Can. J. Biochem.* **59**, 430–5.

Cavalieri, S.J. and Snyder, I.S. (1982a). Effect of *Escherichia coli* alpha-hemolysin on human peripheral leukocyte viability in vitro. *Infect. Immun.* **36**, 455–61.

Cavalieri, S.J. and Snyder, I.S. (1982b). Effect of *Escherichia coli* alpha-hemolysin on human peripheral leukocyte function in vitro. *Infect. Immun.* **37**, 966–74.

Cavalieri, S.J. and Snyder, I.S. (1982c). Cytotoxic activity of partially purified *Escherichia coli* alpha haemolysin. *J. Med. Microbiol.* **15**, 11–21.

Cavalieri, S.J., Bohach, G.A. and Snyder, I.S. (1984). *Escherichia coli* α-hemolysin: characteristics and probable role in pathogenicity. *Microbiol. Rev.* **48**, 326–43.

Chakraborty, T., Huhle, B., Bergbauer, H. and Goebel, W. (1986). Cloning, expression, and mapping of the *Aeromonas hydrophila* aerolysin gene determinant in *Escherichia coli* K-12. *J. Bacteriol.* **167**, 368–74.

Chakraborty, T., Huhle, B., Hof, H., Bergbauer, H. and Goebel, W. (1987). Marker exchange mutagenesis of the aerolysin determinant in *Aeromonas hydrophila* demonstrates the role of aerolysin in *A. hydrophila*-associated systemic infections. *Infect. Immun.* **55**, 2274–80.

Chakraborty, T., Husslein, V., Huhle, B., Bergbauer, H., Jarchau, T., Hof, H. and Goebel, W. (1988). Genetics of hemolysins in *Aeromonas hydrophila*. In: *Bacterial Protein Toxins* (eds F. Fehrenbach, J.E. Alouf, P. Falmagne, W. Goebel, J. Jeljaszewicz, D. Jurgen and R. Rappuoli) Zbl. Bakt. Suppl., **17**, pp.215–22. Gustav Fischer, Stuttgart and New York.

Chakraborty, T., Schmid, A., Notermans, S. and Benz, R. (1990). Aerolysin of *Aeromonas sobria*: evidence for formation of ion-permeable channels and comparison with alpha-toxin of *Staphylococcus aureus*. *Infect. Immun.* **58**, 2127–32.

Champsaur, H., Andremont, H., Mathieu, D., Rottmann, E. and Anzepy, P. (1982). Cholera-like illness due to *Aeromonas sobria*. *J. Infect. Dis.* **145**, 248–54.

Chang, Y.-F., Young, R., Post, D. and Struck, D.K. (1987). Identification and characterization of the *Pasteurella haemolytica* leukotoxin. *Infect. Immun.* **55**, 2348–54.

Chang, Y.-F., Young, R., Moulds, T.L. and Struck, D.K. (1989a). Secretion of the *Pasteurella* leukotoxin by *Escherichia coli*. *FEMS Microbiol. Lett.* **60**, 169–74.

Chang, Y.-F., Young, R. and Struck, D.K. (1989b). Cloning and characterization of a hemolysin gene from *Actinobacillus* (*Haemophilus*) *pleuropneumoniae*. *DNA* **8**, 635–47.

Clinkenbeard, K.D., Mosier, D.A. and Confer, A.W. (1989). Transmembrane pore size and role of cell swelling in cytotoxicity caused by *Pasteurella haemolytica* leukotoxin. *Infect. Immun.* **57**, 420–5.

Coleman, K., Dougan, G. and Arbuthnott, J.P. (1983). Cloning, and expression in *Escherichia coli* K-12, of the chromosomal hemolysin (phospholipase C) determinant of *Pseudomonas aeruginosa*. *J. Bacteriol.* **153**, 909–15.

Colman, A. and Robinson, C. (1986). Protein import into organelles: hierarchical targeting sequences. *Cell* **46**, 321–2.

Confer, D.L. and Eaton, J.W. (1982). Phagocyte impotence caused by an invasive bacterial adenylate cyclase. *Science* **217**, 948–50.

Coutinho, I.R., Berk, R.S. and Mammen, E. (1988). Platelet aggregation by a phospholipase C from *Pseudomonas aeruginosa*. *Thromb. Res.* **51**, 495–505.

Cross, M.A., Koronakis, V., Stanley, P.L.D. and Hughes, C. (1990). HlyB-dependent secretion of hemolysin by uropathogenic *Escherichia coli* requires conserved sequences flanking the chromosomal *hly* determinant. *J. Bacteriol.* **172**, 1217–24.

Daily, O.P., Joseph, S.W., Coolbaugh, J.C., Walker, R.I., Merrell, B.R., Rollins, D.M., Seidler, R.J., Colwell, R.R. and Lissner, C.R. (1981). Association of *Aeromonas sobria* with human infection. *J. Clin. Microbiol.* **13**, 769–77.

DeBoy II, J.M., Wachsmuth, I.K. and Davis, B.R. (1980). Hemolytic activity in enterotoxigenic and non-enterotoxigenic strains of *Escherichia coli*. *J. Clin. Microbiol.* **12**, 193–8.

De la Cruz, F., Zabala, J.C. and Ortiz, J.M. (1979). Incompatibility among α-hemolytic plasmids studied after inactivation of the α-hemolysin gene by transposition of Tn802. *Plasmid* **2**, 507–19.

De la Cruz, F., Müller, D., Ortiz, J.M. and Goebel, W. (1980). Hemolysis determinant common to *Escherichia coli* hemolytic plasmids of different incompatibility groups. *J. Bacteriol.* **143**, 825–33.

De la Cruz, F., Zabala, J.C. and Ortiz, J.M. (1983). Hemolysis determinant common to *Escherichia coli* strains of different O serotypes and origins. *Infect. Immun.* **41**, 881–7.

Delepelaire, P. and Wandersman, C. (1989). Protease secretion by *Erwinia chrysanthemi*. Proteases B and C are synthesized and secreted as zymogens without a signal peptide. *J. Biol. Chem.* **264**, 9083–89.

Delepelaire, P. and Wandersman, C. (1990). Protein secretion in Gram-negative bacteria: the extracellular metalloprotease B from *Erwinia chrysanthemi* contains a C-terminal secretion signal analogous to that of *Escherichia coli* α-hemolysin. *J. Biol. Chem.* **265**, 17118–25.

Delisse-Gathoye, A.-M., Locht, C., Jacob, F., Raaschou-Nielsen, M., Heron, I., Ruelle, J.-L., de Wilde, M. and Cabezon, T. (1990). Cloning, partial sequence, expression, and antigenic analysis of the filamentous hemagglutinin gene of *Bordetella pertussis*. *Infect. Immun.* **58**, 2895–905.

De Moor, C.E. (1963). A non-haemolytic vibrio. *Trop. Geogr. Med.* **15**, 97–107.

De Pauw, A.P., Gill, W.B. and Fried, F.A. (1971). Etiology of pyelonephritis: renal lysosome disruption by hemolytic *Escherichia coli*. *Invest. Urol.* **9**, 230–3.

Devenish, J., Rosendal, S., Johnson, R. and Hubler, S. (1989). Immunoserological comparison of 104-kilodalton proteins associated with hemolysis and cytolysis in *Actinobacillus pleuropneumoniae*, *Actinobacillus suis*, *Pasteurella haemolytica*, and *Escherichia coli*. *Infect. Immun.* **57**, 3210–13.

Ding, J., Lory, S. and Tai, P.C. (1985). Orientation and expression of the cloned hemolysin gene of *Pseudomonas aeruginosa*. *Gene* **33**, 313–21.

Dudgeon, L.S., Wordley, E. and Bawtree, F. (1921). On *Bacillus coli* infections of the urinary tract especially in relation to haemolytic organisms. *J. Hyg.* **20**, 137–64.

Economou, A., Hamilton, W.D.O., Johnston, A.W.B. and Downie, J.A. (1990). The *Rhizobium* nodulation gene *nodO* encodes a Ca^{2+}-binding protein that is exported without N-terminal cleavage and is homologous to haemolysin and related proteins. *EMBO J.* **9**, 349–54.

Erb, K., Vogel, M., Wagner, W. and Goebel, W. (1987). Alkaline phosphatase which lacks its own signal sequence becomes enzymatically active when fused to N-terminal sequences of *Escherichia coli* haemolysin (HlyA). *Molec. Gen. Genet.* **208**, 88–93.

Farmer III, J.J. (1979). *Vibrio (Beneckea")* *vulnificus*, the bacterium associated with sepsis, septicaemia, and the sea. *Lancet* **ii**, 903.

Felmlee, T. and Welch, R.A. (1988). Alterations of amino acid repeats in the *Escherichia coli* hemolysin affect cytolytic activity and secretion. *Proc. Natl. Acad. Sci. USA* **85**, 5269–73.

Felmlee, T., Pellett, S., Lee, E.-Y. and Welch, R.A. (1985a). *Escherichia coli* hemolysin is released extracellularly without cleavage of a signal peptide. *J. Bacteriol.* **163**, 88–93.

Felmlee, T., Pellett, S. and Welch, R.A. (1985b). Nucleotide sequence of an *Escherichia coli* chromosomal hemolysin. *J. Bacteriol.* **163**, 94–105.

Forestier, C. and Welch, R.A. (1990). Nonreciprocal complementation of the *hlyC* and *lktC* genes of the *Escherichia coli* hemolysin and *Pasteurella haemolytica* leukotoxin determinants. *Infect. Immun.* **58**, 828–32.

Freij, B.J. (1984). *Aeromonas*: biology of the organism and diseases of children. *Pediatr. Infect. Dis.* **3**, 164–75.

Frey, J. and Nicolet, J. (1988a). Purification and partial characterization of a hemolysin produced by *Actinobacillus pleuropneumoniae* type strain 4074. *FEMS Microbiol. Lett.* **55**, 41–6.

Frey, J. and Nicolet, J. (1988b). Regulation of hemolysin expression in *Actinobacillus pleuropneumoniae* serotype 1 by Ca^{2+}. *Infect. Immun.* **56**, 2570–5.

Fried, F.A., Vermeulen, C.W., Ginsburg, M.J. and Cone, C.M. (1971). Etiology of pyelonephritis: further evidence associating the production of experimental pyelonephritis with hemolysis in *Escherichia coli*. *J.Urol.* **106**, 351–4.

Friedman, E., Farfel, Z. and Hanski, E. (1987). The invasive adenylate cyclase of *Bordetella pertussis*. Properties and penetration kinetics. *Biochem. J.* **243**, 145–51.

Fry, T.L., Fried, F.A. and Goven, B.A. (1975). Pathogenesis of pyelonephritis: *Escherichia coli*-induced renal ultrastructural changes. *Invest. Urol.* **13**, 47–51.

Fujita, K., Akino, T. and Yoshioka, H. (1988). Characteristics of heat-stable extracellular hemolysin from *Pseudomonas aeruginosa*. *Infect. Immun.* **56**, 1385–7.

Gadeberg, O.V. and rskov, I. (1984). In vitro cytotoxic effect of α-hemolytic *Escherichia coli* on human blood granulocytes. *Infect. Immun.* **45**, 255–60.

Gadeberg, O.V., rskov, I. and Rhodes, J.M. (1983). Cytotoxic effect of an alpha-hemolytic *Escherichia coli* strain on human blood monocytes and granulocytes in vitro. *Infect. Immun.* **41**, 358–64.

Gaillard, J.L., Berche, P., Mounier, J., Richard, S. and Sansonetti, P. (1987). In vitro model of penetration and intracellular growth of *Listeria monocytogenes* in the human erythrocyte-like cell line Caco-2. *Infect. Immun.* **55**, 2822–9.

Gallut, J. (1971). La septieme pandemie cholerique. *Bull. Soc. Pathol. Exot. (Paris)* **64**, 551–60.

Gallut, J. (1974). The cholera vibrios. In: *Cholera* (eds D. Barua and W. Burrows), pp. 17–40. W.B. Saunders, Philadelphia.

Garland, W.J. and Buckley, J.T. (1988). The cytolytic toxin aerolysin must aggregate to disrupt erythrocytes, and aggregation is stimulated by human glycophorin. *Infect. Immun.* **56**, 1249–53.

George, W.L., Nakata, M.M., Thompson, J. and White, M.L. (1985). *Aeromonas*-related diarrhea in adults. *Arch. Intern. Med.* **145**, 2207–11.

Gerlach, J.H., Endicott, J.A., Juranka, P.F., Henderson, G., Sarangi, F., Deuchars, K.L. and Ling, V. (1986). Homology between P-glycoprotein and a bacterial haemolysin transport protein suggests a model for multi-drug resistance. *Nature* **324**, 485–9.

Glaser, P., Ladant, D., Sezer, O., Pichot, F., Ullmann, A. and Danchin, A. (1988a). The calmodulin-sensitive adenylate cyclase of *Bordetella pertussis*: cloning and expression in *Escherichia coli. Molec. Microbiol.* **2**, 19–30.

Glaser, P., Sakamoto, H., Bellalou, J., Ullmann, A. and Danchin, A. (1988b). Secretion of cyclolysin, the calmodulin-sensitive adenylate cyclase-haemolysin bifunctional protein of *Bordetella pertussis. EMBO J.* **7**, 3997–4004.

Goebel, W. and Hedgpeth, J. (1982). Cloning and functional characterization of the plasmid-encoded hemolysin determinant of *Escherichia coli. J. Bacteriol.* **151**, 1290–8.

Goebel, W. and Schrempf, H. (1971). Isolation and characterization of supercoiled circular deoxyribonucleic acid from beta-hemolytic strains of *Escherichia coli. J. Bacteriol.* **106**, 311–17.

Goebel, W., Royer-Pokora, B., Lindenmaier, W. and Bujard, H. (1974). Plasmids controlling synthesis of hemolysin in *Escherichia coli*: molecular properties. *J. Bacteriol.* **118**, 964–73.

Goldberg, S.L. and Murphy, J.R. (1984). Molecular cloning of the hemolysin determinant from *Vibrio cholerae* El Tor. *J. Bacteriol.* **160**, 239–44.

Goldberg, S.L. and Murphy, J.R. (1985). Cloning and characterization of the hemolysin determinants from *Vibrio cholerae* RV79(Hly+), RV79 (Hly−), and 569B. *J. Bacteriol.* **162**, 35–41.

Gonzalez-Carrero, M.I., Zabala, J.C., de la Cruz, F. and Ortiz, J.M. (1985). Purification of α-hemolysin from an overproducing *E. coli* strain. *Molec. Gen. Genet.* **199**, 106–10.

Gracey, M., Burke, V. and Robinson, J. (1982). *Aeromonas*-associated gastroenteritis. *Lancet* **ii**, 1304–6.

Gray, G.L., Berka, R.M. and Vasil, M.L. (1981). A *Pseudomonas aeruginosa* mutant non-derepressible for orthophosphate-regulated proteins. *J. Bacteriol.* **147**, 675–8.

Gray, L.D. and Kreger, A.S. (1985). Purification and characterization of an extracellular cytolysin produced by *Vibrio vulnificus. Infect. Immun.* **48**, 62–72.

Gray, L., Mackman, N., Nicaud, J.-M. and Holland, I.B. (1986). The carboxy-terminal region of haemolysin 2001 is required for secretion of the toxin from *Escherichia coli. Molec. Gen. Genet.* **205**, 127–33.

Green, C.P. and Thomas, V.L. (1981). Hemagglutination of human type O erythrocytes, hemolysin production, and serogrouping of *Escherichia coli* isolates from patients with acute pyelonephritis, cystitis, and asymptomatic bacteriuria. *Infect. Immun.* **31**, 309–15.

Green, M.J. and Buckley, J.T. (1990). Site-directed mutagenesis of the hole-forming toxin aerolysin: studies on the roles of histidines in receptor binding and oligomerization of the monomer. *Biochemistry* **29**, 2177–80.

Grünig, H.-M. and Lebek, G. (1988). Haemolytic activity and characteristics of plasmid and chromosomally borne *hly* genes isolated from *E. coli* of different origin. *Zbl. Bakt. Hyg.* **A267**, 485–94.

Grünig, H.-M., Rutschi, D., Schoch, C. and Lebek, G. (1987). The regulation of the plasmid encoded haemolysin secretion by the chromosomally encoded *fur* gene. *Zbl. Bakt. Hyg.* **A266**, 231–8.

Guzzo, J., Duong, F., Wandersman, C., Murgier, M. and Lazdunski, A. (1991). The secretion genes of *Pseudomonas aeruginosa* alkaline protease are functionally related to those of *Erwinia chrysanthemi* proteases and *E. coli* α-haemolysin. *Molec. Microbiol.*, in press.

Gygi, D., Nicolet, J., Frey, J., Cross, M., Koronakis, V. and Hughes, C. (1990). Isolation of the *Actinobacillus pleuropneumoniae* haemolysin gene and the activation and secretion of the prohaemolysin by the HlyC, HlyB and HlyD proteins of *Escherichia coli. Molec. Microbiol.* **4**, 123–8.

Hacker, J. and Hughes, C. (1985). Genetics of *Escherichia coli* hemolysin. *Curr. Top. Microbiol. Immunol.* **118**, 139–62.

Hacker, J., Hughes, C., Hof, H. and Goebel, W. (1983a). Cloned hemolysin genes from *Escherichia coli* that cause urinary tract infection determine different levels of toxicity in mice. *Infect. Immun.* **42**, 57–63.

Hacker, J., Schröter, G., Schrettenbrunner, A., Hughes, C. and Goebel, W. (1983b). Hemolytic *Escherichia coli* strains in the human fecal flora as potential urinary pathogens. *Zbl. Bakt. Hyg.*, I. Abt. Orig. **A254**, 370–8.

Hall, R.H. and Dvasar, B.S. (1990). *Vibrio cholerae* HlyA hemolysin is processed by proteolysis. *Infect. Immun.* **58**, 3375–9.

Hanski, E. (1989). Invasive adenylate cyclase toxin of *Bordetella pertussis. TIBS* **14**, 459–63.

Härtlein, M., Schießl, S., Wagner, W., Rdest, U., Kreft, J. and Goebel, W. (1983). Transport of hemolysin by *Escherichia coli. J. Cell. Biochem.* **22**, 87–97.

Von Heijne, G. (1986). Mitochondrial targeting sequences may form amphiphilic helices. *EMBO J.* **5**, 1335–42.

Heinrich, S. and Pulverer, G. (1959). Zur Ätiologie und Mikrobiologie der Aktinomykose. III. Die pathogene Bedeutung des *Actinobacillus actinomycetem-comitans* unter den Begleitbakterien" des *Actinomyces israeli. Zentralbl. Bakteriol. Parasitenkd. Infektionskr. Hyg.*, Abt. 1, Orig. **176**, 91–101.

Hess, J., Wels, W., Vogel, M. and Goebel, W. (1986). Nucleotide sequence of a plasmid-encoded hemolysin determinant and its comparison with a corresponding chromosomal hemolysin sequence. *FEMS Microbiol. Lett.* **34**, 1–11.

Hewlett, E.L., Urban, M.A., Manclark, C.R. and Wolff, J. (1976). Extracytoplasmic adenylate cyclase of *Bordetella pertussis. Proc. Natl. Acad. Sci. USA* **73**, 1926–30.

Hewlett, E.L., Gordon, V.M., McCaffery, J.D., Sutherland, W.M. and Gray, M.C. (1989). Adenylate cyclase toxin from *Bordetella pertussis*. Identification and purification of the holotoxin molecule. *J. Biol. Chem.* **264**, 19379–84.

Higgins, C.F., Hiles, I.D., Salmond, G.P.C., Gill, D.R., Downie, J.A., Evans, I.J., Holland, I.B., Gray, L., Buckel, S.D., Bell, A.W. and Hermodson, M.A. (1986). A family of related ATP-binding subunits coupled to many distinct biological processes in bacteria. *Nature* **323**, 448–50.

Höhne, C. (1973). Hly-Plasmide in R-Plasmide tragenden Stämmen von *Escherichia coli. Z. Allg. Mikrobiol.* **13**, 49–53.

Honda, T. and Finkelstein, R.A. (1979). Purification and characterization of a hemolysin produced by *Vibrio cholerae* biotype El Tor: another toxic substance produced by cholera vibrios. *Infect. Immun.* **26**, 1020–7.

Honda, T., Nishibuchi, M., Miwatani, T. and Kaper, J.B. (1986). Demonstration of a plasmid-borne gene encoding a thermostable direct hemolysin in *Vibrio cholerae* non-O1 strains. *Appl. Environ. Microbiol.* **52**, 1218–20.

Honda, T., Ni, Y. and Miwatani, T. (1988). Purification and characterization of a hemolysin produced by a clinical isolate of Kanagawa phenomenon-negative *Vibrio parahaemolyticus* and related to the thermostable direct hemolysin. *Infect. Immun.* **56**, 961–5.

Howard, S.P. and Buckley, J.T. (1982). Membrane glycoprotein receptor and hole-forming properties of a cytolytic protein toxin. *Biochemistry* **21**, 1662–7.

Howard, S.P. and Buckley, J.T. (1985a). Protein export by a gram-negative bacterium: production of aerolysin by *Aeromonas hydrophila. J. Bacteriol.* **161**, 1118–24.

Howard, S.P. and Buckley, J.T. (1985b). Activation of the hole-forming toxin aerolysin by extracellular processing. *J. Bacteriol.* **163**, 336–40.

Howard, S.P. and Buckley, J.T. (1986). Molecular cloning and expression in *Escherichia coli* of the structural gene for the hemolytic toxin aerolysin from *Aeromonas hydrophila. Molec. Gen. Genet.* **204**, 289–95.

Howard, S.P., Garland, W.J., Green, M.J. and Buckley, J.T. (1987). Nucleotide sequence of the gene for the hole-forming toxin aerolysin of *Aeromonas hydrophila. J. Bacteriol.* **169**, 2869–71.

Hughes, C., Müller, D., Hacker, J. and Goebel, W. (1982a). Genetics and pathogenic role of *Escherichia coli* hemolysin. *Toxicon* **20**, 247–52.

Hughes, C., Phillips, R. and Roberts, A.P. (1982b). Serum resistance among *Escherichia coli* strains causing urinary tract infection in relation to O type and the carriage of hemolysin, colicin, and antibiotic resistance determinants. *Infect. Immun.* **35**, 270–5.

Hughes, C., Hacker, J., Roberts, A. and Goebel, W. (1983). Hemolysin production as a virulence marker in symptomatic and asymptomatic urinary tract infections caused by *Escherichia coli. Infect. Immun.* **39**, 546–51.

Hull, S.I., Hull, R.A., Minshew, B.H. and Falkow, S. (1982). Genetics of hemolysin of *Escherichia coli. J. Bacteriol* **151**, 1006–12.

Husslein, V., Huhle, B., Jarchau, T., Lurz, R., Goebel, W. and Chakraborty, T. (1988). Nucleotide sequence

and transcriptional analysis of the *aerCaerA* region of *Aeromonas sobria* encoding aerolysin and its regulatory region. *Molec. Microbiol.* **2**, 507–17.

Issartel, J.-P., Koronakis, V. and Hughes, C. (1991). Activation of *Escherichia coli* prohaemolysin to the mature toxin by acyl carrier protein-dependent fatty acylation. *Nature* **351**, 759–61.

Iwase, M., Lally, E.T., Berthold, P., Korchak, H.M. and Taichman, N.S. (1990). Effects of cations and osmotic protectants on cytolytic activity of *Actinobacillus actinomycetemcomitans* leukotoxin. *Infect. Immun.* **58**, 1782–8.

Janda, J.M., Bottone, E.J., Skinner, C.V. and Calcaterra, D. (1983). Phenotypic markers associated with gastrointestinal *Aeromonas hydrophila* isolates from symptomatic children. *J. Clin. Microbiol.* **17**, 588–91.

Jarvis, F.G. and Johnson, M.J. (1949). A glyco-lipide produced by *Pseudomonas aeruginosa. J. Am. Chem. Soc.* **71**, 4124–6.

Johnson, M.K. and Boese-Marrazzo, D. (1980). Production and properties of heat-stable extracellular hemolysin from *Pseudomonas aeruginosa. Infect. Immun.* **29**, 1028–33.

Juarez, A. and Goebel, W. (1984). Chromosomal mutation that effects excretion of hemolysin in *Escherichia coli. J. Bacteriol.* **159**, 1083–5.

Juarez, A., Hughes, C., Vogel, M. and Goebel, W. (1984). Expression and regulation of the plasmid-encoded hemolysin determinant of *Escherichia coli. Molec. Gen. Genet.* **197**, 196–203.

Kaehler, K.L., Markham, R.J.F., Muscoplat, C.C. and Johnson, D.W. (1980a). Evidence of cytocidal effects of *Pasteurella haemolytica* on bovine peripheral blood mononuclear leukocytes. *Am. J. Vet. Res.* **41**, 1690–3.

Kaehler, K.L., Markham, R.J.F., Muscoplat, C.C. and Johnson, D.W. (1980b). Evidence of species specificity in the cytocidal effects of *Pasteurella haemolytica. Infect. Immun.* **30**, 615–16.

Kaper, J.B., Bradford, H.B., Roberts, N.C. and Falkow, S. (1982). Molecular epidemiology of *Vibrio cholerae* in the U.S. gulf coast. *J. Clin. Microbiol.* **16**, 129–34.

Kaper, J.B., Campen, R.K., Seidler, R.J., Baldini, M.M. and Falkow, S. (1984). Cloning of the thermostable direct or Kanagawa phenomenon-associated hemolysin of *Vibrio parahaemolyticus. Infect. Immun.* **45**, 290–2.

Kayser, H. (1903). Ueber Bakterienhämolysine, im Besonderen das Colilysin. *Z. Hyg. Infectionskr.* **42**, 118–38.

Knapp, S., Hacker, J., Then, I., Müller, D. and Goebel, W. (1984). Multiple copies of hemolysin genes and associated sequences in the chromosomes of uropathogenic *Escherichia coli* strains. *J. Bacteriol.* **159**, 1027–33.

Knapp, S., Then, I., Wels, W., Michel, G., Tschäpe, H., Hacker, J. and Goebel, W. (1985). Analysis of the flanking regions from different haemolysin determinants of *Escherichia coli. Molec. Gen. Genet.* **200**, 385–92.

Knapp, S., Hacker, J., Jarchau, T. and Goebel, W.

(1986). Large, unstable inserts in the chromosome affect virulence properties of uropathogenic *Escherichia coli* O6 strain 536. *J. Bacteriol.* **168**, 22–30.

Kolodrubetz, D., Dailey, T., Ebersole, J. and Kraig, E. (1989). Cloning and expression of the leukotoxin gene from *Actinobacillus actinomycetemcomitans*. *Infect. Immun.* **57**, 1465–9.

König, W., Faltin, Y., Scheffer, J., Shöffler, H. and Braun, V. (1987). Role of cell-bound hemolysin as a pathogenicity factor for *Serratia* infections. *Infect. Immun.* **55**, 2554–61.

Koronakis, V. and Hughes, C. (1988). Identification of the promotors directing in vivo expression of hemolysin genes in *Proteus vulgaris* and *Escherichia coli*. *Molec. Gen. Genet.* **213**, 99–104.

Koronakis, V., Cross, M., Senior, B., Koronakis, E. and Hughes, C. (1987). The secreted hemolysins of *Proteus mirabilis*, *Proteus vulgaris*, and *Morganella morganii* are genetically related to each other and to the alpha-hemolysin of *Escherichia coli*. *J. Bacteriol.* **169**, 1509–15.

Koronakis, V., Cross, M. and Hughes, C. (1988a). Expression of the *E. coli* hemolysin secretion gene *hly*B involves transcript anti-termination within the *hly* operon. *Nucleic Acids Res.* **16**, 4789–800.

Koronakis, V., Koronakis, E. and Hughes, C. (1988b). Comparison of the haemolysin secretion protein HlyB from *Proteus vulgaris* and *Escherichia coli*; site-directed mutagenesis causing impairment of export function. *Molec. Gen. Genet.* **213**, 551–5.

Koronakis, V., Cross, M. and Hughes, C. (1989a). Transcription antitermination in an *Escherichia coli* haemolysin operon is directed progressively by cis-acting DNA sequences upstream of the promoter region. *Molec. Microbiol.* **3**, 1397–1404.

Koronakis, V., Koronakis, E. and Hughes, C. (1989b). Isolation and analysis of the C-terminal signal directing export of *Escherichia coli* hemolysin protein across both bacterial membranes. *EMBO J.* **8**, 595–605.

Kothary, M.H. and Kreger, A.S. (1985). Purification and characterization of an extracellular cytolysin produced by *Vibrio damsela*. *Infect. Immun.* **49**, 25–31.

Kraig, E., Dailey, T. and Kolodrubetz, D. (1990). Nucleotide sequence of the leukotoxin gene from *Actinobacillus actinomycetemcomitans*: homology to the alpha-hemolysin/leukotoxin gene family. *Infect. Immun.* **58**, 920–9.

Kreger, A.S. (1984). Cytolytic activity and virulence of *Vibrio damsela*. *Infect. Immun.* **44**, 326–31.

Kreger, A.S., Bernheimer, A.W., Etkin, L.A. and Daniel, L.W. (1987). Phospholipase D activity of *Vibrio damsela* cytolysin and its interaction with sheep erythrocytes. *Infect. Immun.* **55**, 3209–12.

Kuhn, M., Kathariou, S. and Goebel, W. (1988). Hemolysin supports survival but not entry of the intracellular bacterium *Listeria monocytogenes*. *Infect. Immun.* **56**, 79–82.

Ladant, D., Brezin, C., Alonso, J.-M., Crenon, I.

and Guiso, N. (1986). *Bordetella pertussis* adenylate cyclase. Purification, characterization, and radioimmunoassay. *J. Biol. Chem.* **261**, 16264–9.

Lally, E.T., Golub, E.E., Kieba, I.R., Taichman, N.S., Rosenbloom, J., Rosenbloom, J.C., Gibson, C.W. and Demuth, D.R. (1989). Analysis of the *Actinobacillus actinomycetemcomitans* leukotoxin gene. Delineation of unique features and comparison to homologous toxins. *J. Biol. Chem.* **264**, 15451–6.

Lalonde, G., McDonald, T.V., Gardner, P. and O'Hanley, P.D. (1989). Identification of a hemolysin from *Actinobacillus pleuropneumoniae* and characterization of its channel properties in planar phospholipid bilayers. *J. Biol. Chem.* **264**, 13559–64.

Lebek, G. and Grünig, H.-M. (1985). Relation between the hemolytic property and iron metabolism in *Escherichia coli*. *Infect. Immun.* **50**, 682–6.

Le Minor, S. and Le Coueffic, E. (1975). Studies on hemolysins of *Enterobacteriaceae*. *Ann. Microbiol. (Paris)* **126**, 313–32.

Letoffe, S., Delepelaire, P. and Wandersman, C. (1990). Protease secretion by *Erwinia chrysanthemi*: the specific secretion functions are analogous to those of *Escherichia coli* α-haemolysin. *EMBO J.* **9**, 1375–82.

Letoffe, S., Delepelaire, P. and Wandersman, C. (1991). Cloning and expression in *Escherichia coli* of the *Serratia marcescens* metalloprotease gene: secretion of the protease from *E. coli* in the presence of the *Erwinia chrysanthemi* protease secretion functions. *J. Bacteriol.* **173**, 2160–6.

Leusch, M.S., Paulaitis, S. and Friedman, R.L. (1990) Adenylate cyclase toxin of *Bordetella pertussis*: production, purification, and partial characterization. *Infect. Immun.* **58**, 3621–6.

Linggood, M.A. and Ingram, P.L. (1982). The role of alpha haemolysin in the virulence of *Escherichia coli* for mice. *J. Med. Microbiol.* **15**, 23–30.

Liu, P.V. (1959). Studies on the hemolysin of *Vibrio cholerae*. *J. Infect. Dis.* **104**, 238–59.

Liu, P.V. (1974). Extracellular toxins of *Pseudomonas aeruginosa*. *J. Infect. Dis.* **130** (Suppl.), S94–S99.

Liu, P.V. (1979). Toxins of *Pseudomonas aeruginosa*. In: *Pseudomonas aeruginosa: Clinical Manifestations of Infection and Current Therapy* (ed. R.G. Doggett), pp. 63–88. Academic Press, New York.

Ljungh, A. and Wadström, T. (1983). Toxins of *Vibrio parahaemolyticus* and *Aeromonas hydrophila*. *J. Toxicol. Toxin Rev.* **1**, 257–307.

Lo, R.Y.C., Shewen, P.E., Strathdee, C.A. and Greer, C.N. (1985). Cloning and expression of the leukotoxin gene of *Pasteurella haemolytica* A1 in *Escherichia coli* K-12. *Infect Immun.* **50**, 667–71.

Lo, R.Y.C., Strathdee, C.A. and Shewen, P.E. (1987). Nucleotide sequence of the leukotoxin genes of *Pasteurella haemolytica* A1. *Infect. Immun.* **55**, 1987–96.

Lory, S. and Tai, P.C. (1983). Characterization of the phospholipase C gene of *Pseudomonas aeruginosa* cloned in *Escherichia coli*. *Gene* **22**, 95–101.

Love, M., Teebken-Fisher, D., Hose, J.E., Farmer III,

J.J., Hickman, F.W. and Fanning, G.R. (1981). *Vibrio damsela*, a marine bacterium, causes skin ulcers on the damselfish *Chromis punctipinnis*. *Science* **214**, 1139–40.

Ludwig, A., Vogel, M. and Goebel, W. (1987). Mutations affecting activity and transport of haemolysin in *Escherichia coli*. *Molec. Gen. Genet.* **206**, 238–45.

Ludwig, A., Jarchau, T., Benz, R. and Goebel, W. (1988). The repeat domain of *Escherichia coli* haemolysin (HlyA) is responsible for its Ca^{2+}-dependent binding to erythrocytes. *Molec. Gen. Genet.* **214**, 553–61.

Ludwig, A., Schmid, A., Benz, R. and Goebel, W. (1991). Mutations affecting pore formation by haemolysin from *Escherichia coli*. *Molec. Gen. Genet.* **226**, 198–208.

Mackman, N. and Holland, I.B. (1984a). Secretion of a 107 K Dalton polypeptide into the medium from a haemolytic *E. coli* K12 strain. *Molec. Gen. Genet.* **193**, 312–15.

Mackman, N. and Holland, I.B. (1984b). Functional characterization of a cloned haemolysin determinant from *E.coli* of human origin, encoding information for the secretion of a 107 K polypeptide. *Molec. Gen. Genet.* **196**, 129–34.

Mackman, N., Nicaud, J.-M., Gray, L. and Holland, I.B. (1985a). Genetical and functional organisation of the *Escherichia coli* haemolysin determinant 2001. *Molec. Gen. Genet.* **201**, 282–8.

Mackman, N., Nicaud, J.-M., Gray, L. and Holland, I.B. (1985b). Identification of polypeptides required for the export of haemolysin 2001 from *E. coli*. *Molec. Gen. Genet.* **201**, 529–36.

Mackman, N., Nicaud, J.-M., Gray, L. and Holland, I.B. (1986). Secretion of haemolysin by *Escherichia coli*. *Curr. Top. Microbiol. Immunol.* **125**, 159–81.

Mackman, N., Baker, K., Gray, L., Haigh, R., Nicaud, J.-M. and Holland, I.B. (1987). Release of a chimeric protein into the medium from *Escherichia coli* using the C-terminal secretion signal of haemolysin. *EMBO J.* **6**, 2835–41.

Maki, D.G., Hennekens, C.G., Philips, C.W., Shaw, W.V. and Bennet, J.V. (1973). Nosocomial urinary tract infection with *Serratia marcescens*. *J. Infect. Dis.* **128**, 579–87.

Mandell, R.L. and Socransky, S.S. (1981). A selective medium for *Actinobacillus actinomycetemcomitans* and the incidence of the organism in juvenile periodontitis. *J. Periodontol.* **52**, 593–8.

Manning, P.A., Brown, M.H. and Heuzenroeder, M.W. (1984). Cloning of the structural gene (*hly*) for the haemolysin of *Vibrio cholerae* El Tor strain 017. *Gene* **31**, 225–31.

Markham, R.J.F. and Wilkie, B.N. (1980). Interaction between *Pasteurella haemolytica* and bovine alveolar macrophages. Cytotoxic effect on macrophages and impaired phagocytosis. *Am. J. Vet. Res.* **41**, 18–22.

Marre, R., Hacker, J., Henkel, W. and Goebel, W. (1986). Contribution of cloned virulence factors from

uropathogenic *Escherichia coli* strains to nephropathogenicity in an experimental rat pyelonephritis model. *Infect. Immun.* **54**, 761–7.

Masure, H.R. and Storm, D.R. (1989). Characterization of the bacterial cell associated calmodulin-sensitive adenylate cyclase from *Bordetella pertussis*. *Biochemistry* **28**, 438–42.

Masure, H.R., Au, D.C., Gross, M.K., Donovan, M.G. and Storm, D.R. (1990). Secretion of the *Bordetella pertussis* adenylate cyclase from *Escherichia coli* containing the hemolysin operon. *Biochemistry* **29**, 140–5.

Mercurio, A. and Manning, P.A. (1985). Cellular localization and export of the soluble haemolysin of *Vibrio cholerae* El Tor. *Molec. Gen. Genet.* **200**, 472–5.

Miller, V.L., Taylor, R.K. and Mekalanos, J.J. (1987). Cholera toxin transcriptional activator ToxR is a transmembrane DNA binding protein. *Cell* **48**, 271–9.

Minshew, B.H., Jorgensen, J., Counts, G.W. and Falkow, S. (1978). Association of hemolysin production, hemagglutination of human erythrocytes, and virulence for chicken embryos of extraintestinal *Escherichia coli* isolates. *Infect. Immun.* **20**, 50–4.

Miwatani, T. and Takeda, Y. (1976). *Vibrio parahaemolyticus: a Causative Bacterium of Food Poisoning.* Saikon, Tokyo.

Miyake, M., Honda, T. and Miwatani, T. (1988). Purification and characterization of *Vibrio metschnikovii* cytolysin. *Infect. Immun.* **56**, 954–60.

Miyake, M., Honda, T. and Miwatani, T. (1989). Effects of divalent cations and saccharides on *Vibrio metschnikovii* cytolysin-induced hemolysis of rabbit erythrocytes. *Infect. Immun.* **57**, 158–63.

Miyamoto, Y., Kato, T., Obara, Y., Akiyama, S., Takizawa, K. and Yamai, S. (1969). In vitro hemolytic characteristics of *Vibrio parahaemolyticus*: its close correlation with human pathogenicity. *J. Bacteriol.* **100**, 1147–9.

Monti-Bragadin, C., Samer, L., Rottini, G.D. and Pani, B. (1975). The compatibility of Hly factor, a transmissible element which controls α-haemolysin production in *Escherichia coli*. *J. Gen. Microbiol.* **86**, 367–9.

Morris, J.G. Jr. and Black, R.E. (1985). Cholera and other vibrioses in the United States. *New Engl. J. Med.* **312**, 343–50.

Morris, J.G. Jr., Wilson, R., Hollis, D.G., Weaver, R.E., Miller, H.G., Tacket, C.O., Hickman, F.W. and Blake, P.A. (1982). Illness caused by *Vibrio damsela* and *Vibrio hollisae*. *Lancet* i, 1294–7.

Morris, J.G. Jr., Wright, A.C., Roberts, D.M., Wood, P.K., Simpson, L.M. and Oliver, J.D. (1987). Identification of environmental *Vibrio vulnificus* isolates with a DNA probe for the cytotoxin-hemolysin gene. *Appl. Environ. Microbiol.* **53**, 193–5.

Müller, D., Hughes, C. and Goebel, W. (1983). Relationship between plasmid and chromosomal hemolysin determinants of *Escherichia coli*. *J. Bacteriol.* **153**, 846–51.

Nakahama, K., Yoshimura, K., Marumoto, R., Kikuchi, M., Sook Lee, I., Hase, T. and Matsubura,

H. (1986). Cloning and sequencing of *Serratia* protease gene. *Nucleic Acids Res.* **14**, 5843–55.

Nicaud, J.-M., Mackman, N., Gray, L. and Holland, I.B. (1985a). Characterisation of HlyC and mechanism of activation and secretion of haemolysin from *E. coli* 2001. *FEBS Lett.* **187**, 339–44.

Nicaud, J.-M., Mackman, N., Gray, L. and Holland, I.B. (1985b). Regulation of haemolysin synthesis in *E. coli* determined by HLY genes of human origin. *Molec. Gen. Genet.* **199**, 111–16.

Nishibuchi, M. and Kaper, J.B. (1985). Nucleotide sequence of the thermostable direct hemolysin gene of *Vibrio parahaemolyticus. J. Bacteriol.* **162**, 558–64.

Nishibuchi, M., Ishibashi, M., Takeda, Y. and Kaper, J.B. (1985). Detection of the thermostable direct hemolysin gene and related DNA sequences in *Vibrio parahaemolyticus* and other *Vibrio* species by the DNA colony hybridization test. *Infect. Immun.* **49**, 481–6.

Nishibuchi, M., Hill, W.E., Zon, G., Payne, W.L. and Kaper, J.B. (1986). Synthetic oligodeoxyribonucleotide probes to detect Kanagawa phenomenon-positive *Vibrio parahaemolyticus. J. Clin. Microbiol.* **23**, 1091–5.

Nishibuchi, M., Doke, S., Toizumi, S., Umeda, T., Yoh, M. and Miwatani, T. (1988). Isolation from a coastal fish of *Vibrio hollisae* capable of producing a hemolysin similar to the thermostable direct hemolysin of *Vibrio parahaemolyticus. Appl. Environ. Microbiol.* **54**, 2144–6.

Nishibuchi, M., Taniguchi, T., Misawa, T., Khaeomanee-iam, V., Honda, T. and Miwatani, T. (1989). Cloning and nucleotide sequence of the gene (*trh*) encoding the hemolysin related to the thermostable direct hemolysin of *Vibrio parahaemolyticus. Infect. Immun.* **57**, 2691–7.

Nishibuchi, M., Khaeomanee-iam, V., Honda, T., Kaper, J.B. and Miwatani, T. (1990). Comparative analysis of the hemolysin genes of *Vibrio cholerae* non-O1, *V. mimicus*, and *V. hollisae* that are similar to the *tdh* gene of *V. parahaemolyticus. FEMS Microbiol. Lett.* **67**, 251–6.

Noegel, A., Rdest, U., Springer, W. and Goebel, W. (1979). Plasmid cistrons controlling synthesis and excretion of the exotoxin α-haemolysin of *Escherichia coli. Molec. Gen. Genet.* **175**, 343–50.

Noegel, A., Rdest, U. and Goebel, W. (1981). Determination of the functions of hemolytic plasmid pHly152 of *Escherichia coli. J. Bacteriol.* **145**, 233–41.

Nomura, S., Fujino, M., Yamakawa, M. and Kawahara, E. (1988). Purification and characterization of salmolysin, an extracellular hemolytic toxin from *Aeromonas salmonicida. J. Bacteriol.* **170**, 3694–702.

Notermans, S., Havelaar, A., Jansen, W., Kozaki, S. and Guinee, P. (1986). Production of Asao toxin" by *Aeromonas* strains isolated from feces and drinking water. *J. Clin. Microbiol.* **23**, 1140–2.

Oropeza-Wekerle, R.L., Speth, W., Imhof, B., Gentschev, I. and Goebel, W. (1990). Translocation and compartmentalization of *Escherichia coli* hemolysin (HlyA). *J. Bacteriol.* **172**, 3711–17.

Ostroff, R.M. and Vasil, M.L. (1987). Identification of a new phospholipase C activity by analysis of an insertional mutation in the hemolytic phospholipase C structural gene of *Pseudomonas aeruginosa. J. Bacteriol.* **169**, 4597–601.

Ostroff, R.M., Wretlind, B. and Vasil, M.L. (1989). Mutations in the hemolytic-phospholipase C operon result in decreased virulence of *Pseudomonas aeruginosa* PAO1 grown under phosphate-limiting conditions. *Infect. Immun.* **57**, 1369–73.

Ostroff, R.M., Vasil, A.I. and Vasil, M.L. (1990). Molecular comparisons of a nonhemolytic and a hemolytic phospholipase C from *Pseudomonas aeruginosa. J. Bacteriol.* **172**, 5915–23.

Parker, M.T. (1984). Septic infections due to gram-negative aerobic bacilli. In: *Principles of Bacteriology, Virology and Immunity*, Vol.3. *Bacterial Diseases* (eds G. Wilson, A. Miles, M.T. Parker and G.R. Smith), pp.279–310. Edward Arnold, London.

Pellett, S., Boehm, D.F., Snyder, I.S., Rowe, G. and Welch, R.A. (1990). Characterization of monoclonal antibodies against the *Escherichia coli* hemolysin. *Infect. Immun.* **58**, 822–7.

Poole, K. and Braun, V. (1988). Iron regulation of *Serratia marcescens* hemolysin gene expression. *Infect. Immun.* **56**, 2967–71.

Poole, K., Schiebel, E. and Braun, V. (1988). Molecular characterization of the hemolysin determinant of *Serratia marcescens. J. Bacteriol.* **170**, 3177–88.

Pritchard, A.E. and Vasil, M.L. (1986). Nucleotide sequence and expression of a phosphate-regulated gene encoding a secreted hemolysin of *Pseudomonas aeruginosa. J. Bacteriol.* **167**, 291–8.

Rader, A.E. and Murphy, J.R. (1988). Nucleotide sequences and comparison of the hemolysin determinants of *Vibrio cholerae* El Tor RV79 (Hly⁺) and RV79 (Hly⁻) and classical 569B (Hly⁻). *Infect. Immun.* **56**, 1414–19.

Richardson, K., Michalski, J. and Kaper, J.B. (1986). Hemolysin production and cloning of two hemolysin determinants from classical *Vibrio cholerae. Infect. Immun.* **54**, 415–20.

Rogel, A., Schultz, J.E., Brownlie, R.M., Coote, J.G., Parton, R. and Hanski, E. (1989). *Bordetella pertussis* adenylate cyclase: purification and characterization of the toxic form of the enzyme. *EMBO J.* **8**, 2755–60.

Roland, F.B. (1977). Interaction of blood with *Enterobacteriaceae*. Hemolysis, hemagglutination, fibrino lysis. *Am. J. Clin. Pathol.* **67**, 260–3.

Roy, C. and Mukerjee, S. (1962). Variability in the haemolytic power of El Tor vibrios. *Ann. Biochem. Exp. Med.* **22**, 295–6.

Royer-Pokora, B. and Goebel, W. (1976). Plasmids controlling synthesis of hemolysin in *Escherichia coli*. II. Polynucleotide sequence relationship among hemolytic plasmids. *Molec. Gen. Genet.* **144**, 177–83.

Ruan, Y. and Braun, V. (1990). Hemolysin as a marker for *Serratia. Arch. Microbiol.* **154**, 221–5.

Sakazaki, R., Gomes, C.Z. and Sebald, M. (1967).

Taxonomical studies of the so-called NAG vibrios. *Jpn J. Med. Sci. Biol.* **20**, 265–80.

Sakazaki, R., Tamura, K., Kato, T., Obara, Y., Yamai, S. and Hobo, K. (1968). Studies on the enteropathogenic, facultatively halophilic bacteria, *Vibrio parahaemolyticus*. III. Enteropathogenicity. *Jpn J. Med. Sci. Biol.* **21**, 325–31.

Sansonetti, P.J., Ryter, A., Clerc, P., Maurelli, A.T. and Mounier, J. (1986). Multiplication of *Shigella flexneri* within Hela cells: lysis of the phagocytic vacuole and plasmid-mediated contact hemolysis. *Infect. Immun.* **51**, 461–9.

Scheffer, J., König, W., Hacker, J. and Goebel, W. (1985). Bacterial adherence and hemolysin production from *Escherichia coli* induces histamine and leukotriene release from various cells. *Infect. Immun.* **50**, 271–8.

Schiebel, E. and Braun, V. (1989). Integration of the *Serratia marcescens* haemolysin into human erythrocyte membranes. *Molec. Microbiol.* **3**, 445–53.

Schiebel, E., Schwarz, H. and Braun, V. (1989). Subcellular location and unique secretion of the hemolysin of *Serratia marcescens. J. Biol. Chem.* **264**, 16311–20.

Shattuck, R.L. and Storm, D.R. (1985). Calmodulin inhibits entry of *Bordetella pertussis* adenylate cyclase into animal cells. *Biochemistry* **24**, 6323–8.

Shen, B.-F., Tai, P.C., Pritchard, A.E. and Vasil, M.L. (1987). Nucleotide sequences and expression in *Escherichia coli* of the in-phase overlapping *Pseudomonas aeruginosa plcR* genes. *J. Bacteriol.* **169**, 4602–7.

Shewen, P.E. and Wilkie, B.N. (1982). Cytotoxin of *Pasteurella haemolytica* acting on bovine leukocytes. *Infect. Immun.* **35**, 91–4.

Shirai, H., Ito, H., Hirayama, T., Nakamoto, Y., Nakabayashi, N., Kumagai, K., Takeda, Y. and Nishibuchi, M. (1990). Molecular epidemiologic evidence for association of thermostable direct hemolysin (TDH) and TDH-related hemolysin of *Vibrio parahaemolyticus* with gastroenteritis. *Infect. Immun.* **58**, 3568–73.

Shope, R.E. (1964). Porcine contagious pleuropneumonia. I. Experimental transmission, etiology and pathology. *J. Exp. Med.* **119**, 357–68.

Slots, J. and Genco, R.J. (1984). Black-pigmented *Bacteroides* species, *Capnocytophaga* species, and *Actinobacillus actinomycetemcomitans* in human periodontal disease: virulence factors in colonization, survival, and tissue destruction. *J. Dent. Res.* **63**, 412–21.

Slots, J., Reynolds, H.S. and Genco, R.J. (1980). *Actinobacillus actinomycetemcomitans* in human periodontal disease: a cross-sectional microbiological investigation. *Infect. Immun.* **29**, 1013–20.

Smith, H.W. (1963). The haemolysins of *Escherichia coli. J. Pathol. Bacteriol.* **85**, 197–211.

Smith, H.W. and Halls, S. (1967). The transmissible nature of the genetic factor in *Escherichia coli* that controls haemolysin production. *J. Gen. Microbiol.* **47**, 153–61.

Smith, H.W. and Huggins, M.B. (1985). The toxic

role of alpha-haemolysin in the pathogenesis of experimental *Escherichia coli* infection in mice. *J. Gen. Microbiol.* **131**, 395–403.

Sokolovič, Z. and Goebel, W. (1989). Synthesis of listeriolysin in *Listeria monocytogenes* under heat shock conditions. *Infect. Immun.* **57**, 295–8.

Spitznagel, J. Jr., Kraig, E. and Kolodrubetz, D. (1991). Regulation of leukotoxin in leukotoxic and nonleukotoxic strains of *Actinobacillus actinomycetemcomitans. Infect. Immun.* **59**, 1394–1401.

Stamm, W.E., Martin, S.M. and Bennett, J.V. (1977). Epidemiology of nosocomial infections due to gramnegative bacilli: aspects relevant to development and use of vaccines. *J. Infect. Dis.* **136**, S151–S160.

Stark, J.M. and Shuster, C.W. (1982). Analysis of hemolytic determinants of plasmid pHly185 by Tn5 mutagenesis. *J. Bacteriol.* **152**, 963–7.

Stark, J.M. and Shuster, C.W. (1983). The structure of cloned hemolysin DNA from plasmid pHly185. *Plasmid* **10**, 45–54.

Stinson, M.W. and Hayden, C. (1979). Secretion of phospholipase C by *Pseudomonas aeruginosa. Infect. Immun.* **25**, 558–64.

Strathdee, C.A. and Lo, R.Y.C. (1987). Extensive homology between the leukotoxin of *Pasteurella haemolytica* A1 and the alpha-hemolysin of *Escherichia coli. Infect. Immun.* **55**, 3233–6.

Strathdee, C.A. and Lo. R.Y.C. (1989a). Cloning, nucleotide sequence, and characterization of genes encoding the secretion function of the *Pasteurella haemolytica* leukotoxin determinant. *J. Bacteriol.* **171**, 916–28.

Strathdee, C.A. and Lo, R.Y.C. (1989b). Regulation of expression of the *Pasteurella haemolytica* leukotoxin determinant. *J. Bacteriol.* **171**, 5955–62.

Swihart, K.G. and Welch, R.A. (1990a). The HpmA hemolysin is more common than HlyA among *Proteus* isolates. *Infect. Immun.* **58**, 1853–60.

Swihart, K.G. and Welch, R.A. (1990b). Cytotoxic activity of the *Proteus* hemolysin HpmA. *Infect. Immun.* **58**, 1861–9.

Tacket, C.O., Hickman, F., Pierce, G.V. and Mendoza, L.F. (1982). Diarrhea associated with *Vibrio fluvialis* in the United States. *J. Clin. Microbiol.* **16**, 991–2.

Taichman, N.S. and Wilton, J.M.A. (1981). Leukotoxicity of an extract from *Actinobacillus actinomycetemcomitans* for human gingival polymorphonuclear leukocytes. *Inflammation* **5**, 1–12.

Taichman, N.S., Dean, R.T. and Sanderson, C.J. (1980). Biochemical and morphological characterization of the killing of human monocytes by a leukotoxin derived from *Actinobacillus actinomycetemcomitans. Infect. Immun.* **28**, 258–68.

Takeda, Y., Taga, S. and Miwatani, T. (1978). Evidence that thermostable direct hemolysin of *Vibrio parahaemolyticus* is composed of two subunits. *FEMS Microbiol. Lett.* **4**, 271–4.

Taniguchi, H., Ohta, H., Ogawa, M. and Mizuguchi, Y. (1985). Cloning and expression in *Escherichia coli* of *Vibrio parahaemolyticus* thermostable direct hemolysin and thermolabile hemolysin genes. *J. Bacteriol.* **162**, 510–15.

Tanner, A.C.R., Haffer, C., Bratthall, G.T., Visconti, R.A. and Socransky, S.S. (1979). A study of the bacteria associated with advancing periodontitis in man. *J. Clin. Periodontol.* **6**, 278–307.

Tison, D.L. and Kelly, M.T. (1984). *Vibrio* species of medical importance. *Diagn. Microbiol. Infect. Dis.* **2**, 263–76.

Tsai, C.-C., McArthur, W.P., Baehni, P.C., Hammond, B.F. and Taichman, N.S. (1979). Extraction and partial characterization of a leukotoxin from a plaque-derived gram-negative microorganism. *Infect. Immun.* **25**, 427–39.

Tsai, C.-C., Shenker, B.J., DiRienzo, J.M., Malamud, D. and Taichman, N.S. (1984). Extraction and isolation of a leukotoxin from *Actinobacillus actinomycetemcomitans* with polymyxin B. *Infect. Immun.* **43**, 700–5.

Tschäpe, H. and Rische, H. (1974). Die Virulenzplasmide der *Enterobacteriaceae*. *Z. Allg. Mikrobiol.* **14**, 337–50.

Tsunasawa, S., Sugihara, A., Masaki, T., Narita, K., Sakiyama, F., Takeda, Y. and Miwatani, T. (1983). The primary structure of the toxin protein-hemolysin-*V. parahaemolyticus*. *Jap. Biochem. Soc.* **55**, 807 (in Japanese).

Uphoff, T.S. and Welch, R.A. (1990). Nucleotide sequencing of the *Proteus mirabilis* calcium-independent hemolysin genes (*hpm*A and *hpm*B) reveals sequence similarity with the *Serratia marcescens* hemolysin genes (*shl*A and *shl*B). *J. Bacteriol.* **172**, 1206–16.

Van den Bosch, J.F., Postma, P., de Graaff, J. and MacLaren, D.M. (1981). Haemolysis by urinary *Escherichia coli* and virulence in mice. *J. Med. Microbiol.* **14**, 321–31.

Van den Bosch, J.F., Emödy, L. and Ketyi, I. (1982a). Virulence of haemolytic strains of *Escherichia coli* in various animal models. *FEMS Microbiol. Lett.* **13**, 427–30.

Van den Bosch, J.F., Postma, P., Koopman, P.A.R., de Graaff, J., MacLaren, D.M., van Brenk, D.G. and Guinee, P.A.M. (1982b). Virulence of urinary and faecal *Escherichia coli* in relation to serotype, haemolysis and haemagglutination. *J. Hyg.* **88**, 567–77.

Vandepitte, J., de Geest, H. and Jousten, P. (1977). Subacute bacterial endocarditis due to *Actinobacillus actinomycetemcomitans*. Report of a case with a review of the literature. *J. Clin. Pathol.* **30**, 842–6.

Vasil, M.L., Berka, R.M., Gray, G.L. and Nakai, H. (1982). Cloning of a phosphate-regulated hemolysin gene (phospholipase C) from *Pseudomonas aeruginosa*. *J. Bacteriol.* **152**, 431–40.

Vogel, M., Hess, J., Then, I., Juarez, A. and Goebel, W. (1988). Characterization of a sequence (*hly*R) which enhances synthesis and secretion of hemolysin in *Escherichia coli*. *Molec. Gen. Genet.* **212**, 76–84.

Von Mechow, S., Vaidya, A.B. and Bramucci, M.G. (1985). Mapping of a gene that regulates hemolysin production in *Vibrio cholerae*. *J. Bacteriol.* **163**, 799–802.

Waalwijk, C., van den Bosch, J.F., MacLaren, D.M. and de Graaff, J. (1982). Hemolysin plasmid coding for the virulence of a nephropathogenic *Escherichia coli* strain. *Infect. Immun.* **35**, 32–7.

Waalwijk, C., MacLaren, D.M. and de Graaff, J. (1983). In vivo function of hemolysin in the nephropathogenicity of *Escherichia coli*. *Infect. Immun.* **42**, 245–9.

Waalwijk, C., de Graaff, J. and MacLaren, D.M. (1984). Physical mapping of hemolysin plasmid pCW2, which codes for virulence of a nephropathogenic *Escherichia coli* strain. *J. Bacteriol.* **159**, 424–6.

Wagner, W., Vogel, M. and Goebel, W. (1983). Transport of hemolysin across the outer membrane of *Escherichia coli* requires two functions. *J. Bacteriol.* **154**, 200–10.

Wagner, W., Kuhn, M. and Goebel, W. (1988). Active and inactive forms of hemolysin (HlyA) from *Escherichia coli*. *Biol. Chem. Hoppe-Seyler* **369**, 39–46.

Wall, V.W., Kreger, A.S. and Richardson, S.H. (1984). Production and partial characterization of a *Vibrio fluvialis* cytotoxin. *Infect. Immun.* **46**, 773–7.

Wandersman, C. and Delepelaire, P. (1990). TolC, an *Escherichia coli* outer membrane protein required for hemolysin secretion. *Proc. Natl. Acad. Sci. USA* **87**, 4776–80.

Weiss, A.A. and Hewlett, E.L. (1986). Virulence factors of *Bordetella pertussis*. *Ann. Rev. Microbiol.* **40**, 661–86.

Weiss, A.A., Hewlett, E.L., Myers, G.A. and Falkow, S. (1983). Tn5-induced mutations affecting virulence factors of *Bordetella pertussis*. *Infect. Immun.* **42**, 33–41.

Weiss, A.A., Hewlett, E.L., Myers, G.A. and Falkow, S. (1984). Pertussis toxin and extracytoplasmic adenylate cyclase as virulence factors of *Bordetella pertussis*. *J. Infect. Dis.* **150**, 219–22.

Welch, R.A. (1987). Identification of two different hemolysin determinants in uropathogenic *Proteus* isolates. *Infect. Immun.* **55**, 2183–90.

Welch, R.A. (1991). Pore-forming cytolysins of Gram-negative bacteria. *Molec. Microbiol.* **5**, 521–8.

Welch, R.A., and Falkow, S. (1984). Characterization of *Escherichia coli* hemolysins conferring quantitative differences in virulence. *Infect. Immun.* **43**, 156–60.

Welch, R.A. and Pellett, S. (1988). Transcriptional organization of the *Escherichia coli* hemolysin genes. *J. Bacteriol.* **170**, 1622–30.

Welch, R.A., Dellinger, E.P., Minshew, B. and Falkow, S. (1981). Haemolysin contributes to virulence of extra-intestinal *E. coli* infections. *Nature* **294**, 665–7.

Welch, R.A., Hull, R. and Falkow, S. (1983). Molecular cloning and physical characterization of a chromosomal hemolysin from *Escherichia coli*. *Infect. Immun.* **42**, 178–86.

Wickboldt, L.G. and Sanders, C.V. (1983). *Vibrio vulnificus* infection. Case report and update since 1970. *J. Am. Acad. Dermatol.* **9**, 243–51.

Wolff, J., Cook, G.H., Goldhammer, A.R. and

Berkowitz, S.A. (1980). Calmodulin activates prokaryotic adenylate cyclase. *Proc. Natl. Acad. Sci. USA* **77**, 3841–4.

Wright, A.C. and Morris, J.G. Jr. (1991). The extracellular cytolysin of *Vibrio vulnificus*: inactivation and relationship to virulence in mice. *Infect. Immun.* **59**, 192–7.

Wright, A.C., Morris, J.G. Jr., Maneval, D.R. Jr., Richardson, K. and Kaper, J.B. (1985). Cloning of the cytotoxin-hemolysin gene of *Vibrio vulnificus*. *Infect. Immun.* **50**, 922–4.

Yamamoto, K., Ichinose, Y., Nakasone, N., Tanabe, M., Nagahama, M., Sakurai, J. and Iwanaga, M. (1986). Identity of hemolysins produced by *Vibrio cholerae* non-O1 and *V. cholerae* O1, biotype El Tor. *Infect. Immun.* **51**, 927–31.

Yamamoto, K., Wright, A.C., Kaper, J.B. and Morris, J.G. Jr. (1990). The cytolysin gene of *Vibrio vulnificus*: sequence and relationship to the *Vibrio cholerae* El Tor hemolysin gene. *Infect. Immun.* **58**, 2706–9.

Yoh, M., Honda, T. and Miwatani, T. (1986). Purification and partial characterization of a non-O1 *Vibrio cholerae* hemolysin that cross-reacts with thermostable direct hemolysin of *Vibrio parahaemolyticus*. *Infect. Immun.* **52**, 319–22.

Young, L.S. (1980). The role of exotoxins in the pathogenesis of *Pseudomonas aeruginosa*. *J. Infect. Dis.* **142**, 626–30.

Zabala, J.C., de la Cruz, F. and Ortiz, J.M. (1982). Several copies of the same insertion sequence are present in alpha-hemolytic plasmids belonging to four different incompatibility groups. *J. Bacteriol.* **151**, 472–6.

Zabala, J.C., Garcia-Lobo, J.M., Diaz-Aroca, E., de la Cruz, F. and Ortiz, J.M. (1984). *Escherichia coli* alpha-haemolysin synthesis and export genes are flanked by a direct repetition of IS91-like elements. *Molec. Gen. Genet.* **197**, 90–7.

Zambon, J.J., DeLuca, C., Slots, J. and Genco, R.J. (1983). Studies of leukotoxin from *Actinobacillus actinomycetemcomitans* using the promyelocytic HL-60 cell line. *Infect. Immun.* **40**, 205–12.

8

The Family of the Antigenically-Related, Cholesterol-Binding ('Sulphydryl-Activated') Cytolytic Toxins*

Joseph E. Alouf and Christiane Geoffroy

Bacterial Antigens Unit, CNRS-URA 557, Institut Pasteur, 75724 Paris Cédex 15, France

Introduction

The so-called sulphydryl-activated (also known as oxygen-labile) toxins are a family of 50- to 60-kDa single-chain bacterial proteins sharing the following characteristics: (i) they are lethal (cardiotoxic) to animals (and probably humans); (ii) they are highly potent lytic agents toward eukaryotic cells including erythrocytes (hence the name 'haemolysins' often used for these toxins); (iii) their lytic and lethal properties are suppressed by sulphydryl-group blocking agents and restored by thiols or other reducing agents; (iv) crude or partially purified toxin preparations stored (in the cold) in the presence of air or oxygen progressively lose their lytic properties which are immediately restored by reducing agents, suggesting an 'oxygen-labile' character of the toxins; (v) the toxins are antigenically related (common epitopes) as evidenced by the elicitation in humans or immunized animals of neutralizing and precipitating cross-reacting antibodies; (vi) the lethal and lytic properties are irreversibly lost in the presence of very low (nanomolar) concentrations of cholesterol and other related 3β-hydroxysterols; (vii) membrane cholesterol is thought to be the toxin-binding site at the surface of eukaryotic cells.

Among the various structurally and functionally related toxin families (some of them are reviewed in other chapters), the SH-activated toxin group is the largest. Presently 19 taxonomically different species of Gram-positive aerobic or anaerobic, sporulating or non-sporulating bacteria from the genera *Streptococcus*, *Bacillus*, *Clostridium* and *Listeria* are known to produce the toxins (Table 1). New members may remain to be discovered.

Except for pneumolysin, which is an intracytoplasmic toxin (Johnson, 1977), all the other toxins are secreted in the extracellular medium during bacterial growth. Among the toxin-producing bacteria, only *Listeria* species are intracellular pathogens which grow and release their toxins in the phagocytic cells of the host (Mackaness, 1962; Gaillard *et al.*, 1987).

A number of general, specific or limited reviews have been published in the past 15 years (Table 2). However, since the comprehensive reviews of Bernheimer (1976) and Smyth and Duncan (1978), no new overview covering the substantial information on this toxin family accumulated over the last decade has been provided. In this chapter we attempted to offer an account of the impressive progress realized during this period in an increasing number of laboratories. This review will particularly focus on (i) the characterization and the molecular genetics of the toxins which is presently an intensively active area of investigation, (ii) the mechanism(s) of toxin-induced damage, (iii)

---

* This chapter is dedicated to the memory of Daniel Prigent (1947–86).

SOURCEBOOK OF BACTERIAL PROTEIN TOXINS
ISBN 0-12-053078-3

Table I. The family of the antigenically-related, cholesterol-binding ('sulphydryl-activated') cytolytic toxins

Bacterial genus	Species	Toxin name	Synonymic name
Streptococcus	S. pyogenes[a]	Streptolysin O*	
	S. pneumoniae	Pneumolysin*	
Bacillus	B. cereus	Cereolysin O*	
	B. alvei	Alveolysin*	
	B. thuringiensis	Thuringiolysin O*	
	B. laterosporus	Laterosporolysin	
Clostridium	C. tetani	Tetanolysin*	
	C. botulinum	Botulinolysin	
	C. perfringens	Perfringolysin O*	θ-toxin
	C. septicum	Septicolysin O	δ-toxin
	C. histolyticum	Histolyticolysin O	ϵ-toxin
	C. oedematiens[b]	Oedematolysin O	δ-toxin
	C. chauvoei	Chauveolysin	δ-toxin
	C. bifermentans	Bifermentolysin	
	C. sordellii	Sordellilysin	
	C. caproicum[c]	Caproiciolysin	
Listeria	L. monocytogenes	Listeriolysin O*	
	L. ivanovii	Ivanolysin*	
	L. seeligeri	Seeligerolysin	

[a]Some strains of streptococci of groups C and G also produce streptolysin O.
[b]Type A.
[c]It is still not clear whether this organism is a taxonomically defined species.
*Toxins reported to be purified to apparent homogeneity (see text).
General references: Bernheimer (1976), Alouf (1977, 1980), Smyth and Duncan (1978), Stephen and Pietrowski (1986), Geoffroy et al. (1989), Hatheway (1990).

structure–activity relationship, (iv) pathophysiological aspects of toxin effects.

The problems of toxin nomenclature

The toxins are known in the literature under various denominations: oxygen-labile, (oxygen-sensitive), thiol- or sulphydryl-activated (or -dependent) toxins, lysins, cytolysins or haemolysins ('haemotoxins' in earlier publications). On the basis of our present knowledge, most of these denominations are inappropriate and misleading. For example, the term haemolysin has been and is still often used because toxin identification or detection is in most cases based on the lysis of red blood cells. However, the toxins are not only lytic to erythrocytes but lyse or damage many other eukaryotic cells (all cells so far tested) (see later, and reviews of Bernheimer, 1976; Alouf, 1977, 1980; Smyth and Duncan, 1978). Therefore, the term cytolysin (cytolytic toxin) should be preferably used instead of the restrictive term haemolysin (Bernheimer and Rudy, 1986).

'Oxygen-labile/oxygen-sensitive' toxins

These denominations, which are frequently used though to a lesser extent presently, stem from early observations (see Neill and Mallory, 1926; Todd, 1934; Herbert and Todd, 1941) that the

Table 2. Reviews on the antigenically related cholesterol-binding (SH-activated) cytolytic toxins published from 1976 to 1989

Whole toxin family

Comprehensive reviews	Bernheimer (1976), Smyth and Duncan (1978)
Workshop review	Alouf and Geoffroy (1984)
Short reviews[a]	Freer and Arbuthnott (1976), Alouf (1977)
	Bernheimer and Rudy (1986)
	Stephen and Pietrowski (1986)
	Freer (1986, 1988)

Streptolysin O

Comprehensive review	Alouf (1980)
Short reviews[a]	Wannamaker (1983)
	Bhakdi and Tranum-Jensen (1988)
	Wannamaker and Schlievert (1988)

Perfringolysin O

Comprehensive review	Sato (1986)
Short review[a]	McDonel (1980)

Cereolysin O

Short review[a]	Turnbull (1981)

Listeriolysin O

Short review[b]	Cossart and Mengaud (1989)

[a]Part of general reviews on other cytolytic or non-cytolytic toxins.
[b]Part of a general review on the pathogenicity of *L. monocytogenes*.

haemolytic activities of crude or partially purified preparations of pneumolysin, tetanolysin, perfringolysin O (θ-toxin) and streptolysin O progressively disappeared on standing in air and less rapidly in the absence of oxygen, but were restored to the original level by reducing agents. These observations were confirmed for most unpurified preparations of the toxins listed in Table 1 (see Guillaumie, 1950; Njoku-Obi *et al.*, 1963; Halbert, 1970; Lemeland *et al.*, 1974; Alouf *et al.*, 1977; Sato, 1986; Suzuki *et al.*, 1988; Stevens *et al.*, 1988 and the general reviews mentioned above). The loss of the lytic activity was attributed (without direct chemical proof) to the oxidation of toxin molecules by O_2. However, this hypothesis was not supported by the finding that highly purified preparations of streptolysin O (SLO), alveolysin (ALV) and perfringolysin O (PFO) did not lose their haemolytic activities by flushing air or O_2 in

their solution for several hours (see Alouf, 1980; Sato, 1986). The loss observed in crude preparations is probably due to a kinetically slow O_2-favoured blockade of toxin SH-group by undefined molecules from culture media or bacterial metabolism. Therefore, the term oxygen-labile should be abandoned.

'Sulphydryl-activated' toxins

This denomination or its equivalents (thiol-activated/-dependent toxins) currently used are based on the observations that various thiols and other reducing agents restore the lytic and lethal activities of the toxins previously suppressed by interaction with sulphydryl-group blocking reagents (see Bernheimer, 1976; Smyth and Duncan, 1978; Johnson and Aultman, 1977; Mitsui and Hase, 1979; Alouf and Geoffroy, 1984; Geoffroy *et al.*,

1987). It has been generally assumed, as first hypothesized by Smythe and Harris (1940) and Herbert and Todd (1941) that the oxidation–reduction state involved shifts between an oxidized form with a disulphide bond between two cysteinyl residues, and reduced form with free SH-groups. The disulphide-bridge cleavage hypothesis of activation (see Smyth and Duncan, 1978; Alouf and Geoffroy, 1984) appeared to be supported by the report of the occurrence of more than one cysteinyl residue in cereolysin (Cowell *et al.*, 1976), alveolysin (Geoffroy *et al.*, 1981) and pneumolysin (Alouf and Geoffroy, 1984). In the case of cereolysin, it was reported that toxin activation (reduction) produced a net change of the molecule reflected by differences in the electrophoretic migration of the oxidized and reduced forms in polyacrylamide gels.

In contrast to these data, Mitsui *et al.* (1973) did not detect cysteine residues in PFO which was later shown (Yamakawa *et al.*, 1977; Ohno-Iwashita *et al.*, 1986) to contain only one Cys residue in the performic acid-oxidized derivative of this toxin. This finding is in agreement with the recent finding of one Cys residue deduced from the nucleotide sequences of the structural genes of SLO, PFO, pneumolysin, listeriolysin O and alveolysin (see Figs 1 and 2). This could also be the same for the other 'SH-activated' toxins.

The finding of a single Cys in the toxins appeared to support the concept of a functionally 'essential' sulphydryl group (Geoffroy *et al.*, 1981; Alouf and Geoffroy, 1984; Walker *et al.*, 1987). However, this contention is no longer true. As indicated later, the replacement of the Cys residue in SLO, pneumolysin and listeriolysin O by other amino acids by site-directed mutagenesis generated haemolytically active recombinant molecules. Thereby, the appellation 'sulphydryl-activated toxin' is a misnomer. However, it has been suggested by Pinkney *et al.* (1989) to maintain it because it is a well-established name.

Proposal of a new generic denomination

We think that a generic name reflecting a common, specific, unchallengeable intrinsic property of these membrane-damaging toxins should be preferably used. We propose the denomination 'cholesterol-binding cytolytic toxins (or cytolysins)'. The term

'antigenically related' could be added when necessary for differentiation from other cholesterol-inhibitable cytolysins such as *Pseudomonas pseudomallei* haemolysin (Ashdow and Koehler, 1990). The specific names of each individual member of the toxin family (Table 1) coined by or according to Bernheimer (1976) should be maintained including the appended letter O (reflecting the supposed O_2-lability) due to its wide usage and for differentiation from other homonymic toxins.

Characterization and purification

Among the toxins listed in Table 1, only nine (designated by an asterisk) have been purified to apparent homogeneity to allow a sufficient knowledge of their properties. The other toxins of the group have been practically disregarded since their discovery (see Guillaumie, 1950; Oakley *et al.*, 1947; Moussa, 1958; Rutter and Collee, 1969; Hatheway, 1990). Since they are still poorly characterized these toxins will not be considered here.

This section will deal with the recent progress in the purification and characterization of the nine toxins investigated in this respect in the past decade. Earlier works will be briefly mentioned when necessary. Detailed information can be found in the reviews listed in Table 2. For the historical background and relevant literature concerning the discovery or preliminary characterization of these toxins, the reader is referred to the publications mentioned in Table 3.

Before describing the recent methods used for toxin purification we have to mention that there is still no agreement among the various laboratories for a unified standard technique for the determination of the haemolytic activity of the toxins expressed as haemolytic units (HU). Therefore, the specific activities of the purified toxins (HU/mg of protein) reported in the literature are not always comparable.

Streptolysin O (SLO)

This toxin (M_r 60 151) is produced in culture media by most strains of group A streptococci (*Streptococcus pyogenes*) and some strains of groups C and G streptococci, particularly those causing

Table 3. Early records on the existence or characterization of the toxins purified to date

	Number of publications	References
Streptolysin O	31 (1895–1938)	1–4
Pneumolysin	6 (1905–42)	5–7
Cereolysin O	6 (1930–67)	7, 8
Alveolysin	1 (1967)	7
Thuringiolysin O	1 (1973)	9
Tetanolysin	7 (1898–1927)	10–12
Perfringolysin O	13 (1917–41)	12–15
Listeriolysin O	6 (1934–63)	16, 17
Ivanolysin	4 (1954–78)	17–19

(1) Neill and Mallory (1926); (2) Hewitt and Todd (1941); (3) Halbert (1970); (4) Alouf (1980); (5) Neill (1926a); (6) Shumway and Klebanoff (1971); (7) Bernheimer (1976); (8) Turnbull (1981); (9) Pendleton et al. (1973); (10) Neill (1926c); (11) Fleming (1927); (12) Willis (1969); (13) Neill (1926b); (14) Ispolatovskaya (1971); (15) McDonel (1980); (16) Sword and Kingdon (1971); (17) Seeliger et al. (1982); (18) Vazquez-Boland et al. (1989b); (19) Geoffroy et al. (1989).

human infections (see Herbert and Todd, 1941; Halbert, 1970; Alouf, 1980; Tiesler and Trinks, 1982; Suzuki et al., 1988). Whether the SLO produced by the last two groups is identical to that released by group A strains is still not documented since neither toxin purification, nor gene cloning have been so far reported for groups C and G toxin.

Toxin production in vivo following streptococcal infections in humans is reflected by the occurrence of anti-SLO antibodies in blood plasma (Halbert, 1970; Watson and Kerr, 1985) the determination of which is of great clinical importance (Ferrieri, 1986). A direct evidence of toxin production in vivo was reported by Duncan (1983) who implanted in mice peritoneal cavities micropore filter chambers into which streptococci were grown. Bacterial growth of several different strains, SLO production and antibody response were investigated.

Appropriate culture media have been reported for the production of high amounts of SLO (see Alouf, 1980; Bhakdi et al., 1984; Alouf and Geoffroy, 1988). Individual group A strains vary considerably in the amounts of toxin produced under the same culture conditions (see Halbert, 1970; Tiesler and Trinks, 1979). A list of the strains used for optimal toxin production (500–1000 HU/ml in appropriate media) has been published (Alouf, 1980). The variability of SLO production by the various group A strains tested suggests the involvement of regulation mechanisms

of toxin synthesis and excretion at both genetic and environmental levels which still await investigation. It has been shown, for example, that at concentrations of antibiotics too low to affect bacterial growth, SLO production was suppressed by lincomycin and clindamycin but not by chloramphenicol or erythromycin (Gemmel and Abdul-Amir, 1979).

Several procedures for SLO purification have been reported in the past ten years. An immuno-affinity chromatographic technique taking advantage of the antigenic relationship between SLO and the other related toxins was reported by Linder (1979). The purification process involved culture supernatant precipitation by 80% $(NH_4)_2SO_4$, AH-Sepharose chromatography, immunochromatography on antitetanolysin coupled to CNBr-activated Sepharose 4B and acid elution. The purified SLO material was 80% homogeneous and had a specific activity of 108 771 HU/mg of protein. SDS–PAGE showed a major band corresponding to a molecular weight of 56 000 and two minor bands. A better homogeneity could be obtained by using additional steps. Similar results were obtained by the same technique for the isolation of B. cereus cereolysin (Linder, 1979).

SLO was purified by a classical method described by Bhakdi et al. (1984) involving ammonium sulphate and polyethylene glycol precipitation, DEAE–ion exchange chromatography, preparative isoelectric focusing and chromatography on

Sephacryl S-300. Two forms of the toxin possessing similar haemolytic activity were isolated: a native form of 69 000 Da with a pI at pH 6.0–6.4 and a sedimentation coefficient of 3.9S and a proteolysed form of 57 000 Da (pI 7.0–7.5). The molecular weights were determined at sedimentation equilibrium in an ultracentrifuge. The two forms appeared homogeneous by SDS–PAGE (the calculated M_r values were 80 and 72 kDa respectively) and their specific activity was 8×10^5 HU/mg protein according to the definition proposed by Alouf (1980).

A procedure based on the selective separation of SH-proteins by thiol-disulphide covalent chromatography on thiopropyl Sepharose (Alouf, 1980) allowed also the purification of SLO to apparent electrophoretic homogeneity (5×10^5 HU/mg protein) as described in detail by Alouf and Geoffroy (1988). The determination by Edman degradation of the sequence of the first four amino acids of the purified proteins (NH$_2$-Ser–Asp–Glu–Asp) reported by Alouf et al. (1986) was in agreement with that deduced by Kehoe et al. (1987) from the nucleotide sequence of the slo gene. The purification process involved the following steps: concentration of the culture fluid by ultrafiltration, batchwise adsorption of the retentate on calcium phosphate gel, elution with sodium phosphate, thiol–Sepharose 6B column chromatography, gel filtration on Sephacryl S-200, preparative isoelectric focusing (if necessary) and gel filtration on Biogel P100. The apparent molecular weight of SLO determined by SDS–PAGE ranged from 55 to 60 kDa depending on the batches processed. The purification of alveolysin (Geoffroy and Alouf, 1983), perfringolysin O (Alouf et al. unpublished; Tweten, 1988a), thuringiolysin O (Alouf and Rakotobe, unpublished), pneumolysin (Kanclerski and Möllby, 1987), listeriolysin O (Geoffroy et al., 1987) and ivanolysin (Vazquez-Boland et al., 1989b) have also been realized by chromatography on thiopropyl–Sepharose 6B.

Pneumolysin (PLY)

Few data are available in the literature about the frequency with which Streptococcus pneumoniae strains produce pneumolysin (M_r 52 800) but most, if not all, clinical isolates appear to do so in vivo and in vitro (see Kanclerski and Möllby,

1987; Kalin et al., 1987; Kanclerski et al., 1987, for data and references and also Johnson et al., 1982a). As mentioned above, this toxin is intracellular in contrast to the other related toxins but due to autolysis it is often detected extracellularly. The kinetics of toxin production and bacterial growth has been investigated by Kanclerski and Möllby (1987). The optimal intracellular pneumolysin titre (3000 HU/ml) was reached between 5 and 8 h.

Several procedures for PLY purification have been reported in the last two decades. All techniques used bacterial extracts as crude material. A preparation purified to apparent homogeneity by SDS–PAGE exhibiting a specific activity of 5.5×10^5 HU/mg of protein was first reported by Shumway and Klebanoff (1971). The bacteria grown in a fermenter were disrupted on ice by ultrasonic vibration. The crude extract was then submitted to ammonium sulphate precipitation, ion-exchange chromatography, gel filtration on Sephadex G-100 and preparative acrylamide gel electrophoresis. A similar purification process, except the last step, was used by Johnson (1972) after cell disruption in a French press. The preparation obtained (4.5×10^5 HU/mg of protein) showed a single band by SDS–PAGE. It was later shown by Johnson et al. (1982b) that pneumolysin as well as alveolysin, cereolysin and SLO were hydrophobic in character. This property was used by these authors to purify crude PLY material by fractionation on phenyl-Sepharose followed by chromatography on hydroxyapatite. The specific activity of the toxin obtained was 9.6×10^5 HU/mg of protein. High-pressure hydrophobic interaction chromatography on a TSK-phenyl 5PW column was also used by Mitchell et al. (1989) for the purification of recombinant PLY expressed in E. coli. The specific activity found was 2.4×10^6 HU/mg of protein. Similarly, recombinant PLY expressed in B. subtilis was partially purified on phenyl-Sepharose column (Taira et al., 1989).

Paton and Ferrante (1983) reported toxin purification from S. pneumoniae crude material obtained by cell autolysis with deoxycholate submitted to (NH$_4$)$_2$SO$_4$ salting-out, ion-exchange chromatography and gel filtration on Sephacryl G-200. The toxin (specific activity 5×10^5 HU/mg of protein) was homogeneous by SDS–PAGE and had an apparent molecular weight of 52 000. This procedure was improved later in the same laboratory by Lock *et al.* (1988) who proceeded first to a 38-fold concentration of pneumococcal culture on hollow-fibre cartridges prior to deoxycholate lysis followed by minor modifications. The specific activity of the toxin obtained was *ca.* 1.2×10^6 HU/mg of protein.

Kanclerski and Möllby (1987) described a 10-litre scale fermenter production and purification procedure involving 10-fold culture concentration before centrifugation, cell disruption in an X-press, ion-exchange chromatography on DEAE–Sepharose C16B, covalent thiopropyl-Sepharose 6B chromatography and gel filtration on Sephacryl S-200. The toxin exhibited a single band in SDS gel. The apparent molecular weight was 53 000 and the pI 5.2. The specific activity was 1.4×10^6 HU/mg of protein. The yield was 66% with respect to the original activity of the starting material which is much higher than that reported by the other authors.

Cereolysin O (CLO)

The haemolytic activity of *Bacillus cereus* cultures was known by 1930 (see Turnbull, 1981) and further work pointed to the production of more than one lytic protein (Coolbaugh and Williams, 1978). Presently, three different extracellular cytolysins are known: cereolysin first characterized and designated under this name by Bernheimer and Grushoff (1967a) as an SLO-like cholesterol-inhibitable toxin, cereolysin AB a bifactorial enzymatic component (Gilmore *et al.*, 1989) and another bicomponent haemolysin designated BL (Beecher and Macmillan, 1990). The last two lytic proteins are unrelated to cereolysin and to the other toxins reviewed here. On the basis of the nomenclature proposed by Bernheimer (1976) we propose the term cereolysin O for its differentiation from the other cereolysins.

CLO purification to homogeneity was reported by Cowell *et al.* (1976). A 'secondary' culture supernatant resulting from a bacterial suspension obtained in a preliminary culture was precipitated by ammonium sulphate and the solubilized material obtained was submitted to isoelectric focusing and to gel filtration. The purified CLO (specific activity *ca.* $4–5 \times 10^5$ HU/mg of protein) was characterized

as a single-chain polypeptide with a pI of 6.5–6.6 in the 'reduced' form and a molecular weight of 55 000 as found by SDS–PAGE and gel filtration. Amino acid analysis showed that the toxin contained 518 residues. The calculated molecular weight was 55 636.

Alveolysin (ALV)

This name has been attributed by Bernheimer (1976) to the 'SH-activated' toxin produced by *Bacillus alvei* as first reported by Bernheimer and Grushoff (1967b). No further characterization was done until the systematic study of this toxin by Alouf and his co-workers. *B. alvei* is a soil organism which is often found in honeybee larvae suffering from European foulbrood (Gordon *et al.*, 1973). It is not considered as pathogenic in vertebrates. However, three cases of human opportunistic infection have been reported to date (Reboli *et al.*, 1989).

Alveolysin was purified to homogeneity by Alouf *et al.* (1977) by ammonium sulphate precipitation, gel filtration, isoelectric focusing and ion-exchange chromatography. Toxin production was recently shown to be enhanced by *ca.* 10-fold (2×10^4 HU/ml) in *B. alvei* shaken cultures in the presence of activated charcoal (Geoffroy and Alouf, unpublished data). An improved purification method based on thiol–disulphide interchange chromatography was reported (Geoffroy and Alouf, 1983). The recovery was 40%. The purified toxin (specific activity 10^6 HU/mg of protein) had an apparent molecular weight of 63 000 and a pI of 5.1 (Geoffroy *et al.*, 1981). The sequence of the 28 first N-terminal residues determined by Alouf *et al.* (1986) was in agreement (except for residue 24) with that of the primary structure of the toxin deduced from the nucleotide sequence of the gene (Geoffroy *et al.*, 1990) which showed that alveolysin was a 51 766-Da protein containing only one Cys residue. The previously reported finding of four Cys residues (Geoffroy *et al.*, 1981) is likely due to contaminating thiols.

Thuringiolysin O (TLO)

This toxin produced by *Bacillus thuringiensis* was first characterized as a SLO-like toxin by Pendleton *et al.* (1973) who obtained a partially purified preparation by precipitation of culture supernatants by ammonium sulphate followed by isoelectric focusing. Thuringiolysin O was shown to focus into two distinct peaks of haemolytic activity at pI 6.0 and pI 6.5. The estimated molecular weight was 47 000. This toxin has been recently purified to homogeneity by Rakotobe and Alouf (unpublished) by ultrafiltration of the culture filtrates of strain H1-30 serotype 1 followed by chromatography on thiopropyl–Sepharose 6B column and ion-exchange chromatography on DEAE–cellulose column. The specific activity of the purified toxin was 7×10^5 HU/mg of protein. A single band was observed by SDS–PAGE corresponding to a molecular weight of 50 000 Da. The toxin shared all the properties of the other 'classical' antigenically related cholesterol-binding cytolysins.

Tetanolysin (TLY)

This toxin was first recognized in 1898 by Ehrlich (see Neill, 1926c; Willis, 1969) who clearly differentiated it from the 'true killing toxin' of *C. tetani* (tetanus neurotoxin) as established later in more detail by Fleming (1927) and Hardegree (1965). Little information is available on tetanolysin production by *C. tetani* strains. Lucain and Piffaretti (1977) reported a comparative study of the tetanolysins produced by nine different Tulloch serotypes of *C. tetani* as regards molecular weights, pI values and electrophoretic mobilities. All tetanolysin preparations were similar in these respects. The molecular weights measured by gel filtration were $47 000 \pm 3000$ and the pI values determined by isoelectric focusing in sucrose gradient were 6.54 for a major form and 6.15 and 5.84 for minor forms. Tetanolysin purification from culture filtrates of Massachusetts C2 strain and Harvard A-47 strain was reported by Alving *et al.* (1979) and Mitsui *et al.* (1980) respectively. The procedure used by the former group involved ammonium sulphate precipitation, gel filtration and hydroxylapatite chromatography. This procedure was modified by Rottem *et al.* (1990) who passed on a Mono-Q column the material eluted from the Sephadex G-100 column. In both procedures, the specific activity of the purified toxin was 10^6 HU/mg. The molecular weight was 45 000. Two forms of pI 6.1 and 6.4 were found (Blumenthal and Habig, 1984).

Tetanolysin purification by Mitsui *et al.* (1980) involved ammonium sulphate salting-out, acetone precipitation and two successive gel filtrations on Sephadex G-100. The toxin preparation (specific activity 5×10^5 HU/mg of protein) occurred as a mixture of four haemolytic entities (48–53 kDa). Whether this heterogeneity corresponds to proteolytic nicking is not known. A high molecular weight tetanolysin (≥ 100 kDa) was later characterized by Mitsui *et al.* (1982b). This form eluted from the void volume of Sepharose 6B was shown by electron microscopy as a particulate material which is thought to be constituted by the 50-kDa tetanolysin associated to cytoplasmic membrane fragments.

Perfringolysin O (PFO)

This denomination is that coined by Bernheimer (1976) for the toxin previously known as θ-toxin produced by all types (A to E) of *Clostridium perfringens* (see Willis, 1969; Ispolatovskaya, 1971; McDonel, 1980 for reviews). The toxin (M_r 52 469) was first described by Wuth (1923) and clearly identified by Neill (1926b) as the 'oxidizable hemotoxin of the Welch bacillus' on the basis of earlier observations of haemolytic properties of this microorganism. Several attempts to purify the toxin were reported by various authors since 1941 (see Smyth, 1975 for earlier works). Almost all reported purifications were undertaken on culture fluids from type A strains, particularly the PB6K strain (Mitsui *et al.*, 1973; Smyth, 1975; Yamakawa *et al.*, 1977; Ohno-Iwashita *et al.*, 1986; Alouf and Geoffroy, unpublished). The toxin was also purified in our laboratory from a type C strain (Hauschild *et al.*, 1973).

The procedure described by Smyth (1975) involved precipitation by ammonium sulphate of culture fluid from pH-controlled cultures in fermenters followed by two successive isoelectric focusings in narrow pH (5–8) gradients. Four electrophoretically distinguishable components of pI values between 5.7 and 6.9 were separated with specific activities ranging from 4×10^5 to 1.2×10^6 HU/mg of protein. Disc-gel electrophoresis revealed minor contaminants. The molecular weights determined ranged from 59 000 to 62 000 for each component.

A simple method which allowed toxin purification to electrophoretic and immunological homogeneity with a recovery of 50% was reported by Yamakawa *et al.* (1977). The culture filtrate was submitted to two successive chromatographic separations in DEAE–Sephadex followed by gel filtration on Sephadex G-150. The protein obtained showed a molecular weight of 51 000 and contained a single residue in very good agreement with the data deduced from gene sequencing (see Table 4). The specific activity of the toxin was of the same order as that reported by the authors mentioned above.

The procedure described for alveolysin and SLO (Geoffroy and Alouf, 1983; Alouf and Geoffroy, 1988) using affinity chromatography on thiopropyl-Sepharose 6B was successively used for the purification of perfringolysin O to homogeneity (Alouf and Geoffroy, unpublished data). The specific activity was *ca.* 10^6 HU/mg of protein. The determination by Edman degradation of the sequence of the first 20 amino acid residues reported by Alouf *et al.* (1986) was practically identical to that deduced by Tweten (1988b) from the nucleotide sequence of *pfo* gene which was also in agreement with that of the first 17 residues determined on the toxin purified from *C. perfringens* culture fluid (Tweten, 1988a). It is also interesting to note that the N-terminal residue of the toxin purified from type C strain culture by Hauschild *et al.* (1973) was also lysine.

Ohno-Iwashita *et al.* (1986) purified the toxin to homogeneity with a 64% overall yield by an original procedure based on its cholesterol-binding properties. Ammonium sulphate was added to a final concentration of 1 M to the crude toxin and this solution was then applied to a column of Sepharose 6B on which cholesteryl hemisuccinate was immobilized as a ligand. The toxin was eluted with a linear gradient of a buffer containing 40% ethylene glycol and the resulting pool was fractionated by hydrophobic chromatography on phenyl-Sepharose CL-4B column. The specific activity of the purified toxin was 4.8×10^6 HU/mg of protein. The determination of the amino acid composition showed a single Cys residue. The molecular weight was 54 000. The sequence of the first 17 amino acid residues was partially similar to that reported by Alouf *et al.* (1986) and Tweten (1988b).

This author reported toxin purification from a 6-litre culture fluid concentrated about 100-fold in a tangential flow Millipore concentrator. The concentrate was applied to a solid-matrix Zeta-chrome anion-exchange column and then eluted with a linear gradient of NaCl solution. The appropriate effluent was concentrated, reduced with DTT and passed on a thiopropyl–Sepharose column. The haemolytic fraction was then applied to a mono Q column leading to the purified toxin.

Recombinant PFO expressed in *E. coli* was also purified from *E. coli* extracts (Tweten, 1988a). About 20 mg of toxin were produced (in the periplasm) per litre of culture induced with IPTG. The toxin material was released by osmotic shock, concentrated by tangential-flow ultrafiltration, applied to a column packed with an anion-exchange matrix (Accel QMA) followed by elution. The haemolytic fraction was then purified to homogeneity by high-resolution Superose-12 gel filtration followed by chromatography on a mono Q column. The specific activities of both cloned gene product or authentic PFO were about 10^6 HU/mg of protein. Both products had an estimated molecular weight of 54 000. The calculated M_r deduced from the nucleotide sequence of *pfo* gene was 52 469 (Tweten, 1988b).

Listerial toxins: listeriolysin O, ivanolysin and seligerolysin

Among the seven taxonomically characterized *Listeria* species (Jones, 1988; Rocourt, 1988), only *L. monocytogenes*, *L. ivanovii* and *L. seeligeri* produce cholesterol-binding cytolytic toxins (Leimeister-Wächter and Chakraborty, 1989; Geoffroy *et al.*, 1989; Vazquez-Boland *et al.*, 1989b). *L. monocytogenes* and *L. ivanovii* are pathogenic but not the other *Listeria* species (see Berche *et al.*, 1988). *L. monocytogenes* causes listeriosis in man and other animal species (Seeliger, 1988). Its virulence is attributed to its ability to survive and multiply inside host macrophages (Mackaness, 1962; Berche *et al.*, 1988; Portnoy *et al.*, 1988). All haemolytic (listeriolysin O-producing) strains are virulent in the mouse model (Rocourt *et al.*, 1983; Hof, 1984). Non-haemolytic strains are avirulent. *L. ivanovii*, classified until 1984 as *L. monocytogenes* serovar 5, predominantly infects animals (especially ovine species) and very rarely humans. The toxins produced by these two species but not that produced by *L. seeligeri* have been purified to date. Recent comprehensive information on the microbiological, biological and pathophysiological aspects of these microorganisms is provided in the special issue 'Listeria and Listeriosis' of *Infection*, 1988, Vol. 16 (supplement 2).

Listeriolysin O (LLO)

The 'SLO-like' toxin produced by *L. monocytogenes* is presently known under the following names: *L. monocytogenes* haemolysin (Cluff and Ziegler, 1987; Portnoy *et al.*, 1988), listeriolysin (Goebel *et al.*, 1988; Kathariou *et al.*, 1990; Sokolovic and Goebel, 1989), α-listeriolysin (Parrisius *et al.*, 1986) and listeriolysin O (Geoffroy *et al.*, 1987). The latter denomination, which is also used by other authors (Cossart and Mengaud, 1989; Bielecki *et al.*, 1990; Mounier *et al.*, 1990), is most appropriate and in agreement with the nomenclature of this family of toxins as discussed earlier.

Following the first isolation of *L. monocytogenes* by Murray *et al.* (1926), which was shown to produce β-haemolysis on blood agar. Burn (1934) and Harvey and Faber (1941) reported the presence in culture media of a soluble haemolysin characterized as an oxygen-labile product by Njoku-Obi *et al.* (1963). Its similarity with SLO and related toxins was established on the basis of the increase of the haemolytic activity of crude or partially purified toxin preparations by reducing agents (Jenkins *et al.*, 1964) and its inhibition by cholesterol (see Sword and Kingdon, 1971). Antisera raised against the haemolysin neutralized SLO activity (Jenkins *et al.*, 1964).

Between 1963 and 1974 a number of authors attempted toxin purification from culture filtrates (see, for references, Sword and Kingdon, 1971; Parrisius *et al.*, 1986; Geoffroy *et al.*, 1987). We presently know on the basis of various criteria that the preparations obtained were not purified to homogeneity, as is also inferred from the wide dispersion of the values of the molecular weights reported (10^4–1.7×10^5) and the presence of lipolytic material in certain preparations. The failure of previous authors to purify the toxin is particularly due to the low levels of listeriolysin O (20–200 HU/ml) released in conventional culture media, which also vary from strain to strain.

The purification of LLO to homogeneity was realized by Geoffroy *et al.* (1987) by using a selected strain (EGD, serovar 1/2a from Trudeau Institute) which was grown in a Chelex-treated proteose peptone broth. About 1500 HU/ml are released as compared to 64 HU/ml in the non-treated medium. This enhancing effect is related to the lowering of iron levels in culture media through chelation by Chelex or charcoal in agreement with the observations (see Cowart, 1987) that LLO production is optimal at low iron levels and declines as a function of increasing iron concentrations. The enhancement of toxin production by about 30-fold was later reported for 12 strains of *L. monocytogenes* and strains of *L. seeligeri* cultivated in charcoal-containing broth (Geoffroy *et al.*, 1989). In contrast, haemolysin production by *L. ivanovii* was similar (1500 HU/ml) in the presence or absence of charcoal.

LLO was purified from the culture fluid of the Chelex-treated medium concentrated on Amicon hollow fibre followed by SH-Sepharose chromatography, then by gel filtrations on Sephacryl S-200, Biogel P-100 and Fractogel HW. The purified toxin (specific activity 10^6 HU/mg of protein) gave a single band after SDS–PAGE (calculated M_r 60 000). It exhibited the same classical properties of the other toxins of the family but differed from them in that its haemolytic activity was optimal at pH 5.5 and practically nil at 7.0. This finding suggests that the *in vivo* lytic activity of the toxin might be better expressed in the acidic macrophage phagolysosome where *L. monocytogenes* bacteria replicate.

Listeriolysin O purification from the culture fluid of Sv4B strain was also realized by chromatography on thiopropyl-Sepharose 6B followed by gel filtration on Biogel P-100 and FPLC Sepharose 6B (Goebel *et al.*, 1988; Kreft *et al.*, 1989). The molecular weight of the toxin was 58 kDa. This toxin was shown (Sokolovic and Goebel, 1989; Sokolovic *et al.*, 1990) to be a major listerial extracellular protein synthesized under heat shock or other stress conditions. LLO was coinduced with at least five other heat-shock proteins.

Anti-listeriolysin O is detected in human sera, indicating LLO production *in vivo* (Berche *et al.*, 1990).

Ivanolysin O (ILO)

This term was coined by Vazquez-Boland *et al.* (1989b) for the LLO-like cytolysin released by

L. ivanovii, the most recently discovered member of the toxin family described in this chapter. This toxin was first investigated under the name of α-listeriolysin O by Parrisius *et al.* (1986) who partially purified it from *L. ivanovii* culture fluids or after extraction from the membranes of erythrocytes previously lysed by this toxin. The methods used were similar to those employed in the same laboratory for SLO purification from culture fluids (Bhakdi *et al.*, 1984) or extraction from erythrocyte ghosts (Bhakdi *et al.*, 1985). This toxin was found to cross-react with SLO. Its molecular weight ranged from 55 000 to 60 000. Surprisingly, by using an antiserum raised against *L. ivanovii* α-listeriolysin, Parrisius *et al.* (1986) found that only 2 out of 28 *L. monocytogenes* clinical isolates released the toxin and therefore that α-listeriolysin is not the cause of the β-haemolysis zone in blood agar but rather an unrelated β-listeriolysin which could be also responsible for the CAMP phenomenon which predominates among both human pathogenic strains and *L. ivanovii* strains (Linder and Bernheimer, 1984; Brzin *et al.*, 1990). The occurrence of two listeriolysins was not supported by the experiments of Leimeister-Wächter and Chakraborty (1989). In addition, Goebel *et al.* (1988) and Geoffroy *et al.* (1989) reported that all virulent *L. monocytogenes* strains examined produced LLO. Lemeland *et al.* (1974), who examined 131 strains, found that 98.5% produced 'SLO-like haemolysin'.

Vazquez-Boland *et al.* (1989b) purified ivanolysin O to homogeneity (M_r 61 000, specific activity 2.6×10^6 HU/mg) by thiol–disulphide exchange chromatography as described by Geoffroy and Alouf (1983) for alveolysin. The eluted fraction was concentrated, ultracentrifuged at 90 000 × *g*, gel-filtered on Ultrogel AcA 34 and on Sephacryl S-200. Antigenic relationship with LLO, SLO, PFO and alveolysin was revealed by Western blot analysis. However, ILO differed from LLO by its optimal pH activity (6.5 instead of 5.5) and by differences in migration in SDS–PAGE. A 27-kDa cytolytic protein was found by the authors to copurify with ILO. By using selective sequestration of the latter with cholesterol, the other cytolysin was characterized as a haemolytic sphingomyelinase C involved in the CAMP reaction with *Rhodococcus equi*. Similarly, ivanolysin purification reported by Goebel *et al.* (1988) and Kreft *et al.* (1989) was

also shown to yield a 58-kDa listeriolysin O-like toxin associated to a 24-kDa protein which remained associated under non-denaturing conditions. The 24-kDa protein also exhibited a CAMP-like haemolytic effect. The sequences of the first 23 amino acid residues of ILO and of the 19 amino acid residues of the latter protein were determined by Kreft *et al.* (1989).

Seeligerolysin (SGL)

This toxin, which has not yet been purified, was identified by its haemolytic properties and its cross-reaction with LLO and ILO by immunoblot analysis with anti-LLO serum (Geoffroy *et al.*, 1989; Leimeister-Wächter and Chakraborty, 1989).

Purification of other toxins

Bifermentolysin, sordellilysin and oedematolysin have been purified from culture supernatants by a single-step immunochromatography through columns of anti-pneumolysin monoclonal antibodies coupled to 2C5 or 3H10 Sepharose (Sato, 1986).

Molecular genetics and structural features of the toxins

The structural and functional analysis of the chromosomal region carrying the genes encoding streptolysin O, pneumolysin, perfringolysin O, listeriolysin O, alveolysin and cereolysin O has been reported between 1983 and 1990. This analysis was indeed of the utmost importance for the understanding of the structure–activity relationships of these toxins and of their role in the pathogenicity and virulence of the relevant bacteria. Except for cereolysin O, the nucleotide sequence of the genes encoding these toxins has been determined. Consequently, the sequence of the amino acid residues was deduced (Fig. 1). Some of the physical chemical characteristics of these proteins are shown in Table 4. For the genes encoding the six toxins mentioned above we propose the following specific abbreviations: *slo*, *ply*, *pfo*, *alv*, *llo* and *clo*. The first four symbols have been already used by Kehoe *et al.* (1987), Tweten (1988a,b) and Geoffroy *et al.* (1990). It is preferable to designate listeriolysin O gene as *llo*

Table 4. Some structural characteristics of toxin molecules determined from the nucleotide sequences of toxin genes so far cloned

Toxins	M_r of mature toxins (Da)	AA[a]	N-terminal[b] residue	Cys[c] position	References
Alveolysin	51 766	32 + 469	Ala	429	Geoffroy et al. (1990)
Perfringolysin O	52 469	27 + 472	Lys	429	Tweten (1988a,b)
Pneumolysin	52 800	0 + 471	Ala	428	Walker et al. (1987)
Listeriolysin O	55 842	25 + 504	Asp	459	Mengaud et al. (1988)
Streptolysin O	60 151	33 + 538	Asn	497	Kehoe et al. (1987)
	53 000[d]	471	Ser	430	Kehoe et al. (1987)

[a] Number of amino acid residues of signal peptide + secreted (mature) form (except for pneumolysin which is an intracellular toxin).
[b] Mature form.
[c] Position of the unique cysteine residue of the polypeptide chain starting from the N-terminal residue of the mature form.
[d] Low-molecular-weight active form of SLO resulting from proteolytic cleavage of the mature toxin form subsequent to secretion.

instead of *hlyA* used by Mengaud *et al.* (1987) and Goebel *et al.* (1988) to avoid confusion with the *hlyA* gene of *E. coli* haemolysin.

Gene cloning and sequencing

Streptolysin O

The gene encoding this toxin was cloned and expressed in *E. coli* by Kehoe and Timmis (1984). Two forms of the SLO gene product with M_r of 68 000 and 61 000 were detected in *E. coli* minicells. Hybridization experiments performed at a stringency of *ca.* 80% showed no homology between the cloned *slo* DNA sequences and DNA isolated from bacteria expressing CLO, PLY, LLO, PFO, histolyticolysin O and oedematolysin O. The nucleotide sequence of the *slo* gene and the deduced amino acid sequence of the toxin were reported by Kehoe *et al.* (1987). The primary *slo* gene product consisted of 571 amino acid residues and had a molecular weight of 63 645. The 33 N-terminal residues possessed the feature of a signal peptide suggesting that the excreted toxin (Fig. 1) was a 538-residue polypeptide (M_r 60 151). The polypeptide appeared to undergo a proteolytic cleavage subsequent to secretion removing a *ca.* 7000-Da segment. This low-molecular-weight form of about 53 000 Da is in agreement with the molecular weights of 55 000 reported by Alouf and Raynaud (1973) and 56 000 reported by Linder (1979). The first four N-terminal residues reported by Alouf *et al.* (1986) corresponded to the residues 101 and 104 of the predicted SLO sequence. As a Lys residue occurs at position 100, Kehoe *et al.* (1987) suggested that the low-molecular-weight form was generated by the proteolytic removal of the 67 residues between the end of the predicted signal sequence and the Ser residue at position 101. The data obtained from gene sequencing are in agreement with the high- and low-molecular-weight forms of SLO reported by Bhakdi *et al.* (1984).

Pneumolysin

This toxin is cytoplasmic (Johnson, 1977) and, as expected, no signal sequence was found at its N-terminus, in contrast to the other four toxins so far analysed (Table 4). The structural gene from serotype I strain of *S. pneumoniae* isolated from a patient was cloned and expressed in *E. coli* by Paton *et al.* (1986) who also found by Western blot analysis that two forms of pneumolysin with molecular weights of 54 and 52 kDa were produced. The latter form was indistinguishable from native PLY. The most likely explanation for the expression of the two forms is the existence of two alternative termination sites.

Pneumolysin gene from a serotype II strain was also cloned, expressed in *E. coli* and its nucleotide sequence determined by Walker *et al.* (1987) who also reported the expression of two forms of 56 and 53 kDa. An open reading frame was found that encoded 471 amino acids (Fig. 1). The calculated M_r was 52 800 (Table 4). The amino acid sequence revealed it to be a hydrophobic protein as already found by Johnson *et al.* (1982b). The sequence of the first 16 amino acid residues determined by Walker *et al.* (1987) on purified pneumolysin extracted from *S. pneumoniae* was in total agreement with that deduced from gene sequencing by these authors, and also with that of the sequence of the first eight amino acid residues determined on the recombinant toxin purified from *E. coli* extracts by Mitchell *et al.* (1989). This protein was also identical to the native toxin with respect to specific activity, effect on human polymorphonuclear (PMN) phagocytes and on human complement. The amount of toxin produced in *E. coli* was 5 times higher than that obtained from *S. pneumoniae*. Therefore *E. coli* is a convenient source for the production of pneumolysin for immunization and diagnostic purposes.

Pneumolysin gene was expressed in *Bacillus subtilis* both from its own promoter and as a fusion protein (Taira *et al.*, 1989). The level of expression of pneumolysin from its own promoter was low. The toxin was haemolytically active. A higher level of expression was achieved when either one of two C-terminal fragments (corresponding to amino acids 265–471 or 55–471) or the entire coding part of the gene were fused in the promoter and signal sequence-coding region of the α-amylase of *Bacillus amyloliquefaciens*. The C-terminal fusion peptides remained cell-associated. They were not haemolytic, but reacted with anti-PLY serum. The entire gene fused to the signal sequence yielded an extracellular haemolytically active product which was partially purified.

An approach to the study of the genetic regulation of pneumolysin expression in *S. pneumoniae* was reported by Johnson *et al.* (1984) and Johnson and Alouf (1990) who showed that PLY level was not increased over the wild-type level upon introduction of recombinant shuttle plasmid bearing the *ply* gene into wild-type and PLY-negative strains obtained by nitrosoguanidine mutagenesis (Johnson *et al.*, 1982a). These results may reflect the existence of a negative control determinant on the chromosome which encodes a factor capable of operating in trans. A similar finding was reported in the case of listeriolysin O (Cossart *et al.*, 1989).

Perfringolysin O

The gene encoding this toxin was cloned and expressed in *E. coli* (Tweten, 1988a) and its nucleotide sequence determined (Tweten, 1988b). The product of the cloned gene was purified from *E. coli* extracts and was identical with that determined for perfringolysin O purified from *C. perfringens*. The primary *pfo* gene product consisted of 499 residues including a 27-amino acid signal peptide. The calculated M_r of the mature form of PFO (472 residues) was 52 469 (Table 4). Hydropathy analysis indicated that except for the signal peptide, no major stretches of hydrophobic residues are present. Fairweather *et al.* (1987) reported the cloning of the *pfo* gene and estimated that the gene product identified by immunoreaction with anti-SLO had an M_r of 49 000. This finding was discussed by Tweten (1988a) on the basis of a possible difference between *pfo* genes from various species.

The cloning of the *pfo* gene has been also reported by Shimizu *et al.* (1991) who also cloned and sequenced its regulatory gene (*pfoR*) located upstream and coding for a 343-amino acid protein. The deduced amino acid sequence of *pfoR* product possessed several motifs that are characteristic of DNA-binding proteins. When a region coding for an α-helix turn was deleted within *pfoR*, stimulation was abolished indicating that the regulatory gene positively controls *pfo* expression.

Alveolysin

The structural gene (*alv*) encoding alveolysin production has been cloned, expressed in *E. coli* and its nucleotide sequence determined (Geoffroy *et al.*, 1990). The open reading frame encoding

alveolysin consisted of 501 codons. The first 32 residues of the deduced amino acid sequence corresponded to the signal sequence. The secreted form of the toxin corresponds to a 469-amino acid residue chain (Table 4 and Fig. 1). The calculated molecular weight is 51 766. The previous reported cloning of a gene encoding a haemolytic product presumed to be alveolysin (Geoffroy and Alouf, 1988) appeared by hybridization experiments with *alv* gene to correspond to another haemolytic factor which was also inhibitable by cholesterol.

Listeriolysin O

The identification, cloning and expression in recipient strains of the structural gene encoding this toxin has been reported on the basis of experiments involving recombinant cosmid carrying LLO determinant (Vicente *et al.*, 1985) and transposon-induced mutations into the structural gene leading to non-haemolytic *L. monocytogenes* variants (Gaillard *et al.*, 1986; Kathariou *et al.*, 1987; Mengaud *et al.*, 1987; Portnoy *et al.*, 1988; Goebel *et al.*, 1988). The region of insertion of the transposon used was cloned and sequenced and the deduced amino acid sequence or the open reading frame revealed homology of the truncated protein with SLO and pneumolysin (Mengaud *et al.*, 1987). The homology of the truncated protein with SLO was also detected immunologically (Kathariou *et al.*, 1987).

The complete sequence of the *llo* gene was determined by Mengaud *et al.* (1988) by means of a cosmid vector. It was deduced that this gene encoded a 25-residue signal peptide (Table 4) and the secreted form of the toxin is constituted by a polypeptide of 504 amino acids (Fig. 1). The *llo* gene nucleotide sequence was also reported by Domann and Chakraborty (1989).

Various DNA fragments of *llo* gene and its region were used to probe by hybridization the presence of homologous sequences in the species of the genus *Listeria* and in non-haemolytic *L. monocytogenes* strains. With a specific probe, Mengaud *et al.* (1988) detected the *llo* gene in two non-haemolytic and 13 haemolytic *L. monocytogenes* strains but not in strains from the six *Listeria* species including the LLO-like producing *L. ivanovii* and *L. seeligeri*. Gormley *et al.* (1989), who used probes from different parts of *llo*, detected under low-stringency hybridization conditions sequences homologous to this gene and its 5′ adjacent regions in both *L. ivanovii* and *L. seeligeri* strains. In contrast, the region located downstream from *llo* appeared specific to *L. monocytogenes*. None of the probes spanning the region revealed homologies between *L. monocytogenes* and *L. innocua*, *L. murrayi* and *L. welshimeri*. Leimeister-Wächter and Chakraborty (1989) detected sequences homologous to *llo* gene in both *L. ivanovii* and *L. seeligeri*. These DNA–DNA hybridization results are in agreement with those of Geoffroy *et al.* (1989), who detected immunological cross-reactivity between the haemolysins of *L. monocytogenes*, *L. ivanovii* and *L. seeligeri*.

The *llo* gene was subcloned by Bielecki *et al.* (1990) into an asporogenic mutant of *Bacillus subtilis* under the control of an IPTG-inducible promoter. The gene product secreted by the bacteria transformed was haemolytic and co-migrated with authentic LLO. The haemolytic bacteria incubated with a macrophage-like cell line were internalized. They disrupted the phagosomal membrane and grew rapidly within the cell cytoplasm. These results show that a single gene product is sufficient to convert a common soil bacterium into a parasite that can grow in the cytoplasm of a mammalian cell.

The role of LLO as an essential virulence factor in the *in vivo* infection with *Listeria* in human and animal phagocytes has been intensively investigated at the genetic and cellular levels. This aspect is beyond the scope of this review. For detailed information the reader is referred to the review of Cossart and Mengaud (1989) and the articles of Gaillard *et al.* (1986), Cluff and Ziegler (1987), Portnoy *et al.* (1988), Goebel *et al.* (1988), Cossart *et al.* (1989), Mounier *et al.* (1990).

The genetic regulation of listeriolysin O expression in *L. monocytogenes* was investigated by Mengaud *et al.* (1989) who studied by DNA sequencing the structural organization of a 4.2-kb region containing the 1.5-kb *llo* (*hlyA*) gene. This gene appeared as a monocistronic unit that can be transcribed from two promoters spaced by 10 bp. The gene was found adjacent to another gene ORF U transcribed in the opposite direction. These two transcription units were separated by a perfect palindrome also found in the promoter region of ORF D located downstream for *llo*, suggesting that the three genes are under a similar regulation.

Cossart *et al.* (1989) reported that no increase in haemolytic titre in *L. monocytogenes* cultures was observed when a multicopy plasmid carrying only the *llo* gene was introduced in a wild-type strain as mentioned above for the *ply* gene.

Leimeister-Wächter *et al.* (1990) analysed non-haemolytic variants which revealed the presence of a 450-bp deletion located 1.5 kb upstream of an otherwise intact *llo* gene. These results led to the identification of a gene encoding a 27-kDa polypeptide called *prfA* for positive regulatory factor of listeriolysin expression.

Cereolysin O

The gene encoding the expression of cereolysin O (*clo*) was cloned and expressed in *E. coli* and *B. subtilis* by Kreft *et al.* (1983). Cereolysin O was poorly expressed in *E. coli* whereas the synthesis and transport into the extracellular medium of this toxin took place efficiently in *B. subtilis*. The haemolytic activity found in the culture fluid of this organism was at levels comparable to that obtained in the cultures of the *B. cereus* donor strain. When the recombinant DNA carrying the *clo* gene was used in hybridization experiments with chromosomal DNA preparations carrying *slo* and *llo* genes, no positive hybridization signals were obtained.

Comparative structures of the toxins and their genes

The establishment of the DNA-deduced primary structure of the five toxins shown in Fig. 1 constituted a considerable progress in our knowledge of the cholesterol-binding cytolysins. It enabled for the first time a comparative structural analysis of the toxins at the level of their amino acid sequences and paved the way for a better understanding of their evolutionary relationship, the role of the sulphydryl group, the molecular basis of toxin activity and of the virulence of the corresponding bacteria.

Structural homology of the toxins at the amino acid sequence level

Among the five toxins so far analysed, the smallest polypeptide is alveolysin (469 residues) and the largest is SLO (538 residues for the high-molecular-

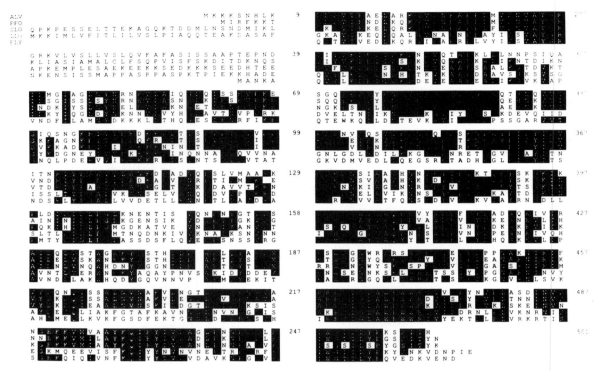

Figure 1. Alignment of the predicted amino acid sequences of alveolysin O (ALV), perfringolysin O (PFO), streptolysin O (SLO), listeriolysin O (LLO) and pneumolysin (PLY). The sequences include the amino acid residues of the signal peptides of ALV (first 32 residues), PFO (first 27 residues) and LLO (first 25 residues). PLY has no signal sequence; its N-terminal residue is Ala. For the sake of simplification, the sequence of SLO starts at residue 16 (Q/Gln) of the high-molecular-weight mature form (see Kehoe *et al.*, 1987 for complete sequence). (Reproduced from Geoffroy *et al.*, 1990 by permission of *J. Bacteriol.*)

weight form) (Table 4). Each polypeptide contains a single cysteine residue towards the C-terminal part of the polypeptide chain (Table 4 and Fig. 1). Interestingly, Iwamoto *et al.* (1987) found by biochemical approaches that the Cys residue of PFO was at a position about 5 kDa from the C-terminus.

The Cys residue lies in a conserved 11-amino acid sequence which represents the largest contiguous region of identity among the five toxins (Fig. 2). When the sequences of the individual toxins are aligned to give maximum similarity (Fig. 1), the common undecapeptide is also aligned. This suggests that this conserved region contributes to a functionally important role of the toxins. That the 'SH-activated toxins' might share a common structure in their C-terminal portion around the Cys residue was predicted by Ohno-Iwashita *et al.* (1986).

A comparison of the sequences of these toxins

revealed homology along the whole molecule of each mature form. However, homology was stronger at the C-terminal end. A match of 42% between the residues of SLO and PLY was found (Kehoe *et al.*, 1987) by using a best-fit program. The homology is higher (60%) when the sequences are aligned by matching structurally similar in addition to identical residues. Mengaud *et al.* (1988) compared SLO, PLY and LLO sequences. The percentage of residue identity between two sequences compared two by two in the common 469-residue region was about 43%. Stronger homologies are found when one compares the sequences in terms of similarity instead of identity. According to Pinkney *et al.* (1989), between 42 and 65% of the residues of SLO, PLY, PFO and LLO match. The comparison of the sequences of SLO, PLY and PFO (Tweten, 1988b) surprisingly shows more homology (65%) of PFO and the low-molecular-weight form of SLO which starts with residue

```
SLO (496-506)
PLY (427-437)
PFO (428-438)     Glu - CYS - Thr - Gly - Leu - Ala - Trp - Glu - Trp - Trp - Arg
LLO (458-468)
ALV (428-438)
```

Figure 2. Conserved 11-amino acid residue sequence encompassing the unique Cys residue in the polypeptide chain of the five toxins of Fig. 1. Numbers in brackets refer to amino acid position in the sequence of mature proteins.

Ser[101] than PLY exhibited with either proteins (42%). If conservative amino acid substitutions are taken into consideration, PFO exhibits 96% homology with SLO (smaller form). No significant homology with either PFO or PLY was detected in the primary sequence of SLO prior to residue 100. The three toxins exhibited similar hydropathic profiles. Those of SLO and PFO were nearly superimposable whereas the PLY profile was slightly different. According to Tweten (1988b), the largest difference resides in residues 158–178 for PLY, 188–208 and 260–280 for PFO and SLO (from the N-terminus of signal peptide) respectively. Both PFO and SLO are relatively hydrophobic in this 20-residue region whereas the same region in PLY is hydrophilic. The first direct evidence of the amphiphilic character of a toxin of the group was shown for alveolysin (Geoffroy and Alouf, 1983). A substantial binding of charged and uncharged detergents by this toxin was detected by charge-shift electrophoresis as found for the unrelated *C. perfringens* delta-toxin and various amphiphilic proteins.

The 67-residue region of SLO between the end of the signal sequence and the beginning of the low-molecular-weight form (Kehoe *et al.*, 1987) is extremely hydrophilic and apparently is not necessary for the *in vitro* activity of SLO since it is as haemolytically active as the heavy form. Since this region has no significant homology with the primary structure of PLY, PFO, LLO and ALV it is likely that it was formed or added after *S. pyogenes* acquired the *slo* gene. Whether this region is present in other SLO or SLO-like toxins produced by group A, C and G streptococci is still unknown.

Structural relationship between toxin genes

Dot-matrix comparison of the nucleotide sequence of the *slo* and *ply* genes led Kehoe *et al.* (1987) to suggest that these genes have undergone a considerable degree of divergence. However,

despite this divergence at the DNA level, a clear homology (42%) exists between the two toxins at the primary structural level.

A similar comparison of the nucleotide sequence of the *pfo*, *slo* and *ply* genes showed 60% homology between *pfo* and *slo* and only 48% between *pfo* and *ply* (Tweten, 1988b). These results explain the non-detection of hybridization at 80% strigency between *slo* and *ply* (Kehoe and Timmis, 1984). The highest homology between *pfo* and *slo* was detected in the coding region for the secreted form of PFO and that for the small form of SLO (Tweten, 1988b). In contrast, little homology was found for the coding region of the hydrophilic 67-residue N-terminal fragment of SLO which is often cleaved after SLO secretion.

Structure–activity relationships

Great progress in our understanding of the relationship between the molecular structure of the toxins and their cytolytic properties, investigated essentially on erythrocytes, has been realized since 1987 by using toxin variants generated by experimental modifications of their structural genes. The most important results obtained concerned the demonstration of the non-essential role of the sulphydryl group of the unique Cys residue of the toxins for cytolytic and other biological activities. The involvement of certain toxin regions, including the common undecapeptide, in the expression of biological activity was also established by both genetic and biochemical approaches.

Role of the sulphydryl group

The finding of a single Cys residue in toxin molecules (Figs 1 and 2) appeared as a strong support for the classical concept of an 'essential' sulphydryl group involved in toxin activity as discussed earlier. This group has been thought to mediate toxin–cholesterol interaction at the surface of target cells (see Smyth and Duncan, 1978 and Bhakdi and Tranum-Jensen, 1988 for reviews). This interaction is the initial step of membrane disruption and subsequent lysis resulting from the aggregation of toxin molecules into oligomers separable from red cell membranes by sucrose density gradient ultracentrifugation (Bhakdi et al., 1985; Boulnois et al., 1990). According to Bhakdi and Tranum-Jensen (1988), the oligomers are transmembrane pores which allow the leakage of cytoplasmic molecules from damaged cells and thereby cytolysis.

To explore the putative functional role of the Cys residue (and also of other residues), mutants of pneumolysin (Saunders et al., 1989; Boulnois et al., 1990), streptolysin O (Pinkney et al., 1989; Pinkney and Kehoe, 1990) and listeriolysin O (Michel et al., 1990) were generated by oligonucleotide-mediated site-directed mutagenesis of toxin genes to systematically change Cys or other residues in the common region. Wild-type and modified toxins expressed in E. coli (and also in L. monocytogenes for llo gene) were purified for examination of haemolytic activity and other biological properties.

The substitution of the Cys residues by Ala and Ser in SLO and LLO and by Ala, Ser, Gly in PLY led to the following results. The Cys → Ala modified SLO or PLY were almost undistinguishable from the wild-type toxins in terms of haemolytic activity (and also inhibitory effects on human PMN leukocytes and complement activation in the case of PLY). In the case of listeriolysin O the Cys → Ala derivative was also haemolytic but its specific activity was 75% that of the wild-type LLO. Clearly, these results indicate that the Cys residue is not essential for toxin activity. The reagents which chemically modify free SH groups caused a marked inhibition of the cytolytic activity of wild-type SLO but predictably did not change that of the Cys-free mutants (Pinkney and Kehoe, 1990). The Cys → Ser derivatives of SLO and LLO were 4- to 5-times less active than the wild types but still considerable (200 000 HU/mg for the SLO mutant). The Cys → Ser and Cys → Gly derivatives of PLY were 6- and 20-times less active on erythrocytes and PMN leukocytes. The reduced activity of the Cys → Ser or Cys → Gly modified toxins might be due to steric factors and or

secondary structure changes which may cause localized distortions of the amino acids surrounding the Cys residue which are functionally important as shown below. The distortion could also involve more distant residues close to the Cys in the 3-dimensional conformation of the molecule. According to Iwamoto et al. (1990), the almost identical activity of both Ala derivative and wild-type toxin may be explained by the similar electronegativity of their respective SH and $=CH_2$ groups. The reduced activity of the Ser derivative could be explained by the larger electronegativity of the OH group as compared to the SH group and the consequent increase of the hydrophilicity of the region.

However, all the modified SLO, PLY and LLO molecules were completely inhibitable by cholesterol and able to bind to cell membrane. The parent and mutant SLO and PLY molecules extracted from lysed erythrocyte membranes had similar sedimentation profiles on sucrose density gradients, suggesting that the absence of Cys residues did not affect their ability to insert into the membranes and to form oligomers. Consequently, the reduced activity of the Cys → Ser and Cys → Gly derivatives could not be explained by a decrease in toxin binding to membranes or its ability to aggregate within membranes. It may be due to the formation of non-functional oligomers according to Saunders et al. (1989), who stated that although Cys is not involved in receptor binding and oligomer formation, it might contribute to the generation of functional pores. The formation of oligomers appears not sufficient by itself for toxin activity.

It has been suggested (Saunders et al., 1989; Pinkney et al., 1989) that the Cys residue may play a critical role in vivo, for example, by stabilizing the toxin through the formation of disulphide bonds with other molecules for its transport to target membranes in this stabilized form where glutathione-like reduction systems release active protein. However, Michel et al. (1990) reported that the Cys residue was not essential for the in vivo virulence of L. monocytogenes mutants producing the Cys → Ala or the Cys → Ser modified toxins. According to Tweten (1988b), the role of the SH group may not be to form a disulphide with other molecules or membrane proteins on target cells but rather it may have some other role such as the formation of a hydrogen bond or hydrophobic interaction.

Role of the common cysteine-containing region

An unusual feature of this region (Fig. 2) is the occurrence of three of the six or seven tryptophan (Trp) residues of toxin molecules. Trp has the largest aromatic sidechain of the amino acids and therefore may confer a local hydrophobic character to the undecapeptide, the structure of which is typically amphiphilic. It is very likely, as shown below, that at least one Trp residue is essential for toxin insertion into the cholesterol/phospholipid bilayer (through hydrophobic interactions). This insertion is probably the mechanism by which the cell membrane is disrupted by the toxins.

The individual substitution by site-directed mutagenesis of the Trp[433], Trp[435] and Trp[436] residues of pneumolysin by phenylalanine and the Trp[466] and Trp[467] residues of listeriolysin O by alanine was reported by Boulnois et al. (1990) and Michel et al. (1990) respectively. Compared to the haemolytic activity of the recombinant wild-type PLY, that of PLY Trp[433] → Phe, Trp[435] → Phe and Trp[436] → Phe derivatives was 0.6%, 13% and 100% indicating that Trp[433] was critical for activity whereas Trp[436] was apparently not. Interestingly, the cholesterol-binding capacity of the Trp[433] → Phe remained identical to that of the wild-type protein. The haemolytic activity of the Trp[466] → Ala and Trp[467] → Ala derivatives of listeriolysin O was severely reduced by 95% and 99% respectively. However, these derivatives retained their capacity to bind cell membranes and to combine with cholesterol similarly to the wild-type toxins. The differences in activity between PLY and LLO mutated at the same Trp positions may be due to steric factors since alanine is a much smaller molecule and less hydrophobic than the bulky Phe or Trp molecules with their phenyl or indole rings which may confer to the undecapeptide a different conformation and higher hydrophobicity than that resulting from alanine substitution.

Another modification of the residues of the common peptide was the substitution of Glu[434] in pneumolysin by asparagine. The activity of the resulting derivative was 25% that of the unsubstituted recombinant toxin.

More information is still required and particularly modification of the undecapeptide residues of the other toxins in order to evaluate more precisely the contribution of the different residues of this region to the lytic activity and other biological properties of the toxins. On the basis of the data hitherto available it appears that despite their pronounced conservation, each of the residues of the undecapeptide region does not appear to be absolutely required individually for activity except probably one of the three Trp residues. This activity appears to depend rather on an appropriate overall structure of the undecapeptide. In addition, the unmodified cholesterol-binding capacity of the haemolytically inactive derivatives suggests that this region is not significantly involved in toxin interaction with its receptor (cholesterol) on target cells. This contention is supported by the finding that a 52-kDa haemolytically inactive truncated fragment of LLO (obtained by transposon mutagenesis) lacking 48 residues of the C-terminal part of the toxin including the Cys region, could be complexed and precipitated by cholesterol (Vazquez-Boland et al., 1989a). The results are in agreement with the finding by Ohno-Iwashita et al. (1988) that the nicking of PFO by limited proteolysis led to an almost non-haemolytic fragment which retained its cholesterol-binding capacity below 20°C. Therefore, different domains appear to be involved in the cytolytic activity and cholesterol-binding function of the toxins. Consequently, the Cys region appears essential for the membrane damage process but not for the cholesterol binding as currently thought. This view is in agreement with the two-site hypothesis of Alouf and Raynaud (1968), who showed by biochemical and immunochemical approaches that two antigenically and topologically distinct sites called f (for fixation) and l (for lytic) are involved in the SLO-induced cytolysis (Alouf, 1977, 1980). The occurrence of these two predicted domains was confirmed immunologically by Watson and Kerr (1985) and is presently supported by the genetic and biochemical approaches mentioned above and in the next section.

Role of other toxin regions in the cytolytic activity

The treatment of pneumolysin with diethylpyrocarbonate (DEPC) abolished the cytolytic activity of the toxin and its ability to bind to erythrocytes (Boulnois *et al.*, 1990). As DEPC modifies primarily histidine (His) residues and since only one His aligned for maximum similarity in the five toxin sequences, the common region encompassing this residue (Fig. 3) located upstream of the Cys-containing undecapeptide might likely be the cholesterol/receptor-binding domain. This domain probably mediates the recognition and binding of the 3β-hydroxyl group of cholesterol. The critical role of the His[367] residue of PLY in this respect is supported by the finding (Boulnois *et al.*, 1990) that its substitution by arginine by site-directed mutagenesis led to a haemolytically inactive derivative. It can be therefore speculated that the Cys region is involved in the lytic process after critical toxin binding via another domain (His-containing region?) and that the Cys region mediates the recognition of the cholesterol isooctyl sidechain as a part of the transmembrane insertion process involving either pore formation or other membrane disorganization. This contention appears in agreement with the study of the haemolytic activity and/or cholesterol-binding properties of SLO and other related toxins in relation to sterol structure. That the cytolytic activity involves different domains is also supported by the data of Taira *et al.* (1989) who showed that the PLY peptides corresponding to the amino acid sequences 55–471 and 265–471, which both contained the Cys region and the critical His region, were haemolytically inactive and that the missing 53 N-terminal fragment was therefore required for the cytolytic activity. The abrogation of the lytic activity of both truncated peptides could be due to the lack of an essential region involved in sterol recognition or to an important change in the overall confor-

mation of these peptides with respect to that of the native toxin molecule.

In the LLO⁻ avirulent mutant of *L. monocytogenes* obtained by transposon mutagenesis (Gaillard *et al.*, 1986), cloning and sequence analysis showed that the transposon was inserted in codon 481 of the *llo* gene three codons upstream from the Cys codon (Mengaud *et al.*, 1987). The resulting truncated protein was devoid of haemolytic activity.

Sterol-binding properties

The most characteristic biochemical property of the 'SH-activated' toxins is the specific and irreversible inhibition of their lytic, lethal and cardiotoxic properties by very low (nanomolar) amounts of cholesterol and certain structurally related sterols (reviewed by Bernheimer, 1976; Alouf, 1977, 1980; Cowell and Bernheimer, 1978; Smyth and Duncan, 1978; McDonel, 1980). The inactivation by cholesterol of the haemolytic activity of tetanolysin was shown as early as 1902 and that of pneumolysin and SLO in 1914 and 1939, respectively (see Bernheimer, 1976). The inactivation of the other toxins by cholesterol and other sterols was subsequently demonstrated (see the above-mentioned references and Geoffroy and Alouf, 1983; Sato, 1986; Geoffroy *et al.*, 1987; Ohno-Iwashita *et al.*, 1988; Vazquez-Boland *et al.*, 1989b).

Structural requirements

The structural and stereospecific features necessary for inhibition were investigated in detail on quantitative bases in the case of PLY, SLO and PFO (see Watson and Kerr, 1974; Prigent and Alouf,

ALV 365-376	---	Lys	Leu	Asp	**HIS**	Ser	Gly	Ala	Tyr	Val	Ala	Gln	Phe ---
PFO 367-378	---	Asn	Leu	Asp	**HIS**	Ser	Gly	Ala	Tyr	Val	Ala	Gln	Phe ---
SLO 433-444	---	Asn	Leu	Ser	**HIS**	Gln	Gly	Ala	Tyr	Val	Ala	Gln	Tyr ---
LLO 395-406	---	Asn	Ile	Asp	**HIS**	Ser	Gly	Gly	Tyr	Val	Ala	Gln	Phe ---
PLY 364-375	---	Leu	Leu	Asp	**HIS**	Ser	Gly	Ala	Tyr	Val	Ala	Gln	Tyr ---

Figure 3. Sequence of the region of 12 residues containing the His residue numbered 400 in the aligned sequence of the five toxins of Fig. 1. The numbers given under the toxin symbols are the position of the residues of the region numbered from the N-terminus of the mature forms.

1976 and the references mentioned above). All inhibitory sterols possessed (i) an OH group in β configuration on carbon-3 of ring A of the cyclopentanoperhydrophenanthrene nucleus, such as is found in cholesterol and kindred sterols; (ii) a lateral aliphatic sidechain of suitable size (isooctyl chain in the case of cholesterol) attached to carbon-17 on the D ring; (iii) the presence of a methyl group at C-20; (iv) an intact B ring (Fig. 4). Neither the saturation state of the B ring nor the positions of the double bonds nor the stereochemical relationships of rings A and B were critical factors. The presence of an α-OH group on C-3 (epicholesterol), esterification of the β-OH group or its substitution with a keto group or with 3β-SH group (thiocholesterol) rendered the sterol inactive (Alouf and Geoffroy, 1979) presumably because such substitutions or orientations preclude correct presentation of the reactive 3β-OH group to the toxins. The usual sterols fulfilling the appropriate criteria for inhibition of toxin activity are: cholesterol, 7-dehydrocholesterol, dihydrocholesterol, stigmasterol, ergosterol, lathosterol and β-sitosterol.

Characteristics of toxin–sterol interaction

The discussion will be restricted here to the interaction of toxin molecules with free cholesterol and other inhibitory sterol molecules dispersed in appropriate buffers. Toxin–sterol interactions with target cells or with artificial membranes (liposomes, lipid films) are considered later. Early experiments in the 1970s suggested the formation of complexes between toxin and sterol molecules (Smyth and Duncan, 1978; Cowell and Bernheimer, 1978). The first direct demonstration of the occurrence of these complexes and thereafter their separation

were provided by Alouf and his co-workers. The complexes were visualized by allowing SLO or alveolysin to diffuse from wells in sterol-containing agar gels (Prigent and Alouf, 1976; Alouf and Geoffroy, 1979; Geoffroy and Alouf, 1983). The complexes formed appeared as white, opaque halos (around toxin wells) constituted by insoluble (hydrophobic) precipitates of irreversibly bound toxin–sterol material stainable by protein dyes. Only inhibitory sterols showed such a pattern.

The separation of toxin–sterol complexes formed in liquid phase was reported by Johnson et al. (1980) who mixed pure pneumolysin, SLO and alveolysin preparations with [³H]cholesterol solutions at concentrations in which this sterol was present in a soluble micellar form. Advantage was taken of the fact that when solutions of cholesterol in phosphate buffer are added to Sephadex on Sephacryl columns, the free cholesterol sticks to the gel and only toxin–sterol complexes are eluted with buffer. The separated complexes were haemolytically inactive and of high molecular weight. The amounts of bound sterol increased linearly with toxin concentration. The reaction was rapid and temperature independent, similar to the binding step of toxin interaction with target cells. The specificity of cholesterol binding was assessed by adding unlabelled inhibitory or non-inhibitory sterols to toxin before adding [³H]cholesterol. Epicholesterol caused only a small decrease of binding whereas 7-dehydrocholesterol inhibited radiolabelled binding to an extent equal to that observed with unlabelled cholesterol. Haemolytically inactive SLO obtained by treatment with oxidized dithiothreitol or by reaction with parahydroxymercuribenzoate showed no decrease in cholesterol-binding activity, whereas the ability of the toxins to bind to erythrocytes was modified by such treatment. This result is in agreement with the further finding that the SH group of cysteine is not involved in toxin interaction with its receptor (cholesterol). The oxidation of the SH group or its substitution very likely produced a steric hindrance to toxin fixation to the cholesterol embedded in the erythrocyte; such a hindrance does not take place in toxin binding to free cholesterol.

The isolation of toxin–sterol complexes in the absence of solid support was reported by Geoffroy and Alouf (1983) who separated and identified

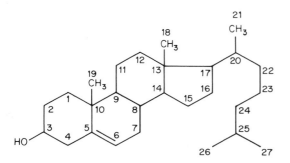

Figure 4. The cholesterol molecule.

[³H]cholesterol–alveolysin complexes by sucrose gradient ultracentrifugation. The complexes formed were relatively heterogeneous in size. Heat-denaturated toxin did not bind cholesterol, indicating that the native structure is essential for binding. Ivanolysin O–cholesterol complexes were also separated by centrifugation at 25 000 *g* by Vazquez-Boland *et al.* (1989a,b).

Stoichiometry

The limited solubility of cholesterol and other sterols in aqueous phase, the non-uniform nature of their dispersions in water and their tendency to stick to solid surfaces, among other reasons, has made it difficult to obtain information on the stoichiometry, affinity and other thermodynamic parameters of toxin–sterol interaction. Cholesterol has a maximum solubility in aqueous solutions of 1.8 µg/ml (4.7 µM) and undergoes a reversible self-association at the critical micellar concentration (CMC) of approximately 27–44 nM at 25°C (see Geoffroy and Alouf, 1983 for references). Such cholesterol aggregates are heterogeneous in size and rod-shaped, each containing *ca.* 260–360 cholesterol molecules (Smyth and Duncan, 1978). Ideally, the study of toxin–cholesterol interactions should be performed at concentrations below the CMC to ensure proper stoichiometry with cholesterol molecules dispersed as monomers rather than aggregates. The latter aggregate state appears to have been the case in the studies on SLO (Prigent and Alouf, 1976; Badin and Denne, 1978), CLO (Cowell and Bernheimer, 1978) and PFO (Hase *et al.*, 1976) as cholesterol concentrations found to inhibit 1 HU ranged from 12 to 25 nmol, corresponding to a molar cholesterol–toxin ratio of *ca.* 500–1000. The data of the literature compiled by Smyth and Duncan (1978) showed that the number of cholesterol molecules required to neutralize a single molecule of toxin ranged from 170 to 1×10^6. By using a dilution technique involving sterol solutions in absolute ethanol, Geoffroy and Alouf (1983) avoided dealing with micellar solutions and were able to determine for the first time a linear inhibition 'titration' curve of alveolysin by cholesterol. The stoichiometry was found practically equimolar (*ca.* 1.6 molecule of cholesterol neutralized by 1 molecule of alveolysin).

Effects of phospholipids on toxin–sterol binding

Phospholipids were shown by Badin and Denne (1978) to greatly affect the inactivation of SLO by cholesterol. Stable pseudosolutions of cholesterol were obtained by the association of polyethyleneglycol (PEG) and phosphatidylcholine (PC) to cholesterol. This pseudosolution was 32 times more active (in terms of the inhibitory effects on SLO) than ethanolic solution of cholesterol dispersed in saline. PC or PEG alone did not inhibit toxin activity. Cholesterol was inactive when mixed with 5 times its weight of PC.

Nature of the binding forces

The physical forces involved in sterol–toxin interaction are poorly understood. The formation of a covalent linkage between the SH group and the 3β-OH of cholesterol is unlikely. London–van der Waals forces and/or hydrophobic interaction probably occur (but not exclusively?) in the interaction. The requirement of both A ring and isooctyl chain of sterol molecules for the inhibition of toxin activity suggest that appropriate aromatic and/or sidechains of amino acid residues of cholesterol-binding domain of the toxins interact with these two separate regions of cholesterol molecule to create the complex. This hypothesis may be clarified if it becomes possible to co-crystallize toxin–cholesterol complexes for X-ray diffraction analysis and the establishment of the 3-dimensional structure of the complex.

Cytolytic and membrane-damaging effects

The most striking biological property of the toxins reviewed here is their potent lytic activity toward all mammalian and other eukaryotic cells so far tested. Lysis results from the disorganization or disruption of the cytoplasmic membrane of target cells. The intracellular organelles are also disrupted. The membrane-damaging effects are also reflected by alterations in the permeability and integrity of artificial model membranes.

Cell lysis markers

The lytic process triggered by the toxins reviewed here and other cytolysins is classically investigated and monitored by the measurement of appropriate intracellular components (from cell cytoplasm or from the disrupted organelles) released in the incubation medium by the toxin-damaged cells (Alouf, 1977; Smyth and Duncan, 1978). The release of radiolabelled, coloured or fluorescent markers entrapped in resealed erythrocyte vesicles (Buckingham and Duncan, 1983) or in liposomes (Duncan, 1984; Menestrina et al., 1990) is also used in the studies of toxin-induced damage of these vesicles.

The spectrophotometric determination of haemoglobin (Hb) is the most usual method employed in the monitoring of erythrocyte lysis (haemolysis) and its kinetics. The release of K^+ or ^{86}Rb from toxin-treated erythrocytes was also used as a marker of cell lysis by tetanolysin (Blumenthal and Habig, 1984) or by SLO (Smyth and Duncan, 1978). A highly sensitive approach is the measurement by a bioluminescence method of the release of ATP from erythrocytes and other cells devised by Fehrenbach et al. (1980) which was used for the study of the kinetics of erythrocyte lysis by SLO (Fehrenbach et al., 1982; Niedermeyer, 1985). A comparative study of Hb and ATP release showed that the latter was a much more sensitive indicator of SLO-induced lysis (10^{-11}–10^{-12} M ATP) than Hb.

Platelet lysis by SLO and alveolysin was followed by the assay of released serotonin, lactate dehydrogenase and other enzymes (Launay and Alouf, 1979; Launay et al., 1984). Membrane damage of human lung fibroblasts by SLO, PFO and ALV was assessed by the leakage of three different-sized cytoplasmic markers (Thelestam and Möllby, 1980; Thelestam et al., 1981). The cells were first preloaded with [^{14}C]amino isobutyric acid (M_r 103) or treated with radiolabelled uridine to obtain either a labelled nucleotide ($M_r < 1000$) or RNA ($M_r > 200\,000$) before toxin challenge. The release of these markers was monitored for the study of the time course of the lytic process and for the estimation of the size of the functional 'pores' (whatever their physical reality) elicited in the damaged membrane (Thelestam and Möllby, 1983). In this respect, reference to 'pores', 'channels' or 'lesions' are meant to indicate functional entities with the potential to allow passage of molecules up to a certain size (Buckingham and Duncan, 1983).

Preloaded [^{3}H]choline or [^{35}S]methionine Lettre cells (murine tumour cells) were used as precursors for labelled intracellular markers for the study of the mechanism of lysis of these cells by PFO and other toxins (Menestrina et al., 1990).

Membrane cholesterol as toxin-binding site and target

It is presently clearly established that cell membrane cholesterol which is located in the two leaflets of the lipid bilayer of the cytoplasmic membrane of eukaryotic cells (Houslay and Stanley, 1982) is the binding site of the 'SH-activated' toxins. This hypothesis is supported by several lines of experimental evidence which can be summarized as follows:

(1) The toxins are inactivated by cholesterol and can no longer bind target cells (Alouf, 1980; Ohno-Iwashita et al., 1986).
(2) The toxins have no lytic effects on prokaryotic cells (bacterial protoplasts and spheroplasts) which lack cholesterol in their cytoplasmic membranes.
(3) Parasitic mycoplasma cells that contain sufficient cholesterol are bound and lysed or damaged by the toxins but not saprophytic mycoplasma cells in which carotenol replaces cholesterol as a membrane constituent (see Smyth and Duncan, 1978; Rottem et al., 1976, 1990).
(4) The lytic activity of the toxins is abolished by incubation with erythrocyte membranes; in membrane extracts only those lipid fractions that contain cholesterol possess the inactivating material (see Hase et al., 1976; Badin and Denne, 1978; Smyth and Duncan, 1978).
(5) Acholeplasma laidlawii cells grown in the presence of cholesterol inhibited the haemolytic activity of cereolysin O, but A. laidlawii cells grown in the absence of cholesterol did not; the incubation of these cells with a cholesterol–Tween mixture re-established their ability to bind CLO (Cowell and Bernheimer, 1978).

(6) The pretreatment of erythrocyte membranes with agents that are known to bind to cholesterol such as polyene antibiotics and alfa alfa saponins prevented the membranes from inhibiting the haemolytic activity of CLO and SLO (Shany *et al.*, 1974).

(7) The partial evulsion of cholesterol (*ca.* 30%) from human erythrocytes decreased perfringolysin O binding and cell susceptibility to lysis (Mitsui *et al.*, 1982a), whereas experimentally cholesterol-enriched erythrocytes exhibited increased sensitivity to lysis by SLO (Linder and Bernheimer, 1984).

(8) The treatment of human erythrocyte membranes and *A. laidlawii* cells containing cholesterol with cholesterol oxidase abolished their ability to bind CLO and to inhibit its haemolytic activity (Cowell and Bernheimer, 1978).

(9) The treatment of nucleated mammalian cells (L cells and HeLa cells) with inhibitors of cholesterol synthesis such as 20α-hydroxysterol or 25-hydroxysterol significantly reduced SLO binding and abolished cell susceptibility to the lytic effect of the toxin; the incubation of refractory cells with serum or cholesterol restored their sensitivity to SLO (Duncan and Buckingham, 1980).

(10) Only cholesterol-containing phospholipid vesicles or films are bound and disrupted by the toxins.

Detailed information about the topological aspects of toxin–membrane cholesterol interaction is still lacking. The experiments of Badin and Denne (1978) led to the conclusion that a small fraction (*ca.* 7%) of human erythrocyte membrane cholesterol was available to SLO and that this sterol is distributed asymmetrically in the membrane cholesterol leaflets, the inner being more rich in available cholesterol.

These data are in agreement with the study of cholesterol binding by PFO and protease-nicked derivatives at the surface of sheep and human erythrocytes and mouse thymocytes by Ohno-Iwashita *et al.* (1988), who reported the occurrence of two classes of toxin-binding sites classified as high-affinity and low-affinity sites for PFO. The

calculated dissociation constants were *ca.* 2.3 nM and 0.2 μM respectively for human erythrocytes and very similar for sheep erythrocytes indicating that the K_D of the low-affinity sites is 100-times that of the high-affinity sites and thereby that the former are far less accessible to the toxin. On the basis of a concentration of 1.24 mg of membrane cholesterol/ml of packed sheep erythrocytes, an apparent K_D of 4.7 μM was calculated by Iwamoto *et al.* (1987) for PFO binding, on the assumption that all membrane cholesterol is available. From these data it was deduced that the number of the high-affinity sites is quite limited (6.2×10^4 per cell). As not more than 10–100 molecules are required for the lysis of one erythrocyte, it was concluded that the lysis by PFO results from toxin binding to the high-affinity sites.

Interesting attempts have been made by certain authors to use the toxins mentioned above as potential probes of cholesterol distribution in eukaryotic cell biomembranes which is still poorly documented. For information concerning this topic which is beyond the scope of this chapter, the reviews of Smyth and Duncan (1978) and Arbuthnott (1982) can be consulted.

Toxin-induced cell damage and lysis

Historical background and bibliographic sources

The lytic process triggered by toxin binding to cell membrane has been intensively investigated since the pioneering studies of Neill and Fleming (1927) and Herbert and Todd (1941) and thereafter by many authors between 1945 and 1980. Most lytic studies during this period were performed on erythrocyte suspensions from various species (generally from sheep, rabbits and humans) because of their easy obtention, manipulation and standardization. In addition, red blood cells are very convenient systems for quantitative studies, particularly for the investigation of the parameters of toxin binding, cell lysis and toxin-induced kinetics (monitored by the measurement of the haemoglobin released). Another advantage over other cells is the better knowledge of erythrocyte membrane composition and structure. However, cell lysis has been also investigated on other blood cells and platelets, human fibroblasts, cardiac cells, HeLa cells, mouse L-M cells, Ehrlich ascites tumour cells, etc. Much of the information and references germane to the work on the membrane-damaging properties and lytic mechanisms of the different toxins of the group until the end of the 1970s can be found in the particularly complete review of Smyth and Duncan (1978) and also in the reviews of Bernheimer (1976), Freer and Arbuthnott (1976), Alouf (1977) and Arbuthnott (1982). An extensive review on the lytic properties of SLO was published by Alouf (1980).

The mechanisms of cell lysis and membrane attack by the toxins has been actively investigated on erythrocytes and other cells during the past 12 years (Table 5). The progress made during this period was reviewed by Linder and Bernheimer (1984), Bernheimer and Rudy (1986), Freer (1986), Bhakdi and Tranum-Jensen (1988).

Table 5. Reports (1978–80) on the mechanisms of cell lysis and membrane attack by the cholesterol-binding toxins

Target cells	Toxin	References
Erythrocytes	Streptolysin O	1–8
	Perfringolysin O	9–13
	Cereolysin O	14
	Ivanolysin O	15
Platelets	Streptolysin O	16, 17
	Alveolysin	16
Human fibroblasts	Streptolysin O	18
	Perfringolysin O	18
	Alveolysin	19
HeLa and L cells	Streptolysin O	20
Myocardial cells	Streptolysin O	21
Mycoplasma cells	Tetanolysin	22

(1) Badin and Denne (1978); (2) Fehrenbach et al. (1982); (3) Buckingham and Duncan (1983); (4) Bhakdi et al. (1984, 1985); (5) Bhakdi and Tranum-Jensen (1985); (6) Niedermeyer (1985); (7) Hugo et al. (1986); (8) Linder and Bernheimer (1984); (9) Mitsui et al. (1979a,b); (10) Mitsui et al. (1982a); (11) Thelestam et al. (1983); (12) Ohno-Iwashita et al. (1986, 1988); (13) Iwamoto et al. (1987); (14) Cowell et al. (1978); (15) Parrisius et al. (1986); (16) Launay and Alouf (1979); (17) Launay et al. (1984); (18) Thelestam and Möllby (1980); (19) Thelestam et al. (1981); (20) Duncan and Buckingham (1980); (21) Fisher et al. (1981); (22) Rottem et al. (1990).

Features of the lytic process

Most of the studies concerning the phenomenological aspects of the lytic process used erythrocytes as target cells (haemolysis). In contrast to many other bacterial cytolysins, erythrocyte lysis occurs in a matter of minutes with SLO, PFO and the other related toxins even as low as 1 HU. This amount of toxin is equivalent to about 1 ng of protein, corresponding to *ca.* 1.1×10^{10} molecules on the basis of an average M_r of 53 000. As 1 HU is the amount of toxin which lyses *ca.* 1×10^8 sheep, rabbit or human erythrocytes, it can be calculated that about 100 molecules of toxin are statistically required for the lysis of one erythrocyte. Human platelets appear more sensitive as about 15 molecules were required for lysis (Launay and Alouf, 1979). According to Kanbayashi *et al.* (1972), only two or three molecules may be required to lyse a single erythrocyte. Comparatively as high as 10^8 molecules of polyene antibiotics, which are also cholesterol-binding lysins, are necessary for the lysis of one red blood cell (Alouf, 1980). All authors agree, as originally put forward in 1968 by Alouf and Raynaud (see Arbuthnott, 1982), that erythrocyte and nucleated cell lysis by the toxins is a multi-hit, two-step process which involves toxin binding to target cells followed by cell disruption as mentioned in the reviews of Smyth and Duncan (1978) and Alouf (1977, 1980) and more recently documented by Fehrenbach *et al.* (1982), Niedermeyer (1985), Ohno-Iwashita *et al.* (1986, 1988), Iwamoto *et al.* (1987), Bhakdi and Tranum-Jensen (1988).

Binding-step. Toxin binding is practically immediate (within less than 2 min), irreversible and independent of pH, ionic strength and temperature. This can be demonstrated by incubating 10–20 HU of toxin with erythrocytes at 0°C where no lysis occurs. However, as shown by Alouf and Raynaud (1968), erythrocyte lysis takes place at this temperature if high amounts (~20 HU) of toxin are incubated with the cells as confirmed by Niedermeyer (1985) and Hugo *et al.* (1986), who observed a similar effect after challenge with 500–1000 HU/7×10^6 cells or 5000–10 000 molecules/cell respectively. If the cells are resuspended in fresh isotonic buffer at 37°C, they lyse completely. At the same time incubation of fresh cells at 37°C with the 0°C supernate produces no lysis.

The ability to separate the initial binding step from the subsequent lytic events made it easy to study the features of the binding step separately from the whole process (see Alouf, 1980; Smyth and Duncan, 1978; Ohno-Iwashita *et al.*, 1986, 1988; Iwamoto *et al.*, 1987). That toxin binding could be reversible was reported (Kanbayashi *et al.*, 1972) on the basis of SLO transferability from erythrocyte to erythrocyte. However, the experiments reported by the other authors indicated the irreversibility of binding of the toxins studied.

It has been clearly established that the abrogation of the lytic activity of the toxins by the blockade of their free SH group with the classical thiol group reagents (Hg salts, p-chloromercuribenzoate, *N*-ethylmaleimide, etc.) is involved at the level of the binding step. As discussed earlier, the free SH group is not essential *per se* for toxin fixation as previously thought. On these grounds, the hypothesis of a possible formation of a disulphide bond or S-acyl linkage with a membrane protein for which no evidence was provided in previous studies (Wannamaker, 1963) could be ruled out. Such a conclusion was also supported by the work of Iwamoto *et al.* (1987) with PFO treated with the thiol-blocking agent 5,5′-dithiobis (2-nitrobenzoic acid). This treatment was shown to affect both toxin binding and the subsequent lytic process.

Our interpretation is that the blockade of the SH group creates a steric hindrance which prevents to various extents the interaction of the cholesterol-binding domain of the toxins with the corresponding fixation site on target cells, depending on the nature and size of the molecules combined to the SH group and that of the local structure, conformation and hydrophobicity of the toxin domain(s) (variable from one toxin to another) involved in their fixation. This view is supported by the new insights provided by Alouf and Geoffroy (1984) who used for the first time tosyl phenylalanine chloromethyl ketone (TPCK) and tosyl lysine chloromethyl ketone (TLCK) in probing toxin activity. These components are known to block histidine residue(s) in certain enzymes and also to alkylate and inactivate thiol-proteinases by combination with their SH groups. The inhibition of the haemolytic activity of SLO, ALV, PLY and PFO by these two reagents showed great differences among these toxins, suggesting variable steric hindrances in spite of the blockade of

their respective SH groups. This variability and differences in the hydrophobicity of the respective domains of these toxins may also explain the differences in the inhibitory activity of various classical SH group reagents examined by Alouf and Geoffroy (1984) and Yamakawa and Ohsaka (1986).

Substantial information on the binding step was provided by Ohno-Iwashita *et al.* (1988) and Iwamoto *et al.* (1987, 1990), who investigated the interaction of intact and nicked perfringolysin O molecules with sheep erythrocytes. PFO derivatives called Cθ and Tθ were obtained by limited proteolysis of this toxin with subtilisin and trypsin respectively. Cθ was a complex of 15-kDa (N-terminal) and 38-kDa (C-terminal) fragments called C2 and C1 respectively, and Tθ a complex of 28-kDa and 25-kDa fragments called T1 and T2 respectively. PFO and Tθ lysed sheep, human and rat erythrocytes over a wide range (4–37°C) of temperatures, whereas Cθ caused almost no haemolysis below 20°C (even after addition of large excess) but retained binding affinity to membrane cholesterol very close to that of PFO. The haemolytic activity of Cθ and Tθ at 37°C was 40% and 27% that of native PFO respectively. Scatchard plot of Cθ binding revealed the presence of low- and high-affinity sites for the toxin on target cells. In contrast to PFO, Cθ bound the erythrocytes in a reversible manner at low temperature. This explains the differences in the lytic properties of PFO and Cθ. Cθ did not cause obvious damage below 20°C as shown by the absence of the arc- and ring-shaped structures visualized by electron microscopy on cell membranes treated with native cholesterol-binding toxins. In addition, Cθ did not show any influx of extracellular Ca^{2+}, in contrast to native PFO (Saito, 1983). Thus, Cθ appears as a convenient probe for the study of toxin binding without the interference of the lytic process.

The separation of fragments T1 and T2 from the Tθ complex allowed an interesting study of toxin–cell interaction (Iwamoto *et al.*, 1990). In contrast to Tθ, neither T1 nor T2 had haemolytic activity. However, T2, which contains the SH group region, exhibited the same potential as native PFO in its binding specificity for cholesterol. Its binding affinity for erythrocytes was lost by treatment with 5,5′-dithiobis (2-nitrobenzoic acid). The incubation of T2 with PFO suppressed the lytic activity of the latter but not its binding to erythrocytes due to abrogation of its oligomerization on the membrane by T2.

Lytic step. This step is well-known and analysed from a phenomenological standpoint but is still poorly understood at the molecular level. It is characterized by the progressive release of intracellular components or markers after a short pre-lytic lag period (30 s to 5 min for erythrocytes challenged with doses between 25 HU and 1 HU per 10^8 cells). It probably corresponds to toxin insertion into the bilayer indicating that this process is rather rapid as compared to many other cytolytic agents. The time course of the lysis of erythrocytes and other cells (platelets, many nucleated cells, mycoplasma cells) follows a non-linear process most often of the sigmoidal type (see Rottem *et al.*, 1976; Smyth and Duncan, 1978; Launay and Alouf, 1979; Alouf, 1977, 1980; Buckingham and Duncan, 1983; Blumenthal and Habig, 1984; Niedermeyer, 1985; Ohno-Iwashita *et al.*, 1988; Iwamoto *et al.*, 1987, 1990).

The lytic step is temperature dependent. It is inhibited by high ionic strength and certain divalent cations, particularly Zn. It is also pH dependent and does not take place at pH below 4.5 and above 9. As shown by Geoffroy *et al.* (1987), great differences in the optimal pH of activity were found for SLO, PLY, ALV, PFO and LLO. That of the latter was the lowest (pH 5.5) and is likely relevant to the intracellular growth of *Listeria monocytogenes* in acidic phagosomes. The pH range of activity was also very narrow for LLO (no activity at pH 7.0) as compared to the other toxins. Certain anti-SLO antibodies have been shown to block significantly the lytic step but not toxin binding to erythrocytes which was inhibited by another type of antibodies (Alouf, 1980; Watson and Kerr, 1985; Hugo *et al.*, 1986) which very likely interact with the cholesterol-binding domain or interfere with toxin access to it by steric hindrance after binding other epitopes.

Some dissimilarities were reported by Thelestam and Möllby (1980) and Thelestam *et al.* (1981) as concerns the lytic effects of SLO, PFO and ALV on human fibroblasts. They noted among other differences that in contrast to SLO, PFO and ALV do not appear to bind irreversibly to target cells but were rather transiently adsorbed. SLO induced

large functional pores (see next section) in contrast to the others which also appeared active at 0°C (marker release). These differences led to the suggestion of a 'hit and run' mechanism for the action of PFO and ALV on fibroblasts. This behaviour was also consistent with the concept of a repair phenomenon of the 'holes' in the toxin-damaged membranes as reflected by the lowered uptake of Trypan Blue by the cells 48 h after challenge with toxins (Thelestam and Möllby, 1983).

Generation of functional pores

The observation of Duncan and associates (see Smyth and Duncan, 1978) of the simultaneous leakage of ^{86}Rb and Hb from SLO-treated erythrocytes led to the concept that the lesions induced by the toxin in target cell membrane were very large. A similar simultaneous release from erythrocytes of ATP and Hb by SLO (Fehrenbach *et al.*, 1982) and K^+ and Hb by tetanolysin (Blumenthal and Habig, 1984) support this contention. The addition of large molecules such as serum albumin was shown to retard the escape of these substances and suggests that the 'functional holes' or lesions have an effective radius greater than 3.5 nm. These results rule out a colloid osmotic process of cell lysis by these toxins, in contrast to that elicited by streptolysin S, staphylococcal α-toxin or complement. In a colloid osmotic mechanism, the lytic agent initially causes a permeability change of small ions or molecules which thereby lose their ability to serve as an osmotic balance against the macromolecules inside the cell, which therefore swells and eventually ruptures. In such a mechanism the egress of small molecules precedes that of large molecules (Blumenthal and Habig, 1984).

The size of the functional holes/pores was shown by Buckingham and Duncan (1983) to depend on the amount of toxin presented to the cells as supported by the study of the egress of various markers entrapped in resealed sheep erythrocyte ghosts challenged with SLO. No release of molecules larger than 72.2 Å in diameter (M_r 68 000, that of Hb) was observed. Larger channels were formed with increased toxin concentrations. The egress of the largest marker molecules was seen after challenge with 300–500 HU of SLO indicating lesions > 128 Å. Higher levels (4000 HU/ml) resulted in even greater release of the largest markers.

The data reported by Menestrina *et al.* (1990) showed that in contrast to SLO the lesions generated by PFO on Lettre cells were rather small as shown by the study of the leakage of monovalent cations, [³H]- and [³⁵S]-labelled material and lactate dehydrogenase (LDH). The amounts of toxins which cause almost complete leakage of the low-molecular-weight compounds caused little leakage of LDH (30 kDa for the monomer, 120 kDa for the tetramer) or other proteins. Divalent cations inhibited the leakage of the internal components ($Zn^{2+} > Ca^{2+} > Mg^{2+}$) by an effect involving pore closure rather than the prevention of pore formation.

The generation of transmembrane diffusion channels in *Mycoplasma gallisepticum* elicited by tetanolysin was inferred by Rottem *et al.* (1990) from the study of the swelling of these wall-less microorganisms in the presence of toxins and various osmotic stabilizers. The postulated water-filled toxin-generated pores allowed the diffusion of hydrophilic molecules into the cells.

Applications of cell permeabilization by the toxins in cell biology research. The poration of various cells by SLO has been recently employed for the introduction of cell effectors for the analysis of a number of eukaryotic cell functions such as intracellular metabolism (Ahnert-Hilger *et al.*, 1989), the phosphorylation of CD3 antigens in T lymphocytes (Alexander *et al.*, 1989), inositol phosphate turnover (Ali *et al.*, 1989) and Ca^{2+} metabolism (Edwardson *et al.*, 1990).

Cholesterol-binding toxins as tools in platelet research. The generation of limited lesions in human platelets by very low amounts of SLO and ALV were applied for the isolation of the dense bodies which are still very poorly investigated and for the study of serotonin compartments (Launay and Alouf, 1979). The same toxins were also used by Launay *et al.* (1984) for the isolation and the study of various platelet enzymes which could not be analysed easily by other methods.

Toxin-induced damage of artificial model membranes

Liposomes and monomolecular films constituted by various phospholipids and cholesterol or other sterols have been used as models for cell membranes in the study of the membrane-damaging effects of various bacterial toxins (see Duncan, 1984 for a review) including the toxins reviewed in this chapter.

Liposomal systems

The first attempts to demonstrate the release of intraliposomal markers from toxin-treated multilamellar vesicles were unsuccessful (Smyth and Duncan, 1978). This failure was attributed to unsuitable physical state of the vesicles and inappropriate phospholipid/cholesterol proportions as first reported by Cowell and Bernheimer (1978). These authors showed that cereolysin O caused the release of 43% of the trapped glucose from liposomes containing 50 mol % of cholesterol, whereas no release was obtained from liposomes containing only 19 mol % of cholesterol. However, the disruption of the high-cholesterol content liposomes required very high amounts of toxin (10 000 HU/ml). Similarly 4500 HU of tetanolysin released 45% of the glucose of vesicles containing 50 mol % cholesterol but only 4% glucose was released from similar vesicles containing 43% cholesterol (Alving *et al.*, 1979).

A detailed study of the binding of SLO by phospholipid–cholesterol vesicles as a function of their respective molar ratio and the nature of phospholipid constituents was reported by Delattre *et al.* (1979). The results of these authors have been summarized by Duncan (1984). Rosenqvist *et al.* (1980) investigated the lytic effect of SLO on large unilamellar vesicles and multilamellar liposomes prepared with varying molar ratios of egg lecithin and cholesterol and loaded with the water-soluble spin label TEMPO-choline chloride. Multilamellar liposomes were not lysed under any condition by SLO (up to 1000 HU/μmol lipid). In contrast, the unilamellar vesicles were damaged provided their cholesterol content was greater than 33 mol %. It was maximal for 50 mol % cholesterol and lower for 67 mol % due to phase separation. Fehrenbach *et al.* (1984) reported the release of entrapped ATP from both multilamellar and monolamellar vesicles treated with low concentrations (24 HU/ml) of SLO. Maximum release was found for 50 mol % of cholesterol. A similar result was found by Alouf and Prigent (see Duncan, 1984).

The damage of cholesterol-containing liposomes by 1500 HU of alveolysin was followed by the kinetics of the release of entrapped [^{14}C]arginine (Geoffroy and Alouf, 1983). Small unilamellar cholesterol–phosphatidyl vesicles entrapping calcein (M_r 700) or fluorescein–dextran (M_r 20 000, estimated radius 8 nm) released these markers when treated with 6 µg/ml (ca. 6000 HU) of perfringolysin O (Menestrina et al., 1990). Calcein release was biphasic and faster at low pH than at neutrality. The leakage was sensitive to inhibition by divalent cations, as in the toxin-treated Lettre cells.

Despite the discrepancies in the minimal amounts of toxin required to damage the liposomes, the results mentioned above clearly show that an equimolar cholesterol/phospholipid ratio is required for vesicle damage to an extent which allows maximal release of entrapped markers.

Monomolecular films

Monomolecular films of phosphatidylcholine (PC) and cholesterol or other sterols have been used by Alouf et al. (1984) as artificial models in the study of toxin–membrane interaction. The films spread at an initial pressure of 25 mN/m, which is close to that found in biomembranes, were shown to be penetrated by SLO, PFO, ALV and PLY in the reduced but not in the SH blocked state. Toxin penetration in the film was reflected by the increase of surface pressure up to 48 mN/m (collapse pressure). Penetration was temperature dependent, proportional to toxin concentration and optimal for 50 mol % cholesterol. No penetration was observed in the absence of cholesterol. For identical toxin concentrations in terms of haemolytic units, the magnitude of pressure increase was in the following order SLO $\geqslant$ PFO $\sim$ ALV > PLY, indicating that film penetration depended on the molecular structure of the toxins. Penetration was observed in films containing other 3β-hydroxysterols with aliphatic sidechains, but was very weak in the films constituted by non-inhibitory sterols.

Planar and lipid bilayers

Cholesterol-containing planar lipid bilayers, also known as black lipid membranes, have been used for the study of the mechanism of the membranolytic properties of tetanolysin

(Blumenthal and Habig, 1984) and perfringolysin O (Menestrina et al., 1990). Toxin interaction with the bilayer is reflected by the pattern of conductance fluctuation. Below 30% of cholesterol, tetanolysin did not show any effect on bilayer conductance. Above 30% the bilayer was disrupted. The characteristics of the conductance pattern and non-linear concentration dependence of toxin-induced conductance led to the inference that the toxin acted by lipid perturbation rather than by the formation of structural channels, such as those generated by gramicidin. In contrast, the results of Menestrina et al. (1990) are in favour of the formation of quite heterogeneously sized pores in the bilayers. Divalent cations decreased the open state probability of the toxin-generated channels, as was found for liposomes and Lettre cells.

Conclusions

Although liposomes appear useful for studying certain aspects of toxin–membrane interactions, they could not be considered as systems which significantly mimic biomembranes in the case of the toxins reviewed in this chapter. The data reported above for tetanolysin and PPO as regards the pore-formation concept versus membrane perturbation or the actual size of the pores clearly show that the extrapolation of data from experiments in artificial systems to intact cells are not always justified.

Ultrastructural changes at the surface of target cells and artificial membranes

Striking ring- and arc-shaped structures lying in the broad plane of erythrocyte membranes lysed with SLO were observed by electron microscopy (negative staining) in the pioneering work of Dourmashkin and Rosse (1966). These remarkable structures were subsequently found on erythrocyte membranes and the surrounding medium after cell lysis with SLO (Duncan and Schlegel, 1975; Bhakdi et al., 1984, 1985; Bhakdi and Tranum-Jensen, 1985; Niedermeyer, 1985), perfringolysin O (Smyth et al., 1975; Mitsui et al., 1979a), cereolysin O (Cowell et al., 1978) and listeriolysin O (Parrisius et al., 1986). The same structures were found on Acholeplasma laidlawii cells treated with CLO (Cowell et al., 1978).

No rings or arcs formed when toxin–erythrocyte mixtures were maintained at 0°C (i.e. under conditions where toxin binding occurs without membrane damage) or with heat-inactivated toxins. These structures were also not visible on erythrocytes treated with small but sufficient lytic amounts of toxin (Cowell *et al.*, 1978; Saito, 1983).

Most studies showed that ring and arc dimensions were heterogeneous. The rings exhibited an internal diameter of 30–50 nm and a border thickness of 5–8 nm. The arcs varied in length from 33 to 170 nm and sometimes more with a border thickness of 6–8 nm. Interestingly, the same structures have been observed at the surface of erythrocytes lysed with metridiolysin (Bernheimer *et al.*, 1979), a cytolytic and lethal toxin produced by the sea anemone *Metridium senile*. Metridiolysin is an 'SH-activated', cholesterol-inhibitable polypeptide with biological effects indistinguishable from the toxins reviewed here. It differs from them in not being inhibited by sterols structurally related to cholesterol and by the absence of antigenic cross-reactivity.

The ring- and arc-shaped structures were also found in the following situations: (i) interaction with erythrocyte ghosts and/or cholesterol-containing liposomes and/or cholesterol dispersions treated with SLO (Duncan and Schlegel, 1975; Delattre *et al.*, 1979; Rosenqvist *et al.*, 1980), PFO (Smyth *et al.*, 1975; Mitsui *et al.*, 1979a), CLO (Cowell *et al.*, 1978) or tetanolysin (Rottem *et al.*, 1982); (ii) concentrated solutions of PFO (Smyth *et al.*, 1975; Mitsui *et al.*, 1979b), CLO (Cowell *et al.*, 1978), tetanolysin (Rottem *et al.*, 1982) and SLO (Niedermeyer, 1985) in the absence of cholesterol and other lipids.

The rings and arcs have been first interpreted as aggregates of toxin–cholesterol complexes initially dispersed throughout the membrane and thereafter reassembled and expelled from the bilayer with subsequent disorganization of cell surface, ending ultimately in lysis (see Duncan, 1984; Freer, 1986). However, as these structures may spontaneously arise in concentrated toxin solutions in the absence of cholesterol or other lipids, it was suggested as first considered by Smyth *et al.* (1975) that the arcs and rings represent polymers of toxin molecules which increase with toxin concentrations and not toxin–cholesterol complexes as originally thought (Freer, 1986; Bhakdi and Tranum-Jensen,

1988). Cholesterol at cell or liposome surfaces or in dispersed form (see Fig. 5) may confer favourable orientation of toxin molecules and thereby enhance the tendency of the toxin to polymerize. This concept is supported by the isolation of purified cholesterol-free toxin oligomers from erythrocyte-bound toxin by solubilization with deoxycholate followed by ultracentrifugation in sucrose gradient. The isolated polymers were seen by electron microscopy as 7–8-nm broad curved rods with a 13- to 16-nm radius of curvature and length ranging from 25 to 100 nm (Bhakdi *et al.*, 1985). On the basis of a molecular volume *ca.* 6000 nm^3 for a typical closed ring (M_r 5 × 10^6), a fully circularized toxin oligomer corresponds to 70–75 protomers of SLO. This structure, which may correspond to a single lesion, is in agreement with the earlier estimate of Alouf and Raynaud (1968) of about 100 molecules necessary for lysis of one erythrocyte. The various sizes of the ring- and arc-shaped structures can be interpreted as different stages in the polymerization of toxin molecules, as shown in the models proposed by Niedermeyer (1985) and Bhakdi *et al.* (1985).

Mechanism of toxin-induced cell lysis

Two different models have been proposed to explain the mechanism whereby erythrocytes and presumably other kinds of cells are lysed by the

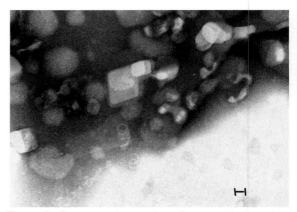

Figure 5. Electron micrography of a negatively stained preparation showing ring- and arc-shaped polymers of alveolysin after interaction with cholesterol dispersions in phosphate-buffered saline (Alouf and Geoffroy, unpublished). Bar = 100 nm.

cholesterol-binding toxins (Bernheimer and Rudy, 1986; Freer, 1986).

First model: Membrane disorganization without transmembrane pore formation

According to this model, the arc- and ring-shaped structures are an incidental by-product, are not transmembrane pores and do not play a role in the lytic process itself. This view is based on the absence of alteration of the internal leaflet of cell membranes in freeze-fracture studies of erythrocytes lysed with CLO (Cowell *et al.*, 1978), PFO (Mitsui *et al.*, 1979a) and SLO (Niedermeyer, 1985). The results obtained with tetanolysin interacting with black lipid membranes are consistent with this view, which is also supported by the finding that PFO, in contrast to the pore-forming staphylococcal α-toxin, was not photolabelled by an apolar photoactivatable hydrophobic probe in the membrane of toxin-treated erythrocytes (Thelestam *et al.*, 1983).

In the non-pore forming model, the current postulate is the sequestration of cholesterol from certain regions of the membranes which causes a lipid phase transition and thereby reduced molecular cohesion and fragilization of the membranes ('functional holes') which impair its permeability (see discussion in Smyth and Duncan, 1978; Niedermeyer, 1985; Freer, 1986, 1988).

Second model: Transmembrane channels

This model, advocated by Bhakdi and co-workers (see Bhakdi and Tranum-Jensen, 1988 for a review), provides compelling evidence that at least at relatively high toxin concentrations the ring and arc structures are indeed transmembrane pores spanning the bilayer and lined with the polymerized toxin. These channels permit direct escape of cell content. In this model (Hugo *et al.*, 1986), the membrane-bound toxin is exclusively present in the monomer form. Membrane damage will ensue only after oligomerization of toxin molecules in the membrane as in the case of the complement attack complex and lymphocyte cytolysins. The oligomerization process takes place by lateral aggregation of toxin monomers which collide with each other, leading to unfolding of polypeptide chains with exposure of hydrophobic domains that enter the membrane and ultimately create

transmembrane channels. The convex side of the oligomeric rod structure carries a strong apolar surface that anchors the toxin complex within the membrane through tight association with membrane lipids. Cholesterol is probably no longer required for this interaction once the oligomer is formed. The concave side of the rod is hydrophilic and, accordingly, repels membrane lipid molecules. If toxin circularization proceeds to completion, the lipids will be fully forced aside and excluded from the pore structure. If circularization is incomplete, a straight free edge of lipid membranes will remain between the two ends of the curved rod to complete the circumference (Fig. 6). The failure of the detection of the inner leaflet bilayer reported by the authors mentioned in the preceding section is explainable by technical reasons of electron microscopical analysis (Bhakdi and Tranum-Jensen, 1988).

The possibility that the mechanisms of membrane damage by cholesterol-binding toxins are not identical for all toxins or under certain conditions cannot be excluded (Sato, 1986; Freer, 1986).

Toxin effects at sublytic doses

Earlier work showed that sublytic concentrations of streptolysin O and perfringolysin O inhibited or suppressed: (i) phagocytosis by macrophages, (ii) mobility and chemotaxis of neutrophils, (iii) lymphocyte transformation by mitogens, and (iv) membrane transport systems in mammalian cells (see for references Smyth and Duncan, 1978; Bhakdi and Tranum-Jensen, 1988). A number of studies have been reported in the past decade on

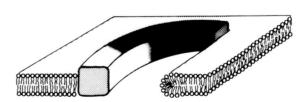

Figure 6. Diagrammatic presentation of the gross principal features proposed for the streptolysin O-induced lesion generated by incompletely circularized toxin oligomers. (Reproduced from Bhakdi *et al.*, 1985 by permission of the senior author and the editor of *Infection and Immunity.*)

the impairment of many cellular functions by sublytic or sublethal doses of toxins.

Alveolysin, PFO and SLO incubated with human polymorphonuclear granulocytes (PMN) increased at sublytic doses the release of leukotrienes, particularly the chemotactic (LTB_4) and spasmogenic (LTC_4, LTD_4 and LTE_4) factors as well as leukotriene-metabolizing enzymes (Bremm et al., 1985, 1987). This is likely related to the inflammatory response in chronic or acute streptococcal and clostridial diseases. The inflammatory potential of SLO was also shown by Bhakdi and Tranum-Jensen (1985) in the context of complement activation and attack on autologous cell membranes induced by nanomolar concentrations of SLO. This toxin and very likely other related toxins may circulate as immune complexes activating the formation of C5b-9 complement attack complex.

Listeriolysin O elicited at sublytic and sublethal concentrations a direct and rapid inflammatory reaction by i.d. injection in mouse footpad (Geoffroy et al., 1987). The toxin also induced delayed hypersensitivity inflammatory reaction in mice infected with as low as 10^3 replicative L. monocytogenes bacteria (Berche et al., 1987a,b).

At concentrations too low to affect PMN, pneumolysin inhibited the chemotaxis and migration of these cells as well as their ability to undergo a respiratory burst in response to zymosan stimulation and to kill opsonized pneumococci (Paton and Ferrante, 1983). The toxin also inhibits human lymphocyte response to mitogens (Ferrante et al., 1984), phospholipid transmethylation and the ability of monocytes to degranulate (Nandoskar et al., 1986). The activation of the classical pathway of complement in an antibody-independent manner (Paton et al., 1984) is very likely relevant to the induction of the inflammatory process and tissue damage. PLY has been shown to cause ciliary slowing and disruption of the respiratory epithelium in organ culture (Steinfort et al., 1989) and to play an important role in ocular infections by pneumococci (Johnson et al., 1990). These various effects of PLY at low concentrations probably compromise the normal non-specific defenses in infected tissues by toxin and interference with the antimicrobial activities of macrophages, PMN and complement. The immunization of mice with PLY significantly increased their survival time after challenge with virulent S.

pneumoniae, indicating the involvement of this toxin in pneumococcal pathogenicity (Lock et al., 1988).

The in vivo cardiotoxic and lethal effects of PFO (Stevens et al., 1988) support the hypothesis of toxin-induced stimulation of the synthesis of platelet-activating factor by human endothelial cells which is known to generate shock and compromise cardiovascular functions in a variety of experimental models.

Concluding remarks and trends for future research

In assessing the past 10 years of research of the 'sulphydryl-activated', cholesterol-binding cytolytic toxins, the greatest progress has been realized in the area of the molecular genetics and the mechanisms of toxin-induced membrane damage. Gene cloning and sequencing technology allowed the knowledge of the primary structure of five toxins which was unknown four years ago. Comparative study of the sequences has led to very important conclusions about the structure–activity relationships. The most important finding is that the unique cysteine residue is not critical per se through its SH group for activity as was thought for over 60 years. A great advance has been also achieved in the study of listeriolysin O, which was poorly documented when the previous comprehensive review was published (Smyth and Duncan, 1978). This toxin has been shown in many laboratories to be an essential virulence factor in the in vivo intracellular development of listeria organisms in host macrophages.

Future perspectives should involve the purification of the ten toxins which have not yet been obtained in the homogeneous state. Gene sequencing of the toxins which have been purified so far and that of the others will greatly contribute to a better understanding of the structural relationships and evolutionary aspects of the toxin family reviewed here and those of their structural genes. The analysis of the genetic regulation of toxin expression will greatly contribute to our understanding of the molecular genetics of the toxins. Toxin crystallization, if successful for the determination of the 3-dimensional structures, will be a great step toward the elucidation of the mechanism of action of the toxin at the molecular level. Finally, the evaluation of the role of the toxins in natural disease, which is still a very difficult question, may contribute to a significant progress in bacterial pathogenicity.

Acknowledgements

The authors would like to thank Miss Luce Cayrol for her excellent and invaluable secretarial assistance in the preparation of this manuscript.

References

Ahnert-Hilger, G., Mach, W., Föhr, K.J. and Gratzl, M. (1989). Poration of alpha-toxin and streptolysin O: an approach to analyze intracellular process. In: *Methods in Cell Biology*, Vol. 31, *Vesicular Transport Part A* (ed. A.M. Tartakoff), pp. 63–90. Academic Press, London.

Alexander, D.R., Hexham, J.M., Lucas, S.C. and Graves, J.D. (1989). A protein kinase C pseudosubstrate peptide inhibits phosphorylation of the CD3 antigen in streptolysin O-permeabilized human T lymphocytes. *Biochem. J.* **260**, 893–901.

Ali, H., Cunha-Melo, J.R. and Beaven, M.A. (1989). Receptor-mediated release of inositol 1,4,5-triphosphate and 1,4 biphosphate in rat basophilic leukemia RBL-2H3 cells permeabilized with streptolysin O. *Biochim. Biophys. Acta* **973**, 88–99.

Alouf, J.E. (1977). Cell membranes and bacterial toxins. In: *The Specificity and Action of Animal, Bacterial and Plant Toxins* (ed. P. Cuatrecasas), pp. 219–70. Chapman and Hall, London.

Alouf, J.E. (1980). Streptococcal toxins (streptolysin O, streptolysin S, erythrogenic toxin). *Pharmacol. Therap.* **11**, 661–717.

Alouf, J.E. and Geoffroy, C. (1979). Comparative effects of cholesterol and thiocholesterol on streptolysin O. *FEMS Microbiol. Lett.* **6**, 413–16.

Alouf, J.E. and Geoffroy, C. (1984). Structure activity relationships in sulfhydryl-activated toxins. In: *Bacterial Protein Toxins* (eds J.E. Alouf, F.J. Fehrenbach, J.H. Freer and J. Jeljaszewicz), pp. 165–71. Academic Press, London.

Alouf, J.E. and Geoffroy, C. (1988) Production, purification and assay of streptolysin O. In: *Microbial Toxins: Tools in Enzymology. Methods in Enzymology*, Vol. 165 (ed. S. Harshman), pp. 52–9. Academic Press, San Diego.

Alouf, J.E. and Raynaud, M. (1968). Some aspects of the mechanisms of lysis of rabbit erythrocytes by streptolysin O. In: *Current Research on Group A Streptococcus* (ed. R. Caravano), pp. 192–206. Excerpta Medica Foundation, Amsterdam.

Alouf, J.E. and Raynaud, M. (1973). Purification and some properties of streptolysin O. *Biochimie* **55**, 1187–93.

Alouf, J.E., Kiredjian, M. and Geoffroy, C. (1977). Purification de l'hémolysine thiol-dépendante extracellulaire de *Bacillus alvei*. *Biochimie (Paris)* **59**, 329–36.

Alouf, J.E., Geoffroy, C., Pattus, F. and Verger, R. (1984). Surface properties of bacterial sulfhydryl-activated cytolytic toxins. Interaction with monomolecular films of phosphatidylcholine and various sterols. *Eur. J. Biochem.* **141**, 205–10.

Alouf, J.E., Geoffroy, C., Gilles, A.M. and Falmagne, P. (1986). Structural relatedness between five bacterial sulfhydryl activated toxins: streptolysin O, perfringolysin O, alveolysin, pneumolysin and thuringiolysin. In: *Bacterial Protein Toxins* (eds P. Falmagne, J.E. Alouf, F.J. Fehrenbach, J. Jeljaszewicz and M. Thelestam), pp. 49–50. Gustav Fischer Verlag, Stuttgart.

Alving, C.R., Habig, W.H., Urban, K.A. and Hardegree, M.C. (1979). Cholesterol-dependent tetanolysin damage to liposomes. *Biochim. Biophys. Acta* **551**, 224–8.

Arbuthnott, J.P. (1982). Bacterial cytolysins (membrane-damaging toxins). In: *Molecular Action of Toxins and Viruses* (eds P. Cohen and S. van Heyningen), pp. 107–29. Elsevier Biomedical Press, Amsterdam.

Ashdow, L.E. and Koehler, J.M. (1990). Production of hemolysin and other extracellular enzymes by clinical isolates of *Pseudomonas pseudomallei*. *J. Clin. Microbiol.* **28**, 2331–4.

Badin, J. and Denne, M.-A. (1978). A cholesterol fraction for streptolysin O binding on cell membranes and lipoproteins – I. Optimal conditions for the determination. *Cell Molec. Biol.* **22**, 133–43.

Beecher, D.J. and MacMillan, J.D. (1990). A novel bicomponent hemolysin from *Bacillus cereus*. *Infect. Immun.* **58**, 2220–7.

Berche, P., Gaillard, J.L. and Sansonetti, P. (1987a). Intracellular growth as a prerequisite for in vivo induction of T-cell immunity. *J. Immunol.* **138**, 2266–71.

Berche, P., Gaillard, J.L., Geoffroy, C. and Alouf, J.E. (1987b). T-cell recognition of listeriolysin O is induced during infection with *Listeria monocytogenes*. *J. Immunol.* **139**, 3813–21.

Berche, P., Gaillard, J.-L. and Richard, S. (1988). Invasiveness and intracellular growth of *Listeria monocytogenes*. *Infection* **16** (suppl. 2), S145–8.

Berche, P., Reich, K.A., Bonnichon, M., Beretti, J.L., Geoffroy, C., Raveneau, J., Cossart, P., Gaillard, J.L., Geslin, P., Kreis, H. and Veron, M. (1990). Detection of anti-LLO for serodiagnosis for human listeriosis. *Lancet* **335**, 624–7.

Bernheimer, A.W. (1976). Sulfhydryl activated toxins. In: *Mechanisms in Bacterial Toxinology* (ed. A.W. Bernheimer), pp. 85–97, John Wiley & Sons, New York.

Bernheimer, A.W. and Grushoff, P. (1967a). Cereolysin:

production, purification and partial characterization. *J. Gen. Microbiol.* **46**, 143–50.

Bernheimer, A.W. and Grushoff, P. (1967b). Extracellular hemolysins of aerobic sporogenic bacilli. *J. Bacteriol.* **93**, 1541–3.

Bernheimer, A.W. and Rudy, B. (1986). Interaction between membranes and cytolytic peptides. *Biochim. Biophys. Acta* **864**, 123–41.

Bernheimer, A.W., Avigad, L.S. and Kim, K.-S. (1979). Comparison of metridiolysin from the sea anemone with thiol-activated cytolysins from bacteria. *Toxicon* **17**, 69–75.

Bhakdi, S. and Tranum-Jensen, J. (1985). Complement activation and attack on autologous cell membranes induced by streptolysin O. *Infect. Immun.* **48**, 713–19.

Bhakdi, S. and Tranum-Jensen, J. (1988). Damage to cell membranes by pore-forming bacterial cytolysins. *Prog. Allergy* **40**, 1–43.

Bhakdi, S., Roth, M., Sziegoleit, A. and Tranum-Jensen, J. (1984). Isolation and identification of two hemolytic forms of streptolysin O. *Infect. Immun.* **46**, 394–400.

Bhakdi, S., Tranum-Jensen, J. and Sziegoleit, A. (1985). Mechanism of membrane damage by streptolysin O. *Infect. Immun.* **47**, 52–60.

Bielecki, J., Youngman, P., Connelly, P. and Portnoy, D.A. (1990). *Bacillus subtilis* expressing a haemolysin gene from *Listeria monocytogenes* can grow in mammalian cells. *Nature* **345**, 175–6.

Blumenthal, R. and Habig, W.H. (1984). Mechanism of tetanolysin-induced membrane damage: studies with black lipid membranes. *J. Bacteriol.* **157**, 321–3.

Boulnois, G.J., Mitchell, T.J., Saunders, F.K., Mendez, F.J. and Andrew, P.W. (1990). Structure and function of pneumolysin, the thiol-activated toxin of *Streptococcus pneumoniae*. In: *Bacterial Protein Toxins* (eds R. Rappuoli, J.E. Alouf, P. Falmagne *et al.*), Zbl. Bakt. Suppl. 19, pp. 43–51. Gustav Fischer Verlag, Stuttgart.

Bremm, K.D., König, W., Pfeiffer, P., Rauschen, I., Theobald, K., Thelestam, M. and Alouf, J.E. (1985). Effect of thiol-activated toxins (streptolysin O, alveolysin and theta toxin) on the generation of leukotrienes and leukotriene-inducing and -metabolizing enzymes from human polymorphonuclear granulocytes. *Infect. Immun.* **50**, 844–51.

Bremm, K.D., König, W., Thelestam, M. and Alouf, J.E. (1987). Modulation of granulocyte functions by bacterial exotoxins and endotoxins. *Immunology* **62**, 363–71.

Brzin, B., Kuhar, N., Naverznick, B. and Vadnjal, A. (1990). Functional similarity of *Listeria ivanovii* and *Staphylococcus aureus* in CAMP test. *Zbl. Bakt.* **273**, 179–83.

Buckingham, L. and Duncan, J.L. (1983). Approximate dimensions of membrane lesions produced by streptolysin S and streptolysin O. *Biochim. Biophys. Acta* **729**, 115–22.

Burn, C.G. (1934). Unidentified gram-positive bacillus associated with meningo-encephalitis. *Proc. Soc. Exp. Biol. Med.* **31**, 1095.

Cluff, C.W. and Ziegler, H.K. (1987). Inhibition of macrophage-mediated antigen presentation by hemolysin-producing *Listeria monocytogenes*. *J. Immunol.* **139**, 3808–12.

Coolbaugh, J.C. and Williams, R.P. (1978). Production and characterization of two hemolysins of *Bacillus cereus*. *Can. J. Microbiol.* **24**, 1289–95.

Cossart, P. and Mengaud, J. (1989). *Listeria monocytogenes*: A model for the molecular study of intracellular parasitism. *Molec. Biol. Med.* **6**, 463–74.

Cossart, P., Vicente, M.F., Mengaud, J., Baquero, F., Perez-Diaz, J.C. and Berche, P. (1989). Listeriolysin O is essential for the virulence of *Listeria monocytogenes*: direct evidence obtained by gene complementation. *Infect. Immun.* **57**, 3629–36.

Cowart, R.E. (1987). Iron regulation of growth and haemolysin production by *Listeria monocytogenes*. *Ann. Inst. Pasteur (Microbiol.)* **138**, 246–9.

Cowell, J.L. and Bernheimer, A.W. (1978). Role of cholesterol in the action of cereolysin membranes. *Arch. Biochem. Biophys.* **190**, 603–10.

Cowell, J.L., Grushoff-Kosyk, P. and Bernheimer, A.W. (1976). Purification of cereolysin and the electrophoretic separation of the active (reduced) and inactive (oxidized) forms of the purified toxin. *Infect. Immun.* **14**, 144–54.

Cowell, J.L., Kim, K. and Bernheimer, A.W. (1978). Alteration by cereolysin of the structure of cholesterol-containing membranes. *Biochim. Biophys. Acta* **507**, 230–41.

Delattre, J., Lesbir, R., Panouse-Perrin, J. and Badin, J. (1979). A cholesterol fraction for streptolysin O binding in cell membranes and lipoproteins – II. Interaction of streptolysin O with phospholipid-cholesterol liposomes of different nature and molar ratios. *Cell. Molec. Biol.* **24**, 157–66.

Domann, E. and Chakraborty, T. (1989). Nucleotide sequence of the listeriolysin gene from a *Listeria monocytogenes* serotype 1/2a strain. *Nucleic Acids Res.* **17**, 6406.

Dourmashkin, R.R. and Rosse, W.F. (1966). Morphologic changes in the membranes of red blood cells undergoing hemolysis. *Am. J. Med.* **41**, 699–710.

Duncan, J.L. (1983). Streptococcal growth and toxin production *in vivo*. *Infect. Immun.* **40**, 501–5.

Duncan, J.L. (1984). Liposomes as membrane models in studies of bacterial toxins. *J. Toxicol. Toxin Rev.* **3**, 1–51.

Duncan, J.L. and Buckingham, L. (1980). Resistance to streptolysin O in mammalian cells treated with oxygenated derivatives of cholesterol: cholestrol content of resistant cells and recovery of streptolysin O sensitivity. *Biochim. Biophys. Acta* **603**, 278–87.

Duncan, J.L. and Schlegel, R. (1975). Effect of streptolysin O on erythrocyte membranes, liposomes and lipid dispersions. *J. Cell Biol.* **67**, 160–73.

Edwardson, J.M., Vickery, C. and Christy, L.J. (1990). Rat pancreatic acini permeabilized with streptolysin

O secrete amylase at Ca^{2+} concentrations in the micromolecular range when provided with ATP and GTP.S. *Biochim. Biophys. Acta* **1053**, 32–6.

Fairweather, N.F., Pickard, D.J., Morrissey, P.M., Lyness, V.A. and Dougan, G. (1987). Genetic analysis of toxin determinants of *Clostridium perfringens*. *Recent Adv. Anaerobic Bacteriol.* **12**, 138–47.

Fehrenbach, F.-J., Huser, H. and Jaschinski, C. (1980). Measurement of bacterial cytolysins with a highly sensitive kinetic method. *FEMS Microbiol. Lett.* **7**, 285–8.

Fehrenbach, F.-J., Schmidt, C.-M. and Huser, H. (1982). Early and long events in streptolysin O-inducing hemolysis. *Toxicon* **20**, 233–8.

Fehrenbach, F.-J., Schmidt, C.-M., Sterzik, B. and Jürgens, D. (1984). Interaction of amphiphilic bacterial polypeptides with artificial membranes. In: *Bacterial Protein Toxins* (eds J.E. Alouf, F.J. Fehrenbach, J.H. Freer and J. Jeljaszewicz), pp. 317–24. Academic Press, London.

Ferrante, A., Rowan-Kelly, B. and Paton, J.C. (1984). Inhibition of *in vitro* human lymphocyte response by the pneumococcal pneumolysin. *Infect. Immun.* **46**, 585–9.

Ferrieri, P. (1986). Immune response to streptococcal infections. In: *Manual of Clinical Laboratory Immunology* (eds N.E. Rose, H. Friedman and J.L. Fahey), pp. 336–41. American Society for Microbiology, Washington, DC.

Fisher, M.H., Kaplan, E.L. and Wannamaker, L.M. (1981). Cholesterol inhibition of streptolysin O for myocardial cells in tissue culture. *Proc. Soc. Exp. Biol. Med.* **163**, 233–7.

Fleming, W.L. (1927). Studies on the oxidation and reduction of immunological substances. VII. The differentiation of tetanolysin and tetanospasmin. *J. Exp. Med.* **46**, 279–90.

Freer, J.H. (1986). Membrane damage caused by bacterial toxins: recent advances and new challenges. In: *Natural Toxins. Animal, Plant and Microbial* (ed. J.B. Harris), pp. 189–211. Clarendon Press, Oxford.

Freer, J.H. (1988). Toxins as virulence factors of Gram-positive pathogenic bacteria of veterinary importance. In: *Virulence Mechanisms of Bacterial Pathogens* (ed. J.A. Roth), pp. 264–88. American Society for Microbiology, Washington, DC.

Freer, J.H. and Arbuthnott, J.P. (1976). Biochemical and morphologic alterations of membranes by bacterial toxins. In: *Mechanisms in Bacterial Toxinology* (ed. A.W. Bernheimer), pp. 169–93. Wiley, New York.

Gaillard, J.L., Berche, P. and Sansonetti, P. (1986). Transposon mutagenesis as a tool to study the role of hemolysin in the virulence of *Listeria monocytogenes*. *Infect. Immun.* **52**, 50–5.

Gaillard, J.L., Berche, P., Mounier, J., Richard, S. and Sansonetti, P. (1987). *In vitro* model of penetration and intracellular growth of *Listeria monocytogenes* in the human enterocyte-like cell line Caco2. *Infect. Immun.* **55**, 2822–9.

Gemmel, C.G. and Abdul-Amir, M.K. (1979). Effect of certain antibiotics on the formation of cellular antigens and extracellular products by group A streptococci. In: *Pathogenic Streptococci* (ed. M.T. Parker), pp. 67–8. Reedbooks, Chertsey, Surrey.

Geoffroy, C. and Alouf, J.E. (1982). Interaction of alveolysin, a sulfhydryl-activated bacterial cytolytic toxin, with thiol group reagents and cholesterol. *Toxicon* **20**, 239–41.

Geoffroy, C. and Alouf, J.E. (1983). Selective purification by thiol-disulfide interchange chromatography of alveolysin, a sulfhydryl-activated toxin of *B. alvei*. *J. Biol. Chem.* **258**, 9968–72.

Geoffroy, C. and Alouf, J.E. (1988). Molecular cloning and characterization of *Bacillus alvei* thiol-dependent cytolytic toxin expressed in *E. coli*. *J. Gen. Microbiol.* **134**, 1961–70.

Geoffroy, C., Gilles, A.M. and Alouf, J.E. (1981). The sulfhydryl groups of the thiol-dependent cytolytic toxin from *B. alvei*: evidence for one essential sulfhydryl group. *Biochem. Biophys. Res. Commun.* **99**, 781–8.

Geoffroy, C., Gaillard, J.L., Alouf, J.E. and Berche, P. (1987). Purification, characterization and toxicity of the sulfhydryl-activated hemolysin listeriolysin O from *Listeria monocytogenes*. *Infect. Immun.* **55**, 1641–6.

Geoffroy, C., Gaillard, J.L., Alouf, J.E. and Berche, P. (1989). Production of thiol-dependent hemolysins by *Listeria monocytogenes* and related species. *J. Gen. Microbiol.* **135**, 481–7.

Geoffroy, C., Mengaud, J., Alouf, J.E. and Cossart, P. (1990). Molecular cloning and complete nucleotide sequence of alveolysin, the thiol-activated toxin of *Bacillus alvei*: It is homologous to listeriolysin O, perfringolysin O, pneumolysin, and streptolysin O and contains a single cysteine. *J. Bacteriol.* **172**, 7301–5.

Gilmore, M.S., Cruz-Rodz, A.L., Leimeister-Wächter, M., Kreft, J. and Goebel, W. (1989). A *Bacillus cereus* cytolytic determinant, cereolysin AB which comprises the phospholipase C and sphingomyelinase genes: nucleotide sequence and genetic linkage. *J. Bacteriol.* **171**, 744–53.

Goebel, W., Kathariou, S., Kuhn, M., Sokolovic, Z., Kreft, J., Köhler, S., Funke, D., Chakraborty, T. and Leimeister-Wächter, M. (1988). Hemolysin from Listeria. Biochemistry, genetics and function in pathogenesis. *Infection* **16** (Suppl. 2), S149–59.

Gordon, R.E., Haynes, W.C. and Pang, C.H.N. (1973). The genus *Bacillus*. US Department of Agriculture. Agricultural Handbook No. 427. US Department of Agriculture, Washington, DC.

Gormley, E., Mengaud, J. and Cossart, P. (1989). Sequences homologous to the listeriolysin O gene region of *Listeria monocytogenes* are present in virulent and avirulent haemolytic species of the genus *Listeria*. *Res. Microbiol.* **140**, 631–43.

Guillaumie, M. (1950). Hémolysines bactériennes et anti-hémolysines. *Ann. Inst. Pasteur* **79**, 661–71.

Halbert, S.P. (1970). Streptolysin O. In: *Microbial Toxins*, Vol. III (eds T.C. Montie, S. Kadis and S.J. Ajl), pp. 69–98. Academic Press, New York.

Hardegree, M.E. (1965). Separation of neurotoxin and hemolysin of *Clostridium tetani*. *Proc. Soc. Exp. Biol. Med.* 119, 405–8.

Harvey, P.C. and Faber, J.E. (1941). Some biochemical reactions of the Listerella group. *J. Bacteriol.* 41, 45–6.

Hase, J., Mitsui, K. and Shonaka, E. (1975). *Clostridium perfringens* exotoxins. III. Binding of θ-toxin to erythrocyte membrane. *Jap. J. Exp. Med.* 45, 433–8.

Hase, J., Mitsui, K. and Shonaka, E. (1976). *Clostridium perfringens* exotoxins. IV. Inhibition of θ-toxin induced hemolysis by steroids and related compounds. *Jap. J. Exp. Med.* 46, 45–50.

Hatheway, C.L. (1990). Toxigenic clostridia. *Clin. Microbiol. Rev.* 3, 66–98.

Hauschild, A.H.W., Lecroisey, A. and Alouf, J.E. (1973). Purification of *Clostridium perfringens* type C theta toxin. *Can. J. Microbiol.* 19, 881–5.

Herbert, D. and Todd, E.W. (1941). Purification and properties of a haemolysin produced by group A haemolytic streptococci (streptolysin O). *Biochem. J.* 35, 1124–39.

Hof, H. (1984). Virulence of different strains of *Listeria monocytogenes* serovar 1/2a. *Med. Microbiol. Immunol.* 173, 207–18.

Houslay, M.D. and Stanley, K.K. (1982). *Dynamics of Biological Membranes*. Wiley, New York.

Hugo, F., Reichweiss, J., Arvand, M., Krämer, S. and Bhakdi, S. (1986). Use of monoclonal antibody to determine the mode of action of transmembrane pore formation by streptolysin O. *Infect. Immun.* 54, 641–5.

Ispolatovskaya, M.V. (1971). Type A *Clostridium perfringens* toxin. In: *Microbial Toxins*, Vol. IIA (eds S. Kadis, T.C. Montie and S.J. Ajl), pp. 109–58. Academic Press, New York.

Iwamoto, M., Ohno-Iwashita, Y. and Ando, S. (1987). Role of the essential thiol group in the thiol-activated cytolysin from *Clostridium perfringens*. *Eur. J. Biochem.* 167, 425–30.

Iwamoto, M., Ohno-Iwashita, Y. and Ando, S. (1990). Effect of isolated C-terminal fragment of θ-toxin (perfringolysin O) on toxin assembly and membrane lysis. *Eur. J. Biochem.* 194, 25–31.

Jenkins, E.M., Njoku-Obi, A.N. and Adams, E.W. (1964). Purification of the soluble hemolysins of *Listeria monocytogenes*. *J. Bacteriol.* 88, 418–24.

Johnson, M.K. (1972). Properties of purified pneumococcal hemolysin. *Infect. Immun.* 6, 755–60.

Johnson, M.K. (1977). Cellular location of pneumolysin. *FEMS Microbiol. Lett.* 2, 243–52.

Johnson, M.K. and Alouf, J.E. (1990). Expression of plasmid-borne pneumolysin gene in wild-type and pneumolysin-negative pneumococci. In: *Bacterial Protein Toxins* (eds R. Rappuoli, J.E. Alouf, P. Falmagne *et al.*), Zbl. Bakt. Suppl. 19, pp. 351–2. Gustav Fischer Verlag, Stuttgart.

Johnson, M.K. and Aultman, K.S. (1977). Studies on the mechanism of action of oxygen-labile haemolysins. *J. Gen. Microbiol.* 101, 237–41.

Johnson, M.K., Geoffroy, C. and Alouf, J.E. (1980). Binding of cholesterol by sulfhydryl-activated cytolysins. *Infect. Immun.* 27, 97–101.

Johnson, M.K., Boeso-Marrazzo, D. and Pierce, W.A. (1981). Effect of pneumolysin on human polymorphonuclear leukocytes and platelets. *Infect. Immun.* 34, 171–6.

Johnson, M.K., Hamon, D. and Drew, G.K. (1982a). Isolation and characterization of pneumolysin-negative mutants of *Streptococcus pneumoniae*. *Infect. Immun.* 37, 837–9.

Johnson, M.K., Knight, R.J. and Drew, G.K. (1982b). The hydrophobic character of thiol-activated cytolysins. *Biochem. J.* 207, 557–60.

Johnson, M.K., Johnson, E.J., Geoffroy, C. and Alouf, J.E. (1984). Physiologic and genetic regulation of the synthesis of sulfhydryl-activated cytolytic toxins. In: *Bacterial Protein Toxins* (eds J.E. Alouf, F.J. Fehrenbach, J.H. Freer and J. Jeljaszewicz), pp. 54–63. Academic Press, London.

Johnson, M.K., Hobden, J.A., Hagenah, M., O'Callaghan, J.O., Hill, J.M. and Chen, S. (1990). The role of pneumolysin in ocular infections with *Streptococcus pneumoniae*. *Current Eye Res.* 9, 1107–14.

Jones, D. (1988). The place of Listeria among gram-positive bacteria. *Infection* 16 (Suppl. 2), S85–8.

Kalin, M., Kanclerski, K., Granström, M. and Möllby, R. (1987). Diagnosis of pneumococcal pneumonia by enzyme-linked immunosorbent assay of antibodies to pneumococcal hemolysin (pneumolysin). *J. Clin. Microbiol.* 25, 226–9.

Kanbayashi, Y., Hotta, M. and Koyama, J. (1972). Kinetic study of streptolysin O. *J. Biochem.* 71, 227–37.

Kanclerski, K. and Möllby, R. (1987). Production and purification of *Streptococcus pneumoniae* hemolysin (pneumolysin). *J. Clin. Microbiol.* 25, 222–5.

Kanclerski, K., Gandström, M. and Möllby, R. (1987). Immunological relation between serum antibodies against pneumolysin and against streptolysin O. *Acta Pathol. Microbiol. Immunol. Scand.*, Sect. B 95, 241–4.

Kathariou, S., Metz, P., Hof, H. and Goebel, W. (1987). Tn916-induced mutations in the hemolysin determinant affecting virulence of *Listeria monocytogenes*. *J. Bacteriol.* 169, 1291–7.

Kathariou, S., Pine, L., George, V., Carlone, G.M. and Holloway, B.P. (1990). Nonhemolytic *Listeria monocytogenes* mutants that are also noninvasive for mammalian cells in culture: evidence for coordinate regulation of virulence. *Infect. Immun.* 58, 3988–95.

Kehoe, M.A. and Timmis, K.N. (1984). Cloning and expression in *Escherichia coli* of the streptolysin O determinant from *Streptococcus pyogenes*: characterization of the cloned streptolysin O determinant and demonstration of the absence of substantial homology with determinants of other thiol-activated toxins. *Infect. Immun.* **43**, 804–10.

Kehoe, M.A., Miller, L., Walker, J.A. and Boulnois, G.J. (1987). Nucleotide sequence of the streptolysin O (SLO) gene: structural homologies between SLO and other membrane, thiol-activated toxins. *Infect. Immun.* **55**, 3228–32.

Kreft, J., Berger, H., Härtlein, M., Müller, B., Weidinger, G. and Goebel, W. (1983). Cloning and expression in *Escherichia coli* and *Bacillus subtilis* of the hemolysin (cereolysin) determinant from *Bacillus cereus. J. Bacteriol.* **155**, 681–9.

Kreft, J.D., Funke, A., Haas, F., Lottspeich, F. and Goebel, W. (1989). Production, purification and characterization of hemolysins from *Listeria ivanovii* and *Listeria monocytogenes* sv4b. *FEMS Microbiol. Lett.* **57**, 197–202.

Launay, J.M. and Alouf, J.E. (1979). Biochemical and ultrastructural study of the disruption of blood platelets by streptolysin O. *Biochim. Biophys. Acta* **556**, 278–91.

Launay, J.M., Geoffroy, C., Costa, J.L. and Alouf, J.E. (1984). Purified SH-activated toxins (streptolysin O, alveolysin): new tools for determination of platelet enzyme activities. *Thromb. Res.* **33**, 189–96.

Leimeister-Wächter, M. and Chakraborty, T. (1989). Detection of listeriolysin O, the thiol-dependent hemolysin in *Listeria monocytogenes, Listeria ivanovii* and *Listeria seeligeri. Infect. Immun.* **57**, 2350–7.

Leimeister-Wächter, M., Haffner, C., Domann, E., Goebel, W. and Chakraborty, T. (1990). Identification of a gene that positively regulates expression of listeriolysin, the major virulence factor of *Listeria monocytogenes. Proc. Natl. Acad. Sci. USA* **87**, 8336–40.

Lemeland, J.-F., Allaire, R. and Boiron, H. (1974). Contribution à l'étude de l'hémolysine de "*Listeria monocytogenes*" (listériolysine). *Pathol. Biol. (Paris)* **22**, 764–70.

Linder, R. (1979). Heterologous immunoaffinity chromatography in the purification of streptolysin O. *FEMS Microbiol. Lett.* **5**, 339–42.

Linder, R. and Bernheimer, A.W. (1984). Action of bacterial cytotoxins on normal mammalian cells and cells with altered membrane lipid composition. *Toxicon* **22**, 641–51.

Lock, R.A., Paton, J.C. and Hansman, D. (1988). Comparative efficacy of pneumococcal neuraminidase and pneumolysin as immunogens protective against *Streptococcus pneumoniae. Microb. Pathogen.* **5**, 461–7.

Lucain, C. and Piffaretti, J.-C. (1977). Characterization of the haemolysins of different serotypes of *Clostridium tetani. FEMS Microbiol. Lett.* **1**, 231–4.

Mackaness, G.B. (1962). Cellular resistance to infection. *J. Exp. Med.* **116**, 382–406.

McDonel, J.L. (1980). *Clostridium perfringens* toxins (types A, B, C, D, E). *Pharmacol. Ther.* **10**, 617–55.

Menestrina, G., Bashford, C.L. and Pasternak, C.A. (1990). Pore-forming toxins: experiments with *S. aureus* α-toxin, *C. perfringens*, θ-toxin and *E. coli*

haemolysin in lipid bilayers, liposomes and intact cells. *Toxicon* **28**, 477–91.

Mengaud, J., Chenevert J., Geoffroy, C., Gaillard J.L. and Cossart, P. (1987). Identification of the structural gene encoding the SH-activated hemolysin of *Listeria monocytogenes*: listeriolysin O is homologous to streptolysin O and pneumolysin. *Infect. Immun.* **55**, 3225–7.

Mengaud, J., Vicente, M.F., Chenevert, J., Moniz-Pereira, J., Geoffroy, C., Gicquel-Sanzey, B., Baquero, F., Perez-Diaz, J.C. and Cossart, P. (1988). Expression in *Escherichia coli* and sequence analysis of the listeriolysin O determinant of *Listeria monocytogenes. Infect. Immun.* **56**, 766–72.

Mengaud, J., Vicente, M.F. and Cossart, P. (1989). Transcriptional mapping and nucleotide sequence of the *Listeria monocytogenes hlyA* region reveal structural features that may be involved in regulation. *Infect. Immun.* **57**, 3695–701.

Michel, E., Reich, K.A., Favier, R., Berche, P. and Cossart, P. (1990). Attenuated mutants of the intracellular bacterium *Listeria monocytogenes* obtained by single amino-acid substitutions in listeriolysin O. *Molec. Microb.* **4**, 2167–78.

Mitchell, T.J., Walker, J.A., Saunders, F.K., Andrew, P.W. and Boulnois, G.J. (1989). Expression of the pneumolysin gene in *Escherichia coli*: rapid purification and biological properties. *Biochim. Biophys. Acta* **1007**, 67–72.

Mitsui, K. and Hase, J. (1979). *Clostridium perfringens* exotoxins. VI. Reactivity of perfringolysin O with thiol and disulfide compounds. *Jap. J. Exp. Med.* **49**, 13–18.

Mitsui, K., Mitsui, N. and Hase, J. (1973). *Clostridium perfringens* exotoxins. II. Purification and some properties of θ-toxin. *Jap. J. Exp. Med.* **43**, 377–91.

Mitsui, K., Sekiya, T., Nozawa, Y. and Hase, J. (1979a). Alteration of human erythrocyte plasma membranes by perfringolysin O as revealed by freeze-fracture electron microscopy: studies on *Clostridium perfringens* exotoxins. V. *Biochim. Biophys. Acta* **554**, 68–75.

Mitsui, K., Sekiya, T., Okamura, S., Nozawa, Y. and Hase, J. (1979b). Ring formation of perfringolysin O as revealed by negative stain electron microscopy. *Biochim. Biophys. Acta* **558**, 307–13.

Mitsui, K., Saeki, Y. and Hase, J. (1982a). Effects of cholesterol evulsion on susceptibility of perfringolysin O of human erythrocytes. *Biochim. Biophys. Acta* **686**, 177–81.

Mitsui, K., Mitsui, N., Kobashi, K. and Hase, J. (1982b). High-molecular-weight hemolysin of *Clostridium tetani. Infect. Immun.* **35**, 1086–90.

Mitsui, N., Mitsui, K. and Hase, J. (1980). Purification and some properties of tetanolysin. *Microbiol. Immunol.* **24**, 575–84.

Mounier, J., Ryter, A., Coquis-Rondon, M. and Sansonetti, P.J. (1990). Intracellular and cell-to-cell spread of *Listeria monocytogenes* involves interaction with F-actin in the enterocytelike cell line Caco-2. *Infect. Immun.* **58**, 1048–58.

Moussa, R.S. (1958). Complexity of toxins from *Clostridium septicum* and *Clostridium chauvoei*. *J. Bacteriol.* **76**, 538–45.

Murray, E.G.D., Webb, R.E. and Swann, M.B.R. (1926). A disease of rabbits characterized by a large mononuclear leucocytosis caused by a hitherto undescribed bacillus *Bacterium monocytogenes* (n. sp.). *J. Pathol. Bacteriol.* **29**, 407–39.

Nandoskar, M., Ferrante, A., Bates, E.J., Hurst, N. and Paton, J.C. (1986). Inhibition of human monocyte respiratory burst, degranulation phospholipid methylation and bactericidal activity by pneumolysin. *Immunology* **59**, 515–20.

Neill, J.M. (1926a). Studies on the oxidation and reduction of immunological substances. I. Pneumococcus hemotoxin. *J. Exp. Med.* **44**, 199–213.

Neill, J.M. (1926b). Studies on the oxidation and reduction of immunological substances. II. The hemotoxin of *B. welchii*. *J. Exp. Med.* **44**, 215–26.

Neill, J.M. (1926c). Studies on the oxidation and reduction of immunological substances. III. Tetanolysin. *J. Exp. Med.* **44**, 227–40.

Neill, J.M. and Fleming, W.L. (1927). Studies on the oxidation of immunological substances, VI. The 'reactivation' of the bacteriolytic activity of oxidized pneumococcus extracts. *J. Exp. Med.* **46**, 263–77.

Neill, J.M. and Mallory, J.B. (1926). Studies on the oxidation and reduction of immunological substances. IV. Streptolysin. *J. Exp. Med.* **44**, 241–60.

Niedermeyer, W. (1985). Interaction of streptolysin-O with biomembranes: kinetic and morphological studies on erythrocytes membranes. *Toxicon* **23**, 425–39.

Njoku-Obi, A.E., Jenkins, E.M., Njoku-Obi, J.C., Adams, J. and Covington, V. (1963). Production and nature of *Listeria monocytogenes* hemolysins. *J. Bacteriol.* **86**, 1–8.

Oakley, C.L., Warrack, G.H. and Clarke, P.H. (1947). The toxins of *Clostridium oedematiens (Cl. novyi)*. *J. Gen. Microbiol.* **1**, 91–107.

Ohno-Iwashita, Y., Iwamoto, M., Mitsui, K., Kawasaki, H. and Ando, S. (1986). Cold-labile hemolysin produced by limited proteolysis of θ-toxin from *Clostridium perfringens*. *Biochemistry* **25**, 6048–53.

Ohno-Iwashita, Y., Iwamoto, M., Mitsui, K., Ando, S. and Nagai, K. (1988). Protease-nicked θ-toxin of *Clostridium perfringens* a new membrane probe with no catalytic effect reveals two classes of cholesterol as toxin-binding sites on sheep erythrocytes. *Eur. J. Biochem.* **176**, 95–101.

Parrisius, J., Bhakdi, S., Roth, M., Tranum-Jensen, J., Goebel, W. and Seeliger, H.P.R. (1986). Production of listeriolysin by beta-hemolytic strains of *Listeria monocytogenes*. *Infect. Immun.* **51**, 314–19.

Paton, J.C. and Ferrante, A. (1983). Inhibition of human polymorphonuclear leukocyte respiratory burst, bactericidal activity and migration by pneumolysin. *Infect. Immun.* **41**, 1212–16.

Paton, J.C., Rowan-Kelly, B. and Ferrante, A. (1984). Activation of human complement by the pneumococcal toxin pneumolysin. *Infect. Immun.* **43**, 1085–7.

Paton, J.C., Berry, A.M., Locke, R.A., Hansmann, D. and Manning, P.A. (1986). Cloning and expression in *Escherichia coli* of the *Streptococcus pneumoniae* gene encoding pneumolysin. *Infect. Immun.* **54**, 50–5.

Pendleton, I.R., Bernheimer, A.W. and Grushoff, P. (1973). Purification and characterization of hemolysin from *Bacillus thuringiensis*. *J. Inverteb. Pathol.* **21**, 131–5.

Pinkney, M. and Kehoe, M. (1990). A thiol group is not required for the *in vitro* activity of the thiol-'activated' toxin streptolysin O. In: *Bacterial Protein Toxins* (eds R. Rappuoli, J.E. Alouf, P. Falmagne *et al.*), Zbl. Bakt. Suppl. 19, pp. 103–104. Gustav Fischer, Stuttgart.

Pinkney, M., Beachey, E. and Kehoe, M. (1989). The thiol-activated toxin streptolysin O does not require a thiol group for cytolytic activity. *Infect. Immun.* **57**, 2553–8.

Portnoy, D., Jacks, P.S. and Hinrichs, D. (1988). Role of hemolysin for the intracellular growth of *Listeria monocytogenes*. *J. Exp. Med.* **167**, 1459–71.

Prigent, D. and Alouf, J.E. (1976). Interaction of streptolysin O with sterols. *Biochim. Biophys. Acta* **443**, 288–300.

Reboli, A.C., Bryan, S.C. and Farrar, W.E. (1989). Bacteremia and hip prosthesis caused by *Bacillus alvei*. *J. Clin. Microbiol.* **27**, 1395–6.

Rocourt, J. (1988). Taxonomy of the genus Listeria. *Infection* **16** (Suppl. 2), S85–S88.

Rocourt, J., Alonso, J.M. and Seeliger, H.P.R. (1983). Virulence comparée des cinq groupes génomiques de *Listeria monocytogenes* (sensu lato). *Ann. Inst. Pasteur (Microbiol.)* **134A**, 354–64.

Rosenqvist, E., Michaelsen, T.E. and Vistnes, A.I. (1980). Effect of streptolysin O and digitonin on egg lecithin/cholesterol vesicles. *Biochim. Biophys. Acta* **600**, 91–102.

Rottem, S., Hardegree, M.C., Grabowski, M.R., Fornwald, R. and Barile, M.F. (1976). Interaction between tetanolysin and mycoplasma cell membrane. *Biochim. Biophys. Acta* **455**, 879–88.

Rottem, G., Cole, R.M., Habig, W.H., Barile, M.F. and Hardegree, M.C. (1982). Structural characteristics of tetanolysin and its binding to lipid vesicles. *J. Bacteriol.* **152**, 2594–9.

Rottem, S., Groover, K., Habig, W.H., Barile, M.F. and Hardegree, M.C. (1990). Transmembrane diffusion channels in *Mycoplasma gallisepticum* induced by tetanolysin. *Infect. Immun.* **58**, 598–602.

Rutter, J.M. and Collee, J.F. (1969). Studies on the soluble antigens of *Clostridium oedematiens (Cl. novyi)*. *J. Med. Microbiol.* **2**, 395–421.

Saito, M. (1983). Activation of the calcium permabilitty of erythrocyte membrane by perfringolysin O. *J. Biochem. (Tokyo)* **94**, 323–6.

Sato, H. (1986). Monoclonal antibodies against *Clostridium perfringens* θ toxin (perfringolysin O). In: *Mono-*

clonal Antibodies against Bacteria, Vol. III (eds A.J.L. Macario and E. Conway de Macario), pp. 203–28. Academic Press, New York.

Saunders, F.K., Mitchell, T.J., Walker, J.A., Andrew, P.W. and Boulnois, G.J. (1989). Pneumolysin, the thiol-activated toxin of *Streptococcus pneumoniae*, does not require a thiol group for *in vitro* activity. *Infect. Immun.* **57**, 2547–52.

Seeliger, H.P.R. (1988). Listeriosis – History and actual development. *Infection* **16** (Suppl. 2), S80–S84.

Seeliger, H.P.R., Schrettenbrunner, A., Pongratz, G. and Hof, H. (1982). Zur Sonderstellung stark hämolysierender Stämme der Gattung Listeria (Special position of strongly haemolytic strains of the genus Listeria). *Zbl. Bakt. Hyg. I. Abt. Orig.* A **252**, 176–90.

Shany, S., Bernheimer, A.W., Grushoff, P.S. and Kim, K.-S. (1974). Evidence for membrane cholesterol as the common binding site for cereolysin, streptolysin O and saponin. *Cell. Molec. Biochem.* **3**, 179–86.

Shimizu, T., Okabe, A., Minami, J. and Hayashi, H. (1991). An upstream regulatory sequence stimulates expression of the perfringolysin O gene of *Clostridium perfringens*. *Infect. Immun.* **59**, 137–42.

Shumway, C.N. and Klebanoff, S.J. (1971). Purification of pneumolysin. *Infect. Immun.* **4**, 388–92.

Smyth, C.J. (1975). The identification and purification of multiple forms of θ-haemolysin (θ-toxin) of *Clostridium perfringens* type A. *J. Gen. Microbiol.* **87**, 219–38.

Smyth, C.J. and Duncan, J.L. (1978). Thiol-activated (oxygen-labile) cytolysins. In: *Bacterial Toxins and Cell Membranes* (eds J. Jeljaszewicz and T. Wadström), pp. 129–83. Academic Press, London.

Smyth, C.J., Freer, J.H. and Arbuthnott, J.P. (1975). Interaction of *Clostridium perfringens* θ-haemolysin a contaminant of commercial phospholipase C with erythrocyte ghost membranes and lipid dispersions. *Biochim. Biophys. Acta* **382**, 479–93.

Smythe, C.V. and Harris, T.N. (1940). Some properties of a hemolysin produced by group A hemolytic streptococci. *J. Immunol.* **38**, 283–300.

Sokolovic, Z. and Goebel, W. (1989). Synthesis of listeriolysin in *Listeria monocytogenes* under heat shock conditions. *Infect. Immun.* **57**, 295–8.

Sokolovic, Z., Fuchs, A. and Goebel, W. (1990). Synthesis of species-specific stress proteins by virulent strains of *Listeria monocytogenes*. *Infect. Immun.* **58**, 3582–7.

Steinfort, C., Wilson, R., Mitchell, T., Feldman, C., Rutman, A., Todd, H., Sykes, D., Walker, J., Saunders, K., Andrew, P.W., Boulnois, G.J. and Cole, P.J. (1989). Effect of *Streptococcus pneumoniae* on human respiratory epithelium *in vitro*. *Infect. Immun.* **57**, 2006–13.

Stephen, J. and Pietrowski, R.A. (1986). *Bacterial Toxins*. Van Nostrand Reinhold, Wokingham, UK.

Stevens, D.L., Troyer, B.E., Merrick, D.T., Mitten, J.E. and Olson, R.D. (1988). Lethal effects and cardiovascular effects of purified α- and θ-toxins from

Clostridium perfringens. *J. Infect. Dis.* **157**, 272–9.

Suzuki, J., Kobayashi, S., Kagaya, K. and Fukazawa, Y. (1988). Heterogeneity of hemolytic efficiency and isoelectric point of streptolysin O. *Infect. Immun.* **56**, 2474–8.

Sword, C.P. and Kingdon, G.C. (1971). *Listeria monocytogenes* toxin. In: *Bacterial Protein Toxins*, Vol. IIA (eds S. Kadis, T. Montie and S.J. Ajl), pp. 357–77. Academic Press, New York.

Taira, S., Jalonen, E., Paton, J.C., Saravas, M. and Runeberg-Nyman, K. (1989). Production of pneumolysin, a pneumococcal toxin, in *Bacillus subtilis*. *Gene* **77**, 211–18.

Thelestam, M. and Möllby, R. (1980). Interaction of streptolysin O from *Streptococcus pyogenes* and thetatoxin from *Clostridium perfringens* with human fibroblasts. *Infect. Immun.* **29**, 863–77.

Thelestam, M. and Möllby, R. (1983). Survival of cultured cells after functional and structural disorganization of plasma membrane by bacterial haemolysins and phospholipases. *Toxicon* **21**, 805–15.

Thelestam, M., Alouf, J.E., Geoffroy, C. and Möllby, R. (1981). Membrane-damaging action of alveolysin from *Bacillus alvei*. *Infect. Immun.* **32**, 1187–92.

Thelestam, M., Jolivet-Reynaud, C. and Alouf, J.E. (1983). Photolabeling of staphylococcal α-toxin from within rabbit erythrocyte membranes. *Biochem. Biophys. Res. Commun.* **111**, 444–9.

Tiesler, E. and Trinks, U. (1979). Die Streptolysin O – Bildung beta-hämolysierender Streptokokken der Gruppe A (The Production of streptolysin O by betahemolytic streptococci of group A). *Zbl. Bakt. Hyg. I Abt. Orig.* A **245**, 17–24.

Tiesler, E. and Trinks, C. (1982). Das Vorkommen extrazellulärer Stoffwechselprodukte bei Streptokokken der Gruppen C und G. (Release of extracellular products by groups C and G streptococci). *Zbl. Bakt. Hyg. I. Abt. Orig.* A **253**, 81–7.

Todd, E.W. (1934). A comparative serological study of streptolysins derived from human and from animal infections, with notes on pneumococcal haemolysin, tetanolysin and staphylococcus toxin. *J. Pathol. Bacteriol.* **39**, 299–321.

Turnbull, P.C.B. (1981). *Bacillus cereus* toxins. *Pharmacol. Ther.* **13**, 453–505.

Tweten, R.K. (1988a). Cloning and expression in *Escherichia coli* of the perfringolysin O (theta toxin) gene from *Clostridium perfringens* and characterization of the gene product. *Infect. Immun.* **56**, 3228–34.

Tweten, R.K. (1988b). Nucleotide sequence of the gene for perfringolysin O (theta toxin) from *Clostridium perfringens*: significant homology with the genes for streptolysin and pneumolysin. *Infect. Immun.* **56**, 3235–40.

Vazquez-Boland, J.A., Dominguez, L., Rodriguez-Ferri, E.F., Fernandez-Garayzabal, J.F. and Suarez, G. (1989a). Preliminary evidence that different domains are involved in cytolytic activity and receptor (cholesterol) binding in listeriolysin O, the *Listeria monocytogenes* thiol-activated toxin. *FEMS Microbiol. Lett.* **65**, 95–100.

Vazquez-Boland, J.A., Dominguez, L., Rodriguez-Ferri, E.F. and Suarez, G. (1989b). Purification and characterization of two *Listeria ivanovii* cytolysins, a sphingomyelinase C and thiol-activated toxin (ivanolysin O). *Infect. Immun.* **57**, 3928–35.

Vicente, M.F., Baquero, F. and Perez-Diaz, J.C. (1985). Cloning and expression of the *Listeria monocytogenes* haemolysin in *Escherichia coli. FEMS Microbiol. Lett.* **30**, 77–9.

Walker, J.A., Allen, R.L., Falmagne, P., Johnson, M.K. and Boulnois, G. (1987). Molecular cloning, characterization and complete nucleotide sequence of the gene for pneumolysin, the sulfhydryl-activated toxin of *Streptococcus pneumoniae. Infect. Immun.* **55**, 1184–9.

Wannamaker, L.W. (1983). Streptococcal toxins. *Rev. Infect. Dis.* **5** (Suppl. 4), S723–32.

Wannamaker, L.W. and Schlievert, P.M. (1988). Exotoxins of group A streptococci. In: *Bacterial Toxins, Handbook of Natural Toxins* (eds M.C. Hardegree and A.T. Tu), pp. 267–95. Marcel Dekker, New York.

Watson, K.C. and Kerr, E.J.C. (1974). Sterol requirements for inhibition of streptolysin O activity. *Biochem. J.* **140**, 95–8.

Watson, K.C. and Kerr, E.J.C. (1985). Specificity of antibodies for T sites and F sites of streptolysin O. *Med. Microbiol.* **19**, 1–7.

Willis, A.T. (1969). *Clostridia of Wound Infections.* Butterworths, London.

Wuth, O. (1923). Serologische und biochemische Studien über das Hämolysin des Fränkelschen Gasbrandbazillus. *Biochem. Zeit.* **142**, 19–28.

Yamakawa, Y. and Ohsaka, A. (1986). Hydrophobic interaction between θ-toxin of *Clostridium perfringens* and erythrocytes membrane. *Jap. J. Med. Sci. Biol.* **39**, 254–5.

Yamakawa, Y., Ito, A. and Sato, H. (1977). Theta-toxin of *Clostridium perfringens*. I. Purification and some properties. *Biochim. Biophys. Acta* **494**, 301–13.

9

Cooperative Membrane-Active (Lytic) Processes

Franz J. Fehrenbach and Dagmar Jürgens

Department of Microbiology, Robert Koch-Institut, Nordufer 20, D-1000 Berlin 65, Germany

Introduction

While most of the contributions of this book deal with the structure, function or genetics of a single or closely related group of bacterial proteins or toxins, this chapter aims to describe a principle by which two or more bacterial products with biological, enzymic or toxic activity, act in a cooperative way on a common target structure *in vivo* or *in vitro*. It is evident that cooperative processes of this kind are difficult to analyse due to the complexity of the physico-chemical reactions that determine the kinetics of these multicomponent systems. In such cases, the course of the reaction is not only dependent on the number of individual reaction steps but also on the time sequence in which these reactions occur.

The term 'membrane-active cooperative processes' will be used in this review to describe all events of a cooperative process, which involve two or more components of bacterial origin acting on a target membrane either at the same time or consecutively. However, it should be stated that the term 'cooperative processes' is not used in the sense of the physico-chemical understanding of cooperativity between lipids.

Bacterial exoenzymes or other metabolites active in cooperative processes interact predominantly with membrane lipid constituents. Natural cells (mainly ruminant red blood cells) (RBC), membrane-bounded subcellular entities or artificial membranes may function as susceptible targets in cooperative processes which result either in lytic disruption of the membrane or in the non-lytic degradation of membrane lipids leading to impaired membrane function.

In the past, cooperative lytic processes leading to RBC lysis have been synonymously designated 'synergistic' lytic processes (Marks and Vaughan, 1950; Marks, 1952; Smith et al., 1964), whereas those capable of blocking the reaction were called 'antagonistic' (Fraser, 1964). Since these designations contribute little to the understanding of the mechanism of cooperativity their use will be avoided.

It was rather by chance that Christie et al. (1944) and Munch-Petersen and Christie (1947) discovered a phenomenon of haemolysis in blood agar plates of sheep red blood cells (SRBC) which obviously was due to the production of two diffusible substances produced by a mixed culture of Staphylococcus aureus and group B streptococci. The phenomenon was explained by the sequential interaction of S. aureus sphingomyelinase (β-toxin) (Doery et al., 1963, 1965; Wiseman, 1965; Wiseman and Caird, 1967) and 'CAMP-factor' of group B streptococci, a polypeptide of non-enzymic activity (Christie et al., 1944; Esseveld et al., 1958, 1975) on sheep erythrocytes (Doery et al., 1963; Colley et al., 1973; Brown et al., 1974; Bernheimer et al., 1979; Sterzik and Fehrenbach, 1985). Further work revealed that an enzymic modification of the SRBC membrane was the activating step of this lytic cooperative process (Wiseman and Caird, 1967). Therefore, 'synergistic lysis' (Marks and Vaughan, 1950; Marks, 1952; Smith et al., 1964; Fraser, 1964) could clearly be distinguished from the action of a single (true) bacterial haemolysin.

The blood agar technique applied by Christie *et al.* (1944) has been used widely thereafter to detect cooperative lytic processes in mixed bacterial cultures using a variety of species and strains (Munch-Petersen and Christie, 1947; Fraser, 1961, 1962, 1964; Gubash, 1978; Bae and Bottone, 1980; Smith and Ngui-Yen, 1980; Barksdale *et al.*, 1981; Hébert and Hancock, 1985; Figura and Guglielmetti, 1987). A notable observation by Christie *et al.* (1944) and Fraser (1961, 1962, 1964) was that the cooperative principle was active in the bacterial-free filtrate, indicating the presence of soluble substances.

The work of Fraser (1961, 1962, 1964) in particular demonstrated that numerous pathogenic microorganisms such as *S. aureus*, streptococci, members of the Enterobacteriaceae, Corynebacteriae and *Bacillus* spp. were capable of producing extracellular substances which could interact cooperatively with the lipid membrane when combined in an appropriate way.

From the study of the reaction mechanism of some of the well-defined cooperative (lytic) systems, it has become evident that phospholipids present in the outer leaflet of the membrane bilayer of the target cell are initially attacked and hydrolysed by bacterial phospholipases or sphingomyelinases in a 'first step' (Table 3) without causing cell lysis. Hydrolysis of these lipid substrates results in the release of the polar head groups, mainly of phosphatidylcholine (PC), sphingomyelin (SPM), and to a lesser extent of phosphatidylethanolamine (PE). PC and SPM are preferentially located in the outer leaflet of the membrane. Non-lytic degradation of membrane phospholipids (PL) leads to impaired membrane function characterized by disturbance of lipid/lipid or protein/lipid interaction, partial phase separation, altered membrane surface pressure, enhanced osmotic lability, and partial loss of ion selectivity. As a consequence of membrane phospholipid hydrolysis in target cells, the 'second step agents' gain access to the membrane and may induce lysis of the cell or further membrane destruction. A diagrammatic representation of the 'first and second step' of the cooperative principle is shown in Fig. 1.

Following these principles one may theoretically combine numerous 'first- and second-step agents' to create a cooperative (lytic) principle active in membrane alteration or disruption.

Studies on the asymmetric distribution of PL in biomembranes (Bretscher, 1972; Verkleij *et al.*, 1973) with phospholipases have contributed enormously to our understanding of membrane structure and function. With a better knowledge of the mode of action of individual phospholipases and their substrate specificities, the 'first- and second-step events' of cooperative processes have become more thoroughly understood.

Since few medically important bacteria possess a pathogenic principle based on the action of a single lethal toxin, interest in understanding multifactorial mechanisms of pathogenicity has steadily increased and has focused on cooperative systems. However, the type of cooperative lysis of the classical CAMP reaction (Christie *et al.*, 1944; Fraser, 1964) observed *in vitro* has tended to be regarded as an epiphenomenon, being of no relevance for *in vivo* pathogenicity. From theoretical grounds it is, nevertheless, conceivable that cooperative lytic or toxic principles might be of relevance *in vivo* and emerge from aerobic or anaerobic mixed infections or simply from the coexistence of different bacteria or microbial species in an appropriate localization such as the oropharynx, the upper respiratory tract, the urogenital tract, or the gut.

Hence, it will be important for future research to develop models that allow evaluation of the possible significance of cooperative processes *in vivo*, especially when the cooperative factors are produced by a single strain. This is necessary because recent work on the regulation of genetic processes shows clearly that phenotypic expression of extracellular proteins can be influenced by environmental factors.

Natural and artificial targets

While the mechanisms of bacterial toxins have been studied in the past with a variety of different target cells (for reviews see Bernheimer, 1976; Alouf, 1977; Jeljaszewicz and Wadström, 1978; Alouf *et al.*, 1984; Falmagne *et al.*, 1986; Fehrenbach *et al.*, 1988a), erythrocytes have been used predominantly to investigate cooperative (lytic) processes. RBC from ox, sheep, horse, guinea-pig, rabbit and fowl were incorporated in agar plates or used in suspension by Christie *et al.* (1944) and Fraser (1964) to study the lytic effect of bacterial

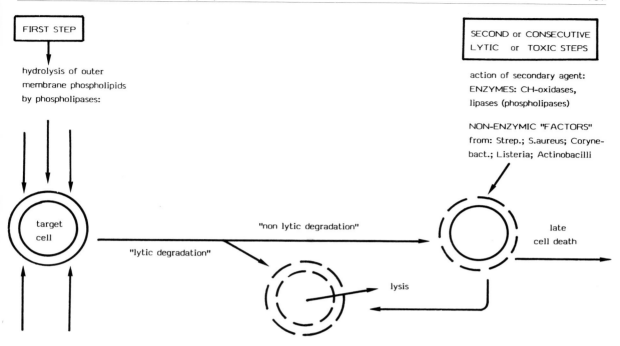

Figure 1. Reaction scheme of cooperative lytic or toxic processes. The pathway of lytic and non-lytic degradation of the target cell membrane is shown. Among the 'second-step agents' cholesterol oxidases, lipases and phospholipases are listed together with the non-enzymic factors of some bacterial species.

products acting cooperatively on target cells (Souček *et al.*, 1967a,b; Součková and Souček, 1972; Gubash, 1978; Skalka *et al.*, 1979; Bae and Bottone, 1980; Smith and Ngui-Yen, 1980; Barksdale *et al.*, 1981; Skalka and Smola, 1982; Figura and Guglielmetti, 1987). Although the authors noticed that RBC from different sources were lysed to different extents, they were unable to explain their observations. Other groups had also recognized differences in the susceptibility of RBC from different species to *S. aureus* α- and β-toxin (Smith and Price, 1938; Bernheimer and Schwartz, 1963; Freer *et al.*, 1968; Jeljaszewicz, 1972; Bernheimer, 1974; Low *et al.*, 1974). Later it was found that *S. aureus* β-toxin was a sphingo-myelinase (Doery *et al.*, 1963, 1965; Wiseman, 1965; Wiseman and Caird, 1967) and that the differences in sensitivity of target cells were due to differences in lipid composition of the membrane and the property of individual enzymes to gain access to their substrates (Coleman *et al.*, 1970; Roelofsen *et al.*, 1971; Woodward and Zwaal, 1972; Laster *et al.*, 1972; Colley *et al.*, 1973; Zwaal *et al.*, 1973, 1975; Gul and Smith, 1974;

Bernheimer *et al.*, 1974; Low *et al.*, 1974; Gazitt *et al.*, 1975; van Deenen, 1981).

The distribution of phospholipids in natural membranes has been studied mainly by the interaction of individual phospholipases with erythrocytes. The lipid composition of different RBC membranes and the asymmetric distribution of PL between the two leaflets of the membrane bilayer is shown in Table 1.

Hence, our present view of membrane structure, as well as the concept of asymmetric distribution of PL in the bilayer is derived mainly from studies on a unique, highly specialized cell, namely the erythrocyte. With few exceptions (Möllby *et al.*, 1974; Thelestam and Möllby, 1975a,b, 1979; Linder and Bernheimer, 1984) all the studies on cooperative lytic or toxic processes have also been performed with erythrocytes as target cells.

Following early studies with intact erythrocytes, procedures were developed for the preparation of erythrocyte stromata and ghosts (Dodge *et al.*, 1963; Bodemann and Passow, 1972; Schwoch and Passow, 1973; Steck and Kant, 1974; Johnson, 1975). These provided further flexibility in mem-

Table I. Lipid composition of red cell membranes from different mammalian species

Species	% of total lipid				Individual phospholipids (% of total)								References
	Neutral lipid	Glycolipid	Ganglio-side	Phospholipid	PA	PE Δ	PS Δ	PI	PC o	SPM o	LPC o	Others	
Rat	24.7	2.0	6.3	67.0	<0.3	21.5	10.8	3.5	47.5	12.8	3.8	—	1
Rabbit	28.9	0.8	4.5	65.8	1.6	31.9	12.2	1.6	33.9	19.0	<0.3	—	1
Pig	26.8	10.1	3.3	59.8	<0.3	29.7	17.8	1.8	23.2	26.5	0.9	—	1
Dog	24.7	10.9	11.8	52.6	0.5	22.4	15.4	2.2	46.9	10.8	1.8	—	1
Horse	24.5	8.0	15.5	52.0	<0.3	24.3	18.0	<0.3	42.4	13.5	1.7	—	1
Sheep	26.5	2.5	7.8	63.2	<0.3	26.2	14.1	2.9	—	51.0	—	4.8	1
Cow	27.5	2.2	5.5	64.8	<0.3	29.1	19.3	3.7	—	46.2	—	1.7	1
Goat	26.2	17.9	5.7	50.2	<0.3	27.9	20.8	4.6	—	45.9	—	0.8	1
Cat	26.8	3.1	8.8	61.3	0.8	22.2	13.2	7.4	30.5	26.1	<0.3	—	1
Guinea-pig	27.0	15.2	2.2	55.6	4.2	24.6	16.8	2.4	41.1	11.1	<0.3	—	1
Mouse					—	15.2	11.2	3.1	46.2	12.1	3.4	—	2
Human					2.2	27.2	13.0	1.3	28.9	26.9	1.1	—	3

Abbreviations: PA, phosphatidic acid; PE, phosphatidylethanolamine; PS, phosphatidylserine; PI, phosphatidylinositol; PC, phosphatidylcholine; SPM, sphingomyelin; LPC, lysophosphatidylcholine.
The asymmetric distribution of phospholipids in erythrocytes is marked with o (outer layer) and Δ (inner layer); Verkleij et al. (1973).
(1) Nelson (1967); (2) Montfoort and Boere (1978); (3) Rouser et al. (1968).

brane studies and were particularly useful techniques in studies of membrane asymmetry. From 1956, artificial membrane models such as liposomes, and bi- or mono-molecular films were developed for investigations of membrane structure and function (Bangham and Dawson, 1962; Miller and Ruysschaert, 1971; Verger and de Haas, 1973; Hendrickson et al., 1974; Demel et al., 1975; Linder and Bernheimer, 1978; Bangham, 1980; Tomita et al., 1983a,b; Thurnhofer et al., 1986).

Liposomes of different lipid composition were used by Linder and Bernheimer (1978), Bernheimer et al. (1979), Fehrenbach et al. (1984) and Sterzik et al. (1986) to investigate cooperative processes. Liposomes are attractive in that lipid components of the bilayer can be chemically defined. The total absence of protein constituents representative of natural membranes provide a simple model system, but this may be unrepresentative in terms of lipid characteristics in natural membranes, because of this apparent advantage.

Modification of target cell membranes: The 'first step'

Biochemical modification

In agreement with the definition given above, cooperative lytic or toxic processes should be considered as reactions in which at least two individual components interact sequentially with a target structure. In this process the initial event, the 'first step', is characterized by an enzymic degradation (non-lytic) of individual PL in either natural or artificial membranes by appropriate phospholipases.

The 'first step' involves binding of an individual phospholipase to the membrane followed by hydrolysis of the respective substrate(s). It should be noticed that only a few bacterial phospholipases are able to hydrolyse their substrates in intact cell membranes.

The site of attack and the potential substrates for phospholipases are shown in Fig. 2.

Not contained in the scheme (Fig. 2) are the sphingomyelinases (EC 3.1.4.3. and EC 3.1.4.12.) which hydrolyse sphingomyelin and play an important role as 'first-step agents' in cooperative processes (Fig. 3).

Due to the asymmetric distribution of phospholipids in natural membranes (Bretscher, 1972; Gordesky and Marinetti, 1973; Zwaal et al., 1973, 1975; Verkleij et al., 1973; van Deenen, 1981) the phospholipids present in the outer leaflet of the bilayer are primarily degraded by phospholipases C and sphingomyelinases (Roelofsen et al., 1971; Colley et al., 1973; Zwaal et al., 1975; van Deenen et al., 1976). Though quantitative differences exist in the composition of membrane lipids for RBC of different species, choline-containing PL are preferentially found in the outer leaflet – whereas amino-phospholipids are mainly present in the inner leaflet of RBC membranes investigated so far (Table 1). Bacterial phospholipases, including sphingomyelinases and phospholipases A, C and D, degrade membrane PL either separately or in combination (Colley et al., 1973). Phospholipases of wider substrate specificity (i.e. phospholipases of C. perfringens or P. aeruginosa) are usually active against a broad variety of RBC. Depending on the enzyme concentration they are either directly lytic or participate, at lower concentrations, in cooperative lytic processes as 'first-step agents'.

Table 2 lists bacterial phospholipases and their substrate specificities. As seen from Table 2 enzymes of absolute specificity, i.e. for PE, PI, SPM or CLP, were found in contrast to those of broad specificity which lyse several substrates such as PC, SPM, PE and PS (e.g. phospholipase C of C. perfringens). Although, enzyme purification methods have vastly improved over the last two decades, it should be recalled that several authors (Avigad, 1976; Freer and Arbuthnott, 1976; Möllby, 1978; Linder, 1984) had argued earlier that stringent criteria of homogeneity for the characterization of phospholipases were often neglected. Hence, some of the data in Table 2 may require revision in the future.

An important prerequisite for the hydrolysis of individual lipid constituents is the accessibility of the substrate for the enzyme in the intact membrane (Verkleij et al., 1973; Zwaal et al., 1973, 1975; Colley et al., 1973). Besides the accessibility, the molar ratio of phospholipids present in the outer leaflet is crucial. Furthermore, the extent of degradation of one or more phospholipid(s) by an enzyme (wide or narrow substrate specificity) plays an important role in the outcome of cooperative phenomena in membranes (Table 2).

Table 2. Phospholipases of bacterial origin

Enzyme	Species	Enzyme preparation	Substrate Phospholipid	Specificity	Activated by	Haemolysis	References
Phospholipase A	E. coli	Membrane-bound	PE, PG	A_1	detergent	–	1
EC 3.1.1.32 A_1		Membrane-bound	PE, PC	A_1, A_2, L	Ca^{2+}, detergent	–	2
EC 3.1.1.4 A_2		Cytoplasm and	PG, CLP		Ca^{2+}	–	3
EC 3.1.1.5 L		membrane-bound	PG, PE, PC, CLP	A_1, A_2, L	Ca^{2+}, detergent	–	
		Supernatant and	PE	A_2	Ca^{2+}	–	
		membrane-bound	PE	A_2	Ca^{2+}, detergent	–	4
	S. typhimurium	Membrane-bound	PE, PG, CLP	A_1, A_2, L	Ca^{2+}, detergent	–	5,6
	B. subtilis	Cytoplasm and	LPG, LPE,	L_1	Ca^{2+}	–	
		membrane-bound	PG, PE	A_1	Ca^{2+}, detergent	–	7–10
	B. megaterium	Spores	PG, PC, PE	A_1	detergent	–	11,12
	Mycobact. phlei	Membrane-bound	PE, PC, LPC	A_1, L_1	Ca^{2+}, detergent	–	13
	V. parahaemolyticus	Membrane-bound	LPC, LPE, LCLP	L	–	–	14
Phospholipase C	Cl. perfringens	Supernatant	PC, SPM, PE, PS		Ca^{2+}, detergent	+	15–22
EC 3.1.4.3	Cl. novyi	Supernatant	PC, PE, PS, PI, SPM		Mg^{2+} or Ca^{2+}, detergent	+	23
EC 3.1.4.10		Supernatant	PI		detergent	–	24
	C. equi	Supernatant	SPM, PC, PE, PS, PA	Not clear	Mg^{2+} or Ca^{2+}, detergent	–	25
	B. cereus	Supernatant	PC, PE, PS, PI, PG		Zn^{2+}	–	26–33
		Supernatant	PC, PI, SPM	3 distinct enzymes	Mg^{2+} or Ca^{2+}	–	34
		Supernatant	PI		–	–	35

B. thuringiensis	Supernatant	PI	—	—	36,37
S. aureus	Supernatant	PI	—	—	38
P. aeruginosa	Supernatant	PC, LPC, SPM	+	—	39–41
Sphingomyelinase C					
EC 3.1.4.3.					
S. aureus	Supernatant	SPM, PC	a	Mg^{2+}	42–46
Cl. perfringens	Supernatant	SPM, PC, LPC	a	Mg^{2+}	47
Leptosp. interrogans	Supernatant	SPM, PC	a	Mg^{2+}	48
Phospholipase D					
EC 3.1.4.4.					
C. ovis	Supernatant	SPM, LPC	—	Mg^{2+}	49–52
C. ulcerans	Supernatant	SPM, LPC	—	Mg^{2+}	53
H. parainfluenzae	Membrane-bound	CLP	—	Mg^{2+}	54,55
E. coli	Cytoplasm	CLP	—	Mg^{2+}	56
V. damsela	Supernatant	SPM, PC, PE	+	—	57
B. subtilis	Supernatant	SPM, PC	—	Ca^{2+} or Zn^{2+}	58
Sphingomyelinase					
EC 3.1.4.12.					
S. aureus	Supernatant	SPM, LPC	a	Mg^{2+}, detergent	59
B. cereus	Supernatant	SPM, LPC	a	Mg^{2+}, detergent	60,61,34

Abbreviations: CLP, cardiolipin; PA, phosphatidic acid; PC, phosphatidylcholine; PE, phosphatidylethanolamine; PG, phosphatidylglycerol; PI, phosphatidylinositol; PS, phosphatidylserine; LCLP, lysocardiolipin; LPC, lysophosphatidylcholine; LPE, lysophosphatidylethanolamine; LPG, lysophosphatidylglycerol; SPM, sphingomyelin.

a Hot-cold lysis.

(1) Scandella and Kornberg (1971); (2) Nishijima et al. (1977); (3) Doi et al. (1972); (4) Bernard et al. (1973); (5) Osborn et al. (1972); (6) Osborn et al. (1974); (7) Kent and Lennarz (1972); (8) Kent et al. (1973); (9) Kennedy et al. (1974); (10) Krag and Lennarz (1975); (11) Yamaguchi and Morishita (1969); (12) Raybin et al. (1972); (13) Nishijima et al. (1974); (14) Misaki and Matsumoto (1978); (15) Macfarlane and Knight (1941); (16) Nameroff et al. (1973); (17) Sabban et al. (1972); (18) Bernheimer et al. (1968); (19) Takahashi et al. (1974a); (20) Zwaal et al. (1975); (21) Yamakawa and Ohsaka (1977); (22) Takahashi et al. (1974b); (23) Taguchi and Ikezawa (1975); (24) Taguchi and Ikezawa (1978); (25) Bernheimer et al. (1980); (26) Zwaal et al. (1971); (27) Zwaal and Roelofsen (1975); (28) Otnaess et al. (1972); (29) Little et al. (1975); (30) Little and Otnaess (1975); (31) Little (1981); (32) Otnaess (1980); (33) Gilmore et al. (1989); (34) Gerasimene et al. (1985); (35) Ikezawa et al. (1976); (36) Taguchi et al. (1980); (37) Ikezawa et al. (1983); (38) Low and Finean (1976); (39) Stinson and Hayden (1979); (40) Berka and Vasil (1982); (41) Pritchard and Vasil (1986); (42) Doery et al. (1963); (43) Doery et al. (1965); (44) Wadström and Möllby (1971a); (45) Wadström and Möllby (1971b); (46) Bernheimer et al. (1974); (47) Pastan et al. (1968); (48) Bernheimer and Bey (1986); (49) Souček et al. (1967a); (50) Souček et al. (1971); (51) Souček and Součková (1974); (52) Bernheimer et al. (1985); (53) Barksdale et al. (1981); (54) Ono and White (1970a); (55) Ono and White (1970b); (56) Benns and Proulx (1974); (57) Kreger et al. (1987); (58) Garutskas et al. (1977); (59) Wiseman and Caird (1967); (60) Ikezawa et al. (1978); (61) Tomita et al. (1982).

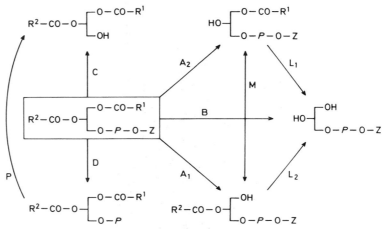

Figure 2. The scheme shows the site of hydrolysis of individual phospholipases attacking diacylphosphoglycerides (boxed) or their split products. Abbreviations: A1, A2, B, C and D designate phospholipases of the respective type. L1, L2: lysophospholipase 1 and 2; M: migratase; P: phosphatidic acid phosphatase (EC 3.1.3.4); Z: designates the residues; choline, ethanolamine and serine. (From O.W. Thiele (1979) *Lipide, Isoprenoide mit Steroiden*, p. 164. Georg Thieme Verlag, Stuttgart.)

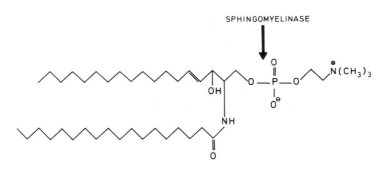

Figure 3. Hydrolysis of sphingomyelin by sphingomyelinase results in the formation of choline and ceramide.

Considering the non-lytic degradation of phospholipids (Colley *et al.*, 1973; Zwaal *et al.*, 1975; Gazitt *et al.*, 1976), e.g. by *S. aureus* sphingomyelinase C or corynebacterial phospholipase D, an extensive breakdown of PL in the outer leaflet may be observed ('first step'). Indeed, *S. aureus* sphingomyelinase C catalyses the breakdown of 50–80% of sphingomyelin in human, swine, bovine and sheep RBC without lysing the cells (Colley *et al.*, 1973). Although the phosphomonoester hydrolases, phosphatidic acid phosphatases, lysophospholipases (L1/L2), and migratases (M) have not, as yet, been demonstrated to be involved in cooperative membrane damage (Fig. 2), it is conceivable that such enzymes may contribute to membrane lesions of this type, either as 'first-, second- or consecutive-step agents'. With knowledge of the phospholipid composition of membranes, cooperative principles can be predicted and combinations of cooperative factors selected which generate membrane disruption in appropriate target cells (Verkleij *et al.*, 1973).

Among the lipid-degrading enzymes which cannot directly interact with intact membranes, cholesterol oxidases (Gottlieb, 1977; McGuinness *et al.*, 1978; Linder and Bernheimer, 1982; Lange *et al.*, 1984; Thurnhofer *et al.*, 1986) and lipases (Christie and Graydon, 1941; Jürgens *et al.*, 1981; Kötting et al., 1985, 1988) should be mentioned since they function as 'second-step agents' in cooperative processes (Table 3).

Unlike the majority of the phospholipases C and D or the sphingomyelinases, phospholipases A of bacterial origin are membrane-bound and not secreted or released (Table 2). Consequently, it is not known whether these enzymes, once liberated through *in vivo* degradation or autolysis of the bacterial cell, will become soluble and catalytically active against target structures. Furthermore, it should be mentioned that phospholipase A, C, D and sphingomyelinases are dependent on divalent cations such as Zn^{2+}, Ca^{2+} and/or Mg^{2+} (Table 2).

Non-enzymic modification of target cell membranes

In their basic study on lipid organization in the human erythrocyte membrane, Zwaal et al. (1975) pointed out that the *in situ* accessibility of membrane phospholipids 'is governed by at least three underlying principles:

- substrate specificity of phospholipases,
- sidedness of phospholipids when only one side of the membrane is exposed to the phospholipase,

and

- compression state of the lipids in the membrane'.

Direct proof of the strong influence of surface pressure or the density of lipid packing on phospholipase activity was obtained by Demel *et al.* (1975), Zwaal *et al.* (1975) and van Deenen *et al.* (1976). Of interest in this context was that phospholipases with direct haemolytic activity (phospholipase C of *S. aureus* and *C. perfringens*) were able to hydrolyse their substrates in mixed lipid monolayers at air–water interfaces above a surface pressure of 31 dynes/cm. This value corresponds approximately to the lipid surface pressure calculated for the intact erythrocyte membrane (31–34.8 dynes/cm).

Changes in lipid packing in the membrane may also be observed when RBC are suspended in

hypotonic buffer solutions. This is indicated by the hydrolysis of PL in intact human RBC (HRBC) membrane by phospholipase C of *B. cereus* (Laster *et al.*, 1972; Woodward and Zwaal, 1972; Zwaal *et al.*, 1973), which cannot hydrolyse its substrate in the intact cell and is non-lytic for HRBC under isotonic conditions (Verkleij *et al.*, 1973). These observations have been explained by the assumption that the enzyme only gains access to its substrate(s) in the osmotically stretched membrane (Woodward and Zwaal, 1972; Zwaal *et al.*, 1973).

Similar observations have been made after addition of sublytic concentrations of detergents such as Triton X-100 or desoxycholate to ·HRBC (Roelofsen *et al.*, 1971). Under these conditions PC, PE and PS were almost completely hydrolysed by phospholipase C of *B. cereus*, although the enzyme gains no access to the membrane phospholipids of intact cells in the absence of detergents. The picture is still incomplete since metabolic and membrane dynamic influences also govern the accessibility of membrane lipids for phospholipases. Gazitt *et al.* (1976) reported that the combined action of phospholipase C of *B. cereus* and sphingomyelinase C of *S. aureus* on ATP-depleted rat-RBC, which caused 43% hydrolysis of phospholipids, was followed by immediate lysis. In contrast, rat RBC with a normal ATP content were insensitive to both enzymes (Frish *et al.*, 1973; Gazitt *et al.*, 1975), whereas a low ATP content in bovine RBC reduced the hydrolysis of sphingomyelin by sphingomyelinase of *B. cereus* (Tomita *et al.*, 1987). It was concluded from these experiments that ATP depletion governs the availability and/or accessibility of membrane phospholipids for individual phospholipases.

Consequences of phospholipid degradation for membrane organization and function

It has been shown in numerous studies (Lenard and Singer, 1968; Ottolenghi and Bowman, 1970; Coleman *et al.*, 1970; Bowman *et al.*, 1971; Roelofsen *et al.*, 1971; Souček *et al.*, 1971; Laster *et al.*, 1972; Součková and Souček, 1972; Woodward and Zwaal, 1972; Zwaal *et al.*, 1973; Verkleij *et al.*, 1973; Low *et al.*, 1973, 1974; Hendrickson *et al.*, 1974; Colley *et al.*, 1973; Bernheimer *et al.*, 1974; Demel *et al.*, 1975; Gazitt *et al.*, 1976; van Deenen *et al.*, 1976; Linder and

Bernheimer, 1978; van Deenen, 1981; Kreger *et al.*, 1987; Allan and Walklin, 1988) that susceptible RBC or RBC ghosts, lipid vesicles and lipid monolayers undergo physico-chemical, morphological and/or functional changes when exposed to the action of particular phospholipases. In line with the concept of phospholipid asymmetry in RBC membranes, three types of reaction may occur between phospholipases and membranes of intact cells:

(a) hydrolysis of phospholipids of the inner and outer half of the membrane bilayer in a direct lytic step (e.g. phospholipase C of *C. perfringens*, *C. novy* and *P. aeruginosa*);

(b) degradation of membrane phospholipids of the outer leaflet of the bilayer in a non-lytic step (e.g. sphingomyelinase of *S. aureus*, *C. perfringens*, and *Leptospira* spp.; phospholipases D of *Corynebacteria*, *V. damsela*, and *B. subtilis*); or

(c) no hydrolysis of phospholipids of intact erythrocytes but hydrolysis of phospholipids of unsealed- or inside-out ghosts (i.e. phospholipase C of *B. cereus*).

For the initiation ('first step') of cooperative lytic processes, the reaction described under (b) is most commonly encountered (Table 3). Additionally, phospholipases which are directly lytic, as described in type (a), may also initiate cooperative processes when acting upon RBC at sublytic concentrations. As a result of the interaction of phospholipase C (*C. perfringens*, *B. cereus*) with erythrocytes, ghosts (Lenard and Singer, 1968; Gordon *et al.*, 1969; Finean and Coleman, 1970), ascites cells (Gordon *et al.*, 1969) and microsomal membranes of muscle cells (Finean and Coleman, 1970) electron-dense droplets associated with the membranes seen by electron microscopy probably represented the split products of phospholipid degradation (Finean and Martonosi, 1965; Ottolenghi and Bowman, 1970; Bowman *et al.*, 1971; Verkleij *et al.*, 1973; Low *et al.*, 1974).

Although it seems unlikely that the authors (Ottolenghi and Bowman, 1970) had access to highly purified phospholipase C (*B. cereus*), their findings (Fig. 4) were later confirmed by the work of Verkleij *et al.* (1973), Low *et al.* (1974), and Bernheimer (1974). Verkleij *et al.* (1973)

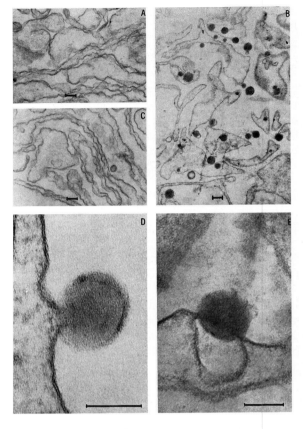

Figure 4. Human red blood cell (HRBC) ghosts fixed with glutaraldehyde and Os-tetroxide and stained with uranyl acetate and lead citrate (original magnification: approx. × 8000). A: Control; B: HRBC ghosts treated with phospholipase C from *B. cereus* showing electron-dense areas. C: Same as B, but after addition of orthophenanthroline showing inhibition of enzyme activity. D–E: Same as B, but electron-dense masses (droplets) shown at higher magnification (× 50000 and 35000). (From A.C. Ottolenghi and H. Bowman (1970). *J. Memb. Biol.* **2**, 186).

demonstrated by freeze-etch electron microscopy that sphingomyelinase treatment (*S. aureus*) of HRBC caused aggregation and clustering of intramembranous particles in both fracture faces together with the formation of small spheres on the outer (EF) and corresponding pits on the inner fracture face (PF). Large membrane-associated droplets of 100–300 nm were also seen in HRBC ghosts after treatment with the same enzyme (Fig. 5).

Similar data have also been reported by Ottolenghi and Bowman (1970), Bowman *et al.* (1971), Coleman *et al.* (1970) and Low *et al.* (1974). There

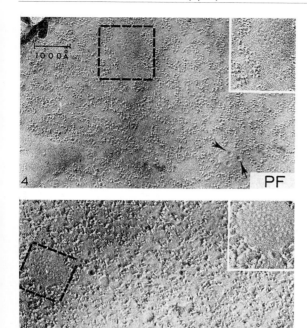

Figure 5. (Upper) Inner fracture face (PF) and (lower) outer fracture face (EF) of human red cell membranes after treatment with sphingomyelinase (S. aureus). Notice the characteristic areas with pits (75 Å in diameter) on the PF (inset upper panel) and the corresponding spheres on the EF in the lower panel. Arrows indicate pits and spheres (200 Å in diameter) on the PF and EF respectively (original magnification × 160 000; inset: × 240 000). (From A.J. Verkleij et al. (1973). Biochim. Biophys. Acta **323**, 178–93.)

were some discrepancies in the estimate of the size of the droplets found by these authors, which ranged from 300 to 1000 nm. With respect to the chemical nature of the 'droplets' or 'dots' observed in RBC following exposure to phospholipases C or sphingomyelinases C (Coleman et al., 1970; Bowman et al., 1971; Colley et al., 1973; Verkleij et al., 1973; Low et al., 1974; Bernheimer et al., 1974) it was suggested that they may represent discrete pools of aggregated diglyceride or ceramide. In fact, this view seemed likely since diglyceride droplets disappeared after digestion with pancreaselipase (Colley et al., 1973). Additionally, the pools incorporated lipophilic dyes (Ottolenghi and Bowman, 1970).

In summary, the non-lytic degradation of phospholipids of intact erythrocytes by phospholipases

C and sphingomyelinases results in the removal of the polar head groups of mainly choline-PL with the concomitant formation of small ceramide or diglyceride droplets within the membrane continuum. Despite extensive degradation (up to 85%) of the membrane sphingomyelin by S. aureus sphingomyelinase, the membrane still retains its barrier function (van Deenen, 1981). Similarly, the function of the bilayer in target liposomes of appropriate lipid composition remains largely intact when exposed to S. aureus sphingomyelinase digestion as judged by retainment of entrapped ATP (Fehrenbach et al., 1984). The process of non-lytic degradation of membrane lipids by individual phospholipases has recently been summarized in a diagrammatic scheme (Fig. 6) by van Deenen (1981). This scheme is based on the work of Ottolenghi and Bowman (1970), Coleman et al. (1970), Bowman et al. (1971), Roelofsen et al. (1977) and Zwaal et al. (1973).

Besides the structural findings mentioned above, other workers have observed membrane invaginations (Ottolenghi and Bowman, 1970; Freer and Arbuthnott, 1976) or internal vesicles after incubation of RBC or ghosts (human, sheep, pig) with phospholipase C (Ottolenghi and Bowman, 1970) or sphingomyelinase C (Low et al., 1974; Bernheimer et al., 1974). Whether these invaginations (Low et al., 1974) correspond to the endovesicles observed recently (Allan and Walklin, 1988) is unclear. These authors reported the formation of endovesicles in human RBC following sphingomyelinase treatment together with a decrease in membrane surface area and discocyte/spherocyte transition. Similar results were reported earlier by Coleman et al. (1970), Bernheimer et al. (1974) and Low et al. (1974). Essentially the same results were obtained by scanning electron microscopy (Fehrenbach et al., 1988a) in an investigation of the morphology of SRBC after interaction with sphingomyelinase (S. aureus) and CAMP factor.

An additional consequence of non-lytic degradation of membrane phospholipids is an increase in osmotic lability of target cells (Laster et al., 1972; Woodward and Zwaal, 1972; Zwaal et al., 1973; Verkleij et al., 1973). Along with this lability, modified target cells exhibit the phenomenon of 'hot–cold lysis' upon chilling (≤ 10°C) (Walbum, 1922; Roy, 1937; Smith and Price, 1938;

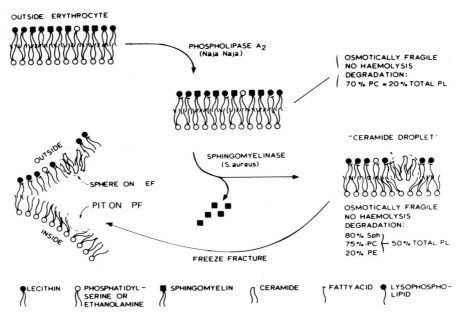

Figure 6. Schematic representation of the non-lytic action of two phospholipases on intact human erythrocytes. Note: the action of these phospholipases on open ghosts produces 100% phospholipid hydrolysis. (From L.L.M. van Deenen (1981) *FEBS Lett.* **123**, 3–14.)

Bernheimer, 1974; Smyth *et al.*, 1975; Linder and Bernheimer, 1982; Allan and Walklin, 1988).

Cooperative lytic systems: The 'second step'

Origin and nature of 'second-step agents'

Since the first report on a cooperative haemolytic system by Christie *et al.* (1944) numerous 'CAMP-like' reactions have been described. In these systems *S. aureus* sphingomyelinase acts in combination with other bacterial phospholipases or with non-enzymic compounds such as bacterial extracellular proteins. Among the 'second-step agents' with enzymic activity are cholesterol oxidases (*C. equi* and *Brevibacterium* spp.) and lipases (*S. aureus*). In addition, there exist a considerable number of soluble bacterial exosubstances active in the 'second step' which are as yet poorly characterized. CAMP factor (Fraser, 1961; Bernheimer *et al.*, 1979; Fehrenbach *et al.*, 1984, 1988b; Jürgens *et al.*, 1985, 1987; Sterzik and Fehrenbach, 1985; Schneewind *et al.*, 1988; Sterzik

et al., 1986; Rühlmann *et al.*, 1988, 1989), delta-toxin (Fitton *et al.*, 1980, 1984) the CAMP-like protein of *A. pleuropneumoniae* (Frey *et al.*, 1989) and *S. aureus* lipase (Jürgens *et al.*, 1981; Kötting *et al.*, 1985, 1988) are the only 'second-step agents' which have either been characterized, cloned or sequenced.

Table 3 lists known cooperative systems together with the bacterial sources from which the 'first- and second-step agents' are derived. Also listed are the target cells sensitive to an individual lytic combination. It is obvious from the Table that RBC of mammalian origin have been predominantly used in such studies.

It is still difficult to classify the 'second-step agents' specified in Table 3, since the cooperative components belong to different classes of proteins, many of which are not fully defined. While the mechanism of membrane disruption by lipolytic enzymes may be explained by the sequential degradation or modification of membrane lipids, little is known about the mode of action of the 'second-step agents' that are non-enzymic proteins.

Consequences of the interaction of 'second-step agents' with modified membranes

While a wealth of data is available describing the morphological changes in membranes following exposure of target cells to phospholipases or sphingomyelinases details of changes in membrane morphology due to the action of 'second-step agents' of the non-enzymic type are scarce. We have reported recently (Fehrenbach *et al.*, 1988a) that the CAMP factor lyses SRBC pretreated with sphingomyelinase (*S. aureus*) by the formation of large membrane defects (Fig. 7).

Moreover, scanning electron microscopy (SEM) shows that sphingomyelinase treatment of SRBC (the 'first step') results in discocyte/spherocyte transition (Low *et al.*, 1974; Allan and Walklin, 1988) and the formation of membrane-associated exocytotic projections or spicules (Fehrenbach *et al.*, 1988a). It remains to be shown that these exocytotic elements on the membrane surface are identical with the membrane-associated 'droplets' seen earlier (Coleman *et al.*, 1970; Bowman *et al.*, 1971; Low *et al.*, 1974).

So far, no evidence has been obtained from SEM as to whether or not hydrolysis of SPM in the SRBC membrane induces endovesiculation as recently described by Allan and Walklin (1988). Furthermore, it is desirable to define any changes in membrane lipid organization which may occur in target cells after exposure to the 'second-step agents' of both types: enzymes and non-enzymic proteins.

Structure/function relationship of 'second-step agents' (CAMP factor and delta-toxin)

Bacterial phospholipases C, cholesterol oxidases and lipases participate in cooperative lytic processes as 'second-step agents' (Table 3). Although these enzymes hydrolyse different substrates, they exhibit the common function of degrading or modifying the membrane lipids of the target cell. They differ from each other in physico-chemical properties such as molecular weight, substrate specificity and structure. However, since they all interact with lipid components in membranes, it may be assumed

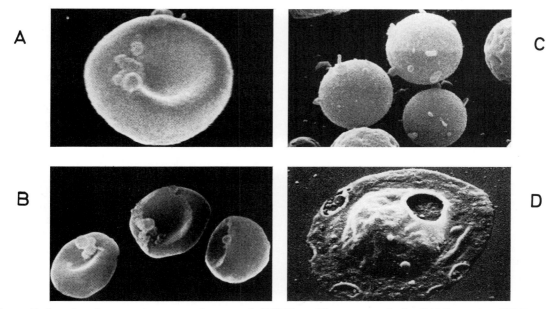

Figure 7. Scanning electron microscopy of target cells (SRBC) at different stages in the CAMP reaction. SRBC control (A); SRBC after addition of CAMP factor (B); after sphingomyelinase treatment SRBC show a spherocyte morphology (C); lysis of sphingomyelinase-treated SRBC by CAMP factor leads to lesions in the cell membrane (D). (From B. Sterzik, Dissertation Freie Universität Berlin, 1988.) Original magnification: × 15 000.

Table 3. Cooperative membrane active processes generated by bacterial products

First step		Second step		Target cells lysed	References
Enzymes acting on membrane lipids	Source of enzyme	Agents reacting with modified membranes	Source of agent		
Sphingomyelinase	S. aureus	Phospholipase C	B. cereus	HRBC	1
	B. cereus	Phospholipase	B. cereus	HRBC	2
	S. aureus	CAMP-factor	S. group B	Ruminant RBC	3
	S. aureus	'CAMP-like factor'	S. uberis		3–6
	S. aureus	'CAMP-like factor'	Actinobacillus pleuropneumoniae	Ruminant RBC	7
	S. aureus	Renalin	C. renale	Ruminant RBC	3
	S. aureus	Delta-toxin	S. aureus	SRBC	8–11
	S. aureus	Oxyd. streptolysin O	S. pyogenes	SRBC	12
	S. aureus	Lipase ?	S. aureus	SRBC	13
	S. aureus	Aeromonas factor ?	A. hydrophila	SRBC	14
	S. aureus	Aeromonas factor ?	A. sobria	SRBC	14
	S. aureus	Vibrio factor ?	Vibrio spec.	SRBC	15,16

Agent 1	Species 1	Agent 2	Species 2	RBC	Ref.
Sphingomyelinase C	S. aureus	Phospholipase C	B. cereus	Ruminant + HRBC	*
	C. perfringens	Phospholipase C	B. cereus	Ruminant + HRBC	*
	Lept. interrogans	Phospholipase C	B. cereus	Ruminant + HRBC	*
Phospholipase D	C. ovis	Cholesterol oxidase	C. equi	SRBC	17
	C. ovis	Cholesterol oxidase	Brevibacterium sp.	SRBC	17
	C. ulcerans	Phospholipase C ?	C. equi	SRBC	18
Phospholipase C	C. perfringens	CAMP factor	S. group B	H/M/R-RBC	19
	C. perfringens	Cholesterol oxidase	Brevibacterium sp.	HRBC	*
	C. perfringens	'Lipase' ?	P. acnes	H/S-RBC	20
	C. perfringens	Lipase	S. aureus (TEN 5)	MRBC, liposomes	u

*Predicted. ? Agent to be characterized. u Fehrenbach et al., unpublished.

Species in bold type indicate 'intrinsic' possession of the cooperative principle.

(1) Colley et al. (1973); (2) Gilmore et al. (1989); (3) Bernheimer and Avigad (1982); (4) Heeschen et al. (1967); (5) Skalka et al. (1980); (6) Skalka and Smola (1981); (7) Frey et al. (1989); (8) Williams and Harper (1947); (9) Marks and Vaughan (1950); (10) Kleck and Donahue (1968); (11) Kreger et al. (1971); (12) Fehrenbach et al. (1988b); (13) Christie and Graydon (1941); (14) Figura and Guglielmetti (1987); (15) Köhler (1988); (16) Lesmana and Rockhill (1985); (17) Linder and Bernheimer (1982); (18) Bernheimer et al. (1980); (19) Fehrenbach et al. (1980); (20) Kar Choudhury (1978).

that they possess amphiphilic structures or domains which allow protein–lipid interaction.

The same may be true for the 'second-step agents' that are non-enzymic proteins, such as CAMP- and CAMP-like factors. While numerous lipolytic bacterial enzymes have been thoroughly investigated, few 'second-step agents' belonging to the class of non-enzymic substances have been purified and chemically defined. Well-defined agents in this group include CAMP factor of group B streptococci (Jürgens *et al.*, 1985, 1987; Sterzik *et al.*, 1986; Rühlmann *et al.*, 1988, 1989), 'renalin' of *C. renale* (Bernheimer and Avigad, 1982) the 'CAMP-like' protein of *A. pleuropneumoniae* (Frey *et al.*, 1989) and delta-toxin of *S. aureus* (Kreger *et al.*, 1971; Fitton *et al.*, 1980, 1984). The latter is active in cooperative processes when used in sublytic concentrations (Williams and Harper, 1947; Marks and Vaughan, 1950; Kleck and Donahue, 1968; Kreger *et al.*, 1971). Interestingly, delta-toxin may also be classed as a 'first-step agent' since cooperativity with sphingomyelinase (*S. aureus*) occurs regardless of the sequence of addition of delta-toxin (Kreger *et al.*, 1971).

The biochemical diversity of the cooperative systems and factors (Table 3) and the different mechanisms of membrane disruption, suggest that little structural homology exists between the individual factors.

Nevertheless, it was of interest to compare the structure of the isofunctional 'second-step agents', namely delta-toxin and CAMP factor, both of which have been sequenced (Fitton *et al.*, 1980, 1984; Rühlmann *et al.*, 1988, 1989).

CAMP factor

CAMP factor is an amphiphilic polypeptide (Jürgens *et al.*, 1985) of known amino acid sequence (Rühlmann *et al.*, 1988). It consists of 226 amino acids, 37% of which are hydrophobic. The latter are evenly distributed throughout the protein and not clustered in a 'hydrophobic domain'. However, hydrophobic domains may result from protein folding in a way which is not predictable from sequence data.

Secondary structure predictions of the CAMP protein reveal the existence of several helical segments. Since amphiphilic helical structures seem to be responsible for the lipid-binding properties of various proteins and peptides (Kaiser and

Kézdy, 1987), these segments have been examined for amphiphilic characteristics by the helical wheel projection (Schiffer and Edmundson, 1967).

Figure 8 shows helical wheel plots of the potential amphiphilic helices of CAMP factor. All four amphiphilic helices extend over a length of 11–12 amino acids which corresponds to the average length of α-helical structures in globular proteins (11 residues or three helical turns; Schulz and Schirmer, 1979) but is not long enough to span a membrane. The hydrophobic domain of these amphiphilic helices occupies almost 180° of the cylindrical surface (Fig. 8).

Interestingly, all the amphiphilic helices (Fig. 8) are located within the N-terminal part of CAMP factor, which exhibits 32% sequence similarity with a segment of the putative lipid-binding domain of human apolipoprotein A-IV (Rühlmann *et al.*, 1989). In addition, Sterzik *et al.* (1986) have shown that the 9-kDa CNBr fragment of CAMP factor, which corresponds almost to this N-terminal part, binds to lipid suspensions of cholesterol and sphingomyelin, whereas the C-terminal part does not bind. These findings led Rühlmann *et al.* (1989) to conclude that the N-terminal part of CAMP factor seems to represent the membrane-binding domain and the four potential amphiphilic helices may be involved in protein/lipid interaction.

Delta-toxin

Delta-toxin is a surface-active directly lytic peptide of 26 amino acids, with a relatively high content of hydrophobic amino acids (42%). Compared with other cytolytic toxins of *S. aureus*, delta-toxin has a rather low lytic potency. Delta-toxin interacts with many types of phospholipid monolayers and artificial membranes and lyses a broad variety of target cells at appropriate concentrations (Freer *et al.*, 1984; Freer and Arbuthnott, 1986). In addition to its direct lytic activity, at sublytic concentrations it acts as a 'second-step agent' in cooperative lytic processes together with sphingomyelinase (Kreger *et al.*, 1971).

The amino acid sequences of the two immunologically distinct forms of delta-toxin from the canine and human strains have been determined (Fitton *et al.*, 1980, 1984). Both delta-toxins exhibit 62% identical residues and if conservative replacements are allowed, the similarity increases to 81%. The sequences exhibit a balanced distribution of

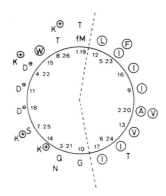

AXIAL PROJECTION OF THE α-HELIX OF DELTA-TOXIN (POS. 1 - 26).

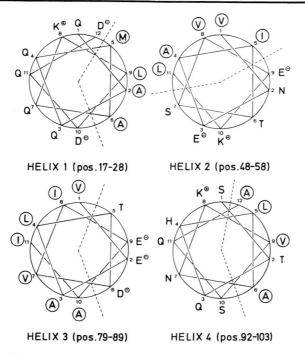

HELIX 1 (pos.17-28)   HELIX 2 (pos.48-58)

HELIX 3 (pos.79-89)   HELIX 4 (pos.92-103)

AXIAL PROJECTION OF THE POTENTIAL AMPHIPHILIC HELICES OF CAMP-FACTOR.

Figure 8. Hydrophobic residues are shown in circles. The dashed lines separate hydrophobic and hydrophilic residues. Helical wheel projections were derived from unpublished work (1989) and kindly supplied by Jörg Rühlmann, Department of Microbiology, Robert Koch-Institut, Berlin, Germany.

charged and polar amino acids separated by hydrophobic residues. If the peptide adopts an α-helical structure, this would result in an amphiphilic distribution of the amino acids with hydrophobic and hydrophilic amino acids on opposite sides of the helix (Freer and Birkbeck, 1982) (Fig. 8 top). Indeed, delta-toxin is predominantly α-helical in aqueous ethanol and methanol (Tappin

et al., 1988) and shows a high α-helical content in water (Colacicco et al., 1977; Fitton, 1981). In contrast to the amphiphilic helices of CAMP factor the length of the delta-toxin helices would suffice to span the membrane. Alouf et al. (1988) have synthesized analogues of delta-toxin with several amino acid substitutions but with preserved amphiphilicity. Interestingly, these analogues still exhibit lytic activity, which may indicate that the amphi-philic helical structure is a requirement for activity. Sequence analysis of both delta-toxin and CAMP factor thus revealed the existence of amphiphilic helical structures (Fig. 8) which seem to be the only common property shared by these agents active in cooperative (haemo-)lysis.

Possible mechanisms of membrane disintegration by cooperative processes

It has been emphasized earlier that the initiation of cooperative processes ('first step') starts with the hydrolysis of membrane phospholipids of the target cell by phospholipases active in hydrolysing their substrates in intact cells. As a consequence of the non-lytic degradation of membrane phospho-lipids in the outer half of the bilayer, structural and functional changes result which obviously provide the conditions for the 'second-step agents' to interact with the modified membrane.

Enzyme hydrolysis of phospholipids not only results in the loss of polar head groups but may also result in accumulation of the split products in discrete pools (ceramide, diglyceride, phosphatidic acid?) and in partial phase separation. The degraded lipids and the split products no longer fully participate in the coordinated lipid/protein or lipid/lipid interaction and thus contribute to the functional disorders observed in the processed membrane. The loss of polar head groups together with the accumulation of neutral lipids or other split products deposited in the membrane may result in altered lipid packing, changes in mem-brane surface pressure and/or membrane fluidity. Additionally, it has been reported that enzymic degradation of phospholipids in one of the leaflets of the bilayer leads to an 'induced asymmetry' which may generate the formation of endovesicles or invaginations and induce considerable shrinking of the membrane.

Therefore, the present data suggest that phos-pholipases create a 'metastable lipid phase' by removing the zwitterionic residues from the outer leaflet, thus, eliminating the cation-mediated inter-action of charged phospholipids at the membrane surface. 'Second-step agents' of the CAMP factor type may then interact with lipids in the metastable membrane and induce phase separation. It has earlier been shown (Fehrenbach et al., 1984) that a suggested receptor-mediated mechanism involving CAMP factor binding to ceramide (Bernheimer et al., 1979) is not required for binding and lysis. The accessibility of the 'second-step agents' to the membrane may be enhanced by the absence of the zwitterionic residues in the outer half of the membrane and/or the reduction in surface pressure which may facilitate insertion or translocation of an amphiphilic domain. The same concept would apply for 'second-step agents' that are phospholipases unable to hydrolyse initially their substrates in the intact cell membrane but are active in a cooperative way.

This does not conflict with the observation that treatment of target cells with phospholipases protects them against subsequent digestion by phospholipases C or sphingomyelinases (Munch-Petersen, 1954; Fraser, 1962, 1964; Zaki, 1965; Souček et al., 1967a,b; 1971; Součková and Souček, 1972; Souček and Součková, 1974; Carne and Onon, 1978; Onon, 1979). This effect, which has been designated 'antagonism' (Fraser, 1964; Linder, 1984), is best explained by the fact that hydrolysis of membrane phospholipids by phospholipase(s) D with specificities for PC (PE) or SPM degrade their substrates by the formation of phosphatidic acid and/or ceramide phosphate. However, ceramide phosphate or phosphatidic acid no longer possess substrate properties for phospholipases C or sphingomyelinases. This hypothesis is sustained by the work of Zaki (1965) and Součkova and Souček (1972).

Inhibition of *S. aureus* alpha-toxin activity on SRBC and rabbit RBC (RRBC) after treatment with phospholipase D (*C. ovis* and *C. equi*) is more difficult to explain. Watanabe et al. (1987) and Bhakdi et al. (1988) suggested that phosphatidyl-choline and/or cholesterol are important in the initial absorption of alpha-toxin to the membrane, although PC is not expected to exhibit receptor function. Absorption or initial binding of alpha-toxin to PC may be blocked in target cells which have been exposed to phospholipase D. Alternatively, the formation of phosphatidic acid in the outer leaflet may considerably influence the coordination of lipids in the outer half of the membrane and interfere with monomer–hexamer transition and generation of the trans-membrane hexameric pores of alpha-toxin (Bhakdi and Tranum-Jensen, 1987; Reichwein et al., 1987).

Comparison of the primary structure of CAMP factor and the delta-toxin did not reveal extended sequence homologies, although both proteins participate in cooperative systems as 'second-step agents'. The helical wheel models (Fig. 8), however, demonstrated that both proteins possess amphiphilic helices which may be in-volved in membrane binding and/or membrane destruction. It should be noted that an amphi-philic domain may not be predicted from secondary structure analysis, but emerges only upon protein folding at the level of the tertiary structure.

Although our knowledge of the molecular mechanisms of membrane disruption by 'second-step agents' that are non-enzymic proteins is still incomplete, some reaction principles may be proposed on recent findings obtained with the CAMP system (Sterzik et al., 1989). Assuming that 'second-step agents' of protein nature all exhibit common amphiphilic properties or struc-tures effective in membrane (lipid) binding and/or membrane translocation, three reaction principles may be proposed (Fig. 9). Using photoaffinity labelling, Sterzik et al. (1989) recently showed that translocation of a domain of the CAMP protein into target liposomes occurs upon bind-ing, suggesting deep penetration into the bilayer.

Besides the lytic principles shown in Fig. 9, one should also consider the cytolytic or cytotoxic mechanisms operative in the direct lysis of target cells by the large number of bacterial cyto- and haemolysins (Wilbrandt, 1942; Bernheimer, 1947; Cooper et al., 1964; Zucker-Franklin, 1965; Freer et al., 1968; Bernheimer, 1974; Fehrenbach et al., 1982; Bhakdi et al., 1984; Bhakdi and Tranum-Jensen, 1987). Cooperative mechanisms in which the lipids in the target cell membrane are sequen-tially hydrolysed may exhibit much more complex reaction patterns since the dynamics of phospho-lipid breakdown, the release of split products giving rise to side reactions and the potential amphiphilic nature of the enzyme proteins have to be considered.

Summary

The list of cooperative (lytic) systems (Table 3) shows that only a few medically important bacteria belong to the group of species that produce either 'first- or second-stage agents'. Furthermore, the majority of species contributing to cooperative systems are not at all or rarely found as infectious agents under normal conditions. Among the few exceptions are S. aureus, corynebacteria and strep-tococci which either colonize the human body or cause severe infections. S. aureus and B. cereus strains are currently the sole species known to produce both the 'first- and second-step agents' and thus possess both factors of the cooperative system (Table 3, bold type). In mono-infections with S. aureus the cooperative principle exhibited by an individual strain may be expected to be of relevance for in vivo pathogenicity. In contrast, the other combinations listed in Table 3 may be of theoretical value only, since there is little clinical evidence of how they occur in mixed infections in nature, except in the cases mentioned below.

While discussing cooperative systems we have focused our attention on those in which 'coopera-tivity' resulted in an easily detectable effect (i.e. haemo- or cytolysis). Moreover, most of these studies have been conducted under in vitro con-ditions and as yet, there is little information from in vivo systems. It is thus possible that the known cooperative systems are irrelevant in vivo and cooperative systems that are generated under natural conditions and may contribute to patho-genicity still remain to be detected. Finally, such in vivo cooperative mechanisms may not be as easily detected as those demonstrated in vitro by virtue of their haemolytic or cytolytic activities, the osmotic lability of target cells or by alterations in membrane functions.

The cooperative principles so far described and the existence of an 'intrinsic' cooperative system, e.g. in S. aureus strains, indicate that they may be more important in host–parasite interaction than presently appreciated. Although this chapter deals mainly with cooperative processes involving two components, it is conceivable that the consecutive action of three or more different factors on individual target structures may initiate early or late events leading to structural and/or functional disorders.

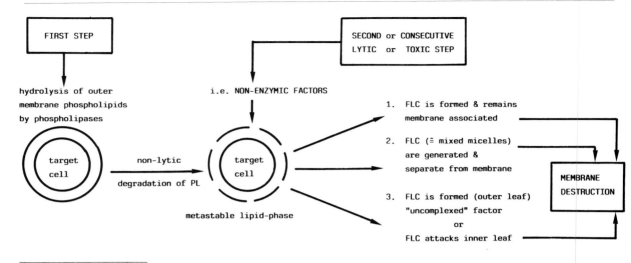

FLC = factor/lipid complex
PL = phospholipids

Figure 9. Bilayer disruption in cooperative processes by non-enzymic bacterial products (possible mechanisms).

Cooperative systems which interact with target structures other than lipids may also exist. These may act through membrane proteins or glycoconjugates to influence regulatory systems of the target cells.

It remains for the future to investigate with suitable animal models the role of cooperative systems in microbial pathogenicity. Additionally, it may be important to extend the search for unknown cooperative factors produced by medically important bacteria coexisting in mixed infections, such as anaerobic bacteria or combinations of bacteria and yeasts which are frequently found in human infections. Hopefully, progress in the understanding of infectious processes and mechanisms of pathogenicity will result from such studies.

findings we would like to acknowledge the help of our colleagues J. Kreft and W. Goebel, R. Lütticken, J. Frey and J. Nicolet. We are grateful for the permission to reproduce material, tables and/or pictures from L.L.M. van Deenen; A.C. Ottolenghi and M.H. Bowman; V.O.W. Thiele; A.J. Verkleij et al.

Thanks are especially due to our secretaries Rita Sprengel and Gabriele Gericke for typing the manuscript. Finally, the authors extend their thanks to the editors J.E. Alouf and J.H. Freer for carefully reading the manuscript.

The work of F.J. Fehrenbech, D. Jürgens, J. Rühlmann and B. Sterzik has been supported by the 'Deutsche Forschungsgemeinschaft', Sfb 9, Technische Universität, Berlin, Germany.

Acknowledgements

The authors express their gratitude to Barbara Sterzik and Jörg Rühlmann for valuable discussions and critical reading of the manuscript. We would like to thank J. Rühlmann especially for providing unpublished material from his own work on CAMP factor and also B. Sterzik for contributing unpublished material to this article. For providing preprints of manuscripts or other unpublished

References

Allan, D. and Walklin, C.M. (1988). Endovesiculation of human erythrocytes exposed to sphingomyelinase C: A possible explanation for the enzyme-resistant pool of sphingomyelin. *Biochim. Biophys. Acta* **938**, 403–10.

Alouf, J.E. (1977). Cell membranes and cytolytic bacterial toxins. In: *The Specificity and Action of Animal, Bacterial and Plant Toxins* (ed. P. Cuatrecasas), pp. 221–64. Chapman and Hall, London.

Alouf, J.E., Fehrenbach, F.J., Freer, J.H. and Jeljaszewicz, J. (eds) (1984). *Bacterial Protein Toxins*, pp. 1–455. Academic Press, London.

Alouf, J.E., Dufourq, J., Siffert, O. and Geoffroy, C. (1988). Comparative properties of natural and synthetic staphylococcal delta toxin and analogues. In: *Bacterial Protein Toxins* (eds F.J. Fehrenbach *et al.*), pp. 39–46. Gustav Fischer, Stuttgart.

Avigad, G. (1976). Microbial phospholipases. In: *Mechanisms in Bacterial Toxinology* (ed. A.W. Bernheimer), pp. 100–67. John Wiley, New York.

Bae, B.H. and Bottone, E.J. (1980). Modified Christie–Atkins–Munch-Petersen (CAMP) test for direct identification of hemolytic and nonhemolytic group B streptococci on primary plating. *Can. J. Microbiol.* **26**, 539–42.

Bangham, A. (1980). Development of the liposome concept. In: *Liposomes in Biological Systems* (eds G. Gregoriadis and A.C. Allison), pp. 1–24. John Wiley, New York.

Bangham, A.D. and Dawson, R.M.C. (1962). Electrokinetic requirements for the reaction between *Cl. perfringens* α-toxin (phospholipase C) and phospholipid substrates. *Biochim. Biophys. Acta* **59**, 103–15.

Barksdale, L., Linder, R., Sulea, I.T. and Pollice, M. (1981). Phospholipase D activity of Corynebacterium pseudotuberculosis (*Corynebacterium ovis*) and *Corynebacterium ulcerans*, a distinctive marker within the genus *Corynebacterium*. *J. Clin. Microbiol.* **13**, 335–43.

Benns, G. and Proulx, P. (1974). The effect of ATP and Mg^{2+} on the synthesis of phosphatidylglycerol in *Escherichia coli* preparations. *Biochim. Biophys. Acta* **337**, 318–24.

Berka, R.M. and Vasil, M.L. (1982). Phospholipase C (heat-labile hemolysin) of *Pseudomonas aeruginosa*: Purification and preliminary characterization. *J. Bacteriol.* **152**, 239–45.

Bernard, M.C., Brisou, J., Dennis, F. and Rosenberg, A.J. (1973). Metabolisme phospholipasique chez les bactéries à gram négatif: Purification et étude cinétique de la phospholipase A2 soluble d'E. coli 0118. *Biochimie* **55**, 377–88.

Bernheimer, A.W. (1947). Comparative kinetics of hemolysis induced by bacterial and other hemolysins. *J. Gen. Physiol.* **30**, 337–53.

Bernheimer, A.W. (1974). Interactions between membranes and cytolytic bacterial toxins. *Biochim. Biophys. Acta* **344**, 27–50.

Bernheimer, A.W. (1976). Sulfhydryl activated toxins. In: *Mechanisms in Bacterial Toxinology* (ed. A.W. Bernheimer), pp. 85–97. John Wiley, New York.

Bernheimer, A.W. and Avigad, L.S. (1982). Mechanism of hemolysis by Renalin, a CAMP-like protein from *Corynebacterium renale*. *Infect. Immun.* **36**, 1253–6.

Bernheimer, A.W. and Bey, R.F. (1986). Copurification of *Leptospira interrogans* serovar pomona hemolysin and sphingomyelinase C. *Infect. Immun.* **54**, 262–4.

Bernheimer, A.W. and Schwartz, L.L. (1963). Isolation and composition of staphylococcal alpha toxin. *J. Gen. Microbiol.* **30**, 455–68.

Bernheimer, A.W., Grushoff, P. and Avigad, L.S. (1968). Isoelectric analysis of cytolytic bacterial proteins. *J. Bacteriol.* **95**, 2439–41.

Bernheimer, A.W., Avigad, L.S. and Kim, K.S. (1974). Staphylococcal sphingomyelinase (beta-hemolysin). *Ann. N.Y. Acad. Sci.* **236**, 292–306.

Bernheimer, A.W., Linder, R. and Avigad, L.S. (1979). Nature and mechanism of action of the CAMP protein of group B streptococci. *Infect. Immun.* **23**, 838–44.

Bernheimer, A.W., Linder, R. and Avigad, L.S. (1980). Stepwise degradation of membrane sphingomyelin by corynebacterial phospholipases. *Infect. Immun.* **29**, 123–31.

Bernheimer, A.W., Campbell, B.J. and Forrester, L.J. (1985). Comparative toxinology of *Loxosceles reclusa* and *Corynebacterium pseudotuberculosis*. *Science* **228**, 590–1.

Bhakdi, S. and Tranum-Jensen, J. (1987). Damage to mammalian cells by proteins that form transmembrane pores. *Rev. Physiol. Biochem. Pharmacol.* **107**, 147–223.

Bhakdi, S., Muhly, M. and Füssle, R. (1984). Correlation between toxin binding and hemolytic activity in membrane damage by staphylococcal alpha-toxin. *Infect. Immun.* **46**, 318–23.

Bhakdi, S., Menestrina, G., Hugo, F., Seeger, W. and Tranum-Jensen, J. (1988). Pore-forming bacterial cytolysins. In: *Bacterial Protein Toxins* (eds F.J. Fehrenbach *et al.*), pp. 71–7. Gustav Fischer, Stuttgart.

Bodemann, H. and Passow, H. (1972). Factors controlling the resealing of the membrane of human erythrocyte ghosts after hypotonic hemolysis. *J. Membr. Biol.* **8**, 1–26.

Bowman, M.H., Ottolenghi, A.C. and Mengel, C.E. (1971). Effects of phospholipase C on human erythrocytes. *J. Membr. Biol.* **4**, 156–64.

Bretscher, M.S. (1972). Asymmetrical lipid bilayer structure for biological membranes. *Nature, New Biol.* **236**, 11–12.

Brown, J., Farnsworth, R., Wannamaker, L.W. and Johnson, D.W. (1974). CAMP factor of group B streptococci: Production, assay, and neutralization by sera from immunized rabbits and experimentally infected cows. *Infect. Immun.* **9**, 377–83.

Carne, M.R. and Onon, E.O. (1978). Action of *Corynebacterium ovis* exotoxin on endothelial cells of blood vessels. *Nature* **271**, 246–8.

Christie, R. and Graydon, J.J. (1941). Observations on staphylococcal haemolysins and staphylococcal lipase. *Aust. J. Exp. Biol. Med. Sci.* **19**, 9–16.

Christie, R., Atkins, N.E. and Munch-Petersen, E. (1944). A note on a lytic phenomenon shown by group B streptococci. *Aust. J. Exp. Biol. Med. Sci.* **22**, 197–200.

Colacicco, G., Basu, M.K., Buckelew, A.R. Jr. and Bernheimer, A.W. (1977). Surface properties of membrane systems transport of staphylococcal delta toxin from aqueous to membrane phase. *Biochim. Biophys. Acta* **465**, 378–90.

Coleman, R. Finean, J.B., Knutton, S. and Limbrick, A.R. (1970). A structural study of the modification of erythrocyte ghosts by phospholipase C. *Biochim. Biophys. Acta* **219**, 81–92.

Colley, C.M., Zwaal, R.F.A., Roelofsen, B. and van Deenen, L.L.M. (1973). Lytic and non-lytic degradation of phospholipids in mammalian erythrocytes by pure phospholipases. *Biochim. Biophys. Acta* **307**, 74–82.

Cooper, L.Z., Madoff, M.A. and Weinstein, L. (1964). Hemolysis of rabbit erythrocytes by purified staphylococcal alpha-toxin. I. Kinetics of the lytic reaction. *J. Bacteriol.* **87**, 127–35.

Demel, R.A., Guerts van Kessel, W.S.M., Zwaal, R.F.A., Roelofsen, B. and van Deenen, L.L.M. (1975). Relation between various phospholipase actions on human red cell membranes and the interfacial phospholipid pressure in monolayers. *Biochim. Biophys. Acta* **406**, 97–107.

Dodge, J.T., Mitchell, C. and Hanahan, D.J. (1963). The preparation and chemical characteristics of hemoglobin-free ghosts of human erythrocytes. *Arch. Biochem. Biophys.* **100**, 119–30.

Doery, H.M., Magnusson, B.J., Cheyne, I.M. and Gulasekharam, J. (1963). A phospholipase in staphylococcal toxin which hydrolyzes sphingomyelin. *Nature* **198**, 1091–2.

Doery, H.M., Magnusson, B.J., Gulasekharam, J. and Pearson, J.E. (1965). The properties of phospholipase enzymes in staphylococcal toxins. *J. Gen. Microbiol.* **40**, 283–96.

Doi, O., Oki, M. and Nojima, S. (1972). Two kinds of phospholipase A and lysophospholipase in *Escherichia coli*. *Biochim. Biophys. Acta* **260**, 244–58.

Esseveld, H., Daniels-Bosman, M.S.M. and Leijnse, B. (1958). Some observations about the CAMP reaction and its application to human β haemolytic streptococci. *Antonie van Leeuwenhoek* **24**, 145–56.

Esseveld, H., Monnier-Goudzwaard, C., van Eijk, H.G. and Soestbergen, M.G. (1975). Chemical nature of a substance isolated from a group B Streptococcus causing the 'CAMP' reaction. *Antonie van Leeuwenhoek* **41**, 449–54.

Falmagne, P., Alouf, J.E., Fehrenbach, F.J., Jeljaszewicz, J. and Thelestam, M. (eds) (1986). *Bacterial Protein Toxins*, pp. 1–397. Gustav Fischer, Stuttgart.

Fehrenbach, F.J., Schmidt, C.M. and Huser, H. (1982). Early and late events in streptolysin-O induced hemolysis. *Toxicon* **20**, 233–238.

Fehrenbach, F.J., Schmidt, C.M., Sterzik, B. and Jürgens, D. (1984). Interaction of amphiphilic bacterial polypeptides with artificial membranes. In: *Bacterial Protein Toxins* (eds J.E. Alouf *et al.*), pp. 317–24. Academic Press, London.

Fehrenbach, F.J., Alouf, J.E., Falmagne, P., Goebel, W., Jeljaszewicz, J., Jürgens, D. and Rappuoli, R. (eds) (1988a). *Bacterial Protein Toxins*, pp. 1–459. Gustav Fischer, Stuttgart.

Fehrenbach, F.J., Jürgens, D., Rühlmann, J., Sterzik, B. and Özel, M. (1988b). Role of CAMP-factor

(protein B) for virulence. In: *Bacterial Protein Toxins* (eds F.J. Fehrenbach *et al.*), pp. 351–7. Gustav Fischer, Stuttgart.

Figura, N. and Guglielmetti, P. (1987). Differentiation of motile and mesophilic *Aeromonas* strains into species by testing for a CAMP-like factor. *J. Clin. Microbiol.* **25**, 1341–2.

Finean, J.B. and Coleman, R. (1970). Integration of structural and biochemical approaches in the study of cell membranes. In: *FEBS Federation of European Societies, Proceedings of the 6th Meeting*, Vol. 20: *Membranes Structure and Function*. Symposium VIII (eds J.R. Villanueva and F. Ponz), pp. 9–16. Academic Press, New York.

Finean, J.B. and Martonosi, A. (1965). The action of phospholipase C on muscle microsomes: A correlation of electron microscope and biochemical data. *Biochim. Biophys. Acta* **98**, 547–53.

Fitton, J.E. (1981). Physicochemical studies on delta haemolysin, a staphylococcal cytolytic polypeptide. *FEBS Lett.* **130**, 257–60.

Fitton, J.E., Dell, A. and Shaw, W.V. (1980). The amino acid sequence of the delta haemolysin of *Staphylococcus aureus*. *FEBS Lett.* **115**, 209–12.

Fitton, J.E., Hunt, D.F., Marasco, J., Shabanowitz, J., Winston, S. and Dell, A. (1984). The amino acid sequence of delta haemolysin purified from a canine isolate of *S. aureus*. *FEBS Lett.* **169**, 25–9.

Fraser, G. (1961). Haemolytic activity of *Corynebacterium ovis*. *Nature* **189**, 246.

Fraser, G. (1962). The effect of staphylococcal toxins on ruminant erythrocytes treated with diffusible substances formed by *Corynebacteria*. *Vet. Rec.* **74**, 753–4.

Fraser, G. (1964). The effect on animal erythrocytes of combinations of diffusible substances produced by bacteria. *J. Pathol. Bacteriol.* **88**, 43–53.

Freer, J.H. and Arbuthnott, J.P. (1986). Toxins of *Staphylococcus aureus*. In: *Pharmacology of Bacterial Toxins* (eds F. Dorner and J. Drews), pp. 581–633. Pergamon Press, Oxford.

Freer, J.H. and Arbuthnott, J.P. (1976). Biochemical and morphologic alterations of membranes by bacterial toxins. In: *Mechanisms in Bacterial Toxinology* (ed. A.W. Bernheimer), pp. 169–93. John Wiley, New York.

Freer, J.H. and Birkbeck, T.H. (1982). Possible conformation of delta-lysin, a membrane-damaging peptide of *Staphylococcus aureus*. *J. Theor. Biol.* **94**, 535–40.

Freer, J.H., Arbuthnott, J.P. and Bernheimer, A.W. (1968). Interaction of staphylococcal alpha-toxin with artificial and natural membranes. *J. Bacteriol.* **95**, 1153–68.

Freer, J.H., Birkbeck, T.H. and Bhakoo, M. (1984). Interaction of staphylococcal δ-lysin with phospholipid monolayers and bilayers – a short review. In: *Bacterial Protein Toxins* (eds J.E. Alouf *et al.*), pp. 181–9. Academic Press, London.

Frey, J., Perrin, J. and Nicolet, J. (1989). Cloning and expression of a cohemolysin, the CAMP factor of

Actinobacillus pleuropneumoniae. Infect. Immun. **57**, 2050–6.

Frish, A., Gazitt, Y. and Loyter, A. (1973). Metabolically controlled hemolysis of chicken erythrocytes. *Biochim. Biophys. Acta* **291**, 690–700.

Garutskas, R.S., Glemzha, A.A. and Kulene, V.V. (1977). Isolation and some properties of phospholipase D from *Bacillus subtilis* G-22. *Biokhimiia* **42**, 1910–18.

Gazitt, Y., Ohad, I. and Loyter, A. (1975). Changes in phospholipid susceptibility toward phospholipases induced by ATP depletion in avian and amphibian erythrocyte membranes. *Biochim. Biophys. Acta* **382**, 65–72.

Gazitt, Y., Loyter, A., Reichler, Y. and Ohad, I. (1976). Correlation between changes in the membrane organization and susceptibility to phospholipase C attack induced by ATP depletion of rat erythrocytes. *Biochim. Biophys. Acta* **419**, 479–92.

Gerasimene, G.B., Makariunaite, I.u.P., Kulene, V.V., Glemzha, A.A. and Ianulaitene, K.K. (1985). Properties of the phospholipase C from *Bacillus cereus. Prikl. Biokhim. Mikrobiol.* **21**, 184–9.

Gilmore, M.S., Cruz-Rodz, A.L., Leimeister-Wächter, M., Kreft, J. and Goebel, W. (1989). A *Bacillus cereus* cytolytic determinant, cereolysin AB, which comprises the phospholipase C and sphingomyelinase genes: Nucleotide sequence and genetic linkage. *J. Bacteriol.* **171**, 744–53.

Gordesky, S.E. and Marinetti, G.V. (1973). The asymmetric arrangement of phospholipids in the human erythrocyte membrane. *Biochem. Biophys. Res. Commun.* **50**, 1027–31.

Gordon, A.S., Wallach, D.F.H. and Straus, J.H. (1969). The optical activity of plasma membranes and its modification by lysolecithin, phospholipase A and phospholipase C. *Biochim. Biophys. Acta* **183**, 405–16.

Gottlieb, M.H. (1977). The reactivity of human erythrocyte membrane cholesterol with a cholesterol oxidase. *Biochim. Biophys. Acta* **466**, 422–8.

Gubash, S.M. (1978). Synergistic hemolysis phenomenon shown by an alpha-toxin-producing *Clostridium perfringens* and streptococcal CAMP factor in presumptive streptococcal grouping. *J. Clin. Microbiol.* **8**, 480–8.

Gul, S. and Smith, A.D. (1974). Haemolysis of intact human erythrocytes by purified cobra venom phospholipase A2 in the presence of albumin and Ca^{2+}. *Biochim. Biophys. Acta* **367**, 271–81.

Hébert, G.A. and Hancock, G.A. (1985). Synergistic hemolysis exhibited by species of staphylococci. *J. Clin. Microbiol.* **22**, 409–15.

Heeschen, W., Tolle, A. and Zeidler, H. (1967). Zur Klassifizierung der Gattung Streptococcus. *Zbl. Bakt. Hyg. I. Abt. Orig. A* **205**, 250–9.

Hendrickson, H.S., Rustad, D.G., Scattergood, E.M. and Engle, D.E. (1974). The action of phospholipase C and lipase on black film bilayer membranes. *Chem. Phys. Lipids* **13**, 63–70.

Ikezawa, H., Yamanegi, M., Taguchi, R., Miyashita, T. and Ohyabu, T. (1976). Studies on phosphatidyl-inositol phosphodiesterase (phospholipase C type) of *Bacillus cereus.* I. Purification, properties and phosphatase-releasing activity. *Biochim. Biophys. Acta* **450**, 154–64.

Ikezawa, H., Mori, M., Ohyabu, T. and Taguchi, R. (1978). Studies on sphingomyelinase of *Bacillus cereus.* I. Purification and properties. *Biochim. Biophys. Acta* **528**, 247–56.

Ikezawa, H., Nakabayashi, T., Suzuki, K., Nakajima, M., Taguchi, T. and Taguchi, R. (1983). Complete purification of phosphatidylinositol-specific phospholipase C from a strain of *Bacillus thuringiensis. J. Biochem. (Tokyo)* **93**, 1717–19.

Jeljaszewicz, J. (1972). Toxins (hemolysins). In: *The Stapylococci* (ed. J.O. Cohen), pp. 249–80. John Wiley, New York.

Jeljaszewicz, J. and Wadström, T. (1978). In: *Bacterial Toxins and Cell Membranes* (eds J. Jeljaszewicz and T. Wadström), pp. 1–432. Academic Press, New York.

Johnson, R.M. (1975). The kinetics of resealing of washed erythrocyte ghosts. *J. Membr. Biol.* **22**, 231–53.

Jürgens, D., Huser, H., Brunner, H. and Fehrenbach, F.J. (1981). Purification and characterization of *Staphylococcus aureus* lipase. *FEMS Microbiol. Lett.* **12**, 195–9.

Jürgens, D., Shalaby, F.Y.Y.J. and Fehrenbach, F.J. (1985). Purification and characterization of CAMP-factor from *Streptococcus agalactiae* by hydrophobic interaction chromatography and chromatofocusing. *J. Chromatogr.* **348**, 363–70.

Jürgens, D., Sterzik, B. and Fehrenbach, F.J. (1987). Unspecific binding of group B streptococcal cocytolysin (CAMP factor) to immunoglobulins and its possible role in pathogenicity. *J. Exp. Med.* **165**, 720–32.

Kaiser, E.T. and Kézdy, F.J. (1987). Peptides with affinity for membranes. *Ann. Rev. Biophys. Biophys. Chem.* **16**, 561–81.

Kar Choudhury, T.K. (1978). Synergistic lysis of erythrocytes by *Propionibacterium acnes. J. Clin. Microbiol.* **8**, 238–41.

Kennedy, M., Krag, S.S. and Lennarz, W.J. (1974). Localization of a phospholipase and its inhibitor in *Bacillus subtilis. Fedn. Proc.* **33**, (Part 2), 1469.

Kent, C. and Lennarz, W.J. (1972). An osmotically fragile mutant of *Bacillus subtilis* with an active membrane-associated phospholipase A_1. *Proc. Natl. Acad. Sci. USA* **69**, 2793–7.

Kent, C., Krag, S.S. and Lennarz, W.J. (1973). Procedure for the isolation of mutants of *Bacillus subtilis* with defective cytoplasmic membranes. *J. Bacteriol.* **113**, 874–83.

Kleck, L.J. and Donahue, J.A. (1968). Production of thermostable hemolysin by cultures of *Staphylococcus epidermidis. J. Infect. Dis.* **118**, 317–23.

Köhler, W. (1988). CAMP-like phenomena of Vibrios. *Zbl. Bakt. Hyg. A* **270**, 35–40.

Kötting, J., Jürgens, D., Schiller, R. and Fehrenbach,

F.J. (1985). Biochemical and biological properties of *Staphylococcus aureus* lipases (E.C. 3.1.1.3). In: *The Staphylococci* (ed. J. Jeljaszewicz), pp. 301–9. Gustav Fischer, Stuttgart.

Kötting, J., Eibl, H. and Fehrenbach, F.J. (1988). Substrate specificity of *Staphylococcus aureus* (TEN5) lipases with isomeric oleoyl-sn-glycerol ethers as substrates. *Chem. Phys. Lipids* **47**, 117–22.

Krag, S.S. and Lennarz, W.J. (1975). Purification and characterization of an inhibitor of phospholipase A$_1$ in *Bacillus subtilis. J. Biol. Chem.* **250**, 2813–22.

Kreger, A.S., Kim, K.S., Zaboretzky, F. and Bernheimer, A.W. (1971). Purification and properties of staphylococcal delta hemolysin. *Infect. Immun.* **3**, 449–65.

Kreger, A.S., Bernheimer, A.W., Etkin, L.A. and Daniel, L.W. (1987). Phospholipase D activity of *Vibrio damsela* cytolysin and its interaction with sheep erythrocytes. *Infect. Immun.* **55**, 3209–12.

Lange, Y., Matthies, H. and Steck, T.L. (1984). Cholesterol oxidase susceptibility of the red cell membrane. *Biochim. Biophys. Acta* **769**, 551–62.

Laster, Y., Sabban, E. and Loyter, A. (1972). Susceptibility of membrane phospholipids in erythrocyte ghosts to phospholipase C and their refractiveness in the intact cell. *FEBS Lett.* **20**, 307–10.

Lenard, J. and Singer, S.J. (1968). Structure of membranes: Reaction of red blood cell membranes with phospholipase C. *Science* **159**, 738–9.

Lesmana, M. and Rockhill, R.C. (1985). A CAMP phenomenon between *Vibrio cholerae* biotype El Tor and staphylococcal β-hemolysin. *Southeast Asian J. Trop. Med. Public Health* **16**, 261–4.

Linder, R. (1984). Alteration of mammalian membranes by the cooperative and antagonistic actions of bacterial proteins. *Biochim. Biophys. Acta* **779**, 423–35.

Linder, R. and Bernheimer, A.W. (1978). Effect on sphingomyelin-containing liposomes of phospholipase D from *Corynebacterium ovis* and the cytolysin from *Stoichactis helianthus. Biochim. Biophys. Acta* **530**, 236–46.

Linder, R. and Bernheimer, A.W. (1982). Enzymatic oxidation of membrane cholesterol in relation to lysis of sheep erythrocytes by corynebacterial enzymes. *Arch. Biochem. Biophys.* **213**, 395–404.

Linder, R. and Bernheimer, A.W. (1984). Action of bacterial cytotoxins on normal mammalian cells and cells with altered membrane lipid composition. *Toxicon* **22**, 641–51.

Little, C. (1981). Phospholipase C from *Bacillus cereus* EC 3.1.4.3 Phosphatidylcholine cholinephosphohydrolase. *Methods Enzymol.* **71**, 725–30.

Little, C. and Otnaess, A.B. (1975). The metal ion dependence of phospholipase C from *Bacillus cereus. Biochim. Biophys. Acta* **391**, 326–33.

Little, C., Aurebekk, B. and Otnaess, A.B. (1975). Purification by affinity chromatography of phospholipase C from *Bacillus cereus. FEBS Lett.* **52**, 175–9.

Low, D.K.R., Freer, J.H., Arbuthnott, J.P., Möllby, R. and Wadström, T. (1974). Consequences of

sphingomyelin degradation in erythrocyte ghost membranes by staphylococcal beta-toxin (sphingomyelinase C). *Toxicon* **12**, 279–85.

Low, M.G. and Finean, J.B. (1976). The action of phosphatidylinositol-specific phospholipases C on membranes. *Biochem. J.* **154**, 203–8.

Low, M.G., Limbrick, A.R. and Finean, J.B. (1973). Phospholipase C *Bacillus cereus* acts only at the inner surface of the erythrocyte membrane. *FEBS Lett.* **34**, 1–4.

Macfarlane, M.G. and Knight, B.C.J.G. (1941). The biochemistry of bacterial toxins. I. The lecithinase activity of *Cl. welchii* toxins. *Biochem. J.* **35**, 884–902.

McGuinness, E.T., Brown, H.D., Chattopadhyay, S.K. and Chen, F. (1978). Cholesterol oxidase: Thermochemical studies and the influence of hydroorganic solvents on enzyme activity. *Biochim. Biophys. Acta* **530**, 247–57.

Marks, J. (1952). Recognition of pathogenic staphylococci with notes on non-specific staphylococcal haemolysin. *J. Pathol. Bacteriol.* **64**, 175–86.

Marks, J. and Vaughan, A.C.T. (1950). Staphylococcal δ-haemolysin. *J. Pathol. Bacteriol.* **62**, 597–615.

Miller, I.R. and Ruysschaert, J.M. (1971). Enzymatic activity and surface inactivation of phospholipase C at the water–air interface. *J. Colloid Interface Sci.* **35**, 340–5.

Misaki, H. and Matsumoto, M. (1978). Purification of lysophospholipase of *Vibrio parahaemolyticus* and its properties. *J. Biochem. (Tokyo)* **83**, 1395–405.

Möllby, R. (1978). Bacterial phospholipases. In: *Bacterial Toxins and Cell Membranes* (eds T. Wadström and J. Jeljaszewicz), pp. 367–424. Academic Press, New York.

Möllby, R., Wadström, T., Smyth, C.J. and Thelestam, M. (1974). The interaction of phospholipase C from *Staphylococcus aureus* and *Clostridium perfringens* with cell membranes. *J. Hyg. Epidemiol. Microbiol. Immunol.* **3**, 259–70.

Montfoort, A. and Boere, W.A.M. (1978). Cholesterol and phospholipid composition of erythroblasts isolated from mouse spleen after Rauscher Leukemia. *Lipids* **13**, 580–7.

Munch-Petersen, E. (1954). A corynebacterial agent which protects ruminant erythrocytes against staphylococcal β toxin. *Aust. J. Exp. Biol.* **32**, 361–8.

Munch-Petersen, E. and Christie, R. (1947). On the effect of the interaction of staphylococcal β toxin and group-B streptococcal substance on red blood corpuscles and its use as a test for the identification of *Streptococcus agalactiae. J. Pathol. Bacteriol.* **59**, 367–71.

Nameroff, M., Trotter, J.A., Keller, J.M. and Munar, E. (1973). Inhibition of cellular differentiation by phospholipase C. I. Effects of the enzyme on myogenesis and chondrogenesis *in vitro. J. Cell Biol.* **58**, 107–18.

Nelson, G.J. (1967). Lipid composition of erythrocytes in various mammalian species. *Biochim. Biophys. Acta* **144**, 221–32.

Nishijima, M., Akamatsu, Y. and Nojima, S. (1974). Purification and properties of a membrane-bound phospholipase A₁ from *Mycobacterium phlei*. *J. Biol. Chem.* **249**, 5658–67.

Nishijima, M., Nakaike, S., Tamori, Y. and Nojima, S. (1977). Detergent-resistant phospholipase A of *Escherichia coli* K-12. Purification and properties. *Eur. J. Biochem.* **73**, 115–24.

Ono, Y. and White, D.C. (1970a). Cardiolipin-specific phospholipase D activity in *Haemophilus parainfluenzae*. *J. Bacteriol.* **103**, 111–15.

Ono, Y. and White, D.C. (1970b). Cardiolipin-specific phospholipase D of *Haemophilus parainfluenzae*. II. Characteristics and possible significance. *J. Bacteriol.* **104**, 712–18.

Onon, O.E. (1979). Purification and partial characterization of the exotoxin of *Corynebacterium ovis*. *Biochem. J.* **177**, 181–6.

Osborn, M.J., Gander, J.E. and Parisi, E. (1972). Mechanism of assembly of the outer membrane of *Salmonella typhimurium*. Site of synthesis of lipopolysaccharide. *J. Biol. Chem.* **247**, 3973–86.

Osborn, M.J., Rick, P.D., Lehmann, V., Rupprecht, E. and Singh, M. (1974). Structure and biogenesis of the cell envelope of gram-negative bacteria. *Ann. N.Y. Acad. Sci.* **235**, 52–65.

Otnaess, A.B. (1980). The hydrolysis of sphingomyelin by phospholipase C from *Bacillus cereus*. *FEBS Lett.* **114**, 202–4.

Otnaess, A.B., Prydz, H., Bjorklid, E. and Berre, A. (1972). Phospholipase C from *Bacillus cereus* and its use in studies of tissue thromboplastin. *Eur. J. Biochem.* **27**, 238–43.

Ottolenghi, A.C. and Bowman, M.H. (1970). Membrane structure morphological and chemical alterations in phospholipase C treated mitochondria and red cell ghosts. *J. Membr. Biol.* **2**, 180–91.

Pastan, I., Macchia, V. and Katzen, R. (1968). A phospholipase specific for sphingomyelin from *Clostridium perfringens*. *J. Biol. Chem.* **243**, 3750–5.

Pritchard, A.E. and Vasil, M.L. (1986). Nucleotide sequence and expression of a phosphate-regulated gene encoding a secreted hemolysin of *Pseudomonas aeruginosa*. *J. Bacteriol.* **167**, 291–8.

Raybin, D.M., Bertsch, L.L. and Kornberg, A. (1972). A phospholipase in *Bacillus megaterium* unique to spores and sporangia. *Biochemistry* **11**, 1754–60.

Reichwein, J., Hugo, F., Roth, M., Sinner, A. and Bhakdi, S. (1987). Quantitative analysis of the binding and oligomerization of staphylococcal alpha-toxin in target erythrocyte membranes. *Infect. Immun.* **55**, 2940–4.

Roelofsen, B., Zwaal, R.F.A., Comfurius, P., Woodward, C.B. and van Deenen, L.L.M. (1971). Action of pure phospholipase A₂ and phospholipase C on human erythrocytes and ghosts. *Biochim. Biophys. Acta* **241**, 925–9.

Rouser, G., Nelson, G.J., Fleischer, S. and Simon, G. (1968). Lipid composition of animal cell membranes, organelles and organs. In: *Biological Membranes, Physical Fact and Function* (ed. D. Chapman), pp. 5–69. Academic Press, London and New York.

Roy, T.E. (1937). The titration of alpha and beta haemolysins in staphylococcal toxin. *J. Immunol.* **33**, 437–69.

Rühlmann, J., Wittmann-Liebold, B., Jürgens, D. and Fehrenbach, F.J. (1988). Complete amino acid sequence of protein B. *FEBS Lett.* **235**, 262–6.

Rühlmann, J., Kruft, V., Wittmann-Liebold, B. and Fehrenbach, F.J. (1989). Sequence similarity between protein B and human apolipoprotein A-IV. *FEBS Lett.* **249**, 151–4.

Sabban, E., Laster, Y. and Loyter, A. (1972). Resolution of the hemolytic and the hydrolytic activities of phospholipase-C preparation from *Clostridium perfringens*. *Eur. J. Biochem.* **28**, 373–80.

Scandella, C.J. and Kornberg, A. (1971). A membrane-bound phospholipase A₁ purified from *Escherichia coli*. *Biochemistry* **10**, 4447–56.

Schiffer, M. and Edmundson, B.M. (1967). Use of helical wheels to represent the structures of proteins and to identify segments with helical potential. *Biophys. J.* **7**, 121–35.

Schneewind, O., Friedrich, K. and Lütticken, R. (1988). Cloning and expression of the CAMP factor of group B streptococci in *Escherichia coli*. *Infect. Immun.* **56**, 2174–9.

Schulz, G.E. and Schirmer, R.H. (1979). Patterns of folding and association of polypeptide chains. In: *Principles of Protein Structure* (ed. C.R. Cantor), pp. 69–70. Springer Verlag, New York.

Schwoch, G. and Passow, H. (1973). Preparation and properties of human erythrocyte ghosts. *Molec. Cell. Biochem.* **2**, 197–218.

Skalka, B. and Smola, J. (1981). Lethal effect of CAMP-factor and UBERIS-factor – a new finding about diffusible exosubstances of *Streptococcus agalactiae* and *Streptococcus uberis*. *Zbl. Bakt. Hyg. A* **249**, 190–4.

Skalka, B. and Smola, J. (1982). Hemolytic properties of exosubstance of serovar 5 *Listeria monocytogenes* compared with beta toxin of *Staphylococcus aureus*. *Zbl. Bakt. Hyg. A* **252**, 17–25.

Skalka, B., Smola, J. and Pillich, J. (1979). Diagnostic utilization of hemolytically active exosubstances of certain gram-positive bacteria. I. Detection of staphylococcal hemolysins with prepurified preparations of staphylococcal beta-toxin and CAMP-factor of *Streptococcus agalactiae*. *J. Hyg. Epidemiol. Microbiol. Immunol.* **23**, 407–16.

Skalka, B., Smola, J. and Pillich, J. (1980). Comparison of some properties of the CAMP-factor from *Streptococcus agalactiae* with the haemolytically latent active exosubstance from *Streptococcus uberis*. *Zentralbl. Veterinärmed. (B)* **27**, 559–66.

Smith, D.C., Folz, V.D. and Lord, T.H. (1964). Demonstration of induced synergistic hemolysis by 'nonhemolytic' *Staphylococcus* species. *J. Bacteriol.* **87**, 188–95.

Smith, J.A. and Ngui-Yen, J.H. (1980). Augmentation

of clostridial partial hemolysis by some bacterial species. *Can. J. Microbiol.* **26**, 839–43.

Smith, M.L. and Price, S.A. (1938). *Staphylococcus* β haemolysin. *J. Pathol. Bacteriol.* **47**, 361–77.

Smyth, C.J., Möllby, R. and Wadström, T. (1975). Phenomenon of hot–cold hemolysis: Chelator-induced lysis of sphingemyelinase-treated erythrocytes. *Infect. Immun.* **12**, 1104–11.

Souček, A. and Součková, A. (1974). Toxicity of bacterial sphingomyelinase D. *J. Hyg. Epidemiol. Microbiol. Immunol.* **18**, 327–35.

Souček, A., Michalec, C. and Součková, A. (1967a). Enzymic hydrolysis of sphingomyelins by a toxin of *Corynebacterium ovis*. *Biochim. Biophys. Acta* **144**, 180–2.

Souček, A., Součková, A. and Patočka, F. (1967b). Inhibition of the activity of alpha-toxin of *Clostridium perfringens* by toxic filtrates of *Corynebacteria*. *J. Hyg. Epidemiol. Microbiol. Immunol.* **11**, 123–4.

Souček, A., Michaelec, C. and Součková A. (1971). Identification and characterization of a new enzyme of the group 'phospholipase D' isolated from *Corynebacterium ovis*. *Biochim. Biophys. Acta* **227**, 116–28.

Součková, A. and Souček, A. (1972). Inhibition of the hemolytic action of α and β lysins of *Staphylococcus pyogenes* by *Corynebacterium hemolyticum*, *C. ovis* and *C. ulcerans*. *Toxicon* **10**, 501–9.

Steck, T.L. and Kant, J.A. (1974). Preparation of impermeable ghosts and inside-out vesicles from human erythrocyte membranes. *Methods Enzymol.* **31**, 172–80.

Sterzik, B. (1988). Untersuchungen zum Mechanismus der CAMP-Reaktion. Dissertation Freie Universität Berlin, 1988.

Sterzik, B. and Fehrenbach, F.J. (1985). Reaction components influencing CAMP factor induced lysis. *J. Gen. Microbiol.* **131**, 817–20.

Sterzik, B., Jürgens, D. and Fehrenbach, F.J. (1986). Structure and function of CAMP factor of *Streptococcus agalactiae*. In: *Bacterial Protein Toxins* (eds P. Falmagne *et al.*), pp. 101–8. Gustav Fischer, Stuttgart.

Sterzik, B., Jürgens, D., Montecucco, C. and Fehrenbach, F.J. (1989). Interaction of protein B (CAMP-factor) with artificial membranes as studied by hydrophobic photolabeling. In: *Bacterial Protein Toxins*, Proceedings of the 4th European Workshop, Urbino/Italy, (eds R. Rappuoli *et al.*), pp. 197–8. Gustav Fischer, Stuttgart.

Stinson, M.W. and Hayden, C. (1979). Secretion of phospholipase C by *Pseudomonas aeruginosa*. *Infect. Immun.* **25**, 558–64.

Taguchi, R. and Ikezawa, H. (1975). Phospholipase C from *Clostridium novyi* type A$_1$. *Biochim. Biophys. Acta* **409**, 75–85.

Taguchi, R. and Ikezawa, H. (1978). Phosphatidyl inositol-specific phospholipase C from *Clostridium novyi* type A. *Arch. Biochem. Biophys.* **186**, 196–201.

Taguchi, R., Asahi, Y. and Ikezawa, H. (1980). Purification and properties of phosphatidylinositol-

specific phospholipase C of *Bacillus thuringiensis*. *Biochim. Biophys. Acta* **619**, 48–57.

Takahashi, T., Sugahara, T. and Ohsaka, A. (1974a). Purification of *Clostridium perfringens* phospholipase C (alpha-toxin) by affinity chromatography on agarose-linked egg-yolk lipoprotein. *Biochim. Biophys. Acta* **351**, 155–71.

Takahashi, T., Sugahara, T. and Ohsaka, A. (1974b). Purification of alpha toxin phospholipase C of *Clostridium perfringens* by affinity chromatography. *Jpn. J. Med. Sci. Biol.* **27**, 89–92.

Tappin, M.J., Pastore, A., Norton, R.S., Freer, J.H. and Campbell, I.D. (1988). High-resolution 1H NMR study of the solution structure of delta-hemolysin. *Biochemistry* **27**, 1643–7.

Thelestam, M. and Möllby, R. (1975a). Determination of toxin-induced leakage of different-size nucleotides through the plasma membrane of human diploid fibroblasts. *Infect. Immun.* **11**, 640–8.

Thelestam, M. and Möllby, R. (1975b). Sensitive assay for detection of toxin-induced damage to the cytoplasmic membrane of human diploid fibroblasts. *Infect. Immun.* **12**, 225–32.

Thelestam, M. and Möllby, R. (1979). Classification of microbial, plant and animal cytolysins based on their membrane-damaging effects of human fibroblasts. *Biochim. Biophys. Acta* **557**, 156–69.

Thiele, O.W. (1979). *Lipide, Isoprenoide mit Steroiden*, p. 164. Georg Thieme Verlag, Stuttgart.

Thurnhofer, H., Gains, N., Mütsch, G. and Hauser, H. (1986). Cholesterol oxidase as a structural probe of biological membranes: Its application to brush-border membrane. *Biochim. Biophys. Acta* **856**, 174–81.

Tomita, M., Taguchi, R. and Ikezawa, H. (1982). Molecular properties and kinetic studies on sphingomyelinase of *Bacillus cereus*. *Biochim. Biophys. Acta* **704**, 90–9.

Tomita, M., Taguchi, R. and Ikezawa, H. (1983a). The action of sphingomyelinase of *Bacillus cereus* on bovine erythrocyte membrane and liposomes. Specific adsorption onto these membranes. *J. Biochem.* **93**, 1221–30.

Tomita, M., Taguchi, R. and Ikezawa, H. (1983b). Adsorption of sphingomyelinase of *Bacillus cereus* onto erythrocyte membranes. *Arch. Biochem. Biophys.* **223**, 202–12.

Tomita, M., Sawada, H., Taguchi, R. and Ikezawa, H. (1987). The action of sphingomyelinase from *Bacillus cereus* on ATP-depleted bovine erythrocyte membranes and different lipid composition of liposomes. *Arch. Biochem. Biophys.* **255**, 127–35.

van Deenen, L.L.M. (1981). Topology and dynamics of phospholipids in membranes. *FEBS Lett.* **123**, 3–15.

van Deenen, L.L.M., Demel, R.A., Guerts van Kessel, W.S.M., Kamp, H.H., Roelofsen, B., Verkleij, A.J., Wirtz, K.W.A. and Zwaal, R.F.A. (1976). Phospholipases and monolayers as tools in studies of membrane structure. In: *The Structural Basis of*

Membrane Function (eds Y. Hatefi and L. Djavadi-Ohaniance), pp. 21–38. Academic Press, New York.

Verger, R. and de Haas, G.M. (1973). Enzyme reactions in a membrane model. Part 1: A new technique to study enzyme reactions in monolayers. *Chem. Phys. Lipids* **10**, 127–36.

Verkleij, A.J., Zwaal, R.F.A., Roelofsen, B., Comfurius, P., Kastelijn, D. and van Deenen, L.L.M. (1973). The asymmetric distribution of phospholipids in the human red cell membrane. A combined study using phospholipases and freeze etch electron microscopy. *Biochim. Biophys. Acta* **323**, 178–93.

Wadström, T. and Möllby, R. (1971a). Studies on extracellular proteins from *Staphylococcus aureus*. VI. Production and purification of β haemolysin in large scale. *Biochim. Biophys. Acta* **242**, 288–307.

Wadström, T. and Möllby, R. (1971b). Studies on extracellular proteins from *Staphylococcus aureus*. VII. Studies on β haemolysin. *Biochim. Biophys. Acta* **242**, 308–20.

Walbum, L.E. (1922). Studien über die Bildung der bakteriellen Toxine. *Biochem. Z.* **129**, 367–443.

Watanabe, M., Tomita, T. and Yasuda, T. (1987). Membrane-damaging action of staphylococcal alpha-toxin on phospholipid-cholesterol liposomes. *Biochim. Biophys. Acta* **898**, 257–65.

Wilbrandt, W. (1942). Osmotische Natur sogenannter nichtosmotischer Hämolysen (Kolloidosmotische Hämolyse). *Pflügers' Arch. ges. Physiol.* **245**, 22–52.

Williams, R.E.O. and Harper, G.J. (1947). Staphylococcal haemolysins on sheep-blood agar with evidence for fourth haemolysin. *J. Pathol. Bacteriol.* **59**, 69–78.

Wiseman, G.M. (1965). Factors affecting the sensitization of sheep erythrocytes to staphylococcal beta lysin. *Can. J. Microbiol.* **11**, 463–71.

Wiseman, G.M. and Caird, J.D. (1967). The nature of staphylococcal beta hemolysin. I. Mode of action.

Can. J. Microbiol. **13**, 369–76.

Woodward, C.B. and Zwaal, R.F.A. (1972). The lytic behavior of pure phospholipase A₂ EC 3.1.1.4 and phospholipase C EC 3.1.4.3 towards osmotically swollen erythrocytes and resealed ghosts. *Biochim. Biophys. Acta* **274**, 272–8.

Yamaguchi, T. and Morishita, M. (1969). The role of *Bacillus megaterium* in the cell membrane. *Proc. Ann. Meeting Agr. Chem. Soc. Japan*, p. 262–3.

Yamakawa, Y. and Ohsaka, A. (1977). Purification and some properties of phospholipase C (alpha-toxin) of *Clostridium perfringens*. *J. Biochem. (Tokyo)* **81**, 115–26.

Zaki, M.M. (1965). Relation between staphylococcal beta-lysin and different *Corynebacteria*. *Vet. Rec.* **77**, 941.

Zucker-Franklin, D. (1965). Electron microscope study of the degradation of polymorphonuclear leukocytes following treatment with streptolysin. *Am. J. Pathol.* **47**, 419–33.

Zwaal, R.F.A. and Roelofsen, B. (1975). Phospholipase C (phosphatidylcholine cholinephosphohydrolase, EC 3.1.4.3) from *Bacillus cereus*. *Methods Enzymol.* **32**, 154–61.

Zwaal, R.F.A., Roelofsen, B., Comfurius, P. and van Deenen, L.L.M. (1971). Complete purification and some properties of phospholipase C from *Bacillus cereus*. *Biochim. Biophys. Acta* **233**, 474–9.

Zwaal, R.F.A., Roelofsen, B. and Colley, C.M. (1973). Localization of red cell membrane constituents. *Biochim. Biophys. Acta* **300**, 159–82.

Zwaal, R.F.A., Roelofsen, B., Comfurius, P. and van Deenen, L.L.M. (1975). Organization of phospholipids in human red cell membranes as detected by the action of various purified phospholipases. *Biochim. Biophys. Acta* **406**, 83–96.

10

Electrophysiological Methods for the Study of Toxin–Membrane Interaction

Gianfranco Menestrina

Dipartimento di Fisica, Università di Trento, I-38050 Povo, Trento, Italy

Introduction

A major aspect of current studies in electrophysiology involves investigation of the ionic channels present in the cell membrane. In this respect there are two ways in which a bacterial protein toxin can interfere with the electrical properties of a cell: (a) the toxin interacts with ionic channels already present in the cell membrane (effect on endogenous pores) and (b) the toxin itself forms an ionic channel in the cell membrane (formation of exogenous pores). Toxins of the first group are usually called neurotoxins whereas toxins of the second group are usually referred to as cytolysins. In the first part of this chapter I will review the experimental techniques which are used in modern electrophysiology to study the permeability properties of cells. Data obtained in recent years on the electrophysiological effects of bacterial toxins will be reviewed in the second part.

Electrophysiological methods

The first electrophysiological studies of biological tissues date back to the end of the last century (Nernst, 1888) but it was only about 40 years ago with the pioneering work of Hodgkin, Huxley and Katz that classical electrophysiology was established (Katz, 1949; Hodgkin and Huxley, 1952a,b). It was devoted to the study of large excitable cells (e.g. squid giant axon) and relied on the development of the voltage-clamp technique.

At that time it became clear that electrical signals travelling between cells, action potentials and resting potentials, had a basis in the ionic permeability of the cell membrane and its changes. The idea of ionic pores or channels soon developed but only later was it realized that these could be membrane-spanning proteins. The first indirect proof of the existence of discrete ionic pores came from noise analysis experiments (Stevens, 1972).

Model systems such as vesicles and planar bilayers were then introduced and it was with such systems that the first proof was obtained that a proteic polypeptide can insert into a lipid bilayer and create an ion-conducting pore (Hladky and Haydon, 1970).

A clear demonstration of the existence of such proteic ionic pores in natural membranes had to await the development of the patch-clamp technique (Neher and Sakmann, 1976). Nowadays a combination of patch electrophysiology on native or host cells, reconstitution in model systems, chemical and genetical manipulation and immunological approaches has provided a successful strategy for the study of endogenous and exogenous pore-forming proteins.

In the first part of this chapter I will provide only a comparatively small amount of reference material, choosing publications which are more general, complete and user-friendly. For a more detailed description of most of the methods presented here the interested reader is referred to some excellent recently published books (Sakmann and Neher, 1983; Hille, 1984; Latorre, 1986; Miller, 1986; Skulachev, 1988).

Electrophysiological studies on cells

Electrophysiology was developed to clarify the functioning of excitable cells and, more recently, its molecular basis. Only in the last few years have these methods been extended to the study of other exogenous 'membrane-seeking' proteins, such as bacterial toxins. Relatively few of the methods which will be reviewed in the following sections have been successfully applied to the action of bacterial toxins. Nevertheless, because of the tremendous wealth of information that these methods can provide, it is clear that the electrophysiological approach will become increasingly important in future studies.

Voltage clamp

This method, aimed at measuring the current flowing through the cell membrane while keeping constant the voltage drop across it, has been the best technique for the study of ionic channels for several decades. It was originally developed by Hodgkin, Huxley and Katz (Katz, 1949; Hodgkin and Huxley, 1952a) and subsequently improved by several researchers, for example see Deck et al. (1964) and Chandler and Meves (1965).

To clamp the transmembrane voltage of a cell at a desired value is not a simple task because of the unpredictable and large voltage drop which develops across the high-impedance electrodes that have to be used to gain access to its interior. The problem was solved using a feedback amplifier which injects into the cell, at any time, exactly the amount of current that is necessary to maintain the voltage at the desired value. A simplified scheme of this apparatus is shown in Fig. 1A.

Typically, in response to a voltage jump, transient currents in the range of nanoamperes (lasting from tenths to hundreds of milliseconds) can be satisfactorily detected by this technique. This current sensitivity, although quite respectable, is inadequate for the resolution of the individual contribution of a single ion channel to the recorded current. Hence, the existence of such discrete entities was based on the intuition of those early researchers (Hodgkin and Huxley, 1952b).

Noise analysis

Historically, noise analysis was the first technique which provided an experimental basis for the hypothesis that the electrical currents in excitable cells were due to the presence of specialized membrane proteins, i.e. ionic channels (Stevens, 1972). It relied on a simple consideration (Neher and Stevens, 1977; DeFelice, 1981; Conti and Wanke, 1975), namely that because ion channels are discrete molecules which open and close stochastically, the number of open channels in any voltage-clamped cell fluctuates, even at equilibrium. Such fluctuations provide an extra noise in the current traces which can be readily distinguished from common sources of electrical noise such as thermal noise (or Johnson), shot noise and $1/f$ noise (or flicker).

In such conditions the conductance of the channel may be estimated from a simultaneous measurement of the mean current and its mean square deviation. Derivation of these values implies the use of some mathematical tools such as the covariance function or the spectral (Fourier) analysis. Interpretation of the results also relies on a number of assumptions (not all of which are always warranted) such as: channels are all equal and independent; the number of channels under observation does not change with time; each channel has only two possible states: open and closed; only the open state is conductive, etc.

Despite its apparent complexity, this technique was successfully applied to most of the channels of physiological importance and provided values for their conductance with surprising accuracy long before they could be measured directly (Colquhoun et al., 1975; Conti et al., 1975; Neher and Stevens, 1977). The reason for this success lies in two favourable circumstances: (a) the number of copies of the ion channel under investigation in each preparation is intrinsically limited to a reasonably small number (less than 10^6) by the physical size of the cell; the amplitude of the fluctuations is thus expected to be at least 0.1% of the steady-state signal (in fact with N fluctuating channels the variance is proportional to $\sqrt{N}$); and (b) each channel, switching a current flow of typically 10^7 ions/s at the expense of the electrochemical gradient, and ultimately of the energy stored in the voltage generator, behaves as a highly efficient amplifier.

Both these factors contribute towards making the current fluctuations large enough to be safely detected above the instrumental noise level in a classical voltage-clamp experiment. The same two reasons, after all, are those which made possible the development of a technique for the direct detection of single-channel currents in cells (known as the patch clamp).

Patch clamp

Early in 1976 Neher and Sakmann presented a new technique designed to elucidate the discrete events involved in the formation of the macroscopic current flowing through a cell (Neher and Sakmann, 1976). They used a glass pipette with a fire-polished tip of a few micrometres in diameter to electrically isolate a piece of plasma membrane from the rest of the cell by pressing the pipette on its surface.

A few years later this method was refined by the same group to improve resolution and achieve what was called the giga-seal (Hamill *et al.*, 1981). The patch clamp takes advantage of the relatively low density of the ionic channels on the cell membrane (typically ranging from 10^{-2} to 10^3 channels/μm^2) which is a consequence of their high efficiency in moving ions through the membrane. Only with such low density (compared, for example, to that of rhodopsin which is present as 5×10^4 molecules/μm^2 in vertebrate rod disks), it was realistic to try the isolation of a few copies of the ion channel of interest within a pipette tip of 1 μm (which approaches the lower limit achievable because of mechanical and electrical problems). A key contribution came from the semiconductor industry which at that time optimized an amplifier (the field-effect transistor) with femtoampere bias current and very low input voltage noise. Current measurements with less than 0.1 pA noise at 2 kHz bandwidth thus became possible.

It should be noted that the achievement of such high resolution of current during the experiment depends solely on the small physical dimensions of the patch isolated by the pipette tip. In fact, this limits the contribution of the cell membrane to the input capacitance to less than 0.1 pF (specific membrane capacitance is 0.8 $\mu F/cm^2$), a value which is well below that of the pipette itself and of the amplifier intrinsic input capacitance (both are typically in the range of 1 pF). It is actually the input capacitance which determines both the bandwidth and the current noise under experimental conditions (Alvarez, 1986).

Single-channel recordings of the acetylcholine-operated channel (the most abundant in excitable tissues) showing rectangular-shaped current fluctuations were first produced by the new technique (Neher and Sakmann, 1976; Sakmann *et al.*, 1980); the sodium channel (Sigworth and Neher, 1980) and potassium-delayed rectifier (Conti and Neher, 1980) soon came. Following these studies almost all physiologically relevant endogenous channels have been investigated.

The advantage of extending this powerful technique to the study of exogenous channels (like those formed by many bacterial toxins) is obvious. It offers a means of investigating the mechanism of cell attack at a molecular level directly on target cells. Actually, a patch-clamp study of the channel-forming effects of the terminal complex of the complement cascade (Jackson *et al.*, 1981) appeared early in 1981. Complement contributes to the humoral immune defence of mammals by an attack mechanism directed against invading cells which is quite similar to the action of many bacterial cytotoxins (Bhakdi and Tranum-Jensen, 1984, 1987). Despite this early report relatively few investigations were done on bacterial toxins. Penner *et al.* (1986) showed with a related technique that tetanus toxin inhibits cellular exocytosis in bovine adrenal medullary chromaffin cells, and quite recently a preliminary study (Dreyer *et al.*, 1990) demonstrated the formation of ionic channels by a cytotoxin from *Pseudomonas aeruginosa* (PACT) in such cultured cells.

The reasons for the paucity of data involving bacterial toxins probably reflects a number of drawbacks of this technique, namely (a) if the toxin is applied extracellularly before patching the cell, the probability of isolating a patch bearing a toxin lesion within the pipette tip is very small (unless non-physiological high concentrations of toxin are used); (b) if the toxin is applied through the pipette, after having already patched the cell, the membrane patch may be destabilized and the seal between the tip and the lipid deteriorated; (c) in most cases all the endogenous channels of the cell have to be blocked to detect those formed by the toxin.

Once these problems, which are not insurmountable, are solved the patch-clamp technique could be applied effectively in research on bacterial

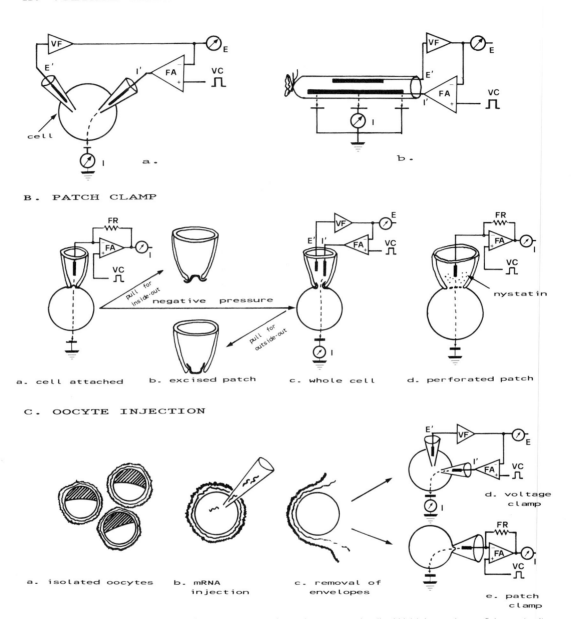

Figure I. Electrophysiological methods for studying ion channels on natural cells. (A) Voltage clamp. Schematic diagram of the voltage clamp apparatus applied either to a small cell (a) or to an excised axon (b). Two microelectrodes are inserted into the cell, a voltage-recording electrode E' and a current-delivering electrode I'. The cell potential E is recorded through a unitary-gain voltage-follower (VF) of high input impedance, and compared at any time to the command voltage pulse VC. The feedback amplifier (FA) injects into the cell exactly the amount of current which is necessary to maintain the voltage at the desired value. The extracellular medium is grounded and the current I flowing to ground is continuously recorded (the route of the current is indicated by a dashed line). (B) Patch clamp. (a) A glass micropipette is pressed on the surface of a cell. A tight seal develops which isolates electrically a piece of plasma membrane from the rest of the cell (cell-attached patch). Because the impedance of the microelectrode is small compared to that of the patch, a single electrode may be used in this case both to inject the current and to monitor

toxins. A few typical configurations used in this approach are reported in Fig. 1B and will be briefly described below.

Cell attached path. This is the classical configuration. The patch is studied in-place on an intact cell. The reference electrode is in the external solution and the method relies on the fact that the permeability of the cell membrane outside the patch is so large (compared to that of the small patch) that it can be safely considered a short circuit.

Excised patch (inside-out or outside-out). Pulling away the pipette from the cell after the giga-seal is achieved (with or without breaking the patch first), produces an outside-out or inside-out excised patch respectively (Fig. 1B). These may be conveniently used to study membrane channels with either their periplasmic or cytoplasmic face exposed to a medium in which all the electrochemical parameters can be controlled and varied at any time. Methods have been developed recently which allow abrupt changes in media (within milliseconds), thereby allowing study of the kinetics of chemical or pharmacological control of channels directly on single molecules (Qin and Noma, 1988; Maconochie and Knight, 1989).

Whole-cell recording. After the giga-seal is established the patch is broken (either by negative pressure, or by a high-voltage pulse or otherwise). A low-resistance access to the interior of the cell is thus gained. The cytoplasmic medium can now be changed but the main problem is that it cannot be prevented from changing, and undesirable

dilution of cell contents into the patching pipette solution occurs. Despite this drawback the method has been used successfully to study a variety of endogenous channels as well as a bacterial pore-forming toxin (PACT) (Dreyer *et al.*, 1990).

Perforated patch. A variant of the previous method is now becoming increasingly popular (Horn and Marty, 1988). In this case the patch is not broken but rather made very permeable by using the channel-forming antibiotic nystatin. Electrical access is gained but (due to the small molecular weight cut-off of the nystatin pores) internal proteins are prevented from escaping from the cell. This method is relevant to the study of bacterial toxins in at least two ways. First, the effects of toxins with an intracellular protein target (e.g. diphtheria toxin or *Pseudomonas* exotoxin A) can be conveniently studied in this way. Second, pore-forming toxins can themselves be used to permeabilize the patch, providing a choice of different pore diameters and characteristics (see Table 2).

Expression in oocytes

Recently Barnard *et al.* (1982) demonstrated that the acetylcholine (ACh) channel can be expressed on the outer membrane of *Xenopus* oocytes injected with mRNA encoding for this protein. Since that time most of the endogenous channels of excitable tissues have been successfully expressed in oocytes in this way, and the resulting voltage- or ligand-controlled pores have been demonstrated by electrophysiological methods, i.e. voltage- or patch-

Figure I. Continued. the voltage. The feedback amplifier converts current to voltage through a feedback resistor FR (ranging from $10^8 \Omega$ to $10^{10} \Omega$). (b) Pulling away the pipette from the cell after the tight seal is achieved, permits the study of an excised patch of membrane. The excised patch is either inside-out or outside-out depending on whether the starting configuration was the cell-attached or the whole cell respectively. (c) By applying a pulse of negative pressure to the pipette the patch is broken whereas the seal is preserved. Recording currents flowing through the whole cell is thus possible. Because of the low cell resistance a voltage-clamp configuration with two microelectrodes (both in the same pipette) is better suited in this case. (d) In an alternative to the classical whole-cell configuration electrical access to the interior of the cell is gained by permeabilizing the patch with nystatin, a pore-forming antibiotic (perforated patch configuration). Because of the small molecular weight cut-off of the nystatin pores, internal proteins and macroions are prevented from escaping from the cell, and thus help to preserve the original 'milieu interieur'. (C) Oocyte injection for expression. (a) Mature *Xenopus* oocytes are dissected from the ovary and stored in sterile medium. (b) Through a glass micropipette the oocyte is injected with the mRNA encoding for a protein, e.g. an ion channel. (c) After a few days the follicular and vitelline envelopes are removed by enzymic cleavage and osmotic shrinkage respectively. Channels expressed by the oocyte on its plasma membrane are studied either by conventional voltage clamp (d) or by patch-clamp (e).

clamp methods (Miledi *et al.*, 1982; Methfessel *et al.*, 1986; Dascal, 1987; Snuthc, 1988) (Fig. 1C).

This technique combines the advantage of expressing the channel on a big cell like the oocyte (which is easy to manipulate) with the possibility of injecting mRNAs encoding either for single subunits or for a number of different subunits (even taken from different species) of the same channel. The injection of genetically engineered mRNA has also been successfully performed. The route to clarification of the structure–function relationship is clearly opened.

The applicability of this approach to exogenous channels formed by bacterial toxins is not at all trivial because these proteins are secreted in a water-soluble form and assume their membrane-competent configuration only later.

A potentially interesting application of this technique would be the expression on the oocyte surface of protein receptors for suitable toxins (Table 1), e.g. (among others) colicins (Cramer *et al.*, 1983; Lazdunski *et al.*, 1988) and probably diphtheria toxin (Olsnes *et al.*, 1988). This should make the cell sensitive (if it was not) to the toxin and would provide an instrument to study the structural basis of the toxin–receptor interaction.

Transmembrane potential determination

Endogenous ion channels contribute via their selectivity to the establishment of the internal cell potential, whereas exogenous pores very often contribute in dissipating it. In both cases it may be interesting to measure experimentally this potential. Besides measuring it directly with a microelectrode (e.g. using a voltage-clamp or a whole-cell patch-clamp configuration) a number of alternative methods have been developed (Skulachev, 1988) suitable for cells which are too small to be punched with an electrode (e.g. platelets or bacteria).

These methods are usually based on the tendency of hydrophobic ions, administered extracellularly, to partition into the plasma membrane and the cytoplasm, according to the internal potential of the cell. The distribution of the hydrophobic ion is then determined by one of a variety of methods. For example, the concentration in the external solution may be directly measured with an ion-selective electrode (Eisenbach *et al.*, 1984) or derived by an NMR spectrum (Kirk *et al.*, 1988).

Alternatively, fluorescent molecules can be used (Loew, 1988) with maxima which differ whether they are in a water phase or in a lipid phase (ANS has been widely employed (Slavik, 1982) as well as cyanine (Waggoner, 1979) and styryl dyes (Grinvald *et al.*, 1982)). Oxonol dyes are also very popular (Bashford and Smith, 1979) because they change their absorbance properties when inserted into a membrane.

All these techniques share several advantages. They are non-invasive, they may be used on small cells, and finally they can be applied to a sample containing a large number of cells. Hence the values they provide are averages from a population of cells and thus they are less prone to the artifactual errors which may occur when a single cell is studied. However, they have many serious drawbacks. Since they are based simply on electrostatic attraction, they cannot in general discriminate between internal potential and surface potential created by the fixed charges on the cell membrane (Robertson and Rottenberg, 1983; Eisenbach *et al.*, 1984), nor between the potential of the cell and that of the subcellular organelles it contains. Furthermore, because of their hydrophobic nature, all of these ionic dyes behave like carriers of counterions and thus increase the permeability of the cell. In the worst case they may even be toxic and interfere with cell metabolism.

At least two recent developments are worth mentioning here. (1) Cytofluorimetric techniques can be applied to mixed populations of cells which are simultaneously scrutinized for their internal potential and for their size (Bashford *et al.*, 1986). Two-dimensional plots are obtained from which the individual contributions of the different cell populations are easily recognized. (2) Enhanced video microscopy (Inouè, 1986) allows observation of single cells at high resolution and monitoring spatial distribution and temporal changes of internal cell potential with approximately a millisecond time constant.

Release (uptake) experiments

The presence of ion channels on the membrane of a cell can, of course, be detected by directly measuring the flux of molecules which can cross them. Several methods are currently employed to measure the release (or uptake) of ions, metabolites, neutral soluble molecules or proteins; most of them

Table I. Properties of bacterial toxins forming ion channels in cells and model membranes

Bacterium	Toxin	Physical state	Lipid receptor	Protein receptor	Channel-former	Channel-forming part	Proposed role for channel	Requirements for channel formation[e]	Comments on channel formation
A/B type									
Vibrio cholerae	Cholera toxin	Heterologous aggregate A + 5 B	GM1 (1)	? (1)	Yes	B-promoter (2)	Translocation of part A (2)	—	B-promoter as effective as whole toxin (2)
Escherichia coli	Heat-labile enterotoxin	Heterologous aggregate A + 5 B	GM1 (1)	Glycoprotein? (1)	—	—	—	—	—
Bordetella pertussis	Pertussis toxin	Heterologous aggregate A+B1, B2, 2B3, B4	Gangliosides? (1)	Glycoprotein? (1)	—	—	—	—	—
Shigella dysenteriae	Shigella toxin	Heterologous aggregate A + 5–7 B	? (1)	Glycoprotein? (1)	—	—	—	—	—
Bacillus anthracis	Anthrax lethal toxin	Complex of 3 proteins PA, EF, LF	—	Yes, unknown (3)	Yes	PA, C-terminal 65-kDa fragment (4)	Translocation of EF and LF (4)	$V_{trans} < 0$ pH < 6.8 (4)	Only the fragment is effective (4)
Corynebacterium diphtheriae	Diphtheria toxin	s.c. (dimer)* (5)	Acidic lipids (5) PI (6)	Glycoprotein? (1) Cl-transporter? (7)	Yes	B fragment (5) N-terminal 45-kDa fragment	Translocation of fragment A (8)	$V_{trans} < 0$ pH < 5 (5,8)	Fragment as effective as whole toxin (5,8)
Pseudomonas aeruginosa	Exotoxin A	s.c.	Acidic lipids (9)	Glycoprotein? (1)	Yes	Domain II (10)	Translocation of domain III (10)	pH < 6 (9)	—
Clostridium tetani	Tetanus toxin	s.c.	GD1b > GT1b (1)	Glycoprotein? (1)	Yes	Heavy chain (11,12) N-terminal 50-kDa fragment	Translocation of light chain (11)	$pH_{cis} < 5$ (11) $pH_{trans} = 7$	Fragment as effective as whole toxin (12)
Clostridium botulinum	Botulinum toxins A, B, Cα, Cβ, D, E, F, G	s.c. (dimer?)* (13)	GT1b > GQ1b > GD1b (1)	Glycoprotein? (1)	Yes	Heavy chain (11) N-terminal 50 kDa fragment	Translocation of light chain (11)	$pH_{cis} < 5$ (11) $pH_{trans} = 7$	Fragment more effective then whole toxin (11)
Thiol-activated									
Streptococcus pyogenes	Streptolysin 0[a]	s.c. large aggregate*Cholesterol (15) (14)	—	Yes	w.t.	Colloid-osmotic shock (15)	Thiol-reduction (15)	Formation of rings and arcs in the EM. i.a. 150–450 nm² (14,15)

Table I. Continued

Bacterium	Toxin	Physical state	Lipid receptor	Protein receptor	Channel former	Channel-forming part	Proposed role for channel	Requirements for channel formation[a]	Comments on channel formation[a]
non A/B enterotoxins									
Clostridium perfringens type A	Enterotoxin	s.c. oligomer?* (16)	—	Yes, 50 kDa (17)	Yes	w.t.	Colloid-osmotic shock (17)	—	—
Staphylococcus aureus	Enterotoxin A, B, C1, C2, D, E	s.c.	—	Yes (18)	—	—	—	—	—
Gram-negative haemolysins									
Escherichia coli	Haemolysin[b]	s.c. monomer?* (19,20) aggregate? (21)	Acidic lipids (22)	—	Yes	w.t.	Colloid-osmotic shock (19)	$V_{trans} < 0$ (20)	Extremely labile (19) protease-sensitive (20)
Pseudomonas aeruginosa	Cytotoxin (PACT)	s.c. oligomer?* (23)	Cholesterol (23)	—	Yes	w.t.	Colloid-osmotic shock (23)	—	—
Aeromonas hydrophila	Aerolysin	s.c.	—	Glycoprotein? (24)	Yes	N-terminal part (25)	Colloid-osmotic shock (24)	—	Extracellularly activated by cleavage at C-terminal (25)
Low molecular weight peptides									
Pseudomonas syringae	Syringomycin[c] E, G, A1	s.c. oligomer?* (26)	—	—	yes	w.t.	Collapse of transmembrane pH and voltage gradient (27)	pH < 8 (28)	—
Gram-positive haemolysins									
Staphylococcus aureus	α-toxin	s.c. hexamer* (29, 30)	PC (31)	? (32)	Yes	w.t.	Colloid-osmotic shock (29)	—	Promoted by cholesterol (33) protease-resistant (29)
Clostridium perfringens	δ-toxin	s.c.	GM2 (34)	—	Yes	w.t.	Colloid-osmotic shock (34)	—	Inhibited by 1 mM Zn^{2+} (35)

Organism	Toxin	Membrane-bound active form	Binding requirement	Channel?	Inserted part	Mechanism	Lipid/pH requirement	Comments
Bacillus thuringiensis	δ-endotoxin	s.c. oligomer?* (36)	—	Yes	N-terminal part (36)	Colloid-osmotic shock (36)	pH > 9 (37)	Extracellularly activated by C-terminal half (36)
Low molecular weight peptides								
Staphylococcus aureus	δ-lysin	s.c. hexamer?* (38, 39)	—	Yes	w.t.	Colloid-osmotic shock (38)	Fluid lipid phase (40)	Insertion of amphiphilic α-helices (38,41)
Streptococcus pyogenes	Streptolysin S	s.c. oligomer?* (42)	Phospholipids (43)	—	w.t.	Colloid-osmotic shock (43)	Fluid lipid phase (43)	Promoted by cholesterol (43)
Bacillus subtilis	Iturin A[d]	s.c. oligomer* (44)	Cholesterol (45)	Yes	w.t.	Membrane (44) destabilization	—	Stabilized by cholesterol (46)
Colicins								
Escherichia coli (different strains)	Colicin A, B, EI, K, Ia, Ib	s.c. monomer?* (47) oligomer?* (48)	Acidic lipids (49) Porin (50,51)	Yes	C-terminal 20-kDa fragment (51,52)	Cell (48–50) depolarization	$V_{trans} < 0$ low pH (49)	—
	Colicin E2, E3, M cloacin		Porin (50)	—	—	—	—	—

References: (1) Eidels et al. (1983); (2) Moss et al. (1976b); (3) Singh et al. (1989); (4) Blaustein et al. (1989); (5) Kagan et al. (1981); (6) Donovan et al. (1982); (7) Olsnes et al. (1988); (8) Donovan et al. (1981); (9) Menestrina et al., unpublished observation; (10) Hwang et al. (1987); (11) Hoch et al. (1985); (12) Boquet and Duflot (1982); (13) Donovan and Middlebrook (1986); (14) Bhakdi and Tranum-Jensen (1987); (15) Bernheimer and Rudy (1986); (16) Sugimoto et al. (1988); (17) McClane et al. (1988); (18) Fraser (1989); (19) Bhakdi et al. (1986); (20) Menestrina et al. (1987); (21) Benz et al. (1989); (22) Menestrina (1988); (24) Howard and Buckley (1982); (25) Howard and Buckley (1985); (26) Ziegler et al. (1984); (27) Zhang and Takemoto (1986); (28) Ziegler et al. (1986); (29) Fussle et al. (1981); (30) Tobkes et al. (1985); (31) Watanabe et al. (1987); (32) Thelestam and Bolmqvist (1988); (33) Forti and Menestrina (1989); (34) Cavaillon et al. (1986); (35) Jolivet-Reynaud and Alouf (1984); (36) Haider and Ellar (1989); (37) Knowles et al. (1989); (38) Freer and Birkbeck (1982); (39) Mellors et al. (1988); (40) Bhakoo et al. (1985); (41) Tappin et al. (1988); (42) Buckingham and Duncan (1983); (43) Duncan and Buckingham (1981); (44) Maget-Dana et al. (1985a); (45) Quentin et al. (1982); (46) Maget-Dana et al. (1985b); (47) Peterson and Cramer (1987); (48) Parker et al. (1989); (49) Schein et al. (1978); (50) Cramer et al. (1983); (51) Lazdunski et al. (1988); (52) Bullock et al. (1983).

Abbreviations: I.a., inner area; s.c., single polypeptide chain; w.t., whole toxin; PA, EF and LF, protective antgen, oedema factor and lethal factor of anthrax toxin. PC, phosphatidylcholine; PS, phosphatidylserine; PI, phosphatidylinositol; GM1, GM2, GT1b, GQ1b, GD1b, gangliosides (see Eidels et al. (1983).

*Indicates the physical state of the membrane-bound active form.

[a]Streptolysin 0 is a prototype for a class of proteins which includes at least 14 other toxins, i.e. Streptococcus pneumoniae pneumolysin, Clostridium perfringens perfringolysin (or θ-toxin), Clostridium tetani tetanolysin, Clostridium botulinum botulinolysin, Clostridium bifermentans bifermentolysin, Clostridium hystolyticum hystolyticolysin, Clostridium novyi oedematolysin, Clostridium septicum septicolysin, Clostridium chauvoei chauveolysin, Bacillus cereus cereolysin, Bacillus alvei alveolysin, Bacillus laterosporus laterosporolysin, Bacillus thuringiensis thuringiolysin, and Listeria monocytogenes listeriolysin.

[b]Escherichia coli haemolysin is a prototype for a class of proteins which includes at least seven other toxins, i.e. Proteus vulgaris haemolysin, Proteus mirabilis haemolysin, Morganella morganii haemolysin, Bordetella pertussis cyclolysin, Pasteurella haemolytica leukotoxin, Serratia marcescens haemolysin, and Actinobacillus pleuropneumoniae haemolysin.

[c]Pseudomonas syringae syringomycin A1 may be considered a prototype for a class of lipodepsinonapeptides which includes at least six other toxins, i.e. Bacillus circulans permetins A and B, polypeptins A and B, and Erwinia herbicola herbicolins A and B.

[d]Bacillus subtilis iturin A is a prototype for a class of lipopeptides which includes at least four other toxins, i.e. Bacillus subtilis bacyllomycin D and L and Nocardia asteroides peptidolipin NA.

[c]cis compartment is that in which the toxin was added; $V_{trans} < 0$ means that a negative voltage in the opposite compartment is required; similarly for the pH requirements indicated above.

make use of radioactively marked molecules but fluorescent dyes have also been used. To review all such methods is outside the scope of this chapter. Using this approach Thelestam and Möllby (1979) measured the functional dimensions of the lesions formed by a variety of bacterial membrane-damaging toxins in nucleated cells. This and other similar studies suggested an interesting application of bacterial toxins such as *S. aureus* α-toxin and streptolysin O. They can be successfully used to permeabilize cells to small molecules (below the cut-off of the pore) but not to large molecules such as cytoplasmic proteins which are above that cut-off (McEwen and Arion, 1985; Bader *et al.*, 1986; Ahnert-Hilger and Gratzl, 1988). For example, the cellular requirements for exocytosis (Ahnert-Hilger *et al.*, 1987) and IgE-induced degranulation (Homan, 1988) have been investigated in this way.

Electrophysiological studies on model systems

After the pioneering work of Hodgkin and Huxley (1952a) not only electrophysiologists but also biochemists and biophysicists became interested in the study of ionic channels. Efforts were made to find simplified systems mimicking the cell membrane where such pores could be studied under fully controlled experimental conditions.

The incorporation of enzymes into unilamellar lipid vesicles was already an acquired technique (Kagawa and Racker, 1971) and an extension of the same approach was used to investigate ionic channels (Miller, 1986). In the mean time a new specific tool was developed giving an unprecedented resolution in the study of the electrical properties of pores: the bilayer (black) lipid film (Mueller *et al.*, 1962). Because of their intrinsic simplicity, these methods found early application in the study of many 'membrane-seeking' bacterial toxins (Latorre and Alvarez, 1981; Blumenthal and Klausener, 1982).

Planar lipid membranes

The first description of a surfactant thinning process, and the intuition that it was due to the formation of a molecular monostrate, dates back to Newton (see Birch, 1757) who correctly estimated the thickness of the film as 50 Å. Nonethe-

less, control of the process of lipid-monolayer formation had to await the pioneering work of Langmuir (1933) several centuries later. The original observation of black holes in soap bubbles was made by Hooke early in 1672 (see Boys, 1959) but the idea that black films of lipid could be formed and used for a biophysical approach to the study of excitatory ionic currents came only about 300 years later and was due to P. Mueller and co-workers (Mueller *et al.*, 1962).

Black lipid membranes. This is the classic black lipid film first described by Mueller *et al.* (1962). A bilayer forms spontaneously when a lipid solution is spread across a small (1–2 mm) hole in a teflon septum separating two aqueous solutions (Fig. 2A). Viewed under reflected light the film shows bright colours during the thinning process but becomes optically black (hence its common name) when the bilayer (about 50 Å thick) is formed. The spreading solution is formed by a lipid (e.g. egg phosphatidylcholine) dissolved in an organic solvent, usually *n*-decane.

The system offered many advantages. First, it was widely accessible, meaning that virtually all the physico-chemical parameters could be varied by the experimenter: lipid composition of the membrane; chemical composition and temperature of the water phase (independently on the two sides); applied voltage and so on. Second was the high resolution with which currents could be resolved. Despite the fact that the relatively large membrane area (contributing an input capacitance of typically 10 nF) limited the resolution with which a square-shaped current jump of 1 pA could be recorded to events lasting at least several seconds, this was unsurpassed performance at that time. After the first polypeptide-induced channels (alamethicin (Mueller and Rudin, 1968), monazomycin (Mueller and Rudin, 1969), and gramicidin (Hladky and Haydon, 1970)) were revealed a wealth of pore-forming substances, including bacterial and animal toxins, were demonstrated and studied in this system (Latorre and Alvarez, 1981; Blumenthal and Klausener, 1982).

Of course, there are also a few major problems with this technique. Probably the most important is that some organic solvent is retained in the lipid film (White, 1986) making it thicker than a biological membrane. Not all membrane proteins are compatible with this residual solvent.

Another problem, as far as endogenous channels are concerned, is that proteins have to cross the water phase before inserting into the bilayer. Accordingly, when trying to incorporate a solubilized intrinsic protein into a black lipid film, detergents have to be used. However, detergents destabilize the bilayer. This problem was finally overcome by a technique pioneered by Miller and Racker (1976): intrinsic proteins were first incorporated into lipid vesicles which were later fused to the black lipid film.

In the case of exogenous channels formed by bacterial toxins, however, this is not a problem because an inherent property of such proteins is the ability to cross a water phase before reaching their final membrane target. This is one reason why studies on such toxins found early success.

Folded bilayers. A major improvement was achieved by Montal and Mueller (1972) who developed a new technique to produce planar bilayers by the apposition of two solvent-free monolayers (Fig. 2B). The resulting bilayer is virtually free of solvent – virtually, because a long-chain organic solvent had to be used (White, 1986) for a proper sealing of the membrane to the supporting septum. Usually *n*-hexadecane is used because it is excluded from the plane of the bilayer and accumulates only at its rim (White, 1986).

The major problem (the presence of residual solvent) was thus solved. But the electrical resolution was improved at the same time because these membranes were successfully apposed on a thin teflon foil bearing a hole typically 0.1 mm in diameter. Membrane area (and thus input capacitance) was decreased by two orders of magnitude compared to black lipid films and, more or less by the same factor, the response time of the I–V converter was improved (Alvarez, 1986). Single channels could be studied at a typical bandwidth resolution of 0.3–1 kHz.

Two more variations of this method are worth mentioning. First, planar bilayers have been formed from apposing monolayers produced by addition of native membrane vesicles in the subphase, thus achieving reconstitution of endogenous channels directly (Schindler, 1980) (Fig. 2B). Second, asymmetrical bilayers in which different lipids were present on the two faces could be

assembled, for a better reproduction of cell membrane polarity (Latorre and Hall, 1976).

Since this approach offers all the advantages described above for the black lipid film, it is in most cases the system of choice.

Bilayers at the tip of a pipette. A further development was recently described which produced planar bilayers at the tip of a glass pipette similar to those used in a patch-clamp (Fig. 2C) (Hanke *et al.*, 1984; Schuerholz and Schindler, 1983; Coronado and Latorre, 1983). Truly solvent-free, stable bilayers were obtained in this way. Because the membrane area was further reduced, the maximal resolution of around 10 kHz (limited by the amplifier noise) was approached in single-channel recording.

The price of this increased sensitivity was that the ready accessibility (discussed above) was lost, at least on one side of the bilayer, i.e. on the inside of the pipette.

Patch clamp of planar bilayers. In a similar approach a patch-clamp apparatus was used to patch a planar bilayer formed by the conventional Mueller technique (Fig. 2C) (Andersen, 1983). The advantage was that simultaneous single-channel recording from the patch pipette and multichannel recording from the black lipid membrane was possible. The drawback was that the chance of getting a patch with a single channel in it was small because the surface density of channels (at most 10^{-1} per μm^2) is very low compared to that of natural cells.

Lipid vesicles

Endogenous channels are intrinsic membrane proteins which can be made water soluble only by detergent treatment. Nonetheless, understanding their properties requires that they are extracted from the native membrane, purified to homogeneity and isolated in a functional state. The problem of preserving their functionality was solved by reconstituting them into a host bilayer, i.e. a lipid vesicle, traditionally by the detergent-removal technique (Miller, 1986).

To demonstrate the functionality of reconstituted channels requires the development of techniques to study ion fluxes in such small objects as lipid vesicles. It is a reflection of the ingenuity of a generation of researchers that so many different

A. BLACK LIPID MEMBRANES

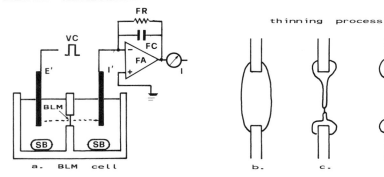

a. BLM cell

thinning process

b.   c.   d.

B. PLANAR LIPID MEMBRANES

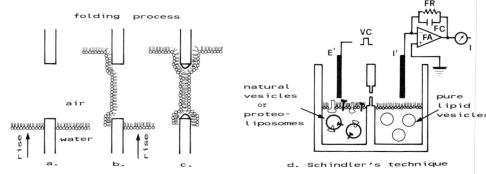

folding process

air

rise   water   rise

a.   b.   c.

natural
vesicles
or
proteo-
liposomes

pure
lipid
vesicles

d. Schindler's technique

C. BILAYERS ON PIPETTES

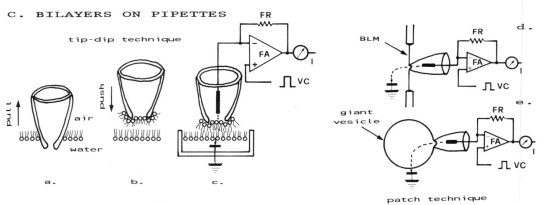

tip-dip technique

pull   air   push

water

a.   b.   c.

BLM   d.

giant
vesicle   e.

patch technique

D. FUSION OF VESICLES TO BLM

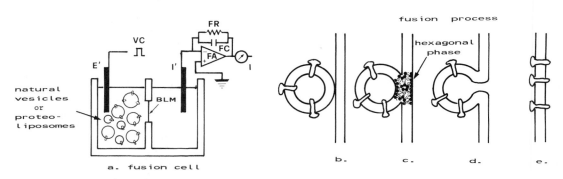

natural
vesicles
or
proteo-
liposomes

BLM

a. fusion cell

fusion process

hexagonal
phase

b.   c.   d.   e.

Figure 2. Electrophysiological methods for studying ion channels on model systems. (A) Black lipid membranes (BLM). (a) Schematic representation of a cell designed for the preparation of black lipid films. Two water-filled compartments are connected through a small (1–2 mm) hole in a teflon septum. A bilayer forms spontaneously when a lipid solution is spread on the hole. Two electrodes are used, one to apply voltage, E', the other to record current, I'. Current is converted to voltage with a feedback amplifier (FA) having a resistance (usually $10^9\ \Omega$) and a capacitance (around 1 pF) in the feedback loop (FR and FC respectively). Buffer solutions are stirred with magnetic spin-bars (SB). The spontaneous thinning process is shown in detail in (b) to (d). Viewed under reflected light the membrane first appears white (b) then brightly coloured with black spots (c) and eventually black when the bilayer (about 50 Å thick) is formed (d). (B) Planar lipid membranes. (a–c) In a cell similar to that used for BLMs two solvent-free lipid monolayers are apposed on a small (0.1–0.2 mm) hole, punched in a thin (12 μm) teflon foil. The folding process, obtained by sequentially raising the water level on each of the two sides of the membrane is shown (a–c). A long-chain organic solvent (which is excluded from the centre of the bilayer) has to be used (White, 1986) to obtain a proper seal between the membrane and the supporting septum. (d) In a similar approach, planar bilayers can be made by apposing monolayers formed by equilibration with lipid vesicles added in the subphase. Pure lipid vesicles, proteoliposomes with reconstituted channels or natural membrane vesicles may be used. (C) Bilayers at the tip of a micropipette. (a–c) A planar lipid bilayer of very small area may be formed on the tip of a glass micropipette by first extracting and then dipping the pipette into a well with a monolayer spread on its surface (tip-dip technique). The electrical configuration for channel recording is the same as in the patch clamp (see Fig. 1). Alternatively planar bilayers (d) or giant vesicles (e) containing pre-inserted channels may be patched with the micropipette. In the planar bilayer configuration simultaneous single-channel recording from the patch pipette and multi-channel recording from the whole bilayer (with the electrical configuration shown in A) is possible. (D) Fusion of vesicles to planar lipid bilayers. (a) Proteoliposomes with reconstituted channels or natural pore-containing membrane vesicles may be incorporated into a pre-existing planar bilayer by membrane fusion (e.g. calcium- or pH-induced). All the channels are transferred to the new host membrane which is easily accessible with electrodes (as shown in A). The fusion process is shown in detail (b–e); it probably goes through formation of a localized hexagonal phase in the contact region. The transmembrane orientation of channels on the vesicle is maintained during fusion.

of the lipid they spontaneously undergo a transition from hydrophilic to amphiphilic which renders them competent to insert into the membrane. At this point most of the techniques developed to study reconstituted endogenous channels can be applied to such toxin-containing vesicles, providing information on the molecular basis of the damaging action of the toxin.

Patch clamp. A very elegant way to demonstrate the incorporation of ion channels in a lipid vesicle and to assess their functionality (or to study their properties), is to apply the patch-clamp technique to the vesicle. However, because of their small dimensions, vesicles produced by traditional techniques are unsuitable for this purpose. A solution was found by inducing such vesicles to repeatedly fuse one with another until giant proteoliposomes resulted (Tank *et al.*, 1982; Criado and Keller, 1987). Such structures can be patch-clamped successfully and the experimental resolution attained is comparable to that of conventional patch-recording. The advantage over cell systems is that proteoliposomes contain only well-defined lipid and protein components.

Fusion to planar lipid bilayers. An alternative approach widely used for studying channels of natural tissues, is to induce the pore-containing vesicle to fuse with a planar lipid bilayer (Miller and Racker, 1976; Cohen, 1986; Hanke, 1986). All channels are thus transferred to the new host membrane which is easily accessible with electrodes (Fig. 2D). The polarized transmembrane orientation of channels (if present on the vesicle) is maintained during fusion (Pasquali *et al.*, 1984). One advantage is that the number of channels present on the vesicles can be estimated by the dimensions of the current jump following the fusion of the vesicle with the bilayer. Furthermore, the planar bilayer can host from a few channel copies up to several thousands, thus providing the opportunity to compare single-channel data with many-channel averages (which are statistically more significant). Unfortunately (as discussed above) the current resolution of single events is much lower than with the patch-recording.

Transmembrane potential determination. Most of the techniques described in the previous section to estimate the internal potential of a cell (and the effects of endogenous or exogenous channels) can, in principle, be applied to a lipid vesicle. The

techniques had been introduced to accomplish this (Miller, 1984; Garcia, 1986).

Exogenous channels, e.g. those formed by bacterial cytotoxins, can be incorporated in lipid vesicles in a simpler way. Soluble toxins are mixed with vesicles in the water phase and in the presence main problem is that the lipid vesicle (lacking ion pumps) has no internal potential. This is overcome by artificially providing the vesicle with one. To do this a gradient of a given ion is generated across the vesicle membrane (by adding it only to the outside medium) and then a specific carrier for that ion is provided (Skulachev, 1988). A potential (quantified by the Nernst equation) establishes through the vesicle membrane and this can be satisfactorily monitored. The depolarizing effect of diphtheria (Shiver and Donovan, 1987) and tetanus (Menestrina et al., 1989) toxin as well as of S. aureus δ-lysin (S. Forti and G. Menestrina, unpublished results) and colicin E1 (Cramer et al., 1983) have been studied in this way.

Release (uptake) experiments. As is the case for whole cells, release or uptake experiments on lipid vesicles can be performed in such a wide variety of ways (Miller, 1984; Garcia, 1986) that it is impossible to satisfactorily review all of them here. Many of these techniques have been fruitfully applied to the study of bacterial protein toxins (see Tables 1 and 2) (Boquet and Duflot, 1982; Kayalar and Düzgünes, 1986; Shone et al., 1987; Menestrina, 1988).

Particularly noteworthy is the possibility of measuring the time course of marker release from a vesicle with ion channels in its membrane using a fast kinetics approach (Garcia, 1986). The number and conductivity of pores can be inferred from these experiments as well as the chemical constants governing a ligand-operated channel (Hess and Udgaonkar, 1987).

Bacterial toxins with electrophysiological effects

Neurotoxins

The term neurotoxin usually applies to protein toxins which directly bind to ionic channels of the nervous system, thus affecting their normal behaviour. Though there are a wealth of examples of neurotoxins produced by higher organisms, e.g. tetrodotoxin, α-bungarotoxin, conotoxins, scorpion and sea-anemone toxins (see Moczydlowski et al., 1986; Hille, 1984), only a few bacterial neurotoxins have been described.

Tetanus and botulinum toxins are usually considered neurotoxins because they interfere with the release of neurotransmitters at the neuromuscular junction, thus indirectly affecting the acetylcholine receptor channel. However, their target is intracellular (and still unknown) and thus they will be included in the next section.

An interesting group of toxins potentially belonging to the neurotoxin class are the heat-stable enterotoxins produced by several strains of E. coli (Holmgren, 1985) (but also by Vibrio cholerae and Yersinia enterocolitica (Takao et al., 1985)). This family of closely related peptides (17–30 residues long) in fact shares a strong homology (up to 54%) with conotoxins, a group of potent neurotoxins produced by the sea animal Conus geographus which block the acetylcholine receptor channel at the neuromuscular junction (Lazure et al., 1983) as well as the sodium channel (Moczydlowski et al., 1986).

Although the strong homology is in the conserved sequence thought to be the active site (Lazure et al., 1983), heat-stable enterotoxins have a cellular target completely different from conotoxins. They provoke severe forms of diarrhoea by stimulating the guanylate cyclase of gut cells after interacting with their membrane in a still unknown manner (Holmgren, 1985). Intracellular cGMP is thus increased, which in turn opens a chloride channel in the apical membrane and a potassium channel in the basolateral membrane (Huott et al., 1988).

Toxins with intracellular targets (A/B type)

This is an important group of toxins characterized by the fact that two domains with different physiological roles (called A and B) can be identified. The domains are either fragments of a unique polypeptide chain or different subunits of the protein toxin. These toxins act intracellularly by an enzymic action promoted by subunit (or fragment) A, whereas subunit (or fragment) B is responsible for binding to the membrane of the target cell and, in some cases, translocation (Eidels et al., 1983; Middlebrook and Dorland, 1984).

A somewhat surprising feature of these toxins is that most of them are able to open ionic channels into cells and model membranes. For this reason they have been included in Tables 1 and 2. It has

Table 2. Properties of the pores formed by bacterial toxins in model lipid membranes

Pore former	Method (reference)	Lipid	Buffer (pH 7 if not specified)	Conductance (pS)	Conductance distribution	Estimated[a] cross-sectional area (Å²)	Rectification of the open channel	On-off gating	Selectivity	Divalent cations permeant	Inhibited by
Cholera toxin	a (1)	GM0 + GMI	0.1 M NaCl	20–40	Narrow	15–30*	No	Yes (v.i.)[b]	Poorly anion selective $P(Na^+) = P(K^+) = 2P(Cl^-)$[c]	—	—
	b (2,3)	Phospholipids + GMI	0.15 M NaCl	—	—	55	—	—	—	—	—
Anthrax toxin PA65 fragment	c (4)	Asolectin or DiPhPC	0.1 M KCl	165	Narrow	> 63	No	Yes (v.d. −)[d]	Almost ideal for univalent cations	No	—
Diphtheria toxin Whole toxin	a (5)	Asolectin or PE	0.2 M NaCl / 1 M NaCl	6.2 / 20	Narrow Narrow	< 20	No	Yes (v.d.)	Anion selective $P(Cl^-)/P(K^+) = 6$	—	—
Fragment B45	a (6)	Soybean phospholipids	0.1 M KCl	10	Narrow	> 250	No	Yes (v.d. −)	—	—	—
P. aeruginosa exotoxin A	a (7)	PC/PS=1/1	0.16 M NaCl	50–500	Broad	> 175	—	Yes (v.d. +)	—	—	—
Tetanus toxin w.t. or B-fragment	a(8)	Asolectin	1 M KCl / 5 mM CaCl₂	45 (pH 7) / 15 (pH 5)	Narrow	> 50	—	Yes (v.d. −)	Poorly cation selective	—	—
	a (9,10)	PS	0.5 M KCl	89 (pH 7) / 30 (pH 4.8)	Narrow	—	Yes	Yes (v.d. ±)	Cation selective	—	—
Botulinum toxin Type C (w.t.)	a (11)	Asolectin	0.1 M NaCl	12 (pH 6.6)	Narrow	—	—	Yes (v.i.)	—	—	—
Type B (h.c.)	a (8)	Asolectin	1 M KCl / 5 mM CaCl₂	15 (pH 7) / 20 (pH 5)	Narrow	> 50	—	Yes (v.i.)	Poorly cation selective	—	—
Type A N-term fragment of h.c.	a (12)	DiPhPC or PE/PC/PS=2/2/1	1 M KCl / 5 mM CaCl₂	15 (pH 7) / 20 (pH 5)	Narrow	—	—	Yes (v.d. −)	—	—	—
	b (13)	Various phospholipids	0.1 M NaCH₃COO	—	—	> 175	—	—	—	—	—
C. perfringens perfringolysin	a (14)	PC/cholesterol 1/1	0.1 M KCl	10×10³–30×10³	Broad	17×10³*	No	No	Non-selective	Yes	Zn^{2+}
S. pyogenes streptolysin O	g (15)	Sheep RBC ghosts	PEM	—	—	> 13×10³	—	—	—	—	—

Table 2. Continued

Pore former	Method (reference)	Lipid	Buffer (pH 7 if not specified)	Conductance (pS)	Conductance distribution	Estimated[a] cross-sectional area (Å²)	Rectification of the open channel	On–off gating	Selectivity	Divalent cations permeant	Inhibited by
C. tetani tetanolysin	b (16)	PC/cholesterol 1/1	0.3 M glucose	—	—	> 55	—	—	—	—	—
C. perfringens enterotoxin	d (17)	Asolectin	0.13 M NaCl 4 mM $CaCl_2$	40–450	Broad	55–91[e]	No	Yes (v.i.)	—	Yes	—
E. coli haemolysin	a (18)	PC/PE=5/1 or POPC	0.1 M NaCl	200	Narrow	310*	No	Yes (v.d.−)	Cation selective $P(Na^+) = P(K^+) = 16 \times P(Cl^-)$	Yes	Zn^{2+}[f]
A. pleuro-pneumoniae haemolysin	d (19)	PC/PE=1/1	PEM	350–400	Narrow	< 380, 50*	No	Yes (v.i.)	—	—	—
P. aeruginosa cytotoxin	b,c (20)	PC/cholesterol	1 M KCl	100–4×10³	Broad	> 450	—	—	Slightly cation-selective	—	—
	e,f (21)	Chromaffin cells	PEM	120	Narrow	90*	No	Yes (v.d.+)	Non-selective	—	—
A. hydrophila aerolysin	b (22)	PC/chol/PA=7/2/1 or PC/PA=9/1	PBS	—	—	< 700	—	—	—	—	—
P. syringae syringomycin	c (23–25)	Soy bean phospholipids or PC/chol=1/1	1 M KCl 5 mM $MgCl_2$	180 (pH 5.5)	Narrow	20*	Yes	Yes (v.d.+)	Anion selective $P(Cl^-)/P(K^+)=8$ pH 6 $P(Cl^-)/P(K^+)=3$ pH 9	No	pH 10
S. aureus α-toxin	a (26)	PC or PC/PS	0.1 M KCl	90	Narrow	95*	Yes	No	Anion selective $P(Cl^-)/P(K^+)=2.3$ pH 5 $P(Cl^-)/P(K^+)=1.6$ pH 7	Yes	Ca^{2+} Zn^{2+}
B. thuringiensis δ-endotoxin	c (27)	POPE	0.3 M KCl	48 (pH 9.5)	Narrow	80–310	No	Yes (v.i.)	Almost ideally cation-selective	—	Ca^{2+} Zn^{2+}
S. aureus δ-lysin	d (28)	DPPC	0.5 M KCl	Small 20–110 large 450[g]	Broad Narrow	58[h]	No	Yes (v.d.±)	Cation selective Small $P(K^+)/P(Cl^-)=9.3$ large $P(K^+)/P(Cl^-)=2.6$	—	—
	b (29)	DPPC	0.1 M PBS	—	—	150	—	—	—	—	—

Organism/toxin	Lipid	Solution	Conductance (pH)	Channel	Area[a]	[e/f]	Gating[d]	Selectivity		
S. pyogenes streptolysin S b (15,30)	Various phospholipids	Diluted PBS 1:10	—	Broad	—	—	—	—	—	—
				Broad						
B. subtilis iturin A c (31,32)	GMO or PC	1 M KCl	6–30	Narrow	1–10*	No	Yes (v.i.)	Slightly anion-selective	—	—
		4 M KCl	6–300							
Colicin A a (33)	Soy bean phospholipids	0.5 M KCl 5 mM CaCl₂	7.5 (pH 6.1)	Narrow	—	No	Yes (v.d. −)	Cation selective $P(Na^+)=P(K^+)=4\times P(Cl^-)$	Yes	—
w.t. or 20-kDa C-term fragment h (34,35)	Soy bean phospholipids	1 M KCl 5 mM CaCl₂	13 (pH 6.2)	Narrow	—	Yes	Yes(v.d. −)	—	—	—
Colicin E1 b (36)	PC/PS or PC/PG	0.1 M NaCl	—	—	150	—	—	—	—	—
w.t. or 20-kDa C-term fragment a (37)	Asolectin	1 M NaCl 3 mM CaCl₂	20 (pH 6)	Narrow	50	Yes	Yes (v.d. −)	Cation selective $P(Na^+)=P(K^+)=4.5\times P(Cl^-)$	—	—
Colicin B c (38)	DiPhPC/PS=4/1	1 M KCl	30 (pH 7)	Narrow	—	—	Yes (v.d. −)	Cation selective	—	—
			14 (pH 5)	Narrow						
Colicin Ia a (39)	Asolectin	0.5 M KCl 5 mM CaCl₂	30 (pH 6.2)	Narrow	—	—	Yes (v.d. −)	—	—	—
Colicin Ib a (40)	Soy bean phospholipids	3 M KCl 5 mM CaCl₂	41 (pH 6.1)	Narrow	—	No	Yes (v.d. −)	—	—	—

References: (1) Tosteson and Tosteson (1978); (2) Moss et al. (1976a); (3) Moss et al. (1976b); (4) Blaustein et al. (1989); (5) Donovan et al. (1981); (6) Kagan et al. (1981); (7) Menestrina et al (unpublished observation); (8) Hoch et al. (1985); (9) Borochov-Neori et al. (1984); (10) Gambale and Montal (1988); (11) Donovan and Middlebrook (1986); (12) Blaustein et al. (1987); (13) Shone et al. (1987); (14) Menestrina et al. (1990); (15) Buckingham and Duncan (1983); (16) Alving et al. (1979); (17) Sugimoto et al. (1988); (18) Menestrina et al. (1987); (19) Lalonde et al. (1989); (20) Weiner et al. (1985); (21) Dreyer et al. (1990); (22) Howard and Buckley (1982); (23) Zegler et al. (1984); (24) Pokornj and Ziegler (1984); (25) Ziegler et al. (1986); (26) Forti and Menestrina (1989); (27) Knowles et al. (1989); (28) Mellors et al. (1988); (29) Yianni et al. (1986); (30) Duncan and Buckingham (1981); (31) Maget-Dana et al. (1985a); (32) Maget-Dana et al. (1985b); (33) Schein et al. (1978); (34) Martinez et al. (1987); (35) Collarini et al. (1987); (36) Kayalar and Duzgunes (1986); (37) Bullock et al. (1983); (38) Pressler et al. (1986); (39) Noguera and Varanda (1988); (40) Weaver et al. (1981).

Abbreviations: w.t., whole toxin; h.c., heavy chain; N-term and C-term, N-terminal and C-terminal; PBS, phosphate-buffered saline; PEM, physiological extracellular medium, i.e. 140 mM NaCl, 2 mM CaCl₂, 2 mM MgCl₂, 3 or 4 mM KCl, pH 7; RBC, red blood cells; GMO, glycerol monooleate; PC, phosphatidylcholine; POPC, palmitoyl-oleoyl-phosphatidylcholine; DPhPC, diphytanoyl-phosphatidylcholine; DPPC, dipalmitoyl-phosphatidylcholine; PE, phosphatidylethanolamine; POPE, palmitoyl-oleoyl-phosphatidylethanolamine; PS, phosphatidylserine; PI, phosphatidylinositol; PG, phosphatidylglycerol; PA, phosphatidic acid; GMI, ganglioside (see Eidels et al. 1983).

Methods: (a) folded planar bilayers; (b) release from lipid vesicles; (c) painted bilayers; (d) bilayers at the tip of a pipette; (e) whole-cell recording; (f) excised patch recording; (g) release from cells; (h) folded bilayers from lipid vesicles.

[a] Cross-sectional area is calculated either from the cut-off molecular weight of molecules released from the channel or it is estimated from its conductance in planar lipid membranes (indicated by an asterisk).

[b] v.i. stands for voltage-independent gating.

[c] P(x) indicates the permeability of the ion species x.

[d] v.d. stands for voltage-dependent gating; the symbol following (when present) indicates the sign of the voltage which has to be applied to the trans compartment in order to keep the channels open.

[e] From McClane (1984).

[f] From Menestrina et al. (1990).

[g] Two kind of channels, small and large, were observed.

[h] This area is estimated theoretically for a pore surrounded by six α-helices.

been proposed that this pore-forming capacity is related to the mechanism by which their enzymically active part is introduced into the target cell (Hoch *et al.*, 1985). Whether ion-channel formation is a necessary step of intoxication or a side-effect is still a matter of debate (Olsnes *et al.*, 1988; Papini *et al.*, 1988).

Toxins increasing intracellular cAMP

This class includes cholera and cholera-like toxins, pertussis toxin and the invasive adenylate cyclase toxins of *Bacillus anthracis* and *Bordetella pertussis* (Middlebrook and Dorland, 1984).

Cholera toxin (and the closely related heat-labile enterotoxins of *Escherichia coli*, *Vibrio mimicus*, *Aeromonas hydrophila*, *Campylobacter jejuni* (Finkelstein *et al.*, 1987)) consists of one A1 subunit linked via an A2 subunit to five B subunits. Binding of the B subunits to the cell is mediated by the ganglioside GM1 on the periplasmic face of the membrane (Eidels *et al.*, 1983), whereas enzymic activity of the A1 subunit is exerted on the G_s protein on the cytoplasmic face of the membrane. A few years ago it was shown that this toxin is able to permeabilize GM1-ganglioside-containing liposomes (Moss *et al.*, 1976a) and also to open pores in a planar lipid bilayer (Tosteson and Tosteson, 1978) (but the later result has never been confirmed). Permeabilization was induced by the B protomer alone (Moss *et al.*, 1976b) and accordingly it was suggested that pore formation was necessary for transmembrane translocation of the A1 subunit. Pertussis toxin is similar to cholera toxin but it has a different G protein as intracellular target and no reported channel-forming activity.

Anthrax toxin is atypical because it consists of a complex of three protein components. One, called protective antigen PA because of its immunogenicity, is the B part of the toxin involved in cell binding and internalization of the A parts. The other two proteins (called oedema factor, EF, and lethal factor, LF) constitute two independent A parts competing for the same B part (Middlebrook and Dorland, 1984). Oedema factor increases intracellular cAMP by a calmodulin-dependent adenylate cyclase activity whereas lethal factor kills the cell in an unknown way. PA (83 kDa) binds to high-affinity cell surface receptors, a 20-kDa N-terminal fragment is then cleaved by cell surface proteases to produce a membrane-

bound form of PA competent for the binding of EF or LF (Singh *et al.*, 1989). It was recently shown that the enzymically-cleaved 65-kDa C-terminal part of the protective antigen is able to open cation-selective pores in a planar lipid bilayer (Blaustein *et al.*, 1989). As yet, pore-forming ability has not been demonstrated for a functionally analogous adenylate cyclase produced by *Bordetella pertussis* (see Chapter 12 of this volume).

Toxins suppressing protein synthesis

Belonging to this group are diphtheria toxin from *Corynebacterium diphtheriae*, exotoxin A from *Pseudomonas aeruginosa*, and shiga toxin from *Shigella dysenteriae* (as well as some shiga-like toxins from *E. coli* (Saxena *et al.*, 1989)). All these toxins probably enter the cell cytosol via receptor-mediated endocytosis (Eidels *et al.*, 1983).

Both diphtheria toxin (Donovan *et al.*, 1981; Papini *et al.*, 1988) and exotoxin A (Menestrina and Gambale, unpublished observation) are able to form ion channels in model membranes. Channel formation is stimulated in both cases by a low pH in the toxin-containing compartment and by the presence of acidic phospholipids in the bilayer (preferentially PI in the case of diphtheria toxin (Donovan *et al.*, 1982) and PG in the case of exotoxin A).

In the case of diphtheria toxin the channel-forming activity is expressed by the C-terminal fragment (Kagan *et al.*, 1981). A channel-forming activity in shiga or shiga-like toxins has not been reported yet.

Toxins inhibiting neurotransmitter release

These consist of tetanus toxin (produced by *C. tetani*) and a number of closely related botulinum toxins (produced by strains of *C. botulinum*) called A, B, C_α, C_β, D, E, F, G. All of these are expressed as a unique polypeptide chain. After secretion the protein is enzymically processed to an active 'nicked' form in which the two fragments A and B (in this case known as light and heavy chain respectively) are linked by a disulphide bridge (Middlebrook and Dorland, 1984). The enzymic activity is exerted intracellularly by the light chain alone (Ahnert-Hilger *et al.*, 1989).

Both tetanus and botulinum toxins are able to form ion pores in model membranes (Borochov-

Neori et al., 1984; Hoch et al., 1985; Donovan and Middlebrook, 1986). As in the case of diphtheria toxin and exotoxin A, channel formation is stimulated by low pH in the toxin-containing solution and by the presence of acidic phospholipids in the bilayer (PI or PS) (Menestrina et al., 1989).

In both cases it has been shown that the channel-forming activity is expressed by the heavy chain and in particular by its N-terminal part (Boquet and Duflot, 1982; Hoch et al., 1985; Blaustein et al., 1987; Shone et al., 1987).

Membrane-damaging toxins

Toxins of this group attack target cells by increasing rather non-specifically their membrane permeability to ions and small molecules. The permeabilization may be lethal if it extends beyond the capacity of the cell to maintain its 'milieu interieur' by active transport.

The number of known toxins belonging to this group is large and still growing very quickly (Bernheimer and Rudy, 1986; Bhakdi and Tranum-Jensen, 1987). In this chapter, I will review only data on those toxins which produce well-defined lesions in the cell membrane, meaning, for example, that the diameter of the lesion may be at least estimated. I will not deal with those toxins which produce enzymic damage to the membranes such as the phospholipases and the sphingomyelinases, or those toxins which have merely a detergent effect on the lipid film.

According to some structural and/or genetic characteristics, several subdivisions may be identified within this group.

Thiol-activated toxins

This group includes at least 15 closely related protein toxins produced by Gram-positive bacteria (Bernheimer and Rudy, 1986). All of these require the presence of cholesterol in the target membrane and are actually inhibited by minute amounts of cholesterol (or closely related sterols) present in solution. They are inactivated by oxidation and reactivated by reduction with thiols. Though their molecular weights range from 48 to 68 kDa and their isoelectric points range from pI 4.9 to pI 7.8, they are clearly antigenically related, indicating at least partial homology.

They cause release of large molecules from different cell types and produce typical ring- or arc-like lesions on the cell surface when observed in the electron microscope (Bernheimer and Rudy, 1986; Bhakdi and Tranum-Jensen, 1987). Such lesions, with an external diameter ranging from 30 to 50 nm, are formed by large toxin aggregates.

At least two of them have been studied in model lipid membranes i.e. tetanolysin (Blumenthal and Habig, 1984; Alving et al., 1979) and perfringolysin (Menestrina et al., 1990). They both produce heterogeneous pores which might be very large. These proteins are also currently used for selective permeabilization of cells (Ahnert-Hilger and Gratzl, 1988).

Non A/B heat-labile enterotoxins

Clostridium perfringens produces a 35-kDa enterotoxin responsible for diarrhoeal-type food poisoning (Matsuda and Sugimoto, 1979). This toxin binds to sensitive cells (which possess a still unknown receptor (McClane, 1984; McClane et al., 1988)), inserts into their membranes and increases their permeability to ions and small molecules (up to molecular weight 200), thus disrupting the colloid-osmotic equilibrium. Subsequent effects depend on this permeability change (Matsuda and Sugimoto, 1979; McClane et al., 1988). Recently it was shown that the toxin forms discrete ion channels in planar bilayers of pure phospholipids (Sugimoto et al., 1988) without requiring any receptor. It was suggested that such pore-forming activity could explain the physiological effects of the toxin on cells.

Staphylococcal food poisoning in turn is due to a family of enterotoxins of molecular weight around 30 kDa called A, B, C_1, C_2, C_3, D, E which cause emesis and diarrhoea (Schmidt and Spero, 1983). Whether their mode of action is similar to that of C. perfringens enterotoxin is still unknown.

Haemolysins from Gram-negative bacteria

High molecular weight proteins. Virulent strains of Escherichia coli produce a haemolysin which may be regarded as the prototype of a group of related toxins. It is the only protein genuinely secreted by E. coli. Export of E. coli haemolysin relies on a secretion machinery composed of two membrane-bound proteins, one of which is closely related to the ATP-activated multidrug transporter (the glycoprotein which confers drug resistance to mammalian cancer cells) (Holland *et al.*, 1989).

This toxin is haemolytic and cell-damaging (Cavalieri *et al.*, 1984; Bhakdi *et al.*, 1988). Its pore-forming ability was first demonstrated in cells (Bhakdi *et al.*, 1986) and then in model systems (Menestrina *et al.*, 1987; Menestrina, 1988). About 15 haemolysins produced by other Gram-negative bacteria are related to *E. coli* haemolysin; they include haemolysins of *Proteus vulgaris*, *Proteus mirabilis*, *Morganella morganii*, *Bordetella pertussis* (cyclolysin), *Pasteurella haemolytica* and *Serratia marcescens* (Koronakis *et al.*, 1987; Lo *et al.*, 1987). Common to all these proteins is a repeated octapeptide motif thought to be involved in Ca^{2+} binding (Ludwig *et al.*, 1988). Some experimental evidence indicates that the toxins form monomeric channels (Jorgensen *et al.*, 1980; Bhakdi *et al.*, 1986; Menestrina *et al.*, 1987; Menestrina, 1988) but the existence of aggregates has also been proposed (Benz *et al.*, 1989).

Other pore-forming cytolysins produced by Gram-negative bacteria include a widely investigated cytotoxin from *Pseudomonas aeruginosa* (Dreyer *et al.*, 1990; Weiner *et al.*, 1985; Lutz *et al.*, 1987), a 107-kDa haemolysin from *Actinobacillus pleuropneumoniae* (Lalonde *et al.*, 1989), aerolysin from *Aeromonas hydrophila* (Howard and Buckley, 1982), and some cytolysins produced by *Vibrio damsela*, *V. vulnificus* and *V. parahaemolyticus* (Bernheimer and Rudy, 1986).

Low molecular weight polypeptides. *Pseudomonas syringae* is a bacterial pathogen of numerous plants. Its virulence is related to the production of some low molecular weight phytotoxins called syringomycins E, G and A1 (Segre *et al.*, 1989). These are cyclic lipodepsinonapeptide antibiotics which open cation-selective channels in model membranes (Ziegler *et al.*, 1984, 1986). The mechanism of action is probably similar to that of other cyclic lipopeptide antibiotics produced by Gram-positive bacteria (see below).

Haemolysins from Gram-positive bacteria

High molecular weight proteins. The most thoroughly studied toxin of this group is probably *Staphylococcus aureus* α-toxin. This toxin oligomerizes on the surface of mammalian cells and

liposomes to form a membrane-embedded hexamer appearing as a hollow cylinder in the electron microscope (Füssle *et al.*, 1981; Tobkes *et al.*, 1985; Thelestam and Bolmqvist, 1988). A pore is formed through which ions and small molecules (up to molecular weight 2000) readily diffuse. The osmotic imbalance causes lysis of erythrocytes whereas nucleated cells can survive its action if adequately protected. Divalent cations are particularly effective in cell protection (Bashford *et al.*, 1986). Pore formation is observed also on model membranes (Menestrina, 1986; Forti and Menestrina, 1989).

A similar pore-forming divalent cation-inhibited action was described for *Clostridium perfringens* δ-toxin (Jolivet-Reynaud and Alouf, 1984).

Another toxin produced by a Gram-positive bacterium known to be a pore-former is δ-toxin from *Bacillus thuringiensis* (Haider and Ellar, 1989; Knowles *et al.*, 1989).

Low molecular weight polypeptides. A number of Gram-positive bacteria produce pore-forming polypeptides, among these are: *Staphylococcus aureus* δ-lysin, *Streptococcus pyogenes* streptolysin S and *Bacillus subtilis* iturins.

S. aureus δ-lysin is a 26-residue polypeptide which can adopt α-helical structure with a highly amphiphilic profile (Freer and Birkbeck, 1982; Tappin *et al.*, 1988). It is haemolytic to a variety of cells at relatively high concentrations and is inhibited by several lipids (Bhakoo *et al.*, 1985). It opens channels in model membranes probably through aggregation of several monomers into the lipid film (Yianni *et al.*, 1986; Mellors *et al.*, 1988).

S. pyogenes streptolysin S is a peptide containing 32 amino acids, but two such chains (complexed through an oligonucleotide) are thought to form the active unit (Bernheimer and Rudy, 1986). In contrast to δ-lysin it is extraordinarily potent in lysing cells and is inhibited by minute amounts of phospholipids. Its mechanism of action involves colloid-osmotic lysis of cells (Duncan and Buckingham, 1981). The toxin also increases the permeability of lipid vesicles by forming pores roughly 4.5 nm in diameter (Buckingham and Duncan, 1983).

Also to this group belongs a family of lipopeptide antibiotics produced by *Bacillus subtilis* called iturins which are studied for their potent antimicotic activity (Maget-Dana *et al.*, 1985a,b). They form ion-conducting pores in planar bilayers, most probably through aggregation of several monomers into the lipid film.

Colicins

Colicins are protein toxins produced by strains of *E. coli* and specifically directed only towards competing bacterial strains. Their mechanism of action is now becoming clear (Cramer *et al.*, 1983; Lazdunski *et al.*, 1988). They are particularly interesting because they share properties with at least two of the toxin groups described above. In fact, because of the molecular organization in functional domains (four domains devoted respectively to binding, translocation, activity and immunity may be distinguished (Lazdunski *et al.*, 1988)) they are similar to toxins of the A/B type. However, only a few of them, i.e. colicins E2, E3 and cloacin, actually have enzymic activity on an intracellular target (Cramer *et al.*, 1983).

All the others exert their killing effect by opening an ionic channel of well-defined size in the cytoplasmic membrane of the competing bacterium. Thus they can be included among the membrane-damaging toxins (Cramer *et al.*, 1983). Even in the absence of their specific protein receptor (i.e. bacterial porins) colicins form channels on model membranes such as vesicles (Davidson *et al.*, 1984; Kayalar and Düzgünes, 1986) and planar bilayers (Schein *et al.*, 1978; Lazdunski *et al.*, 1988). The size of the channel is such that a single hit is sufficient to depolarize and eventually kill a bacterial cell.

Pore-forming activity is limited to a 20-kDa C-terminal domain of the molecule which is easily obtained by mild enzymic digestion. The fragment itself forms pores indistinguishable from those formed by the parent molecule (Cramer *et al.*, 1983; Lazdunski *et al.*, 1988). In the case of colicin A the pore-forming fragment has been crystallized and its water-soluble structure determined (Parker *et al.*, 1989). On this basis it was suggested that aggregation of three molecules is required for pore formation (Parker *et al.*, 1989), though some other experimental evidences point to a monomeric action (Peterson and Cramer, 1987).

Acknowledgements

I wish to thank J. Freer, F. Gambale, and C. Montecucco for helpful comments during the preparation of this manuscript.

References

Ahnert-Hilger, G. and Gratzl, M. (1988). Controlled manipulation of the cell interior by pore-forming proteins. *Trends Pharmacol. Sci.* **9**, 195–7.

Ahnert-Hilger, G., Brautigam, M. and Gratzl, M. (1987). Ca^{++}-stimulated catecholamine release from α-toxin-permeabilized PC12 cells: biochemical evidence for exocytosis and its modulation by protein kinase C and G proteins. *Biochemistry* **26**, 7842–8.

Ahnert-Hilger, G., Weller, U., Dauzenroth, M.-E., Habermann, E. and Gratzl, M. (1989). The tetanus toxin light chain inhibits exocytosis. *FEBS Lett.* **242**, 245–8.

Alvarez, O. (1986). How to set up a bilayer system. In: *Ion Channel Reconstitution* (ed. C. Miller), pp. 115–31. Plenum Press, New York.

Alving, C.R., Habig, W.H., Urban, K.A. and Hardegree, M.C. (1979). Cholesterol-dependent tetanolysin damage to liposomes. *Biochim. Biophys. Acta* **551**, 224–8.

Andersen, O.S. (1983). Ion movement through gramicidin A channels. Single channel measurements at very high potentials. *Biophys. J.* **41**, 119–33.

Bader, M.-F., Thiersé, D., Aunis, D., Ahnert-Hilger, G. and Gratzl, M. (1986). Characterization of hormone and protein release from α-toxin-permeabilized chromaffin cells in primary culture. *J. Biol. Chem.* **261**, 5777–83.

Barnard, E.A., Miledi, R. and Sumikawa, K. (1982). Translation of exogenous messenger RNA coding for nicotinic acetylcholine receptors produces functional receptors in *Xenopus* oocytes. *Proc. R. Soc. Lond. B* **215**, 241–6.

Bashford, C.L. and Smith, J.C. (1979). The use of optical probes to monitor membrane potential. *Methods Enzymol.* **55**, 569–86.

Bashford, C.L., Alder, G.M., Menestrina, G., Micklem, K.J., Murphy, J. and Pasternak, C.A. (1986). Membrane damage by haemolytic viruses, toxins, complement and other agents: a common mechanism blocked by divalent cations. *J. Biol. Chem.* **261**, 9300–8.

Benz, R., Schmid, A., Wagner, W. and Goebel, W. (1989). Pore formation by the *Escherichia coli* haemolysin: evidence for an association–dissociation equilibrium of the pore-forming aggregates. *Infect. Immun.* **57**, 887–95.

Bernheimer, A.W. and Rudy, B. (1986). Interactions between membranes and cytolytic peptides. *Biochim. Biophys. Acta* **864**, 123–41.

Bhakdi, S. and Tranum-Jensen, J. (1984). Mechanism of complement cytolysis and the concept of channel-forming proteins. *Phil. Trans. R. Soc. Lond. B* **306**, 311–24.

Bhakdi, S. and Tranum-Jensen, J. (1987). Damage to mammalian cells by proteins that form transmembrane pores. *Rev. Physiol. Biochem. Pharmacol.* **107**, 147–223.

Bhakdi, S., Mackman, N., Nicaud, J.-M. and Holland, I.B. (1986). *Escherichia coli* haemolysin may damage target cell membranes by generating transmembrane pores. *Infect. Immun.* **52**, 63–9.

Bhakdi, S., Mackman, N., Menestrina, G., Gray, L., Hugo, F., Seeger, W. and Holland, I.B. (1988). The haemolysin of *Escherichia coli*. *Eur. J. Epidemiol.* **4**, 135–43.

Bhakoo, M., Birkbeck, T.H. and Freer, J. (1985). Phospholipid-dependent changes in membrane permeability induced by staphylococcal δ-lysin and bee venom melittin. *Can. J. Biochem. Cell Biol.* **63**, 1–6.

Birch, T. (1757). *History of the Royal Society*, Vol. III, p. 29. A. Millard, London.

Blaustein, R.O., Germann, W.I., Finkelstein, A. and DasGupta, B.R. (1987). The N-terminal half of the heavy chain of botulinum type A neurotoxin forms channels in planar lipid bilayers. *FEBS Lett.* **226**, 115–20.

Blaustein, R.O., Koehler, T.M., Collier, R.J. and Finkelstein, A. (1989). Anthrax toxin: channel-forming activity of protective antigen in planar phospholipid bilayers. *Proc. Natl. Acad. Sci. USA* **86**, 2209–13.

Blumenthal, R. and Habig, W.H. (1984). Mechanism of tetanolysin-induced membrane damage: studies with black lipid membranes. *J. Bacteriol.* **151**, 321–3.

Blumenthal, R. and Klausener, R.D. (1982). The interaction of proteins with black lipid membranes. In: *Membrane Reconstitution.* (eds G. Poste and G.L. Nicolson), pp. 43–82. Elsevier Biomedical Press, Amsterdam.

Boquet, P. and Duflot, E. (1982). Tetanus toxin fragment forms channels in lipid vesicles at low pH. *Proc. Natl. Acad. Sci. USA* **79**, 7614–18.

Borochov-Neori, H., Yavin, E. and Montal, M. (1984). Tetanus toxin forms channels in planar lipid bilayers containing gangliosides. *Biophys. J.* **45**, 83–5.

Boys, C.V. (1959). *Soap Bubbles. Their Colours and the Forces which Mold Them.* Dover Publications, New York.

Buckingham, L. and Duncan, J.L. (1983). Approximate dimensions of membrane lesions produced by streptolysin S and streptolysin O. *Biochim. Biophys. Acta* **729**, 115–22.

Bullock, J.O., Cohen, F.S., Dankert, J.R. and Cramer, W.A. (1983). Comparison of the macroscopic and single channel conductance properties of colicin E1 and its COOH-terminal tryptic peptide. *J. Biol. Chem.* **258**, 9908–12.

Cavalieri, S.J., Bohach, G.A. and Snyder, I.S. (1984). *Escherichia coli* α-haemolysin: characteristics and probable role in pathogenicity. *Microbiol. Rev.* **48**, 326–46.

Cavaillon, J.-M., Jolivet-Reynaud, C., Fitting, C., David, B. and Alouf, J.E. (1986). Ganglioside identification on human monocyte membrane with *Clostridium perfringens* delta-toxin. *J. Leuk. Biol.* **40**, 65–72.

Chandler, W.K. and Meves, H. (1965). Voltage clamp experiments on internally perfused giant axons. *J. Physiol. (Lond.)* **180**, 788–820.

Cohen, F.S. (1986). Fusion of liposomes to planar bilayers. In: *Ion Channel Reconstitution* (ed. C. Miller), pp. 131–9. Plenum Press, New York.

Collarini, M., Amblard, G., Lazdunski, C.J. and Pattus, F. (1987). Gating processes of channels induced by colicin A, its C-terminal fragment and colicin E1 in planar bilayers. *Eur. Biophys. J.* **14**, 147–53.

Colquhoun, D., Dionne, V.E., Steinbach, J.H. and Stevens, C.F. (1975). Conductance of channels opened by acetylcholine-like drugs in muscle end-plate. *Nature (Lond.)* **253**, 204–6.

Conti, F. and Neher, E. (1980). Single channel recordings of K$^+$ currents in squid axons. *Nature (Lond.)* **285**, 140–3.

Conti, F. and Wanke, E. (1975). Channel noise in nerve membranes and lipid bilayers. *Q. Rev. Biophys.* **8**, 451–506.

Conti, F., DeFelice, L. and Wanke, E. (1975). Potassium and sodium ion current noise in the membrane of the squid giant axon. *J. Physiol. (Lond.)* **248**, 45–82.

Coronado, R. and Latorre, R. (1983). Phospholipid bilayers made of monolayers on patch-clamp pipettes. *Biophys. J.* **43**, 231–6.

Cramer, W.A., Dankert, J.R. and Uratani, Y. (1983). The membrane channel-forming bacteriocidal protein, colicin E1. *Biochim. Biophys. Acta* **737**, 173–93.

Criado, M. and Keller, B.U. (1987). A membrane fusion strategy for single-channel recording of membranes usually non-accessible to patch-clamp pipette electrodes. *FEBS Lett.* **224**, 172–6.

Dascal, N. (1987). The use of *Xenopus* oocytes for the study of ion channels. *CRC Crit. Rev. Biochem.* **22**, 317–87.

Davidson, V.L., Brunden, K.R., Cramer, W.A. and Cohen, F.S. (1984). Studies on the mechanism of action of channel-forming colicins using artificial membranes. *J. Memb. Biol.* **79**, 105–18.

Deck, K.A., Kern, R. and Trautwein, W. (1964). Voltage clamp technique in mammalian cardiac fibers. *Pflügers Arch.* **280**, 50–62.

DeFelice, L. (1981). *Introduction to Membrane Noise.* Plenum Press, New York.

Donovan, J.J. and Middlebrook, J.L. (1986). Ion-conducting channels produced by botulinum toxin in planar lipid membranes. *Biochemistry* **25**, 2872–6.

Donovan, J.J., Simon, M.I., Draper, R.K. and Montal, M. (1981). Diphtheria toxin forms transmembrane channels in planar lipid bilayers. *Proc. Natl. Acad. Sci. USA* **78**, 172–6.

Donovan, J.J., Simon, M.I. and Montal, M. (1982).

Insertion of diphtheria toxin into and across membranes: role of phosphoinositide asymmetry. *Nature* **298**, 669–72.

Dreyer, F., Maric, K. and Lutz, F. (1990). The use of patch-clamp technique for the study of the mode of action of bacterial toxins. In: *Bacterial Protein Toxins* (eds R. Rappuoli *et al.*) pp. 227–34. Gustav Fischer Verlag, Stuttgart.

Duncan, J.L. and Buckingham, L. (1981). Effect of streptolysin S on liposomes. Influence of membrane lipid composition on toxin action. *Biochim. Biophys. Acta* **648**, 6–12.

Eidels, L., Proia, R.L. and Hart, D.A. (1983). Membrane receptors for bacterial toxins. *Microbiol. Rev.* **47**, 596–620.

Eisenbach, M., Margolin, Y., Ciobotariu, A. and Rottenberg, H. (1984). Distinction between changes in membrane potential and surface charge upon chemotactic stimulation of *Escherichia coli*. *Biophys. J.* **45**, 463–7.

Finkelstein, R.A., Burks, M.F., Zupan, A., Dallas, W.S., Jacob, C.O. and Ludwig, D.S. (1987). Epitopes of the cholera family of enterotoxins. *Rev. Infect. Dis.* **9**, 544–61.

Forti, S. and Menestrina, G. (1989). Staphylococcal α-toxin increases the permeability of lipid vesicles by a cholesterol and pH dependent assembly of oligomeric channels. *Eur. J. Biochem.* **181**, 767–73.

Fraser, J.D. (1989). High-affinity binding of staphylococcal enterotoxins A and B to HLA-DR. *Nature* **339**, 221–3.

Freer, J.H. and Birkbeck, T.H. (1982). Possible conformation of delta-lysin, a membrane-damaging peptide of *Staphylococcus aureus*. *J. Theor. Biol.* **94**, 535–40.

Füssle, R., Bhakdi, S., Sziegoleit, A., Tranum-Jensen, J., Kranz, T. and Wellensiek, H.J. (1981). On the mechanism of membrane damage by *Staphylococcus aureus* α-toxin. *J. Cell Biol.* **91**, 83–94.

Gambale, F. and Montal, M. (1988). Characterization of the channel properties of tetanus toxin in planar lipid bilayers. *Biophys. J.* **53**, 771–83.

Garcia, A.M. (1986). Methodologies to study channel-mediated ion fluxes in membrane vesicles. In: *Ionic Channels in Cells and Model Systems* (ed. R. Latorre), pp. 127–40. Plenum Press, New York.

Grinvald, A., Hildesheim, R., Farber, I.C. and Anglister, R. (1982). Improved fluorescent probes for the measurement of rapid changes in membrane potential. *Biophys. J.* **39**, 301–8.

Hamill, O.P., Marty, A., Neher, E. and Sakmann, B. (1981). Improved patch-clamp techniques for high resolution current recording from cells and cell-free membrane patches. *Pflügers Arch.* **391**, 85–100.

Haider, M.Z. and Ellar, D.J. (1989). Mechanism of action of *Bacillus thuringiensis* δ-endotoxin: interaction with phospholipid vesicles. *Biochim. Biophys. Acta* **978**, 216–22.

Hanke, W. (1986). Incorporation of ion channels by fusion. In: *Ion Channel Reconstitution* (ed. C. Miller), pp. 141–53. Plenum Press, New York.

Hanke, W., Methfessel, C., Wilmsen, U. and Boheim, G. (1984). Ion channel reconstitution into lipid bilayer membranes on glass pipettes. *Bioelectrochem. Bioenerg.* **122**, 329–39.

Hess, G.P. and Udgaonkar, B. (1987). Chemical kinetic measurement of transmembrane processes using rapid reaction techniques: acetylcholine receptor. *Ann. Rev. Biophys. Biophys. Chem.* **16**, 507–34.

Hille, B. (1984). *Ionic Channels of Excitable Membranes*. Sinauer Associates Publishers, Sunderland, Massachusetts.

Hladky, S.B. and Haydon, P.A. (1970). Discreteness of conductance changes in bimolecular lipid membranes in the presence of certain antibiotics. *Nature (Lond.)* **225**, 451–3.

Hoch, D.H., Romero-Mira, M., Ehrlich, B.E., Finkelstein, A., DasGupta, B.R. and Simpson, L.L. (1985). Channels formed by botulinum, tetanus, and diphtheria toxins in planar lipid bilayers: relevance to translocation of proteins across membranes. *Proc. Natl. Acad. Sci. USA* **82**, 1692–6.

Hodgkin, A.L. and Huxley, A.F. (1952a). Currents carried by sodium and potassium ions through the membrane of the giant axon of Loligo. *J. Physiol. (Lond.)* **116**, 449–72.

Hodgkin, A.L. and Huxley, A.F. (1952b). A quantitative description of membrane current and its application to conduction and excitation in nerve. *J. Physiol. (Lond.)* **117**, 500–44.

Holland, B.I., Wang, R., Seror, S.J. and Blight, M. (1989). Haemolysin secretion and other protein translocation mechanisms in gram-negative bacteria. In: *Society for General Microbiology Symposium, 44* (eds M. Banmberg, I. Hunter and M. Rhodes), pp. 219–54. Cambridge University Press, Cambridge.

Holmgren, J. (1985). Toxins affecting intestinal transport processes. *Spec. Publ. Soc. Gen. Microbiol.* **13**, 177–91.

Homan, R.J. (1988). Aggregation of IgE receptors induces degranulation in rat basophilic leukemia cells permeabilized with α-toxin from *Staphylococcus aureus*. *Proc. Natl. Acad. Sci. USA* **85**, 1624–8.

Horn, R. and Marty, A. (1988). Muscarinic activation of ionic currents measured by a new whole-cell recording method. *J. Gen. Physiol.* **92**, 145–59.

Howard, S.P. and Buckley, J.T. (1982). Membrane glycoprotein receptor and hole-forming properties of a cytolytic protein toxin. *Biochemistry* **21**, 1662–7.

Howard, S.P. and Buckley, J.T. (1985). Activation of the hole-forming toxin aerolysin by extracellular processing. *J. Bacteriol.* **163**, 336–40.

Huott, P.A., McRoberts, J.A., Giannella, R.A. and Dharmsathaphorn, K. (1988). Mechanism of action of *Escherichia coli* heat-stable enterotoxin in a human colonic cell line. *J. Clin. Invest.* **82**, 514–23.

Hwang, J., Fitzgerald, D.J., Adhya, S. and Pastan, I. (1987). Functional domains of *Pseudomonas exotoxin* identified by deletion analysis of the gene expressed in *E. coli*. *Cell* **48**, 129–36.

Inouè, S. (1986). *Video Microscopy*. Plenum Press, New York.

Jackson, M.B., Stephens, C.L. and Lecar, H. (1981). Single channel currents induced by complement in antibody-coated cell membranes. *Proc. Natl. Acad. Sci. USA* **78**, 6421–5.

Jolivet-Reynaud, C. and Alouf, J. (1984). Study of membrane disruption by *Clostridium perfringens* delta-toxin: dissociation of binding and lytic step. In: *Bacterial Protein Toxins* (eds J. Alouf *et al.*), pp. 327–8. Academic Press, New York.

Jorgensen, S.E., Hammer, R.F. and Wu, G.K. (1980). Effect of a single hit from the alpha-haemolysin produced by *Escherichia coli* on the morphology of sheep erythrocytes. *Infect. Immun.* **27**, 988–94.

Kagan, B., Finkelstein, A. and Colombini, M. (1981). Diphtheria toxin fragment forms large pores in phospholipid bilayer membranes. *Proc. Natl. Acad. Sci. USA* **78**, 4950–4.

Kagawa, Y. and Racker, E. (1971). Partial resolution of the enzymes catalyzing oxidative phosphorylation. XXV. Reconstitution of vesicles catalyzing inorganic phosphorus-32-adenosine triphosphate exchange. *J. Biol. Chem.* **246**, 5477–87.

Katz, B. (1949). Les constantes electriques de la membrane du muscle. *Arch. Sci. Physiol.* **2**, 285–99.

Kayalar, C. and Düzgünes, N. (1986). Membrane action of colicin E1: detection of the release of carboxyfluorescein and calcein from liposomes. *Biochim. Biophys. Acta* **869**, 51–6.

Kirk, K., Kuchel, P.W. and Labotka, R.J. (1988). Hypophosphite ion as a ^{31}P nuclear magnetic probe of membrane potential in erythrocyte suspensions. *Biophys. J.* **54**, 241–7.

Knowles, B.H., Blatt, M.R., Tester, M., Horsnell, J.M., Carrol, J., Menestrina, G. and Ellar, D.J. (1989). A cytolytic δ-endotoxin from *Bacillus thuringiensis* var. israelensis forms cation selective channels in planar lipid bilayers. *FEBS Lett.* **244**, 259–62.

Koronakis, V., Cross, M., Senior, B., Koronakis, E. and Hughes, C. (1987). The secreted haemolysins of *Proteus mirabilis*, *Proteus vulgaris* and *Morganella morganii* are genetically related to each other and to the alpha-haemolysin of *Escherichia coli*. *J. Bacteriol.* **169**, 1509–15.

Lalonde, G., McDonald, T.V., Gardner, P. and O'Hanley, P.D. (1989). Identification of a haemolysin from *Actinobacillus pleuropneumoniae* and characterization of its channel properties in planar phospholipid bilayers. *J. Biol. Chem.* **264**, 13559–64.

Langmuir, I. (1933). Oil lenses on water and the nature of monomolecular expanded films. *J. Chem. Phys.* **1**, 756–76.

Latorre, R. (ed). (1986). *Ionic Channels in Cells and Model Systems*. Plenum Press, New York.

Latorre, R. and Alvarez, O. (1981). Voltage dependent channels in planar lipid bilayer membranes. *Physiol. Rev.* **61**, 77–150.

Latorre, R. and Hall, J.E. (1976). Dipole potential measurement in asymmetric membranes. *Nature (Lond.)* **264**, 361–3.

Lazdunski, C.J., Baty, D., Geli, V., Cavard, D., Morlon, J., Lloubes, R., Howard, S.P., Knibiehler, M., Chartier, M., Varenne, S., Frenette, M., Dasseux, J.-L. and Pattus, F. (1988). The membrane channel-forming colicin A: synthesis, secretion, structure, action and immunity. *Biochim. Biophys. Acta* **947**, 445–64.

Lazure, C., Seidah, N.G., Chretien, M., Lallier, R. and St-Pierre, S. (1983). Primary structure determination of *Escherichia coli* heat-stable enterotoxin of porcine origin. *Can. J. Biochem. Cell Biol.* **61**, 287–92.

Lo, R.Y.C., Strathdee, C.A. and Shewen, P.E. (1987). Nucleotide sequence of the leucotoxin genes of *Pasteurella haemolytica* A1. *Infect. Immun.* **55**, 1987–96.

Loew, L.M. (ed.) (1988). *Spectroscopic Membrane Probes*. CRC Press, Boca Raton, Florida.

Ludwig, A., Jarchau, T., Benz, R. and Goebel, W. (1988). The repeat domain of *Escherichia coli* haemolysin (HlyA) is responsible for its Ca^{2+}-dependent binding to erythrocytes. *Molec. Gen. Genet.* **214**, 553–61.

Lutz, F., Maurer, M. and Failing, K. (1987). Cytotoxic protein from *Pseudomonas aeruginosa*: formation of hydrophilic pores in Erlich ascites tumor cells and effect on cell viability. *Toxicon* **25**, 293–305.

McClane, B.A. (1984). Osmotic stabilizers differentially inhibit permeability alterations induced in Vero cells by *Clostridium perfringens* enterotoxin. *Biochim. Biophys. Acta* **777**, 99–106.

McClane, B.A., Wnek, A.P., Hulkower, K.I. and Hanna, P. (1988). Divalent cation involvement in the action of *Clostridium perfringens* type A enterotoxin. Early events are divalent cation-independent. *J. Biol. Chem.* **263**, 2423–35.

McEwen, B.F. and Arion, W.J. (1985). Permeabilization of rat erythrocytes with *Staphylococcus aureus* α-toxin. *J. Cell Biol.* **100**, 1922–9.

Maconochie, D.J. and Knight. D.E. (1989). A method for making solution changes in the sub-millisecond range at the tip of a patch pipette. *Pflügers Arch.* **414**, 589–96.

Maget-Dana, R., Heitz, F., Ptak, M., Peypoux, F. and Guinand, M. (1985a). Bacterial lipopeptides induce ion-conducting pores in planar bilayers. *Biochem. Biophys. Res. Commun.* **129**, 965–71.

Maget-Dana, R., Ptak, M., Peypoux, F. and Michel, G. (1985b). Pore-forming properties of iturin A, a lipopeptide antibiotic. *Biochim. Biophys. Acta* **815**, 405–9.

Martinez, C., Lazdunski, C.J. and Pattus, F. (1987). Isolation, molecular and functional properties of the C-terminal domain of colicin A. *EMBO J.* **2**, 1501–7.

Matsuda, M. and Sugimoto, N. (1979). Calcium-independent and dependent steps in action of *Clostridium perfringens* enterotoxin on HeLa and Vero cells. *Biochem. Biophys. Res. Commun.* **91**, 629–36.

Mellors, I.R., Thomas, D.H. and Sansom, M.S.P. (1988). Properties of ion channels formed by *Staphylococcus aureus* δ-toxin. *Biochim. Biophys. Acta* **942**, 280–94.

Menestrina, G. (1986). Ionic channels formed by *Staphylococcus aureus* alpha-toxin: voltage dependent inhibition by di and trivalent cations. *J. Membr. Biol.* **90**, 177–90.

Menestrina, G. (1988). *Escherichia coli* haemolysin permeabilizes small unilamellar vesicles loaded with calcein by a single hit mechanism. *FEBS Lett.* **232**, 217–20.

Menestrina, G., Mackman, N., Holland, I.B. and Bhakdi, S. (1987). *Escherichia coli* haemolysin forms voltage-dependent channels in lipid membranes. *Biochim. Biophys. Acta* **905**, 109–17.

Menestrina, G., Forti, S. and Gambale, F. (1989). Interaction of tetanus toxin with lipid vesicles. Effects of pH, surface charge and transmembrane potential on the kinetics of channel formation. *Biophys. J.* **55**, 393–405.

Menestrina, G., Bashford, C.L. and Pasternak, C.A. (1990). Pore-forming toxins: experiments with *S. aureus* α-toxin, *C. perfringens* θ-toxin and *E. coli* haemolysin in lipid bilayers, liposomes and intact cells. *Toxicon* **28**, 477–91.

Methfessel, C., Witzeman, V., Takahashi, T., Mishina, M., Numa, S. and Sakmann, B. (1986). Patch clamp measurements on *Xenopus laevis* oocytes: current through endogenous channels and implanted acetylcholine receptor and sodium channels. *Pflügers Arch.* **407**, 577–88.

Middlebrook, J.L. and Dorland, R.B. (1984). Bacterial toxins: cellular mechanisms of action. *Microbiol. Rev.* **48**, 199–221.

Miledi, R., Parker, I. and Sumikawa, K. (1982). Properties of acetylcholine receptors translated by cat muscle mRNA in *Xenopus* oocytes. *EMBO J.* **1**, 1307–12.

Miller, C. (1984). Ion channels in liposomes. *Ann. Rev. Physiol.* **46**, 459–8.

Miller, C. (ed.) (1986). *Ion Channel Reconstitution*. Plenum Press, New York.

Miller, C. and Racker, E. (1976). Calcium induced fusion of fragmented sarcoplasmic reticulum with artificial planar bilayers. *J. Membr. Biol.* **30**, 283–300.

Moczydlowski, E., Uehara, A. and Hall, S. (1986). Blocking pharmacology of batrachotoxin-activate sodium channels. In: *Ion Channel Reconstitution* (ed. C. Miller), pp. 405–28. Plenum Press, New York.

Montal, M. and Mueller, P. (1972). Formation of bimolecular membranes from lipid monolayers and a study of their electrical properties. *Proc. Natl. Acad. Sci. USA* **69**, 3561–6.

Moss, J., Fishman, P.H., Richards, R.L., Alving, C.R., Vaughan, M. and Brady, R.O. (1976a). Choleragen-mediated release of trapped glucose from liposomes containing gangliosides G_{M1}. *Proc. Natl. Acad. Sci. USA* **73**, 3480–3.

Moss, J., Richards, R.L., Alving, C.R. and Fishman, P.H. (1976b). Effect of the A and B promoters of choleragen on release of trapped glucose from liposomes containing or lacking gangliosides G_{M1}. *J. Biol. Chem.* **252**, 797–8.

Mueller, P. and Rudin, D. (1968). Action potentials induced in biomolecular lipid membranes. *Nature (Lond.)* **217**, 713–19.

Mueller, P. and Rudin, D. (1969). Translocators in bimolecular lipid membranes: their role in dissipative and conservative bioenergy transduction. *Curr. Top. Bioenerg.* **3**, 157–249.

Mueller, P., Rudin, D., Tien, H.T. and Westcott, W.C. (1962). Reconstitution of cell membrane structure in vitro and its transformation into an excitable system. *Nature (Lond.)* **194**, 979–81.

Neher, E. and Sakmann, B. (1976). Single-channel currents recorded from membrane of denervated frog muscle fibers. *Nature (Lond.)* **260**, 799–802.

Neher, E. and Stevens, C.F. (1977). Conductance fluctuations and ionic pores in membranes. *Ann. Rev. Biophys. Bioeng.* **6**, 345–81.

Nernst, W. (1888). Zur kinetic der lösung befindlichen körper: theorie der diffusion. *Z. Phys. Chem.* 613–37.

Nogueira, R.A. and Varanda, W.A. (1988). Gating properties of channels formed by colicin Ia in planar lipid bilayer membranes. *J. Membr. Biol.* **105**, 143–53.

Olsnes, S., Moskaug, J.O., Stenmark, H. and Sandvig, K. (1988). Diphtheria toxin entry: protein translocation in the reverse direction. *Trends Biochem. Sci.* **13**, 348–51.

Papini, E., Sandonà, D., Rappuoli, R. and Montecucco, C. (1988). On the membrane translocation of diphtheria toxin: at low pH the toxin induces ion channels on cells. *EMBO J.* **7**, 3353–9.

Parker, M.W., Pattus, F., Tucker, A.D. and Tsernoglou, D. (1989). Structure of the membrane-pore-forming fragment of colicin A. *Nature* **337**, 93–6.

Pasquali, F., Menestrina, G. and Antolini, R. (1984). Fusion of large unilamellar liposomes containing hemocyanin with planar bilayer membranes. *Z. Naturforsch. C* **39**, 147–55.

Penner, R., Neher, E. and Dreyer, F. (1986). Intracellularly injected tetanus toxin inhibits exocytosis in bovine adrenal chromaffin cells. *Nature (Lond.)* **324**, 76–8.

Peterson, A.A. and Cramer, W.A. (1987). Voltage-dependent, monomeric channel activity of colicin E1 in artificial membrane vesicles. *J. Membr. Biol.* **99**, 197–204.

Pressler, U., Braun, V., Wittmann-Liebold, B. and Benz, R. (1986). Structural and functional properties of colicin B. *J. Biol. Chem.* **261**, 2654–9.

Pokornj, J. and Ziegler, W. (1984). Is the syringotoxin-channel permeable to Pr^{3+} ions? *Biologia (Bratislava)* **39**, 701–6.

Qin, D. and Noma, A. (1988). A new oil-gate concentration jump technique applied to inside-out patch-clamp recording. *Am. J. Physiol.* **255**, H980–H984.

Quentin, M.J., Besson, F., Peypoux, F. and Michel,

G. (1982). Action of peptidolipidic antibiotics of the iturin group on erythrocytes. Effect of some lipids on haemolysis. *Biochim. Biophys. Acta* **684**, 207–11.

Robertson, D.E. and Rottenberg, H. (1983). Membrane potential and surface potential in mitochondria. Fluorescence and binding of 1-anilinonaphtalene-8-sulfonate. *J. Biol. Chem.* **25**, 11039–48.

Sakmann, B. and Neher, E. (eds) (1983). *Single Channel Recording*. Plenum Press, New York.

Sakmann, B., Patlak, J. and Neher, E. (1980). Single acetylcholine-activated channels show burst-kinetics in presence of desensitizing concentrations of agonist. *Nature (Lond.)* **286**, 71–3.

Saxena, S.K., O'Brien, A.D. and Ackerman, E.J. (1989). Shiga toxin, Shiga-like toxin II variant, and ricin are all single-site RNA N-glycosidases of 28 S RNA when microinjected into *Xenopus* oocytes. *J. Biol. Chem.* **264**, 596–601.

Schein, S.J., Kagan, B.L. and Finkelstein, A. (1978). Colicin K acts by forming voltage-dependent channels in phospholipid bilayer membranes. *Nature* **276**, 159–63.

Schindler, H. (1980). Formation of planar bilayers from artificial or native membrane vesicles. *FEBS Lett.* **122**, 77–9.

Schmidt, J.J. and Spero, L. (1983). The complete amino acid sequence of staphylococcal enterotoxin C_1. *J. Biol. Chem.* **258**, 6300–6.

Schuerholz, T. and Schindler, H. (1983). Formation of lipid–protein bilayers by micropipette guided contact of two monolayers. *FEBS Lett.* **152**, 187–90.

Segre, A., Bachmann, R.C., Ballio, A., Bossa, F., Grgurina, I., Iacobellis, N.S., Marino, G., Pucci, P., Simmaco, M. and Takemoto, J.Y. (1989). The structure of syringomycins A_1, E and G. *FEBS Lett.* **255**, 27–31.

Shiver, J.W. and Donovan, J.J. (1987). Interaction of diphtheria toxin with lipid vesicles: determinants of ion channel formation. *Biochim. Biophys. Acta* **903**, 48–55.

Shone, C.C., Hambleton, P. and Melling, J. (1987). A 50-kDa fragment from the NH_2-terminus of the heavy subunit of *Clostridium botulinum* type A neurotoxin forms channels in lipid vesicles. *Eur. J. Biochem.* **167**, 175–80.

Sigworth, F.J. and Neher, E. (1980). Single Na^+ channel currents observed in cultured rat muscle cells. *Nature (Lond.)* **287**, 447–9.

Singh, Y., Leppla, S.H., Bhatnagar, R. and Friedlander, A.M. (1989). Internalization and processing of *Bacillus anthracis* lethal toxin by toxin-sensitive and -resistant cells. *J. Biol. Chem.* **264**, 11099–102.

Skulachev, V.P. (1988). *Membrane Bioenergetics*. Springer Verlag, Berlin.

Slavik, J. (1982). Anilinonaphtalene sulfonate as a probe of membrane composition and function. *Biochim. Biophys. Acta* **694**, 1–25.

Snuthc, T.P. (1988). The use of the *Xenopus* oocytes to probe synaptic communication. *Trends Neurosci.* **11**, 250–6.

Stevens, C.F. (1972). Inferences about membrane properties from electrical noise measurements. *Biophys. J.* **12**, 1028–47.

Sugimoto, N., Takagi, M., Ozutsumi, K., Harada, S. and Matsuda, M. (1988). Enterotoxin of *Clostridium perfringens* type A forms ion-permeable channels in a lipid bilayer membrane. *Biochem. Biophys. Res. Commun.* **156**, 551–6.

Takao, T., Shimonishi, Y., Kobayashi, M., Nishimura, O., Arita, M., Takeda, T., Honda, T. and Miwatani, T. (1985). Amino acid sequence of heat-stable enterotoxins produced by *Vibrio cholerae* non-01. *FEBS Lett.* **193**, 250–4.

Tank, D.W., Miller, C. and Webb, W.W. (1982). Isolated-patch recording from liposomes containing functionally reconstituted chloride channels from *Torpedo* electroplax. *Proc. Natl. Acad. Sci. USA* **79**, 7749–53.

Tappin, M.J., Pastore, A., Norton, R.S., Freer, J.H. and Campbell, I.D. (1988). High-resolution ^1H NMR study of the solution structure of δ-haemolysin. *Biochemistry* **27**, 1643–7.

Thelestam, M. and Blomqvist, L. (1988). Staphylococcal α-toxin – recent advances. *Toxicon* **26**, 51–65.

Thelestam, M. and Möllby, R. (1979). Classification of microbial, plant and animal cytolysins based on their membrane-damaging effects on human fibroblasts. *Biochim. Biophys. Acta* **557**, 156–69.

Tobkes, N., Wallace, B.A. and Bayley, H. (1985). Secondary structure and assembly mechanism of an oligomeric channel protein. *Biochemistry* **24**, 1915–20.

Tosteson, M.T. and Tosteson, D.C. (1978). Bilayers containing gangliosides develop channels when exposed to cholera toxin. *Nature* **275**, 142–4.

Waggoner, A.S. (1979). The use of cyanine dyes for the determination of membrane potentials in cells, organelles, and vesicles. *Methods Enzymol.* **55**, 689–95.

Watanabe, M., Tomita, T. and Yasuda, T. (1987). Membrane-damaging action of staphylococcal α-toxin on phospholipid cholesterol liposomes. *Biochim. Biophys. Acta* **898**, 257–65.

Weaver, C.A., Kagan, B.L., Finkelstein, A. and Konisky, J. (1981). Mode of action of colicin Ib, formation of ion-permeable membrane channels. *Biochim. Biophys. Acta* **645**, 137–42.

Weiner, R.N., Schneider, E., Haest, C.W.M., Deuticke, B., Benz, R. and Frimmer, M. (1985). Properties of the leak permeability induced by a cytotoxic protein of *Pseudomonas aeruginosa* (PACT) in rat erythrocytes and black lipid membranes. *Biochim. Biophys. Acta* **820**, 173–82.

White, S.H. (1986). The physical nature of planar bilayer membranes. In: *Ion Channel Reconstitution* (ed. C. Miller), pp. 3–35. Plenum Press, New York.

Yianni, Y.P., Fitton, J.E. and Morgan, C.G. (1986). Lytic effects of melittin and δ-hemolysin from *Staphylococcus aureus* on vesicles of dipalmitoylphosphatidylcholine. *Biochim. Biophys. Acta* **856**, 91–100.

Zhang, L. and Takemoto, J.Y. (1986). Mechanism of

action of *Pseudomonas syringae* phytotoxin, syringomycin. Interaction with the plasma membrane of wild-type and respiratory-deficient strains of *Saccharomyces cerevisiae*. *Biochim. Biophys. Acta* **861**, 201–4.

Ziegler, W., Pavlovkin, J. and Pokornj, J. (1984). Effect of syringotoxin on the permeability of bilayer lipid membranes. *Biologia (Bratislava)* **39**, 693–9.

Ziegler, W., Pavlovkin, J., Remis, D. and Pokornj, J. (1986). The anion/cation selectivity of the syringotoxin channel. *Biologia (Bratislava)* **41**, 1091–6.

11

Leukocidal Toxins

Masatoshi Noda and Iwao Kato

The Second Department of Microbiology, Chiba University School of Medicine, Chiba 280, Japan

Introduction

Leukocidins produced by *Staphylococcus aureus* and *Pseudomonas aeruginosa* which are cytotoxic to polymorphonuclear leukocytes and macrophages but not to erythrocytes and other cells, are called leukocidal toxins. These toxins are clearly distinct from various cytotoxins that possess cytotoxicity not only to phagocytes but also to other cells. Leukocidal toxins are known to be important in the pathogenicity of staphylococcal and pseudomonal infections.

Staphylococcal leukocidin

Purification of leukocidin (Noda *et al.*, 1980a)

The V8 strain of *Staphylococcus aureus* (American Type Culture Collection No. 27733) is believed to produce more leukocidin than most staphylococcal strains. The cocci are grown at 37°C for 22 h on an enriched medium (Table 1) (Gladstone and van Heyningen, 1957) in a reciprocating shaker (120 cycles/min). The culture is then centrifuged at $11\,000 \times g$ for 20 min at 4°C, and the clear supernatant fluid (5000 ml, step I) is purified at 4°C by the following procedure. $ZnCl_2$ (3.7 M) is added dropwise to the supernatant (pH 6.5) until a final concentration of 75 mM is reached. After 30 min at 4°C, a precipitate is formed which is collected by centrifugation at $8\,000 \times g$ for 15 min.

Table 1. Composition of growth medium (pH 7.4)

Component	Amount
Yeast extract	25 g
Casamino acid	20 g
Sodium glycerophosphate	20 g
$Na_2HPO_4 \cdot 12H_2O$	6.25 g
KH_2PO_4	400 mg
$MgSO_4 \cdot 7H_2O$	20 mg
$MnSO_4 \cdot 4H_2O$	10 mg
Sodium lactate, 50%	19.8 ml
$FeSO_4 \cdot 7H_2O$, 0.32% (w/v), plus citric acid, 0.32% (w/v)	2 ml
Distilled water to total	1 litre

The pellet is dissolved gradually in 0.4 M sodium phosphate buffer (pH 6.5). To remove metal ions, the solution is dialysed against 0.05 M sodium acetate buffer (pH 5.2) containing 0.2 M NaCl (buffer A). Solid $(NH_4)_2SO_4$ is then added until saturation is reached, and the solution allowed to stand overnight. The precipitate is collected by centrifugation at $15\,000 \times g$ for 20 min, dissolved in a small volume of buffer A, and then dialysed against buffer A. The dialysate (315×10^5 U, step II) is applied to a column (5 × 90 cm) of carboxymethyl-Sephadex C-50 equilibrated with buffer A. The fractions containing the highest leukocidin activity are eluted with 0.05 M sodium acetate buffer (pH 5.2) containing 1.2 M NaCl. The active fractions are pooled (254 × 10⁵U, step III), saturated with $(NH_4)_2SO_4$ by addition of the salt, and allowed to stand overnight. The precipitate is collected by centrifugation, dissolved in buffer A, and then dialysed against buffer A.

Further purification is achieved by applying the leukocidin preparation to a column (2.5 × 85 cm) of carboxymethyl-Sephadex C-50 equilibrated with buffer A and elution with a linear gradient from 0.2 to 1.2 M NaCl in 0.05 M sodium acetate buffer (pH 5.2). The fractions with F (fast) component activity, which eluted at about 0.45 M NaCl (230 × 10⁵ U, step IV), and those with S component activity, which eluted at about 0.9 M NaCl (235 × 10⁵ U, step IV), are pooled and concentrated by dialysing against saturated $(NH_4)_2SO_4$.

Each concentrated solution is applied to a column (2.5 × 85 cm) of Sephadex G-100 equilibrated with buffer A. The fractions with F or S component activity (F component, 217 × 10⁵ U or S component, 231 × 10⁵ U, step V) are concentrated by dialysing against saturated $(NH_4)_2SO_4$, and then dialysed against 0.05 M veronal buffer (pH 8.6). After dialysis each component is purified by zone electrophoresis on starch (12 mA for 20 h (vessel, 1.5 × 3 × 40 cm)) using 0.05 M veronal buffer (pH 8.6).

Both the F and S fractions show a single protein peak associated with leukocidin activity. Fractions with the highest activity (F component, 210 × 10⁵ U or S component, 209 × 10⁵ U, step VI) are dialysed against buffer A and pooled for crystallization. The recovery and degree of purification at each step are shown in Table 2. The yields of the highly purified F and S components of leukocidin are about 2 mg/l of culture filtrate.

Crystallization of the F and S components of leukocidin

The F and S components of leukocidin are crystallized by dialysing against a saturated $(NH_4)_2SO_4$ solution, pH 7.0, at 4°C for 16 h. Slight opalescence appears in 6 h, along with a white precipitate on the bottom of the cellulose tube. The crystalline precipitate formed is collected by gentle centrifugation at 4°C and dissolved with the same volume of chilled 0.05 M sodium acetate

buffer (pH 5.2) containing 0.2 M NaCl. The crystals are highly soluble in the buffer. Recrystallization of the two components is repeated twice more by dialysing the component solution against a large volume of a 95% saturated $(NH_4)_2SO_4$ solution at 4°C. Microscopic examinations of the white precipitates of the purified F and S components of leukocidin in the dialysis bag reveals crystals in the form of square plates and very fine needles, respectively.

Determination of leukocidin activity

Polymorphonuclear leukocytes are prepared from rabbit peripheral blood on an isokinetic gradient of Ficoll (Pretlow and Luberoff, 1973) and washed with Hanks' solution. The purity of polymorphonuclear leukocyte suspensions averages 98% as judged by examination of Giemsa-stained smears. Viability averages 99% as assayed by nigrosine dye exclusion. Aliquots of 10⁶ cells (10 μl) are placed on glass slides. Since the F and S fractions individually have little or no leukocidin activity, to quantify the components of leukocidin, 6 ng of the S (or F) component of leukocidin is added to serially diluted fractions containing the other component.

To standardize leukocidin, the microscopic slide adhesion method (Gladstone and van Heyningen, 1957) is used. Serial dilutions (10 μl) of each component of leukocidin in 0.1 M phosphate-buffered saline (pH 7.2) containing 0.5% gelatin are incubated at 37°C for 10 min with the other component (6 ng) on a glass slide (total volume, 20 μl) in a moist chamber. After incubation, morphological changes in a slide field (about 1000 cells) are observed by phase-contrast microscopy. The end point is the smallest amount of leukocidin causing about 100% morphological changes in a standard polymorphonuclear leukocyte suspension (10⁶ cells), the number of units in a leukocidin preparation being numerically equal to the dilution at the end point. When 500 pg of purified F component and various amounts (100–700 pg) of purified S component are incubated with suspensions of polymorphonuclear leukocytes (10⁶ cells) on the glass slide (total volume, 20 μl), optimal destruction of all polymorphonuclear leukocytes is obtained at 500 pg of S component (Fig. 1). A linear relationship between polymorphonuclear leukocyte destruction (percentage) and

Table 2. Preparation of Staphylococcal leukocidin

Step		Volume (ml)	Total protein (mg)	Total activity (1×10^{-5}U)	Specific activity (1×10^{-5} U/mg protein)	Recovery of activity (%)	Recovery of protein (%)
I.	Culture filtrate	5 000	13 840	346	0.025	100	100
II.	ZnCl$_2$ precipitation	300	6 330	315	0.050	91.1	45.7
III.	1.2 M NaCl effluent	1 500	1 420	254	0.180	73.4	10.3
IV.	Gradient elution						
	F component	50	138	230[a]	1.67	66.5	1.0
	S component	80	47.1	235[a]	4.99	67.9	0.34
V.	Gel filtration						
	F component	45	39.9	217[a]	5.44	62.7	0.29
	S component	40	42.0	231[a]	5.50	66.8	0.30
VI.	Zone electrophoresis						
	F component	15	12.0	210[a]	17.50	60.7	0.09
	S component	10	11.0	209[a]	19.00	60.4	0.08

[a]For the determination of the activity of each fraction, the F (or S) component (6 ng) was added into the serially diluted S (or F) fraction solution in 20 μl of phosphate-buffered saline (pH 7.2) containing 0.5% gelatin, since each fraction itself has no or little leukocidin activity.

protein concentration is obtained in the range of 300–500 pg.

Results similar to those shown in Fig. 1 are obtained when 500 pg of S component is incubated with various amounts (100–700 pg) of F component and polymorphonuclear leukocytes (10^6 cells). Optimal destruction of all polymorphonuclear leukocytes is obtained at 500 pg of F component. One unit of each component of leukocidin is obtained at 500 pg. When a suspension of rabbit polymorphonuclear leukocytes (10^6 cells) is incubated with 500 pg of both components of leukocidin, it is revealed by light microscopy that, after 2 min of the incubation, the polymorphonuclear leukocytes become round and somewhat swollen; the lobulated nuclei eventually become spherical. The terminal event in leukocidin action is cytoplasmic degranulation, rupture of nuclei and complete cell lysis. These effects are fully developed after 10 min of incubation.

Leukocidin activity is determined also by assaying for ^{86}Rb released from ^{86}Rb-labelled polymorphonuclear leukocytes. Similar data are obtained by examining morphological changes of polymorphonuclear leukocytes. The study of leukocidin effects is confined to the first 10 min of intoxication; maximal effect results from treating 10^6 polymorphonuclear leukocytes per 20 μl with 500 pg of

both components of leukocidin. ^{51}Cr release assay is also reported (Loeffler *et al.*, 1986).

Physico-chemical properties of leukocidin

The F and S components crystallize in the form of plates and needles, respectively. The N-terminal residues of both components are alanine. No detectable half-cystine and free sulphydryl groups are found in either component. The crystallized components migrate as single bands on the SDS–polyacrylamide gels. Molecular weights of crystallized F and S components of leukocidin determined from relative mobilities of marker proteins are 32 000 and 31 000, respectively. The isoelectric points (pI) of F and S components are pI 9.08 ± 0.05 and 9.39 ± 0.05, respectively. The purified F and S components of leukocidin stored at $-80°C$ in 0.1 M phosphate-buffered saline (pH 7.2) containing 0.5% gelatin are stable for at least 3 months.

Mode of action of leukocidin

Staphylococcal leukocidin, known to be important in the pathogenicity of certain staphylococcal diseases (Woodin, 1970; Wenk and Blobel, 1970), consists of two protein components (S and F) that

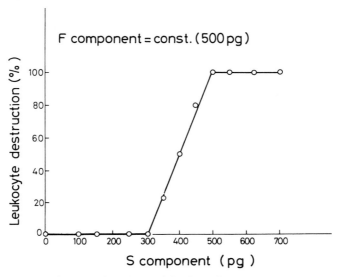

Figure 1. Titration of the optimum leukocytolytic doses of the F and S components. Mixtures of 0.5 ng of F component and varying amounts (0.1–0.7 ng) of S component were incubated at 37°C for 10 min with suspensions of leukocytes (10^6 cells) in 20 μl of 0.1 M phosphate buffer (pH 7.2) containing 0.5% gelatin. Details are described in the text.

act synergistically to induce cytotoxic changes in human and rabbit polymorphonuclear leukocytes and macrophages (Woodin, 1960; Soboll et al., 1973; Noda et al., 1980a; Kato et al., 1988). No other cell type has been found to be susceptible except human promyelocytic leukaemia cell line HL-60 (Morinaga et al., 1987a,b, 1988). These components when tested individually are not cytotoxic. When rabbit polymorphonuclear leukocytes have been preincubated with the S component at 37°C for 10 min, if the F component is added it causes immediate damage to the cells. These findings suggest that the S component is more responsible for the interaction with the leukocytes than the F component. Specific binding of the S component to G_{M1} ganglioside (Noda et al., 1980b) on the cell membrane of leukocidin-sensitive cells induces activation of phospholipid methyltransferases (Noda et al., 1985) and phospholipase A_2 (Noda et al., 1982), leading to an increase in the number of F component-binding sites (Noda et al., 1981). This increased number of binding sites for the F component is dependent on the specific enzymic action of the S component, that is ADP-ribosylation of a 37-kDa membrane protein (Kato and Noda, 1989a,b), and is correlated with increased phospholipid methylation and phospholipase A_2 activation. The F component, bound to leukocidin-sensitive cells in the presence of S component, rapidly caused degradation of the cell membrane, resulting in cell lysis. The F component also possesses ADP-ribosyltransferase and ADP-ribosylates a 41-kDa membrane protein (Kato and Noda, 1989a,b) and stimulates phosphatidylinositol metabolism of leukocidin-sensitive cells (Kato and Noda, 1989a,b). The degradation of the cell membrane is associated with stimulation of phosphatidylinositol metabolism and ouabain-insensitive Na^+, K^+-ATPase activity, and inhibition of cyclic AMP-dependent protein kinase (Noda et al., 1982).

Pseudomonal leukocidin

Purification and crystallization of leukocidin

Pseudomonas aeruginosa 158 is grown on a trypticase soy broth medium containing 0.5% glucose with a reciprocal shaker (120 cycles/min) at 30°C for 12 h.

The culture fluid (10 litres per batch) is then centrifuged at 11 000 × g for 15 min at 4°C, and the bacteria washed and suspended in 1400 ml of phosphate-buffered saline, pH 7.4, containing 0.137 M NaCl, 2.68 mM KCl, 0.5 mM $MgCl_2$, 1.47 mM KH_2PO_4, and 2.9 mM Na_2HPO_4. Autolysis of the bacteria is carried out by shaking incubation at 37°C for 48 h, rotating at 120 cycles/min. The supernatant (1612 ml) after centrifugation of the autolysate at 20 000 × g for 20 min is dialysed against 0.05 M sodium phosphate buffer, pH 7.2. The dialysate is applied to a DEAE–Sephadex A-50 column (5 × 100 cm) equilibrated with 0.05 M sodium phosphate buffer, pH 7.2. After washing the column with the equilibrated buffer (3 700 ml), fractions with leukocytotoxic activity are eluted with 0.5 M sodium phosphate buffer, pH 7.2 (Fig. 2) and are concentrated up to 80 ml by an Amicon ultrafiltration apparatus equipped with a PM-10 membrane. The resulting solution is applied to a Sephadex G-100 column (5 × 100 cm) which was equilibrated with 0.05 M sodium phosphate buffer, pH 7.2. Protein is eluted with the same buffer and is assayed for leukocytotoxic activity. Fractions containing leukocytotoxic activity are collected and concentrated up to 6 ml. This solution, equilibrated with 0.05 M sodium phosphate buffer, pH 8.0, is applied to zone electrophoresis on Pevikon C-870 using the same buffer at 20 mA for 24 h. Active fractions from this final purification step are collected, concentrated by membrane filtration (Diaflo PM-10), crystallized by means of dialysis against ammonium sulphate (Noda et al., 1980a), and stored at −80°C. The recovery and degree of purification at each step are shown in Table 3.

Determination of leukocidin activity

Polymorphonuclear leukocytes are prepared from rabbit peripheral blood on isokinetic Ficoll gradients (Pretlow and Luberoff, 1973). Leukocytotoxic activity is assayed by the microscopic slide adhesion method as described previously (Noda et al., 1980a) except for the use of Ca^{2+}-Dulbecco's phosphate-buffered saline (PBS; 0.01 M sodium–potassium phosphate buffer, pH 7.4, 0.15 M NaCl, 2.5 mM KCl, and 1 mM $CaCl_2$). End points are determined as the smallest amount of the toxin causing morphological changes in 50% of leukocytes in a

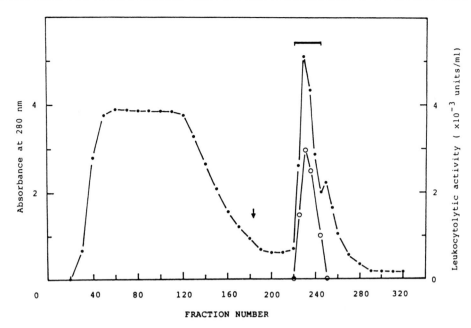

Figure 2. DEAE–Sephadex A-50 column chromatography. The dialysate of bacterial autolysate was applied to the column, equilibrated with 0.05 M sodium phosphate buffer (pH 7.2). The buffer was replaced by 0.5 M sodium phosphate buffer (pH 7.2) at 10°C for elution of leukocidin (indicated by arrow). Fractions were measured for absorbance at 280 nm (●) and leukocytotoxic activity (○) by the microscopic adhesion method. Fractions which were pooled are indicated by a bar.

standard assay suspension (1 × 10⁶ cells/20 μl). The leukocytotoxic activity is expressed in units, the number of units in a toxin preparation being numerically equal to the dilution (in ml) at the point.

Physico-chemical properties of leukocidin
(Hirayama et al., 1983)

The molecular weight of the leukocidin is estimated to be 42 500 by SDS–polyacrylamide gel electrophoresis, 40 000 by gel filtration, and 44 700 by sucrose density gradient centrifugation. The isoelectric point of the leukocidin is estimated to be pI 6.3 by isoelectrofocusing. Three residues of half-cystine are detected. The (Glx + Asx)/(Lys + Arg) ratio is 0.5, in agreement with the acidic nature of the toxin.

Mode of action of leukocidin

Pseudomonas aeruginosa often causes fatal infections in burned and other hospitalized patients (Doggett, 1979; Sabbath, 1980). The importance of leuko-

cidin in the pathogenesis of *Pseudomonas aeruginosa* infections is becoming recognized, since polymorphonuclear leukocytes are known to be one of the most important cell types in the host defence mechanism against pseudomonal infection.

The minimum required doses of leukocidin for destruction of rabbit leukocytes (1 × 10⁶), rabbit blood lymphocytes (1 × 10⁶) and rabbit erythrocytes (1 × 10⁶) are 13 ng, 390 ng and 480 μg, respectively. No morphological changes are observed in rabbit platelets. Hirayama and Kato (1984) reported that the pseudomonal leukocidin specifically bound to a 50 kDa membrane protein of rabbit leukocytes and the subsequent stimulation of the leukocytes by the toxin resulted in a rapid increase in the incorporation of [³²P]Pᵢ into phosphatidyl-4-phosphate (PIP), phosphatidyl-4,5-bisphosphate (PIP₂) and phosphatidic acid (PA). These findings suggested that the leukocidin stimulated the phospholipase C-induced hydrolysis of the inositol phospholipid pathway, which is the general process involved in signal transduction, leading to both activation of protein kinase C and enhancement of Ca²⁺ mobilization. In fact,

Table 3. Purification of leukocidin from the autolysate of *P. aeruginosa* 158

Step		Volume (ml)	Protein (μg/ml)	Total protein (mg)	Total activity (× 10^{-3} U)	Specific activity (U/mg protein)	Recovery of activity (%)
I.	Supernatant of bacterial autolysate	1 612	806	1 300	806	620	100
II.	DEAE–Sephadex A-50 chromatography	520	1 530	796	780	980	96.8
	Concentration by ultrafiltration	80	8 060	645	600	930	74.4
III.	Sephadex G-100 chromatography	162	679	110	486	4 420	60.3
	Concentration by ultrafiltration	6	16 000	96	420	4 380	52.1
IV.	Pevikon zone electrophoresis	12	292	3.5	360	103 000	44.7

Hirayama and Kato (1982, 1983) showed that leukocidal activity of the leukocidin was observed only in the presence of Ca^{2+} and the leukocidin stimulated the activity of Ca^{2+}-dependent protein kinase C, resulting in the phosphorylation of a 28 kDa protein in the membranes of cell surface and lysosome particles from which the leakage of lysosomal enzymes occurred at almost the same time as the leukocyte destruction. Electron microscopic experiments (Hirayama *et al.*, 1983) regarding morphological changes of leukocidin-treated leukocytes indicated an apparent increase in vacuoles resulting from the loss of cytosolic granule contents before leukocyte enlargement.

In conclusion, the early leukocidal action of the toxin might be related to the stimulation of inositol phospholipid-specific phospholipase C inducing activation of Ca^{2+}-dependent protein kinase C. This Ca^{2+}-dependent phosphorylation of a 28-kDa protein in the lysosomal membrane may be involved in the destruction of lysosomal particles in the leukocytes exposed to the leukocidin, but the molecular mechanism remains to be established.

References

Doggett, R.D. (1979). Microbiology of *Pseudomonas aeruginosa*. In: Pseudomonas aeruginosa: *Clinical Manifestations of Infection and Current Therapy* (ed. R.G. Doggett), pp.1–7. Academic Press, New York.

Gladstone, G.P. and van Heyningen, W.E. (1957). Staphylococcal leukocidin. *Br. J. Exp. Pathol.* **38**, 123–37.

Hirayama, T. and Kato, I. (1982). Two types of Ca^{2+}-dependent phosphorylation in rabbit leukocytes. *FEBS Lett.* **146**, 209–12.

Hirayama, T. and Kato, I. (1983). Rapid stimulation of phosphatidylinositol metabolism in rabbit leukocytes by pseudomonal leukocidin. *FEBS Lett.* **157**, 46–50.

Hirayama, T. and Kato, I. (1984). Mode of cytotoxic action of pseudomonal leukocidin on phosphatidylinositol metabolism and activation of lysosomal enzyme in rabbit leukocytes. *Infect. Immun.* **43**, 21–7.

Hirayama, T., Kato, I., Matsuda, F. and Noda, M. (1983). Crystallization and some properties of leukocidin from *Pseudomonas aeruginosa*. *Microbiol. Immunol.* **27**, 575–88.

Kato, I. and Noda, M. (1989a). ADP-ribosylation of cell membrane proteins by staphylococcal α-toxin and leukocidin in rabbit erythrocytes and polymorphonuclear leukocytes. *FEBS Lett.* **255**, 59–62.

Kato, I. and Noda, M. (1989b). ADP-ribosyl transferase activities of staphylococcal cytolytic toxins. In: *ADP-ribosyl Toxins and Target Molecules* (eds I. Kato and T. Uchida), pp.135–61. Saikon, Tokyo.

Kato, I., Morinaga, N. and Muneto, R. (1988). Non-thiol-activated cytolytic bacterial toxin: Current status. *Microbiol. Sci.* **5**, 53–7.

Loeffler, D.A., Schat, K.A. and Norcross, N.L. (1986). Use of ^{51}Cr release to measure the cytotoxic effects of staphylococcal leukocidin and toxin neutralization on bovine leukocytes. *J. Clin. Microb.* **23**, 416–20.

Morinaga, N., Nagamori, M. and Kato, I. (1987a). Suppressive effect of calcium on the cytotoxicity of staphylococcal leukocidin for HL-60 cells. *FEMS Microb. Lett.* **42**, 259–64.

Morinaga, N., Nagamori, M. and Kato, I. (1987b). Stimulation of Ca^{2+}-dependent protein phosphorylation in HL-60 cells by staphylococcal leukocidin. *FEMS Microb. Lett.* **44**, 431–4.

Morinaga, N., Nagamori, M. and Kato, I. (1988). Changes in binding of staphylococcal leukocidin to HL-60 cells during differentiation induced by dimethyl sulfoxide. *Infect. Immun.* **56**, 2479–83.

Noda, M., Hirayama, T., Kato, I. and Matsuda, F. (1980a). Crystallization and properties of staphylococcal leukocidin. *Biochem. Biophys. Acta* **633**, 33–44.

Noda, M., Kato, I., Hirayama, T. and Matsuda, F. (1980b). Fixation and inactivation of staphylococcal leukocidin by phosphatidylcholine and ganglioside G_{M1} in rabbit polymorphonuclear leukocytes. *Infect. Immun.* **29**, 678–84.

Noda, M., Kato, I., Matsuda, M. and Hirayama, T. (1981). Mode of action of staphylococcal leukocidin: Relationship between binding of ^{125}I-labeled S and F components of leukocidin to rabbit polymorphonuclear leukocytes and leukocidin activity. *Infect. Immun.* **34**, 362–7.

Noda, M., Kato, I., Hirayama, T. and Matsuda, F. (1982). Mode of action of staphylococcal leukocidin: Effects of the S and F components on the activities of membrane-associated enzymes of rabbit polymorphonuclear leukocytes. *Infect. Immun.* **35**, 38–45.

Noda, M., Hirayama, T., Matsuda, F. and Kato, I. (1985). An early effect of the S component of staphylococcal leukocidin on methylation of phospholipid in various leukocytes. *Infect. Immun.* **50**, 142–5.

Pretlow, T.G. II and Luberoff, D.E. (1973). A new method for separating lymphocytes and granulocytes from human peripheral blood using programmed gradient sedimentation in an isokinetic gradient. *Immunology* 24, 85–92.

Sabbath, L.D. (ed.) (1980). Pseudomonas aeruginosa: *The Organism, the Diseases it Causes and their Treatment*. Hans Huber, Bern.

Soboll, H., Ito, A., Schaeg, W. and Blobel, H. (1973). Leukozidin von Staphylokokken verschiedener Herkunft. *Zbl. Bakt. Parasitenkd. Infekt. Hyg. Abt. 1 Orig. Reihe* A224, 184–93.

Wenk, K. and Blobel, H. (1970). Untersuchungen und 'Leukozidinen' von Staphylokokken verschiedener Herkunft. *Zbl. Bakt. Parasitenkd. Infekt. Hyg. Abt. 1. Orig.* 213, 479–87.

Woodin, A.M. (1960). Purification of the two components of leukocidin from *Staphylococcus aureus*. *Biochem. J.* 75, 158–65.

Woodin, A.M. (1970). Staphylococcal leukocidin. In: *Microbial Toxins*, Vol.3 (eds T.C. Montie, S. Kadis and S.J. Ajl), pp.327–55. Academic Press, New York.

12
Ciliostatic Toxins

John H. Freer

Department of Microbiology, University of Glasgow, Glasgow G12 8QQ, UK

Introduction

Ciliostatic toxins are active against eukaryotic cells bearing cilia and flagella, causing progressive asynchrony with eventual ciliostasis, usually within a few hours of exposure to active bacterial culture supernates.

These toxins are structurally diverse and the producer organisms are classified in several different bacterial genera, namely *Bordetella*, *Gonococcus*, *Pseudomonas* and *Vibrio*. Ciliostatic activity has been reported in culture supernates of *Haemophilus* species, but as yet the agent responsible remains uncharacterized.

Tracheal cytotoxin in the *Bordetellae*

Background

The agent responsible for ciliostatic activity in exponential-phase culture supernates of virulent strains of *Bordetella pertussis* was designated tracheal cytotoxin (TCT) by Goldman *et al.* (1982). Indeed, our current knowledge of the existence and nature of tracheal cytotoxin (TCT) in members of the genus *Bordetella* is due in very large part to the penetrating studies of Goldman and his colleagues over the last decade (for review see Goldman, 1988). These were facilitated by the earlier technical advance of establishing organ cultures of ciliated epithelia (Hoorn, 1966) and specifically of hamster tracheal rings by Collier (1976), who used them (Collier *et al.*, 1977) to investigate cellular pathology following exposure to virulent *B. pertussis*. The specific destruction of ciliated respiratory epithelial cells was observed, similar to that found in the respiratory tract of man following pertussis infection, and first described in the seminal work on pertussis by Mallory and Horner (1912).

A further advance in enabling technology was the establishment of a more convenient and reproducible *in vitro* system for examining the basis of the specific cellular pathology associated with the interaction of virulent strains of *B. pertussis* with cells of the respiratory tract. The system consisted of cultured hamster tracheal epithelial (HTE) cells (Goldman *et al.*, 1982) which, although non-ciliated, are not transformed, secrete mucus-like glycoprotein, and remain differentiated during 35–40 generations before senescence. In addition, virulent strains of *B. pertussis* adhered to these cells and colonized their surface, a feature which is not, however, peculiar to this cell type (Holt, 1972).

The development and characterization of this assay system allowed Goldman and colleagues (Goldman *et al.*, 1982) to investigate the basis of epithelial cell pathology previously reported by Collier *et al.* (1977). The convenience and reproducibility of this *in vitro* cell assay allowed rapid progress in the identification and purification of a specific low molecular weight component, present in culture supernates of virulent strains of *B. pertussis*, which produced a dose-dependent inhibition of DNA synthesis. While a number of cell lines are colonized by *B. pertussis*, this inhibition of DNA synthesis is a feature not shared by the other cell lines and the activity co-purified with ciliostatic activity demonstrated in the hamster tracheal ring assay. The addition of such fractions to organ cultures induced a slowing of ciliary beat frequency, with eventual cessation of ciliary activity and eversion of ciliated cells from the epithelial layer.

Nature of tracheal cytotoxin

Goldman's search for a biologically active component that might explain the specific cytotoxicity on respiratory epithelia following pertussis infection (Goldman *et al.*, 1982) yielded a biologically active fraction with relatively simple composition. Upon amino acid analysis, it was shown to consist of glutamic acid (5), alanine (5), glycine (2), cysteine (2), diaminopimelic acid (1) (molar ratios are in parenthesis) as well as unspecified amounts of muramic acid and glucosamine. Such an analysis led to the inescapable conclusion by the investigators that at least part of the molecule was related to cell wall peptidoglycan, although the composition and molar ratios indicated that it did not represent a simple precursor or enzymic fragment.

In a subsequent collaborative study (Rosenthal *et al.*, 1987), the similarities in biological activities and chemical composition between extracellular soluble peptidoglycan fragments from *Neisseria gonorrhoeae* and TCT were emphasized and peptidoglycan turnover and release of peptidoglycan fragments during culture were studied in detail, employing methods similar to those used previously for *Neisseria gonorrhoeae*. The results from pulse-chase experiments using [³H] DAP (diaminopimelic acid) and subsequent chromatographic analysis revealed that soluble radiolabelled peptidoglycan was remarkably homogeneous in nature and probably originated from polymeric cell wall material rather than being released as a nucleotide-linked cell wall precursor.

The soluble peptidoglycan fragment shared many properties in common with authentic TCT. It co-migrated with 1,6-anhydro-*N*-acetyl-muramic acid-containing disaccharide peptides on paper chromatography, and was indistinguish-able from authentic TCT by high-voltage paper electrophoresis and by its behaviour in two reversed-phase HPLC systems. The authors (Rosenthal *et al.*, 1987) concluded from this study that TCT was a non-reducing anhydro-muramic acid-containing fragment or possibly a cyclic peptidoglycan derivative. From molecular weight estimations of TCT, it seemed likely that the peptidoglycan fragment was not a simple *N*-acetylglucosaminyl-*N*-acetylmuramyl-alanyl-glutamyl-DAP-alanine (molecular weight 939), but may have consisted of this unit with an additional muramyl dipeptide (*N*-acetylmuramyl-alanyl-glutamic acid) attached (molecular weight 1400) as discussed by Rosenthal *et al.* (1987).

Further purification of TCT activity using more rapid and effective methodologies was reported by Cookson *et al.* (1989a). Employing a combination of solid-phase extraction and reverse-phase HPLC, the active fraction appeared as a single peak and was thus purified to homogeneity in relatively high yield (63% recovery). Amino acid analysis revealed (molar ratios) glucosamine (1), muramic acid (1), alanine (2), glutamic acid (1) and diaminopimelic acid (1), which provided strong supporting evidence for its identity as a peptidoglycan-derived disaccharide tetrapeptide fragment. The molecular weight of this single molecular species was estimated by fast atom bombardment mass spectrometry as 921 (Goldman, 1988).

The unequivocal identity and structure of TCT was reported (Cookson *et al.*, 1989b) after a further detailed study of highly purified and biologically active TCT by fast ion bombardment mass spectrometry. The structure was confirmed as that illustrated in Fig. 1. (*N*-acetyl-glucosaminyl-1, 6-anhydro-*N*-acetylmuramyl-glu-tamyldiaminopimelyl-alanine). Its relationship to peptidoglycan was also unambiguously established, and the authors point out that while the anhydro-muramic acid residue is unusual it is not unique, since such internal glycoside structure is also present as a minor component in peptidoglycan of *N. gonorrhoeae* and *E. coli*. The peptidoglycan of *B. pertussis* is unusual, however, in its uniformity and simplicity of structure with 95% of the monomeric subunits being the disaccharide tetra-peptide (Folkening *et al.*, 1987).

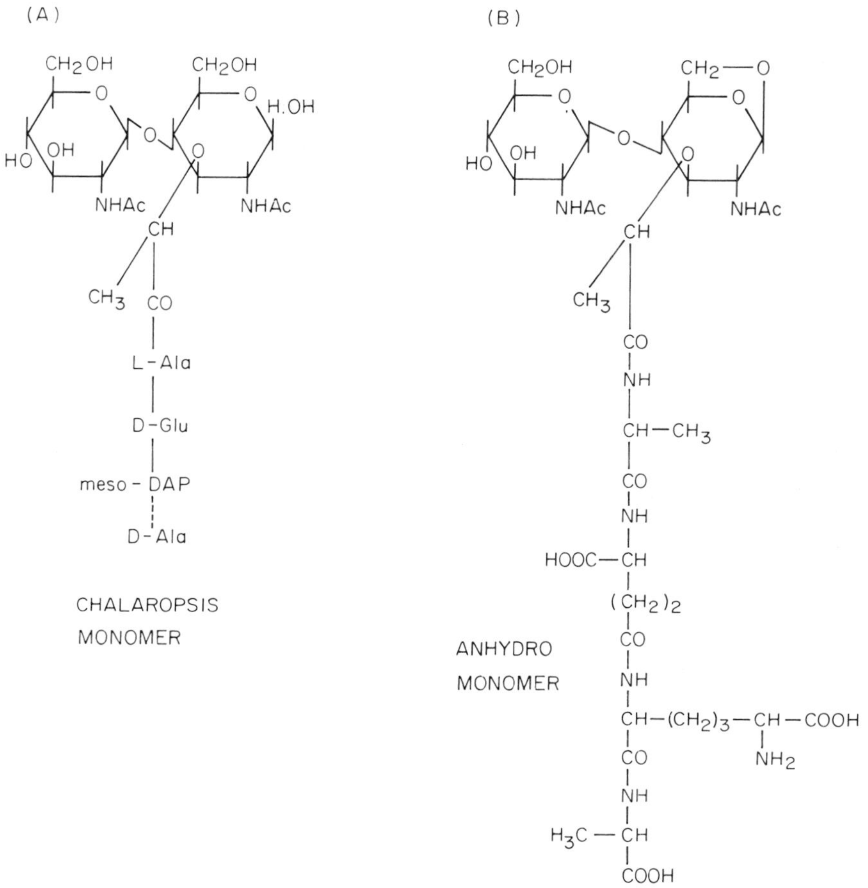

Figure 1. Structure of cell wall-derived disaccharide tetrapeptide fragments with ciliostatic activity. (A) Fragment generated by *Chalaropsis* muramidase contains a reducing muramic acid end, as would fragments generated by the action of host lysozyme on peptidoglycan. (B) Fragment with a non-reducing end generated by the action of endogenous transglycosylase. This fragment corresponds to the tracheal cytotoxin of *Bordetella pertussis*. Disaccharide tri- and tetrapeptides with both reducing and non-reducing ends have been generated from peptidoglycan of *Neisseria gonorrhoeae*. All such fragments caused ciliostasis and epithelial damage in human fallopian tube organ cultures.

Mode of action and possible role in pertussis infections

The site of colonization by virulent bacteria in pertussis infections is the epithelium of the upper respiratory tract of the host species. Depending upon the particular species of *Bordetella*, the host would typically be man in the case of *B. pertussis* or *B. parapertussis*; the dog in the case of *B. bronchiseptica*, or birds, notably turkeys in the case of *B. avium*. All species of the bacterial genus produce TCT in their virulent phase of growth.

Early in the disease, bacteria appear to adhere to and colonize specifically the ciliated cells in the respiratory epithelium. This leads to progressive loss of ciliary beating with eventual eversion of the ciliated cells from the epithelial layer. Reduction of ciliary activity has dire consequences for the muco-ciliary elevator mechanism in the upper respiratory tract. Denuding of the epithelium of ciliated cells results in progressive accumulation of mucus and cell debris, which triggers the cough reflex. Paroxysmal coughing episodes are a cardinal symptom of human pertussis infection. The specific inhibition of DNA synthesis observed in non-ciliated cells *in vitro* (Goldman *et al.*, 1982) is of direct relevance in such infection, since lost ciliated epithelial cells are only replaced by division and subsequent differentiation of cells originating by mitotic activity in the basal cell layer. Inhibition of DNA synthesis in this regenerative cell layer would be likely to prolong the effects of lost ciliary activity by delaying the replacement of everted ciliated epithelial cells. Superimposed on such long-term damage caused by exposure to TCT, are the effects which may result from concomitant release at the epithelial surface of such other toxins as pertussis toxin and adenylate cyclase toxin. Also upon death and lysis of the colonizing bacterial cells, heat-labile toxin and endotoxin would be released. Certainly, the longer term tissue damage resulting from exposure to toxic agents such as these are evident as loss of normal function in the upper respiratory tract (e.g. paroxysmal cough) for some weeks after the disappearance of culturable bacteria from the site of infection.

Ciliostatic activity in *Neisseria gonorrhoeae*

Background

As part of a growing interest in the gonococcal cell surface and its role in the pathogenesis of gonorrhoea, the peptidoglycan component was investigated in a series of studies published in the late 1970s. Particularly noteworthy was the fact that gonococcal peptidoglycan showed substantial and unusually high turnover during exponential growth (Hebeler and Young, 1976a,b; Rosenthal, 1979). It was soon realized that if such soluble peptidoglycan fragments were also released during *in vivo* growth, then they could significantly influence the course of infection, since a number of studies had shown already the immunomodulating activities of various peptidoglycan derivatives and related molecules such as muramyl dipeptide (MDP) (Chedid *et al.*, 1978). For this reason alone, the further characterization of such soluble fragments was considered important (Sinha and Rosenthal, 1980).

Employing different radiolabelled markers, Rosenthal (1979) has shown that the peptidoglycan turned over at a remarkably high rate, equivalent to approximately 35% of the wall polymer per

generation and that soluble fragments released into the medium consisted of four types. In addition to disaccharide peptide monomers and dimers, free disaccharides and free tetrapeptides were also tentatively identified, indicating both hexosaminidase and amidase autolytic activities. In a subsequent study, Sinha and Rosenthal (1980) identified the major component (80%) of the monomeric fraction as *N*-acetylglucosaminyl–β–1,4–1, 6–anhydro-*N*-acetylmuramyl-L-Ala-D-Glu-*meso*-diaminopimelic acid, while the remainder was the corresponding disaccharide tetrapeptide containing a terminal D-alanine. Of particular note was the finding that the glycan was almost exclusively in the anhydro form, indicating an unusual transglycosylase activity. Similar anhydromuramyl moieties were present in the peptide cross-linked dimers and the free disaccharides.

Ciliostatic activity

The successful maintenance of human fallopian tubes in organ culture by McGee *et al.* (1976) provided a very relevant model for the study of gonococcal interaction with its target tissue *in vitro*.

The studies of Hebeler and Young (1976a,b) and of Rosenthal (1979) had already established that gonococci possessed unusual transglycosylase activity, releasing various anhydromuramyl wall fragments into the medium during exponential growth. The structural similarity between these and the fragments released by *B. pertussis* which were shown to have ciliostatic activity (see above) in hamster tracheal organ culture was noted by Melly *et al.* (1984). In this study, the authors showed that either the anhydro monomer or the reducing monomer (*Chalaropsis* monomer)(see Fig. 1) were equally effective in causing eversion of ciliated cells from the epithelial layer, and both tripeptide and tetrapeptide derivatives were equally active, whereas structurally related components such as *N*-acetylglucosamine and muramyl dipeptide (MDP) were inactive at concentrations three times the eliciting dose of active monomers. In addition to sloughing or eversion of the ciliated cells, the authors also noted balloon-like structures formed on the epithelial layer which were associated with loss of mucosal integrity.

Significance of ciliostatic activity in gonococcal infection

Three points which emerge from the results of the study by Melly *et al.* (1984) are worthy of special note: (1) the observed histological changes were specific for the two peptidoglycan monomers; (2) similar histological changes in ciliated epithelium were induced by a peptidoglycan fragment identical to the anhydro monomer and released during exponential phase growth in cultures of *B. pertussis*; and (3) the changes observed in the epithelial layer of fallopian tubes were identical to those seen in organ cultures infected with gonococci and also similar to the damage seen *in vivo* during active gonococcal infections.

Biologically active anhydro monomers and the reducing monomers may be generated by gonococcal transglycosylase or by host lysozyme during infection with gonococcus, and loss of ciliated cell activity in either the fallopian tubes or the upper respiratory tract result in impairment of physiological functions at both these sites. It would be of considerable interest to determine whether the active peptidoglycan fragments from gonococcus also have the ability to inhibit DNA synthesis in HTE cells and whether the fragments from either bacterial species display comparable biological activities in both organ culture systems.

Ciliostatic activity in *Pseudomonas aeruginosa*

Background

Following earlier reports of a factor present in serum from cases of cystic fibrosis which displayed ciliostatic activity in rabbit tracheal explants (Spock *et al.*, 1967), Bowman *et al.* (1969) confirmed the ciliostatic activity in such sera but in a more convenient eukaryotic ciliated assay system, namely segments of oyster gill tissue. They also reported that ciliostatic activity was present in saliva from CF patients, but not normal controls. Both the ciliostatic factors reported by Spock *et al.* (1967) and by Bowman *et al.* (1969) were non-dialysable, heat-labile, high molecular weight components (molecular weight range 75 000–180 000).

Ciliostatic factors active against rabbit ciliated respiratory epithelium were recognized in culture supernates of *Pseudomonas aeruginosa* strain PAKS 1 and several ethyl methane sulphonate-derived mutants by Reimer *et al.* (1980). Interestingly, ciliostatic activity was only partially lost upon heating supernates at 80°C, the remainder of the activity being heat-stable material extractable in chloroform. The chloroform-soluble fraction of culture supernate contained blue phenazine pigment, and upon further purification of this pigment, dose-dependent ciliostatic activity was retained. Haemolysin, partially purified from culture supernate of an EMS-derived hyperproducing derivative of PAKS 1 also showed a dose-dependent ciliostatic activity. However, in both these cases of ciliostatic activity, the nature of the active component was not clearly established and evidence of identity was only circumstantial.

In a later report by Wilson *et al.* (1985), the effect of bacterial culture supernates on ciliary beat frequency was measured by a photometric technique. The bacterial strains were isolated from bronchiectatic patients and the culture supernates assayed against human ciliated epithelial cells obtained by nasal brushing of volunteers. Dose-dependent ciliostatic activity was confirmed in supernates from *Pseudomonas aeruginosa*. However, in this instance, the ciliostatic factor(s) was labile to heat (56°C, 30 min) and relatively unstable at room temperature (activity lost after 120 min at 37°C). The factors were not identified chemically in this study but may equate with the heat-labile fraction of the total ciliostatic activity reported by Reimer *et al.* (1980) in culture supernates of *P. aeruginosa* PAKS 1.

Probable identity of ciliostatic factors

Clarification of the identity of the heat-stable ciliostatic factors produced by *P. aeruginosa* had to await the more penetrating studies of Hingley and colleagues (Hingley *et al.*, 1986a,b). Using the criterion of extractability in chloroform, Hingley and co-workers extracted heat-stable factors with ciliostatic activity against rabbit respiratory cilia from rough variants of *P. aeruginosa* isolated from patients with cystic fibrosis. In an extension of the earlier approach of Reimer *et al.* (1980), they further fractionated the chloroform extracts using thin-layer chromatography and showed that ciliostatic factors correspond to pyo compounds (2-alkyl-4-hydroxyquinolines), to a rhamnolipid haemolysin, and to a phenazine derivative, possibly 1-hydroxyphenazine. They reported a 10-fold difference in ciliostatic activity of supernates from four isolates of *P. aeruginosa*. Typically 50% or more of the total ciliostatic activity was extractable in chloroform, and this fraction was the subject of their study.

Assayed in rabbit tracheal explants, the pyo compounds proved to be the most potent on a weight basis, displaying a ciliostatic ED_{50} (dose required to cause stasis of 50% of cilia within 15 – 30 min) of 50 μg/ml. Purified pyocyanin showed no ciliostatic activity, even at 1 mg/ml, a finding at variance with the earlier report from Reimer *et al.* (1980), who in fact may have been observing the ciliostatic activity of 1-hydroxyphenazine, an oxidation product of pyocyanin, rather than that of pyocyanin. Phenazine derivatives also showed ciliostatic activity, but only at higher conentrations (ED_{50} of approx. 400 μg/ml) and such activity was accompanied by membrane damage. However, ciliary inhibition (decreased beat frequency and/or ciliostasis) by both pyo compounds and the phenazine derivative was partially reversible with exogenous ATP, suggesting intact axoneme structure.

In the case of ciliostasis induced by the rhamnolipid, there was no decreased beat frequency evident before ciliostasis occurred. The ED_{50} in the case of rhamnolipid was much higher than the other components and estimated at 2.8 mg/ml. Ultrastructural studies showed that axoneme structure remained intact after treatment with the pyo and phenazine compounds. Some multilayering of the axoneme membrane was evident after pyo treatment, but complete removal of the axonemal membrane was not observed. In contrast, treatment for 5 min with the rhamnolipid caused significant membrane damage ranging from multilayering of the axonemal membrane to its complete removal. Treatment for 60 min with rhamnolipid resulted in complete removal of axonemal membranes accompanied by removal of outer dynein arms. In addition, because of its detergent-like properties, rhamnolipid disrupted plasma and intracellular epithelial cell membranes, resulting in release of their intracellular constituents at the mucosal surface.

The inferred heat-labile ciliostatic components

in *P. aeruginosa* culture supernates (Reimer *et al.*, 1980; Hingley *et al.*, 1986a) may correspond to proteases such as the pseudomonad elastase and/or alkaline protease which is known to be produced *in vivo* (Jagger *et al.*, 1982; Doring *et al.*, 1983, 1985) and to damage extensively ciliated epithelium (Gray and Kreger, 1979).

Role in respiratory disease

Direct evidence for a role of the heat-stable ciliostatic factors of *Pseudomonas aeruginosa* in disesase is lacking. However, it is certain that release of both heat-stable (pyo compounds and phenazine derivatives) and heat-labile (proteases) ciliostatic factors from bacteria colonizing the upper respiratory tract would contribute directly to the pathogenesis of any subsequent respiratory infection by virtue of their damaging effects on the components of the respiratory elevator and clearance mechanisms. Ciliary damage would likely be exacerbated at a later stage in disease by release of leukocyte elastase at the sites of infiltration during the host inflammatory response (Tegner *et al.*, 1979). The host and bacterial proteases together would be likely to overwhelm the endogenous protease inhibitors of the lung and result in active proteolytic damage to lung mucosa, including ciliated epithelial cells. This pattern of events would be particularly relevant in patients with cystic fibrosis (see Bowman *et al.*, 1969).

Ciliostatic activity in the genus *Vibrio*

Background

Vibrio anguillarum and related marine *Vibrio* species are important pathogens of fish and shellfish and cause the disease vibriosis, which may cause sudden and devastating loss of shellfish larvae stock in hatcheries (Jeffries, 1983). Inhibition of filtration and cessation of feeding in the blue mussel, *Mytilus edulis*, after exposure to *Vibrio anguillarum*, was first reported by McHenery and Birkbeck (1986).

In an earlier study of the virulence factors of *V. anguillarum*, Nottage and Birkbeck (1986) showed that in addition to an extracellular protein toxin, this and some strains of related *Vibrio* species also produced a low molecular weight (<4000 Da) factor, which induced ciliostasis in gill segments of *Mytilus edulis*. In a continuation of this study (Nottage *et al.*, 1989), Birkbeck and colleagues demonstrated production of ciliostatic factors by a number of marine *Vibrio* isolates, most of which were pathogenic for fish and shellfish.

Nature of *Vibrio* ciliostatic toxin

The ciliostatic factor(s) produced by *V. anguillarum*, *V. alginolyticus*, *V. tubiashi* and *Vibrio* spp. NCMB 1338 were partially characterized by Nottage *et al.* (1989). The factor, which they termed VCT (*Vibrio* ciliostatic toxin), was purified by solvent extraction from culture supernates, followed by thin-layer chromatography. It was a heat-stable (retained full activity after 100°C for 10 min), low molecular weight (720 Da by mass spectrometry) molecule which displayed blue fluorescence and reacted with 50% H_2SO_4 and iodine vapour. It had a single absorption peak at 255 nm and was distinct from a second ciliostatic factor detected in culture supernates of *V. alginolyticus* NCMB 1339. This second factor was also haemolytic and was extractable in chloroform as described by Hingley *et al.* (1986b). Whether this factor equates with the rhamnolipid haemolysin described for *Pseudomonas aeruginosa* by Hingley *et al.* (1986b) remains to be established.

Role of ciliostatic toxin in disease

Evidence for a direct role of VCT in the pathogenesis of vibriosis in fish and shellfish is as yet only circumstantial. Nottage *et al.* (1989) showed that the toxin was produced significantly more frequently by fish and bivalve pathogenic isolates than environmental isolates, and that the toxin was produced by bacteria growing in seawater. However, the contribution of ciliostatic factors to the overall virulence of these fish pathogens has yet to be quantified. Also the possible synergistic activity of these ciliostatic factors with bacterial proteases in the overall damage to ciliary function requires further study.

Ciliostatic factors in other bacteria

Ciliostatic factors have been reported in *Haemophilus influenzae* (Wilson *et al.*, 1985), *Aeromonas haemolyticus*, *Aeromonas* spp. and *Moraxella* spp. (Nottage *et al.*, 1989). The factor in *H. influenzae* was present in culture supernates but was not further characterized. Those in *Aeromonas* spp. and *Moraxella* spp. were similar in their heat stability and molecular weight to the VCT reported for *V. anguillarum*, but were not further characterized.

General remarks

The ciliostatic factors of bacterial origin discovered to date represent a structurally diverse group of inimical agents which are likely to be of considerable significance in the pathogenesis of infections caused by these species.

The structures range from low molecular weight metabolites containing aromatic ring structures (e.g. the phenazine and 2-alkyl-4-hydroxyquinolines) found in *Pseudomonas aeruginosa*, through the glycolipid haemolysins (rhamnolipids) found in *P. aeruginosa* and *Vibrio alginolyticus* NCMB 1339, to the peptidoglycan fragments which constitute the ciliostatic toxins in *Bordetella* spp. and *Neisseria gonorrhoeae*.

The latter molecules are particularly fascinating because ciliostasis is but one further biological activity of a family of peptidoglycan-derived muramyl peptides to which they belong. Within this structurally related group of wall peptides are members which induce adjuvanticity, arthritogenicity, cytokine release from leukocytes, and somnogenic activity. Their structure and biological activities have been the subject of a number of fascinating reviews which include those by Adam and Lederer (1984), Kotani *et al.* (1986), and more recently Kreuger (1990). The degree of involvement of such cell wall-derived fragments in the pathogenesis of bacterial infections is not yet fully appreciated, but it is certain that they are of considerable significance *in vivo*.

So far, the anhydromuramyl disaccharide tetrapeptide fragments have been reported in the pathogenic Bordetellae and in *N. gonorrhoeae*, and the possibility that they may constitute a general virulence factor in other respiratory pathogens is under investigation (Cookson *et al.*, 1989b). However, they do not appear to be produced by *Pasteurella haemolytica* strains responsible for respiratory infections in sheep and cattle (unpublished observations).

The somnogenic activity associated with preparations of bacterial cell walls has been recognized for some time (see Toth and Krueger, 1988), as have the specific structural requirements for somnogenic muramyl peptides (Krueger, 1990). These are precise yet different from those required for ciliostatic activity, with the minimal structure for somnogenicity being a muramyl dipeptide. However, an anhydromuramyl residue considerably enhances the activity of the molecule (approx. 10-fold) and the more structurally complex TCT also has somnogenic activity. The interesting hypothesis that muramyl peptides are essential vitamin-like agents for mammals, and that their levels are controlled by activities of macrophages on bacterial substrates has been proposed (Chedid, 1983) and may explain, at least in part, the drowsiness often associated with bacterial infections in man. The somnogenic activity is thought to be a consequence of induction by muramyl peptides of increased production of cytokines, and especially interleukins (IL1).

In conclusion, it seems somewhat paradoxical that the muramyl peptide-based ciliostatic toxins, which are structurally among the most simple toxin molecules produced by bacteria, have activities *in vivo* which involve not only ciliostasis, but also a wide variety of other functions in the host which bear directly on pathogenesis of infection but may also be part of the normal physiological processes of the host. The story of their true role(s) in normal as well as infected hosts is not yet complete.

References

Adam, A. and Lederer, E. (1984). Muramyl peptides: Immunomodulators, sleep factors and vitamins. *Medicinal Res. Rev.* **4** (2) 111–52.

Bowman, B.H., Lockhart, L.H. and McCombs, M.L. (1969). Oyster ciliary inhibition by cystic fibrosis factor. *Science* **164**, 325–6.

Carney, F.E. and Taylor-Robinson, D. (1973). Growth and effect of *Neisseria gonorrhoeae* in organ culture. *Br. J. Vener. Dis.* **49**, 435–40.

Chedid, L. (1983). Muramyl peptides as possible endogenous immunopharmacological mediators. *Microbiol. Immunol.* **27**, 723–32.

Chedid, L., Audibert, F. and Johnson, A.G. (1978). Biological activities of muramyl dipeptide, a synthetic glycopeptide analogous to bacterial immunoregulating agents. *Prog. Allergy* **25**, 63–105.

Collier, A.M. (1976). Techniques for establishing tracheal organ cultures. Procedure 43321 T.C.A. (Tissue Culture Association) *Man* **2**, 333–4.

Collier, A.M., Peterson, L.P. and Baseman, J.B. (1977). Pathogenesis of infection with *Bordetella pertussis* in hamster tracheal organ culture. *J. Infect. Dis.* **136** (Suppl.), S196–S203.

Cookson, B.T. and Goldman, W.E. (1987). Tracheal cytotoxin: a conserved virulence determinant of all *Bordetella* species. *J. Cell. Biochem.* **11B** (Suppl.), 124.

Cookson, B.T., Cho, Hwei-Ling, Herwaldt, L.A. and Goldman, W.E. (1989a). Biological activities and chemical composition of purified tracheal cytotoxin of *Bordetella pertussis*. *Infect. Immun.* **57**, 2223–9.

Cookson, B.T., Nyler, A.N. and Goldman, W.E. (1989b). Primary structure of the peptidoglycan-derived tracheal cytotoxin of *Bordetella pertussis*. *Biochemistry* **28**, 1744–9.

Denny, F. (1974). Effect of toxin produced by *Haemophilus influenzae* on ciliated respiratory epithelium. *J. Infect. Dis.* **129**, 93–100.

Doring, G., Obernesser, H.J., Botzenhart, K., Flehmig, B., Hoiby, N. and Hoffman, A. (1983). Proteases of *Pseudomonas aeruginosa* in patients with cystic fibrosis. *J. Infect. Dis.* **147**, 744–50.

Doring, G., Goldstein, W., Roll, A., Schoitz, P.O., Hoiby, N and Botzenhart, K. (1985). The role of *Pseudomonas aeruginosa* exoenzymes in lung infections of patients with cystic fibrosis. *Infect. Immun.* **49**, 557–62.

Folkening, W.J., Nogami, W., Martin, S.A. and Rosenthal, R.S. (1987). Structure of *Bordetella pertussis* peptidoglycan. *J. Bacteriol.* **169**, 4223–7.

Goldman, W.E. (1986). *Bordetella pertussis* tracheal cytotoxin: Damage to respiratory epithelium. In: *Microbiology – 1986* (ed. L. Leive), pp.65–9. American Society for Microbiology, Washington, DC.

Goldman, W.E. (1988). Tracheal cytotoxin of *Bordetella pertussis*. In: *Pathogenesis and Immunity in Pertussis* (eds A.C. Wardlaw and R. Parton), pp.231–46. J. Wiley & Sons, Oxford.

Goldman, W.E. and Baseman, J.B. (1980a). Selective isolation and culture of a proliferating epithelial cell population from the hamster trachea. *In Vitro* **16**, 313–19.

Goldman, W.E. and Baseman, J.B. (1980b). *In vitro* effect of *Bordetella pertussis* culture supernatant on hamster tracheal epithelial cells. *Abst. Ann. Meet. Amer. Soc. Microbiol.* **80**, 49.

Goldman, W.E. and Herwaldt, L.A. (1985). *Bordetella pertussis* tracheal cytotoxin. In: *Proceedings of the Fourth International Symposium on Pertussis*, Joint

IABS/WHO Meeting, Geneva, 1984. *Devel. Biol. Standard.* **61**, 103–11. S. Karger, Basel.

Goldman, W.E., Klapper, D.G. and Baseman, J.B. (1982). Detection, isolation, and analysis of a released *Bordetella pertussis* product toxic for cultured tracheal cells. *Infect. Immun.* **36**, 782–94.

Gray, L. and Kreger, A. (1979). Microscopic characterisation of rabbit lung damage produced by *Pseudomonas aeruginosa* proteases. *Infect. Immun.* **23**, 150–9.

Hebeler, B.H. and Young, F.E. (1976a). Chemical composition and turnover of peptidoglycan in *Neisseria gonorrhoeae. J. Bacteriol.* **126**, 1180–5.

Hebeler, B.H. and Young, F.E. (1976b). Mechanism of autolysis of *Neisseria gonorrhoeae. J. Bacteriol.* **126**, 1186–93.

Hingley, S.T., Hastie, A.T., Kueppers, F., Higgins, M.L., Weinbaum, G and Shryock, T. (1986a). Effect of ciliostatic factors from *Pseudomonas aeruginosa* on rabbit respiratory cilia. *Infect. Immun.* **51**, 254–62.

Hingley, S.T., Hastie, A.T., Kueppers, F. and Higgins, M.L. (1986b). Disruption of cilia by proteases including those of *Pseudomonas aeruginosa. Infect. Immun.* **54**, 379–85.

Holt, L.B. (1972). The pathology and immunology of *Bordetella pertussis* infection. *J. Med. Microbiol.* **143**, 407–24.

Hoorn, B. (1966). Organ cultures of ciliated epithelium for the study of respiratory viruses. *Acta Path. Microbiol. Scand.*, Suppl. 183.

Hoorn, B. and Lofkvist, T. (1964). The effect of staphylococcal alpha toxin and preparations of staphylococcal antigens on ciliated respiratory epithelium. *Acta Otolaryngol.* **60**, 452–60.

Jagger, K.S., Robinson, D.L., Franz, M.N. and Warren, R.L. (1982). Detection by enzyme-linked immunosorbent assays of antibody specific for *Pseudomonas* proteases and exotoxin A in sera from cystic fibrosis patients. *J. Clin. Microbiol.* **15**, 1054–8.

Jeffries, V. (1983). Three *Vibrio* strains pathogenic to larvae of *Crassostrea gigas* and *Ostrea edulis. Aquaculture* **29**, 145–63.

Kotani, S., Takada, H., Tsujimoto, M., Kubo, T., Ogawa, T., Azuma, I., Ogawa, H., Matsumoto, K., Siddiqui, W.A., Tanaka, A., Nagao, S., Kohashi, S., Kanoh, S., Shiba, T. and Kusumoto, S. (1982) Nonspecific and antigen-specific stimulation of host defence mechanisms by lipophilic derivatives of muramyl dipeptides. In: *Bacteria and Cancer* (eds J. Jeljaszewicz, G. Pulverer and W. Roszkowski) pp.67–107, Academic Press, London.

Kreuger, J.M. (1990). Somnogenic activity of immune response modifiers. *TIPS*, **11** (March 1990) 122–6.

McGee, Z.A., Johnson, A.P. and Taylor-Robinson, D. (1976). Human fallopian tube in organ culture: preparation, maintenance, and quantitation of damage by pathogenic microorganisms. *Infect. Immun.* **13**, 608–18.

McGee, Z.A., Melly, M.A., Gregg, C.R., Horn, R.G., Taylor-Robinson, D., Johnson, A.P. and McCutchan, J.A. (1978). Virulence factors of gonococci: studies using human fallopian tube organ cultures. In: *Immunobiology of Neisseria gonorrhoeae* (eds G.F. Brooks, E.C. Gotschlich, K.K. Holmes, W.D. Sawyer and F.E. Young), pp.258–62. American Society for Microbiology, Washington, DC.

McGee, Z.A., Johnson, A.P. and Taylor-Robinson, D. (1981). Pathogenic mechanisms of *Neisseria gonorrhoeae*: observations on damage to human fallopian tubes in organ culture by gonococci of colony type 1 or type 4. *J. Infect. Dis.* **143**, 423–31.

McHenery, J.G. and Birkbeck, T.H. (1986). Inhibition of filtration in *Mytilus edulis* L. by marine vibrios. *J. Fish Dis.* **9**, 257–61.

Mallory, R.B. and Horner, A.A. (1912). Pertussis: the histological lesion in the respiratory tract. *J. Med. Res.* **145**, 115–23.

Melly, M.A., Gregg, C.R. and McGee, Z.A. (1981). Studies of toxicity of *Neisseria gonorrhoeae* for human fallopian tube mucosa. *J. Infect. Dis.* **143**, 423–31.

Melly, M.A., McGee, Z.A. and Rosenthal, R.S. (1984). Ability of monomeric peptidoglycan fragments from *Neisseria gonorrhoeae* to damage human fallopian tube mucosa. *J. Infect. Dis.* **149**, 378–86.

Muse, K.E., Collier, A.M. and Baseman, J.B. (1977). Scanning electron microscopic study of hamster tracheal organ cultures infected with *Bordetella pertussis. J. Infect. Dis.* **136**, 768–77.

Nottage, A.S. and Birkbeck, T.H. (1986). Toxicity to marine bivalves of culture supernatant fluids of the bivalve-pathogenic *Vibrio* strain NCMB 1338 and other marine vibrios. *J. Fish Dis.* **9**, 249–56.

Nottage, A.L., Sinclair, P.D. and Birkbeck, T.H. (1989). Role of low molecular weight ciliostatic toxins in Vibriosis of bivalve molluscs. *J. Aquatic Animal Health*, **1**, 180–186.

Reimer, A., Klementsson, K., Ursing, J. and Wretlind, B. (1980). The mucociliary activity of the respiratory tract. 1. Inhibitory effects of products of *Pseudomonas aeruginosa* on rabbit trachea *in vitro. Acta Otolaryngol.* **90**, 462–9.

Rosenthal, R.S. (1979). Release of soluble peptidoglycan from growing gonococci: hexaminidase and amidase activities. *Infect. Immun.* **24**, 869–78.

Rosenthal, R.S., Nogami, W., Cookson, B.T., Goldman, W.E. and Folkening, W.J. (1987). Major fragment of soluble peptidoglycan released from growing *Bordetella pertussis* is tracheal cytotoxin. *Infect. Immun.* **55**, 2117–20.

Sinha, R.K. and Rosenthal, R.S. (1980). Release of soluble peptidoglycan from growing gonococci: demonstration of anhydro-muramyl-containing fragments. *Infect. Immun.* **29**, 914–25.

Spock, A., Heick, H.M.C., Cress, H. and Logan, W.S. (1967). Abnormal serum factor in patients with cystic fibrosis of the pancreas. *Pediat. Res.* **1**, 173–9.

Tegner, H., Ohlsson, K., Toremalm, N.G. and von Mecklenburg, C. (1979). Effect of human leukocyte enzymes on tracheal mucosa and its mucociliary activity. *Rhinology* **17**, 199–206.

Toth, L.A. and Kreuger, J.M. (1988). Alteration of sleep in rabbits by *Staphylococcus aureus* infection. *Infect. Immun.* **56**, 1785–91.

Wilson, R., Roberts, D. and Cole. P. (1985). Effect of bacterial products on human ciliary function *in vitro*. *Thorax* **40**, 125–31.

13

Cytolytic Toxins of Pathogenic Marine Vibrios

Arnold S. Kreger

Department of Microbiology and Immunology, The Bowman Gray School of Medicine of
Wake Forest University, Winston-Salem, North Carolina 27103, USA
Present address: The Musculoskeletal Sciences Research Institute, 2190 Fox Mill Road,
Herndon, Virginia 22071, USA

Introduction

Toxins capable of damaging the cytoplasmic membranes of a wide variety of mammalian cells are produced by many prokaryotic and eukaryotic marine organisms (Bernheimer and Rudy, 1986; Shier, 1988). Pathogenic, halophilic *Vibrio* species which produce membrane-damaging toxins (also called cytolytic toxins, cytolysins and haemolysins) and are found in marine or brackish water environments include *V. parahaemolyticus*, *V. hollisae*, *V. vulnificus*, *V. damsela*, *V. metschnikovii*, *V. fluvialis* and *V. anguillarum* (Kodama *et al.*, 1984; Janda *et al.*, 1988; Miyake *et al.*, 1988). This review summarizes current knowledge concerning the purification and characterization of the most intensively studied cytolysins produced by *V. parahaemolyticus*, *V. hollisae*, *V. vulnificus*, *V. damsela* and *V. metschnikovii*. Research has focused on the purification of the toxins and the characterization of their physico-chemical and biological properties and genetic determinants as a prelude to determining (i) the roles of the toxins in the biology of the bacteria and the pathogenesis of infectious diseases, and (ii) whether the toxins are useful as analytical tools to study the structure of biomembranes.

Vibrio parahaemolyticus cytolysins

Many clinical isolates of *V. parahaemolyticus* secrete (i) a relatively heat-resistant, membrane-damaging protein called the thermostable direct haemolysin (TDH) or the Kanagawa phenomenon-associated haemolysin, and/or (ii) a cytolytic protein which is heat-labile but immunologically and genetically related to TDH and is called the TDH-related haemolysin (TRH) (Shirai *et al.*, 1990). In addition, some clinical and environmental isolates of *V. parahaemolyticus* produce a heat-labile cytolysin (called TL haemolysin) which lyses horse erythrocytes (which are resistant to TDH and TRH) and is not genetically related to TDH and TRH (Taniguchi *et al.*, 1985, 1986). Also Taniguchi *et al.* (1990) recently cloned and characterized a gene encoding a thermostable cytolysin (δ-UPH) which is not genetically or immunologically related to the TDH, TRH and TL haemolysin.

TDH of *V. parahaemolyticus*

Several review articles on the purification and characterization of the physico-chemical and biological properties of TDH have been published

(Joseph *et al.*, 1982; Ljungh and Wadström, 1983; Takeda, 1983, 1988).

Purification

Several purification schemes have been described for obtaining highly purified TDH (Sakurai *et al.*, 1973; Honda *et al.*, 1976b, 1986; Miyamoto *et al.*, 1980; Cherwonogrodzky and Clark, 1982a). The most often cited schemes are those of Honda *et al.* (1976b) and Miyamoto *et al.* (1980) which involve sequential ammonium sulphate precipitation, ion-exchange chromatography with DEAE–cellulose, hydroxylapatite adsorption chromatography, and gel filtration with Sephadex G-200 (Honda *et al.*, 1976b), and sequential ion-exchange batch adsorption with DEAE–Sepharose CL-6B, precipitation with acetic acid, gel filtration with Sephadex G-200, and ion-exchange chromatography with DEAE–Sephadex A-50 (Miyamoto *et al.*, 1980). However, the simplest method currently available (Honda *et al.*, 1986) for obtaining the highest recovery of highly purified toxin (60% recovery; approximately 1.7 mg protein from 2 litres of culture) involves ammonium sulphate precipitation and immunoaffinity chromatography with Sepharose 4B coupled with immunoglobulin raised against TDH purified as described by Honda *et al.* (1976b).

Stability and inactivation

The activity of highly purified and crude TDH is not affected by heating at 100°C for 10–60 min; however, the activity of the crude toxin is abolished by heating at 60°C for 10 min (Miwatani *et al.*, 1972; Miyamoto *et al.*, 1980). This effect is thought to be caused by a heat-labile, toxin-inactivating protease contained in crude and partially purified toxin preparations (Takeda *et al.*, 1974, 1975). On the other hand, partial inactivation of highly purified toxin at 60°C for 30 min is believed to be due to aggregation of toxin molecules (Takeda *et al.*, 1975).

TDH is inactivated by pepsin and chymotrypsin (Miyamoto *et al.*, 1980) and by the neuraminidase-sensitive gangliosides GT1 and GD1a (Takeda *et al.*, 1976).

Molecular weight, amino acid composition and isoelectric point

TDH has a molecular weight of 42 000–44 000 (Honda *et al.*, 1976b; Miyamoto *et al.*, 1980) and is composed of two identical subunits (Takeda *et al.*, 1978a). The amino acid composition has been reported in several publications with similar results (Sakurai *et al.*, 1973; Honda *et al.*, 1976b; Miyamoto *et al.*, 1980; Tsunasawa *et al.*, 1987). The N-terminal amino acid is phenylalanine and the toxin contains a high concentration of acidic amino acid residues (approximately 40% of the total residues), which agrees with the estimated isoelectric point values of 4.2 (Honda *et al.*, 1976b) or 4.9 (Miyamoto *et al.*, 1980).

The results of amino acid sequencing studies (Tsunasawa *et al.*, 1987) indicate that the toxin subunit consists of 165 amino acid residues with a disulphide bond between Cys^{151} and Cys^{161}, and the biologically active toxin is formed by non-covalent association of subunits which are not linked together by disulphide bonds. The primary structure of the toxin was essentially the same as that deduced from the reported (Nishibuchi and Kaper, 1985; Taniguchi *et al.*, 1986) nucleotide sequence of the gene encoding the toxin, but differed in nine amino acid residues. The authors (Tsunasawa *et al.*, 1987) proposed that the small discrepancy between the two sequences might result from the presence of two or more closely related genes for the toxin. In fact, *V. parahaemolyticus* WP1, which produces a large amount of TDH, has recently been claimed to carry two chromosomal *tdh* gene copies which share 97.2% DNA homology in their coding regions (Nishibuchi *et al.*, 1990).

Antigenicity

Rabbits vaccinated with TDH produce neutralizing and enzyme-linked immunosorbent assay (ELISA)-reacting antibodies (Sakurai *et al.*, 1975; Miyamoto *et al.*, 1980; Cherwonogrodzky and Clark, 1982b; Honda *et al.*, 1985; Yoh *et al.*, 1988), and elevated anti-TDH antibody titres have been detected in patients convalescing from *V. parahaemolyticus* food poisoning (Miwatani *et al.*, 1976; Miyamoto *et al.*, 1980). Also, the TDH can be detected, by an ELISA, in intestinal loop fluids of rabbits challenged with living *V. parahaemolyticus* (Honda *et al.*, 1985).

TDH is immunologically related to TRH and to cytolysins produced by *V. hollisae* and *V. cholerae* non-01.

Haemolytic and cytotoxic activities

TDH has broad-spectrum haemolytic activity (i.e. is active against erythrocytes from at least nine animal species); however, it is most active against rat and dog erythrocytes and is not active against horse erythrocytes (Zen-Yoji et al., 1971). The cell membrane receptors for the TDH are thought to be neuraminidase-sensitive gangliosides such as GT1 and GD1a, which are absent in the horse erythrocyte membrane (Takeda et al., 1976).

Haemolysis by TDH is temperature dependent (i.e. is not observed at 4°C and is optimal at 37°C) and is thought to be at least a two-step process consisting of a temperature-independent, divalent cation-enhanced, toxin-binding step, followed by a temperature-dependent, membrane-perturbation step(s) that leads to cell lysis (Sakurai et al., 1975).

TDH is cytotoxic for several different types of cultured mammalian cells, such as HeLa cells and L cells (Sakazaki et al., 1974), FL cells (Sakurai et al., 1976; Takeda et al., 1980), human epidermoid carcinoma KB cells, BALB/c 3T3 cells, and mouse melanoma B16 cells (Goshima et al., 1978b), human intestinal CCL 6 cells (Takeda et al., 1980), and mouse and rat myocardial cells (Honda et al., 1976a; Goshima et al., 1977, 1978a,b,c; Takeda et al., 1978b). TDH increases the permeability of the cytoplasmic membrane of cultured myocardial cells to Ca^{2+}, and the pathologic changes observed in the cytoplasm are believed to be caused by excessive uptake of Ca^{2+} by the myocardial cells (Goshima et al., 1978b,c).

In vivo activities

The LD_{50} of purified TDH for 25- to 28-g adult mice is approximately 13 μg by the intraperitoneal route (Miyamoto et al., 1975) and 1.4 μg by the intravenous route (Miyamoto et al., 1980). The LD_{50} for 5- to 6-day-old suckling mice is approximately 30 μg by the oral route, and 6 μg produces diarrhoea in 50% of the mice (Miyamoto et al., 1980).

Toxin (300 ng) injected intradermally into rabbits and guinea-pigs causes a localized increase in vascular permeability (Miyamoto et al., 1980; Yamamoto et al., 1983), and toxin (125 μg) injected into ligated ileal loops of rabbits elicits a positive reaction manifested by distension of the loops with turbid and bloody fluid (Miyamoto et al., 1980). Histopathological examination of intestinal tissues taken from positive ileal loops and from suckling mice challenged with large doses of TDH revealed extensive pathological changes in the tissues.

Studies characterizing the effect of purified TDH on intact rats (Honda et al., 1976a, 1978; Takeda et al. 1978b), on isolated rat and rabbit atrial preparations (Goshima et al., 1977; Seyama et al., 1977), and on mouse and rat myocardial cells in tissue culture (Honda et al., 1976a; Goshima et al., 1977, 1978a,b,c; Takeda et al., 1978b) have documented the cardiotoxicity of the toxin.

Takashi et al. (1982) have reported that intravenous injection of rats with low doses (0.1–0.2 μg/kg) of TDH induces hyperpotassaemia by eliciting potassium leakage from their erythrocytes without lysing the cells.

Cloning and sequencing of the tdh gene

The gene encoding TDH (designated tdh) has been cloned into a phage lambda vector in Escherichia coli (Kaper et al., 1984) and into the pBR322 vector in E. coli (Taniguchi et al., 1985). The gene was subcloned and reported, in two independent studies, to be localized on a 0.9-kb DNA fragment (Taniguchi et al., 1985) and on a 1.3-kb DNA fragment (Nishibuchi et al., 1985). Taniguchi et al. (1985) found, by Southern blot hybridization and colony hybridization experiments, that the tdh gene was present in the chromosomal DNA of 15 Kanagawa phenomenon-positive strains but not in 14 negative strains, and that no homologous DNA sequences were detected in the chromosomes of V. cholerae 01, non-01 V. cholerae, V. vulnificus, and V. anguillarum. However, Nishibuchi et al. (1990) subsequently detected a very homologous sequence in the plasmid-borne structural gene encoding a V. cholerae non-01 cytolysin and in chromosomal genes encoding cytolysins of V. hollisae and V. mimicus.

The nucleotide sequence determined by Nishibuchi et al. (1985) and Taniguchi et al. (1986) indicates that the structural gene encodes a mature TDH subunit of 165 amino acid residues preceded by a putative signal peptide sequence of 24 amino acids.

TRH of *V. parahaemolyticus*

Purification

Highly purified preparations of TRH have been obtained from the culture supernatant fluids of two serotypes of V. parahaemolyticus (03:K6 and 06:K46) isolated from patients with traveller's diarrhoea (Honda *et al.*, 1988, 1989a). The purification schemes were very similar to one another and involved sequential ammonium sulphate precipitation, ion-exchange chromatography with DEAE–cellulose, hydroxylapatite adsorption chromatography, gel filtration with Sepharose 4B (only in Honda *et al.*, 1989a), and fast protein liquid chromatography (FPLC) with a Mono Q (anion-exchange) column. The recovery and specific activity (against a standardized suspension of washed bovine erythrocytes) was approximately 60% (0.2 mg protein from 3 litres of culture) and 4000 haemolytic units (HU)/mg protein, respectively.

Stability and inactivation

The activity of purified TRH, unlike that of TDH, is lost on heating at 60°C or a higher temperature for 10 min (Honda *et al.*, 1988, 1989a).

Molecular weight and isoelectric point

The toxin is slightly larger in size than TDH and is composed of two subunits of about 23 000 Da each (Honda *et al.*, 1988). The isoelectric point of the toxin is approximately 4.6 (Honda *et al.*, 1988).

Antigenicity

The results of (i) cross-neutralization, Ouchterlony double immunodiffusion, and latex agglutination tests with homologous and heterologous antisera (Yoh *et al.*, 1986, 1988; Honda *et al.*, 1988, 1989a), and (ii) ELISA with homologous and heterologous monoclonal antibodies (Honda *et al.*, 1989b) indicate that TRH is immunologically related, but not identical, to TDH and to a *V. hollisae* cytolysin.

Cytolytic activity

The toxin's broad haemolytic spectrum differs from that of the TDH (Honda *et al.*, 1988). Activity against mouse, rabbit, human and guinea-pig erythrocytes is similar to that of TDH and, like TDH, TRH does not lyse horse erythrocytes. However, the TRH is much more active against calf, chicken and sheep erythrocytes than is TDH.

In vivo activity

TRH is lethal for mice after intravenous injection; however, LD_{50} values have not been reported (Honda *et al.*, 1988). The mice died within 1 min after injection of 10 µg of purified TRH or 5 µg of purified TDH.

Cloning and sequencing of the trh gene

The gene encoding TRH (designated *trh*) has been cloned into pBR322 in *E. coli*, and nucleotide sequence analysis of the cloned 1.0-kb DNA fragment revealed that the *trh* gene, like the *tdh* gene, encodes a toxin subunit composed of 189 amino acid residues (Nishibuchi *et al.*, 1989). The *trh* gene had significant nucleotide sequence homology with the *tdh* gene, and the amino acid sequences of the TRH and TDH subunits (deduced from the nucleotide sequences of their genes) were homologous and contained two cysteine residues forming an intrachain disulphide bond at the same positions. Also, hydrophobicity–hydrophilicity analysis and a secondary structure prediction suggested that the conformations of the two toxins are similar. The authors proposed that the *trh* and *tdh* genes may have a common ancestor and may have evolved by single-base changes, so that the fundamental architecture of the toxins was maintained.

TL haemolysin of *V. parahaemolyticus*

The gene encoding the promoter region and structural gene of the TL haemolysin has been cloned into pBR322 in *E. coli* (Taniguchi *et al.*, 1985), and nucleotide sequence analysis of the cloned 1.5-kb DNA fragment revealed (i) no homology between the nucleotide sequences encoding the TDH and the TL haemolysin, and (ii) that the structural gene for the TL haemolysin encodes a mature TL haemolysin subunit of 398 amino acid residues preceded by a putative signal peptide sequence of 20 amino acids (Taniguchi *et al.*, 1986). The gene product has not been purified and characterized as have TDH and TRH; however, maxicell analysis (Taniguchi *et al.*, 1986) revealed that the molecular weight of the TL haemolysin subunit is approximately 45 000 Da (approximately twice the size of the TDH and TRH subunits). Also, the ability of the toxin to lyse horse erythrocytes differentiates it from TDH and TRH (Taniguchi *et al.*, 1985).

Vibrio hollisae cytolysin

Purification

V. hollisae produces an extracellular cytolysin which is immunologically and genetically related to the TDH and TRH of Vibrio parahaemolyticus (Nishibuchi et al., 1985, 1990; Yoh et al., 1986, 1988; Honda et al., 1989b). The toxin has been obtained in a highly purified state by sequential ion-exchange chromatography with DEAE– cellulose and immunoaffinity chromatography with Sepharose 4B coupled with immunoglobulin raised against the V. parahaemolyticus TDH (Yoh et al., 1986). The recovery and specific activity (against a standardized suspension of washed chicken erythrocytes) was approximately 3% (0.59 mg protein from 16 litres of culture) and 1.1×10^5 HU/mg protein, respectively.

Stability and inactivation

The toxin is heat-labile (70°C, 10 min), unlike TDH of V. parahaemolyticus (Yoh et al., 1986).

Molecular weight and amino acid composition

The toxin is slightly smaller in size than the TDH of V. parahaemolyticus and is composed of two subunits of about 20 000–21 000 Da each (Yoh et al., 1986, 1988).

The amino acid composition and sequence are similar to that of TDH but differ in the amounts of serine, alanine, isoleucine, tyrosine and histidine residues (Yoh et al., 1988, 1989).

Antigenicity

The toxin is neutralized by rabbit polyclonal antibodies raised against the TDH of V. parahaemolyticus and a cytolysin produced by a non-01 V. cholerae, and exhibits immunologic cross-reaction (contains both similar and dissimilar antigenic determinants) with TDH and the V. cholerae non-01 cytolysin when examined by Ouchterlony double immunodiffusion (Yoh et al., 1986, 1988). In addition, monoclonal antibodies raised against TDH and TRH of V. parahaemolyticus have recently been tested for their ELISA-reacting activities against the V. hollisae cytolysin (Honda et al., 1989b). Some of the antibodies showed epitopes common to the three cytolysins and some showed the presence of epitopes specific to the V. parahaemolyticus toxins.

Cytolytic activity

The toxin's broad haemolytic spectrum differs from that of the V. parahaemolyticus TDH (Yoh et al., 1986). Activity against rabbit, human, guinea-pig and goose erythrocytes is similar to that of TDH and, like TDH, the V. hollisae cytolysin does not lyse horse erythrocytes. However, the V. hollisae toxin is much more active against chicken, sheep and calf erythrocytes than is TDH.

In vivo activity

The lethal activity of the toxin for mice has been claimed to be similar to that of TDH (Yoh et al., 1988); however, LD_{50} values have not been reported for the V. hollisae cytolysin.

Cloning and comparative analysis of the gene encoding the V. hollisae cytolysin

The gene encoding the cytolysin has recently been cloned into pBR322 in E. coli, on an 8.3-kb fragment of the V. hollisae chromosome (Nishibuchi et al., 1990). A 2.4-kb DNA fragment containing the gene was obtained by subcloning and was compared, by physical mapping and by hybridization with oligodeoxyribonucleotide probes, with the tdh gene and with genes encoding cytolysins produced by V. cholerae non-01 and V. mimicus (Nishibuchi et al., 1990). The nucleotide sequences in the coding regions of the four cytolysin genes (the structural genes) were very homologous and had only minor variations, but the sequences flanking the structural genes were dissimilar. The authors proposed that the cytolysin genes may have a common ancestor that was transferred as a discrete genetic unit from one Vibrio replicon into replicons in other Vibrio species, possibly via plasmids.

Vibrio vulnificus cytolysin

Information pertaining to the purification and characterization of an extracellular *V. vulnificus* cytolysin called vulnificolysin has recently been summarized (Kreger *et al.*, 1988).

Purification

A highly purified preparation of vulnificolysin has been obtained by sequential ammonium sulphate precipitation, gel filtration with Sephadex G-75, hydrophobic interaction chromatography with phenyl-Sepharose CL-4B, and isoelectric focusing in an ethylene glycol density gradient (Gray and Kreger, 1985). The recovery and specific activity (against a standardized suspension of washed mouse erythrocytes) in several different preparations ranged from 20 to 25% (6–7 mg protein from 2.4 litres of culture) and from 70 000 to 90 000 HU/mg protein, respectively.

Stability and inactivation

Gray and Kreger (1985) reported that highly purified vulnificolysin is heat-labile (56°C, 30 min), is inactivated by cholesterol (100 μg/ml), and is partially inactivated by Trypan Blue (25–50 μg/ml) and by various commercially available proteases (pronase, trypsin, chymotrypsin, subtilisin BPN, subtilopeptidase A, papain, thermolysin, and proteinase K; 100 μg/ml). In addition, activity of dilute solutions of the toxin (1–2 μg/ml) was lost at pH 4–10 (4°C, 24 h) unless crystalline bovine albumin was added (1 mg/ml) to the buffers. The toxin was not affected by various phospholipids (cardiolipin, sphingomyelin, phosphatidylcholine, phosphatidylserine, phosphatidylinositol and phosphatidylethanolamine; 100 μg/ml), mixed gangliosides (100 μg/ml), glycophorin (100 μg/ml), dithiothreitol (5 mM), chelating agents (EDTA and EGTA; 1 mM), and divalent cations (Ca^{2+}, Mg^{2+} and Zn^{2+}, 1 mM).

According to Shinoda *et al.* (1985) and Miyoshi *et al.* (1985), haemolytic activity in a partially purified preparation of vulnificolysin is heat-labile and is inhibited by cholesterol (0.1 μg/ml), divalent cations (Ca^{2+}, Mg^{2+} and Zn^{2+}; 25 mM), and dithiothreitol (2.5–10 mM), but is not affected by oxygen, sulphydryl-

blocking agents, and antiserum against streptolysin O. They considered the toxin to be a unique cholesterol-binding but non-thiol-activated cytolysin.

The ability of vulnificolysin to bind to cholesterol was used by Yamanaka *et al.* (1987c) to isolate highly purified toxin bound to cholesterol-containing liposomes, and to prepare monospecific antiserum against the toxin by vaccinating rabbits with the toxin–liposome suspension.

Hydrophobicity

The ability of vulnificolysin to bind to phenyl-Sepharose CL-4B in the presence of 10 mM glycine–NaOH buffer (pH 9.8) containing 25% ethylene glycol (Gray and Kreger, 1985) indicates that the toxin is capable of strong interaction with hydrophobic surfaces.

Molecular weight, amino acid composition and isoelectric point

The molecular weight of vulnificolysin is approximately 56 000 determined by SDS-PAGE. However, Okada *et al.* (1987a,b) reported that some isolates of *V. vulnificus* produce a cytolysin which is antigenically distinct from vulnificolysin and has a molecular weight of approximately 36 000.

The amino acid composition has been determined (Gray and Kreger, 1985) to consist of approximately 13, 22 and 35% basic, acidic and hydrophobic amino acid residues, respectively. However, many of the aspartic acid and glutamic acid residues must occur as asparagine and glutamine because the isoelectric point of the toxin is approximately 7.1 as determined by thin-layer isoelectric focusing (Gray and Kreger, 1985). The presence of eight half-cystine residues suggests that the toxin could have four disulphide bonds. The first 10 N-terminal amino acid residues are Gln–Glu–Tyr–Val–Pro–Ile–Val–Glu–Lys–Pro, and only a single amino acid sequence was detected in purified vulnificolysin (Gray and Kreger, 1985).

Antigenicity

Rabbits vaccinated with vulnificolysin produce neutralizing antibodies (Kreger and Lockwood, 1981; Kreger, 1984), and anti-vulnificolysin antibodies have been detected in the sera of mice and a human convalescing from *V. vulnificus* disease (Gray and Kreger, 1986). Also, ELISA and an indirect immunofluorescence procedure, using polyclonal and monoclonal antibodies specific for the toxin, detected the toxin in mice experimentally infected with *V. vulnificus* (Gray and Kreger, 1989).

The results of cross-neutralization studies with homologous and heterologous antisera (Kreger, 1984) indicate that vulnificolysin is antigenically distinct from the TDH of *V. parahaemolyticus* and from cytolysins produced by *V. damsela* and the El Tor biotype of *V. cholerae*. However, sequencing of the structural gene encoding vulnificolysin (Yamamoto *et al.*, 1990) showed that regions of the gene are homologous to the structural gene for a *V. cholerae* El Tor cytolysin.

Cytolytic activity

Vulnificolysin has broad-spectrum haemolytic activity, i.e. is active against erythrocytes from at least 17 animal species (Gray and Kreger, 1985). The toxin is also lytic for CHO cells in tissue culture (Gray and Kreger, 1985) and for isolated rat mast cells (Yamanaka *et al.*, 1990).

The rate of mouse erythrocyte lysis by the toxin is temperature dependent and is optimal between 30 and 37°C (Gray and Kreger, 1985). Haemolysis is not observed at 4°C and is a multi-hit, at least two-step process consisting of a temperature-independent, cholesterol-dependent, toxin-binding step, followed by a temperature-dependent, divalent cation-inhibitable, membrane-perturbation step(s) that leads to cell lysis (Gray and Kreger, 1985; Shinoda *et al.*, 1985).

Yamanaka *et al.* (1987a,b) studied the action of vulnificolysin on sheep erythrocytes and cholesterol–liposome suspensions and reported that (i) haemolysis involves a colloid-osmotic mechanism, (ii) the temperature dependence of haemolysis is due to the requirement for high temperature which increases the membrane fluidity needed for transmembrane pore formation, (iii) the temperature-dependent cell disruption step consists of two stages: a change in membrane permeability (transmembrane pore formation) and an erythrocyte-bursting stage, and (iv) the effective diameter of the initial membrane pores formed by vulnificolysin is about 3 nm, and large membrane pores (10–40 nm) are formed in the erythrocyte-bursting step.

In vivo activities

The LD_{50} of highly purified vulnificolysin for mice is approximately 3 µg/kg by the intravenous route and 2.2 mg/kg by the intraperitoneal route (Gray and Kreger, 1985). The toxin also produces dermonecrosis and increases vascular permeability in guinea-pig, mouse and rabbit skin (Kreger and Lockwood, 1981; Gray and Kreger, 1985).

Light and electron microscopy of mouse skin damage caused by a single intradermal injection of the toxin revealed acute cellulitis characterized by extensive extracellular oedema, disorganization of collagen bundles, large accumulations of cell debris and plasma proteins, damaged or necrotic fat cells, capillary endothelial cells, and muscle cells, and mild inflammatory cell infiltration (Gray and Kreger, 1987).

Enzyme activity

Partially purified toxin preparations have phospholipase A_2 and lysophospholipase activities; however, both activities are removed by gel filtration with Sephadex G-75 (Testa *et al.*, 1984). Highly purified vulnificolysin does not have detectable phospholipase A, phospholipase C, phospholipase D or lysophospholipase activity (Testa *et al.*, 1984).

Cloning and sequencing of the vulnificolysin gene

The gene(s) coding for the production of vulnificolysin has been cloned in *E. coli* by using the lytic cloning vector λ1059; a 3.2-kb DNA fragment containing the toxin gene was isolated by subcloning in plasmid pBR325 (Wright *et al.*, 1985). Both haemolytic and CHO cell-lytic activities were expressed in *E. coli*, and both activities were neutralized by rabbit antiserum raised against purified vulnificolysin (Wright *et al.*, 1985). When used as a probe, the 3.2-kb DNA fragment detected homologous gene sequences in all the clinical and environmental isolates of *V. vulnificus* examined but did not hybridize with DNA from other *Vibrio* and non-*Vibrio* species under stringent conditions (Wright *et al.*, 1985; Morris *et al.*, 1987).

Sequencing of the DNA encoding cytolytic activity (Yamamoto et al., 1990) revealed that the sequence contained two open reading frames designated vvhA and vvhB. vvhA, the structural gene, encoded a 50-kDa protein preceded by a 20 amino acid signal peptide and containing the N-terminal amino acid sequence previously reported (Gray and Kreger, 1985) for the toxin. Regions of the vvhA gene showed homology to the structural gene for the V. cholerae El Tor cytolysin. At the present time, the vvhB gene product has not been identified and its function is unknown.

Vibrio damsela cytolysin

Information pertaining to the purification and characterization of an extracellular V. damsela cytolysin called damselysin has recently been summarized (Kreger et al., 1988).

Purification

A highly purified preparation of damselysin has been obtained by sequential ammonium sulphate precipitation, gel filtration with Sephadex G-100, and hydrophobic interaction chromatography with phenyl-Sepharose CL-4B (Kothary and Kreger, 1985). The recovery and specific activity (against a standardized suspension of washed mouse erythrocytes) in several different preparations ranged from 42 to 54% (25–30 mg protein from 1.2 litres of culture) and from 1.7×10^6 to 2.0×10^6 HU/mg protein, respectively.

Stability and inactivation

Kothary and Kreger (1985) reported that the activity of dilute solutions (5 µg/ml) of highly purified damselysin is very heat-labile (37°C, 30 min) but is stabilized at 37°C by crystalline bovine albumin (1 mg/ml). The toxin also was unstable at pH 4, 6 and 10 (4°C, 24 h) and was inactivated by pronase and trypsin. Activity was not affected by phospholipids, mixed gangliosides, glycophorin, cholesterol, dithiothreitol, trypan blue, chelating agents (EDTA and EGTA), and divalent cations (Ca^{2+}, Mg^{2+} and Zn^{2+}).

Molecular weight, isoelectric point and amino acid composition

The molecular weight (determined by SDS–PAGE) and isoelectric point (determined by thin-layer isoelectric focusing) of damselysin are approximately 69 000 and 5.5–5.6, respectively (Kothary and Kreger, 1985).

The amino acid composition has been determined (Kothary and Kreger, 1985) to consist of approximately 14, 26 and 35% basic, acidic and hydrophobic amino acid residues, respectively. The presence of four half-cystine residues suggests that the toxin could have two disulphide bonds. The first 10 N-terminal amino acid residues are Phe–Thr–Gln–Trp–Gly–Gly–Ser–Gly–Leu–Thr.

Antigenicity

Rabbits vaccinated with damselysin produce neutralizing antibodies, and the results of cross-neutralization studies with homologous and heterologous antisera indicate that the toxin is antigenically distinct from vulnificolysin, the TDH of V. parahaemolyticus, and a cytolysin produced by the El Tor biotype of V. cholerae (Kreger, 1984).

Cytolytic activity

Damselysin has potent but narrow-spectrum haemolytic activity and is most active against mouse and rat erythrocytes (Kothary and Kreger, 1985). The toxin also is lytic for CHO cells in tissue culture; the minimal toxic dose is approximately 1ng (Kothary and Kreger, 1985).

The rate of mouse erythrocyte lysis by the toxin is temperature and pH dependent and is optimal at 37–47°C and at pH 7–9 (Kothary and Kreger, 1985). Haemolysis is not observed at 4°C and is a multi-hit, at least two-step process consisting of a temperature-independent, toxin-binding step, followed by a temperature-dependent, membrane-perturbation step(s) that leads to cell lysis.

In vivo activities

The LD_{50} of highly purified damselysin for mice is approximately 1 µg/kg by the intraperitoneal route, 2 µg/kg by the intravenous route, and 18 µg/kg by the subcutaneous route. Mice injected subcutaneously with 1 LD_{50} of toxin are lethargic and have ruffled fur, encrustations around their eyelids, and severe local oedema similar to that observed (Kreger, 1984) in mice infected with *V. damsela*.

Phospholipase activity

Damselysin is a phospholipase D which so far is known to be active against sphingomyelin (Kreger *et al.*, 1987), phosphatidylcholine and phosphatidyl-ethanolamine (Daniel *et al.*, 1988), and L-ET-18-OCH$_3$ (an ether-linked phospholipid that exhibits selective toxicity toward several types of tumour cells), platelet-activating factor (PAF) and lyso-PAF (Wilcox *et al.*, 1987). Thus, damselysin appears to be unique in that, to my knowledge, phospholipases D produced by other bacteria have not been reported to possess haemolytic activity.

Treatment of sheep erythrocytes with minute amounts of phospholipase D (sphingomyelinase D) produced by the bacterium *Corynebacterium pseudotuberculosis* and by the brown recluse spider (*Loxosceles reclusa*) is known to protect the cells from lysis by staphylococcal sphingomyelinase C (β-lysin) and helianthin, a non-enzymatic but sphingomyelin-binding cytolysin produced by the sea anemone *Stoichactis helianthus* (Bernheimer *et al.*, 1985). Sheep erythrocytes also can be protected against the lytic activity of these two sphingo-myelin-dependent cytolysins by incubating the cells with subnanogram, sublytic amounts of damselysin (Kreger *et al.*, 1987).

Cloning of the damselysin gene

The damselysin gene (designated *dly*) was cloned in *E. coli*, using plasmid pUC18, on a 23-kb DNA fragment isolated from a highly haemolytic strain of *V. damsela* (Cutter and Kreger, 1990). A 5.9-kb fragment containing *dly* was isolated by subcloning, and transposon mutagenesis using Tn*phoA* localized *dly* to an approximately 1.5-kb region of the fragment and determined the direction of transcription. The fragment was used as a DNA probe to detect homologous sequences in other bacterial isolates. Homology was detected only in other highly haemolytic strains of *V. damsela*; no homology was seen with weakly haemolytic *V.*

damsela isolates or with 318 isolates from eight other haemolytic *Vibrio* species.

Vibrio metschnikovii cytolysin

Purification

A highly purified preparation of an extracellular cytolysin produced by *V. metschnikovii* has been obtained by sequential acid precipitation, hydro-phobic interaction chromatography with phenyl-Sepharose CL-4B, and FPLC with a Mono-Q (anion-exchange) column (Miyake *et al.*, 1988). The recovery and specific activity (against a standardized suspension of washed calf erythro-cytes) was approximately 19% (3.2 mg protein from 6 litres of culture) and 2.4 × 10^5HU/mg protein, respectively.

Stability and inactivation

The toxin is heat-labile (60°C, 5 min) and is inhibited by divalent cations (Zn^{2+} and Cu^{2+}) and cholesterol, but is not affected by air oxidation or activated by dithiothreitol (Miyake *et al.*, 1988).

Molecular weight and isoelectric point

The molecular weight (determined by SDS–PAGE) and isoelectric point (determined by polyacryl-amide gel electrofocusing) of the toxin are approxi-mately 50 000 and 5.1, respectively (Miyake *et al.*, 1988).

Antigenicity

Rabbits vaccinated with the purified toxin produce ELISA-reacting and neutralizing antibodies, and the results of ELISA and cross-neutralization studies indicate that the toxin is antigenically distinct from the TDH of *V. parahaemolyticus* and from cytolysins produced by *V. vulnificus*, *V. fluvialis*, *V. furnissii*, *V. mimicus*, *V. cholerae* 01, and *V. cholerae* non-01 (Miyake *et al.*, 1988).

Cytolytic activity

The toxin has broad-spectrum haemolytic activity (i.e. is active against erythrocytes from at least eight animal species); however, it is most active against calf erythrocytes (Miyake *et al.*, 1988). The cytolysin also lyses cultured mammalian cells, such as Vero and CHO cells. The amounts of toxin required for 50% lysis of Vero and CHO cells were 8 and 1 ng, respectively.

Miyake *et al.* (1988) found that lysis of calf erythrocytes by the toxin is temperature dependent (optimal at 37–43°C), and they suggested that lysis requires interaction of the erythrocyte membrane with more than one molecule of toxin (i.e. is a multi-hit mechanism). They subsequently reported (Miyake *et al.*, 1989) that haemolysis involves a colloid-osmotic mechanism and consists of a sequence of three steps; i.e. (i) a temperature-dependent step binding cytolysin to the erythrocyte membrane, (ii) a temperature-dependent step forming toxin tetramers and transmembrane lesions (pores), and (iii) a temperature-independent lysis step.

In vivo activities

Toxin (2–5 μg) administered intragastrically to infant mice causes fluid accumulation in their intestines, and toxin (200 ng) injected intradermally into rabbits elicits a localized increase in vascular permeability (Miyake *et al.*, 1988). However, the kinetics of the two responses differ from those of *V. cholerae* and *E. coli* enterotoxins.

Future studies

During the past 15 years, a lot of information has been published concerning the purification and characterization of the physico-chemical and biological properties and genetic determinants of membrane-damaging protein toxins produced by pathogenic marine bacteria. However, much remains to be learned about the role and importance of the toxins in the biochemistry of, and the molecular pathogenesis of, diseases caused by the bacteria. For example, why does *V. damsela* produce and secrete large amounts (approximately 50 mg/l) of damselysin, and is the production of this cytolysin/ phospholipase D important in the pathogenesis of human or fish diseases or in maintaining the bacterium in its natural (marine) environment? Experimental approaches to evaluate the importance of the cytolysins in the pathogenesis of disease should include determining (i) whether isogenic mutants defective in toxin production are less virulent than are the wild-type, parental strains, (ii) whether active and passive immunization against the toxins protects laboratory animals challenged with virulent strains of the bacteria against tissue damage and death, (iii) whether the toxins are produced *in vivo* during the development of the infectious disease processes, and (iv) whether the grossly and microscopically observable tissue damage elicited by the toxins mimics that observed during naturally occurring and experimentally induced diseases caused by the bacteria. It should be remembered, however, that the interpretation of some of the results may be complicated because many of the pathogenic marine bacteria produce toxins in addition to cytolysins. In that regard, although *V. parahaemolyticus* and *V. vulnificus* (which produce multiple toxins) produce TDH and vulnificolysin, respectively, *in vivo* (Miwatani *et al.*, 1976; Miyamoto *et al.*, 1980; Gray and Kreger, 1986, 1989) and the cytolysins elicit tissue damage which mimics that caused by the bacteria (Miyamoto *et al.*, 1980; Gray and Kreger, 1987), antiserum against TDH does not prevent accumulation of fluid in rabbit ileal loops challenged with *V. parahaemolyticus* (Honda *et al.*, 1983) and isogenic, vulnificolysin-deficient mutants are as virulent as the cytolysin-positive parental strains (Wright *et al.*, 1991). However, do these observations mean that the two cytolysins do not have a role in the pathogenesis of diseases caused by the bacteria, or that they are only one of several bacterial toxins which cause similar tissue damage during the diseases?

The two pathogenic, marine *Vibrio* species *V. fluvialis* and *V. anguillarum* secrete membrane-damaging toxins which, at present, have been only partially purified and characterized (Munn, 1978, 1980; Lockwood *et al.*, 1982, 1983; Toranzo *et al.*, 1983; Kodama *et al.*, 1984; Moustafa *et al.*, 1984; Wall *et al.*, 1984). Thus, research should continue on the purification and characterization of these toxins and their genetic determinants, as well as on the characterization of the δ-VP and TL haemolysins of *V. parahaemolyticus*, which have not been as intensively studied as the TDH and TRH of the bacterium.

Acknowledgements

The cited studies from my laboratory were supported by Public Health Service grant AI-18184 from the National Institutes of Health of the USA. I thank (i) Dr Alan Bernheimer for introducing me to the study of cytolytic toxins, (ii) Dr David Lyerly for reminding me of the importance of pathogenic marine bacteria, (iii) Frances Hickman-Brenner, Dannie Hollis and Dr Robert Weaver for supplying the *V. vulnificus* and *V. damsela* isolates used in my studies, (iv) the following colleagues for contributing their time and expertise to our studies of vulnificolysin and damselysin: Dr Alan Bernheimer, Patricia Chmielewski, Dr Deborah Cutter, Dr Larry Daniel, Dr George Doellgast, Lori Etkin, Dr Larry Gray, Dr Mahendra Kothary, Dr Jon Lewis, Dr Mark Lively, Dr Donald Lockwood, Dr David Lyerly, Bong Roh and Dr Jacqueline Testa, and (v) Fran Bumgardner for her expertise and patience in typing the manuscript. I acknowledge with thanks the opportunity to collaborate with Dr Donald Lockwood, Dr Stephen Richardson and Victoria Wall during our studies of *V. fluvialis* toxins.

References

Bernheimer, A.W. and Rudy, B. (1986). Interactions between membranes and cytolytic peptides. *Biochim. Biophys. Acta* **864**, 123–41.

Bernheimer, A.W., Campbell, B.J. and Forrester, L.J. (1985). Comparative toxinology of *Loxosceles reclusa* and *Corynebacterium pseudotuberculosis*. *Science* **228**, 590–1.

Cherwonogrodzky, J.W. and Clark, A.G. (1982a). The purification of the Kanagawa haemolysin from *Vibrio parahaemolyticus*. *FEMS Microbiol. Lett.* **15**, 175–9.

Cherwonogrodzky, J.W. and Clark, A.G. (1982b). Production of the Kanagawa hemolysin by *Vibrio parahaemolyticus* in a synthetic medium. *Infect. Immun.* **37**, 60–3.

Cutter, D.L. and Kreger, A.S. (1990). Cloning and expression of the damselysin gene from *Vibrio damsela*. *Infect. Immun.* **58**, 266–8.

Daniel, L.W., King, L. and Kennedy, M. (1988). Phospholipase activity of bacterial toxins. *Methods Enzymol.* **165**, 298–301.

Goshima, K., Honda, T., Hirata, M., Kikuchi, K., Takeda, Y. and Miwatani, T. (1977). Stopping of the spontaneous beating of mouse and rat myocardial cells *in vitro* by a toxin from *Vibrio parahaemolyticus*. *J. Molec. Cell. Cardiol.* **9**, 191–213.

Goshima, K., Honda, T., Takeda, Y. and Miwatani, T. (1978a). Stopping of spontaneous beating of cultured mouse and rat myocardial cells by a toxin (thermostable direct hemolysin) from *Vibrio parahaemolyticus*. In: *Recent Advances in Studies on Cardiac Structure and Metabolism*, Vol. 11 (eds T. Kobayashi, T. Sano and N.S. Dhalla), pp. 615–20. University Park Press, Baltimore.

Goshima, K., Owaribe, K., Yamanaka, H. and Yoshino, S. (1978b). Requirement of calcium ions for cell degeneration with a toxin (vibriolysin) from *Vibrio parahaemolyticus*. *Infect. Immun.* **22**, 821–32.

Goshima, K., Yamanaka, H., Eguchi, G. and Yoshino, S. (1978c). Morphological changes of cultured myocardial cells due to change in extracellular calcium ion concentration. *Devel. Growth Differ.* **20**, 191–204.

Gray, L.D. and Kreger, A.S. (1985). Purification and characterization of an extracellular cytolysin produced by *Vibrio vulnificus*. *Infect. Immun.* **48**, 62–72.

Gray, L.D. and Kreger, A.S. (1986). Detection of anti-*Vibrio vulnificus* cytolysin antibodies in sera from mice and a human surviving *V. vulnificus* disease. *Infect. Immun* **51**, 964–5.

Gray, L.D. and Kreger, A. S. (1987). Mouse skin damage caused by cytolysin from *Vibrio vulnificus* and by *V. vulnificus* infection. *J. Infect. Dis.* **155**, 236–41.

Gray, L.D. and Kreger, A.S. (1989). Detection of *Vibrio vulnificus* cytolysin in *V. vulnificus*-infected mice. *Toxicon* **27**, 459–64.

Honda, T., Goshima, K., Takeda, Y., Sugino, Y. and Miwatani, T. (1976a). Demonstration of the cardiotoxicity of the thermostable direct hemolysin (lethal toxin) produced by *Vibrio parahaemolyticus*. *Infect. Immun.* **13**, 163–71.

Honda, T., Taga, S., Takeda, T., Hasibuan, M.A., Takeda, Y. and Miwatani, T. (1976b). Identification of lethal toxin with the thermostable direct hemolysin produced by *Vibrio parahaemolyticus*, and some physicochemical properties of the purified toxin. *Infect. Immun.* **13**, 133–9.

Honda, T., Goshima, K., Takeda, Y. and Miwatani, T. (1978). A bacterial cardiotoxin: thermostable direct hemolysin produced by *Vibrio parahaemolyticus*. In: *Recent Advances in Studies on Cardiac Structure and Metabolism*, Vol. 11 (eds T. Kobayashi, T. Sano and N.S. Dhalla), pp. 609–14. University Park Press, Baltimore.

Honda, T., Takeda, Y., Miwatani, T. and Nakahara, N. (1983). Failure of antisera to thermostable direct hemolysin and cholera enterotoxin to prevent accumulation of fluid caused by *Vibrio parahaemolyticus*. *J. Infect. Dis.* **147**, 779.

Honda, T., Yoh, M., Kongmuang, U. and Miwatani, T. (1985). Enzyme-linked immunosorbent assays for detection of thermostable direct hemolysin of *Vibrio parahaemolyticus*. *J. Clin. Microbiol.* **22**, 383–6.

Honda, T., Yoh, M., Narita, I., Miwatani, T., Sima, H., Xiegen, Y. and Xiandi, L. (1986). Purification of the thermostable direct hemolysin of *Vibrio parahaemolyticus* by immunoaffinity column chromatography. *Can. J. Microbiol.* **32**, 71–3.

Honda, T., Ni, Y. and Miwatani, T. (1988). Purification and characterization of a hemolysin produced by a clinical isolate of Kanagawa phenomenon-negative *Vibrio parahaemolyticus* and related to the thermostable direct hemolysin. *Infect. Immun.* **56**, 961–5.

Honda, T., Ni, Y. and Miwatani, T. (1989a). Purification of a TDH-related hemolysin produced by a Kanagawa phenomenon-negative clinical isolate of *Vibrio parahaemolyticus* 06:K46. *FEMS Microbiol. Lett.* **57**, 241–6.

Honda, T., Ni, Y., Yoh, M. and Miwatani, T. (1989b). Evidence of immunologic cross-reactivity between hemolysins of *Vibrio hollisae* and *Vibrio parahaemolyticus* demonstrated by monoclonal antibodies. *J. Infect. Dis.* **160**, 1089–90.

Janda, J.M., Powers, C., Bryant, R.G. and Abbott, S.L. (1988). Current perspectives on the epidemiology and pathogenesis of clinically significant *Vibrio* spp. *Clin. Microbiol. Rev.* **1**, 245–67.

Joseph, S.W., Colwell, R.R. and Kaper, J.B. (1982). *Vibrio parahaemolyticus* and related halophilic vibrios. *CRC Crit. Rev. Microbiol.* **10**, 77–124.

Kaper, J.B., Campen, R.K., Seidler, R.J., Baldini, M.M. and Falkow, S. (1984). Cloning of the thermostable direct or Kanagawa phenomenon-associated hemolysin of *Vibrio parahaemolyticus*. *Infect. Immun.* **45**, 290–2.

Kodama, H., Moustafa, M., Ishiguro, S., Mikami, T. and Izawa, H. (1984). Extracellular virulence factors of fish *Vibrio*: relationships between toxic material, hemolysin, and proteolytic enzyme. *Am. J. Vet. Res.* **45**, 2203–7.

Kothary, M.H. and Kreger, A.S. (1985). Purification and characterization of an extracellular cytolysin produced by *Vibrio damsela*. *Infect. Immun.* **49**, 25–31.

Kreger, A.S. (1984). Cytolytic activity and virulence of *Vibrio damsela*. *Infect. Immun.* **44**, 326–31.

Kreger, A. and Lockwood, D. (1981). Detection of extracellular toxin(s) produced by *Vibrio vulnificus*. *Infect. Immun.* **33**, 583–90.

Kreger, A.S., Bernheimer, A.W., Etkin, L.A. and Daniel, L.W. (1987). Phospholipase D activity of *Vibrio damsela* cytolysin and its interaction with sheep erythrocytes. *Infect. Immun.* **55**, 3209–12.

Kreger, A.S., Kothary, M.H. and Gray, L.D. (1988). Cytolytic toxins of *Vibrio vulnificus* and *Vibrio damsela*. *Methods Enzymol.* **165**, 176–89.

Ljungh, Å. and Wadström, T. (1983). Toxins of *Vibrio parahaemolyticus* and *Aeromonas hydrophila*. *J. Toxicol.-Toxin Rev.* **1**, 257–307.

Lockwood, D.E., Kreger, A.S. and Richardson, S.H. (1982). Detection of toxins produced by *Vibrio fluvialis*. *Infect. Immun.* **35**, 702–8.

Lockwood, D.E., Richardson, S.H., Kreger, A.S., Aiken, M. and McCreedy, B. (1983). *In vitro* and *in vivo* biologic activities of *Vibrio fluvialis* and its toxic products. In: *Advances in Research on Cholera and Related Diarrheas* (eds S. Kuwahara and N.F. Pierce), pp. 87–99. KTK Scientific Publishers, Tokyo.

Miwatani, T., Takeda, Y., Sakurai, J., Yoshihara A. and Taga, S. (1972). Effect of heat (Arrhenius effect) on crude hemolysin of *Vibrio parahaemolyticus*. *Infect. Immun.* **6**, 1031–3.

Miwatani, T., Sakurai, J., Takeda, Y., Sugiyama, S. and Adachi, T. (1976). Antibody titers against the thermostable direct hemolysin in sera of patients suffering from gastroenteritis due to *Vibrio parahaemolyticus*. *J. Jpn. Assoc. Infect. Dis.* **50**, 46–51.

Miyake, M., Honda, T. and Miwatani, T. (1988). Purification and characterization of *Vibrio metschnikovii* cytolysin. *Infect. Immun.* **56**, 954–60.

Miyake, M., Honda, T. and Miwatani, T. (1989). Effects of divalent cations and saccharides on *Vibrio metschnikovii* cytolysin-induced hemolysis of rabbit erythrocytes. *Infect. Immun.* **57**, 158–63.

Miyamoto, Y., Obara, Y., Nikkawa, T., Yamai, S., Kato, T., Yamada, Y. and Ohashi, M. (1975). Extraction, purification and biophysico-chemical characteristics of a 'Kanagawa phenomenon'-associated hemolytic factor of *Vibrio parahaemolyticus*. *Jpn. J. Med. Sci. Biol.* **28**, 87–90.

Miyamoto, Y., Obara, Y., Nikkawa, T., Yamai, S., Kato, T., Yamada, Y. and Ohashi, M. (1980). Simplified purification and biophysicochemical characteristics of Kanagawa phenomenon-associated hemolysin of *Vibrio parahaemolyticus*. *Infect. Immun.* **28**, 567–76.

Miyoshi, S., Yamanaka, H., Miyoshi, N. and Shinoda, S. (1985). Non-thiol-activated property of a cholesterol-binding hemolysin produced by *Vibrio vulnificus*. *FEMS Microbiol. Lett.* **30**, 213–16.

Morris, J.G., Jr., Wright, A.C., Roberts, D.M., Wood, P.K., Simpson, L.M. and Oliver, J.D. (1987). Identification of environmental *Vibrio vulnificus* isolates with a DNA probe for the cytotoxin-hemolysin gene. *Appl. Environ. Microbiol.* **53**, 193–5.

Moustafa, M., Kodama, H., Ishiguro, S., Mikami, T. and Izawa, H. (1984). Partial purification of extracellular toxic material of fish *Vibrio*. *Am. J. Vet. Res.* **45**, 2208–10.

Munn, C.B. (1978). Haemolysin production by *Vibrio anguillarum*. *FEMS Microbiol. Lett.* **3**, 265–8.

Munn, C.B. (1980). Production and properties of a haemolytic toxin by *Vibrio anguillarum*. In: *Fish Diseases* (ed W. Ahne), pp. 69–74. Springer-Verlag, Berlin.

Nishibuchi, M. and Kaper, J.B. (1985). Nucleotide sequence of the thermostable direct hemolysin gene of *Vibrio parahaemolyticus*. *J. Bacteriol.* **162**, 558–64.

Nishibuchi, M., Ishibashi, M., Takeda, Y. and Kaper, J.B. (1985). Detection of the thermostable direct hemolysin gene and related DNA sequences in *Vibrio parahaemolyticus* and other *Vibrio* species by the DNA colony hybridization test. *Infect. Immun.* **49**, 481–6.

Nishibuchi, M., Taniguchi, T., Misawa, T., Khaeomanee-Iam, V., Honda, T. and Miwatani, T. (1989). Cloning and nucleotide sequence of the gene (*trh*) encoding the hemolysin related to the thermostable direct hemolysin of *Vibrio parahaemolyticus*. *Infect. Immun.* **57**, 2691–7.

Nishibuchi, M., Khaeomanee-Iam, V., Honda, T., Kaper, J.B. and Miwatani, T. (1990). Comparative analysis of the hemolysin genes of *Vibrio cholerae* non-01, *V. mimicus*, and *V. hollisae* that are similar to the *tdh* gene of *V. parahaemolyticus*. *FEMS Microbiol. Lett.* **67**, 251–6.

Okada, K., Miake, S., Moriya, T., Mitsuyama, M. and Amako, K. (1987a). Variability of haemolysin(s) produced by *Vibrio vulnificus*. *J. Gen. Microbiol.* **133**, 2853–7.

Okada, K., Mitsuyama, M., Miake, S. and Amako, K. (1987b). Monoclonal antibodies against the haemolysin of *Vibrio vulnificus*. *J. Gen. Microbiol.* **133**, 2279–84.

Sakazaki, R., Tamura, K., Nakamura, A., Kurata, T., Ghoda, A. and Kazuno, Y. (1974). Enteropathogenic activity of *Vibrio parahaemolyticus*. In: *International Symposium on Vibrio parahaemolyticus* (eds T. Fujino, G. Sakaguchi, R. Sakazaki and Y. Takeda), pp. 231–5. Saikon Publishing Co, Tokyo.

Sakurai, J., Matsuzaki, A. and Miwatani, T. (1973). Purification and characterization of thermostable direct hemolysin of *Vibrio parahaemolyticus*. *Infect. Immun.* **8**, 775–80.

Sakurai, J., Bahavar, M.A., Jinguji, Y. and Miwatani, T. (1975). Interaction of thermostable direct hemolysin of *Vibrio parahaemolyticus* with human erythrocytes. *Biken J.* **18**, 187–92.

Sakurai, J., Honda, T., Jinguji, Y., Arita, M. and Miwatani, T. (1976). Cytotoxic effect of the thermostable direct hemolysin produced by *Vibrio parahaemolyticus* on FL cells. *Infect. Immun.* **13**, 876–83.

Seyama, I., Irisawa, H., Honda, T., Takeda, Y. and Miwatani, T. (1977). Effect of hemolysin produced by *Vibrio parahaemolyticus* on membrane conductance and mechanical tension of rabbit myocardium. *Jpn. J. Physiol.* **27**, 43–56.

Shier, W.T. (1988). Cytotoxic effect of marine toxins and venoms. In: *Handbook of Natural Toxins*, Vol. 3 (ed. A.T. Tu), pp. 477–91. Marcel Dekker, New York.

Shinoda, S., Miyoshi, S., Yamanaka, H. and Miyoshi-Nakahara, N. (1985). Some properties of *Vibrio vulnificus* hemolysin. *Microbiol. Immunol.* **29**, 583–90.

Shirai, H., Ito, H., Hirayama, T., Nakamoto, Y., Nakabayashi, N., Kumagai, K., Takeda, Y. and Nishibuchi, M. (1990). Molecular epidemiologic evidence for association of thermostable direct hemolysin (TDH) and TDH-related hemolysin of *Vibrio parahaemolyticus* with gastroenteritis. *Infect. Immun.* **58**, 3568–73.

Takashi, K., Nishiyama, M. and Kuga, T. (1982). Hemolysis and hyperpotassemia in rat induced by the thermostable direct hemolysin of *Vibrio parahaemolyticus*. *Jpn. J. Pharmacol.* **32**, 377–80.

Takeda, Y. (1983). Thermostable direct hemolysin of *Vibrio parahaemolyticus*. *Pharmacol. Ther.* **19**, 123–46.

Takeda, Y. (1988). Thermostable direct hemolysin of *Vibrio parahaemolyticus*. *Methods Enzymol.* **165**, 189–93.

Takeda, Y., Hori, Y. and Miwatani, T. (1974). Demonstration of a temperature-dependent inactivating factor of the thermostable direct hemolysin in *Vibrio parahaemolyticus*. *Infect. Immun.* **10**, 6–10.

Takeda, Y., Hori, Y., Taga, S., Sakurai, J. and Miwatani, T. (1975). Characterization of the temperature-dependent inactivating factor of the thermostable direct hemolysin in *Vibrio parahaemolyticus*. *Infect. Immun.* **12**, 449–54.

Takeda, Y., Takeda, T., Honda, T. and Miwatani, T. (1976). Inactivation of the biological activities of the thermostable direct hemolysin of *Vibrio parahaemolyticus* by ganglioside G_{T1}. *Infect. Immun.* **14**, 1–5.

Takeda, Y., Taga, S. and Miwatani, T. (1978a). Evidence that thermostable direct hemolysin of *Vibrio parahaemolyticus* is composed of two subunits. *FEMS Microbiol. Lett.* **4**, 271–4.

Takeda, Y., Takeda, T., Honda, T. and Miwatani, T. (1978b). Comparison of bacterial cardiotoxins: thermostable direct hemolysin from *Vibrio parahaemolyticus*, streptolysin O and hemolysin from *Listeria monocytogenes*. *Biken J.* **21**, 1–8.

Takeda, T., Honda, T., Takeda, Y. and Miwatani, T. (1980). Pathogenesis of *Vibrio parahaemolyticus*. In: *Natural Toxins* (eds D. Eaker and T. Wadström), pp. 251–7. Pergamon Press, New York.

Taniguchi, H., Ohta, H., Ogawa, M. and Mizuguchi, Y. (1985). Cloning and expression in *Escherichia coli* of *Vibrio parahaemolyticus* thermostable direct hemolysin and thermolabile hemolysin genes. *J. Bacteriol.* **162**, 510–15.

Taniguchi, H., Hirano, H., Kubomura, S., Higashi, K. and Mizuguchi, Y. (1986). Comparison of the nucleotide sequences of the genes for the thermostable direct hemolysin and the thermolabile hemolysin from *Vibrio parahaemolyticus*. *Microb. Pathogen.* **1**, 425–32.

Taniguchi, A., Kubomura, S., Hirano, H., Mizue, K. Ogawa, M. and Mizuguchi, Y. (1990). Cloning and characterization of a gene encoding a new thermostable hemolysin from *Vibrio parahaemolyticus*. *FEMS Microbiol. Lett.*, **67**, 339–46.

Testa, J., Daniel, L.W. and Kreger, A.S. (1984). Extracellular phospholipase A_2 and lysophospholipase produced by *Vibrio vulnificus*. *Infect. Immun.* **45**, 458–63.

Toranzo, A.E., Barja, J.L., Colwell, R.R., Hetrick, F.M. and Crosa, J.H. (1983). Haemagglutinating, haemolytic and cytotoxic activities of *Vibrio anguillarum* and related vibrios isolated from striped bass on the Atlantic Coast. *FEMS Microbiol. Lett.* **18**, 257–62.

Tsunasawa, S., Sugihara, A., Masaki, T., Sakiyama, F., Takeda, Y., Miwatani, T. and Narita, K. (1987). Amino acid sequence of thermostable direct hemolysin produced by *Vibrio parahaemolyticus*. *J. Biochem. (Tokyo)* **101**, 111–21.

Wall, V.W., Kreger, A.S. and Richardson, S.H. (1984). Production and partial characterization of a *Vibrio fluvialis* cytotoxin. *Infect. Immun.* **46**, 773–7.

Wilcox, R.W., Wykle, R.L., Schmitt, J.D. and Daniel, L.W. (1987). The degradation of platelet-activating factor and related lipids: susceptibility to phospholipases C and D. *Lipids* **22**, 800–7.

Wright, A.C. and Morris, J.G., Jr. (1991). The extracellular cytolysin of *Vibrio vulnificus*: inactivation and relationship to virulence in mice. *Infect. Immun.* **59**, 192–7.

Wright, A.C., Morris, J.G., Jr., Maneval, D.R., Jr., Richardson, K. and Kaper, J.B. (1985). Cloning of the cytotoxin-hemolysin gene of *Vibrio vulnificus*. *Infect. Immun.* **50**, 922–4.

Yamamoto, K., Honda, T., Takeda, Y. and Miwatani, T. (1983). Production of increased vascular permeability in rabbits by purified thermostable direct hemolysin from *Vibrio parahaemolyticus*. *J. Infect. Dis.* **148**, 1129.

Yamamoto, K., Wright, A.C., Kaper, J.B. and Morris, J.G., Jr. (1990). The cytolysin gene of *Vibrio vulnificus*: sequence and relationship to the *Vibrio cholerae* El Tor hemolysin gene. *Infect. Immun.* **58**, 2706–9.

Yamanaka, H., Katsu, T., Satoh, T., Shimatani, S. and Shinoda, S. (1987a). Effect of *Vibrio vulnificus* hemolysin on liposome membranes. *FEMS Microbiol. Lett.* **44**, 253–8.

Yamanaka, H., Satoh, T., Katsu, T. and Shinoda, S. (1987b). Mechanism of haemolysis by *Vibrio vulnificus* haemolysin. *J. Gen. Microbiol.* **133**, 2859–64.

Yamanaka, H., Satoh, T. and Shinoda, S. (1987c). Preparation of specific antiserum against *Vibrio vulnificus* hemolysin by immunization with hemolysin-bound liposomes. *FEMS Microbiol. Lett.* **41**, 313–16.

Yamanaka, H., Sugiyama, K., Furuta, H. Miyoshi, S. and Shinoda, S. (1990). Cytolytic action of *Vibrio vulnificus* haemolysin on mast cells from rat peritoneal cavity. *J. Med. Microbiol.* **32**, 39–43.

Yoh, M., Honda, T. and Miwatani, T. (1986). Purification and partial characterization of a *Vibrio hollisae* hemolysin that relates to the thermostable direct hemolysin of *Vibrio parahaemolyticus*. *Can. J. Microbiol.* **32**, 632–6.

Yoh, M., Honda, T. and Miwatani, T. (1988). Comparison of hemolysin of *Vibrio cholerae* non-01 and *Vibrio hollisae* with thermostable direct hemolysin of *Vibrio parahaemolyticus*. *Can. J. Microbiol.* **34**, 1321–4.

Yoh, M., Honda, T., Miwatani, T. Tsunasawa, S. and Sakiyama, F. (1989). Comparative amino acid sequence analysis of hemolysin produced by *Vibrio hollisae* and *Vibrio parahaemolyticas*. *J. Bacteriol.* **171**, 6859–61.

Zen-Yoji, H., Hitokoto, H., Morozumi, S. and Le Clair, R.A. (1971). Purification and characterization of a hemolysin produced by *Vibrio parahaemolyticus*. *J. Infect. Dis.* **123**, 665–7.

14

The Anthrax Toxin Complex

Stephen H. Leppla

Laboratory of Microbial Ecology, National Institute of Dental Research, National Institutes of Health, Bethesda, Maryland 20892–0030, USA

Introduction

Anthrax was recognized for many centuries as a serious disease of animals and man, one that inflicted great losses in agricultural economies and caused significant disease in humans. Thus, anthrax was a major concern to the pioneers of microbiology, and its study by Pasteur, Koch, Metchnikoff and others established many of the basic principles of infectious diseases.

In domestic livestock and wild animals, symptoms of infection by *Bacillus anthracis* are rarely evident until the animal becomes lethargic several hours before death. Necropsy shows extensive oedema in many tissues and concentrations of bacteria in blood that may exceed 10^8 per ml (Turnbull, 1990c). Most human cases arise from contact with infected animals or spores present in animal products (wool, leather, bone meal), and begin as cutaneous infection. This is easily treated with antibiotics if correctly diagnosed. The less frequent but more dangerous gastrointestinal form of anthrax is usually contracted by eating contaminated meat; progression of this form is rapid and more difficult to treat.

The virulence of *B. anthracis* for animals and man is attributed to the production of two principal virulence factors, the gamma-linked poly-D-glutamic acid capsule, and the three-component protein exotoxin that is termed anthrax toxin (Smith *et al.*, 1955; Keppie *et al.*, 1963). The capsule appears to protect bacteria from phagocytosis, and therefore plays an essential role during establishment of an infection. The protein exotoxin may also help to establish an infection by incapacitating phagocytes (Keppie *et al.*, 1963; O'Brien *et al.*, 1985; Wade *et al.*, 1985), but its more obvious role is to cause the extensive tissue oedema that appears to be the major cause of death. It is generally accepted that the pathological effects causing death in infected animals are due to the action of the toxin (Smith and Stoner, 1967). A number of reviews have discussed the toxin and its role in pathogenesis and immunity (Stephen, 1981, 1986; Hambleton *et al.*, 1984; Leppla *et al.*, 1985; Turnbull, 1986, 1990a; Leppla, 1988). The published proceedings of an international workshop on anthrax held in 1989 contains a total of 52 papers, of which approximately ten concern some aspects of the toxin (Turnbull, 1990b).

Virulent strains of *B. anthracis* contain two large plasmids, pX01 and pX02. The genes coding for toxin are contained on pX01 (Mikesell *et al.*, 1983; Thorne, 1985), and the genes for capsule are on pX02 (Green *et al.*, 1985; Uchida *et al.*, 1985). Virulence requires the presence of both plasmids, as is evident from comparison of strains which have lost either of the two plasmids. Strains lacking plasmid pX01 do not produce toxin and are essentially avirulent (Ivins *et al.*, 1986; Uchida *et al.*, 1986). Strains lacking pX02 are at least 10^5-fold less virulent than wild-type (Ivins *et al.*, 1986; Welkos and Friedlander, 1988). Although it is possible that these plasmids code for other materials that contribute to virulence, no such materials have yet been identified. *B. anthracis* does produce a number of other secreted and cytosolic materials that are potentially harmful to animal hosts; these include phospholipases, proteases and a thiol-activated haemolysin (unpublished studies of SHL). However, none of these other materials has been shown to contribute to pathogenesis nor to be plasmid-encoded.

The anthrax complex, the principal subject of this chapter, merits study both because it is a principal virulence factor of *B. anthracis*, and because it has several characteristics that make it an attractive model for study of interactions of protein ligands with eukaryotic target cells. Foremost among these is the fact that this toxin complex contains three proteins that are individually non-toxic. These proteins are designated protective antigen (PA), lethal factor (LF), and oedema factor (EF).* Toxic activity is obtained only when the proteins are administered in combinations, and two distinct toxic activities are produced. The existence of two toxic activities, and therefore of two toxins, is emphasized in a recently introduced nomenclature that is recommended for its specificity (Friedlander, 1986). Thus, the combination of PA with LF, which causes rapid death of certain animal species when injected intravenously, is designated 'lethal toxin'. The combination of PA with EF, which causes oedema when injected intradermally, is designated 'oedema toxin'.

Recent progress, to be detailed in this chapter, has shown that the PA component of the toxin binds to receptors on eukaryotic cells and then promotes the internalization of LF and EF to the cytosol. EF is an adenylate cyclase, and converts ATP to cAMP, which accumulates to unphysiologically high concentrations and causes metabolic perturbations (Leppla, 1982, 1984). LF is assumed to be an enzyme that acts in the cytosol, but its mechanism of action remains unknown (Singh *et al.*, 1989a). Because the anthrax toxins act inside cells, the proteins must perform at least three functions – receptor binding, internalization and catalysis. In a number of toxins, the binding and catalytic functions have been localized to specific

domains, protease fragments or subunits. These toxins are then said to fit the A/B model, with the A moiety being the catalytic part, and the B moiety the receptor-binding region (Gill, 1978). Anthrax toxin fits this pattern. The PA protein binds to receptors, and can therefore be considered the B moiety. LF and EF are each alternative A moieties. Anthrax toxin is unusual in that the A and B moieties are separate proteins, whereas in most toxins these functions are performed by separate domains on a single polypeptide. Only a few other toxins are known that have separate protein components. Most analogous in design is the *Clostridium botulinum* toxin C2, in which the receptor recognition protein, after proteolytic activation, binds and promotes the internalization of an actin ADP-ribosylating protein (Ohishi, 1987).

The anthrax oedema toxin is one of two known bacterial 'invasive adenylate cyclases' that were recognized at approximately the same time; the other is the *Bordetella pertussis* cyclase that is the subject of another chapter of this book (Chapter 16). Both adenylate cyclases require calmodulin, a eukaryotic protein, as a cofactor for enzymatic activity. The structural and functional similarities of the two cyclases are discussed in a later section.

Some confusion may arise from use of the designation PA for a component of a toxin. This term arose many years prior to discovery of the toxin, because it was recognized that culture supernatants of *B. anthracis* could be used to immunize animals against infection (Gladstone, 1948). Only after the toxin was discovered did it become clear that the 'protective antigen' in the culture supernates was a component of the toxin. PA remains the principal and essential immunogen in both killed and live anthrax vaccines (Hambleton *et al.*, 1984; Ivins and Welkos, 1988). Therefore, study of anthrax toxin has direct application to development of improved vaccines.

The genetics of toxin and virulence

Plasmids pXOl and pX02

Many pathogenic bacteria have been found to carry the genes for virulence factors on extrachromosomal DNA. The possibility that the genes for *B. anthracis* virulence factors might be extrachromosomal was

---

* The British workers who played a major role in identifying these proteins termed the same proteins as Factors I (=EF), II (=PA), and III (=LF). These numerical designations are no longer favoured, principally because they are less descriptive.

also suggested by the classical studies of Louis Pasteur and Max Sterne, who each isolated variants of *B. anthracis* that had reduced virulence. The ease with which these variants were obtained suggested that the genes for toxin and capsule might be extrachromosomal. After some initial difficulties, plasmids associated with toxin and capsule were discovered (Mikesell *et al.*, 1983; Green *et al.*, 1985; Uchida *et al.*, 1985). Recognition and characterization of the plasmids was delayed because these plasmids are very large, and are destroyed by shear unless special precautions are employed.

The *B. anthracis* plasmids can each be selectively cured, pX01 by repeated passage at 42–43°C, and pX02 by growth in novobiocin. The properties of the variant strains are summarized in Fig. 1. Examination of the variants obtained by curing proved that pX01 was needed for production of toxin, and pX02 for production of capsule (Thorne, 1985). Conjugal transfer of pX01 to plasmid-cured strains of *B. anthracis* confirmed that all the genes necessary for toxin production are contained on the plasmid (Thorne, 1985; Heemskerk and Thorne, 1990). These results appear to provide a basis for explaining the properties and the efficacy of the

anthrax vaccines developed by Louis Pasteur and Max Sterne. It is now well established that *B. anthracis* strains must produce PA in order to induce protective immunity (Ivins and Welkos, 1988). Elimination of the pX01 plasmid therefore yields an avirulent strain, but one which does not induce immunity. In retrospect, it is now evident that Louis Pasteur's attenuation of the virulence of *B. anthracis* cultures by growth at 42°C can be explained as due to partial curing of plasmid pX01. The cultures he used to successfully immunize sheep probably contained a small number of virulent (pX01+, pX02+) bacteria with a larger number of avirulent (pX01−, pX02+) bacteria. Since there is no evidence that capsule or surface antigens play a role in protective immunity, it appears that the efficacy of Pasteur's vaccine depended on the presence of the small number of virulent bacteria, which would induce antibodies to PA. While effective, these vaccines had to be carefully prepared, because they could cause infection if the fraction of virulent organisms was too high.

Max Sterne's important contribution was to carefully analyse the rare, spontaneous non-capsulated variants appearing on agar plates, and to show

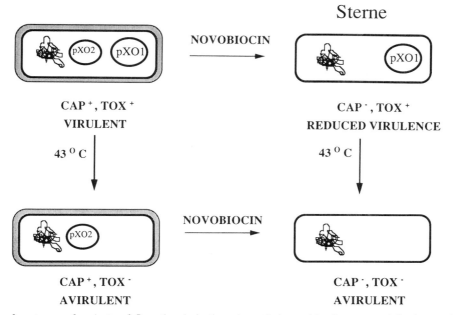

Figure 1. The four types of variants of *B. anthracis*. In the schematic bacterial cell at upper left, the random pattern denotes the chromosomal DNA, labelled ellipses denote plasmids, and the shaded zone at the outer edge denotes the polyglutamate capsule. Labelled arrows identify treatments that lead to loss of each plasmid. Cap+ or Cap− denotes capsule phenotype, and Tox+ or Tox− denotes toxin production phenotype.

that they were greatly reduced in virulence (Sterne, 1937). These variants, now known to have lost the pX02 plasmid, were effective animal vaccines, and did not revert to virulence. The (pX01$^+$, pX02$^-$) 'Sterne' strain was immediately adopted and continues in use today as the preferred live animal vaccine.

Both plasmids have been physically characterized (Kaspar and Robertson, 1987), restriction maps have been constructed, and the origins of replication cloned (Uchida *et al.*, 1987; Robertson *et al.*, 1990). The *B. anthracis* plasmids are very large. Plasmid pX01 has been reported to be 170–185 kbp (Kaspar and Robertson, 1987; Heemskerk and Thorne, 1990; Robertson *et al.*, 1990) while pX02 is 90–95 kbp (Uchida *et al.*, 1985; Robertson *et al.*, 1990). The three toxin genes were localized to specific restriction fragments, showing that the PA and LF genes are separated by about 3 kbp, and the EF gene is located approximately 20 kbp from the PA gene. The location on pX02 of three genes involved in synthesis of the capsule (Makino *et al.*, 1989) has also been deduced (Robertson *et al.*, 1990).

It is not evident why *B. anthracis* maintains these two very large plasmids, pX01 and pX02. The only genes so far identified on the plasmids are the three toxin component genes (discussed below) and the three genes involved in capsule production (Makino *et al.*, 1989). The plasmids must contain a number of other functional genes, but none of these can be essential to growth under laboratory conditions because strains cured of both plasmids grow normally. The presence of some active genes on pX01 is evident from the work of Thorne and colleagues (Robillard *et al.*, 1983; Heemskerk and Thorne, 1990) who found approximately six phenotypic differences between isogenic strain pairs differing only in their pX01 content. An example is that curing strain UM23 of pX01 eliminates the requirement that certain amino acids (but not all) be added to minimal media to obtain growth. Another consequence of the absence of pX01 is that bacteria sporulate at an earlier stage in growth.

Cloning of the toxin genes

The recognition that anthrax toxin was encoded on pX01 facilitated the cloning of the toxin genes

for PA (*pag*), LF (*lef*) and EF (*cya*).* The cloning was also made possible by purification and N-terminal sequencing of the three toxin proteins, and by affinity purification of toxin component-specific goat and rabbit antibodies. Recombinant gene libraries were prepared initially by complete BamH1 digestion of pX01 DNA and ligation to pBR322 (Vodkin and Leppla, 1983), and later by partial MboI digestion and ligation to pUC8 (Robertson and Leppla, 1986).

Immunochemical screening detected PA- and LF-positive colonies, although expression of the proteins from their own promoters was weak in *Escherichia coli*. Immunochemical screening for EF gave very faint reactions, probably due in part to the consistently lower potency of antisera to EF. However, at least one immunoreactive EF clone (pEF42) was identified that was later shown to contain the EF gene (Robertson *et al.*, 1988). The pUC8 library was also screened with oligonucleotide probes based on the N-terminal sequences of the purified proteins; this confirmed the initial selection of positive colonies. The cloning of the individual genes was confirmed by demonstrating that *E. coli* extracts possessed activity in the CHO cell and macrophage assays for toxin (described below).

An initial concern that a toxic entity might be produced by isolation in a single recombinant clone of both the PA gene and either the LF or EF genes was alleviated by showing that none of the clones produced more than one toxin component. Subsequently, the individual genes were mapped on the pX01 plasmid, and found to be non-contiguous (discussed above). Further evidence that the toxin genes are present in single copy and are not contiguous was provided by transduction experiments with bacteriophage CP-51. Transduction of transposon-marked pX01 DNA was expected to yield shortened variants of the toxin plasmid, since the transducing bacteriophage CP-51 can accommodate only 90 kbp of DNA. Among the transductants obtained, several produced only PA, indicating physical separation of the toxin genes (Heemskerk and Thorne, 1990).

EF was independently cloned by Mock *et al.* (1988) using a direct genetic selection for complementation of an adenylate cyclase mutant of *E.*

---

* The gene acronyms are those agreed to by the American researchers in Frederick, MD, Amherst, MA, and Provo, UT.

Table I. Properties of the anthrax toxin proteins

	PA[a]	LF[b]	EF[c]
%G+C in gene	31%	30%	29%
AA residues in signal sequence	29	33	33
AA residues in mature protein	735	776	767
Start codon	ATG	ATG	ATG
Stop codon	TAA	TAA	TAA
Sequence at signal peptide cleavage site[d]	IQA*E	VQG*A	VNA*M
$M_r \times 10^3$			
Calculated from sequence	82.7	90.2	88.8
By SDS electrophoresis[e]	85	83	89
By SDS electrophoresis[f]	85	87	86
Isoelectric point			
Calculated from sequence	5.6	6.1	6.8
Measured[f]	5.5	5.8	5.9
Measured[g]	5.5	5.8	6.4

[a] Except as noted, data are from Welkos et al. (1988).
[b] Except as noted, data are from Bragg and Robertson (1989).
[c] Except as noted, data are from Robertson et al. (1988) and Escuyer et al. (1988). See also text footnote on p. 281.
[d] Asterisk shows site of signal peptide cleavage.
[e] From Leppla (1988).
[f] From Quinn et al. (1988).
[g] Unpublished studies of S.F. Little.

coli, with selection for the ability to ferment maltose. In this case, it was necessary to first insert a separate, compatible plasmid expressing calmodulin, since EF enzymatic activity is strictly dependent on this eukaryotic protein. Positive clones were obtained at a frequency of 1%.

The genes for PA (Welkos et al., 1988), LF (Bragg and Robertson, 1989), and EF (Robertson et al., 1988; Escuyer et al., 1988) have been sequenced.* Each of the genes has a G+C content of about 30% (Table 1), similar to that of the B. anthracis genomic DNA (35% A+T). Similarly, the codon usage patterns were close to those of several other genes sequenced from various bacillus species. Possible promoter sequences were noted in the PA sequence (Welkos et al., 1988), but

transcriptional start sites have not been determined for any of the three genes. Upstream of the ATG start codons in each of the genes is an appropriately located ribosome-binding site, AAAGGAG for the PA and LF genes, and AAAGGAGGT for the EF gene. Each of the three genes shows a typical bacillus signal sequence of 29–33 amino acids, with cleavage occurring after an Ala or Gly. The deduced amino acid sequences beginning at those cleavage sites exactly match the chemically determined sequences of the pure proteins. The putative open reading frames of all three genes end at TAA codons. Following the end of the PA gene is an inverted repeat that may act as a transcriptional stop; no similar structures are present in the LF or EF regions.

To date, most evidence suggests that there is little or no difference in sequence or serological reactivity between the toxin proteins produced by different strains. For example, in preliminary experiments (S.H. Leppla, unpublished work) a

* The DNA sequences for EF determined by Escuyer et al. (1988), and by Robertson et al. (1988) differ by a few nucleotides in four separate regions. After re-examination of the data, it is now accepted that the sequence of Escuyer et al. (1988) is correct for both EF clones (Don Robertson, pers. commun.).

type of peptide mapping was achieved by using immunoblots to compare the patterns of peptide fragments generated when PA in culture supernates becomes degraded by endogenous proteases. Of ten strains compared, nine had identical patterns of more than twelve fragments. One other strain showed a slight shift of a single band.

DNA and amino acid sequence homologies

A principal objective of sequencing the toxin proteins was to discover whether nucleotide or amino acid sequence homology existed with previously characterized prokaryotic or eukaryotic proteins. Therefore, extensive database searches were done; PA was not found to have homology to any sequences in the databases. In the case of the LF gene, the FASTP program (Lipman and Pearson, 1985) has shown weak homology to several microfilament proteins extending over a long region. The best fit (22% homology score over 206 amino acids) was to hamster vimentin. It is not clear whether this homology and the even weaker homology to several other structural proteins reflects a similarity in a structural motif, or whether it is simply fortuitous. Even if this homology is significant, it is not obviously helpful in identifying the putative enzymatic activity of LF, because vimentin appears to serve principally as a structural protein.

Since EF is an adenylate cyclase, it was of immediate interest to compare the sequence to that of other adenylate cyclases. When the sequence was first obtained, the only available cyclase gene sequences were those of *E. coli* and *Saccharomyces cerevisiae* (Aiba *et al.*, 1984; Kataoka *et al.*, 1985). The EF gene was found to have no homology to these genes. Concurrent with sequencing of EF, the *Bordetella pertussis* adenylate cyclase gene was sequenced (Glaser *et al.*, 1988; see also Chapter 16). EF has homology to the *B. pertussis* cyclase in three regions, with the homology being more evident in the amino acid than in the DNA sequence. From studies of the pertussis cyclase, the area of homology to EF is known to be the catalytic domain. The sequence homology suggests that the genes had a common origin, and that mutation then occurred, particularly in the third position of codons, to bring the %G+C closer to that found in each host (31% G+C for *B. anthracis*

and 65% G+C for *B. pertussis*). Details of the amino acid sequence homologies of the two cyclases are discussed later.

Regarding the origins of the two toxin bacterial adenylate cyclases, it was earlier noted (Leppla *et al.*, 1985) that the similarity of EF to bovine brain calmodulin-dependent adenylate cyclase implied its possible derivation from a eukaryotic enzyme, the gene for which was captured and retained because of its ability to enhance virulence. This hypothesis suggested that some sequence homology might exist with eukaryotic adenylate cyclases. With the recent sequencing of an adenylate cyclase from bovine brain (Krupinski *et al.*, 1989), this comparison became possible. No homology was found. However, the *B. pertussis* cyclase has been reported to have serological cross-reactivity to rat brain adenylate cyclase (Monneron *et al.*, 1988), so the question of the relatedness of these proteins remains open.

The final, and perhaps most useful, information to come from comparison of DNA sequences was the recognition that LF and EF share extensive nucleotide homology in the regions coding for the N-terminal 300 amino acids of each protein. It was known that LF and EF compete for interaction with PA, but there was no a priori reason to expect that the PA-binding domains would be located in homologous regions of each polypeptide. The strong homology of the DNA sequences encoding the N-terminal portions of LF and EF, together with the evidence that the catalytic domain of EF is in residues 300–767, strongly suggests that the N-terminus of each protein contains the PA-binding site. The details of the amino acid sequence homology of LF and EF are discussed later.

Genetic transfer and mutagenesis in *B. anthracis*

The availability of the cloned toxin genes makes it possible to consider returning altered genes to *B. anthracis* to study their individual roles in pathogenesis and immunity. Fortunately, the last few years have seen the development of a number of tools for achieving genetic transfers of DNA into *B. anthracis*. These include transduction (Ruhfel *et al.*, 1984) and conjugation (Battisti *et al.*, 1985; Koehler and Thorne, 1987; Heemskerk and Thorne, 1990) and electroporation (Bartkus and Leppla, 1989; Quinn and Dancer, 1990). A conjugational transfer system was used to transfer a mutated PA gene into the Sterne strain, replacing the resident PA gene (Cataldi *et al.*, 1990). Transposon Tn916 has been used to produce aromatic amino acid requiring mutants (Ivins *et al.*, 1988), and the elegant methods developed for use of Tn917 in *Bacillus subtilis* are being extended to *B. anthracis* (Heemskerk and Thorne, 1990).

The proteins

Production of toxin from *B. anthracis*

Study of the anthrax toxin proteins has been greatly facilitated by the relative ease with which they can be prepared in milligram amounts. Extensive work in the period 1940–65 led to development of synthetic media that support good production of PA (Puziss *et al.*, 1963; Haines *et al.*, 1965). Key ingredients of synthetic media are bicarbonate, recently shown to activate toxin gene transcription (Bartkus and Leppla, 1989), and a buffering system that maintains the pH above 7.0. It also appears that particular amino acid combinations promote toxin production, but these effects appear complex, because systematic trials of the effects of individual amino acids failed to show that individual amino acids controlled toxin synthesis (S.H. Leppla, unpublished studies). Later work led to development of a completely synthetic medium, designated R, that was claimed to further increase yields (Ristroph and Ivins, 1983). Subsequently, the R medium was slightly modified, and effective methods were developed for isolation of the toxin proteins from the cell-free supernate of 50-litre fermentor cultures (Leppla, 1988).

Further minor modifications to the medium (S.H. Leppla, unpublished work) that improve the reliability of the published procedure include: (1) reduction of Tris from 75 mM to 50 mM, (2) reduction of $NaHCO_3$ from 0.8% w/v to 0.4% w/v, and (3) increase of phosphate concentration from 3.4 mM to 17 mM. Because the previously used amounts of both the Tris and bicarbonate bordered on inhibitory levels, these changes seem to improve the growth of the bacteria. With these methods, a 50-litre culture typically yields, after purification, 500 mg PA, 100 mg LF and 40 mg EF.

Production from *B. subtilis* and other host systems

More recently, it has been possible to produce native and mutant PA proteins from *B. subtilis* strains containing the PA gene cloned along with its *B. anthracis* 5′ regulatory regions into vectors derived from plasmid pUB110 (Ivins and Welkos, 1986; Singh *et al.*, 1989b). Since pUB110 plasmids exist at high copy number compared to that of pX01, it was thought that production would exceed that in *B. anthracis*. In fact, these strains secrete PA into the culture supernatant at 100 μg/ml when grown in rich media (S.H. Leppla, unpublished studies); this amount of PA is at most three-fold greater than that of the Sterne type *B. anthracis*. Apparently, the control sequences of the *B. anthracis* toxin genes function adequately in *B. subtilis*, but other gene products involved in toxin synthesis and secretion may not be present in *B. subtilis*. Achieving the high levels of PA expression expected from the high plasmid copy number may require cloning and introduction of the putative bicarbonate-activated transcriptional activator gene (Bartkus and Leppla, 1989).

The production of intact PA from these *B. subtilis* strains is made more difficult because *B. subtilis* produces a number of extracellular proteases, and because good production requires aerobic growth in rich media. *B. subtilis* strains from which the two most active proteases have been deleted (Kawamura and Doi, 1984) have been used as hosts in this laboratory, and appear to improve yields. However, even these *B. subtilis* hosts retain some proteases. PA may present a special problem because of its unique sensitivity to proteolysis (discussed below). EF and especially LF appear more resistant to proteolysis, and may prove easier to produce in *B. subtilis*. Further improvements can be anticipated in expression of the anthrax toxin proteins from *B. subtilis* and other bacterial hosts. However, even the current methods of expression in *B. subtilis* yield milligram amounts of the proteins, and are adequate for the many laboratories which would be reluctant to grow *B. anthracis*.

An alternate expression system for PA was proposed in recent work (Iacono-Connors *et al.*, 1990) in which the PA gene was subcloned into baculovirus-derived vectors and the protein expressed in insect cells; cloning into vaccinia virus was also described. Several injections of mice with the infected insect cells or the recombinant vaccinia virus induced high titred antibody to PA. However, production of PA in the insect cells was at relatively low levels, and purification was difficult. While these systems are not likely to replace the less costly bacterial expression systems (Leppla, 1988) for production of PA, the vaccinia recombinants appear to have potential as live vaccines.

Toxin purification

The three toxin proteins together constitute more than 50% of the protein present in *B. anthracis* culture supernates grown in appropriate media. This makes purification of the proteins relatively easy, and several effective methods have been described. Detailed protocols for purification are provided in a recent review (Leppla, 1988). Chromatography on anion-exchange resins (Wilkie and Ward, 1967; Quinn *et al.*, 1988) or hydroxyapatite (Leppla, 1988) separates all three components. Hydrophobic interaction chromatography on phenyl-Sepharose was used to achieve a final purification (Quinn *et al.*, 1988). Although calmodulin–agarose was effective for purifying the *B. pertussis* adenylate cyclase (Ladant, 1988) and would be expected to bind the EF protein, conditions were not found that caused adsorption of EF to this resin (Quinn *et al.*, 1988). Immunoadsorbants have also been used to purify PA and LF (Machuga *et al.*, 1986; Larson *et al.*, 1988a,b), and can be successful if the proteases in the culture supernatants are inhibited. The immmunoadsorbant based on monoclonal antibody 3B6 (Little *et al.*, 1988) is used routinely in the author's laboratory for purification of PA (Singh *et al.*, 1989b).

Heterogeneity within purified toxin components

When modern methods of purification were applied to anthrax toxin, each of the three components was obtained as a homogeneous protein (Leppla, 1988). The extensive work on toxin purification prior to 1965 had suggested the presence of aggregates and multiple species, but this can now be attributed to proteolytic degradation. The ease with which each protein can be obtained in a homogeneous form contrasts with the difficulties experienced in characterization of the *B. pertussis* adenylate cyclase (Chapter 16), where extensive effort was needed to purify and identify the cell-invasive species, and where it remains difficult to obtain milligram amounts of pure protein. While the established methods for production of the anthrax toxin components consistently yield proteins that are homogeneous in size, several types of heterogeneity have been detected; these are discussed below.

When the major *B. anthracis* metalloprotease is inhibited with chelating agents during chromatographic purifications, LF and EF each behave as monodisperse species, eluting in a single peak. The final purified LF and EF protein preparations show a single charge species comprising at least 80% of the total protein when analysed on non-denaturing or isoelectric focusing gels. In contrast, PA shows greater charge heterogeneity. After an initial separation of the three toxin components on hydroxyapatite, the PA fraction is further purified on DEAE–Sepharose, and up to four peaks are sometimes observed. Analysis on native or isoelectric focusing gels shows that these species comprise a series of charge isomers, each differing by a single charge from the next species in the series. These isomers were separated by ion-exchange chromatography (Mono-Q resin) using shallow salt gradients, and preparations greatly enriched in each isomer were obtained (S.H. Leppla, unpublished work). Toxicity tests in macrophages showed that the species with the highest pI (most positive net charge) was more toxic than the other species. The increased negative charge of the less toxic species may be due to hydrolysis of glutamine and asparagine sidechains, possibly by amidases secreted from *B. anthracis*. Alternately, removal of residues at the C-terminus might increase the net negative charge, since the last 15 residues contain four lysines and one glutamate. Recognition that increased negative charge is associated with loss of toxicity makes it possible to estimate the quality and potency of a PA preparation by electrophoresis rather than by quantitative toxicity testing.

The other type of heterogeneity often found in purified PA results from a cryptic proteolytic cleavage near the middle of the protein, observed by the appearance of fragments of 47 and 37 kDa after electrophoresis in SDS (S.H. Leppla, unpublished work). Fragments of this size are obtained after incubation of purified PA with *B. anthracis* culture supernates, so it is evident that the nicking is caused by a *B. anthracis* protease. Fragments of the same size can be produced intentionally by specific cleavage with low concentrations (approx. 1 µg/ml) of chymotrypsin or thermolysin, or higher concentrations of several bacterial metalloproteases (e.g. *Bacillus polymyxa* protease). The fragments produced by cleavage in this region remain strongly associated, and can be separated only after denaturation. Chromatography on Mono S cation-exchange columns (Pharmacia) at pH 5.0 or 6.5 in 7 M urea was used to separate the fragments, and their N-terminal sequences were determined. Comparison to the deduced amino acid sequence showed that cleavage by chymotrypsin occurs C-terminal to the paired phenylalanine residues 313–314, while cleavage by thermolysin occurs N-terminal to these two residues (J. Schmidt and S.H. Leppla, unpublished work). Sequence analysis of the 37-kDa thermolysin fragment showed that it had also been cleaved N-terminal to the paired leucine residues 8–9. The cleavage of residues 1–9 from the 37-kDa fragment was not evident from analysis on SDS gels. The great sensitivity of residues 8–9 and 313–314 probably results from the presence of the two adjacent hydrophobic residues (the sidechain type recognized by these proteases), and the probable exposure of these residues on the protein's surface. While cleavage at residue 9 has neither a predicted nor a known effect on function, cleavage at residue 313–314 inactivates PA, as is discussed in a later section.

Structural features common to the three toxin components

All three of the anthrax toxin proteins are very similar in size and charge (Table 1). The masses calculated from the DNA sequences agree reasonably well with those estimated in several different laboratories by SDS electrophoresis. It is not evident why the proteins should be similar in size, although it can be noted that a relatively large size might be expected for proteins that must perform several functions, which in the case of EF include binding to PA, penetration of a vesicle membrane, binding to calmodulin, and catalysis of ATP cyclization.

A notable feature of all three proteins is their lack of cysteines, and therefore of disulphide bonds. This bias would not be predicted for a protein designed to be active in extracellular spaces, because many secreted eukaryotic proteins (e.g. lysozyme, serum albumin, insulin) appear to employ disulphide bonds to enhance their stability. For these proteins, folding and correct pairing of cysteines requires accessory proteins located in compartments involved in secretion (Golgi etc.). However, in Gram-positive bacteria, secretion to extracellular space occurs coincident with synthesis, and proper disulphide pairing would have to be specified by the primary sequence of the protein. If an adequate structure could be achieved without disulphide bonds, then genetic pressure might operate to eliminate cysteines, because the reactivity of the sulphydryl group makes it susceptible to side reactions such as oxidation, heavy-metal chelation, or reaction with other sulphydryl groups. It was noted some years ago that extracellular bacterial proteins generally have a low cysteine content (Pollack and Richmond, 1962). This generalization appears to hold for a number of other secreted bacterial toxins. Perhaps most striking is the complete absence of cysteines in the *Bordetella pertussis* adenylate cyclase, a protein of 1706 residues (Glaser *et al.*, 1988), and in two of the closely homologous haemolysins of the RTX class, the *E. coli* haemolysin (Felmlee *et al.*, 1985) and the *Pasteurella haemolytica* haemolysin (Lo *et al.*, 1987).

The deduced amino acid sequences of all three anthrax toxin proteins were analysed by the several different algorithms for strongly hydrophobic regions that might interact with target cell membranes. No such sequences were detected. Apparently this type of analysis cannot detect certain types of membrane-active regions, since some other toxins that must cross membranes (e.g. *Pseudomonas* exotoxin A) also lack obviously hydrophobic regions. The hydrophobicity profiles do show that the LF and EF proteins have similar structures in their N-terminal 300 amino acids (Bragg and Robertson, 1989), the regions in which amino acid sequence homology occurs. This similarity supports the belief that this region contains the binding site for PA.

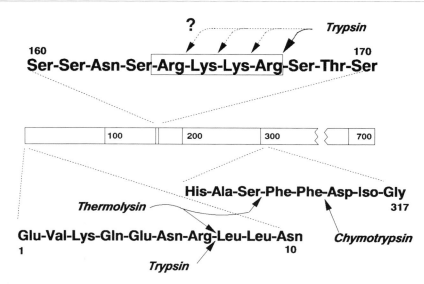

Figure 2. Protease-sensitive sites of protective antigen. Polypeptide bonds selectively cleaved by proteases are identified by the solid arrows, and dotted arrows identify other possible cleavage sites.

PA structure and function

The three functional domains of PA

The PA protein has three regions to which different functions can be assigned. These three regions are defined by two sites that are uniquely sensitive to cleavage by proteases (Fig. 2). Cleavage at residues 313–314 was discussed above. The other protease-sensitive site consists of a cluster of four basic amino acids, Arg^{164}–Lys^{165}–Lys^{166}–Arg^{167}. As might be expected, this site is extremely sensitive to cleavage by trypsin and several related enzymes. Thus, PA in solution at 1 mg/ml is completely cleaved in 30 min when treated with 0.1 μg of trypsin/ml.

In the author's experience, the nicked protein does not immediately dissociate into the separate, soluble fragments of 20 and 63 kDa (designated PA20 and PA63, respectively),* especially at pH > 8.5. However, at pH 7.5, concentrated solutions of the nicked PA incubated for several hours form a precipitate, which can be shown by SDS gel electrophoresis to contain PA63. After extensive trials, a method was found that allows separation

of PA20 and PA63 without obvious denaturation. This was accomplished by chromatography of trypsin-treated PA on the Mono Q anion-exchange resin (Pharmacia) using a pH 9.0 aminoethanol buffer system and a NaCl gradient. Several closely spaced peaks eluting early in the salt gradient were shown by SDS gels to contain fragments of 15–20 kDa. Eluting much later in a single peak was PA63. N-terminal sequencing showed that PA20 began with residue 8, indicating that a second cleavage had occurred at Arg^7 (Fig. 2). PA63 began with residue 168, showing that cleavage had occurred at the cluster of four basic residues. It is possible that an initial cleavage occurs at any of the four basic amino acids, and that subsequent cleavage removes the other residues, leaving the N-terminal Ser^{168} detected by sequencing. The size heterogeneity of the 15–20 kDa fragments indicated that additional cleavages had probably occurred in the N-terminal region.

PA63 purified on Mono Q resin remains soluble indefinitely if kept at pH 9.0, and can be frozen and thawed without precipitating. While it was initially thought that this fragment was monomeric, subsequent analysis by non-denaturing gel electrophoresis and gel filtration chromatography showed it to be approximately 300 kDa, suggesting that it is a pentamer or hexamer of the 63-kDa peptide.

---

* The 63-kDa fragment was initially estimated from SDS gels to be 65 kDa, and was so named in several publications. The availability of DNA sequence showed that the correct size is 63 kDa, and it is this designation, or the abbreviation PA63, which is now recommended.

This oligomer is extremely stable, giving discrete bands and peaks when subjected to separations at pH 9.0. It appears that this oligomer is formed only by exposure to the Mono Q resin at pH 9.0, since incubation of trypsin-nicked PA in buffer at pH 9.0 does not produce significant amounts of this species. Perhaps the high density of quaternary positive charges on the Mono Q resin tightly binds the 63-kDa fragment, displacing the less acidic 20-kDa fragment, which then elutes from the resin. As the NaCl concentration increases during the gradient elution, PA63 may elute and its newly-exposed hydrophobic surfaces self-associate to produce the oligomer. It is not known whether PA63 can also exist in solution as a monomer. It would be advantageous for functional studies if conditions could be found that keep PA63 in a native conformation and soluble at neutral pH.

The functional importance of the protease-sensitive site at residues 163–167 became evident when it was noted that radiolabelled PA incubated with cells becomes nicked at this site (Leppla et al., 1988). Only PA63 remained bound to cells; PA20 could be detected in the supernate if this was concentrated before analysis. Cleavage occurred at 4°C, and also on cells that had been fixed chemically, situations where PA would be restricted to the cell surface. All cells tested to date have the ability to cleave PA, indicating that the cell surface protease is widely distributed. The protease does not appear to be derived from serum, since procedures designed to deplete cells of serum-derived proteins did not decrease nicking. Attempts to identify or classify the cell surface protease using inhibitors have been unproductive; none of the common peptide or active site reactive protease inhibitors block nicking.

The other experiment which revealed the essential functional role of PA63 was the demonstration that this fragment could substitute for PA in combining with LF to produce lethal toxin activity, as measured in the macrophage lysis assay. The PA63 preparation tested was produced by Mono Q chromatography and was completely free of intact (83-kDa) PA. The PA63 had 20% of the potency of intact PA, as might be expected from its oligomeric structure and limited solubility. However, its ability to substitute for PA proves that the N-terminal, 20-kDa domain is not required for binding and internalization of LF.

A separate line of research yielded the other type of evidence needed to construct a consistent model for interaction of the toxin components. Measurements of the interactions of the toxin components in solution done by sedimentation equilibrium in an air-driven ultracentrifuge showed that radiolabelled LF was binding to some protein species in PA preparations that were considered to be homogeneous. However titrations showed that only a few per cent of the PA molecules in the sample had binding activity. When the PA was repurified on the Mono Q column at pH 9.0, a treatment known to remove any PA molecules nicked at the trypsin-sensitive site, the LF-binding activity was lost. Conversely, when trypsin-nicked PA was used, strong binding activity was observed. Competition experiments showed that EF was binding to the same site as LF. These experiments showed that cleavage of PA at residues 163–167 allowed exposure of a site to which LF bound very tightly.

The interaction of LF with PA63 was further characterized using a soluble competitive binding assay in which PA63 was precipitated with mouse monoclonal antibody 3B6 (described below). Binding of trace amounts of labelled LF to PA63 was blocked by non-radioactive LF and EF, and both were equally effective as competitors. A dissociation constant of 10 pM was calculated. Several other methods have been used to characterize the interaction of PA63 with LF and EF. Perhaps the most informative method has been electrophoresis on non-denaturing 5% polyacrylamide gels at pH 8.5 in the presence of detergents. Under these conditions, LF has high electrophoretic mobility, followed by PA, and then the PA63 oligomer. The PA63 moves as a sharp band, much slower than PA. In mixtures of PA63 and LF, several very closely spaced bands migrating even more slowly than PA63 are seen. These are believed to represent the oligomeric PA63 species with increasing numbers (perhaps 1–6) of bound LF molecules. The fact that distinct species are seen suggests that the rate of dissociation of LF from the complex is very low, and that the binding is effectively irreversible. This is consistent with the dissociation constant of 10 pM measured by the soluble binding assay described above.

The protease sensitivity of the phenylalanine residues at positions 313–314 has allowed discrimination of two functional regions within the 63-kDa domain. Several types of experiments using PA nicked at this site showed that residues 168–312 comprise part of the binding site for LF and EF, while residues 315–735 contain the cell receptor-binding site. For example, when receptor-binding studies were done with radiolabelled PA that was partially nicked at residues 313–314, the cell-bound material contained 63- and 47-kDa fragments. This result showed that the cell recognition domain is entirely contained in the 47-kDa fragment, residues 315–735. Evidence that residues 168–312 are needed for LF binding was obtained in experiments with PA that was fully nicked by chymotrypsin; this material was found to be completely inactive in the macrophage cytotoxicity assay. When cells are incubated with this nicked PA at 4°C, they are found to be unable to support subsequent binding of labelled LF.

PA activation and internalization of LF and EF

The data discussed above have been interpreted as supporting the model depicted in Fig. 3 (Leppla et al., 1988). PA binds to cell surface receptors and is cleaved by a cell surface protease with release of the 20-kDa N-terminal fragment. This event exposes a site on the receptor-bound PA fragment, and LF or EF then bind with very high affinity. The complex is internalized by endocytosis, and acidification of the vesicle causes transfer of LF or EF across the membrane to the cytosol.

In support of this model, recent work has provided genetic evidence that cleavage at residues 164–167 is an obligatory step in toxicity (Singh et al., 1989b). The PA gene was placed with its 5′ regulatory region into a shuttle vector able to replicate in both E. coli and B. subtilis. The six codons specifying amino acid residues 163–168 were deleted, and the mutant PA was purified from the supernate of a protease-deficient B. subtilis. The 'deleted PA' could still bind to cellular receptors, but it was not cleaved by trypsin or by the cell surface protease, and was not toxic in macrophages or rats when mixed with LF. Reactivity with monoclonal antibodies and resistance to proteases showed that the deleted PA retained a native conformation, and was not inactive due simply to a loss of normal structure.

Recent evidence has suggested an alternative

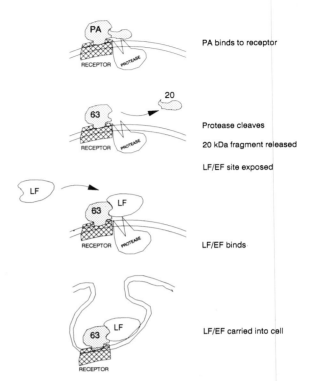

Figure 3. Anthrax toxin binding and internalization. A schematic that summarizes the interaction of the anthrax toxin proteins on the surface of sensitive eukaryotic cells.

mode for proteolytic activation of PA (Ezzell et al., 1990). The sera of guinea-pigs dying of anthrax infection were analysed by immunoblots and a majority of the PA was found to be cleaved at the trypsin site, and to be associated with LF. Furthermore, a Ca^{2+}-dependent proteolytic activity able to cleave PA was found in the plasma of many different animal species. These data suggest that some portion of the PA might be proteolytically activated prior to binding to cells. However, this analysis was done late in infection, at a time when PA had probably saturated its cellular receptors, so it is difficult to estimate the extent and significance of the observed PA cleavage in plasma.

Although the model in Fig. 3 implies that a monomeric, receptor-bound PA63 fragment can bind LF or EF, the data do not rule out the possibility that oligomerization of PA63 is required for either LF binding or membrane penetration. The consistently linear Scatchard analyses and normal dose–response curves argue against cooperative binding. However, it may be relevant that one model for cholera toxin suggests that the pentameric B subunit is needed to insert the single A subunit into membranes. Based on this example, it seems possible that an obligatory oligomerization of PA63 may occur after acidification in endocytic vesicles, and that the oligomeric PA63 causes LF to be inserted across the vesicle membrane. Oligomerization occurring inside cells might not be reflected in the shape of receptor binding and dose–response curves.

Regarding the mechanism by which the toxin components cross membranes, it was shown that PA63 but not intact PA can form ion-conductive channels in artificial membranes (Blaustein et al., 1989). Several other toxins also form channels, especially at low pH, and this ability is believed to correspond to the natural process of membrane penetration from acidic vesicles. However, PA63 was able to insert in membranes at neutral pH, except in the presence of low salt and EDTA, where insertion required low pH. These unusual effects of EDTA and salt will need to be studied further before the significance of the membrane insertion becomes clear.

PA epitopes defined by monoclonal antibodies

Analysis of monoclonal antibodies to PA yielded additional support for the assignment of distinct functions to particular regions of PA. A set of 36 monoclonal antibodies was shown to define approximately 23 different epitopes (Little et al., 1988). The epitope defined by reaction with antibody 3B6 and two other antibodies appears to involve the portion of PA that reacts with cellular receptors. Thus, 3B6 blocks binding of radiolabelled PA to cells and neutralizes the toxin if preincubated with PA, but does not bind to PA that has already bound to cells. Antibody 3B6 reacts with PA63 and with the 47-kDa fragment, confirming the assignment of the receptor-binding region to the C-terminal region (S. Little and S.H. Leppla, unpublished work). Antibody 3B6 appears to react with a conformationally determined epitope, which is destroyed by denaturation and by cyanogen bromide digestion. In recent work, it was shown that 3B6 does not react with a truncated PA that has lost the C-terminal 65 amino acids (Y. Singh and S.H. Leppla, unpublished work),

suggesting that the receptor domain may be near the C-terminus.

The model presented above predicts that monoclonal antibodies might also be found that neutralize toxin by blocking LF and EF binding to PA63. Preliminary studies have identified several anti-PA antibodies that may be of this type (S.F. Little et al., unpublished studies). Because LF and EF have very high affinity for PA63, only those antibodies with extremely high affinity would be expected to compete successfully, especially in experimental protocols where binding was allowed to reach equilibrium. Therefore, antibodies binding at the LF/EF site might be most easily detected in kinetic experiments where PA63 was preincubated with the antibodies prior to addition of LF or EF.

Remarkably, of more than 35 monoclonal antibodies induced following immunization with PA, none was reactive with PA20. This could be explained in part as due to cleavage of PA on the surface of cells, freeing PA20 as a soluble peptide that might be poorly immunogenic. In contrast, the cell-bound PA63 may be efficiently presented to antibody-producing cells. Monoclonal antibodies to PA20 were later induced at lower frequency by immunizing mice with purified PA20 (S. Little and S.H. Leppla, unpublished work).

EF structure and function

EF domain structure

Relatively little experimental work has been done to characterize the protein structure of EF. Therefore, our knowledge of the functional domains of EF is derived largely from comparison of the protein sequence to the sequences of LF and of the B. pertussis adenylate cyclase.

The N-terminal 300 amino acids of LF and EF have substantial homology, and are considered to be the domains that bind to PA63. The hydrophilicity plots of the two proteins in this region also show some similarity, consistent with a shared structure (Bragg and Robertson, 1989) (Fig. 4). The most notable feature of the hydrophilicity plots is a strongly hydrophilic region at the extreme N-terminus of the mature proteins. This region contains a high proportion of charged amino acids, with more than half of residues 1–28 of EF and the homologous residues 11–38 of LF being charged. This suggests that interactions of

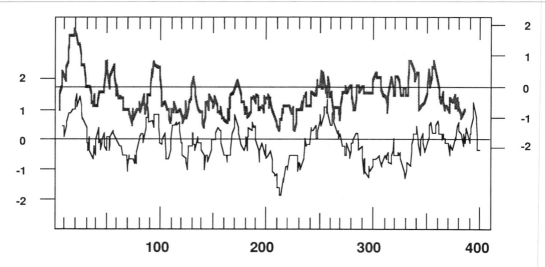

Figure 4. Hydrophilicity plots by the method of Hopp and Woods of the N-terminal regions of lethal factor and oedema factor, aligned to show similarities. Top line and scale on right are for LF, bottom line and left scale are for EF.

charged residues may contribute to the binding to PA.

Comparison of the amino acid sequences of EF and of *B. pertussis* adenylate cyclase shows at least three regions having substantial homology (Fig. 5). Because the functional similarity of the proteins is confined to the shared catalytic activity, it was expected that the region of sequence homologies would identify the catalytic domain. The first of these homologies, denoted region 1 in Fig. 5, contains the sequence GVATKGXXXXGKS (residues 309–321) in EF and GVATKGXX-XXAKS (residues 54–66) in the *B. pertussis* cyclase. The EF sequence exactly matches the consensus sequence, GXXXXGKS, recognized in a large number of nucleoside triphosphate-binding proteins (Higgins *et al.*, 1986). Mutagenesis in this region to replace Lys[320] with Asn in EF (unpublished work cited in Robertson, 1988) and to replace either Lys[58] or Lys[65] with Gln in the pertussis cyclase (Glaser *et al.*, 1989) caused more than a 10-fold loss of catalytic activity. These data support designation of these residues as part of the catalytic centres of the enzymes.

The second and third regions of high homology designated in Fig. 5 have no proven functions, but can be assumed to be part of the active site, or to be involved in calmodulin binding. Prediction of

calmodulin-binding sites from amino acid sequence has been difficult (Bennett and Kennedy, 1987), and no consensus sequences have been found among calmodulin-binding proteins. It has been suggested that calmodulin-binding sites often consist of amphiphilic helices, in which positively charged residues lie on one side of the helix and hydrophobic residue on the other (Cox *et al.*, 1985). Analysis of the EF sequence showed that residues 280–290 (region a in Fig. 5) have a high probability of assuming such a structure (Robertson *et al.*, 1988). However, Glaser *et al.* (1988) predicted that the sequence denoted region b in Fig. 5 would constitute the calmodulin-binding site of pertussis cyclase, and supported this assignment by mutagenizing Trp[242] and showing loss of calmodulin binding. However, subsequent analysis of the *B. pertussis* cyclase was interpreted as showing that two separate regions interact with calmodulin (Ladant, 1988). Since an extended surface of the cyclases may be interacting with calmodulin, extensive mutagenesis studies may be needed to fully define the residues involved.

As might be predicted from the amino acid sequence homology of the two cyclases, there is also some serological cross-reactivity (Mock *et al.*, 1988).

```
EF    255      VEKDRIDVLKGEKALKASGLVPEHADAFKKIARELNTYILFRPVNKLA        302
               .:: :.   ::  |  :.||:.:.   |::| :|:| |. ::|| ||. .
BPAC   1       MQQSHQAGYANAADRESGIPAAVLDGIKAVAKEKNATLMFRLVNPHS         47
                                    region a  (CaM?)


      303      TNLIKSGVATKGLNVHGKSSDWGPVAGYIPFDQDLSKKHGQQLAV-EKGNLE    353
               |.|| .||||||||:|||||| ||||:.:||| |...| .::| .
       48      TSLIAEGVATKGLGVHAKSSDWGLQAGYIPVNPNLSKLFGRAPEVIARADND    99
                    region 1   (ATP site)


      354      NKKSITEHEGEIGKIPLKLDHLRIEELKENGIILKGKKEIDNGKKYYLLE      403
               ..|:.. ...::   |.|.. |:: |::.|::    .:: .:.    .
      100      VNSSLAHGHTAVD---LTLSKERLDYLRQAGLVTGMADGVVASN-----H      141


      404      SNNQVYEFRI---SDENNEVQYKTKEGKITVLGEKFNWRNIEVMAKNVEG      450
               .. : :|||:    ||:. .|||: |:|.      :: .:.|:: |..|
      142      AGYEQFEFRVKETSDGRYAVQYRRKGGD--------DFEAVKVIG-NAAG      182


      451      VLKPLTADYDLFALAPSLTEIKK------------QIPQKEWDKVVNTP       487
               :  |||||.|:||: |  |.::::.        :. .:.:  . :..
      183      I--PLTADIDMFAIMPHLSNFRDSARSSVTSGDSVTDYLARTRRAASEAT      230
                    region 2


      488      NSLEKQKGVTNLLIKYGIERKPDSTKGTLSNWQKQMLDRLNEAV------      531
               .:|:::: .:|| |.. : |.| :|:. :  . :| :|
      231      GGLDRER--IDLLWKIARA-GARSAVGTEARRQFRYDGDMNIGVITDFEL     277
                    region b  (CaM?)


      532      --------KYTGYTGGDVVNHGTEQDNEEFPEKDNEIFIINPEGEFILTK      573
               :   :  .:.:|||.||||||:| .||| |:.||::::.|| |.
      278      EVRNALNRRAHAVGAQDVVQHGTEQNN-PFPEADEKIFVVSATGESQMLT      326
                    region 3


      574      NWEMTGRFIEKNITGKDYLYYFNRSYNKIAPGN-KAYIEWTDPITKAKIN     622
               . ::.: :|:.. |.:|:| ||.|. .:  .. :: ...:.:: .
      327      RGQLKE-YIGQQ-RGEGYVFYENRAYGVAGKSLFDDGLGAAPGVPSGRSK      374


      623      TIPTSAEFIKNLSSIRRSSNVGVYKDSGDKDEFAKKESVKKIAGYLSDYY     672
               |.   |  |:.. ..:|||.| .:| :|:.:: |.:. :|  .  | :||
      375      FSPDVLETVPASPGLRRPSLGAVERQDSGYDSLDGVGSRSFSLGEVSDM-     423


      673      NSANHIFSQEKKRKISIFRGIQAYNEIE       700
               .| .   . |..|.:  :  : |. .| :
      424      -AAVEAAELEMTRQVLHAGARQDDAEPG        450
```

Figure 5. Amino acid sequence homologies of the *B. anthracis* oedema factor and the *Bordetella pertussis* adenylate cyclase. The top line in each set is the sequence of oedema factor (EF), and the bottom line is the sequence of the pertussis cyclase (BPAC). Residue numbers for EF are those of the mature protein, unlike the numbers in the similar figure in Robertson (1988), where numbering began at the start of the 33-residue signal peptide. Sequences were aligned by the GAP program of the Genetics Computer Group, University of Wisconsin. In lines listing amino acid sequence, every tenth residue is shown in bold type, and '—' indicates a gap introduced to maximize alignment of other residues. Symbols vertically connecting homologous residues in the two sequences are '|' for identical residues, ':' for highly similar residues, and '.' for less similar residues. In region a, '■' is above residues constituting the consensus ATP-binding site.

EF catalytic activity

The enzymatic properties of the *B. anthracis* adenylate cyclase have not been extensively studied (Leppla, 1984). This is surprising because the enzyme can be obtained more easily than other bacterial and eukaryotic adenylate cyclases (Leppla, 1988), and because it has high catalytic activity and considerable potential as a pharmacological tool for transiently increasing intracellular cAMP concentrations in eukaryotic cells.

EF has high catalytic activity, with a V_{max} = 1.2 mmol cAMP/min/mg protein (see Fig. 7 of Leppla, 1984), similar to the values of 1.6 mmol cAMP/min/mg protein reported for the *B. pertussis* cyclase (Rogel *et al.*, 1988; Glaser *et al.*, 1988). The K_m for ATP in the presence of Mg^{2+} is 0.16 mM. The enzyme activity is very sensitive to Ca^{2+}, showing optimum activity at 0.2 mM and inhibition at higher concentrations. Apparently a low concentration of Ca^{2+} is needed to activate calmodulin, while a higher concentration directly inhibits the catalytic centre. Because Mn^{2+} can substitute for Ca^{2+} in activation of calmodulin, and does not cause inhibition at the catalytic centre, its use as the only added divalent ion makes the enzyme activity insensitive to small variations in free Ca^{2+}. One reported property of the enzyme, its inhibition by inorganic phosphate (Leppla, 1988), was not confirmed during careful studies by Roger Johnson (pers. commun.), using EF supplied by this author, and therefore the previous report is considered to have been incorrect.

The enzyme activity of EF has an absolute requirement for calmodulin. In the presence of 50 μM Ca^{2+}, the concentration of calmodulin giving half-maximal activity is 2.0 nM. When the Ca^{2+} is chelated by excess EGTA, calmodulin still can activate EF, but 5 μM was needed to get equivalent activity. Thus, as is true also of the *B. pertussis* cyclase, calmodulin can activate even when it contains no bound Ca^{2+}. The *B. anthracis* cyclase does appear to differ from the pertussis enzyme in not having variant peptide species that are active in the absence of calmodulin. A number of calmodulin-dependent eukaryotic enzymes can be activated by proteolytic removal of an inhibitory domain (Kennelly *et al.*, 1987), and this may also be true of the *B. pertussis* cyclase, since several early reports described calmodulin-independent

forms. In retrospect, these were probably fragments produced by proteolysis of the 177-kDa primary gene product. It was thought that the same phenomenon might occur with EF. However, preliminary work in the author's laboratory did not identify protease treatments that produced a calmodulin-independent form of EF. Furthermore, *E. coli* strains containing the EF gene but not calmodulin produced no detectable cAMP and did not have any adenylate cyclase activity (Mock *et al.*, 1988), in spite of the probable presence of breakdown fragments of the EF protein. The details of how calmodulin interacts with EF to activate the catalytic function may become clear once defined regions of the EF protein are expressed and purified from recombinant plasmids.

LF structure and function

Some of the structural characteristics of LF deduced from the nucleotide sequence are summarized in Table 1. Of the three anthrax toxin proteins, the least is known about the structure of LF. This is because less effort has been applied to LF, and because conditions have not been found in which proteases produce limited cleavages that would allow isolation of large peptide fragments.

The evidence that the N-terminal 300 amino acids constitutes the PA-binding domain was discussed earlier. By implication, the putative enzymatic activity lies in the C-terminal region. Between these two domains lies a unique structural feature of LF – a sequence of 19 amino acids that is repeated four times in the region of residues 308 and 383 (Bragg and Robertson, 1989). The repeats are imperfect, and have only about 60% homology (Fig. 6). It is unlikely that this region is part of a catalytic domain. Instead, the repeats may serve either as a structural element that evolved to increase the physical separation of the two domains it connects, or as a site that interacts, perhaps in a multivalent manner, with a protein or lipid surface. A precedent for the latter exists in the RTX family of haemolysins and the *B. pertussis* adenylate cyclase, where a long series of imperfect repeats of 8–9 amino acids is believed to be involved in binding to target cell membranes (Felmlee and Welch, 1988).

It has been widely accepted in the older anthrax toxin literature that LF and EF are serologically

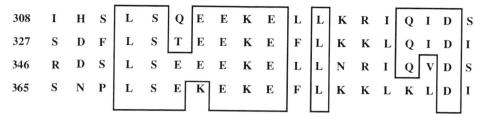

308	I	H	S	L	S	Q	E	E	K	E	L	L	K	R	I	Q	I	D	S
327	S	D	F	L	S	T	E	E	K	E	F	L	K	K	L	Q	I	D	I
346	R	D	S	L	S	E	E	K	E	L	L	N	R	I	Q	V	D	S
365	S	N	P	L	S	E	K	E	K	E	F	L	K	K	L	K	L	D	I

Figure 6. Homologies between the repeated regions in lethal factor. Amino acid residues 308–383 of the lethal factor were aligned to show the homologies.

distinct; no cross-reactivity was detected in immunodiffusion, and this was recently confirmed with purified components (Quinn *et al.*, 1988). Even using the more sensitive ELISA and immunoblot procedures, antisera raised against each component were found to be specific for the immunizing protein. Thus it was surprising to find the substantial regions of sequence homology in the N-terminal 300 amino acids that are assumed to be involved in binding of LF and EF to PA. Since LF and EF bind to PA63 competitively, it would seem probable that similar structures are displayed on the surface of the LF and EF molecules. Whatever these structures are, they seem to have evolved to share function without sharing antigenic structure. Perhaps it has been advantageous for these proteins to evolve so as to avoid having shared epitopes, since shared epitopes would increase the effectiveness of an animal's immune response to the proteins.

Cross-reaction of polyclonal antibodies to LF and EF at a few epitopes would be difficult to detect unless sensitive techniques were employed. Therefore, it would be expected that such shared epitopes would be more readily found using monoclonal antibodies. A set of 61 monoclonal antibodies to LF was characterized recently and three were found to cross-react with EF (Little *et al.*, 1990). Eight of the LF monoclones could neutralize lethal toxin in either *in vivo* or *in vitro* tests, and six of these were shown to do so by preventing LF binding to PA63. Surprisingly, only two of the six that blocked LF binding cross-reacted with EF. Again, this suggsts that the sequences shared by LF and EF and presumably involved in binding to PA63 are not strongly immunogenic.

In spite of considerable interest and effort, no enzymatic activity has been discovered for LF. The rapid action of lethal toxin in rats and macrophages makes it unlikely that LF could act by inhibiting synthesis of proteins or nucleic acids, so interest has focused on other possible activities. The high potency of LF (ED_{50} 10 ng/ml in macrophages) suggests but does not prove that it has catalytic activity. The workers who first purified the three proteins screened them for 11 different enzymes, and found no activity (Stanley and Smith, 1963). In the author's laboratory, LF was screened for modification of nucleotides, ADP-ribosylation, protease activity, phospholipase, and other activities, without success. In some cases, these experiments included pretreatments of LF that might activate a latent activity, since a number of bacterial toxins require activation.

In order to determine whether LF acts enzymatically in the cytosol of eukaryotic cells, several methods have been used to introduce it artificially, bypassing the need for PA binding to receptors. Both microinjection and osmotic lysis of pinosomes (Singh *et al.*, 1989a, 1990) have been used. When introduced by the latter technique, LF was toxic to macrophages even in the absence of PA, although high concentrations were needed. This proved that LF is toxic even in the absence of PA.

Cellular mechanisms of action

From evidence presented above, it was concluded that the toxic actions of both the lethal and oedema toxins result from delivery of LF and EF to the cytosol of target cells. The role of PA is to cause binding of LF and EF to the cell surface, so that they will be internalized by endocytosis. The effect of EF delivery to the cytosol is predictable from its adenylate cyclase activity and the well-known effects of cAMP and of other toxins that elevate cytosolic cAMP concentrations. LF is assumed to possess enzymatic activity, but attempts to identify this have not yet been successful. Some speculations about its mechanism of action can be made based on the cytotoxic action on cells.

Cell specificity and receptors

The available evidence suggests that nearly all types of eukaryotic cells possess receptors for PA. Binding isotherms are easily performed and interpreted because non-specific binding of radioiodinated PA is consistently less than 20% of the total binding. Removal of Ca^{2+} decreases binding of PA by about 50% (Bhatnagar et al., 1989). Analysis of the binding data by the LIGAND program (Munson and Rodbard, 1980) shows that cells possess a single class of high-affinity receptors, having association constants of approximately 10^{-9} M (Singh et al., 1990). Most types of cells have between 5000 and 50000 receptors. The modest number of receptors, the low non-specific binding, and the shape of the binding curves suggests that the receptor is a protein. If this is true then it should prove possible to identify and isolate the receptor using standard techniques. Identification of the receptor will be valuable because it may suggest how the anthrax toxins evolved, as well as suggesting the identity of the protease that activates PA.

All cells so far tested have the ability to proteolytically nick PA. It was initially thought that the resistance of most cell types to the lethal toxin might be due to a failure to bind or nick PA. However, resistant cells are able to perform both these activities (Singh et al., 1989a, 1990), so resistance must be due to processes subsequent to endocytosis.

Internalization and processing

The anthrax toxins are considered to enter cells by receptor-mediated endocytosis and pass through acidic vesicles, because the effects of the toxins are blocked by pharmacological agents that block this process. Thus, the effects of oedema toxin are blocked by cytochalasin D and by amines (Gordon et al., 1988, 1989), and the effects of lethal toxin are blocked by amines (Friedlander, 1986).

Evidence that LF and EF have the ability to insert into membranes from acidic vesicles was inferred from experiments in which cells having surface-bound toxin (PA and either LF or EF) were briefly exposed to acidic buffers under conditions where normal endocytic processes were blocked by cytochalsin D or amines. As in the case of other toxins that pass through acidic vesicles (e.g. diphtheria toxin), this procedure induced toxicity (Friedlander, 1986; Gordon et al., 1988), apparently by causing direct transfer of the toxin into the cytosol. This activity may relate to the ability of PA63 to insert into artificial lipid membranes (Blaustein et al., 1989), discussed above.

Intracellular mechanisms of action

EF action in the cytosol

Oedema toxin treatment of nearly all cell types causes an increase in cAMP. This provides additional evidence that most cell types possess receptors and the protease that activates PA. The plateau levels of cAMP reached in treated cells differ greatly, from levels that barely exceed the basal concentration of about 2 μmol cAMP per mg cell protein to levels of 2000 μmol cAMP per mg protein (Leppla, 1982, 1984; Gordon et al., 1988, 1989). At these latter concentrations, it can be calculated that 20–50% of the ATP has been converted to cAMP. It is not known why different cell types show large differences in response to oedema toxin, but this may be due to inherent differences in cAMP metabolism as much as to differences in toxin binding and internalization.

The normal cellular response to elevated cytosolic cAMP, which is mediated solely through activation of cAMP-dependent protein kinase, is fully activated by modest increases of cAMP above basal levels. The much higher cAMP concentrations in some types of cells exposed to oedema toxin appear to have no phsyiological effect, except possibly through depletion of ATP. Thus, even very high cAMP concentrations are not directly or immediately cytocidal to cultured cells, and toxin-treated cells can recover after removal of toxin. Unlike cholera toxin, the effects of oedema toxin are rapidly reversed, apparently because EF is unstable in the cytosol (Leppla, 1982).

Several cells have been identified that show only a small response to the oedema toxin. Most striking of these is the A/J primary mouse macrophage, which shows only a 10-fold rise in cAMP concentration (A.M. Friedlander *et al.*, unpublished work). Since A/J cells possess PA receptors and proteolytically nick PA, it is likely that the defect in these cells involves intracellular routing, or that the cells have efficient mechanisms to destroy or secrete cAMP. The latter possibilities could be examined by comparing the action of other toxins or agents that elevate cAMP. Because A/J macrophages are also resistant to lethal toxin, it will be useful to understand the causes of their low response.

LF action in the cytosol

In contrast to the broad range of cells sensitive to oedema toxin, only one cell type responds in an acute manner to lethal toxin. Treatment of mouse and rat macrophages with concentrations of 1 μg/ml each of PA and LF leads to complete cell lysis within 2 h. Both resident and elicited primary macrophages are susceptible, as are certain macrophage-like cell lines, including J774A.1. When certain other types of cultures cells are exposed to the lethal toxin, a growth inhibitory effect can be demonstrated; however, the effect was observed only when cells were plated at low densities and growth measured over a period of 3–7 days. For example, when BHK cells are tested in this manner, a 50% inhibition of growth is observed after 3 days (S.H. Leppla, unpublished work).

Comparison of sensitive and resistant cells has provided some clues as to the mechanism of action of LF. It was first asked whether resistant cells lack the intracellular target of LF, and this was addressed by measuring the toxicity of LF introduced artificially into the cytosol (Singh *et al.*, 1989a). A macrophage-like line, IC-21, that is resistant to lethal toxin was chosen for comparison to J774A.1. The method of osmotic lysis of pinosomes was optimized in J774A.1 cells using gelonin. When LF was internalized in this way, only those macrophage-like cells that are sensitive to the lethal toxin applied extracellularly (e.g. J774A.1) were killed; IC-21 cells were unaffected. This suggests that resistant cells may lack the intracellular target of LF. However, it is also possible that the resistant cells lack an enzymatic cofactor of LF, are unable to correctly activate LF, or fail to direct the internalized LF to the site containing the sensitive target. Although LF

introduced by osmotic lysis was active, very high concentrations (50 μg/ml) were required to obtain toxicity. This suggests that LF does not act by a simple enzymatic mechanism in the cytosol, like EF, but may require some combination of correct intracellular targeting and activation.

The role of calcium in the action of LF has been studied in J774A.1 cells (Bhatnagar *et al.*, 1989). Removal of extracellular calcium after LF was allowed to enter the cells prevented lysis. ^{45}Ca was used to show that a large influx of calcium occurred 90 min after toxin addition, prior to lysis at 120 min. Addition of 100 μM verapamil, a calcium channel blocker, prevented both the calcium influx and lysis. These data show that the induced calcium influx is required to cause lysis, but do not prove that calcium is needed for the earlier steps in the action of LF. However, several calmodulin antagonists, trifluoperazine and W-7, also protected cells, arguing for a role of calcium or calmodulin in the lethal toxin action. It will be important to determine whether irreversible damage is occurring in the cytosol of verapamil-protected cells, damage that would lead to lysis were calcium allowed to enter.

In vivo actions

The best-characterized action of the lethal toxin is its rapid toxicity in rats, where it causes overwhelming pulmonary oedema and death in as little as 38 min (Haines *et al.*, 1965; Ezzell *et al.*, 1984). The Fischer 344 rat is particularly sensitive to the toxin. The oedematous fluid in the lungs of the rats has the protein composition of serum, implying a cytotoxic effect that leads to failure of the endothelial barrier, rather than stimulation of fluid secretion. However, when the lethal toxin was tested on cultured endothelial cells, no cytotoxic effects were seen.

The extensive studies done prior to 1965 on the pathophysiological effects of toxin and infection in animals have been critically reviewed (Stephen, 1986). Unfortunately, those early studies employed toxin preparations which by today's standards must be considered to be poorly characterized. The separated toxin components used in some studies probably contained traces of the other toxin components, and perhaps other toxic materials from *B. anthracis*. Furthermore, those early studies were mostly descriptive, and on re-examination provide few insights into the cellular systems that are damaged by toxin *in vivo*. However, it does appear that the effects of lethal toxin administration match those observed in animals dying from infection with virulent *B. anthracis*, and therefore it remains accepted that lethal toxin is the major cause of tissue damage and death. It would be useful to repeat studies of anthrax lethal toxin action in rats now that much more discriminating methods are available to characterize the effects.

How can the demonstrated action of lethal toxin in cultured cells be used to explain the rapid lethal action *in vivo*? In particular, it is not evident that the lysis of macrophages, which occurs in cell cultures only after 120 min, can account for the death of Fischer 344 rats in only 40 min. It is possible that macrophages *in vivo* are more sensitive, and may lyse more quickly; this might be evident if careful *in vivo* studies were performed. Alternatively, macrophages may release toxic substances long before lysis, and these would be detectable only by specific assays. Of course it is also possible that the rapid death of Fischer rats, which is not seen in other rat strains, results from toxin action on other, unidentified cell types. Because a variety of reagents are available to manipulate macrophages *in vivo*, and to analyse their products both *in vivo* and *in vitro*, it should be possible to determine whether macrophages play an essential role in causing the death of lethal toxin-treated rats.

The contribution of the oedema toxin to virulence and pathogenesis is not well understood. In the case of the *B. pertussis* adenylate cyclase, transposon mutagenesis was used to prove that mutants lacking the enzyme are at least 10 000-fold less virulent (Weiss *et al.*, 1984). It will be important to do the analogous experiments with *B. anthracis*. Mutants of *B. anthracis* lacking PA have been made and were shown to be avirulent (Cataldi *et al.*, 1990), and the same procedure could be used to produce EF mutants.

Although thorough histopathological studies are lacking, it appears that the oedema toxin by itself does not cause major tissue damage. This is consistent with the relatively low potency of cholera toxin when administered intravenously (LD_{50} of 2 μg in mice), and with studies in cultured cells showing that high intracellular cAMP concen-

trations are cytostatic but not lethal in most cell types. Evidence does exist that oedema toxin may have an important role early in infection by incapacitating the immune system. Thus, O'Brien *et al.* (1985) showed that low concentrations of PA with EF could inhibit phagocytosis of spores by human polymorphonuclear leukocytes. Later studies suggested a role for oedema toxin in altering the chemotactic response of PMNs to several stimuli (Wright and Mandell, 1986; Wright *et al.*, 1988). In recent work it has been shown that certain inbred mouse strains having a lower natural resistance to *B. anthracis* infection are slower to mobilize phagocytes to the site of infection (Welkos *et al.*, 1990). This argues that the early mobilization and activation of phagocytes is critical for clearance of *B. anthracis*, and supports the view that the oedema toxin may play an essential role at this stage in infection.

Conclusions and summary

It is clear from the data presented above that the anthrax toxin is the principal protein virulence factor of *B. anthracis*. Strains unable to produce the toxin are avirulent, and animals are protected against infection only if they possess antibodies to the toxin. Other bacterial pathogens depend on protein exotoxins for their virulence, but in most cases the toxins have a less complex design. The unique feature of the anthrax toxins is the segregation of the cell receptor and effector functions on separate gene products. This design is shared by the C2 toxin of *Clostridium botulinum*, which also has a receptor-binding protein that requires proteolytic activation for binding of the enzymatic protein. It will be useful to compare the structural features of these two 'binary' toxins.

Certainly one of the most intriguing questions regarding the anthrax toxins is how this system evolved. What features in its design caused its selection and maintenance? While a complete answer will only come when more is known about the toxin and its receptor, some speculations can be offered based on general principles. It was previously suggested that EF might have evolved from a recombinational event in which *B. anthracis* 'captured' a eukaryotic gene (Leppla, 1982). However, there is no firm basis for arguing that the toxin genes are derived from eukaryotic rather than prokaryotic origins, and both possibilities must be considered.

Since the present forms of PA, LF or EF have no toxic activity in the absence of the others, it is difficult to imagine how selective pressures would have operated to cause creation and maintenance of a single factor in the absence of the others. Therefore, it seems more likely that one of the components existed or arose initially in a form that had some toxic activity by itself, and it was therefore maintained. The other factor might then have arisen in a single step, being selected because it greatly increased the potency of the pre-existing factor.

If one factor developed in the absence of the other, it is easier to imagine that this would have been LF or EF. The ancestral forms of these components may have been variants of enzymes involved in normal bacterial metabolism. Phagocytosis followed by lysis of some of the bacteria would have released these catalytic activities into the cytosol of the phagocytes, causing perturbation of the phagocyte metabolism. In this way, catalytic activities of the bacteria might act as virulence factors, and therefore confer a selective advantage. Presence in the bacterium's genome of genes coding for these 'toxic' enzymes would provide a substrate for further recombination and selection of more potent forms.

It is more difficult to imagine how an ancestral form of PA might have arisen from a bacterial protein in the absence of LF or EF, especially when the existence of the signal peptide and the requirement for proteolytic activation is considered. While bacteria make a number of cell surface or secreted proteins that are involved in adhesion or invasion of eukaryotic cells, and these might be candidates for ancestors of PA, none of these appear to undergo proteolytic nicking as an inherent part of their interaction with eukaryotic cells. Therefore, it may be more likely that PA arose from a eukaryotic protein.

Based on these arguments, it seems easiest to imagine that PA and the N-terminal, PA-binding domains of LF or EF are derived from several ancestral proteins (designated B and A2, respectively, in the following discussion) that bound to the surface of eukaryotic cells as part of their normal function. These proteins might have been either prokaryotic or eukaryotic. Fusion of the gene for a catalytic domain such as that of EF (call it A1), with the gene for A2 would then have produced in a single event a system that delivers a toxic enzymatic activity to the cytosol of eukaryotic cells. If the affinity of B and A2 was very high, as is the affinity of PA63 for LF and EF, then this fusion product would have been toxic even if it were expressed at low levels, and this toxicity would have provided the selective advantage needed for its maintenance.

Once either LF or EF had evolved, it is easy to explain the selection of the other enzymatic component as due to a single recombination event. The existence of both LF and EF suggests that the N-terminal 300 amino acids of these proteins (A2) exists as a separate domain that may tolerate attachment to a variety of catalytic domains, being limited only by the possible requirement for other, so far unrecognized, structures that allow membrane penetration and escape from acidic vesicles, and by the need for the fused domains to be assembled so as to be relatively resistant to proteases. Fusion of the gene for the N-terminal 300 amino acids (A2) to a sequence coding for another catalytic domain produces in a single step a gene that contains effective transcriptional and translational sequences and which produces a protein having a signal peptide for secretion from *B. anthracis* and the ability to bind to PA63 and be internalized by eukaryotic cells.

Clearly, if natural selection can produce two proteins using this design, it is certain that modern methods of genetic manipulation can create additional variations on this design. In particular, it should be possible to fuse the N-terminal 300 amino acids of LF or EF with other proteins so as to cause their internalization by eukaryotic cells. Fusion to enzymes such as peroxidase, phosphatase, luciferase or chloramphenicol acetyl transferase would provide reagents for study of endocytic mechanisms, while fusion to various biologically active proteins (lymphokines, hormones, toxins, etc.) would offer another way to internalize these proteins so as to study their metabolic activity.

Given that genetic selection has led to evolution of the design of anthrax toxin, one can ask what features of the design favoured its selection? Did the anthrax toxin arise as a result of a single event or through a series of evolutionary steps driven by selection for features that confer significant efficiencies and economies, ones that can be recognized in the existing proteins described in this chapter.

It does seem evident that the very high affinity of the PA63-binding site for LF and EF makes it possible for the toxin to act even when the concentrations of LF and EF are very low. Because each PA molecule that is proteolytically activated will bind an LF or EF molecule, all of the available LF and EF will be utilized, assuming their amounts do not exceed that of PA. The amount of LF and EF delivered to the interior of cells will be dependent only on the amount of cell-bound PA that is proteolytically activated. Thus, only the PA protein needs to be expressed at high levels by the bacterium. When grown *in vitro*, *B. anthracis* does appear to produce more PA than LF or EF, although this is dependent on the growth medium used. This bias may also exist *in vivo*, since immunization with the Sterne strain invokes higher antibody titres to PA than to LF or EF (Ivins and Welkos, 1988). This sytem provides for very efficient use of LF and EF, and makes it unnecessary for them to be produced at high levels.

Another probable advantage of a design that depends on a high-affinity site for interaction of proteins is that antibodies are less likely to successfully compete with LF or EF at this site so as to neutralize the toxin. As discussed above, the anti-PA monoclonal antibodies that neutralize toxin most effectively are those that block PA binding to receptor. Preliminary tests suggest that several other neutralizing anti-PA monoclonal antibodies may react at the LF-binding site, since they partially inhibit LF binding (S.F. Little *et al.*, unpublished). Of the eight anti-LF antibodies shown to neutralize toxin, six block binding of LF to PA63 (Little *et al.*, 1990). Neutralization does not appear to occur when antibodies bind to most other regions of PA or LF, perhaps because the complex of PA, LF and antibodies is internalized to endosomes, and the antibodies then dissociate when the endosomes are acidified. For those anti-PA and anti-LF monoclonal antibodies that block interaction of PA63 and LF or EF, rather high concentrations were needed to neutralize. This is expected, since antibodies typically have dissociation constants >100 pM, whereas the binding of LF to PA63 has a dissociation constant of 1 pM. Consistent with the difficulty of blocking binding at this site is the evidence that immunization with LF or EF provides only marginal protection against infection by *B. anthracis*; antibodies to PA are much more effective (Ivins and Welkos, 1988).

Another feature of the toxin's design that appears functionally important is the production of PA with a dispensable 20-kDa N-terminal peptide. This domain is reminiscent of the N-terminal 'propeptides' present in the precursors of many Gram-positive secreted proteases (Wandersman, 1989). The propeptides, which can contain 15–300 amino acids, are postulated to be involved in secretion and proper folding of the polypeptides, as well as in repressing enzymatic activity. In the case of PA, it seems likely that one function of the 'propeptide' is to maintain the solubility of the 63-kDa functional domain. Certainly it is the author's experience that PA63 is poorly soluble; this predicts that if it were synthesized as the PA63 peptide, it would be too hydrophobic to remain in the circulation of an infected animal long enough to reach receptors in sensitive tissues. Perhaps the very high-affinity site on PA to which LF and EF bind cannot exist in a soluble protein without causing non-specific binding to other materials, either cell surfaces or proteins.

While it is intriguing to consider how the anthrax toxin evolved, we look forward to replacing these speculations with explicit knowledge about the structure and function of the toxin proteins, and their relationships and interactions with the eukaryotic cells. The important questions that require and hopefully will attract the attention of researchers include: (1) the identity of the PA receptor, (2) the identity of the cell surface nicking protease, (3) the stoichiometry of the toxin components interacting at the receptor, (4) the internalization mechanism and intracellular routing of the toxin components, (5) the structures of the toxin and the cell membranes involved in escape to the cytosol, and (6) the enzymatic mechanism of LF. It is probable that resolution of these questions will provide new insights into biological mechanisms, and that the knowledge gained will find application in cell biology and medicine.

References

Aiba, H., Mori, K., Tanaka, M., Oci, T., Roy, A. and Danchin, A. (1984). The complete nucleotide sequence of the adenylate cyclase gene of *Escherichia coli*. *Nucleic Acids Res.* **12**, 9427–40.

Bartkus, J.M. and Leppla, S.H. (1989). Transcriptional regulation of the protective antigen gene of Bacillus anthracis. *Infect. Immun.* **57**, 2295–300.

Battisti, L., Green, B.D. and Thorne, C.B. (1985). Mating system for transfer of plasmids among *Bacillus anthracis*, *Bacillus cereus* and *Bacillus thuringiensis*. *J. Bacteriol.* **162**, 543–50.

Bennett, H.K. and Kennedy, M.B. (1987). Deduced primary structure of the subunit of brain type II Ca^{2+}/calmodulin-dependent protein kinase determined by molecular cloning. *Proc. Natl. Acad. Sci. USA* **84**, 1794–8.

Bhatnagar, R., Singh, Y., Leppla, S.H. and Friedlander, A.M. (1989). Calcium is required for the expression of anthrax lethal toxin activity in the macrophage-like cell line J774A.1. *Infect. Immun.* **57**, 2107–14.

Blaustein, R.O., Koehler, T.M., Collier, R.J. and Finkelstein, A. (1989). Anthrax toxin: Channel-forming activity of protective antigen in planar phospholipid bilayers. *Proc. Natl. Acad. Sci. USA* **86**, 2209–13.

Bragg, T.S. and Robertson, D.L. (1989). Nucleotide sequence and analysis of the lethal factor gene (*lef*) from *Bacillus anthracis*. *Gene* **81**, 45–54.

Cataldi, A., Labruyere, E. and Mock, M. (1990). Construction and characterization of a protective antigen-deficient *Bacillus anthracis* strain. *Mol. Microbiol.* **4**, 1111–7.

Cox, J.A., Comte, M., Fitton, J.E. and DeGrado, W.F. (1985). The interaction of calmodulin with amphiphilic peptides. *J. Biol. Chem.* **260**, 2527–34.

Escuyer, V., Duflot, E., Sezer, O., Danchin, A. and Mock, M. (1988). Structural homology between virulence-associated bacterial adenylate cyclases. *Gene* **71**, 293–8.

Ezzell, J.W., Ivins, B.E. and Leppla, S.H. (1984). Immunoelectrophoretic analysis, toxicity, and kinetics of in vitro production of the protective antigen and lethal factor components of *Bacillus anthracis* toxin. *Infect. Immun.* **45**, 761–7.

Ezzell, J. W., Jr., Abshire, T.G. and Brown, C. (1990). Analyses of *Bacillus anthracis* vegetative cell surface antigens and of serum protease cleavage of protective antigen. *Salisbury Med. Bull.* **68**, Spec. Suppl., 43–4.

Felmlee, T. and Welch, R.A. (1988). Alteration of amino acid repeats in *Escherichia coli* hemolysin affect cytolytic activity and secretion. *Proc. Natl. Acad. Sci. USA* **85**, 5268–73.

Felmlee, T., Pellett, S. and Welch, R.A. (1985). Nucleotide sequence of an *Escherichia coli* chromosomal hemolysin. *J. Bacteriol.* **163**, 94–105.

Friedlander, A.M. (1986). Macrophages are sensitive to anthrax lethal toxin through an acid-dependent process. *J. Biol. Chem.* **261**, 7123–6.

Gill, D.M. (1978). Seven toxin peptides that cross cell membranes. In: *Bacterial Toxins and Cell Membranes* (eds J. Jeljaszewicz and T. Wadstrom), pp. 291–332. Academic Press, New York.

Gladstone, G.P. (1948). Immunity to anthrax. Production of the cell-free protein antigen in cellophane sacs. *Br. J. Exp. Pathol.* **29**, 379.

Glaser, P., Ladant, D., Sezer, O., Pichot, F., Ullmann, A. and Danchin, A. (1988). The calmodulin-sensitive adenylate cyclase of *Bordetella pertussis*: cloning and expression in *Escherichia coli*. *Molec. Microbiol.* **2**, 19–30.

Glaser, P., Elmaoglou-Lazaridou, A., Krin, E., Ladant, D., Barzu, O. and Danchin, A. (1989). Identification of residues essential for catalysis and binding of calmodulin in *Bordetella pertussis* adenylate cyclase by site-directed mutagenesis. *EMBO J.* **8**, 967–72.

Gordon, V.M., Leppla, S.H. and Hewlett, E.L. (1988). Inhibitors of receptor-mediated endocytosis block the entry of *Bacillus anthracis* adenylate cyclase toxin but not that of *Bordetella pertussis* adenylate cyclase toxin. *Infect. Immun.* **56**, 1066–9.

Gordon, V.M., Young, W.W., Jr., Lechler, S.M., Leppla, S.H. and Hewlett, E.L. (1989). Adenylate cyclase toxins from *Bacillus anthracis* and *Bordetella pertussis*: Different processes for interaction with and entry into target cells. *J. Biol. Chem.* **264**, 14792–6.

Green, B.D., Battisti, L., Koehler, T.M., Thorne, C.B. and Ivins, B.E. (1985). Demonstration of a capsule plasmid in *Bacillus anthracis*. *Infect. Immun.* **49**, 291–7.

Haines, B.W., Klein, F. and Lincoln, R.E. (1965). Quantitative assay for crude anthrax toxins. *J. Bacteriol.* **89**, 74–83.

Hambleton, P., Carman, J.A. and Melling, J. (1984). Anthrax: the disease in relation to vaccines. *Vaccine* **2**, 125–32.

Heemskerk, D.D. and Thorne, C.B. (1990). Genetic exchange and transposon mutagenesis in *Bacillus anthracis*. *Salisbury Med. Bull.* **68**, Spec. Suppl., 63–7.

Higgins, C.F., Hiles, I.D., Salmond, G.P.C., Gill, D.R., Downie, J.A., Evans, I.J., Holland, I.B., Gray, L., Buckel, S.D., Bell, A.W. and Hermodson, M.A. (1986). A family of related ATP-binding subunits coupled to many distinct biological processes in bacteria. *Nature* **323**, 448.

Iacono-Connors, L.C., Schmaljohn, C.S. and Dalrymple, J.M. (1990). Expression of the *Bacillus anthracis* protective antigen gene by baculovirus and vaccinia virus recombinants. *Infect. Immun.* **58**, 366–72.

Ivins, B.E. and Welkos, S.L. (1986). Cloning and expression of the *Bacillus anthracis* protective antigen gene in *Bacillus subtilis*. *Infect. Immun.* **54**, 537–42.

Ivins, B.E. and Welkos, S.L. (1988). Recent advances in the development of an improved, human anthrax vaccine. *Eur. J. Epidemiol.* **4**, 12–19.

Ivins, B.E., Ezzell, J.W., Jr., Jemski, J., Hedlund,

K.W., Ristroph, J.D. and Leppla, S.H. (1986). Immunization studies with attenuated strains of *Bacillus anthracis*. *Infect. Immun.* **52**, 454–8.

Ivins, B.E., Welkos, S.L., Knudson, G.B. and Leblanc, D.J. (1988). Transposon Tn916 mutagenesis in *Bacillus anthracis*. *Infect. Immun.* **56**, 176–81.

Kaspar, R.L. and Robertson, D.L. (1987). Purification and physical analysis of *Bacillus anthracis* plasmids pX01 and pX02. *Biochem. Biophys. Res. Commun.* **149**, 362–8.

Kataoka, T., Broek, D. and Wigler, M. (1985). DNA sequence and characterization of *S. cervisiae* gene encoding adenylate cyclase. *Cell* **43**, 493–505.

Kawamura, F. and Doi, R.H. (1984). Construction of a *Bacillus subtilis* double mutant deficient in extracellular alkaline and neutral proteases. *J. Bacteriol.* **160**, 442–4.

Kennelly, P.J., Edelman, A.M., Blumenthal, D.K. and Krebs, E.G. (1987). Rabbit skeletal muscle myosin light chain kinase. *J. Biol. Chem.* **262**, 11958–63.

Keppie, J., Harris-Smith, P.W. and Smith, H. (1963). The chemical basis of the virulence of *Bacillus anthracis*. IX. Its aggressins and their mode of action. *Br. J. Exp. Pathol.* **44**, 446–53.

Koehler, T.M. and Thorne, C.B. (1987). *Bacillus subtilis* (*natto*) plasmid pLS20 mediates interspecies plasmid transfer. *J. Bacteriol.* **169**, 5271–8.

Krupinski, J., Coussen, F., Bakalyar, H.A., Tang, W.-J., Feinstein, P.G., Orth, K., Slaughter, C., Reed, R.R. and Gilman, A.G. (1989). Adenylyl cyclase amino acid sequence: possible channel- or transporter-like structure. *Science* **244**, 1558–64.

Ladant, D. (1988). Interaction of *Bordetella pertussis* adenylate cyclase with calmodulin: identification of two separate calmodulin-binding domains. *J. Biol. Chem.* **263**, 2612–18.

Larson, D.K., Calton, G.J., Little, S.F., Leppla, S.H. and Burnett, J.W. (1988a). Separation of three exotoxic factors of *Bacillus anthracis* by sequential immunosorbent chromatography. *Toxicon* **26**, 913–21.

Larson, D.K., Calton, G.J., Burnett, J.W., Little, S.F. and Leppla, S.H. (1988b). Purification of *Bacillus anthracis* protective antigen by immunosorbent chromatography. *Enzyme Microb. Technol.* **10**, 14–18.

Leppla, S.H. (1982). Anthrax toxin edema factor: a bacterial adenylate cyclase that increases cyclic AMP concentrations of eukaryotic cells. *Proc. Natl. Acad. Sci. USA* **79**, 3162–6.

Leppla, S.H. (1984). *Bacillus anthracis* calmodulin-dependent adenylate cyclase: chemical and enzymatic properties and interactions with eucaryotic cells. In: *Advances in Cyclic Nucleotide and Protein Phosphorylation Research*, Vol. 17 (ed. P. Greengard), pp. 189–98. Raven Press, New York.

Leppla, S.H. (1988). Production and purification of anthrax toxin. In: *Methods in Enzymology*, Vol. 165 (ed. S. Harshman), pp. 103–16. Academic Press, Orlando, Florida.

Leppla, S.H., Ivins, B.E. and Ezzell, J.W., Jr. (1985). Anthrax toxin. In: *Microbiology – 1985* (eds L. Leive,

P.F. Bonventre, J.A. Morello, S. Schlessinger, S.D. Silver and H.C. Wu), pp. 63–6. American Society for Microbiology, Washington, DC.

Leppla, S.H., Friedlander, A.M. and Cora, E. (1988). Proteolytic activation of anthrax toxin bound to cellular receptors. In: *Bacterial Protein Toxins* (eds F. Fehrenbach, J.E. Alouf, P. Falmagne, W. Goebel, J. Jeljaszewicz, D. Jurgen and R. Rappouli), pp. 111–12. Gustav Fischer, New York.

Lipman, D.J. and Pearson, W.R. (1985). Rapid and sensitive protein similarity searches. *Science* **227**, 1435–41.

Little, S.F., Leppla, S.H. and Cora, E. (1988). Production and characterization of monoclonal antibodies to the protective antigen component of *Bacillus anthracis* toxin. *Infect. Immun.* **56**, 1807–13.

Little, S.F., Leppla, S.H. and Friedlander, A.M. (1990). Production and characterization of monoclonal antibodies against the lethal factor component of *Bacillus anthracis* lethal toxin. *Infect. Immun.* **58**, 1606–13.

Lo, R.Y.C., Strathdee, C.A. and Shewen, P.E. (1987). Nucleotide sequence of the leukotoxin genes of *Pasteurella haemolytica* A1. *Infect. Immun.* **55**, 1987–96.

Machuga, E.J., Calton, G.J. and Burnett, J.W. (1986). Purification of *Bacillus anthracis* lethal factor by immunosorbent chromatography. *Toxicon* **24**, 187–95.

Makino, S., Uchida, I., Terakado, N., Sasakawa, C. and Yoshikawa, M. (1989). Molecular characterization and protein analysis of the cap region, which is essential for encapsulation in *Bacillus anthracis*. *J. Bacteriol.* **171**, 722–30.

Mikesell, P., Ivins, B.E., Ristroph, J.D. and Dreier, T.M. (1983). Evidence for plasmid-mediated toxin production in *Bacillus anthracis*. *Infect. Immun.* **39**, 371–6.

Mock, M., Labruyere, E., Glaser, P., Danchin, A. and Ullmann, A. (1988). Cloning and expression of the calmodulin-sensitive *Bacillus anthracis* adenylate cyclase in *Escherichia coli*. *Gene* **64**, 277–84.

Monneron, A., Ladant, D., d'Alayer, J., Bellalou, J., Barzu, O. and Ullmann, A. (1988). Immunological relatedness between *Bordetella pertussis* and rat brain adenylyl cyclases. *Biochemistry* **27**, 536–9.

Munson, P.J. and Rodbard, D. (1980). Ligand: a versatile computerized approach for characterization of ligand-binding systems. *Analyt. Biochem.* **107**, 220–39.

O'Brien, J., Friedlander, A., Dreier, T., Ezzell, J. and Leppla, S. (1985). Effects of anthrax toxin components on human neutrophils. *Infect. Immun.* **47**, 306–10.

Ohishi, I. (1987). Activation of botulinum C2 toxin by trypsin. *Infect. Immun.* **55**, 1461–5.

Pollack, M.R. and Richmond, M.H. (1962). Low cysteine content of bacterial extracellular proteins: its possible physiological significance. *Nature* **194**, 446–9.

Puziss, M., Manning, L.C., Lynch, J.W., Barclay, E.,

Abelow, I. and Wright, G.G. (1963). Large-scale production of protective antigen of *Bacillus anthracis* in anaerobic cultures. *Appl. Microbiol.* **11**, 330–4.

Quinn, C.P. and Dancer, B.N. (1990). Transformation of vegetative cells of *Bacillus anthracis* with plasmid DNA. *J. Gen. Microbiol.* **136**, 1211–5.

Quinn, C.P., Shone, C.C., Turnbull, P.C. and Melling, J. (1988). Purification of anthrax-toxin components by high-performance anion-exchange, gel-filtration and hydrophobic-interaction chromatography. *Biochem. J.* **252**, 753–8.

Ristroph, J.D. and Ivins, B.E. (1983). Elaboration of *Bacillus anthracis* antigens in a new, defined culture medium. *Infect. Immun.* **39**, 483–6.

Robertson, D.L. (1988). Relationships between the calmodulin-dependent adenylate cyclases produced by *Bacillus anthracis* and *Bordetella pertussis*. *Biochem. Biophys. Res. Comm.* **157**, 1027–32.

Robertson, D.L. and Leppla, S.H. (1986). Molecular cloning and expression in *Escherichia coli* of the lethal factor gene of *Bacillus anthracis*. *Gene* **44**, 71–8.

Robertson, D.L., Tippetts, M.T. and Leppla, S.H. (1988). Nucleotide sequence of the *Bacillus anthracis* edema factor gene (cya): a calmodulin-dependent adenylate cyclase. *Gene* **73**, 363–71.

Robertson, D.L., Bragg, T.S., Simpson, S., Kaspar, R., Xie, W. and Tippetts, M.T. (1990). Mapping and characterization of *Bacillus anthracis* plasmids pX01 and pX02. *Salisbury Med. Bull.* **68**, Spec. Suppl., 55–8.

Robillard, N.J., Koehler, T.M., Murray, R. and Thorne, C.B. (1983). Effects of plasmid loss on the physiology of *Bacillus anthracis*. *Abstr. Ann. Meet. Am. Soc. Microbiol. 1983*, 115 (Abstract).

Rogel, A., Farfel, Z., Goldschmidt, S., Shiloach, J. and Hanski, E. (1988). *Bordetella pertussis* adenylate cyclase: identification of multiple forms of the enzyme by antibodies. *J. Biol. Chem.* **263**, 13310–16.

Ruhfel, R.E., Robillard, N.J. and Thorne, C.B. (1984). Interspecies transduction of plasmids among *Bacillus anthracis*, *B. cereus*, and *B. thuringiensis*. *J. Bacteriol.* **157**, 708–11.

Singh, Y., Leppla, S.H., Bhatnagar, R. and Friedlander, A.M. (1989a). Internalization and processing of *Bacillus anthracis* lethal toxin by toxin-sensitive and - resistant cells. *J. Biol. Chem.* **264**, 11099–102.

Singh, Y., Chaudhary, V.K. and Leppla, S.H. (1989b). A deleted variant of *Bacillus anthracis* protective antigen is non-toxic and blocks anthrax toxin action *in vivo*. *J. Biol. Chem.* **264**, 19103–7.

Singh, Y., Leppla, S.H., Bhatnagar, R. and Friedlander, A.M. (1990). Basis of cellular sensitivity and resistance to anthrax lethal toxin. *Salisbury Med. Bull.* **68**, Spec. Suppl., 46–8.

Smith, H. and Stoner, H.B. (1967). Anthrax toxic complex. *Fedn. Proc.* **26**, 1554–7.

Smith, H., Keppie, J. and Stanley, J.L. (1955). The chemical basis of the virulence of *Bacillus anthracis*. V. the specific toxin produced by *B. anthracis in vivo*. *Br. J. Exp. Pathol.* **36**, 460.

Stanley, J.L. and Smith, H. (1963). The three factors of anthrax toxin: their immunogenicity and lack of demonstrable enzymic activity. *J. Gen. Microbiol.* **31**, 329–37.

Stephen, J. (1981). Anthrax toxin. *Pharmacol. Ther.* **12**, 501–13.

Stephen, J. (1986). Anthrax toxin. In: *Pharmacology of Bacterial Toxins* (eds F. Dorner and J. Drews), pp. 381–95. Pergamon Press, Oxford.

Sterne, M. (1937). Variation in *Bacillus anthracis*. *Onderstepoort J. Vet. Sci. Anim. Ind.* **8**, 271–349.

Thorne, C.B. (1985). Genetics of *Bacillus anthracis*. In: *Microbiology – 85* (eds L. Lieve, P.F. Bonventre, J.A. Morello, S. Schlessinger, S.D. Silver and H.C. Wu), pp. 56. American Society for Microbiology, Washington, DC.

Turnbull, P.C.B. (1986). Thoroughly Modern Anthrax. *Abst. Hyg. Trop. Med.* **61**, R1-R13.

Turnbull, P. (1990a). Anthrax. In: *Bacterial Diseases. Topley and Wilson's Principles of Bacteriology, Virology and Immunity. Vol 3* (eds G.R. Smith and C.R. Easmon), pp. 364–377. Edward Arnold, Sevenoaks, UK.

Turnbull, P.C.B. (1990b). Proceedings of the International Workshop on Anthrax. *Salisbury Med. Bull.* **68**, Spec. Suppl.

Turnbull, P.C.B. (1990c). Proceedings of the International Workshop on Anthrax. *Salisbury Med. Bull.* **68**, Spec. Suppl., p. 53–5.

Uchida, I., Sekizaki, T., Hashimoto, K. and Terakado, N. (1985). Association of the encapsulation of *Bacillus anthracis* with a 60 megadalton plasmid. *J. Gen. Microbiol.* **131**, **131**, 363–7.

Uchida, I., Hashimoto, K. and Terakado, N. (1986). Virulence and immunogenicity in experimental animals of *Bacillus anthracis* strains harbouring or lacking 110MDa and 60MDa plasmids. *J. Gen. Microbiol.* **132**, 557–9.

Uchida, I., Hashimoto, K., Makino, S., Sasakawa, C., Yoshikawa, M. and Teradado, N. (1987). Restriction map of a capsule plasmid of *Bacillus anthracis*. *Plasmid* **18**, 178–81.

Van Ness, G.B. (1971). Ecology of anthrax. *Science* **172**, 1303–6.

Vodkin, M.H. and Leppla, S.H. (1983). Cloning of the protective antigen gene of *Bacillus anthracis*. *Cell* **34**, 693–7.

Wade, B.H., Wright, G.G., Hewlett, E.L., Leppla, S.H. and Mandell, G.L. (1985). Anthrax toxin components stimulate chemotaxis of human polymorphonuclear neutrophils. *Proc. Soc. Exp. Biol. Med.* **179**, 159–62.

Wandersman, C. (1989). Secretion, processing and activation of bacterial extracellular proteases. *Molec. Microbiol.* **3**, 1825–31.

Weiss, A.A., Hewlett, E.L., Myers, G.A. and Falkow, S. (1984). Pertussis toxin and extracytoplasmic adenylate cyclase as virulence factors of *Bordetella pertussis*. *J. Infect. Dis.* **150**, 219–22.

Welkos, S.L. and Friedlander, A.M. (1988). Comparative safety and efficacy against *Bacillus anthracis* of protective antigen and live vaccines in mice. *Microb. Pathogen.* **5**, 127–39.

Welkos, S.L., Lowe, J.R., Eden-McCutchan, F., Vodkin, M., Leppla, S.H. and Schmidt, J.J. (1988). Sequence and analysis of the DNA encoding protective antigen of *Bacillus anthracis*. *Gene* **69**, 287–300.

Welkos, S., Becker, D., Friedlander, A. and Trotter, R. (1990). Pathogenesis and host resistance to *Bacillus anthracis*: a mouse model. *Salisbury Med. Bull.* **68**, Spec. Suppl., 49–52.

Wilkie, M.H. and Ward, M.K. (1967). Characterization of anthrax toxin. *Fedn. Proc.* **26**, 1527–31.

Wright, G.G. and Mandell, G.L. (1986). Anthrax toxin blocks priming of neutrophils by lipopolysaccharide and by muramyl dipeptide. *J. Exp. Med.* **164**, 1700–9.

Wright, G.G., Read, P.W. and Mandell, G.L. (1988). Lipopolysaccharide releases a priming substance from platelets that augments the oxidative response of polymorphonuclear neutrophils to chemotactic peptide. *J. Infect. Dis.* **157**, 690–6.

15

Molecular Biology of Clostridial Neurotoxins

Heiner Niemann

Institute for Microbiology, Federal Research Centre for Virus Diseases of Animals, PO Box 1149, D7400 Tübingen, Germany

Introduction

Historical and clinical aspects

Tetanus toxin (TeTx) and botulinum neurotoxins (BoNTs) are produced by rod-like, spore-forming anaerobic bacilli of the genus *Clostridia* (Greek: *kloster* = spindle). The seven serologically different types of BoNTs, designated BoNT/A through G, have been discovered through botulism outbreaks in humans (types A, B, E and F), birds (type C1), cattle (type D), or through experimental isolation from soil (type G). In contrast, there is only a single type of TeTx, and the strains that produce it are phenotypically similar (Hatheway, 1989).

The clinical manifestations of tetanus (Greek: *tetanos* = tension) have been known since the beginning of medical documentation. Hippocrates described the onset and course of the disease as follows:

'The master of a large ship crushed the index finger of his right hand with the anchor. Seven days later a somewhat foul discharge appeared; then trouble with his tongue – he complained that he could not speak properly. . .his jaws became depressed together, his teeth were locked, then symptoms appeared in his neck; on the third day opisthotonus appeared with sweating. Six days after the diagnosis was made he died.'

Aretaios from Kapadokia, a Greek physician from the third century, used more impressive words: 'An inhuman calamity! An unseemly sight! A spectacle even painful to the beholder! An incurable malady! With them, then who are overpowered by the disease, the physician can merely sympathize. He has no means to preserve the patient's life, to correct the luxations. Should he try to straighten the patient's body, he would have to dissect or break it. This is the great agony of the physician' (Ackland, 1959).

The nature of the disease, however, remained obscure until Carle and Rattone (1884) detected its transmissibility to animals. They injected rabbits with suspensions of a patient's acne pustule, the starting point of a fatal tetanus attack, and showed that the disease could then be transmitted to other rabbits by injections of a suspension of nerve tissue.

The disorder begins with quite unspecific symptoms such as headache, pain of the neck and back, high fever (especially observed with children) and muscle stiffness. The generalized form of the disease proceeds with the specific manifestations that can be ascribed to an inhibition of release of GABAergic neurotransmitters from synapses in the central nervous sytem: risus sardonicus, lockjaw (trismus), opisthotonus, and generalized persistent reflex spasms caused by any kind of acoustic, visual or motorial stimulation (Fig. 1). (For a recent review on the clinical aspects of tetanus see Bleck, 1989.)

Edsall (1976) called tetanus an 'inexcusable disease', because a safe and potent vaccine based on formaldehyde-treated toxin is available and today constitutes the most frequently administered

Figure 1. Soldier suffering from generalized tetanus. Etching by an unknown physician during the Napoleonic Wars.

vaccine throughout the world: Soon after Kitasato (1889) had accomplished isolation of a pure culture of the TeTx producing organism under anaerobic conditions from pus, Behring and Kitasato (1890) succeeded in the production of an effective anti-tetanus antiserum that could neutralize the toxin (the intravenous LD_{50} for a mouse is 6pg) and protect animals injected with it.

Active immunization with formaldehyde-treated TeTx was commenced by Ramon (1925). Repeated injections with such toxoid yields efficient protection of immunocompetent hosts with failure rates of less than 4 per 100 million persons (Band and Bennett, 1983). Although vaccination has almost eliminated the disease from the developed countries, the reported mortality figures per 100 000 inhabitants, 28 in Africa and 15 in Asia, indicate that even today tetanus remains a challenge (Cvjetanovic, 1981). About 80% of the mortality caused by tetanus in the Third World must be ascribed to neonatal tetanus. A survey, performed in the United States for the years 1982–84, revealed 0.036 reported cases per 100 000; the mortality rate of those affected was 26% (CDC, 1985).

In contrast to the terrifying and extremely painful symptoms of a generalized tetanus, the clinical features of botulism (Latin: *botulus* = sausage) are much less dramatic and may in earlier centuries even have escaped detection. Probably the first documentation of botulinum toxin stems from an Indian textbook on poisons. The author Schanaq described the production of a highly potent toxin as follows: 'Collect blood from the left vein of the neck of a black bull, fill it into an unrinsed sheep gut, seal it tightly and dry the content in the shade of a mulberry tree. The powdered residue intermingled with food will lead to death within three days.'

The identification of the toxin-producing organism was only achieved in 1897, when van Ermengem investigated the contaminated ham that had affected 30 members of a music club and killed three of them. From the ham, as well as from the bowel of one of the victims an anaerobic bacillus was isolated.

The initial symptoms of the disease involve diarrhoea, nausea, dry mouth and visual problems such as diplopia, followed by muscle weakness and rapidly progressive, flaccid neuromuscular paralysis leading to respiratory depression and death (Tacket and Rogawski, 1989; Colebatch *et al.*, 1989).

While in the past food-borne botulism, due to insufficient hygiene during food production, was the major source of toxification (MacDonald *et al.*, 1986), infant botulism, affecting babies up to 6 months of age, accounted for 59% of the 133 cases reported in the United States in 1983 (CDC, 1984). In contrast to food-borne botulism, in infant botulism an enteric infection with *Clostridium botulinum* takes place. The reason for the age-dependent susceptibility is unknown, but could be related to an immature gut physiology or bacteriology (Arnon, 1980). It is important to note that except for the few cases of wound botulism (MacDonald *et al.*, 1985), the oral route constitutes the major uptake route. Thus, the toxin has to survive the acidic and protease-rich environment within the gastric juice. For this purpose, botulinal toxins are complexed with a haemagglutinin (Tsuzuki *et al.*, 1990) and a poorly characterized non-toxic component (Sakaguchi, 1983). The toxin is absorbed mostly from the upper intestine from where it appears in the lymphatics (May and Whaler, 1958). Both tetanus and botulinum toxins, however, enter into the neuronal tissue preferentially at the motoneuronal end-plate. The clinical manifestations of botulism are due to inhibition of peripheral cholinergic synapses. Tetanus toxin, when present in high doses, may also act at the motoneuronal junction close to the portal of entry of *C. tetani* (Harvey, 1939; Göpfert and Schäfer, 1941). Botulism-like muscle weakness and decreased muscle tone are the consequences (Davies and Wright, 1955; Miyasaki *et al.*, 1967). Transsynaptic migration from the peripheral α-motoneurons into interneurons is accompanied by tetanus-specific symptoms such as increased resting muscle tone and hyperactive reflexes. In generalized tetanus, the toxin has arrived in the central nervous system. Inhibition of central GABAergic presynaptic neurons leads to the specific clinical manifestations described above.

Clostridial neurotoxins versus ADP-ribosylating enzymes

Before we turn to a more specific description of origin, molecular cloning and structure of the neurotoxin genes, we should dispose of reports in literature suggesting that ADP-ribosylation was the molecular basis of blocking neurotransmitter release. Indeed such ADP-ribosylating activities are found in *C. botulinum* strains of types C and D (see Chapter 2). It must be stressed, however, that the cytotoxin C2 wich ADP-ribosylates non-muscle actin (Simpson, 1987; Weigt *et al.*, 1989; Ohishi *et al.*, 1990), and exoenzyme C3 which modifies the Rho group of GTP-binding proteins (Narumiya *et al.*, 1988; Rubin *et al.*, 1988) influence neurotransmitter release by mechanisms that are completely different from those of clostridial neurotoxins.

(1) The neurotoxic and ADP-ribosylating components from *C. botulinum* type D have been separated by chromatographic procedures (Rösener *et al.*, 1987; Aktories *et al.*, 1988).

(2) As demonstrated by Adam-Vizi *et al.* (1988) in bovine chromaffin cells, ADP-ribosylation

caused by exoenzyme C3 and inhibition of regulated exocytosis caused by botulinal neurotoxin D are unrelated events and each of the two processes can be selectively inhibited by application of the corresponding antibody. More recent studies came to similar conclusions (Morii *et al.*, 1990; Hoshijima *et al.*, 1990).

(3) The amino acid sequence of exoenzyme C3 has been determined by sequencing the structural gene (Popoff *et al.*, 1990). No similarities with the sequences of the L chains of the clostridial neurotoxins types C1 and D were detected (Hauser *et al.*, 1990; Binz *et al.*, 1990b).

Scope of this article

In spite of the diverse clinical features of tetanus and botulism and the different uptake routes of the toxins or the toxin-producing organisms, TeTx and the seven serologically distinct botulinal neurotoxins are closely related regarding their biosynthesis, structure and mode of action. The structural model of TeTx (Fig. 2) relies on numerous experiments performed in the 1970s designed to isolate antigenically active subdomains either by cleavage through endogenous proteases and subsequent reduction or by *in vitro* teatment with exogenous proteases (Matsuda, 1989). Figure 2 also depicts the different nomenclature applied in literature for the subdomains of TeTx. A similarly confusing, but again different set of terms could be listed for botulinal toxins. Since the amino acid sequences of TeTx and BoNTs, as deduced from the DNA sequences (Fig. 4), support earlier findings on the close structural relatedness of individual subfragments, a common nomenclature should be used for both types of neurotoxins and will be applied in this chapter (as under-lined in Fig. 2).

Three-step models have been proposed for botulinum and tetanus neurotoxins to account for the inhibition of transmitter release (Simpson, 1980, 1981; Schmitt *et al.*, 1981). This model follows the general principles of cellular uptake of lipid-containing viruses and protein toxins with intracellular targets, such as ricin, exotoxin A of *Pseudomonas aeruginosa* or diphtheria toxin. The three sequential steps involve:

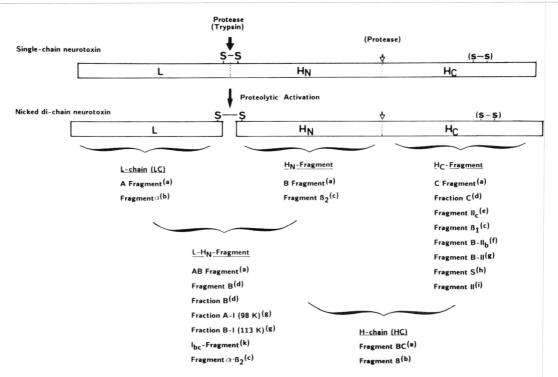

Figure 2. Structural model of clostridial neurotoxins. The single-chain polypeptide produced in *Clostridia is proteolytically nicked by endogenous or exogenous proteases to generate the active di-chain neurotoxin in which the L- and the H-chains remain linked to each other by a single disulphide bond. The open arrow indicates a proteolytic cleavage site that may be recognized in vitro* either by papain (TeTx) or trypsin (BoNT/A) or by endogenous proteases upon longer storage of some toxin preparations. The intrachain disulphide bridge in the H$_C$ region (in brackets) has been mapped for TeTx. The underlined nomenclature is used in this article for both TeTx and BoNTs. This nomenclature has been generally adopted at the Fifth European Workshop on Bacterial Protein Toxins (Veldhoven, July 1991). In addition, the confusing nomenclature of TeTx, as previously used in literature is depicted: (a) Matsuda *et al.* (1983); (b) Matsuda and Yoneda (1975); (c) Matsuda and Yoneda (1977); (d) Helting and Zwisler (1977); (e) Bizzini *et al.*, (1981); (f) Bizzini *et al.*, (1977); (g) Bizzini and Raynaud (1974); (h) Sato *et al.*, (1979); (i) Robinson *et al.*, (1978); (k) Bizzini *et al.*, (1980b). (Modified from Matsuda, 1989).

(1) Binding to cell surface receptor(s) on nerve terminals via the H chain.
(2) Internalization and intraneuronal targeting to the site of action, and
(3) The poisoning step after translocation of the individual L chain into the cytosol.

Although this model has been generally accepted, many of the molecular mechanisms underlying these three steps are still controversial or remain completely unclear: (1) the receptor(s) for the neurotoxins have not been unequivocally defined and characterized at the molecular level; (2) the intraneuronal sorting mechanism(s) that target the neurotoxins within the motoneuron are still unknown; (3) the mechanism(s) by which TeTx

(and botulinal toxins) leave the motoneuron again are not understood; (4) the translocation step which allows the L chains to enter the cytosol awaits elucidation at the molecular level; (5) finally, the molecular mechanism(s) by which neurotransmitter release is blocked and the nature of the target(s) still remain obscure.

An excellent book, edited by Lance L. Simpson (1989) covered the state of the art of clostridial neurotoxins of about 1988 with particular emphasis on pharmacological and toxicological, as well as immunological and clinical aspects. The present chapter will describe the molecular approach which began when research on these neurotoxins also included the search for and deciphering of their structural genes. This approach not only provided

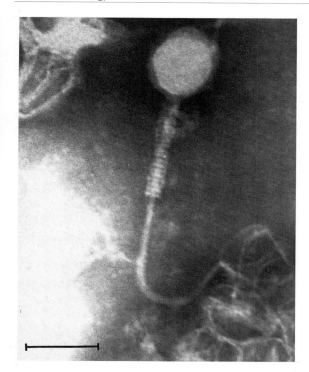

Figure 3. Electron micrograph of a tox⁺ bacteriophage encoding BoNT/CI isolated from *C. botulinum* strain Seddon 460. The bar represents 100nm. (By courtesy of M. Eklund, Northwest Fisheries Center, Seattle).

the first complete amino acid sequences of clostridial neurotoxins, but it also furnished a safe procedure to study the structure–function relationship of specific subdomains of the toxins.

Origin and structure of the neurotoxin genes

General remarks

Spores of *C. tetani* and *C. botulinum* have been isolated from a variety of specimens such as soil and street dust, marine sediments, or faeces of normal farm animals (Hatheway, 1989). The question of why some strains of *Clostridia* produce neurotoxin cannot be answered firmly: there is no indication that producer strains grow any better in culture systems than non-producer strains. On the other hand, the amount of neurotoxin produced may change significantly with the composition of

the growth medium (Hatheway, 1989). What purpose does the neurotoxin have for the bacteria? Apparently, it serves no function other than producing an anaerobic fermenter in the cadaver of the affected animal, thus allowing the production of an enormous amount of progeny spores. Reminiscent of bacterial genes that code for functions important for the survival of the organism in their ecological habitat, the neurotoxin genes are remarkably mobile and have been identified on plasmids and in the genomes of bacteriophages. Even in those instances where the neurotoxin genes are considered to be chromosomally integrated, they seem to be relatively easily transmitted from one bacterium to another, as indicated by the following two lines of evidence:

(1) The production of BoNTs does not necessarily correlate with the physiological properties of the corresponding organisms, as judged by growth conditions, metabolic end products, or proteolytic and lipolytic activities. Each of the four culturally distinct groups of *C. botulinum* (I – IV, Table 1) is related to a non-toxigenic *Clostridium* species that may be known by a distinct species name (Smith, 1977; Suen *et al.*, 1988a,b). BoNT/E or BoNT/F production has recently been detected in unrelated strains such as *C. butyricum* (Aureli *et al.*, 1986; McCroskey *et al.*, 1986) or *C. baratii* (Hall *et al.*, 1985; Suen *et al.*, 1988b), respectively. Another conclusion from these findings is that grouping of *Clostridium* species should not be solely based on neurotoxin production, but should be supplemented by more refined methods such as multi-locus enzyme electrophoresis and DNA-hybridization (Altwegg and Hatheway, 1988).

(2) Some *C. botulinum* strains produce combinations of A and B, A and F, or B and F (Giménez and Ciccarelli, 1970; Poumeyrol *et al.*, 1983; Giménez, 1984; Sakaguchi *et al.*, 1986; Hatheway and McCroskey, 1987).

Taken together, these findings suggest that the neurotoxin genes originally entered the individual strains on extrachromosomal vehicles, such as plasmids or bacteriophages.

Cloning of the neurotoxin genes was facilitated

by the pioneering studies on the distribution of plasmids and bacteriophages in toxigenic and non-toxigenic strains (Eklund *et al.*, 1989). The construction of gene libraries from these extrachromosomal elements drastically accelerated the identification of the structural genes of TeTx, BoNT/C1 and BoNT/D (see below). It should be kept in mind that clostridial genes have a strong bias for AT-rich codons (Marmur *et al.*, 1963). The coding region of the TeTx gene contained 72.1% [A+T] (84.5% in the third position; Eisel *et al.*, 1986). Oligonucleotide probes which reflected this codon usage and thus contained merely A or T residues in the wobble positions were successfully applied to identify gene fragments encoding BoNT/A, BoNT/D and BoNT/E (Binz *et al.*, 1990a,b; Hauser *et al.*, 1990). In order to exclude any potential biohazards, isolation and subcloning of chromosomal fragments smaller than 2.0 kb were performed under L3-B1 containment. Toxicity tests with cell lysates of recombinant *E. coli* clones indicated that none of the newly established clones was toxic in mice.

The following section summarizes major contributions that have led to the characterization of neurotoxin genes.

The extrachromosomal localization of the structural gene encoding TeTx was originally reported by Hara *et al.* (1977). Laird and co-workers (1980, 1981) confirmed that toxigenic strains of diverse origin contained a single large plasmid. Non-toxigenic derivatives were isolated from these strains, and in each case the loss of toxigenicity correlated with loss of the plasmid. Finn *et al.* (1984) provided a first restriction map of this plasmid and localized the L chain encoding sequence on a 16.5 kb *EcoRI* fragment using a mixture of synthetic oligonucleotides. The complete sequence of the *TeTx* gene was established independently by two groups (Eisel *et al.*, 1986; Fairweather and Lyness, 1986). Primer extension studies on mRNA from *C. tetani* indicated that transcription is initiated at the tip of a stem-loop structure 127 nucleotides upstream from the translation start codon (Niemann *et al.*, 1988). Palindromic sequences present in the 3'-non-coding region of the *TeTx* gene are likely to constitute transcription termination signals (Eisel *et al.*, 1986). Taken together, these data suggest that the *TeTx* gene is not part of an operon and

that TeTx is translated from a monocistronic mRNA. The strength of the *TeTx* promoter was tested in *E. coli* using the promoter test vector of Brosius and Holy (1984). The *TeTx* promoter was found to be 3.5-fold weaker in *E. coli* than the *tac* promoter. More detailed studies in our laboratory have indicated, however, that any [A+T]-rich DNA sequence, such as random fragments from the coding region of the *TeTx* gene will function as a weak promoter (Thomas Binz and Heiner Niemann, unpublished data).

Weickert *et al.* (1986) compared the plasmid patterns of 12 **BoNT/A**-producing strains with those of six-non-neurotoxic variants generated by cultivation in the presence of desoxycholate. Each of the strains contained plasmids, but no correlation between neurotoxicity and the presence of a specific plasmid could be made. Eklund *et al.* (1989) showed that a type A strain that was cured of four different bacteriophages, and thus became sensitive to reinfection, continued to produce BoNT/A throughout the experiment. Recently, the BoNT/A gene was cloned and sequenced independently by two groups using chromosomal gene libraries (Binz *et al.*, 1990a; Thompson *et al.*, 1990). Although the sequence was established from two different strains, the entire coding region (3888 bp) contained only two nucleotide exchanges leading to amino acid changes (Pro^2–Gln^2 and Val^{27}–Ala^{27}). Interestingly, the identity of the two sequences extends in the 3'-non-coding regions, but ends 100 nucleotides downstream from the termination codon at the tip of a palindrome (dG(25°C) = 8.8 kcal) that is thought to function as a transcriptional termination signal according to Binz *et al.* (1990a). These authors mapped the initiation of transcription to a site 188 nucleotides upstream from the ATG translational start codon.

The production of **BoNT/B** has been detected in several proteolytic as well as in non-proteolytic strains, all of which carried plasmids of various size (Strom *et al.*, 1984). In addition, organisms have been isolated which simultaneously produce minor amounts of BoNT/A (Giménez and Ciccarelli, 1970) or of BoNT/F (Hatheway and McCroskey, 1987). The involvement of the plasmids in toxin production is not clear. Based on partial amino acid sequences of the L and H chains of the proteolytic strain B657 determined by Das-Gupta and Datta (1988), synthetic oligonucleotides

Table I. Characteristics of the four physiologic groups of *C. botulinum*

Group	Type of toxin	Glucose fermentation	Milk digestion	Lipase	Metabolic products[a] Volatile	Metabolic products[a] Non-volatile	Related Clostridium
I	A,B,F	+	+	+	A, B, iB, iV	PP	*C. sporogenes*
II	B,E,F	+	–	+	A, B		[b]
III	C,D	+	+/–	+	A, P, B		*C. novyi*
IV	G	–	+	–	A, B, iB, iV		*C. argentinense*

[a] Metabolic end-products produced in peptone–yeast extract–glucose medium: A, acetic acid; B, butyric acid; iB, isobutyric acid; iV, isovaleric acid; PA, phenylacetic acid; PP, phenylpropionic acid.
Compiled from Hatheway (1989) and Altwegg and Hatheway (1988).
[b] Non-toxigenic organisms similar to group II are encountered but have not been given species names.

were used successfully to identify and partially sequence the corresponding gene fragments in chromosomal DNA-libraries (Binz *et al.*, 1990a; Kurazono *et al.*, 1991a). It remains to be shown whether BoNT/B from proteolytic and non-proteolytic strains is completely identical.

The structural genes of **BoNT/C1** and **BoNT/D** have been localized on the genomes of bacteriophages. The first electron microscropic characterization of a bacteriophage in lysates of *C. botulinum* type C (Vinet *et al.*, 1968), was soon supplemented by studies by Inoue and Iida (1968, 1970), Eklund *et al.* (1969), and Dolman and Chang (1972). These studies showed that many strains of *C. botulinum* types A through F contained bacteriophages and that strains of type C or D that had lost their bacteriophages always ceased to produce neurotoxin. Reinfection of such non-neurotoxigenic variants with type C- or D-specific phages generated toxin producing strains of the corresponding serotype (for a recent review, see Eklund *et al.*, 1989).

Two additional important results were obtained from such conversion studies: (a) The structural gene of the binary toxin C2 is not located on the bacteriophage genome, since non-neurotoxigenic variants of strains C (strain 153) or D (strain 1873), e.g. strains that were cured of the corresponding C or D bacteriophages, continued to produce C2 binary toxin (Eklund and Poysky, 1974). (b) The non-neurotoxigenic variants of strains C and D were indistinguishable from each other and the high frequency with which such loss of bacteriophages was observed indicated that the strains were not lysogenic (Eklund *et al.*, 1989). This is further supported by recent findings that type C1- and D-specific oligonucleotide probes reacted exclusively with bacteriophage DNA and not with chromosomal DNA (Hauser *et al.*, 1990, Binz *et al.*, 1990b). Interconversion of a bacteriophage-cured type C strain into a type D strain was accomplished by infection with the D-specific bacteriophage and vice versa. In addition, cells infected with either one of the two types of bacteriophage became immune to superinfection with the other phage (Eklund and Poysky, 1974).

Oguma *et al.* (1976a,b) distinguished two groups of converting type C bacteriophages on the basis of their conversion spectra and immunogenicity. Further characterization of the genomes of the two phage groups by restriction maps supported this grouping and indicated that the genome sizes were in the range of 110 – 150 kb (Fujii *et al.*, 1988).

The similarity of the BoNT/C and D viruses is suggested by recent findings that both bacteriophages harbour the structural gene for the exoenzyme C3, the sequene of which was found to be completely identical in both phages (Popoff *et al.*, 1990). It was, therefore, not surprising that the degree of identity found between the genes encoding BoNT/C1 (Hauser *et al.*, 1990; Oguma *et al.*, 1990) or BoNT/D (Binz *et al.*, 1990b) was significantly higher than compared to other genes (for a comparison at the protein level see Table 2).

Experiments designed to link the production of **BoNT/E** to the presence of specific bacteriophages revealed that both the neurotoxigenic and non-toxigenic strains carried one and the same bacteriophage (Eklund *et al.*, 1989). A partial amino acid sequence has been deduced from a cloned *EcoRI* fragment by the group of K. Wernars (Bilthoven, The Netherlands). Again, chromosomal DNA served as the source for the library supporting the idea that the structural gene of BoNT/E is present in the bacterial chromosome (Binz *et al.*, 1990a). A BoNT/E-related neurotoxin was identified in two clinical isolates of *Clostridium butyricum* from patients with infant botulism (Aureli *et al.*, 1986; McCroskey *et al.*, 1986). A partial N-terminal sequence of the neurotoxin isolated from one of the cases has recently been determined by Edman degradation. The sequence found was nearly identical to that derived from BoNT/E (Giménez *et al.*, 1988).

The first known strain of *C. botulinum* type F was isolated by Moller and Scheibel (1960). Dolman and Murakami (1961) identified the strain as proteolytic and saccharolytic and thus belonging to group II. A phage-sensitive derivative was isolated from a non-proteolytic *C. botulinum* type F strain which continued to produce **BoNT/F**. The genetic vehicles by which toxin production is passed amongst individual strains have not been identified. Strom *et al.* (1984) reported the presence of a single 11.5 MDa plasmid in five different proteolytic type F strains, and the presence of a single 2.2 MDa plasmid in four of six different non-proteolytic type F strains. However, it remains to be shown whether these plasmids carry the neurotoxin genes. Some proteolytic and saccharolytic group I strains (Table 1) produce simul-

Table 2. Degree of sequence identity and similarity between individual types of clostridial neurotoxins

A: L chain

	TeTx	BoNT/A	BoNT/B	BoNT/C1	BoNT/D	BoNT/E	
TeTX	100	52.82	69.64	55.51	54.88	61.93	% S I M I L A R I T Y
BoNT/A	32.05	100	53.38	52.64	54.38	53.24	
BoNT/B	51.60	32.40	100	55.63	54.02	58.64	
BoNT/C1	34.83	34.48	33.56	100	60.90	53.38	
BoNT/D	34.47	34.10	34.48	46.52	100	55.01	
BoNT/E	43.86	33.33	36.50	34.06	36.43	100	

%IDENTITY

B: H$_N$ Region

	TeTx	BoNT/A	BoNT/B	BoNT/C1	BoNT/D	BoNT/E	
TeTx	100	59.30		58.46	60.46		% S I M I L A R I T Y
BoNT/A	38.95	100		56.97	56.49		
BoNT/B			100				
BoNT/C1	41.02	34.14		100	84.58		
BoNT/D	41.58	36.06		72.53	100		
BoNT/E						100	

%IDENTITY

C: H$_C$ Region

	TeTx	BoNT/A	BoNT/B	BoNT/C1	BoNT/D	BoNT/E	
TeTx	100	54.50		52.38	50.76		% S I M I L A R I T Y
BoNT/A	34.75	100		55.92	55.14		
BoNT/B			100				
BoNT/C1	28.32	33.00		100	61.63		
BoNT/D	27.92	32.58		41.83	100		
BoNT/E						100	

%IDENTITY

taneously type A and F neurotoxins (Giménez and Ciccarelli, 1970; Sugiyama et al., 1972) or B and F neurotoxins (Hatheway and McCroskey, 1987). These strains were classified as A$_f$ or B$_f$, respectively.

Strom et al. (1984) studied the prevalence of plasmids in various *C. botulinum* G strains, and found that six toxigenic strains all harboured a single 81-MDa plasmid. Eklund et al. (1988) cured these strains of their plasmid by daily passages at 44°C. Plasmid and toxicity analyses indicated that the production of **BoNT/G** ceased whenever the plasmid was lost. Apparently this plasmid also encoded the structural genes mediating production and immunity to a bacteriocin, designated boticin G.

At present nothing is known about the genetic relationship between type C and D bacteriophages on the one hand and the plasmids encoding TeTx or BoNT/G on the other. In particular, there is only little information on the structure and function of additional gene products encoded by the phages or the plasmids. While it is clear that the type C and D bacteriophages also harbour the genes encoding the haemagglutinin (Oguma et al., 1976b; Tsuzuki et al., 1990) and the aforementioned exoenzyme C3 (Popoff et al., 1990), it remains to be shown, whether the E88 plasmid encoding TeTx also contains the information for other cytotoxic factors such as the haemolysin. Such information could provide additional help to establish an evolutionary tree of the individual neurotoxins.

Take-home messages from the amino acid sequences

Searches of the GenBank™/EMBL Data Bank (March 1991) containing 48 288 nucleotide sequences and 8 702 different protein sequences were a great disappointment: No significant similarities to other proteins were found. Fig. 4 shows an alignment of the deduced amino acid sequences of the L chains of TeTx and botulinal toxins type A, B, C1, D and E. Without going into a detailed discussion on computer-based analyses a few conclusions drawn from such an alignment seem worth mentioning:

(1) Similarity and identity analyses (Table 2) indicate that the degree of identity between the L chains of TeTx and the individual BoNTs decreases in the order: BoNT/B, BoNT/E, [BoNT/C1; BoNT/D], BoNT/A. It remains to be shown whether this relatedness can be assigned to an evolution of the botulinal genes from one ancestral TeTx gene. Comparisons at the DNA level are difficult to interpret, since DNA with more than 73% [A+T] will have a significant basal level of relatedness.

(2) On the basis of this alignment at least three regions of enhanced similarity can be detected within the individual L chains: An N-terminal region (involving about the first 120 residues of the individual sequences) contains 37% identical residues. The central segment (extending from residues 120 to almost 300) shows only 19% of

identity, but contains a conserved motif 'DPhhnLhHELnHnnHxLYG' in which 'h' constitutes hydrophobic, 'n' uncharged and 'x' any type of residues. Secondary structure analyses predict an α-helical structure for five out of the six motifs. A helical arrangement would place the imidazole side groups of the three histidines on the same side of the helix. Residues 300 to about 400 again constitute a related region (38% identity). With the exception of BoNT/C1, this region exhibits a significantly increased bias in the use of Lys over Arg, indicating that this domain could represent a catalytic domain, as previously postulated for other dichain toxins (London and Luongo, 1989; Singh, 1990). The remaining residues of the L chains are largely dissimilar in the individual sequences. Recent studies have shown that these residues can be deleted from the L chains of TeTx and BoNT/A without altering their intracellular toxifying activity (Kurazono et al., 1991a).

(3) The cysteine residues involved in the disulphide linkage between the L and H chains can be aligned. Analyses of the complete protein sequences for surface probability support the experimental finding that the nicking regions (between the disulphide-linked cysteines) are highly exposed and can thus be attacked by a variety of proteases (Weller et al., 1988). The length of the loops varies between 8 in BoNT/B and 27 residues in TeTx. All clostridial neurotoxins, with the exception of BoNT/E, undergo proteolytic cleavage by an endogenous protease. A protease that nicks single-chain BoNT/A has recently been purified and characterized (Dekleva and DasGupta, 1990). This protease cleaves specifically arginyl-amide bonds and consists of two subunits of 16.5 and 48 kDa, respectively, which can be separated from each other only in the presence of sodium dodecyl sulphate. Proteolytic activation has been reported for BoNT/B (Ohishi and Sakaguchi, 1977; Kozaki et al., 1985), BoNT/E (Ohishi et al., 1975; Kozaki et al., 1985; Yokosawa et al., 1986; Bittner et al., 1989b; Poulain et al., 1989b), BoNT/F (Ohishi and Sakaguchi, 1977) and TeTx

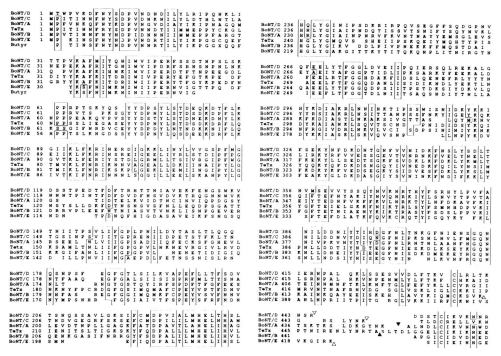

Figure 4. Alignment of the L chain amino acid sequences of BoNT/D (Binz *et al.*, 1990b), BoNT/C1 (Hauser *et al.*, 1990); BoNT/A (Binz *et al.*, 1990a); TeTx (Eisel *et al.*, 1986); BoNT/B (Kurazono *et al.*, 1991b); and BoNT/E (Kurazono *et al.*, 1991b). Identical and equivalent amino acid residues are boxed. The proteolytic cleavage sites between the L and H chains are indicated by filled arrowheads. Open arrowheads indicate putative cleavage sites that have not been verified by protein sequencing.

(Matsuda and Yoneda, 1974; Bergey *et al.*, 1986). More recently, Weller *et al.* (1988) showed that several different proteases are capable of activating single-chain TeTx. The TeTx isotoxins obtained in this manner were all cleaved in the authentic nicking or hinge region and were found to be 5 – 12 times more potent than single-chain TeTx. Results obtained with BoNT/B by Evans *et al.* (1986) indicate that such proteolytic cleavage does not alter the binding characteristics of the neurotoxin.

(4) The sequence of the first 100 N-terminal residues of the H$_N$ regions is quite dissimilar in the individual neurotoxins, suggesting that their presence merely serves a spacer function, perhaps to allow for a translocation of the L chains through a pore formed by the highly conserved portion between residues 550 to 840. For BoNT/C1 and BoNT/D, the degree of identity exceeds 70% within this region (Table 2). A channel

formation has been demonstrated for TeTx (Boquet and Duflot, 1982; Boquet *et al.*, 1984), BoNT/A (Shone *et al.*, 1987), BoNT/B (Hoch *et al.*, 1985) or BoNT/C1 (Donovan and Middlebrook, 1986). Two main peptides derived from the H$_N$ region of TeTx, 27 000 and 21 000 in molecular weight, were protected against pepsin proteolysis after low pH-induced insertion of TeTx into asolectin vescicles (Roa and Boquet, 1985).

(5) The H$_C$ region harbours the ganglioside-binding site of the neurotoxins, as demonstrated for TeTx (Morris *et al.*, 1980; Goldberg *et al.*, 1981; Halpern *et al.*, 1990) or BoNT/A (Kozaki *et al.*, 1989). Binding of gangliosides occurs at a 1:1 ratio, and isolated H$_C$ fragments inhibit to a varying extent binding of the holotoxins (Simpson, 1984, 1985). Binding of holotoxin or of H$_C$ fragment can be inhibited by pretreatment of cells with neuraminidase, and can be partially restored by exogenously added

gangliosides (Kozaki *et al.*, 1989). The sequence alignment of the H_C fragments of TeTx, BoNT/A, BoNT/C1, and BoNT/D indicates that approximately 175 C-terminal residues are dissimilar (Fig. 4). However, each of the sequences contains at least two tryptophan residues within the last 40 amino acids. Residues Cys^{1077} and Cys^{1093} of TeTx, reported to be disulphide-linked in TeTx (Krieglstein *et al.*, 1990), are not conserved in any of the known sequences of botulinal toxins.

Binding, internalization and intraneuronal sorting: only a receptor problem?

Numerous studies in the past have addressed the question of how TeTx and BoNTs bind to the presynaptic membranes, how they then become internalized and directed to their site(s) of action (for reviews see Habermann and Dreyer, 1986; Montecucco, 1986; Critchley *et al.*, 1988; Middlebrook, 1989). Any model that attempts to describe the above sequence of events for TeTx or BoNTs at a molecular level, will have to integrate answers to the following questions:

(1) What factors determine the neuroselectivity and what is the molecular nature of the receptors? Are these receptors identical at the motoneuronal end-plate and at central presynaptic membranes?
(2) Are there different cellular uptake mechanisms for TeTx and BoNTs or at peripheral and central neuronal membranes?
(3) By what mechanism(s) are retrogradely transported neurotoxins released from the α-motoneurons?
(4) Why are BoNTs, although transported retrogradely, not active in the central nervous sytem?
(5) What is the subcellular compartment at which toxification takes place?
(6) What are the cellular targets for the

individual neurotoxins and by what molecular mechanism are these targets altered?

So, let us recollect the published information and follow TeTx and BoNTs on their way into the neuronal cells.

Gangliosides versus proteins as binding sites for neurotoxins

There are numerous biological examples demonstrating that contacts between two partners are initially reversible and in fact take place in two discrete steps. For instance *Klebsiella* bacteriophages make first contact to a susceptible host cell in an enzymatic reaction in which the bacterial polysaccharide capsule (K-antigen) is degraded into oligosaccharide repeats. Only then can the bacteriophage bind irreversibly to a second receptor present in the cell wall (Geyer *et al.*, 1983). Influenza viruses bind via the HA1 subunits of their haemagglutinin spikes to sialic acid-containing glycoproteins. In addition, the virus particle contains a neuraminidase. While this enzyme may serve other essential functions, it makes the initial binding step reversible, thus allowing the virus to stroll along the host cell surface until it gets entrapped by a clathrin-coated pit (Murphy and Webster, 1985).

The neuroselective binding of TeTx was first detected at the end of the last century (Wassermann and Takaki, 1898; Marie 1898). Binding studies performed at the molecular level were stimulated by the now classical findings that gangliosides of the G_{1b} series bind to both TeTx (van Heyningen and Miller, 1961; van Heyningen and Mellanby, 1973) and BoNTs (Simpson and Rapport, 1971). Such gangliosides are highly enriched in neuronal tissue (Wiegandt, 1985). Over the years, a large body of information regarding the interaction of clostridial neurotoxins with gangliosides and putative protein receptors has accumulated, but even today the relevance of the individual receptors is debated controversially (Critchley *et al.*, 1988; Middlebrook, 1989; Simpson, 1989). From the enormous amount of information it becomes evident that binding characteristics determined for a neurotoxin with particulate preparations such as brain synaptosomes or primary tissue culture provide little conclusive evidence regarding the uptake mechanism by which the same toxin enters neuronal tissue at the neuromuscular junction or by which it is internalized at central synapses after systemic administration.

The aim of this section is the proposal of a model that could explain why TeTx and the individual botulinal toxins exhibit identical retrograde transport properties, but have different sites of action. This model is an extension of the dual receptor model originally proposed by Montecucco (1986) in which gangliosides and putative protein receptors act together to control binding and internalization. According to our model, the presence of the specific protein receptor within peripheral nerve endings and central presynaptic membranes then determines by means of the cellular sorting machinery, whether retrograde axonal transport or translocation of the L chains into the cytosol and toxification of the cell takes place.

Gangliosides as receptors

There is absolutely no doubt that TeTx and BoNTs bind to gangliosides via their H_C fragments (see above) under a variety of conditions. Binding of neurotoxins was examined with gangliosides fixed onto solid supports (van Heyningen and Miller, 1961; Mellanby and Whittaker, 1968; Simpson and Rapport, 1971; Helting et al. 1977; Holmgren et al., 1980; Kitamura et al. 1980; Critchley et al., 1985; Habermann and Tayot, 1985; Kamata et al., 1986, 1988; Ochandra et al., 1986; Lazarovici et al., 1987; Yavin et al., 1987), gangliosides embedded into lipid vesicles of defined composition (Lazarovici et al., 1987; Montecucco et al., 1986, 1988, 1989), or present in membranes of primary neuron cultures (Kurokawa et al., 1987; Parton et al., 1987); PC12 cells (Sandberg et al., 1989a; Fujita et al., 1990), neuroblastoma cells (Wendon and Gill, 1982; Staub et al., 1986; Yokosawa et al., 1989), brain membrane preparations (Rogers and Snyder, 1981; Agui et al., 1983; Critchley et al., 1986; Lazarovici and Yavin, 1986; Pierce et al., 1986; Kozaki et al., 1989; Wadsworth et al., 1990). Other studies on binding and/or uptake used even non-neuronal cells such as liver cells (Montesano et al., 1982), secretory cells such as bovine adrenal medullary cells (Marxen et al., 1989) or macrophages (Ho and Klempner, 1985, 1986; Blasi et al., 1990), or membrane preparations

from rabbit kidney (Habermann and Albus, 1986) or thyroid tissue (Ledley et al., 1977).

The major criticism against the concept of gangliosides as the exclusive receptors of clostridial neurotoxins is based on the notion that binding is best under ionic strength and pH conditions that are far from physiologic (i.e. 25–50 mM Tris/acetate, pH $\leq$ 6.0). In addition, even under such optimized binding conditions, TeTx showed only a 350-fold binding preference for GT_{1b} over GM_1 (Kitamura et al., 1980) and the binding affinities to other gangliosides decreased in the order $GT_{1b} > GQ_{1b} > GD_{1b} \gg GM_1 > GD_{1a}$ (Critchley et al., 1988). In this respect, TeTx has been compared with cholera toxin which exhibits a binding preference for GM_1 over other gangliosides by at least three orders of magnitude (van Heyningen and Mellanby, 1973; van Heyningen 1974). At more physiologic ionic strength and pH, binding of TeTx to gangliosides is reduced (Weller et al., 1986). Similar pH-dependent binding properties have also been reported for botulinal neurotoxins (Kamata et al., 1988) and, apparently, individual serotypes have different binding characteristics (Kozaki et al., 1978, 1986; Agui et al., 1983; Kozaki and Sakaguchi, 1982). Finally, Lazarovici et al. (1987, 1989a) reported the existence of two subpopulations of TeTx molecules which differ in their amino acid composition and affinity for gangliosides. At present, however, it is unclear (a) what molecular mechanism(s) induce such different affinities, and (b) whether, and if, under what conditions, a conversion from one species into the other takes place in a host. In addition, since the amino acid compositions of the two affinity-purified TeTx preparations (Lazarovici et al., 1987) showed significant deviations from that deduced from the TeTx DNA sequence (Eisel et al., 1986), impurities altering the binding characteristics of either preparation cannot be excluded.

For reasons which will be discussed in greater detail below, polysialylated gangliosides, however, are sufficient to mediate binding to neuronal cells, internalization, transient presence in the neuron and retrograde export from it. With other words, internalization will not necessarily lead to toxification. According to our model, toxification of cells depends upon the presence of a second proteinaceous receptor and subsequent specific intraneuronal sorting. Toxification taking place in the absence of such a protein receptor should require much higher concentrations of toxins and may be interpreted as a collapse of the cellular sorting mechanism.

Indeed, there is solid evidence that binding of isolated fragment H_C of TeTx and that of holotoxin are indistinguishable in terms of energy dependence and sensitivity to methylamine hydrochloride or ammonium chloride (Simpson, 1985). Recently, Halpern et al. (1990) have shown that the H_C fragment of TeTx, as expressed in E. coli under control of the tac promoter, was able to compete in a dose-dependent manner with ^{125}I-labelled TeTx for binding to GT_{1b}. In addition, the recombinant H_C fragment retained the neurospecificity of TeTx, as demonstrated by binding to foetal mouse spinal cord cells (Halpern et al., 1990). Similar findings concerning the neuroselective binding of the H_C fragment to preganglionic nerve terminals have been reported by Meckler et al. (1990). Manning et al. (1990) studied the anterograde and retrograde transport properties of isolated H_C fragment of TeTx in the rabbit and rat visual systems and showed that H_C exhibited exclusively retrograde transneuronal transport.

Proteins as receptors

Let us begin this section with the general statement that there is as of yet no absolute and direct proof of the existence of proteinaceous receptor for clostridial neurotoxins. Previous experimental approaches such as detection on Western blots with radiolabelled toxin or affinity chromatography of neuronal cell extracts on columns carrying covalently attached H chains have failed (Critchley et al., 1986). However, there is sound evidence that each of the clostridial neurotoxins has a second proteinaceous receptor: Pierce and co-workers (1986) re-evaluated the binding of TeTx to rat brain membranes under various conditions. Under physiological pH and salt conditions, binding was inhibited only by large concentrations ($IC_{50} > \mu M$) of gangliosides, but continued to be sensitive to pretreatment of membranes with neuraminidase. In addition, binding was largely inhibited by heating (100°C, 5 min) or preincubation of membranes with proteases such as trypsin or S. griseus protease. A protease-sensitive receptor was also postulated by Lazarovici and Yavin (1986), who showed that pretreatment of guinea-pig brain

synaptosomes with trypsin significantly inhibited binding of TeTx, but not cholera toxin, under physiological conditions.

Several studies came to the conclusion that a protein receptor should also be involved in the binding of BoNTs to neuronal synaptosomes (Agui et al., 1983, 1985; Evans et al., 1986; Williams et al., 1983) or neuroblastoma cells (Yokosawa et al., 1989). Taken together, these data strongly suggest that brain or spinal cord membrane preparations exhibit under physiological conditions a small number of high affinity protein receptors for both TeTx and BoNTs, while gangliosides constitute only low-affinity binding sites (reviewed by Critchley et al., 1988). Once again, the notion that such putative protein receptors exist in particulate preparations does not necessarily mean that these receptors can indeed be physically reached by a neurotoxin on their retrograde route. On the contrary, since pathological effects of botulinal toxins in the brain have never been observed in vivo after systemic administration, we may conclude that the putative protein receptors for botulinal toxins are either absent from those presynaptic membranes that make the contacts along the retrograde transport route, or, alternatively, that these receptors are for some unknown reasons not efficiently internalized (Black and Dolly, 1987).

A model for binding, uptake and sorting

Both TeTx and BoNTs bind to polysialylated gangliosides via their H_C-regions. As we have discussed above, however, this binding is weak under physiological conditions. Nevertheless, the high concentration of gangliosides of the $G1_b$ series in presynaptic membranes (Wiegandt, 1985) would cause an enrichment of the neurotoxins at the cell surface. At the motoneuronal junction, demonstration of direct binding of TeTx by immunohistochemical methods is apparently much more difficult to achieve than specific staining for acetylcholinesterase (McMahan et al., 1972) or demonstration of binding of α-latrotoxin (Valtorta et al., 1984) or α-bungarotoxin (Anderson and Cohen, 1974; Fertuck and Salpeter, 1974). Using ^{125}I-labelled TeTx and mouse skeletal muscle, Wernig et al. (1977) demonstrated that the toxin was enriched in end-plate regions. Labelling of end-plates could be verified by autoradiography only when cryosections of soleus muscle and high concentrations of ^{125}I-labelled TeTx were used. While it has been discussed that access of exogenously added TeTx to its receptor could be physically obstructed (Dreyer, 1989), these observations also allow the conclusion that, due to a low affinity of TeTx to its ganglioside receptors, previously bound TeTx elutes again during preparation of the specimens for electron microscopy.

In contrast, binding of ^{125}I-labelled BoNT/A to phrenic nerve terminals was demonstrated without major experimental difficulties by Dolly et al. (1984). Interestingly, TeTx could inhibit binding of BoNT/A only partially, indicating that a subpopulation of the BoNT/A molecules had bound to a receptor that could not be recognized by TeTx. Using BoNT/A and BoNT/B in binding competition studies, Black and Dolly (1986a,b) found that the nerve terminals of the mouse hemidiaphragm and frog cutaneous pectoris muscle contained two different types of binding sites for the two serotypes: a single nerve terminal contained about four times more receptors for BoNT/B than for BoNT/A. Dolly et al. (1984) determined that the BoNT/A receptors present in the mouse hemidiaphragm are distributed on unmyelinated areas at an average density of 150 to 500 sites per μm^2.

Taken together, these findings are compatible with a dual receptor model (Montecucco, 1986) in which initial binding of both TeTx and BoNT/A to the motoneuronal terminals is mediated by gangliosides. After lateral movement along the membrane, a subpopulation of BoNT but not of the TeTx molecules binds to as yet unspecified protein receptors which may differ for the individual neurotoxins. The distribution of these receptors should be restricted to the motoneuronal end-plate and cholinergic synapses. This conclusion can be drawn from previous results demonstrating that the BoNT/A does not act on noradrenergic synapses (MacKenzie et al., 1982) at concentrations sufficient to block release of acetylcholine at nerve terminals (Simpson, 1982). TeTx apparently lacks such a second proteinaceous receptor in the periphery and binds merely to gangliosides.

The subsequent internalization of BoNT/A into nerve terminals of the mouse diaphragm was inspected by electron microscopy (Black and Dolly, 1986b). The toxin was found initially in clathrin coated pits and later in smooth walled vesicles that were indistinguishable from endosomes. Binding of all neurotoxins is energy independent, while internalization required energy as evidenced by its sensitivity to sodium azide (Dolly et al., 1984). Simpson (1982) showed that the paralytic action of both BoNT/A and BoNT/B on the mouse phrenic nerve-hemidiaphragms was inhibited by the lysomotropic reagent chloroquine, suggesting that for both toxins receptor-mediated endocytosis was involved in the internalization process (Hubbard, 1989).

It is still not clear whether internalization occurs for both botulinal toxins and TeTx via clathrin-coated pits, or whether, in addition, a second type of uptake mechanism exists for ganglioside-associated toxins. Studies by Schwab and Thoenen (1978) used ^{125}I-labelled TeTx–gold conjugates to analyse the binding at the anterior eye chamber of adult rats. One hour after injection, TeTx–gold particles were selectively associated with membranes of nerve terminals and preterminal axons. Clathrin-coated pits, although detected frequently, were unlabelled and the authors state: 'It could not be decided unequivocally whether the uptake of the toxin–gold complexes by nerve terminals occurred by means of coated vesicles or smooth membrane structures.' Montesano et al. (1982) reported that internalization of TeTx, prebound to hepatocytes at low temperature, occurred via non-coated vesicles. However, in these experiments, the first electron micrographs were taken only ten minutes after temperature upshift, a point at which the toxin-containing vesicles could already have been uncoated by ATP-driven heat shock proteins (Keen, 1990). Endocytosis of cholera toxin (which binds to the ganglioside GM_1), apparently does not occur through coated pits (Joseph et al., 1979; Parton et al., 1987). Therefore, it is possible that the internalization of ganglioside-bound TeTx and those BoNT molecules which have not found the putative protein receptors occurs via smooth-walled vesicles. It is obvious that such vesicles could be processed differently from clathrin-coated vesicles and later on discharge their content, without being exposed to an acidic pH, into transport vesicles which are involved in retrograde axonal transport along microfilaments (alternative pathway in Fig. 5). As stated above, the binding characteristics of TeTx and its isolated H_C fragment, as assessed at the cholinergic neuromuscular

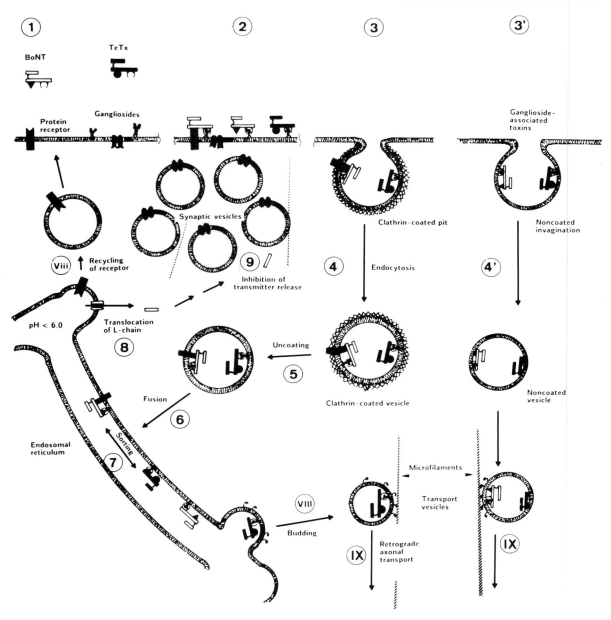

Figure 5. A model for binding and uptake of BoNTs and TeTx at peripheral cholinergic synapses (for details see text).

junction, were completely identical (Simpson, 1985). Furthermore, internalization of both derivatives, as determined by their disappearance from the cell surface or inaccessibility to antibodies, was promoted by nerve stimulation, but antagonized by ammonium chloride and methylamine hydrochloride. These findings suggest that, if presented in larger quantities, even ganglioside-associated neurotoxin derivatives are targeted into vacuoles with an acidic internal environment.

Differential sorting of ganglioside-bound neurotoxins anchored merely in one leaflet of the membrane from the subpopulation of BoNT molecules that have found a transmembrane protein receptor could take place either close to the presynaptic membrane or in an *endosomal reticulum* (Fig. 5, compare steps 3, 3′ and 7). This endosomal compartment has only recently been detected and may be identical with the extreme trans-Golgi compartment previously termed GERL (Golgi-endoplasmic reticulum-lysosomal complex) or trans-Golgi network (Griffith and Simons, 1986). Video recordings made from living cells by confocal laser scan microscopy have suggested that the early part of the endocytic pathway is a network of interconnected tubules (Hopkins *et al.*, 1990; Cooper *et al.*, 1990). Using different fluorescent tags to follow the recycling of both the transferrin and epidermal growth factor receptor, Hopkins *et al.* (1990) showed that the transferrin receptor moved through the tubular network at speeds of up to 1 μm/s. The epidermal growth factor receptor moved more slowly (0.05 μm/s) through these tubes in multivesicular bodies. The advantages of a tubular network system for a differential sorting are obvious. One can imagine that the ganglioside-bound toxins are sorted into different sub-compartments as compared to toxin that has bound to a transmembrane protein receptor which could serve a specific function at the motoneuronal junction, perhaps being directly involved in release of acetylcholine. Membrane turnover through exocytosis and internalization through clathrin coated pits after binding of BoNT followed by recycling through acidic intracellular compartments could be a continuous process (Morris and Saelinger, 1989).

The model would also explain why TeTx at high concentrations can also act in the periphery: the sorting machinery in the motoneuron no longer performs accurately, a 'spill-over' taking place by which ganglioside-associated TeTx is swept into more acidic subcompartments. Here, the H chains of both BoNTs and TeTx would integrate into the membrane and translocation of the L chain into the cytosol could take place. Indeed, injection of large doses of TeTx into mice causes death due to flaccid paralysis (Laird, 1982; Matsuda *et al.*, 1982). As of yet, this translocation event is speculative and based on indirect evidence.

Recently the group of van Heyningen has reported that the H chain of TeTx could mediate entry of cytotoxic gelonin, a cell-impermeant derivative of the A chain of ricin, into intact human colonic carcinoma cells (Johnstone *et al.*, 1990). Gelonin substituted the L chain of TeTx, being linked to the H chain by a disulphide bond. It remains to be shown, however, whether gelonin was indeed actively translocated through a pore generated by the H_N domain or by some other unknown mechanism.

Furthermore, our model is compatible with previous reports demonstrating that BoNT/A is also transported retrogradely to reach the ventral horn of the spinal cord (Habermann, 1974; Wiegand *et al.*, 1976): ganglioside receptors are likely to outnumber the putative protein receptors of BoNT. Therefore, a considerable amount of toxin will enter the cell being merely attached to gangliosides. There is no reason to assume that this fraction should be processed differently from ganglioside-attached TeTx once it has entered the motoneuron. Although it remains to be clarified whether sequestration into retrograde transport vesicles occurs close to the plasma membrane or through a budding process from the endosomal reticulum, ganglioside binding is apparently sufficient to guarantee retrograde intraaxonal transport. This was demonstrated convincingly with the native H_C fragment of TeTx by the group of Bizzini (Büttner-Ennever *et al.*, 1981a,b; Bizzini *et al.*, 1980a,b; Ugolini *et al.*, 1984), and with recombinant H_C (Halpern *et al.*, 1990; Manning *et al.*, 1990).

How then can TeTx and botulinal toxins traverse the synaptic cleft from motoneurons to interneurons and central neurons? Very little is known about transynaptic migration at the molecular level. However, since retrogradely transported neurotoxins are considered to be merely attached to gangliosides, they could easily dissociate again from their receptors upon fusion of the transport vesicles at the synaptic cleft and exposure to physiologic pH. Per diffusion they could then reach the opposite presynaptic membrane of an interneuron, bind again to gangliosides and, where available, to protein receptors. We would then have to postulate that central presynaptic membranes (and perhaps already interneurons) contain a specific protein receptor for TeTx. Furthermore, since BoNTs are inactive when they reach central synapses through retrograde axonal transport, we would have to conclude that the high affinity protein receptor(s) which bind the botulinal toxins at peripheral terminals are either absent from those central synapses or that, for some unknown reasons, these receptors are not readily internalized. Too much speculation in a situation where a protein receptor has been identified neither for BoNTs nor for TeTx? Bigalke *et al.* (1985) demonstrated that neuraminidase treatment of primary nerve cell cultures abolishes their sensitivity to BoNT/A while that against TeTx is only reduced (Bigalke *et al.*, 1985). More recently, Takano *et al.* (1989) reported that fragment H_N-L of TeTx blocked excitatory and inhibitory synapses of the spinal motoneuron of the cat while the same preparation injected intramuscularly had no effects on neuromuscular transmission. Furthermore, Poulain *et al.* (1990) showed that in the central cholinergic neurons of *Aplysia californica*, BoNT/A H_N-L fragment could efficiently block binding of BoNT/A holotoxin. In addition, the H_N fragment of BoNT/A was capable of internalizing the L chain of TeTx into the presynaptic neuron of the cholinergic synapse. Subsequent blockade of transmitter release was stronger than that obtained after extracellular application of complete TeTx or H_N-L of TeTx indicating that uptake through the botulinal H_N domain was more efficient. This finding is consistent with the observation of Rabasseda *et al.* (1988) who demonstrated that toxification of cholinergic synaptosomes derived from *Torpedo marmorata* required larger TeTx than BoNT/A concentrations. Taken together, we may conclude that (1) the H_N domains of BoNT/A and TeTx indeed bind to different protein receptors; (2) the H_N domain alone is sufficient to mediate translocation of the L chains into the cytosol; (3) both TeTx and BoNT/A retain a specificity for non-cholinergic or cholinergic neurons, respectively (Poulain *et al.*, 1990).

Reassessing the model

The above model is still highly speculative. At this moment, we can think of three experimental approaches which could confirm the model and provide additional clues on the molecular nature of the putative protein receptors:

(a) Clearly, similar studies to those of Hopkins *et al.* (1990) should be performed on the same preparation with TeTx and BoNTs carrying either different fluorescent tags, different oligonucleotides (that could be detected by *in situ* hybridization), gold particles of distinct size, or different marker enzymes. It is obvious that such experiments should be extended on the isolated subfragments H_N and H_C of the individual neurotoxins.

(b) Perhaps this latter approach is more easily achieved at the genetic level: hybrid genes can be constructed that encode individual subdomains of the neurotoxins fused in frame to the genes of marker enzymes such as alkaline phosphatase or firefly luciferase. The two enzymes have the advantage of being active as monomeric species. The resulting recombinant proteins could add new information on the binding properties of the toxin-specific moieties, on their intracellular targeting and *in vivo* transport characteristics. In addition, this approach could lead to the development of different neuron-specific transporters that allow the specific introduction of biologically and pharmacologically relevant compounds into the peripheral or central nervous system. With respect to a potential therapeutic application of such transporters in humans, it would be of advantage to construct such transporters on the basis on botulinal sequences, in order to rule out immunological complications. Work in this field is currently in progress in our laboratory.

(c) Direct identification of protein receptors by biochemical or genetic methods. Unfortunately, there is only a limited number of suitable permanent tissue culture systems that allow reproducible binding and release studies. In addition, such cell lines generally express their sensitivity towards individual neurotoxins only at specific differentiation stages induced by exogenous factors such as nerve growth factor (in the rat pheochromocytoma cell line PC12; Sandberg *et al.*, 1989a,b), or by interferon (in human macrophages; Blasi *et al.*, 1990). PC12 and several mouse neuroblastoma cell lines have been successfully applied to study the effects of TeTx (Walton *et al.*, 1988; Sandberg *et al.*, 1989a,b; Wellhöner and Neville, 1987;

Considine *et al.*, 1990) or BoNTs (Yokosawa *et al.*, 1989). The differentiation-dependent sensitivity of PC12 cells to TeTx made by Sandberg *et al.* (1989b) has recently been ascribed to the increased formation of polysialylated gangliosides (Fujita *et al.*, 1990), suggesting that these gangliosides mediate an enhanced uptake of TeTx. It would still be interesting to see whether such cell lines alter their sensitivity to clostridial neurotoxins after pretreatment with processing inhibitors of N-linked carbohydrate sidechains, such as castanospermine. This inhibitor of the RER-associated glucosidase I is virtually non-toxic and acts specifically on glycoproteins by inducing the accumulation of mannose-rich carbohydrate sidechains on the glycoproteins that cannot be further processed by the Golgi-associated mannosidases, glucosaminyl-, galactosyl- and sialyltransferases.

Finally, molecular biology should provide access to the receptor molecule(s): Since the putative protein receptor for TeTx is considered to be absent from peripheral cholinergic and present in central non-cholinergic synapses, expression-cDNA libraries could be generated from either source and screened with recombinant H_N or H_N-L fragments for expression of corresponding receptor molecules.

Towards a molecular understanding of transmitter release: potential modes of action of clostridial neurotoxins

Through the work of Penner *et al.* (1986) it has become clear that the toxins display their activity in the cytosol. These authors used microinjection of TeTx to block exocytosis from bovine adrenal chromaffin cells. A large number of reports have searched for mechanism(s) underlying the toxification process, and hypotheses such as blockade Ca^{2+} influx or changes in intracellular handling of Ca^{2+} (Molgo and Thesleff, 1984), disruption of protein kinase C metabolism (Ho and Klempner, 1988; Aguilera and Yavin, 1990; Augilera *et al.*, 1990; Considine *et al.*, 1990), changes in membrane phosphorylation (Guitart *et al.*, 1987), effects on

cGMP-phosphodiesterase (Sandberg *et al.*, 1989b) or, as stated already in the introduction, ADP-ribosylation of GTP-binding proteins (Matsuoka *et al.*, 1987) have been published. Impurities in the toxin preparations on the one side have led to false conclusions. On the other side, in most of the studies it remains uncertain whether the observed effects are actually due to primary or secondary alterations of the cell metabolism. In fact, most of the postulated cellular targets have been ruled out by subsequent studies performed on preparations from different species or different neuronal tissue. Whereas cGMP-analogues and inhibitors of cGMP-degrading phosphodiesterases rapidly reversed the effects of TeTx in PC12 cells (Sandberg *et al.*, 1989b), altering of intrasynaptosomal concentrations of cAMP and cGMP or 8-bromo-cGMP in the presence of inhibitors of cGMP-phosphodiesterases failed to affect the BoNT/A- or BoNT/B-induced reduction of evoked transmitter release from cerebrocortical synaptosomes (Dolly *et al.*, 1990). The same authors showed that activators of protein kinase C, shown to induce neurotransmitter release from synaptosomes of *Torpedo* electric organ (Guitart *et al.*, 1990), were unable to perturb the effects of BoNTs on cerebrocortical synaptosomes. The matter becomes even more complex, since individual neurotoxins apparently exhibit different sensitivities to chemical drugs that were tested as potential antagonists (Dreyer *et al.*, 1987; Gansel *et al.*, 1987; Simpson, 1989). Data from these studies allow the conclusion that the intra-cellular target of BoNT/A could differ from that of TeTx and BoNT/B and the other clostridial neurotoxins (Simpson, 1988). Furthermore, microtubule-depolymerizing agents such as colchicine, nocodazole or griseofulvin attenuated the inhibition normally produced by BoNT/B whilst inhibition by BoNT/A was unaffected (Dolly *et al.*, 1990).

Let us briefly summarize the intraneuronal cascade of events that is believed to precede vesicular transmitter release. According to textbook knowledge (Shepherd, 1988), the following sequential steps are involved: Presynaptic depolarization (1) causes Ca^{2+} influx by opening specific Ca^{2+} channels (2), leading directly or indirectly to priming of the nerve terminals (3), fusion of the vesicles (4) and subsequent release of the neurotransmitters (5).

Today, it is widely accepted that the neurotoxins act at a late stage in the above cascade and that they do not primarily interfere with (1) the generation and entry of an action potential into the nerve terminal (Harris and Miledi, 1971); (2) the opening of Ca^{2+} channels and increase of intracellular calcium (Gundersen et al., 1982; Dreyer et al., 1983); (3) synthesis, uptake or storage of neurotransmitters in synaptic vesicles (Collingridge et al., 1980). The characterization of the constituents of synaptic vesicles has made rapid progress during the last few years. The following section does not attempt to provide a complete survey of the integral proteins found in neuro-transmitter-containing vesicles. It merely touches upon some of the best defined polypeptides in order to attract the reader's interest in his search for potential targets.

Synaptic vesicles have a membrane composition different from that of the nerve terminal plasma membrane. The neuron contains at least two different types of secretory vesicles involved in regulated secretion (for reviews see De Camilli and Jahn, 1990; Burgoyne, 1990): (1) small electron-translucent vesicles which contain classical non-peptide neurotransmitters. These vesicles are related to the microvesicles of endocrine cells; (2) large, dense-core vesicles which accumulate neuroactive peptides and catecholamines. Such vesicles have a more scattered distribution in the axon, while the small synaptic vesicles are clustered at synaptic junctions. Little is known about the molecular events underlying the final steps that finally lead to quantal release of neuro-transmitters from chemical synapses.

A current working hypothesis (Bähler et al., 1990) suggests that in the resting state of the nerve terminal, the small neurotransmitter-containing vesicles are divided into two different pools (Fig. 6b): a 'releasable pool' of vesicles and a 'reserve pool'. The releasable pool occupies specific, but as yet uncharacterized 'docking' sites at the cytoplasmic face of the presynaptic membrane. Apparently, these releasable vesicles are sitting at 'active zones', ready to fuse, just waiting for an as yet unknown final stimulus. In contrast, vesicles of the 'reserve pool' are attached to actin filaments via synapsin I molecules (Fig. 6a). Such association has been demonstrated by quick-freezing deep-etching techniques (Hirokawa et al., 1989).

Synapsins represent a family of four closely related proteins (synapsin Ia,b; synapsin IIa,b) generated by different splicing and regulated by several protein kinases (Südhof et al., 1989; De Camilli and Jahn, 1990). Synapsin I molecules contain three phosphorylation sites, designated P1 (located within the N-terminal head region; regulated by the cAMP-dependent protein kinase and the calcium/calmodulin-dependent protein kinase I), P2, and P3 (both located in the C-terminal tail region, regulated by the calcium/calmodulin-dependent protein kinase II) (Ueda et al., 1973; Nairn and Greengard, 1987). These latter two sites are absent from synapsin II molecules. Isolated synapsin I was shown to interact with F actin and to bundle actin filaments in a phosphorylation-dependent manner (Hirokawa et al., 1989; Petrucci and Morrow, 1987). In vivo, non-phosphorylated synapsin I molecules are associated with the cytoplasmic vesicle surface, probably due to an interaction of the head region with charged phospholipids and of the tail region with vesicle protein(s) (Benfenati et al., 1989). A calculation of the number of synapsin molecules per small synaptic vesicle suggested that a large area of the vesicle should be covered by synapsin I molecules surrounding the vesicle like a cage (Bähler et al., 1990). Injection of dephospho-synapsin I, but not of phospho-synapsin I inhibits neurotransmitter release at the squid giant synapse (Llinás et al., 1985). Phosphorylation of synapsin I at P1 reduces and phosphorylation in the tail region virtually abolishes the F-actin bundling activity (Hirokawa et al., 1989). As a consequence, the vesicles dissociate from the actin filaments to enter the releasable pool. In agreement with this hypothesis, agents that inhibit actin polymerization such as cytochalasin or the binary botulinum toxin C2 (Ohishi et al., 1990) enhance transiently neurotransmitter release by increasing the pool size of releasable vesicles (Bernstein and Bamburg, 1989; Matter et al., 1989). After fusion with the presynaptic membrane, the specific constituents of the synaptic vesicle have to undergo a rapid retrieval and recycling process in order to guarantee continuous transmitter release. At this stage, synapsin I could facilitate reassembly and direct reutilization of vesicle-specific components. Koenig and Ikeda (1989) recently dissected this retrieval process using the temperature-sensitive mutant of Drosophila,

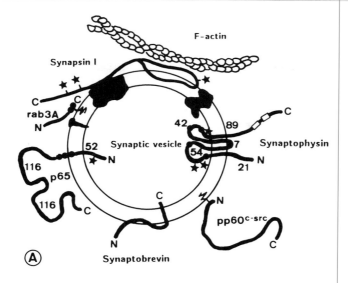

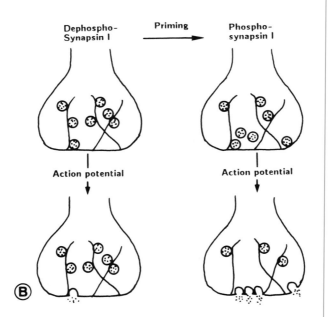

Figure 6. Composition of small translucent synaptic vesicles (A) and their putative arrangement in a nerve terminal (B). (A) Membrane topology of some of the constituent proteins. The positions of the N-terminal (N) and C-terminal ends (C) are indicated. Filled circles indicate cysteine residues of potential importance. N-glycosylation sites are marked by an asterix. Note that acylation of rab3A and pp60^{c-src} alone is not sufficient to induce membrane association but requires, in addition, the interaction with protein(s). For further details see text.

shibire[ts-1] in which endocytosis is reversibly blocked at non-permissive temperature: synaptic vesicle membranes disassembled at the time of transmitter release and reassembled along the plasma membrane. Membranes were then internalized in the form of large cisternae, from which new vesicles were formed. Torri-Tarelli *et al.* (1990) studied the redistribution of synaptophysin and synapsin I during the vesicle recycling process at the neuromuscular junction. The authors found that retrieval was a highly efficient process which did not involve extensive intermixing of vesicle and plasma membrane. Rapidly recycling vesicles continued to contain the bulk of synapsin I associated with their membranes.

The phospholipid-binding protein **p65** is one of the most recent discoveries (Perin *et al.*, 1990). Its intravesicular N-terminal domain (52 residues) is followed by a single transmembrane segment (27 residues) which contains several cysteine residues close to the cytoplasmic border region. Cysteine clusters in such positions are often acylated in viral proteins that exhibit fusion activity (White, 1990). However, at this moment it is not known whether this protein undergoes such a modification. Within the large cytoplasmic portion p65 contains two repeats, each 116 amino acid residues in length, that share 41% identity and, most interestingly, show a similar degree of identity to the C2 regulatory region of protein kinase C (Nishizuka, 1988). The authors demonstrate that a recombinant derivative containing these domains but lacking the N-terminal and the membrane anchor domain binds strongly radiolabelled phosphatidylserine and polysialylated gangliosides, but not phosphatidylcholine. Furthermore, the recombinant molecular species agglutinated trypsin-treated rabbit erythrocytes at concentrations less than 10 nM, indicating the capability of p65 for bivalent interaction with two separate membranes. The notion that haemagglutination was not inhibited by lipids containing only one acyl chain further suggests that the protein makes contact with the hydrophobic portion of the phospholipids and perhaps inserts partially into the bilayer. Together, these results suggest a role for this protein in synaptic vesicle exocytosis, perhaps in the binding to phospholipids present at the cytoplasmic face of the plasma membrane.

The synaptic vesicles contain a number of *GTP-binding proteins* (Ngsee *et al.*, 1990; Fischer von Mollard *et al.*, 1990). This is of particular interest, because GTP-binding proteins such as sec4 and YPT1 play a role in regulating constitutive vesicle traffic between individual compartments of the eukaryotic cell (Zahraoui *et al.*, 1989; Niemann *et al.*, 1989). The GTP-binding protein rab3A is exclusively associated with synaptic vesicles in neuronal tissue and co-localizes in adrenal medulla with a microvesicle population that is distinct from chromaffin granules. The protein quantitatively dissociates from the synaptic vesicles during Ca^{2+}-stimulated exocytosis and reassociates again at later stages during recovery (Fischer von Mollard *et al.*, 1991). Taken together, these findings provide direct evidence for an association–dissociation cycle of GTP-binding proteins during vesicular transmitter release and retrieval of membranes. It will be interesting to see the effects of GTP S, a poorly hydrolysable analogue of GTP on the binding characteristics of rab3A. Unlike fusion of viral and cellular membranes, where the fusion event can be ascribed to particular fusogenic proteins (White, 1990), fusion of such transport vesicles with target membranes involves multisubunit fusion complexes that seem to be assembled at the junction sites (Orci *et al.*, 1989; Weidman *et al.*, 1989). In the Golgi, sequentially coated and uncoated intermediates of vesicles mediate the intercisternal traffic. The uncoating step of such non-clathrin-coated vesicles has to occur prior to fusion and is inhibited by GTPgammaS, again suggesting the involvement of GTP-binding proteins (Pfanner *et al.*, 1990). Studies involving permeabilized rat mast cells also provided evidence that late stages of exocytosis are regulated by GTP-binding proteins (for review see Gomperts, 1990).

Could GTP-binding proteins such as rab3A be targets for clostridial neurotoxins? At present it is unclear whether structural modifications and/or associations with additional proteins which could regulate the dissociation–association cycle of the G proteins are affected in cells treated with clostridial neurotoxins. A Ca^{2+}-independent stimulatory effect on secretion by GTP-analogues was reported for rat insulinoma cells (Vallar *et al.*, 1987) and bovine chromaffin cells in primary culture (Bittner *et al.*, 1986). In electrically permeabilized bovine chromaffin cells or in staphylococcal α-toxin permeabilized PC12 cells, on the other hand,

GTPgammaS inhibits Ca^{2+}-induced secretion (Knight and Baker, 1985; Ahnert-Hilger et al., 1987) indicating that the effects of GTP analogues on regulated exocytosis are inconsistent in individual cell types.

Another constituent of the small synaptic vesicle membrane is the glycoprotein *synaptophysin*, a calcium-binding protein (M_r 38 000) which spans the synaptic membrane four times (Jahn et al., 1985). Both the N-terminal and the C-terminal end of the protein are located cytoplasmically (Johnstone and Südhof, 1990). The C-terminal tail contains a pentapeptide repeat which binds Ca^{2+} and undergoes tyrosine phosphorylation by endogenous and exogenous tyrosine kinases (Pang et al., 1988; Rehm et al., 1986). Guitart et al. (1987) reported that BoNT/A inhibited depolarization-stimulated protein phosphorylation. Although not identified as synaptophysin by antisera or characterized by phosphoamino acid analysis, a 37 kDa species was amongst those proteins the phosphorylation of which was specifically reduced by BoNT/A. Barnekow et al. (1990) suggested that tyrosine phosphorylation of synatophysin was catalysed both *in vivo* and *in vitro* by the proto-oncogene product pp60[c-src]. This tyrosine-specific protein kinase has previously been reported to be associated with synaptic vesicles (Hirano et al., 1988). Clearly, the potential regulatory role of pp60[c-src] *in vivo* still needs more thorough experimental evidence; at any rate, the *in vitro* tyrosine phosphorylation reaction of synaptophysin was neither inhibited by TeTx nor by its isolated L chain (H. Niemann and A. Barnekow, unpublished).

Another protein that undergoes tyrosine phosphorylation and shares epitopes with synaptophysin is **p29**, a non-glycosylated species (M_r 29 000) found in small synaptic vesicles and synaptic-like microvesicles of endocrine cells (Baumert et al., 1990). It is clear that p29 also constitutes an integral membrane protein; its precise membrane topology, however, has not yet been determined.

Cross-linking and sedimentation analyses indicated that synaptophysin is a hexameric homo-oligomer (Thomas et al., 1988) and the transmembrane topology as well as its putative quarternary structure are reminiscent of connexon, the unit element of gap junctions. Like connexon, the hexameric synaptophysin appears to form high conductance channels in planar phospholipid bi-layers (Thomas et al., 1988). Recently, **synaptoporin** (M_r 37 000), another transmembrane protein of synaptic vesicles exhibiting 58% amino acid identity to synaptophysin and a similar membrane topology, was characterized (Knaus et al., 1990). It is tempting to speculate that the synaptophysin/synaptoporin homo-oligomers catalyse binding of the synaptic vesicle to a receptor present at the cytoplasmic face of the presynaptic membrane (Fig. 5). What type of stimulus induces such docking process and whether initially transient pores are generated remain open questions. At low levels of stimulation such membrane-associated vesicle could flicker, thus releasing only minute quantities of neurotransmitters. This would explain why under such conditions the constituents of the synaptic vesicle membrane and the presynaptic membrane do not intermingle (Valtorta et al., 1988). Indeed, the sudden formation of a fusion pore is the earliest detectable event when exocytosis is triggered (Almers, 1990). The demonstration of defined conductances in isolated vesicle preparations from *Torpedo* electric organ (Rahamimoff et al., 1988, 1989) suggests a functional importance of vesicular channel proteins related to that of gap-junction proteins (Betz, 1990).

Is there any direct interaction between clostridial neurotoxins and synaptic vesicles? Lazarovici et al. (1989b) reported binding of TeTx to a G1b type glycoconjugate present at the cytoplasmic face of isolated granules from bovine adrenal chromaffin cells. We have used a different approach to obtain clues to this question: we translated TeTx L chain-specific mRNA in reticulocyte lysate in the presence of [^{35}S] methionine and incubated the radiolabelled L chain together with affinity chromatography-purified vesicles under a variety of salt and pH conditions. Subsequent sedimentation analyses of the mixture on sucrose gradients, however, did not indicate an association of the TeTx L chain with the vesicles. We cannot exclude, however, that such binding would take place in the presence of specific cytosolic cofactors (U. Eisel, R. Jahn, and H. Niemann, unpublished results).

It has been suggested (Aunis and Bader, 1988) that the final trigger for fusion and transmitter release is the sudden removal of cytoskeletal barriers to exocytosis, rather than an active fusogenic mechanism. Recent reports have yielded significant insights into two critical features: the molecular rearrangements during membrane fusion and the role of the cytoskeleton in regulating the exocytotic event. Could the neurotoxins interfere with such late events? Marsal and co-workers (1989) used quick-freezing and freeze etching techniques to answer this question by using pure cholinergic synaptosomes prepared from the *Torpedo* electric organ. The authors found that within 6 min of exposure to BoNT/A, the potassium-induced release of acetylcholine, but not that of ATP, was inhibited in a dose-dependent manner. During this process no leakage of lactate dehydrogenase activity was observed, indicating that the synaptosomes remained sealed. In the absence of BoNT/A, KCl-stimulated exocytosis causes a rearrangement of intramembrane particles detectable in each of the two leaflets of the presynaptic membrane after etching: in the resting state the inner leaflet contains a large number of intramembranal particles of two different size classes, whereas the outer leaflet contains less of these particles. Upon chemical stimulation the number of smaller-sized particles decreased at the inner leaflet and simultaneously the number of larger-sized particles present in the outer leaflet increased. This observation suggests some transfer from the inner to the outer leaflet of the presynaptic membrane during exocytosis. Such translocation, however, was blocked when the synaptosomes were toxified with BoNT/A (Marsal *et al.*, 1989) or with TeTx (Egea *et al.*, 1990) prior to potassium stimulation. We may argue that this translocation event is too close to the final step on the ladder leading to vesicular transmitter release and already above the point at which the neurotoxins act. Nevertheless, it is worth while to approach the unknown inhibitory step from the other end of the ladder by characterizing the intramembrane particles at the molecular level. Do they contain any proteins present in synaptic vesicles? This question could be approached with the increasing set of monoclonal antibodies prepared against vesicular proteins (De Camilli and Jahn, 1990). Does the distribution of such particles correlate with attachment sites of actin filaments? Marxen and Bigalke (1991) used bovine chromaffin cells to follow the F-actin bundling and rearrangement after nicotinic stimulation in normal and BoNT/A-treated cells. When cells were stimulated with carbachol, F-actin formed patches at sites close

to the plasma membrane. Withdrawal of the secretagogue in the presence of Ca^{2+} restored the original phenotype within 30 min. This actin rearrangement was blocked in cells pretreated either with TeTx or BoNT suggesting that the neurotoxins could act either directly on F-actin, on actin-severing proteins or at any earlier stage of the secretory cascade but of course, as stated before, posterior to Ca^{2+} influx.

From the neurotoxin genes to biological activities

In the search for a safe way leading from the cloned neurotoxin genes back to systems that allowed functional analysis of individual subdomains we had to take into account two kinds of concerns:

First, facing the situation that neither the molecular mechanism of toxification nor the distribution of the target molecule(s) within the organism are known, we did not want to clone the entire L chain gene under control of a eukaryotic promoter.

Second, we had no idea about the biological stability of an mRNA that contained more than 70% [A+U] in an eukaryotic environment. Analyses for splice donor and acceptor sites suggested that it was highly unlikely that intact mRNA molecules would be exported from the nucleus. Furthermore, because of the high [A+U] content, neurotoxin-specific mRNAs contain numerous of the 5'AUU-UA3'-motifs known to downregulate the stability of several lymphokines and proto-oncogene mRNAs in eukaryotic cells (Shaw and Kamen, 1986).

For these reasons we decided to pursue an alternative route based on *in vitro* synthesis of biologically stable, i.e. 5'-capped and 3'-polyadenylated mRNA. This mRNA was then translated into the corresponding peptides to map epitopes of monoclonal antibodies. In addition, we and others have expressed non-toxigenic gene fragments in bacteria to test their immunological and biological properties. To define the structural requirements for toxicity, we microinjected the mRNA into vertebrate and non-vertebrate target cells including bovine medullary chromaffin cells or into the identified presynaptic neuron of the cholinergic synapse present in the buccal ganglia of *Aplysia californica*.

The action of clostridial neurotoxins in bovine adrenal chromaffin cells

Adrenomedullary chromaffin cells originate from the neuronal crest and can develop several neuronal characteristics: (1) They form cholinergic synapses in culture when exposed to nerve growth factor (Unsicker *et al.*, 1978). (2) The newly formed synapses are functionally active and the synaptic vesicles respond to K^+ in the culture medium (Ogawa *et al.*, 1984). (3) The release of catecholamine is inhibited by exposure to high concentrations of botulinal toxins (Knight *et al.*, 1985; Knight, 1986). In contrast, the cells are not inhibited by externally applied TeTx. This latter observation must be ascribed to the absence of GT_{1b} gangliosides and/or a TeTx-specific protein receptor, because subsequent experiments have demonstrated that TeTx also blocks secretion totally when administered intracellularly. This was achieved either by microinjection (Penner *et al.*, 1986) or by diffusion into the cell through pores produced by digitonin (Bittner and Holz, 1988; Bittner *et al.*, 1989a; McInnes and Dolly, 1990) or by streptolysin O (Ahnert-Hilger *et al.*, 1989a). Cells permeabilized with streptolysin O or digitonin retain their exocytotic capacity for a short period of time (Howell and Gomperts, 1987). However, the pores have a diameter that allows entry of the pore forming agents into the cytosol and, therefore, under most experimental conditions intracellular compartments may also incorporate the pore forming agent. Furthermore, leakage of cellular compounds essential for secretion constitutes a serious disadvantage and, therefore, experiments are generally hampered by the narrow time window in which exocytosis can be measured before the toxin has already blocked it. Ahnert-Hilger *et al.* (1987) have recently improved the streptolysin permeabilization technique by applying low concentrations of the pore-former at low temperature ($0°C$), i.e. under conditions where only binding but no membrane integration and pore formation takes place. The cells can then be washed to remove unbound streptolysin prior to temperature upshift and generation of pores. Under these conditions entry of streptolysin can be avoided. The pores have a size large enough to allow diffusion of macromolecules such as antibodies (Bhakdi and Tranum-Jensen, 1987).

Two important findings evolved from the studies on permeabilized cells: (1) The L chains of both TeTx and BoNT/A alone are sufficient to block secretion (Bittner *et al.*, 1989a, 1989b; Ahnert-Hilger *et al.*, 1989b; Stecher *et al.*, 1989a,b). (2) Reductive chain separation of BoNT/A was found to be prerequisite for its inhibitory action on exocytosis (Stecher *et al.*, 1989a). For reasons discussed further below, however, this finding does not simply imply that the newly generated cysteine of the L chain reacts with an intracellular target, since this residue may be deleted without changing toxicity.

However, the compelling advantage of cell permeabilization which allows access to a large number of cells under a variety of experimental conditions is compensated by the fact that the cells are leaky and that studies on secretion can be performed only over a period of about 30 to 60 min. Furthermore, extensive controls are required to define whether an exogenous compound acts exclusively intracellularly. Here, microinjection of mRNA into viable cells offers a clear advantage, since *de novo* protein synthesis has to take place in the cytosol, before the corresponding protein can elicit its effects. Binz *et al.* (1990c) have microinjected mRNA encoding the authentic L chain of TeTx into bovine chromaffin cells. 5'-Capped and 3'-polyadenylated mRNA was transcribed *in vitro* and introduced into the cells by a patch electrode. The electrode was removed, the cells were incubated for about 5h at 37°C and analysed electrophysiologically for Ca^{2+} induced capacitance changes applying patch-clamp techniques and the whole cell configuration. Control cells respond to the Ca^{2+}-stimulus with fusion of vesicular membranes and the plasma membrane. The increase in membrane area can be measured as increase of electrical membrane capacitance. Such increase was also observed when cells injected with TeTx L chain-specific mRNA were incubated throughout the experiment with cycloheximide, an inhibitor of protein synthesis. In contrast, five out of seven cells injected with L chain-specific mRNA showed no change in membrane capacitance indicating the block in secretion. The data of Binz *et al.* (1990c) fully supported the results obtained previously with the isolated L chain. Furthermore, this approach opened, for the first time, the possibility to study the effects of specific mutations of the L chain.

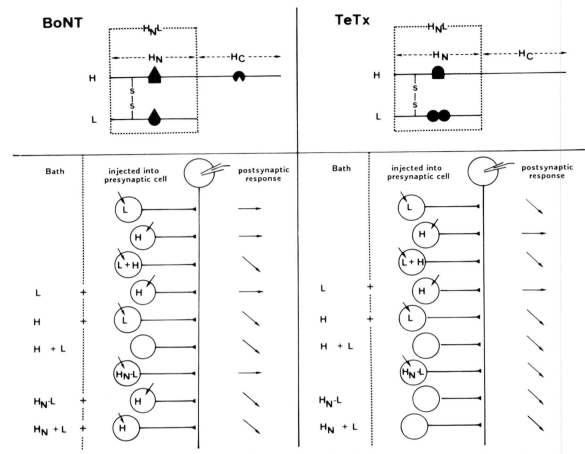

Figure 7. Differential effects of tetanus and botulinum type A toxins on neurotransmitter release from *Aplysia* neurons. The symbol within the H$_C$-fragment of BoNT/A indicates its ability to synergistically contribute to the toxifying action mediated by the L chain. In contrast, the L chain of TeTx is sufficient to inhibit neuronal transmission. The term "bath" indicates that the compound was applied exracellularly; the arrows characterize the postsynaptic response; falling: depression of evoked release; horizontal: no effect. (adapted from Poulain et al., 1990).

As of yet, the small number of microinjected cells does not allow biochemical studies to detect target proteins or metabolic changes following translation of the mRNA. Preliminary experiments based on electroporation and liposome fusion as a means to introduce the mRNA into the cell, however, suggest that even this problem may be solved in the near future.

Lessons from injecting neurons of *Aplysia californica*

It may be questioned to what extent secretion from bovine chromaffin cells follows precisely the same principles as neurotransmitter release from peripheral or central nerve terminals. *Aplysia californica* contains in its central nervous system several separated ganglia. These ganglia consist of many neurons that vary in size, position, shape, firing pattern and the chemical substances by which they transmit information to other cells. Some of the neurons make well-characterized synapses. In particular, synapses in the **buccal ganglion** and in the **cerebral ganglion** have been quite useful in studies on clostridial neurotoxins:

In the **buccal ganglion** recordings can be made from two identified *cholinergic* neurons afferent to the very same postsynaptic neuron making a well-known chloride-dependent inhibitory synapse (see Fig. 8) (Tauc *et al.*, 1974; Poulain *et al.*, 1986). This synapse is almost as sensitive to botulinal toxins as the vertebrate motoneuronal junction (Poulain *et al.*, 1988a,b, 1989a,b; Maisey *et al.*, 1988). The cholinergic synapse was considered an ideal preparation for microinjection experiments, because the

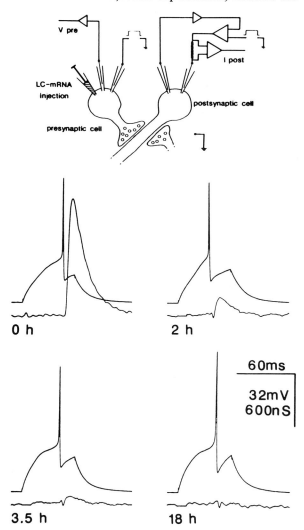

Figure 8. The identified cholinergic synapse of the buccal ganglion of *Aplysia californica* and effects of presynaptic injection of TeTx L chain specific mRNA. Uppper panel: The identified pre- and postsynaptic neurons were impaled with two glass micropipettes for current- and voltage-clamp measurements. Lower panels: Acetylcholine release was monitored by the amplitude of the postsynaptic response (lower traces; expressed as membrane conductance, nS) elicited by an evoked action potential in the presynaptic neuron (upper traces) (From Mochida *et al.*, 1990a).

presynaptic neurons have large bodies, 150–300 μm in diameter and the distances to the synaptic terminals are short (<500 μm). In the experimental setup, acetylcholine release from either of the presynaptic neurons can be induced electrically by evoking presynaptic action potentials (once a minute) and the postsynaptic responses can be recorded as current changes and expressed as membrane conductance (Fig. 9) (Poulain *et al.*, 1986).

In the **cerebral ganglion**, *non-cholinergic* neurons from cluster A make cationic excitatory neuronal synapses on neurons from cluster B (Gaillard and Carpenter, 1986). Here, the neuronal somata are about 75–125 μm in diameter. Post-synaptic responses due to presynaptic stimulation and subsequent release of an undefined neurotransmitter (it is definitely not acetylcholine) are recorded as voltage changes (Poulain *et al.*, 1991).

Experiments with clostridial neurotoxin proteins and toxin subfragments

It is to the merit of the group of Ladislav Tauc at the CNRS (Gif-sur-Yvette) that they exploited the two preparations of *Aplysia californica* as tools to analyse the effects of clostridial neurotoxins. Being continuously superfused with artificial seawater, both preparations (especially the cholinergic synapse) are stable over a period of several hours, thus allowing detailed studies on binding, internalization and toxification at specific temperatures. Furthermore, the availability of sufficiently pure constituent chains or subfragments of the toxins (Weller *et al.*, 1988, 1989; Shone *et al.*, 1985; Maisey *et al.*, 1988; Poulain *et al.*, 1988a, 1989a,b) allowed the generation of hybrid toxins in which the L chain of TeTx or BoNT/A was associated with a heterologous H chain.

These studies yielded a number of important results some of which, however, are in clear contrast to observations made with vertebrate synapses (Fig. 7): (1) The most striking difference is that in *Aplysia* neurons the intracellular administration of the botulinal L chains is insufficient to inhibit neuronal transmission (Poulain *et al.*, 1988a, 1989a; Mochida *et al.*, 1990a,b), whilst the L chain of TeTx behaves in *Aplysia* as in vertebrate secretory cells and is capable of inhibiting neurotransmitter release in the absence of the H chain (Fig. 8; Mochida *et al.*, 1989). No matter what experimental approach to introduce an L chain of

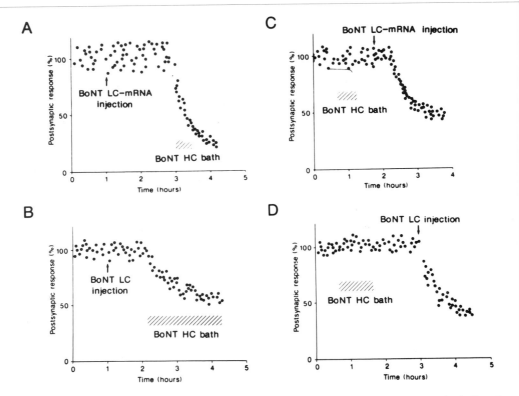

Figure 9. Injection of BoNT/A L chain-specific mRNA (A,C) or native L chain (B,D) into presynaptic cholinergic neurons of the buccal ganglion of *Aplysia californica*. The amplitude of postsynaptic responses is expressed as percentage control before the intracellular injections (indicated by the arrows). Concentration of L chain-specific mRNA: 0.5 μg/μl. Note that depression of postsynaptic responses starts only after addition of the H chain and at least 40 min after injection of the mRNA (C). Hatched areas mark the time points and duration of presence of BoNT/A H chain (10 − 50 nM) in the bath. (From Mochida *et al.*, 1990a).

BoNTs was chosen (microinjection of the L chain, extracellular or intracellular application of H_N-L (Fig. 7) or microinjection of L chain-specific mRNA (Fig. 9)), toxicity was observed only in the presence of the H chain or, to be more precise, when both the L and the H chain were present in the cytosol of one and the same presynaptic neuron (Fig. 9B,D). From these data it may be concluded that the mysterious synergistic effect of the botulinal H chain must be assigned to the H_C-region of the H chain (Poulain *et al.*, 1989b). Although the mechanism of such interaction between the L- and H chain is not understood, the requirement of the H chain for inhibition of neuronal transmission appears to be truly restricted to the *Aplysia* system. Dayanithi *et al.* (1990) reported that the release of vasopressin from digitonin-permeabilized nerve endings from the rat neurol lobe was inhibited by the isolated L chain of BoNT/A.

(2) In *Aplysia* the H_N-L fragments of TeTx and BoNTs are efficiently internalized and block transmitter release, of course, in the latter case only when the H chain was previously microinjected or internalized (Fig. 7) (Poulain *et al.*, 1989b). Furthermore, TeTx and BoNT/A or their isolated H_N-L fragments exhibited a remarkable specificity for non-cholinergic and cholinergic nerve terminals, respectively, when administered extracellularly (Mochida *et al.*, 1989; Poulain *et al.*, 1991). Thus, at the cholinergic synapse of the buccal ganglion BoNT/A was 100-fold more powerful in blocking neurotransmitter release than TeTx. This situation was reversed at the non-cholinergic synapse of the cerebral ganglion. The selectivity is specified by the individual H_N regions, since chimaeric proteins consisting of the H chain or the H_N fragment of BoNT/A and the L chain of TeTx are efficiently internalized into the pre-synaptic cholinergic neu-

ron and are equally potent inhibitors of transmission like extracellularly administered BoNT/A (Poulain *et al.*, 1991). Taken together, these data again support the model (Fig. 5) that different protein receptors are recognized by the H_N regions of TeTx and BoNTs.

As of yet, demonstration of toxicity of mixed-type neurotoxins was restricted to *Aplysia*. Reconstituted chimeric type A and B botulinal toxins are fully toxic in the buccal ganglion, but are relatively inactive on mammalian nerve terminals (Maisey *et al.*, 1988). This may be explained by the notion that, based on a similar overall tertiary structure, heterologous L and H chains form complexes primarily due to hydrophobic and ionic interactions, while homologous L and H chains in addition form oxidized disulphide-linked di-chain species (Maisey *et al.*, 1988; Poulain *et al.*, 1991). It is obvious that the above research area can now be extended by studies directly employing the H_C fragments of TeTx and BoNTs. To analyse the contribution of the individual subfragments, the binding and internalization steps can be further assessed at different temperatures and by competition tests involving subfragments from different serotypes. All these studies, however, depend on the assumption that the isolated subfragments retain their original properties as exhibited in the native di-chain neurotoxins. Furthermore, this approach remains vulnerable to traces of contamination and it ultimately reaches its limits when subdomains smaller than the L, H_N and H_C fragments have to be characterized.

Microinjection of L chain-specific mRNAs into Aplysia neurons

In addition to the physical stability over a long time period, the identified presynaptic cholinergic neurons of the buccal ganglion of *Aplysia californica* offer yet another advantage for the microinjection of mRNA: they allow the injection of large volumes; up to 10% (<0.4 nl) of the cell body volume can be injected. This allowed us to inject concentrated *in vitro* transcription mixtures without previous removal of the plasmid-DNA (Mochida *et al.*, 1990a,b).

Figure 8 shows a schematic drawing of the cholineric synapse in which two presynaptic cells make contact to one postsynaptic cell. One of the presynaptic cells receives mRNA by pressure injection. Evoked action potentials in the presynaptic neuron (Fig. 8, upper curves in the four panels) cause release of acetylcholine and, after a short delay, a postsynaptic response (lower traces). Concomitant with the translation of the injected mRNA, transmitter release becomes inhibited, first detectable after 35–45 min (Fig. 9C). As a consequence, communication with the postsynaptic neuron is eventually blocked and the presynaptically evoked action potential no longer creates a postsynaptic response. Note that at this stage any detrimental effect on the postsynaptic neuron can be excluded by stimulating the non-injected presynaptic cell. Figure 9 shows that the pivotal role of the BoNT/A-chain for inhibition of acetylcholine release can also be demonstrated by microinjection of mRNA (Mochida *et al.*, 1990a). L chain-specific mRNA was injected either before (Fig. 9A) or after bath application of the H chain (Fig. 9C). In agreement with the results obtained with the isolated L chain of BoNT/A (Fig. 9B and D), inhibition of transmitter release occurred only in the presence of the H chain.

To learn more about the specific role of particular subdomains of the L chains of TeTx and BoNT/A, we have generated a series of mutants by site-directed mutagenesis, exonuclease III digestion, or by applying the polymerase chain reaction (Kurazono *et al.*, 1991a,b). The mutations clustered in three regions:

(1) the conserved His-rich region;
(2) the C-terminal regions containing the Cys residue involved in the linkage of the L chain to the H chain; and
(3) the highly conserved N-terminal domain.

As with the wild-type genes, mutant mRNA was prepared by transcription *in vitro* and tested for its biological activity in the presynaptic cholinergic neuron of *Aplysia californica*. The results are summarized in Table 3.

Let us first discuss the effects of specific exchanges of individual His residues in the conserved motif 'DPhhLhHELnHnnHxLYG' in which 'h' constitutes hydrophobic, 'n' uncharged and 'x' any type of amino acid. It has been reported previously that a treatment of BoNT/A and BoNT/E with diethylpyrocarbonate, a reagent specifically carboxyethylating histidine residues, detoxified both toxins (DasGupta and Rasmussen,

1984; Binz *et al.*, 1990a). As stated above, in most of the known L chain sequenced the His-rich motif is part of a predicted α-helix in which the histidine's imidazole sidegroups would be located at the same face of the helix. Since the motif was found in identical positions of all known L chain sequences, it was surprising that each of the three His residues of the TeTx L chain could be replaced by other amino acids, without altering its toxicity (Table 3). Only one of the nine mutants, carrying a proline residue instead of His^{237} was found completely non-toxic in *Aplysia*.

These findings could be explained in several ways, but without further experimental evidence each of the interpretations would be highly speculative. It seems possible, however, that the His-rich motif plays more than just one role: (1) it could be important for an interaction with the H chain during the translocation process; (2) the presence of the α-helix in the centre of the L chains could be important to stabilize the tertiary structure. A proline residue would break the α-helix which again could lead to a different folding of the L chain. However, these hypotheses need to be validated by double and triple mutations involving also botulinal toxins. In addition, they should be examined by an *in vitro* translocation assay.

The examination of N-terminal and C-terminal deletion mutants was performed to identify the minimal essential sequences required for toxicity of the L chains of TeTx and BoNT/A. As shown in Table 3, L chain derivatives of TeTx or BoNT/A that lacked 65 or 32 C-terminal residues, respectively, retained toxicity, whereas those which lacked 68 or 57 residues were non-toxic. The important finding here was that the cysteine residues originally participating in the linkage to the H chains can be deleted, indicating that they are neither part of an active centre nor substrate for cytoplasmic acylation or isoprenylation, as observed in a variety of biologically active molecules including GTP-binding proteins (Hancock *et al.*, 1989) and lamins (Krohne *et al.*, 1989). A similar type of modification could have been postulated for the neurotoxin L chains on the grounds of recent reports by Stecher *et al.* (1989a) and Schiavo *et al.* (1990). The alignment of the amino acid sequences (Fig. 4) reveals a low degree of identity within the 55 C-terminal amino acids of the

individual L chains. Since the first 100 residues of the individual H chains are also quite dissimilar, it is tempting to speculate that the non-essential C-terminal portion of the L chains merely serves a spacer function to provide flexibility for the postulated translocation of the L chains through an H-chain-specified pore (Kurazono *et al.*, 1991b).

A deletion of the N-terminal 44 residues from the L chain of the BoNT/A abolishes toxicity in *Aplysia* neurons (Table 3; Kurazono *et al.*, 1991b). Data obtained with monoclonal antibodies suggest that a deletion of this region which is highly conserved in all serotypes, leads to a significant alteration of the tertiary structure of the residual L chain of BoNT/A.

Molecular biology and immunology: joint efforts in the establishment of structure–function relationships and therapeutic tools

Tetanus toxoid is worldwide the most frequently administered vaccine and commercial vaccines are inexpensive and safe. Nevertheless, tetanus remains a threat, especially in those countries which fail to fully implement the programmes for prevention of tetanus (Stanfield and Galazka, 1984). For these reasons, the immunological features of TeTx have always attracted an enormous scientific attention which would well be worth an independent review. The use of monoclonal antibodies against botulinal toxins has been reviewed only recently (Kozaki *et al.*, 1989). In this section, I shall restrict the discussion to the following immunological aspects which, in my opinion, should be pursued in combination with gene technological methods:

(1) Mapping of B cell epitopes and application of monoclonal antibodies in analyses of structure–function relationships.
(2) Mapping of T cell epitopes and characterization of T cell populations.
(3) Generation of recombinant peptides for the use as second-generation vaccines or therapeutic tools.

Mapping of B cell epitopes

Monoclonal antibodies can provide unique opportunities for generating clinically useful antitoxins

Table 3. Characterization of mutants of the L chains of TeTx and BoNT/A

Derivative[a]	Terminal Sequence[b]	MW[c]	Toxicity[d]	Immunoprecipitable with MAB[e]			
				E2[f]	A109[g]	A113[h]	A115[h]
TeTx-LC	..NRTA^{457}	52412	+	+	−	−	−
TeTx-H^{233}→L^{233}	..L^{233}ELIHVLH..	52388	+	n.t.	−	−	−
TeTx-H^{233}→V^{233}	..V^{233}ELIHVLH..	52374	+	n.t.	−	−	−
TeTx-H^{237}→D^{237}	..HELID^{237}VLH..	52390	+	n.t.	−	−	−
TeTx-H^{237}→G^{237}	..HELIG^{237}VLH..	52332	+	n.t.	−	−	−
TeTx-H^{237}→V^{237}	..HELIV^{237}VLH..	52374	+	n.t.	−	−	−
TeTx-H^{237}→P^{237}	..HELIP^{237}VLH..	52372	−	n.t.	−	−	−
TeTx-H^{240}→D^{240}	..HELIHVLD^{240}..	52390	+	n.t.	−	−	−
TeTx-H^{240}→R^{240}	..HELIHVLR^{240}..	52431	+	n.t.	−	−	−
TeTx-H^{240}→A^{240}	..HELIHVLA^{240}..	52346	+	n.t.	−	−	−
TeTx3'413	..EYKG^{413}P	47597	+	n.t.	−	−	−
TeTx3'398	..NDTE^{398}AA	45945	+	n.t.	−	−	−
TeTx3'392	..LDDT^{392}LN	45294	+	n.t.	−	−	−
TeTx3'389	..PNLL^{389}I	44849	−	n.t.	−	−	−
TeTx3'381	..MNHD^{381}A	43982	−	n.t.	−	−	−
TeTx3'356	..SIMY^{356}RLIN	41399	−	n.t.	−	−	−
BoNT/A-LC+	..EENI^{494}IDEFELGTTLIN	57824	+	+	+	+	+
BoNT/A5'45	MI^{45}PER..	46337	−	+	−	−	−
BoNT/A5'71	MS^{71}YYD..	43465	−	+	−	−	−
BoNT/A5'121	MS^{121}TID..	37698	−	+	−	−	−
BoNT/A5'173	ML^{173}NLT..	31963	−	−	−	−	−
BoNT/A3'175	..VLNL^{175}A	19888	−	−	−	−	−
BoNT/A3'188	..YIRF^{188}RLIN	21859	−	+	+	−	−
BoNT/A3'207	..TNPL^{207}FA	23707	−	+	+	−	−
BoNT/A3'254	..YEMS^{254}PLN	28938	−	+	+	+	−
BoNT/A3'315	..LQYM^{315}A	35453	−	+	+	+	+
BoNT/A3'384	..KVNY^{384}PLN	44247	−	+	+	+	+
BoNT/A3'406	..GQNT^{406}RLIN	46887	+	+	+	+	+
BoNT/A3'420	..KNFT^{420}RLIN	48583	+	+	+	+	+
BoNT/A3'425	..LFEF^{425}RLIN	49177	+	+	+	+	+
BoNT/A3'447	..KGYN^{447}PLN	51487	+	+	+	+	+

[a] 3'-deletions were introduced by exonuclease III digestion; 5'-deletions were generated by the polymerase chain reaction. mRNA was injected into the identified presynaptic neuron.

[b] All clones were characterized by direct sequencing. Numbers specify the last and the first authentic amino acid residue of the individual L chains. Dots indicate that the protein continues with the authentic N-terminal or C-terminal sequences.

[c] Theoretical molecular weights including the start methionine as deduced from the DNA-sequence.

[d] Depression (+) or failure of depression (−) of evoked postsynaptic responses were determined in at least three separate experiments in which after mRNA was microinjected into the identified presynaptic cholinergic neuron of the buccal ganglia of Aplysia californica.

[e] The corresponding mRNA was translated in reticulocyte lysate in the presence of radiolabelled methionine and subjected to immunoprecipitation with monoclonal antibodies as indicated. All monoclonal antibodies recognized the L chain of BoNT/A on Western blots.

[f] The antibody E2 was originally raised against BoNT/E. In addition the antibody precipitates the L chains of TeTx, BoNT/A, BoNT/B, BoNT/C1 and BoNT/D.

[g] The antibody A109 has BoNT/A-neutralizing activity (Kozaki et al., 1989).

[h] The antibodies A113 and A115 have similar binding characteristics to those A109, however, they do not neutralize BoNT/A.

and for studying the structure–function relationship of the neurotoxins, as long as the epitopes of such antibodies have been mapped. The production of neutralizing monoclonal antibodies against TeTx and BoNTs has been reported by many groups, but only in a few instances have attempts been made to localize the binding sites precisely. This can be achieved easily with genetically engineered deletion mutants synthesized *in vitro*, provided that the epitopes consist of linear peptide sequences (Andersen-Beckh *et al.*, 1989). We have recently mapped the epitopes of four monoclonal antibodies, designated E2, A109, A113 and A115, all of which recognized the L chain of BoNT/A on Western blots (Kozaki *et al.*, 1989). Monoclonal antibody E2 (Table 3), originally raised against BoNT/E, recognized in addition the L chains of TeTx, BoNT/A, BoNT/B, BoNT/C1, and BoNT/D, but had no toxin-neutralizing activity (Kurazono *et al.*, 1991b). An antibody with similar properties has been reported previously (Tsuzuki *et al.*, 1988). Of the three remaining antibodies, only A109 exhibited BoNT/A-neutralizing activity in mice (Kozaki *et al.*, 1989) and in *Aplysia* (Kurazono *et al.*, 1991b), although the three antibodies exhibited similar binding characteristics.

The results summarized in Table 3 indicate that the E2-antibody binds to a linear epitope between residues 120 and 188 of the BoNT/A L chain, a sequence which is not significantly conserved in other serotypes (Fig. 4) with the exception of a rather short hydrophobic stretch ($N^{150}LVIIGP^{156}$). It will be of interest to see whether a synthetic peptide harbouring this sequence will bind to E2. At any rate, the data suggest that this common epitope is located within the above peptide which in BoNT/A and in other serotypes should be exposed at the surface of the individual toxin molecules.

The other three antibodies, including the BoNT/A-neutralizing A109, clearly recognized tertiary epitopes, as evidenced by the finding that each of them recognized large peptides with different but well-defined C-terminal ends, while none of them precipitated L chain variants which lacked the N-terminal 44 amino acids (Table 3).

Let us reconsider these results in connection with previous data obtained through microinjection of mutant mRNAs into *Aplysia* neurons. Several questions come immediately into our mind: How is it possible that a neutralizing monoclonal antibody requires only the N-terminal 188 residues of BoNT/A for binding, while toxicity in *Aplysia* requires the presence of almost the entire L chains of BoNT/A or TeTx (Kurazono *et al.*, 1991a,b)? Furthermore, what is the molecular mechanism by which the single point mutant carrying a proline residue within a predicted α-helix becomes nontoxic? It is intriguing to speculate that the L chains actually contain two functionally important subdomains, domain I being formed by the 200 N-terminal residues, and domain II by residues 300 – 400. The subdomains could serve different functions, such as providing a substrate binding site and an active centre. Finally, in the Pro^{237} mutant an essential interaction of the two subdomains could be irreversibly blocked by a change in the tertiary structure. These are intriguing hypotheses and we will try our best to confirm them.

Mapping of T cell epitopes and characterization of T-cell population

As we have seen above, the precise mapping of B cell epitopes by the genetic approach may be hampered when epitopes are not linear but determined by the tertiary structure of the native protein. In this respect mapping of T cell epitopes appears to be easier: T lymphocytes recognize antigenic peptides, 10 – 15 amino acids in length, in context with the polymorphic class I and class II major histocompatibility complex (MHC) molecules. These are found on antigen-presenting cells like macrophages and B cells (Lanzavecchia, 1985). Such cells bind complex antigens, internalize and proteolytically degrade them in the cytosol and/or the endoplasmic reticulum, and finally present individual peptides bound to the MHC molecules at their cell surface. The crystal structure of the class I MHC molecules has recently been determined (Björkman *et al.*, 1987a,b; Garrett *et al.*, 1990). Binding of the antigenic peptides is mediated by two almost parallel α-helices and a β-sheet trapping the antigen in their middle. Due to the polymorphism of the α-helices, different MHC molecules bind different antigens. According to this model, the T cell receptor recognizes some sidechains of the α-helices in conjunction with some sidechains of the antigen. For several reasons TeTx has become an ideal antigen to study the cascade of events from antigen binding to presentation of specific T cell epitopes: (1) Virtually every human being in the Western world is immunized against TeTx and thus carries TeTx-specific B and T cells in his blood. (2) The TeTx molecule, 1314 amino acids in length, is large enough to allow representative processing studies and, because its primary sequence is known, mapping of T cell epitopes by bacterially expressed subfragments or synthetic peptides is facilitated (Demotz et al., 1989a,b,c). (3) The defined T cell epitopes can be grouped into different classes on the basis of their binding characteristics to MHC molecules, for instance by measuring the persistence on the antigen-presenting cells. These epitope groups can then be further analysed for the presence of common sidechains with the final goal to eventually predict the quality of putative T cell epitopes from a given primary sequence (Corradin and Lanzavecchia, 1990).

Recent studies (Demotz et al., 1989a,b,c; Panina-Bordignon et al., 1989; Corradin et al., 1990) have indicated that TeTx-specific T cell epitopes can indeed be distinguished on the basis of their interaction with different class II allelic MHC molecules. Furthermore, frequency analyses indicated that particular T cell epitopes of TeTx (for instance Q^{830}YIKANSKFIGITE^{843} and F^{947}NNFTVSFWLRVPKVSASHLE^{967}) were recognized by peripheral blood lymphocytes of all individuals tested (Corradin et al., 1990). This finding raises the question, as to whether fusion of such common T cell epitopes to foreign antigens of viral or bacterial origin could stimulate or improve the T cell response in an immunized host. The genetic construction of such mixed-type T cell epitopes can easily be achieved by means of synthetic oligonucleotides. Tetanus toxoid has been used in an adjuvans-like manner for many years and immunizations with conjugates composed of toxoid and weak antigens such as bacterial capsular polysaccharides have successfully been applied as vaccines (Lagergard et al., 1990; Paoletti et al., 1990). Lagergard et al. (1990) reported recently that such conjugates indeed stimulated the T cell response to Streptococcus type III capsular polysaccharides.

Development of recombinant vaccines and therapeutic tools

The development of a new generation of vaccines would have obvious advantages, because antigen production would be absolutely safe and because new routes of immunization could be explored. At this moment, however, it would have the disadvantage that it would be much more expensive than the formaldehyde-treated toxin currently administered. The generation of a recombinant TeTx vaccine based on bacterially expressed fragment H_C was pursued by the group of Neil Fairweather. In a series of studies, this group has gone all the way from cloning, sequencing and expression of fragment H_C in Escherichia coli to application of recombinant peptides as second generation vaccines (Fairweather et al., 1986, 1987; Fairweather and Lyness, 1986; Makoff et al., 1989). Recently, an attenuated Salmonella typhimurium aroA mutant was used as a live carrier to orally immunize mice against tetanus (Fairweather et al., 1990). The H_C fragment was expressed under control of the tac promoter. However, no modifications of the protein were made which would lead to expression of the recombinant peptide at the bacterial surface. Nevertheless, eight out of ten mice which had received a single oral dose of 2 x 10^9 colony-forming units survived the challenge with 500 50% lethal doses of TeTx and oral immunization with two doses gave complete protection (Fairweather et al., 1990). It remains to be shown whether the initial doses for immunization can be reduced when the H_C fragment is presented as a bacterial surface antigen. In addition, it is not known what Ig classes are induced in the immunized animal.

The application of fragment H_C as a neuroselective marker and transporter has been proposed in the past repeatedly by several groups (Bizzini et al., 1980a,b; Simpson, 1981). Halpern et al. (1990) have demonstrated that, indeed, recombinant fragment H_C, as expressed in E. coli, retains its original binding specificity for GT_{1b} ganglioside and neuronal tissue. However, as of yet no chimaeric proteins have been generated by gene fusion. The aforementioned experiments with reconstituted toxins containing L and H chains from different serotypes (Maisey et al., 1988; Poulain et al., 1991), as well as results obtained by Johnstone et al.

(1990) with derivatives containing gelonin and the H chain of TeTx, suggest that such chimaeric proteins could be developed as useful tools. We can think of a large variety of such fusion proteins in which marker enzymes, peptide hormones, growth factors, or even the antigen-binding domain of the IgG molecule take the position of the L chain. Indeed, the development of human monoclonal antibodies which neutralize TeTx (Kamei et al., 1990; Simpson et al., 1990) and the relative ease with which a desired antigen-binding fragment can be amplified from a human cDNA library (Mullinax et al., 1990) make us hope that such chimaeric antibodies against neurotoxins will be available in the near future. Furthermore, some neurotropic viruses, such as Borna disease virus, are capable of persisting in neuronal tissue for a long period of time, being inaccessible to the host's immune defence system. It is an intriguing task to redesign parts of the TeTx or BoNT molecules and apply them as tools which could overcome the limitations of traditional antibody therapy.

Concluding remarks

During the last couple of years, research on clostridial neurotoxins has received some new stimulation through the elucidation of the amino acid sequences. In this chapter the L chain sequences of five neurotoxins are presented. However, these sequences failed to provide any direct clues regarding the mechanism(s) by which the neurotoxins block transmitter release. Unfortunately, the somewhat frustrating conclusion that the neurotoxins are related to each other but that they differ from any other known protein, will not help much at all to end the speculations regarding their intracellular targets and mode(s) of action. Even today, research in this field appears to be limited to a rather descriptive approach in which individual steps of regulated secretion in the neurotoxin-treated cell are grouped as 'still normal' or 'inhibited'. However, the joint biochemical, molecular biological, and electrophysiological efforts are beginning to narrow down the number of possibly involved reactions, as more and more details of the secretory pathway become unravelled. A safe way back from the neurotoxins genes to biologically active peptides has been established,

the first genetically engineered L chain mutants have been constructed and characterized by patch-clamp technology. It seems to be merely a matter of time until the three functional subdomains involved in binding, translocation and toxification are characterized in greater detail. At that stage, a turning point has been reached, where we can genetically redesign these powerful neurotoxins into their opposite: beneficial tools to fight disorders of the central nervous systems.

Acknowledgements

I thank all colleagues who have made results of their studies available to me prior to publication and who supported me with reprints and discussion. I am especially grateful to Sumiko Mochida, Bernard Poulain, and Ladislav Tauc (Gif-sur-Yvette); Michel Popoff (Institut Pasteur, Paris); Karel Wernars (Bilthoven); Shunji Kozaki and Genji Sakaguchi (Osaka Prefecture University, Osaka); Mel Eklund (Northwest Fisheries Center, Seattle); Antonio Lanzavecchia (Basel Institute of Immunology, Basel), and Giampiedro Corradin (Institut de Biochemie, Lausanne). This work was supported by grant Nie 175/5-2 from Deutsche Forschungsgemeinshaft.

References

Ackland, T.H. (1959). Tetanus prophylaxis. Med. J. Australia Vol.I-46th year, 185–8.

Adam-Vizi, V., Rosener, S., Aktories, K. and Knight, D.E. (1988). Botulinum toxin-induced ADP-ribosylation and inhibition of exocytosis are unrelated events. FEBS Lett. 238, 277–80.

Agui, T., Syuto, B., Oguma, K., Iida, H. and Kubo, S. (1983). Binding of clostridium botulinum type C neurotoxin to rat brain synaptosomes. J. Biochem. 94, 521–7.

Agui, T., Syuto, B., Oguma, K., Iida, H. and Kubo, S. (1985). The structural relation between the antigenic determinants to monoclonal antibodies and binding sites to rat brain synaptosomes and GT_{1b} ganglioside in Clostridium botulinum type C neurotoxin. J. Biochem. 97, 213–8.

Aguilera, J. and Yavin, E. (1990). In vivo translocation and down-regulation of protein kinase C following intraventricular administration of tetanus toxin. J. Neurochem. 54, 339–42.

Aguilera, J., Lopez, L.A. and Yavin, E. (1990). Tetanus

toxin-induced protein kinase C activation and elevated serotonin levels in the perinatal rat brain. *FEBS Lett.* **263**, 61–5.

Ahnert-Hilger, G., Bräutigam, M. and Gratzl, M. (1987). Ca²⁺-stimulated catecholamine release from a-toxin-permeabilized PC²⁺ cells: Biochemical evidence for exocytosis and its modulation by protein kinase C and G proteins. *Biochemistry* **26**, 7842–8.

Ahnert-Hilger, G., Bader, M.-F., Bhakdi, S. and Gratzl, M. (1989a). Introduction of macromolecules into bovine adrenal medullary chromaffin cells and rat pheochromocytoma cells (PC12) by permeabilization with streptolysin O: Inhibitory effect of tetanus toxin on catecholamine secretion. *J. Neurochem.* **52**, 1751–8.

Ahnert-Hilger, G., Weller, U., Dauzenroth, M.-E., Habermann, E. and Gratzl, M. (1989b). The tetanus toxin light chain inhibits exocytosis. *FEBS Lett.* **242**, 245–8.

Aktories, K., Rösener, S., Blaschke, U. and Chhatwal, G.S. (1988). Botulinum ADP-ribosyltransferase C3. Purification of the enzyme and characterization of the ADP-ribosylation reaction in platelet membranes. *Eur. J. Biochem.* **172**, 445–50.

Almers, W. (1990). Exocytosis. *Annu. Rev. Physiol.* **52**, 607–24.

Altwegg, M. and Hatheway, C. (1988). Multilocus enzyme electrophoresis of *Clostridium argentinense* (*Clostridium botulinum* toxin type G) and phenotypically similar asaccharolytic Clostridia. *J. Clin. Microbiol.* **26**, 2447–9.

Andersen-Beckh, B., Binz, T., Kurazono, H., Mayer, T., Eisel, U. and Niemann, H. (1989). Expression of tetanus toxin subfragments in vitro and characterization of epitopes. *Infect. Immunity* **57**, 3498–505.

Anderson, M.J. and Cohen, M.W. (1974). Fluorescent staining of acetylcholine receptors in vertebrate skeletal muscle. *J. Physiol. (London)* **237**, 385–400.

Bähler, M., Benfenati, F., Valtorta, F. and Greengard, P. (1990). The synapsins and the regulation of synaptic function. *BioEssays* **12**, 259–263.

Band, J.D. and Bennett, J.V. (1983). Tetanus. In: *Infectious Diseases* (P.D. Hoeprich, ed.), 3rd Ed. pp. 1107–1114. Harper and Row, Philadelphia.

Barnekow, A., Jahn, R. and Schartl, M. (1990). Synaptophysin: a substrate for the protein tyrosine kinase pp60ᶜ⁻ˢʳᶜ in intact synaptic vesicles. *Oncogene* **5**, 1019–1024.

Baumert, M., Takei, K., Hartinger, J., Burger, P. M., Fischer von Mollard, G., Maycox, P. R., DeCamilli, P. and Jahn, R. (1990). P29: a novel tyrosine-phosphorylated membrane protein present in small clear vesicles of neurons and endocrine cells. *J. Cell Biol.* **110**, 1285–1294.

Behring, E. and Kitasato, S. (1890). Über das Zustandekommen der Diphtherie-Immunität und der Tetanus-Immunität bei Thieren. *Dtsch. Med. Wochenschr.* **16**, 1113–1114.

Benfenati, F., Bähler, M., Jahn, R. and Greengard, P. (1989). Interaction of synapsin with small synaptic vesicles: distinct sites in synapsin I bind to vesicle phospholipids and vesicle proteins. *J. Cell Biol.* **108**, 1863–1872.

Bergey, G. K., Habig, W. H. and Lin, C. (1986). Nicking of tetanus toxin increases activity. *Ann. Neurol.* **20**, 137–138.

Bernstein, B. W. and Bamburg, J. R. (1989). Cycling of actin assembly in synaptosomes and neurotransmitter release. *Neuron* **3**, 257–265.

Betz, H. (1990). Homology and analogy in transmembrane channel design: lessons from synaptic membrane proteins. *Biochemistry* **29**, 3591–3599.

Bhakdi, S. and Tranum-Jensen, J. (1987). Damage to mammalian cells by proteins that form transmembrane pores. *Rev. Physiol. Biochem. Pharmacol.* **107**, 147–223.

Bigalke, H., Dreyer, F. and Bergey, G. (1985). Botulinum A neurotoxin inhibits non-cholinergic synaptic transmission in mouse spinal cord neurons in culture. *Brain Res.* **360**, 318–324.

Binz, T., Kurazono, H., Wille, M., Frevert, J., Wernars, K. and Niemann, H. (1990a). The complete sequence of botulinum neurotoxin type A and comparison with other clostridial neurotoxins. *J. Biol. Chem.* **265**, 9153–9158.

Binz, T., Kurazono, H., Popoff, M., Eklund, M.W., Sakaguchi, G., Kozaki, S., Krieglstein, K., Henschen, A., Gill, D.M. and Niemann, H. (1990b). Nucleotide sequence of the gene encoding *Clostridium botulinum* neurotoxin type D. *Nucl. Acids. Res.* **18**, 5556–5556.

Binz, T., Eisel, U., Kurazono, H., Binschek, T., Bigalke, H. and Niemann, H. (1990c). Tetanus and botulinum A toxins: Comparison of sequences and microinjection of tetanus toxin A subunit-specific mRNA into bovine chromaffin cells. In: *Bacterial Protein Toxins* (R. Rappuoli *et al.*, eds). *Zbl. Bakt. Suppl. 19* pp. 105–108. Gustav Fischer Verlag, Stuttgart.

Bittner, M.A. and Holz, R.W. (1988). Effects of tetanus toxin on catecholamine release from intact and digitonin-permeabilized chromaffin cells. *J. Neurochem.* **51**, 451–456.

Bittner, M.A., Holz, R.W. and Neubig, R.R. (1986). Guanine nucleotide effects on catecholamine secretion from digitonin-permeabilized adrenal chromaffin cells. *J. Biol. Chem.* **261**, 10182–10188.

Bittner, M.A., Habig, W.H. and Holz, R.W. (1989a). Isolated light chain of tetanus toxin inhibits exocytosis: studies in digitonin-permeabilized cells. *J. Neurochem.* **53**, 966–968.

Bittner, M.A., DasGupta, B.R. and Holz, R.W. (1989b). Isolated light chains of botulinum neurotoxins inhibit exocytosis. *J. Biol. Chem.* **264**, 10354–10360.

Bizzini, B. and Raynaud, M. (1974). Etude immunologique et biologiqudee sous-unités de la toxin tétanique. *C.R. Acad. Sci. (Paris)* **279**, 1809–1812.

Bizzini, B., Stoeckel, K. and Schwab, M. (1977). An antigenic polypeptide fragment isolated from tetanus toxin: Chemical characterization, binding to gangliosides and retrograde axonal transport in various neuron systems. *J. Neurochem.* **28**, 529–542.

Bizzini, B., Grob, P., Glicksman, M.A. and Akert, K. (1980a). Use of B-11$_b$ tetanus toxin derived fragment as a specific pharmacological transport agent. *Brain Res.* **193**, 221–227.

Bizzini, B., Grob, P., Glicksman, M.A. and Akert, K. (1980b). Preparations of conjugates using two tetanus toxin-derived fragments: Their binding to gangliosides and isolated synaptic membranes and their immunological properties. *Toxicon* **18**, 561–572.

Bizzini, B., Grob, P. and Akert, K. (1981). Papain derived fragment II$_c$ of tetanus toxin: Its binding to isolated synaptic membranes and retrograde axonal transport. *Brain Res.* **210**, 291–299.

Bjorkman, P.J., Saper, M.A., Samraoui, B., Bennett, W.S., Strominger, J.L. and Wiley, D.C. (1987a). Structure of the human histocompatibility antigen, HLA-A2. *Nature* **329**, 506–512.

Bjorkman, P.J., Saper, M.A., Samraoui, B., Bennett, W.S., Strominger, J.L. and Wiley, D.C. (1987b). The foreign antigen binding site and T cell recognition regions of class I histocompatibility antigens. *Nature* **329**, 512–518.

Black, J.D. and Dolly, J.O. (1986a). Interaction of ^{125}I-labeled botulinum neurotoxins with nerve terminals. I. Ultrastructural autoradiographic localization and quantitation of distinct membrane acceptors for types A and B on motor nerves. *J. Cell Biol.* **103**, 521–534.

Black, J.D. and Dolly, J.O. (1986b). Interaction of ^{125}I-labeled botulinum neurotoxins with nerve terminals. II. Autoradiographic evidence for its uptake into motor nerves by acceptor-mediated endocytosis. *J. Cell Biol.* **103**, 535–544.

Black, J.D. and Dolly, J.O. (1987). Selective location of acceptors for botulinum neurotoxin A in the central and peripheral nervous systems. *Neuroscience* **23**, 767–779.

Blasi, E., Pitzurra, L., Burhan Fuad, A.M., Marconi, P. and Bistoni, F. (1990). Gamma interferon-induced specific binding of tetanus toxin on the GG2EE macrophage cell line. *Scand. J. Immunol.* **32**, 289–292.

Bleck, T.P. (1989). Clinical Aspects of Tetanus In: Botulinum Neurotoxin and Tetanus Toxin (L.L. Simpson, Ed.) pp. 379–398. Academic Press, New York.

Boquet, P. and Duflot, E. (1982). Tetanus toxin fragment forms channels in lipid vesicles at low pH. *Proc. Natl. Acad. Sci. USA* **79**, 7614–7618.

Boquet, P., Duflot, E. and Hauttecoeur, B. (1984). Low pH induces a hydrophobic domain in tetanus toxin molecule. *Eur. J. Biochem.* **144**, 339–44.

Brosius, P. and Holy, A. (1984). Regulation of the ribosomal RNA-promoters with a synthetic lac operon. *Proc. Natl. Acad. Sci. USA*, **81**, 6929–6933.

Büttner-Ennever, J.A., Grob, P., Akert, K. and Bizzini, B. (1981a). A transsynaptic autoradiographic study of the pathways controlling the extraocular eye muscles, using (^{125}I)B-II$_b$ fragment. *Ann. N.Y. Acad. Sci.* **374**, 157–170.

Büttner-Ennever, J.A., Grob, P., Akert, K. and Bizzini, B. (1981b). Transsynaptic retrograde labeling in the oculomotor system of the monkey with (^{125}I) tetanus toxin B-11$_b$ fragment. *Neurosci. Lett.* **26**, 233–238.

Burgoyne, R.D. (1990). Secretory vesicle-associated proteins and their role in exocytosis. *Annu. Rev. Physiol.* **52**, 647–659.

Carle, A. and Rattone, C. (1884). Studio sperimentale sull'etiologia del tetano. *G. Accad. med. Torino* **32**, 174–180.

Centers for Disease Control (1984). Morbidity and Mortality Weekly Reports, Summary 1983. **35**, 490.

Centers for Disease Control (1985). Tetanus - United States, 1982–1984. Morbidity and Mortality Weekly Reports **34**, 601–611.

Colebatch, J.G., Wolff, A.H., Gilbert, R.J., Mathias, C.J., Smith, S.E., Hirsch, N., Wiles, C.M. (1989). Slow recovery from severe foodborne botulism. *Lancet* **1989**, 1216–7.

Collingridge, G.L., Collins, G.G.S., Davies, J., James, T.A.S., Neal, M.J. and Tongroach, P. (1980). Effect of tetanus toxin from substantia nigra and striatum in vitro. *J. Neurochem.* **34**, 540–7.

Considine, R.V., Bielicki, J.K., Simpson, L.L. and Sherwin, J.R. (1990). Tetanus toxin attenuates the ability of phorbol myristate acetate to mobilize cytosolic protein kinase C in NG-108 cells. *Toxicon,* **28**, 13–9.

Cooper, M.S., Cornell-Bell, A.H., Chernjavsky, A., Dani, J.W. and Smith, S.J. (1990). Tubovesicular processes emerge from trans-Golgi cisternae, extend along microtubules, and interlink adjacent trans-Golgi elements into a reticulum. *Cell* **61**, 135–45.

Corradin, G. and Lanzavecchia, G. (1990). Chemical and functional analysis of MHC class II-restricted T cell epitopes. *Intern. Rev. Immun.* **7**, 139–47.

Corradin, G., Cordey, A.S., Niemann, H., Matricardi, P., Panina, P., Lanzavecchia, A. and Demotz, S. (1990). Degradation and presentation of tetanus toxin to specific human T cell clones. In: *Bacterial Protein Toxins* (eds R. Rappuoli et al.). *Zbl. Bakt. Suppl.* **19** pp. 431–6. Gustav Fischer Verlag, Stuttgart.

Critchley, D.R., Nelson, P.G., Habig, W.H. and Fishman, P.H. (1985). Fate of tetanus toxin bound to the surface of primary neurons in culture: evidence for rapid internalization. *J. Cell Biol.* **100**, 1499–507.

Critchley, D.R., Habig, W.H. and Fishman, P.H. (1986). Reevaluation of the role of gangliosides as receptors for tetanus toxin. *J. Neurochem.* **47**, 213–22.

Critchley, D.R., Parton, R.G., Davidson, M.D. and Pierce, E.J. (1988). Characterization of tetanus toxin binding by neuronal tissue. In: *Neurotoxins in Neurochemistry* (ed. J.O. Dolly). pp. 109–122. Ellis Harward Series in Biotechnology. J. Wiley and Sons.

Cvjetanovic, B. (1981). Public health aspects in tetanus control. In: *Tetanus, Important New Concepts* (ed. R. Veronesi), pp 1–7. Excerpta Medica, Amsterdam.

DasGupta, B.R. and Rasmussen, S. (1984). Effect of diethylpyricarbonate on the biological activities of botulinum neurotoxin types A and E. *Arch. Biochem. Biophys.* **232**, 172–8.

DasGupta, B.R. and Datta, A. (1988). Botulinum neurotoxin type B (strain 657): partial sequence and similarity with tetanus toxin. *Biochimie* **70**, 811–7.

Davies, J.R. and Wright, E.A. (1955). The specific precocious protective action of tetanus toxoid. *Br. J. Exp. Path.* **36**, 487–93.

Dayanithi, G., Ahnert-Hilger, G., Weller, U., Nordmann, J.J. and Gratzl, M. (1990). Release of vasopressin from isolated permeabilized neurosecretory nerve terminals is blocked by the light chain of botulinum A toxin. *Neuroscience* **39**, 711–5.

De Camilli, P. and Jahn, R. (1990). Pathways to regulated exocytosis in neurons. *Ann. Rev. Physiol.* **52**, 625–45.

Dekleva, M.L. and DasGupta, B.R. (1990). Purification and characterization of a protease from *Clostridium botulinum* type A that nicks single-chain type A botulinum neurotoxin into the di-chain form. *J. Bacterol.* **172**, 2498–503.

Demotz, S., Lanzavecchia, A., Eisel, U., Niemann, H., Widman, C. and Corradin, G. (1989a). Delineation of several DR-restricted tetanus toxin T cell epitopes. *J. Immunol.* **142**, 394–402.

Demotz, S., Matricardi, P.M., Irle, C., Panina, P., Lanzavecchia, A. and Corradin, G. (1989b). Processing of tetanus toxin by human antigen-presenting cells. Evidence for donor and epitope-specific processing pathways. *J. Immunol.* **143**, 3881–6.

Demotz, S., Matricardi, P., Lanzavecchia, A. and Corradin, G. (1989c). A novel and simple procedure for determining T cell epitopes in protein antigens. *J. Immun. Meth.* **122**, 67–72.

Dolly, J.O., Black, J., Williams, R.S. and Melling, J. (1984). Acceptors for botulinum neurotoxin reside on motor nerve terminals and mediate its internalization. *Nature (London)* **307**, 457–60.

Dolly, J.O., Lande, S. and Wray, D.W. (1987). The effects of *in vitro* application of purified botulinum neurotoxin at mouse motor nerve terminals. *J. Physiol. (London)* **386**, 475–84.

Dolly, J.O., Ashton, A.C., McInnes, C., Wadsworth, J.D.F., Poulain, B., Tauc, L., Shone, C.C. and Melling, J. (1990). Clues to the multiphase inhibitory action of botulinum neurotoxins on release of transmitters. *J. Physiol. (Paris)* **84**, 237–46.

Dolman, C.E. and Murakami, L. (1961). *Clostridium botulinum* type F with recent observations on other types. *J. Infect. Dis.* **109**, 107–28.

Dolman, C.E. and Chang, E. (1972). Bacteriophages of *Clostridium botulinum*. *Can. J. Microbiol.* **18**, 67–76.

Donovan, J.J. and Middlebrook, J.L. (1986). Ion-conducting channels produced by botulinum toxin in planar lipid membranes. *Biochemistry* **25**, 2872–6.

Dreyer, F. (1989). Peripheral actions of tetanus toxin. In: *Botulinum Neurotoxin and Tetanus Toxin* (ed. L.L. Simpson) pp. 179–202, Academic Press, New York.

Dreyer, F., Mallart, A. and Brigant, J.L. (1983). Botulinum A toxin and tetanus toxin do not effect presynaptic membrane currents in mammalian nerve endings. *Brain Res.* **270**, 373–5.

Dreyer, F., Rosenberg, F., Becker, C., Bigalke, H. and Penner, R. (1987). Differential effects of various secretagogues on quantal transmitter release from mouse nerve terminals treated with botulinum A and tetanus toxin. *Naunyn Schmiedeberg's Arch. Pharmacol.* **335**, 1–7.

Edsall, G. (1976). The inexcusable disease *J. Am. Med. Assoc.* **235**, 62–3.

Egea, G., Rabasseda, X., Solsona, C., Marsal, J. and Bizzini, B. (1990). Tetanus toxin blocks potassium-induced transmitter release and rearrangement of intramembrane particles at pure cholinergic synaptosomes. *Toxicon* **28**, 311–8.

Eisel, U., Jarausch, W., Goretzki, K., Henschen, A., Engels, J., Weller, U., Hudel, M., Habermann, E. and Niemann, H. (1986). Tetanus toxin: primary structure, expression in *E. coli*, and homology with botulinum toxins. *EMBO J.* **5**, 2495–502.

Eklund, M.W., Poysky, F.T. and Boatman, E.S. (1969). Bacteriophages of *Clostridium botulinum* types A, B, E and F and nontoxigenic strains resembling type E. *J. Virol.* **3**, 270–4.

Eklund, M.W. and Poysky, F.T. (1974). Interconversion of type C and D strains of *Clostrium botulinum* by specific bacteriophages. *Appl. Microbiol.* **27**, 251–8.

Eklund, M.W., Poysky, F.T., Meyers, J.A. and Pelroy, J.A. (1974). Interspecies conversion of *Clostridium botulinum* type C to *Clostridium novyi* by bacteriophage. *Science* **186**, 456–8.

Eklund, M.W., Poysky, F.T., Mseitif, M.L. and Strom, M.S. (1988). Evidence for plasmid-mediated toxin and bacteriocin production in *Clostridium botulinum* type G. *Appl. Environ. Microbiol.* **54**, 1405–8.

Eklund, M.W., Poysky, F.T. and Habig, W.H. (1989). Bacteriophages and plasmids in *Clostridium botulinum* and *Clostridium tetani* and their relationship to production of toxins. In: *Botulinum Neurotoxin and Tetanus Toxin* (ed. L.L. Simpson) pp. 25–51, Academic Press, New York.

Evans, D.M., Williams, R.S., Shone, C.S., Hambleton, P., Melling, J. and Dolly, J.O. (1986). Botulinum neurotoxin type B. Its purification, radioiodination and interaction with rat-brain synaptosomal membranes. *Eur. J. Biochem.* **154**, 409–16.

Fairweather, N.F. and Lyness, V.A. (1986). The complete nucleotide sequence of tetanus toxin. *Nucleic Acids Res.* **14**, 7809–12.

Fairweather, N.F., Lyness, V.A., Pickard, D.J., Allen, G. and Thomson, R.O. (1986). Cloning, nucleotide sequencing and expression of tetanus toxin fragment C in Escherichia coli. *J. Bacteriol.* **165**, 21–7.

Fairweather, N.F., Lyness, V.A. and Maskell, D.J. (1987). Immunization of mice against tetanus with fragments of tetanus toxin synthesized in Escherichia coli. *Infect. Immun.* **55**, 2541–5.

Fairweather, N.F., Chatfield, S.N., Makoff, A.J., Strugnell, R.A., Bester, J., Maskell, D.J. and Dougan, G. (1990). Oral vaccination of mice against tetanus by use of a live attenuated salmonella carrier. *Infect. Immun.* **58**, 1323–6.

Fertuck, H.C. and Salpeter, M.M. (1974). Localization of acetylcholine receptor by ^{125}I-a-bungarotoxin binding at mouse motor end plates. *Proc. Natl. Acad. Sci. USA* **71**, 1376–8.

Finn, C.W. Jr., Silver, R.P., Habig, W.H., Hardegree, M.C., Zon, G. and Garon, C.F. (1984). The structural gene for tetanus neurotoxin is on a plasmid. *Science* **224**, 881–4.

Fischer von Mollard, G., Mignery, G.A., Baumert, M., Perin, M.S., Hanson, T.J., Burger, P.M., Jahn, R. and Südhof, T.C. (1990). rab3 is a small GTP-binding protein exclusively localized to synaptic vesicles. *Proc. Natl. Acad. Sci. USA* **87**, 1988–92.

Fischer von Mollard, G., Südhof, T.C. and Jahn, R. (1991). A small GTP-binding protein dissociates from synaptic vesicles during exocytosis. *Nature* **349**, 79–81.

Fujii, N., Oguma, K., Yokosawa, N., Kimura, K. and Tsuzuki, K. (1988). Characterization of bacteriophage nucleic acids obtained from *Clostridium botulinum* types C and D. *Appl. Environ. Microbiol.* **54**, 69–73.

Fujita, K., Guroff, G., Yavin, E., Gopin, G., Orenberg, R. and Lazarovici, P. (1990). Preparation of affinity-purified, biotinylated tetanus toxin and characterization and localization of cell surface binding sites on nerve growth factor-treated PC12 cells. *Neurochem. Res.* **15**, 373–83.

Gaillard, W.D. and Carpenter, D.O. (1986). On the transmitter at the A-to-B cell in *Aplysia californica*. *Brain Res.* **373**, 311–5.

Gansel, M., Penner, R. and Dreyer, F. (1987). Distinct sites of action of clostridial neurotoxins revealed by double-poisoning of mouse motor nerve terminals. *Pflügers Arch.* **409**, 533–9.

Garrett, T.P.J., Saper, M.A., Bjorkman, P.J., Strominger, J.L. and Wiley, D.C. (1990). Specificity pockets for the side chains of peptide antigens in HLA-Aw68. *Nature* **342**, 692–6.

Geyer, H., Himmelspach, K., Kwiatkowsky, B., Schlecht, S. and Stirm, S. (1983). Degradation of bacterial surface carbohydrates by virus-associated enzymes. *Pure Appl. Chem.* **55**, 637–53.

Giménez, D.F. (1984). *Clostridium botulinum* subtype B$_a$. *Zentralbl. Bakteriol. Hyg.* **A257**, 68–72.

Giménez, D. and Ciccarelli, A. (1970). Another type of *Clostridium botulinum*. *Zentralbl. Bakteriol.* **215**, 221–4.

Giménez, J., Foley, J. and DasGupta, B.R. (1988). Neurotoxin type E from *Clostridium botulinum* and *C. butyricum*; Partial sequence and comparison. *FASEB J.* (Abstract 8447) A1750.

Göpfert H. and Schäfer, H. (1941). Über die Mechanik des Wundstarrkrampfes. Naunyn Schmiedeberg's *Arch. Exp. Pathol. Pharmacol.* **197**, 93–122.

Goldberg, R.L., Costa, T., Habig, W.H., Kohn, L.D. and Hardegree, M.C. (1981). Characterization of fragment C and tetanus toxin binding to rat brain membranes. *Mol. Pharmacol.* **20**, 565–70.

Gomperts, B.D. (1990). G$_E$: a GTP-binding protein mediating exocytosis. *Annu. Rev. Physiol.* **52**, 591–606.

Griffith, G. and Simons, K. (1986). The trans-Golgi network: Sorting at the site of the Golgi complex. *Science* **234**, 438–43.

Guitart, X., Egea, G., Solsona, C. and Marsal, J. (1987). Botulinum neurotoxin inhibits depolarization-stimulated protein phosphorylation in pure cholinergic synaptosomes. *FEBS Lett.* **219**, 219–23.

Guitart, X., Marsal, J. and Solsona, C. (1990). Phorbol esters induce neurotransmitter release in cholinergic synaptosomes from torpedo electric organ. *J. Neurochem.* **55**, 468–72.

Gundersen, C.B., Katz, B. and Miledi, R. (1982). The antagonism between botulinum toxin and calcium in motor nerve terminals. *Proc. R. Soc. London (B)* **216**, 369–76.

Habermann, E. (1974). ^{125}I-labeled neurotoxin from *Clostridium botulinum* A: Preparation, binding to synaptosomes and ascent to the spinal cord. *Naunyn Schmiedeberg's Arch. Pharmacol.* **281**, 47–56.

Habermann, E. and Albus, U. (1986). Interaction between tetanus toxin and rabbit kidney: A comparison with rat brain preparations. *J. Neurochem.* **46**, 1219–26.

Habermann, E. and Dreyer, F. (1986). Clostridial neurotoxins: Handling and action at the cellular and molecular level. *Curr. Top. Microbiol. Immunol.* **129**, 93–176.

Habermann, E. and Tayot, J.L. (1985). Interaction of solid phase gangliosides with tetanus toxin and tetanus toxoid. *Toxicon* **23**, 913–20.

Hall, J.D., McCroskey, L.M., Pincomb, B.J. and Hatheway, C.L. (1985). Isolation of an organism resembling *Clostridium baratii* which produced type F botulinal toxin from an infant with botulism. *J. Clin. Microbiol.* **21**, 654–5.

Halpern, J.L., Habig, W.H., Neale, E.A. and Stibitz, S. (1990). Cloning and expression of functional fragment C of tetanus toxin. *Infect. Immun.* **58**, 1004–9.

Hancock, J.F., Magee, A.I., Childs, J.E. and Marshall, C.J. (1989). All ras proteins are polysioprenylated but only some are palmitoylated. *Cell* **57**, 1167–77.

Hara, T., Matsuda, M. and Yoneda, M. (1977). Isolation and some properties of nontoxigenic derivatives of a strain of Clostridium tetani. *Biken J.* **20**, 105–15.

Harris, A.J. and Miledi, R. (1971). The effect of type D botulinum toxin on frog neuromuscular junctions. *J. Physiol. (London)* **217**, 497–515.

Harvey, A.M. (1939). The peripheral action of tetanus toxin. *J. Physiol. (London)* **96**, 348–65.

Hatheway, C.L. (1989). Bacterial sources of clostridial neurotoxins. In: *Botulinum Neurotoxin and Tetanus Toxin* (ed. L.L. Simpson) pp. 3–24. Academic Press, New York.

Hatheway, C.L. and McCroskey, L.M. (1987). Examination of feces and serum for diagnosis of infant botulism: Experience in 336 patients. *J. Clin. Microbiol.* **25**, 2334–8.

Hauser, D., Eklund, M.W., Kurazono, H., Binz, T., Niemann, H., Gill, D.M., Boquet, P. and Popoff, M.R. (1990). Nucleotide sequence of *Clostridium botulinum* C1 neurotoxin. *Nucl. Acids Res.* **18**, 4924–4924.

Helting, T.B. and Zwisler, O. (1977). Structure of tetanus toxin. I. Breakdown of the toxin molecule and discrimination between polypeptide fragments. *J. Biol. Chem.* **252**, 187–93.

Helting, T.B., Zwisler, O. and Wiegandt, H. (1977). Structure of tetanus toxin. II. Toxin binding to Gangliosides. *J. Biol. Chem.* **253**, 194–8.

Hirano, A.A., Greengard, P., Huganir, R.L. (1988). Protein tyrosine kinase activity and its endogenous substrates in rat brain: a subcellular and regional survey. *J. Neurochem.* **50**, 1447–55.

Hirokawa, N., Sobue, K., Kanda, K., Harada, A. and Yorifuji, H. (1989). The cytoskeletal architecture of the presynaptic terminal and molecular structure of synapsin I. *J. Cell. Biol.* **108**, 111–26.

Ho, J.L. and Klempner, M.S. (1985). Tetanus toxin inhibits secretion of lysosomal contents from human macrophages. *J. Infect. Dis.* **152**, 922–9.

Ho, J.L. and Klempner, M.S. (1986). Inhibition of macrophage secretion by tetanus toxin is not directly linked to cytosolic calcium homeostasis. *Biochem. Biophys. Res. Commun.* **135**, 16–24.

Ho, J.L. and Klempner, M.S. (1988). Diminished activity of protein kinase C in tetanus toxin-treated macrophages and in the spinal cord of mice manifesting generalized tetanus intoxication. *J. Infect. Dis.* **157**, 925–33.

Hoch, D.H., Romero-Mira, M., Ehrlich, B.E., Finkelstein, A., DasGupta, B.R. and Simpson L.L. (1985). Channels formed by botulinum, tetanus, and diptheria toxins in planar lipid bilayers: relevance to translocation of proteins across membranes. *Proc. Natl. Acad. Sci. USA* **82**, 1692–1696.

Holmgren, J., Elwing, H., Fredman, P. and Svennerholm, L. (1980). Polystyrene-adsorbed gangliosides for investigation of the structure of the tetanus toxin receptor. *Eur. J. Biochem.* **106**, 371–9.

Hopkins, C.R., Gibson, A., Shipman, M. and Miller, K. (1990). Movement of internalized ligand-receptor complexes along a continuous endosomal reticulum. *Nature* **340**, 335–9.

Hoshijima, M., Kondo, J., Kikuchi, A., Yamamoto, K. and Takai, Y. (1990). Purification and characterization from bovine brain membranes of a GTP-binding protein with a M_r of 21,000, ADP-ribosylated by an ADP-ribosyltransferase contaminated in botulinum toxin type C1- Identification as the *rhoA* gene product. *Mol. Brain Res.* **7**, 9–16.

Howell, T.W. and Gomperts, B.D. (1987). Rat mast cells permeabilized with streptolysin O secrete histamine in response to Ca^{2+} at concentration buffered in the micromolar range. *Biochim. Biophys. Acta* **927**, 177–83.

Hubbard, A.L. (1989). Endocytosis. *Curr. Opinions Cell Biol.* **1**, 675–83.

Inoue, K. and Iida, H. (1968). Bacteriophages of *Clostridium botulinum*. *J. Virol.* **2**, 537–40.

Inoue, K. and Iida, H. (1970). Conversion to toxigenicity in *Clostridium botulinum* type C. *Jpn. J. Microbiol.* **14**, 87–9.

Jahn, R., Schiebler, W., Ouimet, C. and Greengard, P. (1985). A 38,000-dalton membrane protein (p38) present in synaptic vesicles. *Proc. Natl. Acad. Sci. USA* **82**, 4137–41.

Johnston, P.A. and Südhof, T.C. (1990). The multisubunit structure of synaptophysin. *J. Biol. Chem.* **265**, 8869–73.

Johnstone, S.R., Morrice, L.M. and Heyningen van, S. (1990). The heavy chain of tetanus toxin can mediate the entry of cytotoxic gelonin into intact cells. *FEBS Lett.* **265**, 101–3.

Joseph, K.C., Stieber, A. and Gonatas, N.K. (1979). Endocytosis of cholera toxin in GERL-like structures of murine neuroblastoma cells treated with GM_1 gangliosides. *J. Cell Biol.* **81**, 543–54.

Kamata, Y., Kozaki, S., Sakaguchi, G., Iwamori, M. and Nagai, Y. (1986). Evidence for direct binding of Clostridium botulinum type E derivative toxin and its fragments to gangliosides and free fatty acids. *Biochem. Biophys. Res. Commun.* **140**, 1015–9.

Kamata, Y., Kozaki, S. and Sakaguchi, G. (1988). Effects of pH on the binding of *Clostridium botulinum* type E derivative toxin to gangliosides and phospholipids. *FEMS Microbiol. Lett.* **55**, 71–6.

Kamei, M., Hashizume, S., Sugimoto, N. and Matsuda, M. (1990). Stable hybridomas producing anti-tetanus antibodies of IgG with high neutralizing activity. Abstracts of the IUMS Congress: bacteriology and Mycology, PS059–01; Osaka, 16th - 22.9.1990.

Kitamura, M., Iwamori, M. and Nagai, Y. (1980). Interaction between *Clostridium botulinum* neurotoxin and gangliosides. *Biochim. Biophys. Acta* **628**, 328–35.

Keen, J.H. (1990). Clathrin and associated assembly and disassembly proteins. *Annu. Rev. Biochem.* **59**, 415–38.

Kitasato, S. (1889). Über den Tetanusbazillus. *Z. Hyg. Infektionskr.* **7**, 225–30.

Knaus, P., Marquèze-Pouey, B., Scherer, H. and Betz, H. (1990). Synaptoporin, a novel putative channel protein of synaptic vesicles. *Neuron* **5**, 453–62.

Knight, D.E. and Baker, P.F. (1985). Guanine nucleotides and Ca^{2+}-dependent exocytosis. Studies on two adrenal cell preparations. *FEBS Lett.* **189**, 345–9.

Knight, D.E., Tonge, D.A. and Baker, P.F. (1985). Inhibition of exocytosis in bovine adrenal medullary cells by botulinum toxin type D. *Nature* **317**, 719–21.

Knight, D.E. (1986). Botulinum toxin types A, B and D inhibit catecholamine secretion from bovine adrenal medullary cells. *FEBS Lett.* **207**, 222–6.

Koenig, J.H. and Ikeda, K. (1989). Disappearance and reformation of synaptic vesicle membrane upon transmitter release observed under reversible blockage of membrane retrieval. *J. Neuroscience* **9**, 3844–60.

Kozaki, S. and Sakaguchi, G. (1982). Binding to mouse brain synaptosomes of *Clostridium botulinum* type E derivative toxin before and after tryptic activation. *Toxicon* **20**, 841–6.

Kozaki, S., Miyazaki, S. and Sakaguchi, G. (1978). Structure of *Clostridium botulinum* type B derivative toxin: inhibition with a fragment of toxin from binding to synaptosome fraction. *Jpn. J. Med. Sci. Biol.* **31**, 163–6.

Kozaki, S., Oga, Y., Kamata, Y. and Sakaguchi, G. (1985). Activation of *Clostridium botulinum* type B and E derivative toxins with lysine-specific proteases. *FEMS-Microbiol. Lett.* **27**, 149–54.

Kozaki, S., Kamata, Y., Nagai, Y., Ogasawara, J. and Sakaguchi, G. (1986). The use of monoclonal antibodies to analyze the structure of *Clostridium botulinum* type E derivative toxin. *Infect. Immun.* **52**, 786–91.

Kozaki, S., Miki, A., Kamata, Y., Ogasawara, J., Sakaguchi, G. (1989). Immunological characterization of papain-induced fragments of clostridium botulinum type A neurotoxin and interaction of the fragments with brain synaptosomes. *Infect. Imm.* **57**, 2634–9.

Krieglstein, K., Henschen, A., Weller, U. and Habermann, E. (1990). Arrangement of disulfide bridges and positions of sulphydryl groups in tetanus toxin. *Eur. J. Biochem.* **188**, 39–45.

Krohne, G., Waizenegger, I. and Höger, T.H. (1989). The conserved carboxy-terminal cystein of nuclear lamins in essential for lamin association with the nuclear envelope. *J. Cell. Biol.* **109**, 2003–11.

Kurazono, H., Mochida, S., Binz, T., Eisel, U., Poulain, B., Tauc, L. and Niemann, H. (1991a). A novel approach to test the biological properties of the light chains of tetanus toxin and botulinum neurotoxin type A: Analyses of carboxyl-terminal deletion mutants. *J. Biol. Chem.* submitted.

Kurazono, H., Mochida, S., Kozaki, S., Binz, T., Grebenstein, O., Poulain, B., Tauc, L. and Niemann, H. (1991b). Two subdomains within the L-chains of tetanus and botulinum A toxin? Lessons from monoclonal antibodies and electrophysiological properties of various deletion mutants. *J. Biol. Chem.* submitted.

Kurokawa, Y., Oguma, K., Yokosawa, N., Skuto, B., Fukatsu, R. and Yamashita, I. (1987). Binding and cytotoxic effects of clostridium botulinum type A, C₁ and E toxins in primary neuron cultures from foetal mouse brains. *J. Gen. Microbiol.* **133**, 2647–57.

Lagergard, T., Shiloach, J., Robbins, J.B. and Schneerson, R. (1990). Synthesis and immunological properties of conjugates composed of group B streptococcus type III capsular polysaccharide covalently bound to tetanus toxoid. *Infect. Immun.* **58**, 687–94.

Laird, W.J. (1982). Botulinum toxin-like effects of tetanus toxin. *Proc. VI Int. Conf. Tetanus. Dec. 3-5*, 33–34.

Laird, W.J., Aaronson, W., Silver, R.P., Habig, W.H., Hardegree, M.C. (1980). Plasmid-associated toxigenicity in Clostridium tetani. *J. Infect. Dis.* **142**, 623.

Laird, W.J., Aaronson, W., Habig, W.H., Hardegree, M.C., Silver, R.P. (1981). Proceedings of the VIth international conference on tetanus toxin. Foundation Marcel Mérieux, Lyon, France. 9–20.

Lanzavecchia, A. (1985). Antigen-specific interaction between T and B cells. *Nature* **314**, 537–9.

Lazarovici, P. and Yavin, E. (1986). Affinity purified tetanus neurotoxin interaction with synaptic membranes: Properties of a protease-sensitive receptor component. *Biochemistry* **25**, 7047–54.

Lazarovici, P., Yanai, P. and Yavin, E. (1987). Molecular interactions between micelar polysialogangliosides and affinity-purified tetanotoxins in aqueous solution. *J. Biol. Chem.* **262**, 2645–51.

Lazarovici, P., Yavin, E., Bizzini, B. and Fedinec, A. (1989a). Retrograde transport in sciatic nerves of ganglioside-affinity-purified tetanus toxins. In: *Neurotoxins as Tools in Neurochemistry* (ed. J.O. Dolly). Ellis Horwood Limited, Chichester.

Lazarovici, P., Fujita, K., Contreras, M.L., DiOrio, J.P. and Lelkes, P.I. (1989b). Affinity purified tetanus binds to isolated chromaffin granules and inhibits catecholamine release in digitonin-permeabilized chromaffin cells. *FEBS Lett.* **253**, 121–8.

Ledley, F.D., Lee, G., Kohn, L.D., Habig, W.H. and Hardegree, M.C. (1977). Tetanus toxin interactions with thyroid plasma membranes. Implications for structure and function of tetanus toxin receptors and potential pathophysiological significance. *J. Biol. Chem.* **252**, 4049–55.

Llinás, S., McGuiness, T.L., Leonard, C.S., Sugimori, M. and Greengard, P. (1985). Intra-terminal injection of synapsin I or calcium/calmodulin-dependent protein kinase II alters neurotransmitter release at the squid giant synapse. *Proc. Natl. Acad. Sci. USA* **82**, 3035–9.

London, E. and Luongo, C.L. (1989). Domain specific bias in arginine/lysine usage by protein toxins. *Biochem. Biophys. Res. Commun.* **160**, 333–9.

MacDonald, K.L., Rutherford, G.W., Friedman, S.M., Dietz, J.R., Kaye, B.R., McKinley, G.F., Tenney, J.F. and Cohen, M.L. (1985). Botulism and botulism-like illness in chronic drug abusers. *Ann. Intern. Med.* **102**, 616–8.

MacDonald, K.L., Cohen, M.L. and Blake, P.A. (1986). The changing epidemiology of adult botulism in the United States. *Am. J. Epidemiol.* **124**, 794–9.

MacKenzie, I., Burnstock, G. and Dolly, J.O. (1982). The effects of purified botulinum neurotoxin type A on cholinergic, adrenergic and non-adrenergic, atropine-resistant autonomic neuromuscular transmission. *Neuroscience* **7**, 997–1006.

Maisey, E.A., Wadsworth, J.D.F., Poulain, B., Shone, C.C., Melling, J., Gibbs, P., Tauc, L. and Dolly, J.O. (1988). Involvement of the constituent chains of botulinum neurotoxins A and B in the blockade of neurotransmitter release. *Eur. J. Biochem.* **177**, 683–91.

Makoff, A.J., Ballantine, S.P., Smallwood, A.E. and Fairweather, N.F. (1989). Expression of tetanus toxin fragment C in E. coli: Its purification and potential use as a vaccine. *Biotechn.* **7**, 1043–6.

Manning, K.A., Erichsen, J.T. and Evinger, C. (1990). Retrograde transneuronal transport of fragment C of tetanus toxin. *Neuroscience* **34**, 251–63.

Marie, A. (1898). Recherches sur les propriétés antitetaniques des centres nerveux de l'animal sain. *Ann. Inst. Pasteur.* 91–95.

Marmur, J., Falkow, S. and Mandel, M. (1963). New approaches to bacterial taxonomy. *Ann. Rev. Microbiol.* **17**, 329–72.

Marsal, J., Egea, G., Solsona, C., Rabasseda, X. and Blasi, J. (1989). Botulinum toxin type A blocks the morphological changes induced by chemical stimulation on the presynaptic membrane of torpedo synaptosomes. *Proc. Natl. Acad. Sci. USA* **86**, 372–6.

Marxen, P. and Bigalke, H. (1991). Tetanus and botulinum A toxins inhibit stimulated F-actin rearrangement in chromaffin cells. *NeuroReport* **2**, in press. Month of appearance: January.

Marxen, P., Fuhrmann, U. and Bigalke, H. (1989). Gangliosides mediate inhibitory effects of tetanus and botulinum a neurotoxins on exocytosis in chromaffin cells. *Toxicon* **27**, 849–59.

Matsuda, M. (1989). The structure of tetanus toxin. In: *Botulinum Neurotoxin and Tetanus Toxin* (ed. L.L. Simpson). pp. 69–92. Academic Press, New York.

Matsuda, M. and Yoneda, M. (1974). Dissociation of tetanus neurotoxin into two polypeptide fragments. *Biochem. Biophys. Res. Commun.* **57**, 1257–62.

Matsuda, M. and Yoneda, M. (1975). Isolation and purification of two antigenically active, "complementary" polypeptide fragments of tetanus neurotoxin. *Infect. Immun.* **12**, 1147–53.

Matsuda, M. and Yoneda, M. (1977). Antigenic substructure of tetanus neurotoxins. *Biochem. Biophys. Res. Commun.* **77**, 268–74.

Matsuda, M., Sugimoto, N., Ozutsumi, K. and Toshiro, H. (1982). Acute botulinum-like intoxication by tetanus neurotoxin in mice. *Biochem. Biophys. Res. Commun.* **104**, 799–805.

Matsuda, M., Makinaga, G. and Hirai, T. (1983). Studies on the antibody composition and neutralizing activity of tetanus antitoxin sera from various species of animals in relation to the antigenic substructure of the tetanus toxin molecule. *Biken J.* **26**, 133–43.

Matter, K., Dreyer, F. and Aktories, K. (1989). Actin involvement in exocytosis from PC12 cells: Studies on the influence of botulinum C2 toxin on stimulated noradrenaline release. *J. Neurochem.* **52**, 370–6.

Matsuoka, I., Syuto, B., Kurihara, K. and Kubo, S. (1987). ADP-ribosylation of specific membrane proteins in pheochromocytoma and primary-cultured brain cells by botulinum neurotoxins types C and D. *FEBS Lett.* **216**, 295–9.

May, A.J. and Whaler, B.C. (1958). The adsorption of *Clostridium botulinum* type A from the alimentary canal. *Br. J. Exp. Path.* **39**, 307–16.

McCroskey, L.M., Hatheway, C.E., Fenicia, L., Pasolini, B. and Aureli, P. (1986). Characterization of an organism that produces type E botulinal toxin that resembles *Clostridium butyricum* from feces of an infant with botulism. *J. Clin. Microbiol.* **23**, 201–2.

McInnes, C. and Dolly, J.O. (1990). Ca^{2+}-dependent noradrenaline release from permeabilized PC12 cells is blocked by botulinum neurotoxin A or its light chain. *FEBS Lett.* **261**, 323–6.

McMahan, U.J., Spitzer, N.C. and Peper, K. (1972). Visual identification of nerve terminals in living isolated skeletal muscle. *Proc. R. Soc. London B.* **181**, 421–30.

Meckler, R.L., Baron, R., McLachlan, E.M. (1990). Selective uptake of C-fragment of tetanus toxin by sympathetic preganglionic nerve terminals. *Neuroscience* **36**, 823–9.

Mellanby, J. and Whittaker, V.P. (1968). The fixation of tetanus toxin by synaptic membranes. *J. Neurochem.* **15**, 205–8.

Middlebrook, J.L. (1989). Cell surface receptors for protein toxins. In: *Botulinum Neurotoxin and Tetanus Toxin* (ed. L.L. Simpson). pp. 95–119. Academic Press, New York.

Miyasaki, S., Okada, K., Muto, S., Itokazu, T., Matsui, M., Ebisawa, I., Kagabe, K. and Kimura, T. (1967). On the mode of action of tetanus toxin in rabbit. I. Distribution of tetanus toxin in vivo and development of paralytic signs under some conditions. *Jpn. J. Exp. Med.* **37**, 217–25.

Mochida, S., Poulain, B., Weller, U., Habermann, E. and Tauc, L. (1989). Light chain of tetanus toxin intracellularly inhibits acetylcholine release at neuroneuronal synapses, and its internalization is mediated by heavy chain. *FEBS Lett.* **253**, 47–51.

Mochida, S., Poulain, B., Eisel, U., Binz, T., Kurazono, H., Niemann, H. and Tauc, L. (1990a). Exogenous mRNA encoding tetanus or botulinum neurotoxins expressed in Aplysia neurons. *Proc. Natl. Acad. Sci. USA* **87**, 7844–8.

Mochida, S., Poulain, B., Eisel, U., Binz, T., Kurazono, H., Niemann, H. and Tauc, L. (1990b). Molecular biology of clostridial toxins: Expression of mRNA encoding tetanus and botulinum neurotoxins in Aplysia neurons. *J. Physiol. (Paris)* **84**, 278–84.

Molgo, J. and Thesleff, S. (1984). Studies on the mode of action of botulinum toxin type A at the frog neuromuscular junction. *Brain Res.* **297**, 309–16.

Moller, V. and Scheibel, I. (1960). Preliminary report on the isolation of an apparently new type of *Cl. botulinum. Acta Path. Microbiol. Scand.* **48**, 80.

Montecucco, C. (1986). How do tetanus and botulinum toxins bind to neuronal membranes? *TIBS* **11**, 314–7.

Montecucco, C., Shiavo, G., Brunner, J., Duflot, E., Boquet, P. and Roa, M. (1986). Tetanus toxin is labeled with photoactivatable phospholipid at low pH. *Biochemistry* **25**, 919–24.

Montecucco, C., Shiavo, G., Gao, Z., Bauerlein, E., Boquet, P., and DasGupta, B.R. (1988). Interaction of botulinum and tetanus toxins with the lipid bilayer surface. *Biochem. J.* **251**, 379–83.

Montecucco, C., Shiavo, G. and DasGupta, B.R. (1989). Effect of pH on the interaction of botulinum neurotoxins A, B and E with liposomes. *Biochem. J.* **259**, 47–53.

Montesano, R., Roth, J., Robert, A. and Orci, L. (1982). Non-coated membrane invaginations are involved in binding and internalization of cholera and tetanus toxin. *Nature* **296**, 651–3.

Morii, N., Ohashi, Y., Nemoto, Y., Fujiwara, M., Ohnishi, Y., Nishiki, T., Kamata, Y., Kozaki, S., Narumiya, S., Sakaguchi, G. (1990). Immunochemical identification of the ADP-ribosyltransferase in Botulinum C1 Neurotoxin as C3 exoenzyme-like molecule. *J. Biochem.* **107**, 769–75.

Morris, N.P., Consiglio, E., Kohn, L., Habig, D. and Hardegree, W.H. (1980). Interaction of fragment-B and fragment-C of tetanus toxin with neural and thyroid membranes and with gangliosides. *J. Biol. Chem.* **255**, 6071–6.

Morris, R.E. and Saelinger, C.B. (1989). Entry of bacterial toxins into mammalian cells. In: *Botulinum Neurotoxin and Tetanus Toxin* (ed. L.L. Simpson). pp. 121–152. Academic Press. New York.

Mullinax, R.L., Gross, E.A., Amberg, J.R., Hay, B.N., Hogrefe, H.H., Kubitz, M.M., Greener, A., Alting-Mees, M., Ardourel, D., Short, J.M., Sorge, J.A. and Shopes, B. (1990). Identification of human fragment clones specific for tetanus toxoid in a bacteriophage lambda immunoexpression library. *Proc. Natl. Acad. Sci. USA* **87**, 8095–9.

Murphy, B.R. and Webster, R.G. (1985). Influenza viruses. In: *Virology* (eds Fields *et al.*) pp. 1179–239. Raven Press, New York.

Nairn, A.C. and Greengard, P. (1987). Purification and characterization of Ca^{2+}/calmodulin-dependent protein kinase I from bovine brain. *J. Biol. Chem.* **262**, 7273–81.

Narumiya, S., Sekine, A. and Fujiwara, M. (1988). Substrate for botulinum ADP-ribosyltransferase, Gb, has an amino acid sequence homologous to a putative rho gene product. *J. Biol. Chem.* **263**, 17255–7.

Ngsee, J.K., Miller, K., Wendland, B. and Scheller, R.H. (1990). Multiple GTP-binding proteins from cholinergic synaptic vesicles. *J. Neuroscience* **10**, 317–22.

Niemann, H., Andersen-Beckh, B., Binz, T., Eisel, U., Demotz, S., Mayer, T. and Widmann, C. (1988). Tetanus toxin: Evaluation of the primary sequence and potential applications. *Zbl. Bakt. Suppl.* **17**, 29–38.

Niemann, H., Mayer, T. and Tamura, T. (1989). Signals for membrane-associated transport in eukaryotic cells. *Subcellular Biochemistry* **15**, 303–61.

Nishizuka, Y. (1988). The molecular heterogeneity of protein kinase C and its implications for cellular regulation. *Nature* **334**, 661–5.

Ochandra, J.O., Syuto, B., Ohishi, I., Naiki, M. and Kubo, S. (1986). Binding of *Clostridium botulinum* neurotoxin to gangliosides. *J. Biochem. (Tokyo)* **100**, 27–33.

Ogawa, M., Ishikawa, T. and Irimajiri, A. (1984). Adrenal chromaffin cells form functional cholinergic synapses in culture. *Nature* **307**, 66–8.

Oguma, K., Iida, H., Shiozaki, M. and Inoue, K. (1976a). Antigenicity of converting phages obtained from *Clostridium botulinum* tapes C and D. *Infect. Immun.* **13**, 855–60.

Oguma, K., Iida, H. and Shiozaki, M. (1976b). Phage conversion to hemagglutinin production in *Clostridium botulinum* types C and D. *Infect. Immun.* **14**, 597–602.

Oguma, K., Kimura, K., Fujii, N., Tsuzuki, K., Murakami, T., Indoh, T. and Yokosawa, N. (1990). Cloning and whole nucleotide sequence of *Clostridium botulinum* type C toxin gene. Abstracts of the IUMS Congress: Bacteriology and Mycology (Osaka, 16.-22.9.1990) PS59-04, p157.

Ohishi, I. and Sakaguchi, G. (1977). Activation of botulinum toxins in the absence of nicking. *Infect. Immun.* **17**, 402–7.

Ohishi, I., Okada, T. and Sakaguchi, G. (1975). Response of *Clostridium botulinum* type B and E progenitor toxins to some clostridial sulfhydryl-dependent proteases. *Jpn. J. Med. Sci. Biol.* **28**, 157–64.

Ohishi, I., Morikawa, Y. and Baba, T. (1990). ADP-ribosylation of nonmuscle actin by component of botulinum C2 Toxin activates the ability to interact with unmodified actin. *J. Biochem.* **107**, 420–5.

Orci, L., Malhotra, V., Amherdt, M., Serafini, T. and Rothman, J.E. (1989). Dissection of a single round of vesicular transport: sequential coated and uncoated intermediates mediate intercisternal movement in the Golgi stack. *Cell* **56**, 357–68.

Pang, D., Wang, J., Valtorta, F., Benfenati, F. and Greengard, P. (1988). Protein tyrosine phosphorylation in synaptic vesicles. *Proc. Natl. Acad. Sci. USA* **85**, 762–6.

Panina-Bordignon, P., Tan, A., Termijtelen, A., Corradin, G. and Lanzavecchia, A. (1989). Universally immunogenic T cell epitopes: promiscuous binding of human MHC class II and promiscuous recognition by T cells. *Eur. J. Immunol.* **19**, 2237–42.

Paoletti, L.C., Kasper, D.L., Michon, F., DiFabio, J., Holme, K., Jennings, H.J. and Wessels, M.R. (1990). An oligosaccharide-tetanus toxoid conjugate vaccine against type III group B streptococcus. *J. Biol. Chem.* **265**, 18278–283.

Parton, R.G., Ockleford, C.D. and Critchley, D.R. (1987). A study of the mechanism of internalization of tetanus toxin by primary mouse spinal cord cultures. *J. Neurochem.* **49**, 1057–68.

Penner, R., Neher, E. and Dreyer, F. (1986). Intracellularly injected tetanus toxin inhibits exocytosis in bovine adrenal chromaffin cells. *Nature* **324**, 76–8.

Perin, M.S., Fried, V.A., Mignery, G.A., Jahn, R. and Südhof, T.C. (1990). Phospholipid binding by a synaptic vesicle protein homologous to the regulatory region of protein kinase C. *Nature* **345**, 260–3.

Petrucci, T.C. and Morrow, J.S. (1987). Synapsin I: An actin-bundling protein under phosphorylation control. *J. Cell Biol.* **105**, 1355–63.

Pfanner, N., Glick, B.S., Arden, S.R. and Rothman, J.E. (1990). Fatty acylation promotes fusion of transport vesicles with Golgi cisternae. *J. Cell Biol.* **110**, 955–61.

Pierce, E.J., Davison, M.D., Parton, R.G., Habig, W.H. and Critchley, D.R. (1986). Characterization of tetanus toxin binding to rat brain membranes. *Biochem. J.* **236**, 845–52.

Popoff, M., Boquet, P., Gill, D.M. and Eklund, M.W. (1990). DNA sequence of exoenzyme C3, an ADP-ribosyltransferase encoded by Clostridium botulinum C and D phages. *Nucleic Acids Res.* **18**, 1291.

Poulain, B., Baux, G. and Tauc, L. (1986). Presynaptic transmitter content controls the number of quanta released at a neuro-neuronal cholinergic synapse. *Proc. Natl. Acad. Sci. USA* **83**, 170–3.

Poulain, B., Fossier, P., Baux, G. and Tauc, L. (1987). Hemicholinum-3 facilitates the release of acetylcholine by acting on presynaptic nicotinic receptors at a central synapse in *Aplysia. Brain Res.* **435**, 63–70.

Poulain, B., Tauc, L., Maisey, E.A., Wadsworth, J.D.F., Mohan, P.M. and Dolly, J.O. (1988a). Neurotransmitter release is blocked intracellularly by botulinum neurotoxin, and this requires uptake of both toxin polypeptides by a process mediated by the larger chain. *Proc. Natl. Acad. Sci. USA* **85**, 4090–4.

Poulain, B., Tauc, L., Maisey, E.A. and Dolly, J.O. (1988b). Les synapses ganglionnaires d'Aplysie comme modèle d'étude du mode d'action des neurotoxines botuliques. *C.R. Acad. Sci. Paris* **306**, 483–8.

Poulain, B., Wadsworth, J.D.F., Maisey, E.A., Shone, C.C., Melling, J., Tauc, L. and Dolly, J.O. (1989a). Inhibition of transmitter release by botulinum neurotoxin A. *Eur. J. Biochem.* **185**, 197–203.

Poulain, B., Wadsworth, J.D.F., Shone, C.C., Mochida, S., Lande, S., Melling, J., Dolly, J.O. and Tauc, L. (1989b). Multiple domains of botulinum neurotoxin contribute to its inhibition of transmitter release in aplysia neurons. *J. Biol. Chem.* **264**, 21928–33.

Poulain, B., Mochida, S., Wadsworth, J.D.F., Weller, U., Habermann, E., Dolly, J.O. and Tauc, L. (1990). Inhibition of neurotransmitter release by botulinum neurotoxins and tetanus toxin at *Aplysia* synapses: Role of the constituent chains. *J. Physiol. (Paris)* **84**, 247–61.

Poulain, B., Mochida, S., Weller, U., Högy, B., Habermann, E., Wadsworth, J.D.F., Shone, C.C., Dolly, J.O. and Tauc, L. (1991). Heterologous combinations of heavy and light chains from botulinum neurotoxin A and tetanus toxin inhibit neurotransmitter release in *Aplysia. J. Biol. Chem.* in press.

Poumeyrol, M., Billon, J., DeLille, F., Haas, C., Marmonier, A. and Sebold, M. (1983). Intoxication botylique due à une souche de Clostridium botulinum de type AB. *Med. Malad. Infect.* **13**, 750–4.

Rabasseda, X., Blasi, J., Marsal, J., Dunant, Y.,

Casanova, A. and Bizzini, B. (1988). Tetanus and botulinum toxins block the release of acetylcholine from slices of rat striatum and from the isolated electric organ of *Torpedo* at different concentrations. *Toxicon* **26**, 329–36.

Rahamimoff, R., DeRiemer, S.A., Sakmann, B., Stadler, H. and Yakir, N. (1988). Ion channels in synaptic vesicles from *Torpedo* electric organ. *Proc. Natl. Acad. Sci. USA* **85**, 5310–4.

Rahamimoff, R., DeRiemer, S.A., Ginsburg, S., Kaiserman, I., Sakmann, B., Shapira, R., Stadler, H. and Yakir, N. (1989). Ion channels in synaptic vesicles: are they involved in transmitter release? *Quart. J. Exp. Physiol.* **74**, 1019–31.

Ramon, G. (1925). Sur la production des antitoxines. *C.R. Acad. Sci. (Paris)* **181**, 157.

Rehm, H., Weidermann, B. and Betz, H. (1986). Molecular characterization of synaptophysin, a mayor calcium-binding protein of the synaptic vesicle membrane. *EMBO J.* **5**, 535–41.

Roa, M. and Boquet, P. (1985). Interaction of tetanus toxin with lipid vesicles at low pH. *J. Biochem.* **260**, 6827–35.

Robinson, J.P., Chen, H.C.J., Hash, J.H. and Puett, D. (1978). Enzymatic fragmentation of tetanus toxin. Identification and characterization of an atoxic, immunogenetic fragment. *Mol. Cell. Biochem.* **5**, 23–31.

Rösener, S., Chhatwal, G.S. and Aktories, K. (1987). Botulinum ADP-ribosyl-transferase C3 but not botulinum neurotoxins C1 and D ADP-ribosylates low molecular weight GTP-binding proteins. *FEBS Lett.* **224**, 38–42.

Rogers, T.B. and Snyder, S.H. (1981). High affinity binding of tetanus toxin to mammalian brain membranes. *J. Biol. Chem.* **256**, 2402–7.

Rubin, E.J., Gill, D.M., Boquet, P. and Popoff, M.R. (1988). Functional modification of a 21-kilodalton G protein when ADP-ribosylated by exoenzyme C3 of *Clostridium botulinum. Mol. Cell. Biol.* **8**, 418–26.

Sakaguchi, G. (1983). *Clostridium botulinum* toxins. *Pharmac. Ther.* **19**, 165–94.

Sakaguchi, G., Sakaguchi, S., Kozai, S. and Takahashi, M. (1986). Purification and some properties of *Clostridium botulinum* type AB toxin. *FEMS Microbiol. Lett.* **33**, 23–9.

Sandberg, K., Berry, C.J. and Rogers, T.B. (1989a). Studies on the intoxication pathway of tetanus toxin in the rat pheochromocytoma (PC12) cell line. *J. Biol. Chem.* **264**, 5679–86.

Sandberg, K., Berry, C.J., Eugster, E. and Rogers, T.B. (1989b). A role for cGMP during tetanus toxin blockade of acetylcholine release in the rat pheochromocytoma (PC12) cell line. *J. Neuroscience* **9**, 3946–54.

Sato, H., Ito, A., Yamakawa, Y. and Murata, R. (1979). Toxin-neutralizing effect of antibody against subtilisin-digested tetanus toxin. *Infect. Immun.* **24**, 958–61.

Schiavo, G., Papini, E., Genna, G. and Montecucco,

C. (1990). An intact disulfide bond is required for the neurotoxicity of tetanus toxin. *Infect. Immun.* **58**, 4136–41.

Schmitt, A., Dreyer, F. and John, C. (1981). At least three sequential steps are involved in the tetanus toxin-induced block of neuromuscular transmission. *Naunyn-Schmiedeberg's Arch. Pharmacol.* **317**, 326–30.

Schwab, M.E. and Thoenen, H. (1978). Selective binding, uptake, and retrograde transport of tetanus toxin by nerve terminals in the rat iris. *J. Cell Biol.* **77**, 1–13.

Shaw, G. and Kamen, R. (1986). A conserved AU sequence from the 3'-untranslated region of GM-CSF mRNA mediates selective mRNA degradation. *Cell* **46**, 659–67.

Shepherd, G.M. (1988). *Neurobiology.* Oxford University Press, Oxford.

Shone, C.C., Hambleton, P. and Melling, J. (1985). Inactivation of *Clostridium botulinum* type A neurotoxin by trypsin and purification of the two tryptic fragments. Proteolytic action near the COOH-terminus of the heavy subunit destroys toxin-binding activity. *Eur. J. Biochem.* **151**, 75–82.

Shone, C.C., Hambleton, P. and Melling, J. (1987). A 50kDa fragment from the NH$_2$-terminus of the heavy subunit of *Clostridium botulinum* type A neurotoxin forms channels in lipid vesicles. *Eur. J. Biochem.* **167**, 175–180.

Simpson, L.L. (1980). Kinetic studies on the interaction between botulinum toxin type A and the cholinergic neuromuscular junction. *J. Pharmacol. Exp. Ther.* **212**, 16–21.

Simpson, L.L. (1981). The origin, structure, and pharmacological activity of botulinum toxin. *Pharmacol. Rev.* **33**, 155–88.

Simpson, L.L. (1982). The interaction between aminoquinolines and presynaptically acting neurotoxins. *J. Pharmacol. Exp. Ther.* **222**, 43–8.

Simpson, L.L. (1983). Ammonium chloride and methylamine hydrochloride antagonize clostridial neurotoxins. *J. Pharmacol. Exp. Ther.* **225**, 546–52.

Simpson, L.L. (1984). Fragment C of tetanus toxin antagonizes the neuromuscular blocking properties of native tetanus toxin. *J. Pharmacol. Exp. Ther.* **228**, 600–4.

Simpson, L.L. (1985). Pharmacological experiments on the binding and internalization of the 50,000 Dalton carboxyterminus of tetanus toxin at the cholinergic neuromuscular junction. *J. Pharmacol. Exp. Ther.* **234**, 100–105.

Simpson, L.L. (1987). The pathophysiological actions of the binary toxin produced by Clostridium botulinum. In: *Avian Botulism: An International Perspective* (eds M.W. Eklund and V.R. Dowell, Jr.). pp. 249–264. Charles C. Thomas, Springfield, IL.

Simpson, L.L. (1988). The use of pharmacological antagonists to deduce commonalities of biological activity among clostridial neurotoxins. *J. Pharmacol. Exp. Ther.* **245**, 867–72.

Simpson, L.L. (ed) (1989). *Botulinum Neurotoxin and*

Tetanus Toxin. Academic Press, New York.

Simpson, L.L. and Rapport, M.M. (1971). Ganglioside inactivation of botulinum toxin. *J. Neurochem.* **18**, 1341–3.

Simpson, L.L., Lake, P. and Kozaki, S. (1990). Isolation and characterization of a novel human monoclonal antibody that neutralizes tetanus toxin. *J. Pharmacol. Exp. Ther.* **254**, 98–103.

Singh, B.R. (1990). Identification of specific domains in botulinum and tetanus neurotoxins. *Toxicon* **28**, 992–6.

Smith, L. (1977). *Botulism: The Organism, its Toxins, the Disease.* Charles C. Thomas, Publisher, Springfield, Illinois, USA.

Stanfield, J.P. and Galazka, A. (1984). Neonatal tetanus in the third world today. *Bull. W.H.O.* **62**, 710–4.

Staub, G.C., Walton, K.M., Schnaar, R.L., Nichols, T., Baichwal, R., Sandberg, K. and Rogers, T.B. (1986). Characterization of the binding and internalization of tetanus toxin in a neuroblastoma hybrid cell line. *J. Neuroscience* **6**, 1443–51.

Stecher, B., Gratzl, M. and Ahnert-Hilger, G. (1989a). Reductive chain separation of botulinum A toxin—a prerequisite to its inhibitory action on exocytosis in chromaffin cells. *FEBS Lett.* **248**, 23–7.

Stecher, B., Weller, U., Habermann, E., Gratzl, M. and Ahnert-Hilger, G. (1989b). The light chain but not the heavy chain of botulinum A toxin inhibits exocytosis from permeabilized adrenal chromaffin cells. *FEBS Lett.* **255**, 391–4.

Stoeckel, K., Schwab, M. and Thoenen, H. (1975). Comparison between the retrograde axonal transport of cholera toxin and tetanus toxin as compared to nerve growth factor and wheat germ agglutinin. *Brain Res.* **132**, 273–85.

Strom, M.S., Eklund, M.W. and Poysky, F.T. (1984). Plasmids in *Clostridium botulinum* and related *Clostridium* species. *Appl. Environ. Microbiol.* **48**, 956–63.

Südhof, T.C., Czernik, A.J., Kao, H.-T., Takei, K., Johnston, P.A., Horiuchi, A., Kanazir, S.D., Wagner, M.A., and Perin, M.S., De Camilli, P. and Greengard, P. (1989). Synapsins: mosaics of shared and individual domains in a family of synaptic vesicle phosphoproteins. *Science* **245**, 1474–80.

Suen, J.C., Hatheway, C.L., Steigerwalt, A.G. and Brenner, D.J. (1988a). *Clostridium argentinense* sp. nov.: a genetically homogenous group of all strains of *Clostridium botulinum* toxin type G and some nontoxigenic strains previously identified as *Clostridium subterminale* and *Clostridium hastiforme*. *Int. J. Syst. Bacteriol.* **38**, 375–81.

Suen, J.C., Hatheway, C.L., Steigerwalt, A.G. and Brenner, D.J. (1988b). Genetic confirmation of identities of neurotoxigenic *Clostridium baratii* and *Clostridium butyricum* implicated as agents of infant botulism. *J. Clin. Microbiol.* **26**, 2191–2.

Sugiyama, H., Mizutani, D. and Yang, K.W. (1972). Basis of type A and F toxicities of *Clostridium botulinum* strain 84. *Proc. Soc. Exp. Biol. Med.* **191**, 1063–7.

Tacket, C.O. and Rogawski, M.A. (1989). Botulism, In: *Botulinum Neurotoxin and Tetanus Toxin* (ed. L.L. Simpson) pp. 351–378. Academic Press, New York.

Takano, K., Kirchner, F., Gremmelt, A., Matsuda, M., Ozutsumi, N. and Sugimoto, N. (1989). Blocking effects of tetanus toxin and its fragment (A-B) on excitatory and inhibitory synapses of the spinal motoneurone of the cat. *Toxicon* 27, 385–92.

Tauc, L., Hoffmann, A., Tsujii, S., Hinzen, D.H. and Faille, L. (1974). Transmission abolished on a cholinergic synapse after injection of AChE into the presynaptic neuron. *Nature* 250, 496–8.

Thomas, L., Hartung, K., Langosch, D., Rehm, H., Bamberg, E., Franke, W.W. and Betz, H. (1988). Identification of synaptophysin as a hexameric channel protein of the synaptic vesicle membrane. *Science* 242, 1050–3.

Thompson, D.E., Brehm, J.K., Oultram, J.D., Swinfield, T.-J., Shone, C.C., Atkinson, T., Melling, J. and Minton, N.P. (1990). The complete amino acid sequence of the clostridium botulinum type A neurotoxin deduced by nucleotide sequence analysis of the encoding gene. *Eur. J. Biochem.* 189, 73–81.

Torri-Tarelli, F., Villa, A., Valtorta, F., De Camilli, P., Greengard, P. and Ceccarelli, B. (1990). Redistribution of synaptophysin and synapsin I during a-Latrotoxin-induced release of neurotransmitter at the neuromuscular junction. *J. Cell. Biol.* 110, 449–59.

Tsuzuki, K., Yokosawa, N., Syuto, B., Ohishi, I., Fujii, N., Kimura, K. and Ogume, K. (1988). Establishment of a monoclonal antibody recognizing an antigenic site common to *Clostridium botulinum* types B, C1, D and E toxins and tetanus toxin. *Infect. Immun.* 56, 898–902.

Tsuzuki, K., Kimura, K., Fujii, N., Yokosawa, N., Indoh, T., Murakami, T. and Oguma, K. (1990). Cloning and complete sequence of the gene for the main component of hemagglutinin produced by *Clostridium botulinum* type C. *Infect. Immun.* 58, 3173–7.

Ueda, T., Maeno, H. and Greengard, P. (1973). Regulation of endogenous phosphorylation of specific proteins in synaptic membrane fractions from rat brain by adenosine-3'.5'-monophosphate. *J. Biol. Chem.* 248, 8295–305.

Ugolini, G., Kuypers, H.G.J.M. and Bizzini, B. (1984). Retrograde labelling of motoneurons by means of tetanus toxin fragment II_c, demonstrated immunohistologically. *Neurosci. Lett. Suppl.* 18, S339.

Unsicker, K., Krisch, B., Otten, U. and Thoenen, H. (1978). Nerve growth factor-induced fiber outgrowth from isolated rat adrenal chromaffin cells: Impairment by glucocorticoids. *Proc. Natl. Acad. Sci. USA* 75, 3498–502.

Vallar, L., Biden, T.J. and Wollheim, C.B. (1987). Guanine nucleotides induce Ca^{2+}-independent insulin secretion from permeabilized RINm5F cells. *J. Biol. Chem.* 262, 5049–56.

Valtorta, F., Mededdu, L., Meldolesi, J. and Ceccarelli, B. (1984). Specific localization of the a-latrotoxin

receptor in the nerve terminal plasma membrane. *J. Cell. Biol.* 99, 124–32.

Valtorta, F., Jahn, R., Fesce, R., Greengard, P. and Ceccarelli, B. (1988). Synaptophysin (p38) at the frog neuromuscular junction: its incorporation into the axolemma and recycling after intense quantal secretion. *J. Cell. Biol.* 107, 2717–27.

van Ermengem, E. (1897). Über einen neuen anaeroben Bacillus und seine Beziehungen zum Botulismus. *Z. Hyg. Infektionskr.* 26, 1–56.

van Heyningen, W.E. (1974). Gangliosides as membrane receptors for tetanus toxin, cholera toxin and serotonin. *Nature* 249, 415–7.

van Heyningen, W.E. and Mellanby, J. (1973). A note on the specific fixation, specific deactivation and nonspecific inactivation of bacterial toxins by gangliosides. *Naunyn Schmiedebergs Arch. Pharmacol.* 276, 297–302.

van Heyningen, W.E. and Miller, P.A. (1961). The fixation of tetanus toxin by ganglioside. *J. Gen. Microbiol.* 24, 107–119.

Vinet, G., Berthiaume, L. and Fredette, V. (1968). Un bacteriophage dans une culture de *C. botulinum* C. *Rev. Can. Biol.* 27, 73–4.

Wadsworth, J.D.F., Desai, M., Tranter, H.S., King, H.J., Hambleton, P., Melling, J., Dolly, J.O. and Shone, C.C. (1990). Botulinum type F neurotoxin. *Biochem. J.* 268, 123–8.

Walton, K.M., Sandberg, K., Rogers, T.B. and Schnaar, R.L. (1988). Complex ganglioside expression and tetanus toxin binding by PC12 pheochromocytoma cells. *J. Biol. Chem.* 263, 2055–63.

Wassermann, A. and Takaki, I. (1898). Über tetanusantitoxische Eigenschaften des normalen Centralnervensystems. *Klin. Wochenschr.* 35, 5–6.

Weickert, M.J., Chambliss, G.H. and Sugiyama, H. (1986). Production of toxin by *Clostridium botulinum* type A strains cured of plasmids. *Appl. Environ. Microbiol.* 51, 52–6.

Weidman, P.J., Melancon, P., Block, M.R. and Rothman, J.E. (1989). Binding of an NEM-sensitive fusion protein to Golgi membranes requires both a soluble protein(s) and integral membrane receptor. *J. Cell Biol.* 108, 1589–96.

Weigt, C., Just, I., Wegner, A. and Aktories, K. (1989). Nonmuscle actin ADP-ribosylated by botulinum C2 toxin caps actin filaments. *FEBS Lett.* 246, 181–4.

Weller, U., Taulor, C.F. and Habermann, E. (1986). Quantitative comparison between tetanus toxin, some fragments and toxoid for binding and axonal transport in the rat. *Toxicon* 24, 1055–63.

Weller, U., Mauler, F. and Habermann, E. (1988). Tetanus toxin: Biochemical and pharamacological comparison between its protoxin and some isotoxins obtained by limited proteolysis. *Naunyn-Schmiedeberg's Arch. Pharmacol.* 338, 99–106.

Weller, U., Dauzenroth, M.-E., Meyer zu Heringdorf, D. and Habermann, E. (1989). Chains and fragments of tetanus toxin. *Eur. J. Biochem.* 182, 649–56.

Wellhöner, H.H. and Neville, D.M., Jr. (1987). Tetanus

toxin binds with high affinity to neuroblastoma x glioma hybrid cells ND-108-15 and impairs their stimulated acetylcholine release. *J. Biol. Chem.* **262**, 17374–8.

Wendon, L.M.B. and Gill, D.M. (1982). Tetanus toxin action on nerve cells. Does it modify a neuronal protein? *Brain Res.* **238**, 292–7.

Wernig, A., Stover, H. and Tonge, D. (1977). The labelling of motor end-plates in skeletal muscle of mice with ^{125}I-tetanus toxin. *Naunyn Schmiedeberg's Arch. Pharmacol.* **298**, 37–42.

White, J.M. (1990). Viral and cellular membrane fusion proteins. *Annu. Rev. Physiol.* **52**, 675–97.

Wiegand, H., Erdmann, G. and Wellhöner, H.H. (1976). ^{125}I-labeled botulinum A neurotoxin: pharmacokinetics in cats after intramuscular injection. *Naunyn Schmiedeberg's Arch. Pharmacol.* **292**, 161–5.

Wiegandt, H. (1985). *Glycolipids.* Elsevier Science Publishers, Amsterdam.

Williams, R.S., Tse, C.-K., Dolly, J.O., Hambleton, P. and Melling, J. (1983). Radioiodination of botulinum neurotoxin type A with retention of biological activity

and its binding to brain synaptosomes. *Eur. J. Biochem.* **131**, 437–45.

Yavin, E., Lazarovici, P. and Nathan, A. (1987). Molecular interactions of ganglioside receptors with tetanotoxin on solid supports, aqueous solutions and natural membranes. In: *Membrane Receptors, Dynamics, and Eragetics* (ed. K.W.A. Wirtz). *Natio Asi Series,* Plenum Press **133**, 135–47.

Yokosawa, N., Tsuzuki, K., Syuto, B. and Oguma, K. (1986). Activation of clostridium botulinum type E toxin purified by two different procedures. *J. Gen. Microbiol.* **132**, 1981–8.

Yokosawa, N., Kurokawa, Y.Y., Tsuzuki, K., Syuto, B., Fujii, N., Kimura, K. and Oguma, K. (1989). Binding of *Clostridium botulinum* type C neurotoxin to different neuroblastoma cell lines. *Infect. Immun.* **57**, 272–7.

Zahraoui, A., Touchot, N., Chardin, P. and Tavitian, A. (1989). The human rab genes encode a family of GTP-binding proteins related to the yeast YPT1 and sec4 products involved in secretion. *J. Biol. Chem.* **264**, 12394–401.

16

Bordetella pertussis Adenylate Cyclase Toxin

Emanuel Hanski[1] and John G. Coote[2]

[1]Department of Medical Microbiology, The Hebrew University – Hadassah Medical School, Jerusalem 91010, Israel and [2]Department of Microbiology, The University of Glasgow, Glasgow G12 9QQ, UK

Introduction

The Gram-negative, small coccobacillus *Bordetella pertussis* is the aetiologic agent of pertussis, or whooping cough, in humans. The disease has been clinically recognized since the sixteenth century (Cone, 1970) and the organism was first isolated in 1906 (Bordet and Gengou, 1906). Pertussis is considered to be a disease of infants and young children (Olson, 1975), but adults may also be infected and develop a persistent cough (Robertson *et al.*, 1987). In infants, pertussis, at its paroxysmal stage, is characterized by frequent episodes of violent coughing, associated with a typical 'whoop' – a forced inspiration of air over a partially closed glottis (Olson, 1975). Until the wide-scale introduction of the whole-cell vaccine, whooping cough was a major cause of childhood morbidity and mortality (Cherry, 1984). It still remains so in many developing countries (Muller *et al.*, 1984). Recent concern over adverse reactions to the whole-cell vaccine decreased its acceptance by the public, and caused increased incidence of the disease (Centers for Disease Control, 1984; Romanus *et al.*, 1987). This has stimulated the development of a safer acellular vaccine.

The pathogenesis of pertussis is complex and involves the activity of a variety of components implicated as virulence factors (Weiss and Hewlett, 1986). These include: pertussis toxin, filamentous haemagglutinin, dermonocretic toxin, lipopolysaccharide, tracheal cytotoxin, haemolysin and adenylate cyclase (AC). The existence of adenylate cyclase activity in *B. pertussis* was discovered in the early 1970s (Fishel *et al.*, 1970; Wolff and Cook, 1973). It took a decade to realize that *B. pertussis* AC has rather unusual properties and is, in fact, a toxin. Most bacteria produce AC (Ide, 1971), however, only in *B. pertussis* (Hewlett *et al.*, 1976), *Bacillus anthracis* (see Chapter 14) and possibly *Yersinia pestis* (Asseeva *et al.*, 1987) is the location of AC extracytoplasmic. In contrast to the AC of other bacteria which are either stimulated or inhibited by alpha-keto acids (Hirata and Hayaishi, 1967; Ide, 1969), *B. pertussis* AC is resistant (Hewlett and Wolff, 1976). Wolff *et al.* (1980) discovered that *B. pertussis* AC is greatly stimulated by the Ca^{2+}-binding protein calmodulin (CaM), which is absent in bacteria.

Confer and Eaton (1982) were the first to demonstrate that *B. pertussis* acts as a toxin which may suppress host defences. They showed that exposure of polymorphonuclear leukocytes and macrophages to *B. pertussis* extract containing AC led to the generation of uncontrolled levels of cAMP and concomitantly to inhibition of phagocytic functions. More recently, the potential importance of *B. pertussis* AC as a virulence factor was emphasized by the finding that a Tn5 insertion mutant deficient in AC was avirulent in an infant mouse model of the infection (Weiss *et al.*, 1984). Introduction of the cloned AC gene into this mutant strain restored its virulence (Brownlie *et al.*, 1988).

The objectives of this chapter are to describe

the biochemical nature of *B. pertussis* AC toxin and the gene structure encoding it, to discuss its mechanism of penetration and its fate within the host, and to evaluate its role in the pathogenesis of whooping cough.

Production and localization of AC

High levels of AC activity were detected in all virulent strains of *B. pertussis* and in fresh isolates from cases of whooping cough (Wolff, 1985). *B. pertussis* is a slow-growing, fastidious organism and *in vitro* passaging or alteration in growth conditions tends to promote the loss of virulence-associated properties including AC activity (Leslie and Gardner, 1931; Parton and Durham, 1978; Peppler, 1982; Goldman *et al.*, 1984). There are three other members of the genus *Bordetella*; *B. parapertussis* and *B. bronchiseptica* produce high levels of the enzyme (Endoch *et al.*, 1980), but *B. avium* had no detectable AC activity (Gentry-Weeks *et al.*, 1988) and hybridization analysis showed that it did not possess the AC genetic determinant found in the other *Bordetella* (Brownlie *et al.*, 1988).

B. pertussis AC activity can be detected in both whole cells and in culture medium. The enzyme was originally characterized as an extracellular enzyme; the majority of its cell-associated activity was readily assayed by supplying exogenous ATP to intact cells and was exquisitely sensitive to trypsin (Hewlett and Wolff, 1976; Hewlett *et al.*, 1976). These data suggested that the catalytic domain of the enzyme was localized on the outer surface of the bacterial cell. Recently it was shown that release of the contents of the periplasm by formation of spheroplasts did not liberate a significant amount of the cell-associated AC, suggesting that the enzyme may be associated with the inner membrane (Masure and Storm, 1989).

The amount of AC activity secreted into the culture medium during bacterial growth may vary among different wild-type strains. Hewlett *et al.*, (1976) and Endoch *et al.* (1980) found that the cell-associated AC activity in *B. pertussis* was higher than the activity in the supernatant. In contrast, McPheat *et al.* (1983), Brownlie *et al.* (1985a) and Masure and Storm (1989) found a greater proportion of the AC activity in the culture supernatant. The distribution of *B. pertussis* AC

during growth in various bacterial compartments should be further studied using direct visualization techniques. It would appear that *B. pertussis* AC is extruded directly to the medium without the formation of a periplasmic intermediate.

Purification of AC

The AC activity of *B. pertussis*, either cell-bound or released into the culture medium, has been associated with several polypeptides of different molecular mass (Masure *et al.*, 1987). Until recently it was not clear whether the larger forms of the enzyme represented aggregates of the catalytic subunit, either with itself or with other polypeptides, or whether different molecular weight (M_r) forms of the same enzyme existed.

Purification of AC from culture supernatant was first reported by Hewlett and Wolff (1976). The enzyme they obtained was characterized as a single polypeptide of M_r 70 000, with a specific activity of 0.14 μmol/min/mg protein, but was poorly activated by CaM. Kessin and Franke (1986) purified two forms of CaM-sensitive AC from culture medium of high (700 000) and low (50 000–65 000) apparent M_r. It was shown that the high-M_r form could be decomposed and partially converted to the low-M_r form. This study suggested that the minimal M_r of the enzyme in the culture medium was 50 000, and it was able to aggregate with itself or with other proteins.

The first substantial purification of the extracellular AC was achieved by Shattuck *et al.* (1985). The enzyme was purified about 670-fold to a specific activity of 608 μmol/min/mg protein. Since CaM was added during the purification, the purified preparation did not respond to CaM. Analysis of this preparation on SDS–PAGE revealed numerous protein bands including two major bands of M_r 45 000 and 64 000. The M_r of the enzyme calculated according to independent hydrodynamic measurements was 44 300. It was assumed therefore, that the 45 000-band represented AC (Shattuck *et al.*, 1985).

Unambiguous identification of *B. pertussis* AC from bacterial culture medium was recently achieved by Ladant *et al.* (1986). These investigators used a *B. pertussis* strain which secreted high levels of AC into the culture medium. The enzyme was concentrated from culture medium by absorption to a filter membrane, desorbed using detergent and subsequently purified by two different procedures. The first involved affinity chromatography on immobilized CaM. *B. pertussis* AC bound strongly to CaM and could be eluted only by 8.8 M urea. The second procedure involved sequential chromatography on DEAE–Sephacel, first in the absence and then in the presence of CaM. The enzyme obtained by these procedures had a specific activity of 1600 μmol/min/mg protein and existed as three forms of M_r 43 000, 45 000 and 50 000. By limited proteolysis it was shown that all these forms of the enzyme were structurally related (Ladant *et al.*, 1986; Ladant, 1988).

Wolff *et al.* (1984) were the first to claim the purification of AC from whole cells. They reported the isolation of a protein of M_r 47 000 which retained CaM sensitivity and had a high specific activity, but no further details such as extent of purification, recovery and SDS–PAGE data were provided. Friedman (1987) claimed to have achieved a 104-fold purification of AC from bacterial extract and the isolation of a major band of M_r 60 000 and a minor band of M_r 200 000. In this case, however, the specific activity obtained was very low compared to the AC purified from bacterial culture medium. In addition, this enzyme preparation was only activated only two-fold in the presence of CaM. Rogel *et al.* (1988) achieved a significant purification from bacterial extracts of 200 000 and 47 000 forms of AC. By using specific antibodies raised against the 47 000 form they showed that the different forms of *B. pertussis* AC are related. Gel filtration of an extract of concentrated bacterial cells resulted in the separation of two major peaks of AC activity; one peak of an aggregated material, and a second peak of apparent M_r of 50 000 (Hanski and Farfel, 1985; Rogel *et al.*, 1988). By applying the two resolved peaks to CaM–agarose, two forms of AC (M_r 200 000 and 47 000) were purified. They had specific activities of 470 and 1690 μmol/min/mg protein respectively and were activated 50- to 200-fold by CaM. Polyclonal antibodies raised against the purified 47 000 polypeptide specifically recognized both the 200 000 and 47 000 polypeptides on immunoblots. Furthermore, they inhibited the AC activity of all forms of the enzyme found in bacterial extract, suggesting that they are all

related and presumably derived from a common polypeptide (Rogel *et al.*, 1988). It was recently shown that the 200 000 form can be spontaneously converted to the 47 000 form in sonicated spheroplasts (Masure and Storm, 1989).

In summary, it may be concluded that *B. pertussis* AC is synthesized in the bacterium as a polypeptide of 200 000 which is proteolytically degraded and appears in the medium in a lower M_r form. The structure of the toxic form of *B. pertussis* AC and whether the degradation of the enzyme has any relevance to the formation of the toxic form or to its secretion, will be discussed below.

Regulation of AC Activity by Calmodulin

The first clue indicating the existence of a potent activator of *B. pertussis* AC was suggested by the fact that organisms grown on blood agar plates exhibited much higher AC activity than those grown on synthetic media (Hewlett *et al.*, 1978). Furthermore, addition of red-cell lysates caused a dramatic increase (up to 1000-fold) in the activity of the enzyme. The widespread distribution of the activator led eventually to its identification as CaM (Wolff *et al.*, 1980) which is a contaminant of many commercial protein preparations (Goldhammer *et al.*, 1981).

Many enzymes are regulated by CaM (Klee and Vanaman, 1982). The characteristics of CaM regulation vary considerably from one protein to another with respect to the dose required for activation and the participation of Ca^{2+} in this process. The activation of *B. pertussis* AC displayed unusual properties: it occurred over a wide concentration range (Goldhammer and Wolff, 1982) and Ca^{2+} was not absolutely required (Greenlee *et al.*, 1982; Kilhoffer *et al.*, 1983). Two types of activation-related binding were identified. The first one occurred at nanomolar concentrations of CaM, required Ca^{2+} and was inhibited by EGTA. The second one occurred at micromolar concentrations, did not require Ca^{2+} and was unaffected by the presence of EGTA. Since both modes of activation were displayed using pure preparations of the enzyme (Ladant, 1988; Rogel *et al.*, 1988), one can conclude that they are intrinsic properties of the CaM interaction with *B. pertussis* AC. The absence of an absolute Ca^{2+} requirement for activation suggested that *B. pertussis* AC may be less stringent than other CaM-activated enymes in its requirement for the intact CaM molecule. A C-terminal fragment of CaM (residues 78–148), containing only two Ca^{2+}-binding sites, could fully activate the enzyme with 10–15% of the potency of CaM. Like intact CaM, this fragment could fully activate the enzyme in the presence of EGTA, but the activation required concentrations 20- to 60-fold higher (Wolff *et al.*, 1986). Thus, it may be concluded that the two C-terminal Ca^{2+}-binding sites on the CaM molecule can account for most of the stimulation of the enzyme by CaM, and their occupancy by Ca^{2+} merely alters the affinity of CaM towards the enzyme.

The *B. pertussis* AC can also be stimulated by other tryptic fragments of CaM (amino acids 1–77, 1–90 and 107–148), but at much lower potency (Wolff *et al.*, 1986). In spite of the apparently less stringent structural requirements for the activation of *B. pertussis* AC, the enzyme is unresponsive to other Ca^{2+}-binding proteins, such as troponin C and parvalbumin which share many structural features with CaM (Wolff *et al.*, 1980; Kilhoffer *et al.*, 1983). It is possible that subtle structural differences around the Ca^{2+}-binding domains between these Ca^{2+}-binding proteins and CaM (Herzberg and James, 1985) can account for the lack of activation by the former.

Proteolytic degradation by trypsin of the 43 000 form of *B. pertussis* AC in the presence of CaM resulted in the generation of two fragments of M_r of 25 000 and 18 000 (Ladant *et al.*, 1986; Ladant, 1988). Initial biochemical studies located the CaM-binding site on the C-terminal 18 000 fragment and the catalytic domain on the N-terminal 25 000 fragment (Ladant, 1988). More precise analysis showed that the CaM-binding domain was located on the N-terminal part of the 18 000 fragment. This domain included tryptophan which could explain its cross-linking with azido-^{125}I-CaM (Ladant *et al.*, 1989). The amino acid sequence situated around this tryptophan has features characteristic of other CaM-binding domains, such as the presence of basic and hydrophobic amino acids forming an amphiphilic α-helix structure (Glaser *et al.*, 1989, see next section). Furthermore, a synthetic peptide produced according to the nucleotide sequence of this region inhibited the

CaM-mediated activation of the enzyme and bound CaM in a Ca^{2+}-dependent manner (Ladant *et al.*, 1989). Resolved M_r 25 000 and 18 000 fragments could be reassociated in the presence of CaM and formed a CaM-sensitive AC which had a similar structure to the native enzyme (Ladant *et al.*, 1989).

Gene Structure and Function Relationship

Cloning and expression

Two major difficulties were encountered in the cloning of the genetic determinant of *B. pertussis* AC in *E. coli*. The first was that the expression of *B. pertussis* virulence-associated genes requires the action of a *trans*-acting regulatory factor encoded by a separate *B. pertussis* locus termed *vir* (Weiss and Falkow, 1984; Stibitz *et al.*, 1988) which is not present in *E. coli*. The second was that in the absence of CaM the level of cAMP in *B. pertussis* AC-producing *E. coli* clones would be negligible, not allowing their identification according to their cAMP content. Two different strategies were employed for the cloning of the *B. pertussis* AC gene: the first involved the complementation of an AC-deficient *B. pertussis* mutant and the second complementation of an AC-defective *E. coli* strain expressing CaM. Transposon Tn5 mutagenesis of *B. pertussis* produced several strains that were defective in their ability to produce various virulence factors (Weiss *et al.*, 1983). One of these, mutant BP348 (AC⁻,HLY⁻), was deficient both in AC and haemolytic activities. Another Tn5-induced non-haemolytic mutant, BP349 was also impaired in the synthesis of AC, suggesting that the genetic determinants for these two properties were closely linked (Weiss *et al.*, 1983). Brownlie *et al.* (1986) constructed a gene library of *B. pertussis* DNA in *E. coli* using the broad-host-range cosmid vector pLAFR1. Recombinant clones were transferred from *E. coli* to BP348 (AC⁻,HLY⁻) and one recombinant cosmid, pRMB1 (Fig. 1), restored haemolytic activity to this strain, as detected by screening conjugated clones on blood agar (Brownlie *et al.*, 1986). BP348 harbouring pRMB1 also expressed high levels of CaM-dependent AC activity (Brownlie *et al.*, 1988). A

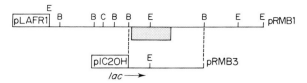

Figure 1. Plasmids containing the *B. pertussis* AC genetic determinant (redrawn from Brownlie *et al.* 1988). The hatched area represents the AC coding region, positioned according to the sequence data of Glaser *et al.* (1988a). The arrow indicates the direction of transcription from the *lac* promoter contained in pIC20H. Restriction sites are shown for *Bam*HI (B), *Cla*I (C), and *Eco*RI (E).

10-kb (kilobase pair) *Bam*H1 fragment of the 33-kb insert in pRMB1, when subcloned into the vector pIC20H (Marsh *et al.*, 1984), was able to complement a strain of *E. coli* carrying a deletion of its AC gene. High-level expression of CaM-dependent AC activity was obtained from this plasmid, designated pRMB3, which was dependent on the *lac* promoter contained in the vector (Brownlie *et al.*, 1988; Fig. 1). The ACs purified from BP348 containing pRMB1 and from an *E. coli* Lon⁻ protease defective strain harbouring pRMB3 both had an M_r of 200 000 and shared similar biochemical and immunological properties (Rogel *et al.*, 1989).

Glaser *et al.* (1988a) cloned the *B. pertussis* AC gene in an *E. coli* (*cya-*) strain harbouring the plasmid pVUC-1, which expresses high levels of CaM. By this elegant procedure clones complementing the *E. coli cya* defect were identified on MacConkey-maltose plates as red colonies, indicating maltose degradation by enzymes induced by cAMP formed by the cloned AC. The level of expression of AC was about 1000-fold lower than in wild-type *B. pertussis*, indicating that some expression of the enzyme can be obtained in *E. coli*, even in the absence of the *vir* regulatory product. The smallest fragment encoding a CaM-sensitive AC was 2.1 kb located at the 5'-end of the open reading frame (ORF) (Fig. 1).

Nucleotide and predicted amino acid sequence

Sequencing of the *B. pertussis* AC determinant revealed a single ORF, GC-rich (66.6% G+C ratio), of 1740 codons (Glaser *et al.*, 1988a). Two overlapping promoter-like regions, located

upstream of the ATG translational start codon, contained sequences matching in part the *E. coli* −35 and −10 consensus sequences. Comparison of the putative AC gene with other *B. pertussis* genes which have been sequenced indicates that the coding regions share a similar codon usage. In most cases the ATG codon is preceded by a region with homology to the *E. coli* ribosomal binding site and, except where the gene may represent part of an operon, regions similar to the *E. coli* −35 and −10 promoter sequences have been found (Locht and Keith, 1986; Nicosia *et al.*, 1986; Livey *et al.*, 1987; Glaser *et al.*, 1988a,b; Maskell *et al.*, 1988; Charles *et al.*, 1989; Relman *et al.*, 1989).

The ORF assigned to the *B. pertussis* AC structural gene has the potential to encode a polypeptide of 1706 amino acids, having a calculated M_r of 177 000 (Glaser *et al.*, 1988a). The M_r 200 000 forms of AC isolated by us (Rogel *et al.*, 1988, 1989) probably represent the full length of the transcript. Discrepancies between M_r values predicted according to sequence data and determined by SDS–PAGE have often been observed. Translation of the complete ORF yields a protein much longer than the major form of *B. pertussis* AC detected in the culture medium which has an M_r of 45 000. This represents the first 450 N-terminal amino acids, and as noted earlier, biochemical analysis of this form indicated that the ATP-binding site was located on the N-terminal domain (residues 1–235/237), and the CaM-binding site on the C-terminal domain (residues 236/238–399) (Ladant *et al.*, 1989). The function of the remainder of the protein is not known, but it carries the Hly activity of *B. pertussis* and may represent a region necessary for interaction of the enzyme with the host cell prior to penetration (see below).

Bacillus anthracis produces AC which is CaM-dependent and enters eukaryotic cells when combined with a separate protein termed protective antigen (see Chapter 14). This AC determinant has recently been cloned and sequenced (Mock *et al.*, 1988; Robertson *et al.*, 1988). *B. anthracis* AC gene encodes a 800-amino acid protein, but the overall homology of *B. anthracis* AC with *B. pertussis* AC is low. However, three highly conserved amino acid domains were identified, one of which contains consensus sequences present in many eukaryotic and prokaryotic ATP-binding proteins (Escuyer et al., 1988; Robertson, 1988). This homologous domain is located on B. pertussis AC between amino acid 54 and 70, and has the following sequence: ^{54}GVATKGLGVHAKSSDWG^{70}. Site-directed mutagenesis of Lys^{58} or Lys^{65} drastically reduced the catalytic activity of B. pertussis AC without having pronounced effects on its CaM binding (Glaser et al., 1989; Au et al., 1989). This is in agreement with previous studies suggesting the involvement of a basic amino acid in the catalysis of cAMP production (Eckstein et al., 1981; Gerlt et al., 1980).

The location of the CaM-binding site on B. pertussis AC was predicted to reside around Trp^{242}, based on slight primary sequence homology to other CaM-stimulated enzymes (Glaser et al., 1988a) and was supported by the biochemical results mentioned earlier. However, comparison with B. anthracis AC did not display any significant primary structure similarity in the corresponding region, instead the region was bracketed by two regions of strong homology between the two enzymes (Escuyer et al., 1988). Usually, the degree of primary sequence homology between CaM-binding domains is low, although these domains form amphiphilic α-helix type structures (Cox et al., 1985). Site-directed mutagenesis of Trp^{242} with amino acids such as glycine and aspartate, considerably reduced the affinity of the enzyme for CaM without affecting its catalytic potency, thus establishing the importance of Trp^{242} in the binding of CaM to B. pertussis AC (Glaser et al., 1989). The fact that the conservation of primary structure was in regions flanking Trp^{242} indicates that it is secondary structure, probably in the form of an α-helix, that is important for CaM binding.

Analysis of the sequence of B. pertussis AC according to the hydrophobicity plot (Eisenberg et al., 1984) predicted the presence of four possible membrane-spanning regions within a 200-residue segment starting at amino acid 530 (Fig. 2). This segment was suggested to play a role in the penetration of B. pertussis AC into target cells (Hanski, 1989). Glaser et al. (1988b) found that the 1250 C-terminal amino acids of B. pertussis AC display significant degrees of homology with E. coli α-haemolysin (25%) and P. hemolytica leukotoxin (22%). The homology was particularly high between residues 640–910 of B. pertussis AC and corresponding internal domains of the other two

toxins. In addition, a repetitive motif of nine residues, with glycine predominant, which appeared 47 times in the 700 C-terminal amino acids of B. pertussis AC was also found similarly placed with regard to the C-terminus in α-haemolysin (11 repeats) and leukotoxin (6 repeats). The presence of this repeat motif in toxins from a variety of species has prompted the term RTX cytotoxins (for repeats in the structural toxin) to be used to describe this family of virulence factors (Felmlee and Welch, 1988; Strathdee and Lo, 1989). It was suggested that the homology has a functional significance which is exhibited in the mechanism of secretion of B. pertussis AC, and its ability to act as a haemolysin (Glaser et al., 1988b, see below). A schematic representation of the key features of the AC protein predicted from the gene structure is shown in Fig. 2.

The Structure of the Toxic Form

The structure of the toxic form of the enzyme, i.e. the form capable of penetrating eukaryotic cells and catalysing the formation of intracellular cAMP, has been a subject of some controversy. Until recently, none of the extensively purified forms of the enzyme was found to be toxic. Gel filtration of dialysed urea extract of wild-type B. pertussis bacteria showed that the toxic activity migrated as a distinct peak of a minor AC activity and had an apparent size of 200 000 (Fig. 3A, see also Hanski and Farfel, 1985; Weiss et al., 1986; Rogel et al., 1988). However, most of the non-toxic AC activity was associated with proteins of M_r 200 000 and 47 000 which migrated as one peak of aggregated material (Fig. 3A, peak I) and a second peak of apparent size of 50 000 (Fig. 3A, peak III). Attempts to purify the toxic form were frustrated by its minute quantity and the fact that it was probably masked by various non-toxic forms of AC having different M_r values (Fig. 3A, inset). BP348 harbouring pRMB1 (Fig. 1) over-produced two forms of AC, one of which was toxic (Fig. 3B, peak II and Rogel et al., 1989). These two forms of AC, both of M_r 200 000 (Fig. 3B, inset) were resolved and purified to homogeneity. They had similar biochemical and immunological properties. Yet one form migrated on gel filtration as an

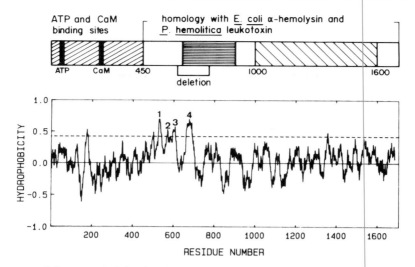

Figure 2. Key features of *B. pertussis* AC related to the gene structure. The location of the ATP- and the CaM-binding domains is discussed in detail in the text. The first 450 N-terminal amino acids represent the form of the enzyme which is detected in *B. pertussis* culture medium. The area of the horizontal bars is highly homologous with internal parts of *E. coli* α-haemolysin and *P. hemolytica* leukotoxin (Glaser *et al.*, 1988b). A C-terminal domain from amino acids 1000 to 1600 contains repeats of nine amino acids having a typical sequence of GGDGDDTLX. The hydrophobicity plot was produced as described by Eisenberg *et al.* (1984) using a 'window' length of 21 residues.

aggregated protein (Fig. 3B, peak I) and was non-toxic, whereas the other migrated as non-aggregated protein (Fig. 3B, peak II), and generated high levels of intraceullar cAMP (Rogel *et al.*, 1989).

The 10-kb insert in pRMB3 (Fig. 1) contains the entire AC ORF and, when transferred to a Lon⁻ *E. coli* strain to minimize proteolytic degradation, expressed an AC protein of M_r 200 000, which was non-toxic and migrated as an aggregate on gel filtration (Rogel *et al.*, 1989). Although the AC purified from *E. coli* had no penetrative activity, it nonetheless displayed similar biochemical and immunological properties to the enzyme purified from BP348 harbouring pRMB1 (Rogel *et al.*, 1989). It was concluded that both the enzymatic and the toxic properties of *B. pertussis* AC reside on the same macromolecule, and no protein entity other than that of the M_r 200 000 polypeptide is required for its penetration. In addition, it was proposed that post-translational modification, which occurs in *B. pertussis* but not in *E. coli*, confers upon the M_r 200 000 protein the toxic structure. A precedent for such a mechanism of toxin activation occurs with *E. coli* α-haemolysin, which shares structural and functional homology with *B. pertussis* AC (Glaser *et al.*, 1988b).

The production of biologically active haemolysin requires a separate gene (*hlyC*). Its product probably acts by modifying the product of the structural gene (*hlyA*), rendering it haemolytically active (Nicaud *et al.*, 1985). As in the case of *B. pertussis* AC, active and non-active forms of *E. coli* haemolysin having identical M_r values have been detected (Wagner *et al.*, 1988).

Others, however, have claimed that the M_r 45 000 form of *B. pertussis* is able, by itself or in combination with other polypeptides, to invade target cells (Masure *et al.*, 1988; Masure and Storm, 1989). The 45 000 AC is the major form of the enzyme which appears in the culture medium. It is possible that the toxic 200 000 form of the enzyme is released first to the medium, but is rapidly degraded and therefore its quantity is very small and barely detectable. In fact, extensive purification of the 45 000 AC yielded a CaM-sensitive enzyme which lacked toxic activity (Masure *et al.*, 1988). Furthermore, it was recently shown that inclusion of red blood cells or bovine serum albumin in a synthetic growth medium for *B. pertussis* resulted in the secretion of a 200 000 form of the enzyme which could penetrate target cells (Bellalou *et al.*, 1990). It appears, therefore,

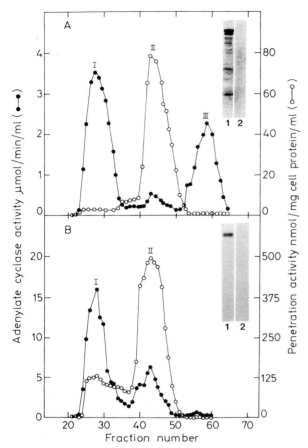

Figure 3. Gel filtration of *B. pertussis* extracts. (A) Dialysed urea extract of the wild-type strain 165 of *B. pertussis* was subjected to chromatography on an AcA 34 column. The determinations of enzymatic and penetration activities, as well as the Western blot analysis were performed as described by Rogel *et al.* (1988). Lanes 1 and 2 represent incubation of the extracts with immune and non-immune sera respectively. (B) Dialysed urea extract of the recombinant strain of *B. pertussis* (BP348pRMB1) was subjected to chromatography and the activity determinations and Western blot analysis were performed as described in A.

that the proteolytic degradation of the 200 000 polypeptide to the 45 000 form may not be relevant to the secretion and production of the toxic form of the enzyme and may not happen under physiological conditions within the host.

Secretion and haemolytic activity

The secretion of proteins by Gram-negative bacteria usually requires their synthesis with an N-terminal

extension, termed the signal or leader peptide. Proteins possessing this leader peptide are secreted in a two-step procedure. In the first step, they are translocated into the periplasm and the leader peptide is cleaved. Then they are transferred across the outer membrane by a specific export mechanism (Pugsley and Schwartz, 1985; Hirst and Welch, 1988). *E. coli* α-haemolysin has no leader peptide and is secreted by a rather unusual mechanism which does not involve the formation of a periplasmic intermediate (Mackman *et al.*, 1986; Koronakis *et al.*, 1989). The sequence predicted for *B. pertussis* AC did not display the presence of a leader sequence in its N-terminal part, as was found, for instance, for the subunits of pertussis toxin (Locht and Keith, 1986; Nicosia *et al.*, 1986). Because of the extensive sequence homology to *E. coli* α-haemolysin, it was postulated that the secretion of *B. pertussis* AC may proceed via a similar mechanism (Glaser *et al.*, 1988b). This was also supported by the observation of Masure and Storm (1989) that little AC activity was detected in the periplasm of *B. pertussis*.

The *E. coli* α-haemolysin system is organized in an operon of four genes, *hlyC*, *hlyA*, *hlyB* and *hlyD* (see Fig. 4). Both *hlyC* and *hlyA*, together, are required for the production of haemolytic activity, whereas *hlyB* and *hlyD*, but not *hlyC*, are essential for secretion of the HlyA gene product (Mackman *et al.*, 1986). Glaser *et al.* (1988b) found downstream of the AC structural gene (*cyaA*) three additional genes that were essential for the secretion of AC. The first two had predicted amino acid sequences very similar to those of *hlyB* and *hlyD*, and were therefore termed *cyaB*

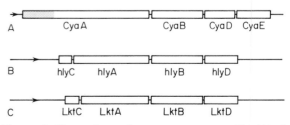

Figure 4. Comparison of operon structures for (A) *B. pertussis* AC, (B) *E. coli* α-haemolysin and (C) *P. hemolytica* leukotoxin. The open boxes represent the length of the protein coding regions for each gene. The hatched area in A represents the catalytic and calmodulin-binding region of the *cyaA* AC gene product. The arrow in each case represents the direction of transcription.

cyaD. The third gene, *cyaE*, produced a protein of unknown function that did not show a significant sequence similarity to any known proteins of *E. coli*. Based on the structural similarity with α-haemolysin it was proposed that the four *cya* genes are organized in a single operon, as depicted in Fig. 4 (Glaser *et al.*, 1988b). The structurally similar operon for leukotoxin secretion in *Pasteurella hemolytica* is also included in Fig. 4. As with the α-haemolysin of *E. coli*, the LktA gene product requires the action of the LktC product to create leukotoxic activity and the *lktB* and *lktD* genes are required for secretion of the toxin (Lo *et al.*, 1987; Strathdee and Lo, 1989; Highlander *et al.*, 1989).

Secretion of HlyA strongly depends on the amino acid sequence in its C-terminal end which may form a secretion signal for direct translocation of the toxin through a trans-envelope channel formed by H1yB and HlyD (Gray *et al.*, 1986; Ludwig *et al.*, 1987; Koronakis *et al.*, 1989). The C-terminal ends of HlyA, LktA and CyaA are poorly conserved, and as the role of CyaE in the secretion of AC is unknown, it is possible that the mechanisms of secretion of these three proteins may differ substantially.

Previous studies using transposon mutagenesis in *B. pertussis* suggested close proximity between the genes for AC and Hly (Weiss *et al.*, 1983). Furthermore, recombinant cosmid pRMB1 restored both AC and Hly activities to BP348 (AC⁻, Hly⁻) (Brownlie *et al.*, 1988). Glaser *et al.* (1988b) showed that restoration of haemolytic activity to BP348 and secretion of AC required the expression of *cyaA*, *B*, *D* and *E* genes, suggesting that *cyaA* may encode a single bifunctional protein carrying both activities. This notion was supported by generation of an in-frame deletion in a region conserved between *B. pertussis* AC, *E. coli* α-haemolysin and *P. hemolytica* leukotoxin (Fig. 2). The protein produced as a result of this deletion possessed enzymatic activity but lacked toxic and haemolytic activities. Mutations affecting haemolytic activity in *E. coli* have also been located to this homologous region (Ludwig *et al.*, 1987).

Properties of cellular penetration

Evidence of entry

Many eukaryotic cells respond to *B. pertussis* AC by extensive production of intracellular cAMP. It was first proposed by Confer and Eaton (1982) that the accumulation of cAMP reflects the entry of the bacterial enzyme rather than activation of the intrinsic membrane-bound eukaryotic AC. Direct evidence for physical penetration of the bacterial enzyme was produced by showing dose-, time- and temperature-dependent accumulation of AC activity in human lymphocytes despite the presence of trypsin in the medium (Hanski and Farfel, 1985; Friedman *et al.*, 1987a). This intracellular activity displayed properties which were similar to those of the bacterial enzyme, but were completely different from those of the intrinsic membrane-bound AC (Friedman *et al.*, 1987a).

Kinetics and route of penetration

Analysis of the kinetics of *B. pertussis* AC entry into mammalian cells showed that the activity appeared in the cells immediately after cell exposure and reached a constant level within 10–40 min (Friedman *et al.*, 1987a; Farfel *et al.*, 1987). This level was maintained provided that the enzyme was present in the incubation medium. Upon transfer of cells into enzyme-free medium a rapid decrease in the intracellular AC activity was observed ($t_{\frac{1}{2}}$ = 15 min). This decrease reflected intracellular inactivation of the bacterial enzyme and did not result from the release of the enzyme back to the incubation medium (Friedman *et al.*, 1987a; Farfel *et al.*, 1987).

The exact mechanism by which *B. pertussis* AC enters target cells is as yet unknown. Nevertheless, several lines of evidence support the notion that the enzyme penetrates plasma membranes of target cells directly and does not enter these cells from endocytic vesicles. (1) The accumulation of the enzyme within the cells proceeds without any noticeable lag period (Friedman *et al.*, 1987a; Farfel *et al.*, 1987). (2) Agents that interfere with endocytosis have no effect on the penetration of the enzyme (Hanski and Farfel, 1985; Gentile *et al.*, 1988; Gordon *et al.*, 1988). (3) Fractionation of human lymphocytes and erythrocytes showed that the *B. pertussis* AC activity was always detected in a particulate fraction, identified as plasma membrane, irrespective of the time of the cells exposure to the enzyme (Farfel *et al.*, 1987).

Comparison of penetration of Chinese hamster ovary (CHO) cells by *B. pertussis* AC and *B. anthracis* AC clearly demonstrated that these two AC toxins enter the same cell using different mechanisms. The entry of *B. anthracis* was by means of receptor-mediated endocytosis into acidic compartments, whereas the entry of *B. pertussis* AC occurred probably directly via the plasma membrane (Gordon *et al.*, 1988). The different routes of cell entry may be attributed to the different structures of the two toxins. Analysis of the sequence of *B. pertussis* AC according to the hydrophobicity plot (Eisenberg *et al.*, 1984) predicts the presence of four possible membrane-spanning regions within a 200-residue segment starting at amino acid 530 (Fig. 2). Deletion of 157 codons in the AC ORF from amino acid 623 to 779 obliterated haemolysin activity without affecting AC activity or secretion (Glaser *et al.*, 1988b), which may indicate that at least the fourth hydrophobic segment (Fig. 2) may be essential for the penetration of *B. pertussis* AC into target cells. Relevant to this is the report of Raptis *et al.* (1989a) who showed that by selective proteolysis of *B. pertussis* AC one can completely inhibit the penetration capacity of the enzyme without affecting its enzymatic activity. The hydrophobicity plot of *B. anthracis* AC sequence detected no potential membrane spanning regions in its whole ORF, consistent with its penetration from acidic vesicles.

The target cell receptor for AC toxin is not known, but Gable *et al.* (1985) showed inhibition of AC toxin penetration of human neutrophils by bovine brain gangliosides or by treatment of neutrophils with neuraminidase to remove sialic acid residues. They therefore suggested that the mammalian cell receptor for AC toxin may be a ganglioside. This notion was further supported by the studies of Gordon *et al.* (1989) who showed that purified *B. pertussis* AC enzyme specifically penetrates liposomes prepared from a mixture of gangliosides and phospholipids.

Washed, intact *B. pertussis* cells were able to increase intracellular cAMP accumulation to the same level as bacterial cell extracts (Hewlett *et al.*, 1987) in macrophage-like cells. This led to the suggestion that the toxin may be delivered by direct interaction of the bacterium with the target cell (Hewlett *et al.*, 1987). Relevant to this is the report of Raptis *et al.* (1989b) who showed inhibition of AC toxin activity by polycations. They suggested that the target cell surface has an anionic receptor site necessary for binding of AC which is irreversibly blocked by such cationic molecules. Similar observations were reported for diphtheria toxin (Eidels and Hart, 1982).

The effects of Ca^{2+} and CaM on penetration

B. pertussis AC-induced cAMP accumulation in target cells requires extracellular Ca^{2+} at physiological concentrations (Confer *et al.*, 1984; Hanski and Farfel, 1985). Since the Ca^{2+}-mediated stimulation of the enzyme occurs at micromolar concentrations (Masure *et al.*, 1988), this Ca^{2+} requirement may reflect an effect on the configuration of the enzyme (Masure *et al.*, 1988) or on its process of entry. There is some evidence to suggest that Ca^{2+} might be involved in the attachment of *B. pertussis* toxin to target cells. *E. coli* α-haemolysin requires extracellular Ca^{2+} for its action (Springer and Goebel, 1980). Recently, it has been discovered that the 11 tandemly repeated sequences of nine amino acids are responsible for binding of α-haemolysin to erythrocyte plasma membranes (Ludwig *et al.*, 1988) and may represent a recognition site for a receptor on the erythrocyte membrane. The proper conformation of this repeat region may in turn depend on the presence of Ca^{2+}. Similar nanomers were identified on *B. pertussis* AC (Glaser *et al.*, 1988b, Fig. 2).

Addition of CaM to *B. pertussis* AC prevented the formation of cAMP catalysed by the enzyme in some target cells (Shattuck and Storm, 1985; Gentile *et al.*, 1988). The CaM-mediated inhibition of the enzyme penetration was dose dependent and inversely proportional to the CaM-dependent stimulation of the catalytic activity (Rogel *et al.*, 1988). This may suggest that binding of CaM to the enzyme induces a conformational change that disturbs the structure of the toxic form. Indeed, binding of CaM to *B. pertussis* AC stabilized it against proteolysis (Ladant *et al.*, 1986).

Intracellular inactivation

Recently, it was discovered that the inactivation of *B. pertussis* AC in intact cells and cell lysates is dependent on ATP (Gilboa-Ron *et al.*, 1989). Depletion of cellular ATP by treatment of human lymphocytes with metabolic inhibitors completely blocked the intracellular inactivation of the enzyme. The inactivation of *B. pertussis* AC *in vitro* by lymphocyte lysate was associated with ATP-mediated degradation of ^{125}I-labelled enzyme. In this case, ATP was required for degradation but not as an energy source, since non-hydrolysable ATP analogues supported both inactivation and degradation of the enzyme. Although *B. pertussis* AC toxin is short lived, it generates high levels of intracellular cAMP due to its high turnover number. Exposure of *B. anthracis* AC to lymphocyte lysates resulted in the loss of a significant part of its activity. The rate of *B. anthracis* AC inactivation was enhanced by the presence of ATP. In contrast, neither the human lymphocyte AC nor the soluble rat CaM-stimulated brain AC were inactivated by lymphocyte lysates in the presence and absence of ATP (Gilboa-Ron *et al.*, 1989). These results suggest that *B. pertussis* AC, as well as *B. anthracis* AC, are likely to be degraded by the same host protease, supporting the hypothesis that selective proteolysis is a major mechanism of intracellular inactivation of the AC toxins.

In summary, it appears that *B. pertussis* AC binds to the target cell through a Ca^{2+}-dependent process and then directly penetrates the host cell plasma membrane. The enzyme remains associated with the host plasma membrane, with the ATP- and CaM-binding sites exposed to the cytosol. After its activation by CaM it generates high levels of cAMP and is rapidly inactivated. A simplified view of the events which occur upon the interaction of the toxin with a target cell is presented in Fig. 5.

Pathophysiological effects and immunogenicity of AC

Immune effector cells are assumed to be the primary target of AC toxin since cAMP accumulation is known to inhibit the functions of these cells (Bourne *et al.*, 1974). Confer and Eaton (1982) showed that neutrophils and macrophages have impaired chemotaxis, a reduced oxidative response and decreased killing capacity upon their exposure to *B. pertussis* extract containing AC. These observations were later extended to monocytes (Pearson *et al.*, 1987), where it was also demonstrated that under certain conditions it is possible to block oxidative responses without affecting phagocytosis (Symes *et al.*, 1983; Friedman *et al.*, 1987b; Galgiani *et al.*, 1988). Further studies showed that the cytotoxicity of natural killer cells can be blocked by their treatment with *B. pertussis* AC. This effect was synergisitic with that caused by pertussis toxin (Hewlett *et al.*, 1983).

An effect on the mucosa, to produce the mucus secretion which is a feature of pertussis has also been suggested (Hewlett and Gordon, 1988). Brezin *et al.* (1987) and Guiso *et al.* (1989) have associated AC activity with acute haemorrhagic alveolitis following intranasal infection of mice. Delivery of AC by direct interaction of the bacterium and the target cell would predict that the toxin has a localized site of action in the respiratory tract and is not a systemic toxin like pertussis toxin. Disruption of immune cell clearance mechanisms and perhaps damage to respiratory epithelial cells, where the haemolytic properties of AC toxin may be important, are thought to be the main contributions of the toxin to pathogenesis. The observation that the oxidative burst, which is associated with intracellular killing by phagocytes, is inhibited without an effect on phagocytes might indicate that AC toxin may play a role in intracellular survival of *B. pertussis*. Early reports indicated that *B. pertussis* can establish an intracellular habitat (Crawford and Fishel, 1959; Gray and Cheers, 1969) and this has been corroborated by more recent work (Ewanowich *et al.*, 1989; Pileri *et al.*, 1990).

It is important to emphasize that AC is one of a range of virulence factors coordinately regulated by the action of the *vir* locus. Spontaneous phase variants that have simultaneously lost the expression of all virulence factors and which no longer cause disease in mice, arise at high frequency (one per $10^3 - 10^6$ cells). Reversion back to the virulent phase has been shown to occur at a lower frequency (Weiss and Falkow, 1984). The basis of this phase variation has been explained, at least for one strain of *B. pertussis*, by the occurrence of

PENETRATION

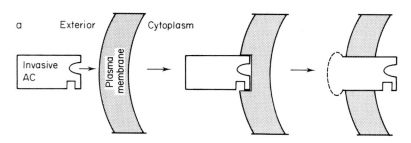

ACTIVATION AND cAMP GENERATION

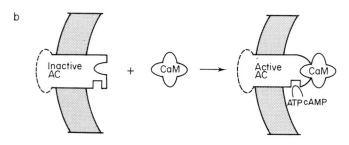

INACTIVATION AND DEGRADATION

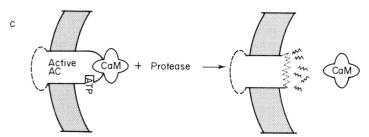

Figure 5. A simplified view of the molecular events which occur upon invasion of target cells by *B. pertussis* AC. (a) *B. pertussis* AC penetrates the host cell plasma membrane with its putative hydrophobic domains, exposing the catalytic and CaM-binding sites to the cytosol. (b) Upon interaction with CaM the enzyme is greatly stimulated and converts intracellular ATP to cAMP. (c) The enzyme within the cell is inactivated by a protease.

spontaneous frame-shift mutations in the *vir* locus (Stibitz *et al.*, 1989). In addition, the virulence genes of *B. pertussis* are also regulated by a process called antigenic modulation (Lacey, 1960), whereby none of the virulence factors are produced at low temperature (<28°C) or when the growth medium is supplemented with high $MgSO_4$ or nicotinic acid concentrations. The effects of these environmental stimuli on virulence gene expression are again regulated through the action of the *vir* locus in an,

as yet, unknown way (Stibitz *et al.*, 1988). There is evidence that both processes may occur *in vivo* (Kasuga *et al.*, 1954; Lacey, 1960) and may help *B. pertussis* to establish a carrier state, whereby it no longer expresses its virulence factors and so remains undetected by the immune system and does not cause disease in the infected individual. Such hosts might also serve as a reservoir for infection because lack of expression, for example, of attachment factors would allow organisms to be expelled more easily from an infected individual for transmission to a new host (Robinson *et al.*, 1986; Wardlaw and Parton, 1988).

There is no evidence to suggest that *B. pertussis* AC, via its product cAMP, acts to regulate gene expression as it does in many other prokaryotes (Botsford, 1981). During antigenic modulation AC was lost concomitantly with other virulence factors. Inducers of modulation, such as $MgSO_4$, has no inhibitory effect on AC activity; neither did extracts of modulated cells. Antiserum against purified *E. coli* catabolite repressor protein gave no reaction with *B. pertussis* (Brownlie *et al.*, 1985b). Monneron *et al.* (1988) reported that *B. pertussis* AC (M_r 43 000–50 000) was immunologically related to rat brain AC catalytic component, but not to *E. coli* AC. As calmodulin is not known to occur in bacteria, they suggested that the bacterial enzyme may have arisen by transfer from an eukaryotic source. Recently, the eukaryotic AC was cloned and sequenced (Krupinski *et al.*, 1989). Its overall amino acid sequence homology with *B. pertussis* AC is very low. However, both enzymes, as expected, display sequence homology with consensus sequences proposed for ATP-binding domains derived from many different ATP-binding proteins (E. Hanski, unpublished results). Indeed, antibodies raised against a synthetic peptide covering the ATP-binding domain in *B. pertussis* AC cross-reacted with the eukaryotic brain AC on Western blots (Goyard *et al.*, 1989).

The immunogenic properties of *B. pertussis* AC have not been fully evaluated, mainly due to the difficulty, until recently, of obtaining sufficient quantities of purified protein. Antibody to AC purified from the culture medium of *B. pertussis* passively protected suckling mice from an otherwise lethal intranasal challenge (Brezin *et al.*, 1987) and in another report active immunization with AC purified from the culture medium protected mice against a lethal respiratory challenge (Guiso *et al.*, 1989). We have found that the 200 000 AC purified from *E. coli* expressing *B. pertussis* AC from plasmid pRMB3 offered partial protection in mice vaccinated at 3 weeks and challenged intranasally at 5 weeks (Brownlie *et al.*, 1990). AC purified from *B. pertussis* urea extracts is also able to partially protect mice from intranasal challenge (I. Livey, pers. commun.). Anti-pertussis rabbit

serum has a high anti-AC neutralizing titre (Gentile *et al.*, 1988) and anti-AC titres were found to be 100-fold higher in whole-cell-vaccinated than in non-vaccinated children (E. Hanski, unpublished results). This latter observation corroborates the report of Hewlett *et al.* (1977) who detected AC enzymic activity in some whole-cell *B. pertussis* vaccines. Taken together these results indicate that AC, or a toxoided derivative of it, could be a candidate antigen for inclusion in an acellular vaccine. The reported immunological cross-reaction between *B. pertussis* AC and rat brain AC (Monneron *et al.*, 1988) should not constitute a serious problem. This can be overcome by using as a vaccine those regions of the protein not associated with AC catalytic activity and which, because they allow interaction with the host target cell, might be more likely to harbour protective epitopes.

Concluding remarks

The accumulated data suggest that *B. pertussis* AC toxin is composed of a single polypeptide that possesses both the binding and the toxic functions. The AC toxin penetrates the host cell directly through the plasma membrane and, due to its own highly active catalyst, it causes a dramatic increase in intracellular cAMP. In contrast to the long-lasting signals produced by the toxins acting by means of ADP-ribosylation, the toxic signal produced by *B. pertussis* AC is transient and depends on the prolonged presence of the invasive enzyme. Removal of the toxin from the target cell medium results in a rapid loss of its intracellular activity, leading to a decay in the level of intracellular cAMP. The exact physiological role and the specificity of the putative host protease responsible for the inactivation of *B. pertussis* AC is not known. Nevertheless, it may represent a system by which the host cell protects itself against foreign invasive proteins.

Why has *B. pertussis* devised two different mechanisms for elevating cAMP in the host? It is possible that pertussis toxin and AC toxin are affecting different types of target cells. Alternatively, it is clear that the elevation of cAMP by AC toxin is immediate whereas pertussis toxin is much slower acting. Thus, it is possible that AC toxin may be responsible for the short-term increase in cAMP while pertussis toxin acting in concert amplifies the cAMP signal thereafter. Many details about the unusual *B. pertussis* AC toxin remain to be clarified: (1) The exact nature of the toxic form, and how it penetrates the plasma membrane. (2) The relationship between the toxic and the haemolytic activities of the enzyme. It is possible, for example, that the process of entry may leave the AC catalytic moiety associated with the target cell membrane and release the remainder of the protein as a haemolysin, or penetration damages the membrane which in the case of red cells is manifest as haemolysis. (3) The mechanism of toxin secretion. (4) The role of the toxin in whooping cough pathogenesis.

Further understanding of *B. pertussis* AC action will contribute to the knowledge of how polypeptide molecules cross the cell membrane and may provide new ideas for delivery of proteins into target cells. In addition, these studies may give insight into the pathophysiology of pertussis infection and into the host–pathogen relationship. Furthermore, they may contribute to a better diagnosis and prevention of whooping cough by development of a safer vaccine.

Acknowledgements

We are grateful to Rona Levin for the excellent word processing of this manuscript. Work in Israel was supported in part by grants from the US–Israel Cooperative Development Programme and the US-Binational Science Foundation. Work in UK was funded by the Medical Research Council, UK. Equipment was made available by the Wood Boyd Fund of the University of Glasgow.

References

Aseeva, L.E., Shevchenko, L.A., Shimaniuk, N.I., Rubler, B.D. and Mishankin, B.N. (1987). Assessment of the modulating effect of adenylate cyclase of the plague microbe on guinea pig peritoneal leukocytes with the aid of chemiluminescence. *Z. Mikrobiol. Epidemiol. Immunobiol.* **7**, 59–63.

Au, D.C., Masure, H.R. and Storm, D.R. (1989). Site directed mutagenesis of lysine 58 in a putative-ATP-binding domain of the calmodulin-sensitive adenylate cyclase from *Bordetella pertussis* abolishes catalytic activity. *Biochemistry* **28**, 2772–6.

Bellalou, J., Geoffroy, C., Ladant, D. and Ullmann, A. (1990). Secretion of *B. pertussis* adenylate cyclase haemolysin bifunctional protein. In: *Fourth European Workshop on Bacterial Protein Toxins* (eds R. Rappuoli *et al.*) pp. 397–8. Gustav Fischer, Stuttgart.

Bordet, J. and Gengou, O. (1906). L'endotoxine coqueluche. *Ann. Inst. Pasteur. (Paris)* **23**, 415–741.

Botsford, J.C. (1981). Cyclic nucleotides in prokaryotes. *Microbiol. Rev.* **45**, 620–42.

Bourne, H.R., Lichtenstein, L.M., Melmon, K.L., Henney, C.S., Weinstein, Y. and Shearer, G.M. (1974). Modulation of inflammation and immunity by cyclic AMP. *Science* **184**, 19–28.

Brezin, C., Guiso, N., Ladant, D., Djavadi-Ohaniance, L., Megret, F., Onyeocha, I. and Alonso, J.M. (1987). Protective effects of anti-*Bordetella pertussis* adenylate cyclase antibodies against lethal respiratory infection of the mouse. *FEMS Microbiol. Lett.* **42**, 75–80.

Brownlie, R.M., Parton, R. and Coote, J.G. (1985a). The effect of growth conditions on adenylate cyclase activity and virulence-related properties of *Bordetella pertussis*. *J. Gen. Microbiol.* **131**, 17–25.

Brownlie, R.M., Coote, J.G. and Parton, R. (1985b). Adenylate cyclase activity during phenotypic variations of *Bordetella pertussis*. *J. Gen. Microbiol.* **131**, 27–38.

Brownlie, R.M., Coote, J.G. and Parton, R. (1986). Complementation of mutations in *Escherichia coli* and *Bordetella pertussis* by *B. pertussis* DNA cloned in broad-host-range cosmid vector. *J. Gen. Microbiol.* **132**, 3221–9.

Brownlie, R.M., Coote, J.G., Parton, R., Schultz, J.E., Rogel, A. and Hanski, E. (1988). Cloning of adenylate cyclase genetic determinant of *Bordetella pertussis* and its expression in *Escherichia coli* and *B. pertussis*. *Microb. Pathogen.* **4**, 335–44.

Brownlie, R.M., Coote, J.G., Parton, R., Rogel, A., Goldschmidt, A. and Hanski, E. (1990). Properties of *Bordetella pertussis* adenylate cyclase expressed in *E. coli*. In: *Fourth European Workshop on Bacterial Protein Toxins* (eds R. Rappuoli *et al.*) pp. 403–4, Gustav Fischer, Stuttgart.

Centers for Disease Control (1984). Pertussis – Washington. *Morbid. Mortal. Weekly Rep.* **34**, 390–400.

Charles, I.G., Dougan, G., Pickard, D., Chatfield, S., Smith, M., Novotny, P., Morrissey, P. and Fairweather, N.F. (1989). Molecular cloning and characterization of protective outer membrane protein P.69 from *Bordetella pertussis*. *Proc. Natl. Acad. Sci. USA* **86**, 3554–8.

Cherry, J.D. (1984). The epidemiology of pertussis and pertussis immunization in the United Kingdom and the United States: A comparative study. *Curr. Prob. Pediatr.* **14**, 1–78.

Cone, T.E. (1970). Whooping cough is first described

as a disease sui generis by Bajallou in 1640. *Pediatrics* **46**, 522.

Confer, D.L. and Eaton, J.W. (1982). Phagocyte impotence caused by the invasive bacterial adenylate cyclase. *Science* **217**, 948–50.

Confer, D.L., Slungaard, A., Graf, E., Panter, S.S. and Eaton, J.W. (1984). *Bordetella* adenylate cyclase toxin: Entry of bacterial adenylate cyclase into mammalian cells. *Adv. Cyclic Nucleotide Res.* **17**, 183–7.

Cox, J.A., Comte, M., Fitton, J.E. and DeGrado, W.F. (1985). The interaction of calmodulin with amphiphilic peptides. *J. Biol. Chem.* **260**, 2527–34.

Crawford, J.G. and Fishel, C.W. (1959). Growth of *Bordetella pertussis* in tissue culture. *J. Bacteriol.* **77**, 465–74.

Eckstein, F., Romaniuk, P.J., Heideman, W. and Storm, D.R. (1981). Stereochemistry of mammalian adenylate cyclase reaction. *J. Biol. Chem.* **256**, 9118–20.

Eidels, L. and Hart, D.A. (1982). Effect of polymers of L-lysine on the cytotoxic action of diphtheria toxin. *Infect. Immun.* **37**, 1054–8.

Eisenberg, D., Schwartz, E., Komaromy, M. and Wall, R. (1984). Analysis of membrane and surface protein sequences with the hydrophobic moment plot. *J. Molec. Biol.* **179**, 125–42.

Endoch, M., Takezawa, T. and Nakase, Y. (1980). Adenylate cyclase activity of *Bordetella* organism, I. Its production in liquid medium. *Microbiol. Immunol.* **24**, 95–104.

Escuyer, V., Duflot, E., Sezer, O., Danchin, A. and Mock, M. (1988). Structural homology between virulence-associated bacterial adenylate cyclases. *Gene* **71**, 293–8.

Ewanowich, C.A., Melton, A.Z., Weiss, A.A. and Sherbe, R.K. (1989). Invasion of HeLa cells by virulent *Bordetella pertussis*. *Infect. Immun.* **57**, 2699–717.

Farfel, Z., Friedman, E. and Hanski, E. (1987). The invasive adenylate cyclase of *Bordetella pertussis*: Intracellular localization and kinetics of penetration into various cells. *Biochem. J.* **243**, 153–8.

Felmlee, T. and Welch, R.A. (1988). Alterations of amino acid repeats in *Escherichia coli* hemolysin affect cytolytic activity and secretion. *Proc. Natl. Acad. Sci. USA* **85**, 5268–73.

Fishel, C.W., O'Bryan, B.S., Smith, D.L. and Jewel, J.W. (1970). Conversion of ATP by mouse lung preparations and by an extract of *Bordetella pertussis*. *Int. Arch. Allergy* **38**, 457–62.

Friedman, R.L. (1987). *Bordetella pertussis* adenylate cyclase: Isolation and purification by calmodulin-sepharose 4B chromatography. *Infect. Immun.* **55**, 129–34.

Friedman, E., Farfel, Z. and Hanski, E. (1987a). The invasive adenylate cyclase of *Bordetella pertussis*: Properties and penetration kinetics. *Biochem. J.* **243**, 145–51.

Friedman, R.L., Fiederlein, R.L., Glasser, L. and

Galgiani, J.N. (1987b). *Bordetella pertussis* adenylate cyclase: Effects of affinity-purified adenylate cyclase on human polymorphonucelar leukocyte functions. *Infect. Immun.* **55**, 135–40.

Gable, P., Eaton, J.W. and Confer, D.C. (1985). Intoxication of human phagocytes by *Bordetella* adenylate cyclase toxin: Implication of a ganglioside receptor. *Clin. Res.* **33**, 844A.

Galgiani, J.N., Hewlett, E.L. and Friedman, R.L. (1988). Effects of adenylate cyclase toxin from *Bordetella pertussis* on human neutrophil interactions with *Coccidioides immitis* and *Staphylococcus aureus*. *Infect. Immun.* **56**, 751–5.

Gentile, F., Raptis, A., Knipling, L.G. and Wolff, J. (1988). *Bordetella pertussis* adenylate cyclase-penetration into host cells. *Eur. J. Biochem.* **175**, 447–53.

Gentry-Weeks, C.R., Cookson, B.T., Goldman, W.E., Rimler, R.B., Porter, S.B. and Curtiss, R. (1988). Dermonectrotic toxin and tracheal cytotoxin, putative virulence factors of *Bordetella avium*. *Infect. Immun.* **56**, 1698–707.

Gerlt, J.A., Coderre, J.A. and Wolin, M.S. (1980). Mechanism of the adenylate cyclase reaction. *J. Biol. Chem.* **255**, 331–4.

Gilboa-Ron, A., Rogel, A. and Hanski, E. (1989). *Bordetella pertussis* adenylate cyclase inactivation by the host cell. *Biochem. J.* **262**, 25–31.

Glaser, P., Ladant, D., Sezer, O., Pichot, F., Ullman, A. and Danchin, A. (1988a). The calmodulin-sensitive adenylate cyclase of *Bordetella pertussis*: Cloning and expression in *Escherichia coli*. *Molec. Microbiol.* **2**(1), 19–30.

Glaser, P., Sakamoto, H., Bellalou, J., Ullmann, A. and Danchin, A. (1988b). Secretion of cyclolysin, the calmodulin-sensitive adenylate cyclase-hemolysin bifunctional protein of *Bordetella pertussis*. EMBO J. **7**, 3997–4004.

Glaser, P., Elmaoglou-Lazaridou, A., Krin, E., Ladant, D., Barzu, O. and Danchin, A. (1989). Identification of residue essential for catalysis and binding of calmodulin in *Bordetella pertussis* adenylate cyclase by site directed mutagenesis. *EMBO J.* **8**, 967–72.

Goldhammer, A. and Wolff, J. (1982). Assay of calmodulin with *Bordetella pertussis* adenylate cyclase. *Analyt. Biochem.* **124**, 45–52.

Goldhammer, A.R., Wolff, J., Cook, G.H., Berkowitz, S.A., Klee, C.B., Manclark, C.R. and Hewlett, E.L. (1981). Spurious protein activators of *Bordetella pertussis* adenylate cyclase. *Eur. J. Biochem.* **115**, 605–9.

Goldman, S., Hanski, E. and Fish, F. (1984). Spontaneous phase variation in *Bordetella pertussis* is a multistep non-random process. *EMBO J.* **3**(6), 1353–6.

Gordon, V.M., Leppla, S.H. and Hewlett, E.L. (1988). Inhibitors of receptor-mediated endocytosis block the entry of *Bacillus anthracis* adenylate cyclase toxin but not that of *Bordetella pertussis* adenylate cyclase toxin. *Infect. Immun.* **56**, 1066–9.

Gordon, V.M., Young, W.W., Lechler, S.M., Gray,

M.C., Leppla, S.H. and Hewlett, E.L. (1989). Adenylate cyclase toxins from *Bacillus anthracis* and *Bordetella pertussis*: Different processes for interaction with and entry into target cells. *J. Biol. Chem.* **264**, 14792–6.

Goyard, S., Orlando, C., Sabatier, J.-M., Labruyere, E., d'Alayer, J., Fontan, G., van Rietschaten, J., Mock, M., Dancin, A. and Monneron, A. (1989). Identification of a common domain in calmodulin-activated eukaryotic and bacterial adenylate cyclases. *Biochemistry* **28**, 1964–7.

Gray, D.F. and Cheers, C. (1969). The sequence of enhanced cellular activity and protective humoral factors in murine pertussis immunity. *Immunology* **17**, 889–96.

Gray, L., Mackman, N., Nicaud, J.-M. and Holland, I.B. (1986). The carboxy-terminal region of haemolysin 2001 is required for secretion of the toxin from *Escherichia coli*. *Molec. Gen. Genet.* **205**, 127–33.

Greenlee, D.V., Andreasen, T.J. and Storm, D.R. (1982). Calcium-independent stimulation of *Bordetella pertussis* adenylate cyclase by calmodulin. *Biochemistry* **21**, 2759–64.

Guiso, N., Rocancourt, M., Szatanik, M. and Alonso, J.M. (1989). *Bordetella* adenylate cyclase is a virulence associated factor and an immunoprotective antigen. *Microb. Pathogen.* **7**, 373–80.

Hanski, E. (1989). Invasive adenylate cyclase toxin of *Bordetella pertussis*. *TIBS* **14**, 459–63.

Hanski, E. and Farfel, Z. (1985). *Bordetella pertussis* invasive adenylate cyclase: Partial resolution and properties of cellular penetration. *J. Biol. Chem.* **260**, 5526–32.

Herzberg, O. and James, M.N.G. (1985). Structure of the calcium regulatory muscle protein troponin-C at 2.8A resolution. *Nature (Lond.)* **313**, 653–9.

Hewlett, E.L. and Gordon, V.M. (1988). Adenylate cyclase toxin of *Bordetella pertussis*. In: *Pathogenicity and Immunity in Pertussis* (eds A.C. Wardlaw and R. Parton), pp. 193–209. Wiley, Chichester and New York.

Hewlett, E. and Wolff, J. (1976). Soluble adenylate cyclase from the culture medium of *Bordetella pertussis*: Purification and characterization. *J. Bacteriol.* **127**, 890–8.

Hewlett, E.L., Urban, M.A., Manclark, C.R. and Wolff, J. (1976). Extracytoplasmic adenylate cyclase of *Bordetella pertussis*. *Proc. Natl. Acad. Sci. USA* **73**, 1926–30.

Hewlett, E.L., Manclark, C.R. and Wolff, J. (1977). Adenyl cyclase in *Bordetella pertussis* vaccines. *J. Infect. Dis.* **136**, 5216–19.

Hewlett, E.L., Wolff, J. and Manclark, C.R. (1978). Regulation of *Bordetella pertussis* extracytoplasmic adenylate cyclase. *Adv. Cyclic. Nucleotide Res.* **9**, 621–8.

Hewlett, E.L., Smith, D.L., Myers, G.A., Pearson, R.D. and Kay, H.D. (1983). Inhibition of *in vitro* natural killer (NK) cell cytotoxicity by *Bordetella* cyclase and *pertussis* toxin. *Clin. Res.* **31**, 365a.

Hewlett, E.L., Gray, M.C. and Pearson, R.D. (1987). Delivery of *Bordetella pertussis* adenylate cyclase toxin to target cells by intact bacteria. *Clin. Res.* **35**, 477A.

Highlander, S.K., Chidambaram, M., Engler, M.J. and Weinstock, G.M. (1989). DNA sequence of the *Pasteurella haemolytica* leukotoxin gene cluster. *DNA* **8**, 15–28.

Hirata, M. and Hayaishi, O. (1967) Adenylate cyclase of *Breivbacterium liquefaciens*. *Biochim. Biophys. Acta* **149**, 1–11.

Hirst, T.H. and Welch, R.A. (1988). Mechanisms for secretion of extracellular protein by gram-negative bacteria. *TIBS* **13**, 265–9.

Ide, M. (1969). Adenyl cyclase of *Escherichia coli*. *Biochem. Biophys. Res. Commun.* **36**, 42–6.

Ide, M. (1971). Adenyl cyclase of bacteria. *Arch. Biochem. Biophys.* **144**, 262–8.

Kasuga, T., Nakase, Y., Ukishima, K. and Takatsu, K. (1954). Studies on *Haemophilus pertussis*. V. Relation between the phase of bacilli and the progress of the whooping cough. *Kitasato Arch. Exp. Med.* **27**, 57–62.

Kessin, R.H. and Franke, J. (1986). Secreted adenylate cyclase of *Bordetella pertussis*: Calmodulin requirements and partial purification of two forms. *J. Bacteriol.* **166**, 290–6.

Kilhoffer, M.-C., Cook, G.H. and Wolff, J. (1983). Calcium-independent activation of adenylate cyclase by calmodulin. *Eur. J. Biochem.* **133**, 11–15.

Klee, C.B. and Vanaman, T.C. (1982). Calmodulin. *Adv. Protein Chem.* **35**, 213–321.

Koronakis, V., Koronakis, E. and Hughes, C. (1989). Isolation and analysis of the C-terminal signal directing export of *E. coli* hemolysin protein across both bacterial membranes. *EMBO J.* **8**, 595–605.

Krupinski, J., Coussen, F., Bakalyar, H.A., Tang, W.-J., Feinstein, P.G., Orth, K., Slaughter, C., Reed, R.R. and Gilman, A.G. (1989). Adenylyl cyclase amino acid sequence: Possible channel- or transporter like structure. *Science* **244**, 1558–64.

Lacey, B.W. (1960). Antigenic modulation of *Bordetella pertussis*. *J. Hyg. Camb.* **58**, 57–93.

Ladant, D. (1988). Interaction of *Bordetella pertussis* adenylate cyclase with calmodulin: identification of two separated calmodulin-binding domains. *J. Biol. Chem.* **263**, 2612–18.

Ladant, D., Brezin, C., Alonso, J.-M., Crenon, I. and Guiso, N. (1986). *Bordetella pertussis* adenylate cyclase, purification, characterization and radioimmunoassay. *J. Biol. Chem.* **261**, 16264–9.

Ladant, D., Michelson, S., Sarfati, R., Gilles, A.-M., Predeleanu, R. and Barzu, O. (1989). Characterization of the calmodulin-binding and the catalytic domains of *Bordetella pertussis* adenylate cyclase. *J. Biol. Chem.* **264**, 4015–20.

Leslie, P.H. and Gardner, A.D. (1931). The phases of *Haemophilus pertussis*. *J. Hyg.* **3**, 423–34.

Livey, I., Duggleby, C.J. and Robinson, A. (1987). Cloning and nucleotide sequence analysis of the serotype α finbrial subunit gene of *Bordetella pertussis*. *Molec. Microbiol.* **1**(2), 203–9.

Locht, C. and Keith, J.M. (1986). Pertussis toxin gene: Nucleotide sequence and genetic organization. *Science* **232**, 1258–63.

Lo, R.Y.C., Strathdee, C.A. and Shewen, P.E. (1987). Nucleotide sequence of the leukotoxin genes of *Pasteurella haemolytica* A1. *Infect. Immun.* **55**, 1987–96.

Ludwig, A., Vogel, M. and Goebel, W. (1987). Mutations affecting activity and transport of haemolysin in *Escherichia coli*. *Molec. Gen. Genet.* **206**, 238–45.

Ludwig, A.T., Jarchau, T., Benz, R. and Goebel, W. (1988). The repeat domain of *E. coli* hemolysin (HlyA) is responsible for its Ca^{2+}-dependent binding to erythrocytes. *Molec. Gen. Genet.* **214**, 553–61.

Mackman, N., Nicaud, J.-M., Gray, L. and Holland, I.B. (1986). Secretion of haemolysin by *Escherichia coli*.: *Curr. Topics Microbiol. Immunol.* **125**, 159–81.

Marsh, J.L., Erfle, M. and Wykes, E.J. (1984). The pIC plasmid and phage vectors with versatile cloning sites for recombinant selection by insertional inactivation. *Gene* **32**, 481–5.

Maskell, D.J., Morrissey, P. and Dougan, G. (1988). Cloning and nucleotide sequence of the aroA gene of *Bordetella pertussis*. *J. Bacteriol.* **170**, 2467–71.

Masure, H.R. and Storm, D.R. (1989). Characterization of the bacterial cell associated calmodulin-sensitive adenylate cyclase from *Bordetella pertussis*. *Biochemistry* **28**, 438–42.

Masure, H.R., Shattuck, R.L. and Storm, D.R. (1987). Mechanisms of bacterial pathogenicity that involves production of calmodulin-sensitive adenylate cyclases. *Microbiol. Rev.* **51**, 60–5.

Masure, H.R., Oldenburg, D.J., Donovan, M.G., Shattuck, R.L. and Storm, D.R. (1988) The interaction of Ca^{2+} with the calmodulin-sensitive adenylate cyclase from *Bordetella pertussis*. *J. Biol. Chem.* **263**, 6933–40.

McPheat, W.L., Wardlaw, A. and Novotny, P.C. (1983). Modulation of *Bordetella pertussis* by nicotinic acid. *Infect. Immun.* **41**, 516–522.

Mock, M., Labruyere, E., Glaser, P., Danchin, A. and Ullmann, A. (1988). Cloning and expression of the calmodulin-sensitive *Bacillus anthracis* adenylate cyclase in *Escherichia coli*. *Gene* **64**, 277–84.

Monneron, A., Ladant, D., d'Alayer, J., Bellalou, J., Barzu, O. and Ullmann, A. (1988). Immunological relatedness between *Bordetella pertussis* and rat brain adenylyl cyclases. *Biochemistry* **27**, 536–9.

Muller, A.S., Leeuwenburg, J. and Voorhoeve, A.M. (1984). Pertussis in rural areas of Kenya: Epidemiology and results of a vaccine trial. *Bull. WHO* **62**, 899–908.

Nicaud, J.-M., Mackman, N., Gray, L. and Holland, I.B. (1985). Characterization of HlyC and mechanism of activation and secretion of haemolysin from *E. coli* 2001. *FEBS Lett.* **187**, 339–44.

Nicosia, A., Perugini, M., Franzini, C., Casagli, M.C.,

Giuseppina, M., Borri, M.G., Antoni, G., Almoni, M., Neri, P., Ratti, G. and Rappuoli, R. (1986). Cloning and sequencing of pertussis toxin genes: Operon structure and gene duplication. *Proc. Natl. Acad. Sci. USA* **83**, 4631–5.

Olson, L.C. (1975). Pertussis. *Medicine (Baltimore)* 427–69.

Parton, R. and Durham, J.P. (1978). Loss of adenylate cyclase activity in variants of *Bordetella pertussis*. *FEMS Microbiol. Lett.* **4**, 287–9.

Pearson, R.D., Symes, P., Conboy, M., Weiss, A.A. and Hewlett, E.L. (1987). Inhibition of monocyte oxidative response by *Bordetella pertussis* adenylate cyclase toxin. *J. Immunol.* **139**, 2749–54.

Peppler, M.S. (1982). Isolation and characterisation of isogenic pairs of domes hemolytic and flat non-hemolytic colony types of *Bordetella pertussis*. *Infect. Immun.* **35**, 840–51.

Pileri, P., Peppolini, S., Nuti, S., Tagliabue, A. and Nencioni, L. (1990). Evaluation of eukaryotic cell attachment and internalisation by *Bordetella*. In: *Fourth European Workshop on Bacterial Protein Toxins* (eds R. Rappuoli *et al.*) pp. 391–5. Gustav Fischer, Stuttgart.

Pugsley, A.P. and Schwartz, M. (1985). Export and secretion of proteins by bacteria. *FEMS Microbiol. Rev.* **32**, 3–38.

Raptis, A., Knipling, L.G. and Wolff, J. (1989a). Dissociation of catalytic and invasive activities of *Bordetella pertussis* adenylate cyclase. *Infect. Immun.* **57**, 1725–30.

Raptis, A., Knipling, L.G., Gentile, F. and Wolff, J. (1989b). Modulation of invasiveness and catalytic activity of *Bordetella pertussis* adenylate cyclase by polycations. *Infect. Immun.* **57**, 1066–71.

Relman, D.A., Domenighini, M., Tuomanen, E., Rappuoli, R. and Falkow, S. (1989). Filamentous hemagglutinin of *Bordetella pertussis*: Nucleotide sequence and crucial role in adherence. *Proc. Natl. Acad. Sci. USA* **86**, 2637–41.

Robertson, D.L. (1988). Relationships between the calmodulin-dependent adenylate cyclases produced by *Bacillus anthracis* and *Bordetella pertussis*. *Biochem. Biophys. Res. Commun.* **157**, 1027–32.

Robertson, D.L., Tippetts, M.T. and Leppla, S.H. (1988). Nucleotide sequence of *Bacillus anthracis* edema factor gene (cya): A calmodulin dependent adenylate cyclase. *Gene* **73**, 363–71.

Robertson, P.W., Goldberg, H., Jarvie, B.H., Smith, D.D. and Whybin, L.R. (1987). *Bordetella pertussis* infection: A cause of persistent cough in adults. *Med. J. Aust.* **147**, 522–5.

Robinson, A., Duggleby, C.J., Gorringe, A.R. and Livey, I. (1986). Antigenic variation in *Bordetella pertussis*. In: *Antigenic Variation in Infectious Diseases* (eds T.H. Birkbeck and C.W. Penn), pp. 147–61. IRL Press, Oxford and Washington, DC.

Rogel, A., Farfel, Z., Goldschmidt, S., Shiloach, J. and Hanski, E. (1988). *Bordetella pertussis* adenylate cyclase: Identification of multiple forms of the enzyme by antibodies. *J. Biol. Chem.* **263**, 13310–16.

Rogel, A., Schultz, J.E., Brownlie, R.M., Coote, J.G., Parton, R. and Hanski, E. (1989). *Bordetella pertussis* adenylate cyclase: Purification and characterization of the toxic form of the enzyme. *EMBO J.* **8**, 2755–60.

Romanus, V., Jonsell, R. and Bergquist, S.D. (1987). Pertussis in Sweden after the cessation of general immunization in 1979. *Pediatr. Infect. Dis. J.* **6**, 364–71.

Shattuck, R.L. and Storm, D.R. (1985). Calmodulin inhibits entry of *Bordetella pertussis* adenylate cyclase into animal cells. *Biochemistry* **24**, 6323–5.

Shattuck, R.L., Oldenburg, D.J. and Storm, D.R. (1985). Purification and characterization of a calmodulin-sensitive adenylate cyclase from *Bordetella pertussis*. *Biochemistry* **2**, 6356–62.

Springer, W. and Goebel, W. (1980). Synthesis and secretion of haemolysin of *Escherichia coli*. *Molec. Gen. Genet.* **175**, 343–50.

Stibitz, S., Weiss, A.A. and Falkow, S. (1988). Genetic analysis of region of *Bordetella pertussis* chromosome encoding filamentous hemagglutinin and the pleiotropic regulatory locus VIR. *J. Bacteriol.* **170**, 2904–13.

Stibitz, S., Aaronson, W., Monack, D. and Falkow, S. (1989). Phase variation in *Bordetella pertussis* by frameshift mutation in a gene for a novel two-component system. *Nature, Lond.* **338**, 266–9.

Strathdee, C.A. and Lo, R.Y.C. (1989). Cloning, nucleotide sequence and characterisation of genes encoding the secretion function of the *Pasteurella haemolytica* leukotoxin determinant. *J. Bacteriol.* **171**, 916–28.

Symes, P.H., Hewlett, E.L., Roberts, O., de Sousa, A. and Pearson, R.D. (1983). The effect of *Bordetella* adenylate cyclase on human monocytes: Suppression of the oxidative burst without inhibition of phagocytosis. *Clin. Res.* **31**, 377a.

Wagner, W., Kuh, M. and Goebel, W. (1988). Active and inactive forms of hemolysin from *Escherichia coli*. *Biol. Chem.* Hoppe-Seyler **369**, 39–46.

Wardlaw, A.C. and Parton, R. (1988). The host–parasite relationship in pertussis. In *Pathogenesis and Immunity in Pertussis*. (eds A.C. Wardlaw and R. Parton) pp. 327–52. Wiley, Chichester, UK.

Weiss, A.A. and Falkow, S. (1984). Genetic analysis of phase change in *Bordetella pertussis*. *Infect. Immun.* **43**, 263–9.

Weiss, A.A. and Hewlett, E.L. (1986). Virulence factors of *Bordetella pertussis*. *Ann. Rev. Microbiol.* **40**, 661–86.

Weiss, A.A., Hewlett, E.L., Myers, G.A. and Falkow, S. (1983). Tn5-induced mutations affecting virulence factors of *Bordetella pertussis*. *Infect. Immun.* **42**, 33–41.

Weiss, A.A., Hewlett, E.L., Myers, G.A. and Falkow, S. (1984). Pertussis toxin and extracytoplasmic adenylate cyclase as virulence factors of *Bordetella pertussis*. *J. Infect. Dis.* **150**, 219–22.

Weiss, A.A., Myers, G.A., Crane, J.K. and Hewlett, E.L. (1986). *Bordetella pertussis* adenylate cyclase toxin: Structure and possible function in whooping cough and the pertussis vaccine. In *Microbiology 1986* (ed L. Leive), pp. 70–4. American Society for Microbiology, Washington DC.

Wolff, J. (1985). Extracellular adenylate cyclase of *Bordetella pertussis*. In: *Pertussis Toxin* (eds R.D. Sekura, J. Moss and M. Vaughan), pp. 225–39. Academic Press, New York.

Wolff, J. and Cook, G.H. (1973). Activation of thyroid membrane adenylate cyclase by purine nucleotides. *J. Biol. Chem.* **248**, 350–5.

Wolff, J., Newton, D.L. and Klee, C.B. (1986). Activation of *Bordetella pertussis* adenylate cyclase by the carboxy-terminal tryptic fragment of calmodulin. *Biochemistry* **25**, 7950–95.

Wolff, J., Cook, G.H., Goldhammer, A.R. and Berkowitz, S.A. (1980). Calmodulin activates prokaryotic adenylate cyclase. *Prtoc. Natl. Acad. Sci. USA* **77**, 3840–4.

Wolff, J., Cook, G.H., Goldhammer, A.R., Londos, C. and Hewlett, E.L. (1984). *Bordetella pertussis*: Multiple attacks on host cell cyclic AMP regulation. *Adv. Cyclic. Nucleotide Protein Phosphorylation Res.* **17**, 161–72.

17

The Family of Mitogenic, Shock-inducing and Superantigenic Toxins from Staphylococci and Streptococci

Joseph E. Alouf[1], Heide Knöll[2] and Werner Köhler[2]

[1]Bacterial Antigens Unit, CNRS URA 557, Institut Pasteur, 75724 Paris Cédex 15, France
[2]Institute for Microbiology and Experimental Therapy, Jena, Germany

Introduction

Staphylococcus aureus and *Streptococcus pyogenes* (group A streptococci) produce a wide array of extracellular toxins and enzymes, which are known to be involved in the pathogenesis of a number of diseases elicited by these organisms. The staphylococcal toxins include membrane-damaging agents known as α, β, γ and δ-haemolysins (Freer and Arbuthnott, 1983), Panton-Valentine leukocidin (Grójec and Jeljaszewicz, 1985), epidermolytic toxins (Blomster-Hautamaa and Schlievert, 1988a), enterotoxins (Spero *et al.*, 1988), pyrogenic exotoxins (Schlievert, 1981) and toxic-shock syndrome toxin-1 (TSST-1) (Blomster-Hautamaa and Schlievert, 1988a).

The streptococcal toxins comprise two cytolysins (streptolysin O and streptolysin S) and erythrogenic (pyrogenic) toxins (Alouf, 1980; Wannamaker and Schlievert, 1988).

This chapter will be limited to staphylococcal enterotoxins, TSST-1 and pyrogenic exotoxins and to streptococcal erythrogenic toxins which constitute a family of single-chain polypeptides (Table 4) sharing a number of common pyrogenic, mitogenic, shock-inducing and immunological properties in addition to specific clinical, epidemiological, biological and pathogenic characteristics. An extensive survey of the historical and basic

aspects of these toxins is unwarranted here. They are covered in many excellent reviews in the reference list. This chapter will therefore focus primarily on a critical review of recent experimental work and areas in which significant progress has been made. The main emphasis will be placed on new achievements concerning the structural, genetic and immunological aspects of these toxins and on the mechanisms underlying their interactions with target cells. We will designate collectively these toxins as immunocytotropic toxins, since many of their biologic and toxic effects appear to derive primarily from their interaction with both monocytes, macrophages and T lymphocytes and the subsequent release of various cytokines (Alouf, 1990).

General characteristics

Staphylococcal enterotoxins (SEs)

These toxins have been widely investigated over the past 30 years in large part by the groups of Bergdoll and Spero (see Bergdoll, 1979, 1983; Freer and Arbuthnott, 1983; Spero *et al.*, 1988; Marrack and Kappler, 1990 for reviews).

The enterotoxins are a group of seven structurally related polypeptides differentiated into five serological groups A, B, C, D, E (SEA to SEE).

Serotype C was further subdivided into three types (SEC1, SEC2, SEC3) based on differences in minor epitopes. Still unidentified serotypes probably do exist (Spero et al., 1988). The in vivo production of the toxins during staphylococcal infections is evidenced by the presence of antitoxin antibodies (see Kunstmann et al., 1989) or free toxin(s) (Hosotsubo et al., 1989) in human sera. The toxins are released in vitro during bacterial growth and in S. aureus-contaminated foods. They are the causative agents of staphylococcal food poisoning in humans that is characterized by emesis and diarrhoea which can also be elicited in experimental animals (monkeys, kittens). The term enterotoxin is a misnomer since these emetic proteins do not elicit the fluid accumulation characteristic of enterotoxins in the classic ligated ileal loop (Iandolo and Tweten, 1988). Rather they are presumed to function by affecting local receptors in the digestive tract and generate impulses that reach the subcortical vomiting centre via vagal and sympathic afferents. Indeed the exact manner in which these food-poisoning toxins cause the classic symptoms is unclear but their massive stimulation of T lymphocytes may play a role in the toxic effects by the release of mediators such as tumour necrosis factor, interleukins or leukotrienes (Scheuber et al., 1987; Jupin et al., 1988; Marrack et al., 1990).

The enterotoxins may be involved in other diseases in humans such as staphylococcal pseudomembranous enterocolitis and probably in other staphylococcal diseases (infected wounds, skin infections, osteomyelitis, septicaemias) in conjunction with other toxins such as TSST-1 and the exfoliatins (Bergdoll, 1986).

More than one serotype of enterotoxin (generally two serotypes, rarely three) may be produced simultaneously by certain strains. Enterotoxins A and D or A/D combination appear to cause foodborne intoxications at a greater frequency than do serotypes B and C. SEA occurs with the greatest frequency. SEB is seldom associated with food poisoning but is produced along with SEA or in association with the latter or other serotypes by S. aureus strains of clinical origin. Valuable data and references concerning these epidemiological aspects collected from various countries may be found in the review of Bennett (1986) and in the articles of Mochmann et al. (1981), Melconian et al. (1983a), Mauff et al. (1983) and Humphreys et al. (1989).

Staphylococcal toxic-shock syndrome toxin-1 (TSST-1)

This toxin is recognized as the major factor responsible for toxic-shock syndrome (TSS), an acute, severe and life-threatening multisystem illness aetiologically associated with S. aureus infection (see Chesney et al., 1984; Hirsch and Kass, 1986; Blomster-Hautamaa and Schlievert, 1988a; Arbuthnott, 1988 and International Symposium on TSS, 1989 for general reviews and survey). TSS is characterized by five major clinical and biological criteria: acute onset of high fever (temperature > 39°C), hypotension (blood pressure < 90 mm Hg in adults) or orthostatic dizziness, scarlatiniform erythematous rash, desquamation of the skin (palms and soles upon recovery one or two weeks after onset of illness) and dysfunction of three or more of the following organ systems: gastrointestinal tract (vomiting, diarrhoea), muscular system, mucous membranes (conjunctival, vaginal, oropharyngeal hyperaemia), kidney and liver (impaired functions), haematologic system (thrombocytopaenia), central nervous system (alterations in consciousness or disorientation).

Since the initial description of TSS (Todd et al., 1978) several thousands of cases, mainly in the United States and to a lesser extent in Europe and other countries, have been reported. The overall fatality rate is around 6%. The majority of the cases are epidemiologically associated with tampons used in menstruating young women vaginally infected with S. aureus (Shands et al., 1980; Davis et al., 1980; Hirsch and Kass, 1986; Blomster-Hautamaa and Schlievert, 1988a). However, about 15–30% of cases may occur in other male and female individuals with focal, non-genital S. aureus wounds or soft tissue infections (Schlievert, 1983; Blomster-Hautamaa and Schlievert, 1988a). The toxin is released in vivo as demonstrated by the presence of antitoxin antibodies (see Kunstmann et al., 1989). Illness results if the individual has no or very low antibody titre to the toxin (Crass and Bergdoll, 1986).

The observations that TSS does not require a deep tissue invasive infection (S. aureus remains focally localized and bacteraemia rarely occurs) pointed to the involvement of circulating exotoxin(s) responsible for the widespread systemic multi-organ effects as first suggested by Todd et al. (1978) and later by Shands et al. (1980) and

Davis *et al.* (1980). A new toxin produced by staphylococci isolated from patients with TSS was identified independently by Schlievert *et al.* (1981) under the name of pyrogenic exotoxin C (PEC) because of its pyrogenicity in rabbits and by Bergdoll *et al.* (1981) under the name of enterotoxin F. The latter was further named toxic-shock toxin because the emetic action in monkeys could not be confirmed (Reiser *et al.*, 1983). Barbour (1981) observed that all the TSS isolates and 50% of non-TSS isolates produced a protein with a pI 7.0 and M_r of 22 000 which could be identical with PEC. These values are those established later for purified TSST-1. The two toxins C and F were later shown to be biochemically and immunologically identical (Bonventre *et al.*, 1983; Igarashi *et al.*, 1984; Bergdoll and Schlievert, 1984) and the term toxic-shock syndrome toxin-1 (TSST-1) was adopted. The designation 1 was added because some investigators were concerned that other toxins may also be involved; if they are found they will be designated TSST-2, TSST-3, etc. A 30-kDa protein has been described as TSST-2 (Scott *et al.*, 1986).

TSST-1 administration to rabbits has been shown to cause many of the signs associated with TSS and to be lethal (Arbuthnott, 1988; International Symposium on TSS, 1989). The toxin also produced disorders similar to TSS in baboons (Quimby and Nguyen, 1985). However, there is mounting evidence that TSST-1 is not or not always the only staphylococcal product involved in TSS. Other toxin(s), particularly SEB or SEC or combinations of toxins may trigger shock and the same cascade of tissue-damaging effects as TSST-1 (Garbe *et al.*, 1985; Crass and Bergdoll, 1986; Arbuthnott, 1988; De Azavedo, 1989). A severe shock-like syndrome was induced in monkeys by i.v. injection of relatively large doses (50–100 µg of SEB/kg) (see Van Miert *et al.*, 1984). TSST-1-negative strains producing enterotoxins have been shown to elicit TSS-like illness in animals (De Azavedo, 1989; McCollister *et al.*, 1990) or in humans (Immerman and Greenman, 1987; Rizkallah *et al.*, 1989). Recently a new shock-inducing 26-kDa toxic (lethal, diarrhoeagenic) polypeptide was identified by Hall *et al.* (1989) in the culture fluid of *S. aureus* isolated from a patient with nonmenstrual TSS. This toxin was unrelated to TSST-1. It was also distinct from SEs with, however, a certain relation to them.

Staphylococcal pyrogenic exotoxins

These toxins are two antigenically distinct proteins named A and B (*ca.* 12 and 18 kDa) purified from the culture fluid (Schlievert *et al.*, 1979a; Schlievert, 1980b) of a strain of *S. aureus* isolated from a patient with subcutaneous lymph node syndrome (Kawasaki's disease). Toxins A and B share most of the biological properties of the toxins reviewed in this chapter and particularly those exhibited by streptococcal erythrogenic (pyrogenic) toxins. The toxins produce erythrodermal skin reaction in the rabbit and are thought (Schlievert, 1981) to be involved in the staphylococcal scarlet fever syndrome (Feldman, 1962; McCloskey, 1973) which is similar to the classical streptococcal scarlet fever. These toxins share most of the biological properties common to the immunocytotropic toxins (Table 9) reviewed in this chapter. The pI values of toxins A and B which were estimated to be 5.3 and 8.5 respectively (Schlievert, 1980b) are likely the acidic and basic polyclonal staphylococcal mitogens reported by Kreger *et al.* (1972) and Taranta (1974). Pyrogenic exotoxin A is elaborated by TSS *S. aureus* but the toxin is not specific for TSS strains and its role in the disease is still not evaluated. Type B toxin did not appear to be made by TSS strains of *S. aureus* (Blomster-Hautamaa and Schlievert, 1988a).

Streptococcal erythrogenic toxins (ETs)

A great deal of work has been devoted to these toxins (see Kim and Watson, 1972; Ginsburg, 1972; Alouf, 1980; Wannamaker and Schlievert, 1988) because of their purported role in scarlet fever and other streptococcal diseases (see Parker, 1984). The name erythrogenic toxin, abbreviated here to ET, was coined in 1924 by George and Gladys Dick who discovered it in the culture filtrates of group A streptococci of scarlatinal origin after the failure of many investigators at the turn of the century to identify 'the' streptococcal toxin involved in the pathogenesis of scarlet fever and other streptococcal diseases. The Dicks found that the intradermal injection of the culture filtrate into certain persons elicited a skin erythrematous reaction mimicking the skin rash observed in scarlet fever (Dick test). A local blanching of this rash is observed by injection of anti-ET antitoxin or the serum of a convalescent patient into the rash (Schultz-Charlton test).

The interest in ET fell off in the 1930s, because the results of active immunizations with ET-containing culture filtrates were disapproved. New interest came up when Schwab *et al.* (1953) and later on, Watson detected substances in skin extracts of locally *S. pyogenes*-infected rabbits which were able to enhance the lethal and cardiotoxic activities of endotoxin and streptolysin O. The substances were found to be pyrogenic and were identified by Watson with the scarlet fever group of toxins. He thus introduced the designation 'streptococcal pyrogenic exotoxins' (Cremer and Watson, 1960). This denomination reflects one biological property among many other activities of ET, including T cell mitogenicity. The multiplicity of the biological activities of ET led to the various denominations presently used in the literature: erythrogenic toxins (Dick and Dick, 1924; Hooker and Follensby, 1934; Stock, 1939; Nauciel *et al.*, 1969; Gerlach *et al.*, 1980; Geoffroy and Alouf, 1988; Kamezawa and Nakahara, 1989), streptococcal pyrogenic exotoxins (Kim and Watson, 1970; Schlievert *et al.*, 1979a) or streptococcal exotoxins (Houston and Ferretti, 1981). Other synonymous names are scarlet fever toxin (Dick and Boor, 1935), streptococcal mitogens (Seravalli and Taranta, 1974), extracellular lymphocyte mitogens (Abe and Alouf, 1976; Cavaillon *et al.*, 1979), blastogen A (Gray, 1979; Schlievert and Gray, 1989), delayed skin reaction (Tomar, 1976) and streptococcal (keratinocyte) proliferative factor (Rasmussen and Wuepper, 1981). For historical reasons we prefer the original name erythrogenic toxin until a new commonly accepted denomination is established.

Two immunologically distinct toxins (ETA and ETB) were first differentiated by Hooker and Follensby (1934) who predicted a third serotype effectively characterized and named C by Watson (1960). These three serotypes are produced exclusively by group A streptococci separately, simultaneously, or in various combinations depending on strains (whether associated with scarlet fever or not) and probably on culture conditions (see Table 1). It is likely that the Dicks dealt with ETA. A fourth antigenic type named D was isolated by Schuh *et al.* (1970) from the strain C203U. This new serotype was later characterized by McMillian *et al.* (1987) in the supernate of a strain isolated from a patient.

Until the first half of this century scarlet fever was a severe life-threatening disease with high mortality rates (5–25%) observed in individuals with group A streptococcal pharyngitis or associated with sepsis and shock following skin wounds or surgery (surgical scarlatina) (Cone *et al.*, 1987; Köhler *et al.*, 1987; Stevens *et al.*, 1989; Quinn, 1989). The *in vivo* production of ETs is documented by the presence of antibodies (Knöll *et al.*, 1990; Abe *et al.*, 1990) or the toxin itself (Alouf, unpublished data) in human sera. In the past 40 years a marked decline in the prevalence of scarlet fever, erysipelas, rheumatic fever and other severe infections caused by *S. pyogenes* has been observed in the developed countries with improved socioeconomic conditions and curative or prophylactic antibiotic treatment of these infections (see Quinn, 1989 for a comprehensive review).

However, very recently severe streptococcal infections (over 700 cases) bearing a striking resemblance to staphylococcal TSS characterized by multisystem failure and high mortality rate have been reported in various countries (Stevens *et al.*, 1989; Köhler *et al.*, 1990). Most of the isolates produced ETs (see Table 1). Several lines of evidence point to the involvement of these toxins in this syndrome and in scarlet fever (Willoughby and Greenberg, 1983). Erythrogenic toxins are known to cause substantial tissue injury (Ginsburg, 1972) in addition to their strong mitogenic effects on lymphocytes and the liberation of various cytokines by immune system cells. The cyclic variability of the severity of scarlet fever and related streptococcal infections still requires detailed epidemiological analysis. Several factors and hypotheses discussed by Köhler *et al.* (1987) have been considered, including quantitative aspects of ET production by the strains, genetic evolution of strain virulence and possible variation in immunity to the toxin.

Toxin detection and assay

The methods for the identification and quantitation of the toxins are based on biological and immunological techniques. Only a brief outline is presented here. The reader is referred to the reviews and articles of the authors cited below for detailed

Table I. Production of erythrogenic toxins A, B and C by strains of group A streptococci from various sources (percentage)

Source	No of strains	A	B	C	AB	AC	BC	ABC	Negative	Reference
Laboratory stocks, USA	75	6.7	12	14.7	5.3	6.7	33.3	12	9.3	Schlievert et al. (1979c)
Laboratory stocks, Clin-isolates, USA	42	4.8	50	7.7	9.5	0	9.5	4.8	11.9	Bloomster and Watson (1982)
Type collection	50	22	18	8	6	4	6	4	32	Köhler et al. (1987)
Streptococcal infections, USA	80	—	22.5	45	—	—	26.2	—	6.3	Cone et al. (1987)
Scarlet fever										
1980–83, UK	40	—	47.5	17.5	—	—	22.5	—	12.5	Hallas (1985)
1972, 1982 (epidemics), GDR	73	42.5	15.1	1.4	4.1	5.5	17.8	2.7	10.9	Köhler et al. (1987)
1985, Czechoslovakia	52	17.3	28.8	—	19.2	—	13.5	15.4	5.8	Knöll et al. (1990)
1988, 1989 (epidemics), GDR	268	3.4	59.3	2.6	1.1	—	13.1	0.7	19.9	Knöll, unpublished
Streptococcal toxic shock like syndrome[a]										
USA	10	40	10	—	30	—	—	10	10	Stevens et al. (1989)
USA	26	50	—	11.5	19	—	—	19.5	—	Lee and Schlievert (1989)
France, GDR, Sweden	17	—	11.8	—	—	—	5.9	82.3	—	Köhler et al. (1990)
Total	733									

— not detected.
[a]Single case reports are not included.

description of the techniques and bibliographic references.

Biological assays

Staphylococcal enterotoxins

The only reliable assay is the monkey feeding test (Bergdoll, 1988) which is also the only method for detecting new serotypes. The monkeys are fed through a catheter and the elicitation of emesis within 5 h is observed. This procedure is cumbersome and the expenses for purchasing and maintaining the animals greatly limit the use of this bioassay. Additionally, the animals may become resistant to the enterotoxin with repeated use. The emetic dose 50 (ED_{50}) is about 5–20 µg per animal. Intravenous injections are also used in the testing of purified SEs as well as for the determination of the neutralizing capacity of heterologous antibodies. Among all other animals tested, only kittens were sensitive. Enterotoxin injection (i.v. or i.p.) of cats or kittens will produce an emetic response when 1 µg of toxin is injected. This test is less reliable since these animals are insensitive to SEC and are subject to non-specific reactions.

Toxic-shock syndrome toxin-1

A bioassay has been devised by De Azavedo et al. (1988a) based on the enhancement of chick embryo lethality elicited by TSST-1 in the presence of a fixed sublethal amount of endotoxin (E. coli LPS). Test samples (100 µl) are injected into embryo veins (10 eggs receive the same dose) and the embryos, incubated at 37°C, are examined for viability 18–24 h after injection. This assay is sensitive, reproducible and simple to perform. However, it may not be suitable for crude TSST-1 material since staphylococcal products other than TSST-1 may kill the embryos.

Erythrogenic toxins

Two tests based on the erythematous skin reaction and mitogenic activity are used for the bioassay of ETs (Geoffroy and Alouf, 1988). The skin test is performed in rabbits or guinea-pigs. Toxin samples (100 µl) are injected intradermally. Positive reaction is demonstrated by an erythematous response. The skin is observed at 24 h and 48 h after injection. One skin test dose (STD) is that amount of toxin which elicits an erythema of 10 mm in diameter. It corresponds approximately to 100 ng of toxin in rabbit and guinea-pig systems according to Cunningham et al. (1976), Hříbalová et al. (1980) and to 1 ng in rabbit according to Kamezawa and Nakahara (1989). One STD in humans (Dick test) corresponds to 1 ng (Stock, 1939). This discrepancy is not explained but it may be related to both the complex nature of the reaction and to the host since the skin reactivity appears to reflect Arthus and delayed hypersensitivity reaction in addition to the toxic effects of ETs (Schlievert et al., 1979b). More work is still required for accurate quantitative evaluation of skin reactivity for the three toxin serotypes. The specificity of this reactivity is assessed by its neutralization by preincubation of ETs with corresponding specific antibodies. A skin test in rabbits based on blueing reaction has been described recently (Kamezawa et al., 1990). This test appears to offer advantages over the classical erythematous skin reaction.

The lymphocyte blast transformation test performed on human or rabbit lymphocytes separated from peripheral blood is based on the mitogenic effect of ETs. The results are expressed as the stimulation index, which is the ratio of the counts per minute (cpm) of the toxin-stimulated culture pulsed with [^{3}H]thymidine to the cpm of the non-stimulated control cultures (Cavaillon et al., 1979; Knöll et al., 1983). Inhibition of the lymphocyte mitogenicity of ETA has been used for the assay of antitoxin antibodies in human sera (Abe et al., 1990).

The pyrogenic effect of the toxins has also been used for bioassay (Cunningham et al., 1976; Hříbalová et al., 1980).

Immunological assays

The obvious drawbacks of the biological assays led to the development of immunological techniques for toxin detection and quantitation since the toxins are reasonably good antigens which elicit specific antibodies in various animal species.

The techniques mainly used for toxin detection and quantitation in culture media and purified fractions are, presently, immunoprecipitation in gel, radioimmunoassay (RIA), enzyme-linked immunoassay (ELISA) and immunoblotting methods. These methods have also been applied to SEs in food (reviewed by Bennett, 1986). Monospecific immune sera (rabbit, goat, horse) or purified polyclonal antibodies are still widely employed. More recently monoclonal antibodies have been used for the detection or assay of SEs (Meyer *et al.*, 1984; Thompson *et al.*, 1984, 1986; Edwin *et al.*, 1986; Lapeyre *et al.*, 1987; Lin *et al.*, 1988; Edwin, 1989; Kienle and Buschmann, 1989), TSST-1 (Blomster-Hautamaa *et al.*, 1986b; Wells *et al.*, 1987; Bonventre *et al.*, 1988; Murphy *et al.*, 1988 and many papers in International Symposium on TSS, 1989) and ETs (Bohach *et al.*, 1988b; Knöll and Alouf, unpublished data).

Immunoprecipitation in gel

Various classical qualitative and quantitative techniques based on this reaction have been used. The lowest detectable quantities have been reported to range from 0.1 to 10 μg of toxin (Weckbach *et al.*, 1984; Bennett, 1986; Lee and Schlievert, 1989). The classical Oudin gel tube test and Oakley double diffusion tube test are still useful for detection in purification fractions but not in foods. In the latter case, three different immunodiffusion tests are applicable for SEs: the microslide method, the optimum sensitivity plate and single radial diffusion (Bennett, 1986). Immunodiffusion techniques for TSST-1 assay have been described by Schlievert (1988).

RIA and ELISA

Both techniques offer about 1000-fold higher sensitivity (0.1–0.5 ng of toxin) than the previous procedures (Bennett, 1986; Spero *et al.*, 1988). The various RIA methods involved antibody binding on polystyrene tubes, insoluble adsorbents such as a bromocetyl-cellulose, double antibody technique with anti-rabbit immunoglobulin as co-precipitant and protein A as immunosorbent. The main disadvantages of RIA are the radiolabelling of the toxin and expensive isotope-counting equipment.

The equally sensitive ELISA procedures are obviously more appropriate and have been widely used for the quantitation of SEs (see Bennett, 1986 for a review; Fey *et al.*, 1984; Edwin *et al.*, 1986; Lapeyre *et al.*, 1987; Kienle and Buschmann, 1989). The incorporation of biotin with avidin or streptavidin in the assay (Hahn *et al.*, 1986; Edwin, 1989) increased sensitivity and prevented protein A interference. TSST-1 in cultural supernatants, purification fractions or vaginal washings was also quantitated by ELISA (Parsonnet *et al.*, 1985; Wells *et al.*, 1987; Rosten *et al.*, 1987; Edwin *et al.*, 1988; Schlievert, 1988; and International Symposium on TSS, 1989). This technique has also been used for ET assay (Houston and Ferretti, 1981; Köhler *et al.*, 1987). The presence of large amounts of hyaluronic acid in test material has been reported to interfere with ET assay (Lee and Schlievert, 1989).

Immunoblotting

Polyclonal and monoclonal antibodies have been used in immunoblotting procedure for detection of SEs (Thompson *et al.*, 1984; Edwin *et al.*, 1986; Lapeyre *et al.*, 1987), TSST-1 (Weckbach *et al.*, 1984; Bonventre *et al.*, 1988; Edwin *et al.*, 1988; Murphy *et al.*, 1988; Whiting *et al.*, 1989) and ETs (Hynes *et al.*, 1987; Schlievert and Gray, 1989).

Production and toxinogenesis

All of the three groups of toxins (enterotoxins, TSST-1 and erythrogenic toxins) are extracellular proteins released in culture fluids during the growth cycle. As evidenced from the sequencing of the structural genes of these toxins (see Table 4) all are synthesized as precursor molecules. The mature exotoxins are produced by removal of the signal peptides during the process of toxin passage across the cell membrane.

Staphylococcal enterotoxins

A number of media have been proposed for toxin production (see Bergdoll, 1979; Freer and Arbuthnott, 1983; Bennett, 1986; Iandolo and Tweten, 1988). The most widely used medium is 3% NZ-amine type A or NAK containing 1% yeast extract and two vitamins, niacin and thiamine. The starting pH is always near neutral. The strains generally used for production (Spero *et al.*, 1988; Iandolo and Tweten, 1988) can be obtained from the Food Research Institute (Madison, WI) or from the American type culture collection (Rockville, MD).

A significant improvement in toxin production has occurred through the use of fermenters when pH, temperature and oxygen tension are controlled. Under these conditions, SEB, SEC1 and SEC2 yield will be up to several hundred micrograms per millilitre while that of the other serotypes will be only a few micrograms per millilitre (except for type A mutant 13-2909 which produces about 100 μg of SEA/ml). This large discrepancy in production, particularly between SEA and SEB even in strains that produce both toxin types, remains an intriguing problem in gene regulation that has resisted solution (Iandolo, 1989).

SEs are secreted during the exponential and stationary stages of cell growth (Iandolo and Tweten, 1988; Otero et al., 1990). The highest amounts of SEB are produced in the stationary phase (Gaskill and Khan, 1988). The larger molecular weight precursor forms of SEA and SEB have been isolated from S. aureus membranes by Iandolo and co-workers (see Iandolo and Tweten, 1988) who also found by pulse-chase techniques both the precursor and mature forms of SEB after 5 s. It was concluded that post-translational and cotranslational modification of the precursor were necessary to account for rapid appearance of the mature form.

The regulation of SEB production has been shown to occur at the transcriptional level and that the accessory element agr involved in the regulation of several exoprotein genes in S. aureus including those of α-toxin, TSST-1 and epidermolytic toxin A, is also involved in SEB regulation (Gaskill and Khan, 1988). The role of upstream sequences in the expression of SEB gene was investigated (Mahmood and Khan, 1990).

The biosynthesis of SEA, SEB and SEC is repressed by glucose (see Smith et al., 1986a for references). Glycerol and maltose were also shown to repress SEA synthesis (Smith et al., 1986b). Mutants lacking the phosphoenolpyruvate phosphotransferase system showed considerably less repression by these compounds. SEB does not appear to be regulated by a mechanism analogous to catabolite repression (Iandolo, 1989). It has been postulated that the inhibition of enterotoxin synthesis may involve phosphorylated sugar inter-

mediates by a mechanism which is still not clear (Smith et al., 1986b).

Toxic-shock syndrome toxin-I

Environmental factors

An evaluation of culture conditions required for TSST-1 production may be found in several publications (Poindexter and Schlievert, 1985a; Blomster-Hautamaa and Schlievert, 1988a,b; International Symposium on TSS, 1989; Wong and Bergdoll, 1990). The toxin which accounts for about 1–2% of the total exoproteins released (Kass, 1989) is produced in complex media (beef heart dialysate, brain–heart infusion, Todd-Hewitt broth NZ-amine NAK or trypticase supplemented with yeast extract). Chemically defined media (CDM) also allow good growth and toxinogenesis (Mills et al., 1986; Taylor and Holland, 1988; James et al., 1989; Reeves, 1989). As shown by Reeves (1989) toxin production was generally greater in CDM (50–80 μg/ml of culture fluid) than in brain–heart infusion (ca. 20 μg/ml). About 50 μg/ml were obtained in trypticase medium (Melconian et al., 1983b). TSST-1 production occurs under aerobic conditions but excessive aeration has an adverse effect (Wong and Bergdoll, 1990).

The toxin is not produced under strict anaerobic conditions although staphylococci are aerobes that can grow under anaerobic conditions (see Tierno and Hanna, 1989). Addition of CO_2 enhanced toxin production as well as that of haemolysins and nuclease. The mechanisms of the effect of CO_2 are unknown. The optimal temperature range for production is 37–40°C with slightly more toxin produced at 40°C. Much less toxin is produced at 30–32°C. Maximum production occurs at neutral pH during late exponential phase and early stationary growth phase (Mills et al., 1986). According to Wong and Bergdoll (1990) maximum toxin yield in the fermenter was obtained when cultures were supplied with air (20 cm³/min) and CO_2 (5 cm³/min) via a sintered glass sparger. Cultures with no pH control made more TSST-1 than those maintained at pH 5.5–7.5. Toxin production was unaffected by glucose up to ca. 0.3% in the medium but was repressed by higher concentrations (Poindexter and Schlievert, 1985a; Taylor and Holland, 1988). The membrane over agar method has been used for the enhancement of toxin production (Melconian et al., 1983a; Parsonnet et al., 1987). The growth of S. aureus strains from vaginal isolates and TSST-1 production were investigated in a medium providing a menses-like environment in order to mimic in vivo conditions (Kirkland et al., 1989).

Conflicting results have been reported about the role of Mg^{2+} in toxin production (see Mills et al., 1986; Taylor and Holland, 1988; Reeves, 1989; Kass, 1989; James et al., 1989). All researchers agree that Mg^{2+} is essential for growth of staphylococci and production of TSST-1, but there is disagreement on the effect of increasing the concentration of Mg^{2+} in the medium on TSST-1 production. Certain authors reported that minimal amounts of Mg^{2+} (2–5 µg/ml) resulted in elevated production of TSST-1 while higher amounts (~ 15 µg/ml) decreased toxin production (Mills et al., 1986; Kass, 1989). Reeves (1989) confirmed this finding. They found in a CDM that when individual metals were tested, only Mg^{2+} stimulated cell growth and TSST-1 production. Maximal yield occurred at 0.1 mM and declined for higher concentrations. The stimulating effect of Mg^{2+} on TSST-1 production was significantly enhanced by adding 0.3 mM Zn^{2+} and 0.003 mM Fe^{2+}. The inhibitory effect of Mg^{2+} on TSST-1 production above the low levels reported was not confirmed in other studies (Taylor and Holland, 1988; James et al., 1989). The discrepancy between the data has been attributed to the interference of other ions and organic nutrients, to various experimental parameters (chemostat continuous culture versus batch culture, culture duration, etc.) and to the method of expression of toxin production with respect to bacterial growth.

The study of Mg^{2+} influence on TSST-1 production was directly related to the in vitro effect of tampons on TSST-1 synthesis. It was reported that highly absorbent tampons can stimulate toxin production (see Robbins et al., 1989; Tierno and Hanna, 1989 for references). It was postulated that certain fibres that were used in some of the tampon products act as ion exchangers. They remove Mg from the surrounding environment and thereby favour production of maximal amounts of TSST-1 (Mills et al., 1986; Kass, 1989). However, the various experimentations reported have led to conflicting conclusions as regards the effects of tampons on TSST-1 production (see International Symposium on TSS, 1989). Indeed it appears that the major contribution of tampons is the provision of sufficient air for S. aureus growth and TSST-1 production, along with relatively large surface areas on which the staphylococci can grow. In addition a constellation of microbial and host-dependent factors govern the production of TSST-1 (see Tierno and Hanna, 1989).

Bacterial strains

In vitro production of TSST-1 occurs in about 20% of randomly screened S. aureus clinical isolates, 60–75% of non-menstrual TSS isolates and nearly 100% of vaginal isolates associated with menstrual TSS (see Barbour, 1981; Blomster-Hautamaa and Schlievert, 1988a; Whiting et al., 1989; See and Chow, 1989; International Symposium on TSS, 1989). These data suggest that (1) TSST-1 is not required for normal maintenance of S. aureus and thereby the production of this toxin is an accessory genetic trait, (2) genital and non-genital TSS strains may have different clonal origins and that menstrual strains have a limited clonality whereas that of non-menstrual strains is more heterogeneous (See and Chow, 1989). The phenotypic characteristics of the isolated TSS-associated strains (production of α, β, γ, δ-haemolysins, enterotoxins A–E, other toxins, nucleases and lipases, proteolytic activity, pigmentation, susceptibility to bacteriocins, heavy metal and antibiotic resistance) as well as the genotypic characteristics (particularly the linkage to Trp auxotrophy) have been widely studied (reviewed by Kreiswirth, 1989; See and Chow, 1989; Aliu and Bergdoll, 1989). Clyne et al. (1988) observed that TSS strains predominantly produced γ- and δ-haemolysins. Crass and Bergdoll (1986) found that 60% of TSS-associated strains produce SEA, SEB or SEC and that SEB and TSST-1 expression appear to be mutually exclusive. Strain MN 8, a vaginal isolate from a woman with typical menstrual TSS, has been used in many laboratories as the source of TSST-1 (overnight culture).

TSS strains of S. aureus are significantly more sensitive to group I phage types (particularly phage type 29 alone or in combination with phage type 52). New phage types (UC 21–25) appear uniquely associated with TSS. However 30–40% of TSS-associated *S. aureus* remain non-typable. *S. aureus* strains sensitive to phage group II do not produce TSST-1 (Kreiswirth, 1989; See and Chow, 1989). TSST-1 does not appear to be produced by coagulase-negative strains as shown by Parsonnet *et al.* (1987) who screened 187 strains. As observed by Kreiswirth and co-workers, TSST-1 gene was not detected in such strains by use of a TSST-1 gene-specific probe (See and Chow, 1989). However, TSST-1 has been reported to be produced by some coagulase-negative staphylococci (Crass and Bergdoll, 1986; Aliu and Bergdoll, 1989). Recently, Ho *et al.* (1989) detected a TSST-1 variant produced by non-human-associated *S. aureus* strains from sheep, goats and cows.

Erythrogenic toxins

Environmental factors

These toxins are released as extracellular proteins in culture media. Maximum production occurs during the mid or late exponential phase of growth and no additional toxin is secreted during the stationary phase (Houston and Ferretti, 1981; Lee and Schlievert, 1989). Most culture media presently used for toxin production are diffusates of: (i) beef heart infusion (Fast *et al.*, 1989) according to Stock's procedure (see Stock, 1939) as modified by Watson (1960), (ii) Todd-Hewitt broth (Houston and Ferretti, 1981; Hallas, 1985; Köhler *et al.*, 1987), and (iii) pancreatic and peptic peptones with or without yeast extracts (Gerlach *et al.*, 1983; Geoffroy and Alouf, 1988; Kamezawa and Nakahara, 1989). A valuable antigen-free medium (Holm and Falsen, 1967) prepared from Sephadex gel-filtered trypticase-yeast autolysate allowed heavy growth and production of high amounts of ETs and of SLO. Large-scale production in 50-litre fermenters (at 30°C under constant pH) was undertaken with yeast–peptone solutions dialysed in the fermenter (Gerlach *et al.*, 1980). Chemically defined media were able to allow toxin production by certain strains only. ETA was produced in such media (Hallas, 1985; Kamezawa and Nakahara, 1989) as well as streptococcal proteinase precursor (Ogburn *et al.*, 1958; Cohen, 1969) which was shown later to be identical to ETB (Gerlach *et al.*, 1983).

According to Houston and Ferretti (1981), ETA production in Todd-Hewitt dialysate medium was not affected by adding various metal ions (Mg, Ca, Mn, Fe) or nutrients, suggesting constitutive synthesis of the toxin. In addition, no differences in bacterial growth and ETA yield were observed between 24°C and 37°C. However, both growth and toxin production decreased at 42°C. Optimal production of ETA and ETC requires a slightly alkaline pH ($\sim$ 7.2). Automatically regulated pH is advantageous but this is not absolutely necessary because neither toxin is destroyed by streptococcal proteinase produced at acidic pH in non-regulated media. ETB production required a pH between 5.5 and 6.5 for proteinase production (see Elliott and Liu, 1970). Gerlach *et al.* (1983) found maximal production (145 mg/litre) by strain T19 at pH 5.9 and at 30°C. Growth at a constant pH of 6.3 lowered the production to 10 mg/ml. Neopeptone was reported to prevent proteinase/ETB production (Cohen, 1969). However, Hallas (1985) reported ETB production by strain T19 in Todd-Hewitt neopeptone medium.

Recently ETA structural gene was cloned in *Bacillus subtilis* (Kreiswirth *et al.*, 1987) and in *Streptococcus sanguis* (see Gerlach *et al.*, 1987). In both cases ETA was produced and secreted by these heterologous hosts. The amount of toxin released by *B. subtilis* was 32 times greater than in the streptococcal host. In contrast the amount of ETA produced by *S. sanguis* (1 μg/ml) was one-quarter that released by NY5 strain. However, this toxin was purified to homogeneity from the culture fluid of the recombinant strain (Gerlach *et al.*, 1987). It was antigenically identical to that purified by Gerlach *et al.* (1980) from *S. pyogenes* and had the same mitogenic potency.

Toxin production *in vivo* was investigated by Knöll *et al.* (1982) by the tissue cage model of Holm *et al.* (1978). The bacteria (10^5 per 0.1 ml) are introduced into subcutaneous steel-net cages implanted 6 weeks before. During this time a layer of connective tissue grows into the net creating a sealed tissue chamber and inhibiting the spread of bacteria out of the chamber. By aspiration the tissue chamber fluid can be removed at different times after infection. The average amounts of ETA, ETB and ETC released by the various strains tested ranged from 5 to 10 μg/ml in the cage fluid. This procedure allowed the detection of ETA production *in vivo* by strains which did not show toxin release under *in vitro* conditions (Knöll et al., 1985). The rabbits responded to the infection with a serologically specific antibody response to the type of toxin produced. Using the cage model, Knöll et al. (1988a) showed that rabbits immunized with ETA toxoid evoked a remarkable protection against death from infection by ETA-producing strains introduced in the cage.

Toxinogenic strains

The detection of the toxins in culture supernates and the determination of their serotype require sensitive techniques since the amounts of toxin produced by most clinical isolates are very low (Köhler et al., 1987). Culture filtrates are usually concentrated 100 times by differential precipitation with ethanol at $-20°C$ and resolubilization in acetate-buffered saline according to Kim and Watson (1970). The concentrated material is tested by immunodiffusion in agar (Köhler et al., 1987; Lee and Schlievert, 1989) or by ELISA (Knöll et al., 1990).

The reference strains commonly used for toxin production and purification are listed in Table 2. Eleven strains investigated by Gerlach et al. (1981) produced ETA in quantities ranging from 16 to 0.03 mg/litre. Strain T19 produced up to 145 mg of ETB/litre (Gerlach et al., 1983). The amounts of ETC produced by strains T18, NY5 and AT13 were 1.2, 0.9 and 1 mg/litre respectively. The mean amount of ETA produced by 52 strains isolated in 1987/88 in the GDR from children suffering from scarlet fever was ca. 1 μg/litre (Knöll et al., 1990) whereas the average titre was 68 and 8 μg/litre from the strains isolated in the same country during scarlet fever epidemics in 1972/73 and 1982/83 (Köhler et al., 1987).

In contrast, much higher amounts were produced by 26 strains isolated from patients with streptococcal toxic-shock-like syndrome in 1987/88 (Lee and Schlievert, 1989). The average concentrations of ETA, ETB and ETC released per litre of culture were 3.2, 0.7 and 0.6 mg respectively. However, recently isolated strains from toxic-shock-like syndromes in France, Sweden and Chile were found to produce very low or no detectable amounts of ETs (Köhler et al., 1990; Knöll, unpublished data). The reasons for the variable toxin production are unclear. They are most likely due to regulatory rather than gene dosage effects (Yu and Ferretti, 1989). The cyclic trends in the qualitative and

Table 2. Reference *S. pyogenes* strains used for the production and purification of erythrogenic (pyrogenic) toxins A, B, C and D

Production of ETA	
NY5	Gerlach et al. (1980), Houston and Ferretti (1981), Bohach et al. (1988a), Kamezawa and Nakahara (1989)
594	Nauciel et al. (1969), Hallas (1985)
T25₃	Lee and Schlievert (1989)
C203 S	Gray (1979), Schlievert and Gray (1989)
S 84	Cavaillon et al. (1979), Geoffroy and Alouf (1988)
Production of ETB	
NY5	Stock and Lynn (1961), Barsumian et al. (1978a)
T19	Barsumian et al. (1978a), Gerlach et al. (1983), Hallas (1985)
86-5885	Lee and Schlievert (1989)
Production of ETC	
NY5	Ozegowski et al. (1984), Bohach et al. (1988a)
T 18	Schlievert et al. (1977), Ozegowski et al. (1984), Bohach et al. (1988a), Lee and Schlievert (1989)
AT13	Ozegowski et al. (1984)
86-104	Lee and Schlievert (1989)
Production of ETD	
C203 U	Schuh et al. (1970)
Clinical isolate	McMillian et al. (1987)

quantitative pattern of the erythrogenic toxins over the past 60 years and their relation to fluctuations in the severity of the diseases and mortality rates have been discussed by Hallas (1985), Köhler et al. (1987), Lee and Schlievert (1989) and Yu and Ferretti (1989).

Epidemiological aspects of toxin production

Schlievert *et al.* (1979c) reported that 91% of 75 strains examined were positive for one or more toxin types A, B and C. Strains producing both types B and C were the most common whereas those which produced type A or both A+B or A+C were the least common. A compilation of the data for 733 strains isolated in the past ten years from different clinical sources and geographic areas shows that positive strains varied from 100 to 68% with great differences as regards toxin serotype patterns (Table 2). Recently, Gaworzewska and Hallas (1989) reported an increase in the United Kingdom in the number of *S. pyogenes* isolates from fatal streptococcal toxic-shock-like infections (32 out of 34 patients died). None of the strains produced ETA whereas 72% produced ETB. In contrast, in recent similar syndromes in the United States, Stevens *et al.* (1989) found that 8 of 10 strains isolated from patients produced ETA alone or associated to other serotypes. Similarly, Lee and Schlievert (1989) reported that 13 out of 26 isolates made type A only, 3 made type C and the 10 others made A and B or the three serotypes together (see Table 1).

Recently, Yu and Ferretti (1989) reported a molecular epidemiologic analysis of ETA gene in 446 clinical isolates of *S. pyogenes* by colony hybridization technique employing a specific internal fragment of the gene. The strains were obtained from 11 different countries. Among 300 strains isolated from patients with a variety of diseases except scarlet fever, 15% were found to contain the toxin gene. For 146 strains collected from individuals described as having scarlet fever, only 45% were found to contain the gene. Two possible explanations are provided for this surprising result. First, there may be variations among physicians in the criteria used to diagnose scarlet fever. Second, ETB and ETC may, in the absence of ETA, produce identical clinical symptoms. The frequency of ETB and ETC genes (*speB* and *speC*) investigated among 500 clinical strains (Yu and Ferretti, 1991) was 100% for the former and 60% for the latter. The *speC* gene was found to be more frequently associated with serotypes M2, M4 and M6 strains and less frequently associated with M1, M3 and M49 strains.

Genetic determinants

The field of molecular genetics of SEs, TSST-1 and ETs has seen major advances in the last decade and particularly in the past 5 years that have made possible the cloning and sequencing of structural genes of these toxins and thereby allowed the establishment of their primary structure.

The genetic analysis of the determinants of these toxins is beyond the scope of this chapter. The reader is referred to the articles quoted in Table 4 and to the articles or reviews of Iandolo (1989), Soltis *et al.* (1990), Kreiswirth (1989), Dickgiesser (1988), Blomster-Hautamaa and Schlievert (1988a) and Wannamaker and Schlievert (1988). The structural genes may be chromosomal or located on bacteriophages or plasmids (Table 4). Interesting analyses of bacteriophage involvement in the production of ETA have been reported by Nida and Ferretti (1982), Weeks and Ferretti (1984), Johnson and Schlievert (1984) and Johnson *et al.* (1986b). SEB gene was shown to be associated with discrete genetic elements (John and Khan, 1988) and to be regulated by upstream sequences (Mahmood and Khan, 1990).

Purification

All staphylococcal enterotoxins, TSST-1 and erythrogenic toxins have been purified to apparent homogeneity in the past 20 years (Table 3). Detailed description of the methods used is not possible in this chapter (see references in Table 3). Generally, culture supernates are concentrated by appropriate techniques (salting-out with $(NH_4)_2SO_4$, precipitation with ethanol at $-20°C$, ultrafiltration) followed by the classical protein separation methods (gel filtration, ion-exchange chromatography, isoelectric focusing, chromatofocusing, etc.). Recent new methodologies have also been used such as HPLC or separation by hydrophobic, immuno- or dye-ligand affinity chromatography.

Technical details and useful recommendations allowing reliable and satisfactory purification may be found in recently published methods for SEs (Spero *et al.*, 1988; Iandolo and Tweten, 1988), TSST-1 (Blomster-Hautamaa and Schlievert,

Table 3. Some reported purifications to homogeneity of staphylococcal enterotoxins, TSST-1 and streptococcal erythrogenic (pyrogenic) toxins

Staphylococcal

SEA	Schantz et al. (1972),[a] Robern et al. (1975), Spero and Metzger (1981),[a] Huang et al. (1987), Lapeyre et al. (1987), Reynolds et al. (1988)
SEB	Schantz et al. (1965),[a] Metzger et al. (1972),[a] Ende et al. (1983), Williams et al. (1983), Strickler et al. (1989)
SEC1	Borja and Bergdoll (1967),[a] Spero et al. (1976), Ende et al. (1983), Lapeyre et al. (1987)
SEC2	Avena and Bergdoll (1967),[a] Blomster-Hautamaa and Schlievert (1988b)
SEC3	Reiser et al. (1984)
SED	Chang and Bergdoll (1979),[a] Lapeyre et al. (1987), Lei et al. (1988)
SEE	Borja et al. (1972)[a]
TSST-1	Schlievert et al. (1981), Bergdoll et al. (1981), Reiser et al. (1983), Melconian et al. (1983b), Igarashi et al. (1984), Blomster-Hautamaa et al. (1986a), Reeves et al. (1986), Blomster-Hautamaa and Schlievert (1988b), Rosten et al. (1989)

Streptococcal

ETA	Nauciel et al. (1969), Cunningham et al. (1976), Cavaillon et al. (1979), Gray (1979), Gerlach et al. (1980), Houston and Ferretti (1981), Geoffroy and Alouf (1988), Kamezawa and Nakahara (1989)
ETA (recomb.)	Gerlach et al. (1987)
ETB	Cunningham et al. (1976), Barsumian et al. (1978a), Gerlach et al. (1983)
ETC	Schlievert et al. (1977), Ozegowski et al. (1984)

[a]See Spero et al. (1988) for these references.

1988b) and ETA (Gerlach et al., 1987; Geoffroy and Alouf, 1988; Wannamaker and Schlievert, 1988). We briefly mention below some significant aspects of toxin purification.

Staphylococcal enterotoxins

SEA was purified with an overall yield of 55% in a single step by dye-ligand affinity chromatography with the triazine dye Red A (Reynolds et al., 1988). This is a significant improvement in time, yield and purity over previously published methods. Chromatofocusing method allowed the purification of SEB and SEC1 (Ende et al., 1983) as well as that of SED (Lei et al., 1988). SEB was purified in a two-step procedure with a yield of 35–45% by HPLC (Strickler et al., 1989). Pure SEA, SEC1 and SED were also obtained on affinity columns of Ultrogel coupled with the corresponding monoclonal antibodies (Lapeyre et al., 1987).

Toxic-shock syndrome toxin-1

In the procedure described by Reiser et al. (1983), the toxin present in the supernatant fluid was adsorbed on Amberlite CG-50 resin, then eluted with buffer and then purified by successive ion-exchange and gel filtration procedures. Adsorption on Biorex cation-exchange resin was used in the procedure of Melconian et al. (1983b). A chromatofocusing step was used by Igarashi et al. (1984). This procedure, combined with purification on Affi-gel-10 column coupled with monoclonal antibody to TSST-1, was used by Reeves et al. (1986). Reverse-phase HPLC was applied by Rosten et al. (1989) and isoelectric focusing by Blomster-Hautamaa et al. (1986c).

Streptococcal erythrogenic toxins

After the discovery of the 'scarlet fever toxin' by the Dicks in 1924, many investigators attempted in the following 20 years to purify the toxins A and B known at that time (see Kim and Watson, 1970 for references). The most significant work in this period was done by Stock (1939) and his associates (see Krejci et al., 1942; Stock and Lynn, 1961). Krejci et al. isolated the toxin as a protein having a biological activity of $110–150 \times 10^6$

STD/mg and a molecular weight of 27 000, which is close to that calculated for the mature toxin on the basis of gene sequence (Table 4). The M_r of ETA isolated by ion-exchange chromatography by Nauciel et al. (1969) or by ethanol precipitation by Kim and Watson (1970) was 30 000 and 29 400 respectively. A further purification of ETA by DEAE–Sephadex chromatography by Cunningham et al. (1976) yielded two serologically identical fractions of 8 and 5 kDa, respectively. These data dominated the literature of recent years although they did not coincide with the M_r values of 26–30 kDa in earlier reports or with those similar from other laboratories (Seravalli and Taranta, 1974; Cavaillon et al., 1979; Gerlach et al., 1980, 1987; Rasmussen and Wuepper, 1981; Cavaillon and Alouf, 1982). The conflicting results wre discussed and explained by Gerlach et al. (1986), who reported the copurification of ETA with a biologically active low molecular weight protein (LMP-10K), which was found to be serologically different from ETA but had the same pI (5.25) and the same solubility in acid buffer.

The M_r of the ETA recently purified by Kamezawa and Nakahara (1989) was calculated as 26 330 on the basis of amino acid analysis of the protein and 28 000 as determined by SDS–PAGE analysis. This value is identical to that reported by Gerlach et al. (1980). ETA could be easily purified from the supernatant fluid of S. sanguis (Fig. 1) containing the cloned gene of the toxin (Gerlach et al., 1987). In the method of Geoffroy and Alouf (1988), the obtention of ETA is facilitated by batchwise adsorption of the supernatant fluid on $Ca_3(PO_4)_2$ gel which allows the elimination of streptolysin O and many other extracellular proteins.

ETB purified by Cunningham et al. (1976) was characterized as a 22 000-Da protein which showed some heterogeneity as regards pI (8.0–9.5) value. By using isoelectric focusing for toxin purification, Barsumian et al. (1978a) isolated three antigenically identical fractions with pI values of 8.0, 8.4 and 9.0. The molecular weight of all three fractions was ca. 17 000 but only that of pI 8.4 was biologically active. When the isolated fractions were electrofocused again they appeared heterogeneous, suggesting an instability of the B toxin. ETB from strain T19 was purified by Gerlach et al. (1983) who reported its identity with streptococcal proteinase.

A molecular weight of 13 000 was reported for

type C by Schlievert et al. (1977). This value was not confirmed by Ozegowski et al. (1984) who found an M_r of 25 500, close to the value of 24 354 calculated from the gene sequence (Goshorn and Schlievert, 1988).

Protocols for the purification of ETB and ETC have been reported by Wannamaker and Schlievert (1988) who also provided a method for recombinant ETA purification from B. subtilis supernatant fluid.

ETD was partially purified by McMillian et al. (1987) and characterized as a 13-kDa mitogenic and pyrogenic protein with a pI of 4.6. It may be identical to one of the mitogens of 12 kDa of pI 4.7 reported by Gerlach et al. (1986). A homology of 69% between ETD and S. aureus phosphocarrier protein was recently found by Gerlach (Abstract P83 of the XI Lancefield Symposium on Streptococci, Siena, 1990).

Molecular and antigenic characteristics

The molecular weights and other physical and chemical characteristics of SEs, pyrogenic exotoxins A and B, TSST-1 and ETs are listed in Table 4. All toxins are single-chain polypeptides. The primary structures deduced from structural gene sequencing have been established for all SEs, TSST-1, ETA, ETB and ETC (see Table 4 references). The primary structure of SEA, SEB and SEC1 has also been established by direct determination of the amino acid sequence of the purified toxins (see Huang et al., 1987). The aligned primary sequences of all these toxins are shown in the review of Marrack and Kappler. SEs but not the other toxins have a disulphide loop.

Staphylococcal enterotoxins

Structural characteristics

The seven serotypes are single-chain polypeptides with M_r values ranging from 26 360 to 28 336 (Table 4). The shortest polypeptide is SED with 228 residues and the longest are SEB, SEC1, SEC2 and SEC3 with 239 residues. SEA and SEE contain 233 and 230 residues respectively. The N-terminal residues of SEs are either Ser or Glu (Table 4). All SEs have two half-cystine residues joined to make a loop near the centre of the molecule (Fig. 2). The

Cultivation of *S. pyrogenes* in 50-litre yeast extract–peptone medium for 16 h

Killing of streptococci in the fermenter by addition of H_2O_2 (final concentration 0.1%) or acidification to pH 3.5 for 1 h

Centrifugation

Supernatant: adjust to pH 5.2, addition of NaCl 2.5% (w/v)

Toxin binding onto 1 kg silica gel under stirring for 2 h

Elution of ETA from sedimented silica gel with 5 litres of 1% (w/v) solution of Na_2CO_3

Precipitation of ETA from the eluate by addition of 500 g $(NH_4)_2SO_4$ per litre

Dissolution of precipitate in acidic buffer (0.15 M NaCl; 0.01 M acetate, pH 4.0)

Dialysis against the same buffer, discard precipitated material

Concentration of toxin by a second ammonium sulphate precipitation

Dissolution of precipitate in acidic buffer, dialysis against the same buffer, followed by dialysis against 0.01 M acetate buffer, pH 4.6

Chromatography on a CM-Sepharose 6B column

Washing with 0.01 M acetate buffer pH 4.6

Elution of toxin with 0.05 M acetate buffer, pH 5.2

Precipitation of toxin containing fractions of the second peak with 4 vol. ethanol

Dissolution of precipitate in acidic buffer (pH 4.0), discard in soluble material

Lyophilization

Figure 1. Purification of erythrogenic toxin A (Gerlach *et al.*, 1987).

numbers of residues of the two half-cystine loop are 19 for SEB, 16 for SECs and 9 for SEA, SED and SEE. The function of the disulphide bond is unknown. However, this bond does not appear to be an active site, as shown for SEB which could be reduced and alkylated with no loss of toxicity. It may serve to stabilize the molecule.

Nucleotide sequence homology. A comparison of the *se* gene sequences (see Betley *et al.*, 1990 for gene nomenclature) has revealed different degrees of relatedness among the genes (Betley and Mekalanos, 1988). The relatedness between *sec* genes and *seb* genes is 75%, that of *seb* and *sec* to *sea* (50%) is lower than that of *sea* to *see* which is 84% (Soltis *et al.*, 1990), *sed* is more closely related to *sea* and *see* (*ca.* 54%) than to *seb* and *sec1* (40%). The three *sec* genes share a homology of 98% (Hovde *et al.*, 1990). The TSST-1 gene shares little sequence homology with *se*, in contrast to the ETA and ETC genes. Table 7 summarizes some of these data.

Amino acid sequence homology. The percentage of amino acid sequence homology among the mature forms of SE varies from 82 to 29%. SEA and SEE

are the most closely related (Couch *et al.*, 1988; Singh and Betley, 1989). The cross-reactivity between these two types (see below) reflects their extensive similarity. The homology between SEB and SEC1 is about 65% (Schmidt and Spero, 1983). Both toxins have a similar level of toxicity. Clearly a 35% difference in the amino acid sequence does not affect the biological effects of the toxins. Moreover, SEB and SEC1 have significant differences in the nature of their constituting amino acids. SEB contains more charged and more hydrophobic amino acids than SEC1. The closest homology between the toxins was in the C-terminal regions where 55 of the last 67 residues were identical. In the N-terminal regions 38 of the 57 residues of SEC1 were identical to the corresponding residues in SEB.

A comparison of amino acid sequences of SEA and SEB or SEA and SEC1 reveals less similarity (Huang *et al.*, 1987). In the N-terminal region up to 50 residues, almost no homology is observed between SEA and SEB or SEA and SEC1. As shown in Table 5, the only sequence in which five successive residues are identical is that following the second Cys residue in SEA, SEB, SECs and

Table 4. Some physical, chemical and genetic characteristics of staphylococcal enterotoxins, pyrogenic exotoxins, TSST-I and streptococcal erythrogenic (pyrogenic toxins)

	M_r (mature form)	Number of AA[a]	pI	N-terminal residues	Structural gene localization[b]	References
S. aureus						
SEA	27078	233 + 24	7.3	Ser	C or B	1–4
SEB	28336	239 + 27	8.6	Glu	C or P	1–3, 5
SECI	27496	239 + 27	8.6	Glu	C or P	1–3, 6
SEC2	27589	239 + 27	7.0	Glu	C	1, 2, 7
SEC3	27563	239 + 27	8.1	Glu	C	1, 2, 7, 8
SED	26360	228 + 30	7.4	Ser	P	9, 10
SEE	26425	230 + 27	7.0	Ser	B (?)	2, 3, 11
Exotoxin A	12000		5.3	?	n.d.	12
Exotoxin B	18000		8.5	Ala	n.d.	12
TSST-I	22049	194 + 40	7.08–7.22	Ser	C	3, 13, 14
S. pyogenes						
ETA	25805	220 + 30	5.2	Ser	B	15
	25787	221 + 30		Gln		16, 17
ETB	40314	371 + 27	8.4	Asp	C	18–21
Breakdown product	27588	253				21
ETC	24354	208 + 27	6.8	Asp	B	22, 23
ETD	13000		4.6			24

[a]Number of amino acids of the mature extracellular toxin + that of leader sequence.
[b]B, bacteriophage; C, chromosome; P, plasmid.
(1) Spero et al. (1988); (2) Iandolo and Tweten (1988); (3) Iandolo (1989); (4) Betley and Mekalanos (1988); (5) Jones and Khan (1986); (6) Bohach and Schlievert (1987a,b); (7) Couch and Betley (1989); (8) Hovde et al. (1990); (9) Lei et al. (1988); (10) Bayles and Iandolo (1988); (11) Couch et al. (1988); (12) Schlievert (1980b); (13) Blomster-Hautamaa and Schlievert (1988a); (14) Blomster-Hautamaa et al. (1986a,b); (15) Kreiswirth et al. (1987); (16) Johnson et al. (1986b); (16) Weeks and Ferretti (1986); (17) Gerlach et al. (1980); (18) Barsumian et al. (1978a); (19) Gerlach et al. (1983); (20) Bohach et al. (1988a); (21) Hauser and Schlievert (1990); (22) Ozegowski et al. (1984); (23) Goshorn and Schlievert (1988) and Goshorn et al. (1988); (24) McMillian et al. (1987).

1. Cystine loops of enterotoxins A and E

2. Cystine loop of enterotoxin B

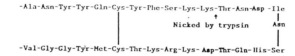

3. Cystine loops of the enterotoxins SEC

4. Cystine loop of enterotoxin D

Figure 2. Cystine loops of staphylococcal enterotoxins.

Table 5. Amino acid sequence homology in staphylococcal enterotoxins

SEA:	Pro–Asn–Lys–Thr–Ala–**Cys–Met–Tyr–Gly–Gly–Val–Thr–**Leu–His–Asp–Asn–Asn–Arg–
SEB:	Asp–Lys–Arg–Lys–Thr–**Cys–Met–Tyr–Gly–Gly–Val–Thr–**Glu–His–Asn–Gly–Asn–Gln–
SECs:	Thr–Gly–Gly–Lys–Thr–**Cys–Met–Tyr–Gly–Gly–**Ile–Thr–Lys–His–Glu–Gly–Asn–His–
SED:	Ile–Asp–Arg–Thr–Ala–**Cys–**Thr–**Tyr–Gly–Gly–Val–Thr–**Pro–His–Glu–Gly–Asn–Lys–
SEE:	Pro–Asn–Lys–Thr–Ala–**Cys–Met–Tyr–Gly–Gly–Val–Thr–**Leu–His–Asp–Asn–Asn–Arg–

[a]Bold, common residues; Bold and underlined, differing residues.
Residue numbers: SEA, 101–118; SEB, 108–125; SECs, 105–122; SED, 96–113; SEE, 98–115.
Data kindly communicated by Dr M. Bergdoll.

SEE (four residues in SED). This common feature might support the hypothesis that this part of the molecule is essential for the emetic and other biological properties of SEs.

The three C serotypes are the most antigenically related and their sequences are almost identical. SEC3 and SEC2 differ by four residues (98.3% homology), SEC1 differs from SEC2 by seven residues (97.1% homology) and from SEC2 by

nine residues (96.2% homology). Overall there are 10 divergent residues among the three types.

Structure–activity relationship

This relationship was investigated by studying the biological and antigenic properties of the toxins submitted to proteolysis or to chemical modifications of specific residues (see Spero *et al.*, 1988). Native SEB and SEC1 are readily nicked by limited tryptic proteolysis while SEA is very resistant. SEB is split between Lys[97] and Thr[98] in the cystine loop and SEC1 is also split in two positions in this loop (Fig. 2) and in two other positions (Lys[57]–Leu[58] and Lys[59]–Asn[60]) outside the loop. Trypsin-cleaved SEB remained emetic and mitogenic even after reduction of the disulphide bond and alkylation of the SH groups. The two peptides remained associated under these conditions. They could be separated by chromatography in 8 M urea and each individual peptide was inactive upon removal of denaturant. Full activity was recovered by reassociation of the two entities.

The study of the nicked SEC1 (Bohach et al., 1989) showed that the mitogenic, pyrogenic and lethal activity of the toxin was associated with the larger C-terminal peptide, which was also shown (see Spero et al., 1988) to provoke diarrhoea in monkeys. The N-terminal fragment of SEC1 was reported to contain the functional site for the mitogenic but not for the emetic activity. Recently Pontzer et al. (1989) confirmed by the synthetic peptide approach and by immune sera raised against the peptides corresponding to the residues 1–27 and 28–45 of SEA that the part of the toxin responsible for T cell proliferation and induction of interferon-γ involves the N-terminal 27 residues.

The chemical modifications of various specific residues of enterotoxins carried out by many investigators gave, according to Spero et al. (1988), very little definitive evidence on the nature of the active site of these toxins, although toxin activity was affected by certain treatments which modified various residues. However, one of the histidine residues appears to be critical. It concerns that present in all enterotoxins in the highly conserved segment following the second cysteine residue (Table 5) proposed as the active site of SEs. This His residue might be the essential residue responsible for the loss of toxicity of SEA and SEB upon alkylation of the His residues of these toxins. The secondary structure of SEB places the His residue of the conserved region in a β-turn and therefore on the surface of the molecule. Streptococcal ETA has a His residue in a comparable sequence segment (Weeks and Ferretti, 1986; Johnson et al., 1986a).

It might be expected that sequences required for specific secondary and tertiary structures as well as amino acid residues critical for biological

function are conserved among the enterotoxins. It therefore appeared to be a good approach for the understanding of structure–activity relationship to investigate the secondary structure of SEs by appropriate methods such as far-UV circular dichroism (CD) spectroscopy. This technique was used by Munoz et al., Middlebrook et al. and Spero between 1976 and 1981 (see Spero et al., 1988) for SEA, SEB and SEC1, and they found an α-helix content for these three toxins with predominant β-forms and aperiodic conformation, which suggested a β-barrel tertiary structure. A more detailed study in terms of α-helix, β-pleated sheets, β-turns and random coil was recently reported for SEB and SEC1 by Singh et al. (1988b) and for SEA and SEE by Singh and Betley (1989). These authors also used fluorescence spectroscopy for these toxins, which contain a single tryptophan residue (Trp[197]) in SEB and SEC1 and two Trp residues in SEA and SEE.

The CD spectra of the four toxins allowed the parameters of their secondary structure to be determined (Table 6). These toxins showed the same basic patterns: low α-helical content varying from 6.5% for SEE to 15% for SEC1 (9.5 and 10.0% for SEB and SEA, respectively) and predominant β-sheets/β-turns structures varying from 63.5% for SEC1 to 84.5% for SEA (71.5 and 79.5 for SEB and SEE respectively). The topography of Trp residues was probed by fluorescence spectroscopy which showed maxima at ~ 342 nm for SEA and SEE and ~ 333 nm for SEB and SEC1. These data indicate that the Trp

Table 6. Secondary structure parameters of enterotoxins, A, B, C1, E and TSST-1 from far UV circular dichroic spectra[a]

Secondary structure	SEA	SEB	SEC1	SEE	TSST-1
% α-helix	10.0	9.5	15.0	6.5	6.25
% β-sheet	84.5	55.0	38.0	78.0	51.25
% β-turns	0.0	16.5	25.5	1.5	9.0
% Random coils	5.5	19.0	9.7	14.0	33.5

[a] Data from Singh et al. (1988a,b) and Singh and Betley (1989).

residue of SEB and SEC1 is in a relatively non-polar environment whereas the Trp residues of SEA and SEE are in a polar environment. The Trp residue in SEB was *ca.* 46% more fluorescent than that in SEC1, indicating a significant difference in the surrounding groups of the Trp of these toxins and thereby different polypeptide folding since the surrounding amino acid residues are identical in both proteins. Similarly, the Trp residues of SEA were 41% more fluorescent than in SEE, also suggesting a different environment for the two toxins. Trp fluorescence quenching of SEB and SEC1 probed by the surface quencher I^- and the neutral quencher acrylamide revealed that the Trp residue was buried in the protein matrix. In contrast, at least one of the two Trp residues of SEA and SEE is located on the outer surface of the proteins.

These spectroscopic data reveal similarities and significant differences between the enterotoxins. The similarities concern common features of their secondary structure irrespective of differences in their sequence. In any case it seems that a higher content of β-pleated sheets is maintained in all SEs and also, as shown below, in TSST-1. The main differences concern different polypeptide folding (tertiary structure) and different environments for Trp residues.

Antigenic structure and cross-reactivity

The elucidation of the primary structure of all SEs and the secondary structure of some of them allows a better understanding of their antigenic features. The cross-reactivity between these toxins has been investigated by immunodiffusion, neutralization, ELISA, RIA and immunoblotting techniques (see Spero *et al.*, 1988 for a review and Meyer *et al.*, 1984; Reiser *et al.*, 1983; Edwin *et al.*, 1986; Lapeyre *et al.*, 1987; Kienle and Buschmann, 1989). Both polyclonal and monoclonal antibodies have been used in recent investigations. Each SE was shown to bear specific epitope(s) not possessed by the others. However, various degrees of antigenic relatedness were demonstrated. As expected from their almost identical primary structure the three SEC subtypes presented the greatest degree of cross-reactivity. They differed by minor epitopes. Cross-reacting determinants were observed between the SEB–SECs group as well as within the SEA–SEE group, reflecting the sequence

homology in each group. No cross-reactivity between SEA and SEB has been demonstrated. SEA, SED and SEE have considerable structural homology. Monoclonal antibodies that cross-reacted with SEA and SED, and SED and SEE, respectively, were obtained, indicating different antigenic relatedness within these three toxins.

The tryptic cleavage of SEB and SEC1 allowed the investigation of the epitope location on the resulting peptide fragments (see Spero *et al.*, 1988; Thompson *et al.*, 1984). There were two epitopes on each toxin capable of reacting with heterologous antibodies: one on the N-terminal peptide and another on the C-terminal peptide.

Seven antigenic determinants for SEB and SEC1 and six determinants for SEA and SEE have been predicted (Singh *et al.*, 1988b; Singh and Betley, 1989) on the basis of hydrophilicity and secondary structure. Three antigenic sites were found similar for SEB and SEC1 and at least four were identical in SEA and SEE. The location of these sites on the molecules were determined and shown to be mainly in the β-turn segments.

Toxic-shock syndrome toxin-1

Structural characteristics

TSST-1 is a single-chain polypeptide constituted by 194 residues on the basis of gene sequencing with a predicted M_r of 22 049 (Blomster-Hautamaa *et al.*, 1986a) which is very close to the value (22 000) determined by SDS–PAGE on the toxin purified by Blomster-Hautamaa *et al.* (1986c). These authors reported that this protein was resolved after isoelectric focusing into two immunologically and functionally identical forms with pI values of 7.08 and 7.22, respectively, which could result either from the presence of a cofactor (AMP, NADH) or from alternative conformations or deamidations during purification. The slight differences in the nature and number of amino acid residues (197 and 199) of the two forms and the discrepancy with that predicted from gene sequencing (see Blomster-Hautamaa and Schlievert, 1988a for a comparative table) remain to be explained. A purified TSST-1 preparation submitted by Eriksson *et al.* (1989) to free zone electrophoresis at pH 8.6 was also resolved into two components with respective mobilities of -2.12×10^{-3} and -3.60×10^{-3} mm^2 V^{-1} s^{-1}.

Charge heterogeneity of SEA, SEC1 and SEC3 was also observed by these authors, who also reported a high surface hydrophobicity of TSST-1 and SEA as probed by high-performance hydrophobic chromatography on octyl-agarose column. This technique also allowed the resolution of TSST-1 into two distinct 22-kDa proteins identical as regards amino acid composition, antigenic reactivity and pI (7.4).

TSST-1 lacks cysteine residues, it is resistant to trypsin but is digestible by pepsin and papain (Reiser et al., 1983; Edwin and Kass, 1989).

Structure–activity relationships

The evaluations of the relationships were based on studies on toxin fragments or on the whole molecule. The cleavage of the two methionine residues of TSST-1 (Met[33] and Met[158]) by cyanogen bromide generates five peptide fragments: CN1 (18 kDa), CN2 (17 kDa), CN3 (14 kDa), CN4 (6–8 kDa) and CN5 (4 kDa) which were isolated and purified (Blomster-Hautamaa et al., 1986a). By using monoclonal antibodies, Blomster-Hautamaa et al. (1986b) demonstrated that the mitogenic and immunosuppressive functions of TSST-1 were located on the central 14-kDA CN3 fragment. Both functions were neutralized by these antibodies. A synthetic decapeptide corresponding to a stretch at the N-terminus of CN3 fragment competitively inhibited the binding to TSST-1 of one of these neutralizing antibodies and its capacity to block the mitogenic activity of the toxin (Murphy et al., 1988). A polyclonal rabbit serum raised against a conjugate of the synthetic decapeptide reacted with TSST-1 and partially neutralized toxin mitogenicity. These results suggest that the epitope recognized by the monoclonal antibody includes the decapeptide sequence.

Papain hydrolysis of TSST-1 generated three fragments identified as 16-, 12- and 10-kDa resulting from the cleavage of peptide bonds between Tyr[52] and Ser[53] and between Gly[87] and Val[88]. The two fragments encompassed between residues 53 and 87 and 88 and 194 were immunologically and biologically active but not the fragment from residues 1 to 52 (Edwin et al., 1988). The mitogenic potency of the 12-kDa fragment was higher than that of the 16-kDa fragment. Monoclonal antibodies developed against segments of

the native toxin depicted two different antigenic and functional segments (Edwin and Kass, 1989). One antibody shown to inhibit the biologic activity of the holotoxin in vitro and in vivo reacted primarily with the 12-kDa fragment. A second antibody that did not neutralize interleukin 1 production reacted primarily with the 16-kDa fragment.

Kokan-Moore and Bergdoll (1989a,b) investigated the role of methionine, histidine and tyrosine residues in the mitogenic effect and antigenic reactivity of TSST-1. Modification of the two Met residues by alkylation with iodoacetic acid did not affect the mitogenic activity and the reaction with polyclonal antibodies. Modification of four of the five His residues with diethylpyrocarbonate inhibited 50% of the mitogenic reaction but did not inhibit toxin precipitation with polyclonal antisera. One of the monoclonal antibodies did not react with the modified toxin. Modification of one or two of the nine Tyr residues did not inhibit the precipitin reaction but did inhibit 85% of the mitogenic activity. The proximity of Tyr-[80] and His-[82] may be essential for the mitogenic activity of the toxin.

A detailed physical study of TSST-1 was carried out by Singh et al. (1988a) by the same techniques used by these investigators for SEs. This study established that: (1) TSST-1 has, similarly to SEs, low α-helix and high β-sheet/β-turn content (Table 6); (2) six of nine Tyr residues are exposed on the surface of the molecule; (3) Trp residues are buried in the hydrophobic protein matrix. Four antigenic sites were predicted, all in β-turn conformation, three of them being located on the central 15-kDa segment. None of the four sites has any similarity with the predicted antigenic sequences of SEA, SEB, SEC1 or SEE. His- and Tyr-modified TSST-1 molecules were analysed for secondary structure parameters by CD measurements (Kokan-Moore and Bergdoll, 1989a). For His-modified molecules there was a gradual increase of the α-helical content with all other parameters staying constant, suggesting that no gross conformational change occurred. In contrast, modifications of Tyr residues resulted in alteration of all secondary parameters and thereby a change in conformation which may explain the loss of mitogenic activity.

Erythrogenic toxins

Structural characteristics of ETA

The nucleotide sequence analysis of the structural gene encoding ETA indicated that the gene product was composed of 251 amino acid residues with a probable signal peptide of 30 residues (Weeks and Ferretti, 1986) or 250 residues including also a 30-residue signal peptide (Johnson et al., 1986a). The calculated molecular weights of the mature protein were 25 787 and 25 805, respectively. The amino acid compositions and sequence reported by the two groups of investigators differed significantly, particularly with regard to the sequence of the N-terminal residues which were Gln–Gln–Asp–Pro–Asp–Pro–Ser–Gln–Leu–His–Arg . . . and Ser–Thr–Arg–Pro–Lys–Pro–Gln–Leu–Arg–Ser . . ., respectively. Other variances occurred throughout the polypeptide chain, including the leader sequence. That reported by Johnson and his co-workers varied from classical prokaryotic signal peptides. The difference between the two sequences is still not explained. It might be due to a frame-shift (Couch et al., 1988).

The amino acid composition of ETA preparations purified by Gerlach et al. (1980) and Kamezawa and Nakahara (1989) from S. pyogenes and that of recombinant ETA purified from S. sanguis (Gerlach et al., 1987) closely paralleled with a few exceptions, that deduced by Weeks and Ferretti. The N-terminal sequence of the first five residues of the ETA obtained by Kamazawa and Nakahara and that of the first nine residues of recombinant ETA were identical to that reported by Weeks and Ferretti, indicating for the latter that ETA is identically processed in S. pyogenes and S. sanguis. An identical N-terminal sequence from residues 2 to 10 was found for blastogen A as reported by Schlievert and Gray (1989) who finally established that this mitogenic factor first reported to be different from ETA (Gray, 1979) was indeed ETA per se. Both preparations showed total immunological cross-reaction by immunodiffusion and Western blot analyses as well as comparable mitogenic properties and molecular weight (28 000).

All data derived from gene sequencing and protein analyses of the recently purified ETA preparations indicated that the toxin is rich in lysine, leucine, glutamic and aspartic acids and contains three cysteine residues located at positions

87, 90 and 98 in the mature protein as deduced from nucleotide sequencing. Whether a disulphide bridge occurs is not known. However, the presence of at least one free sulphydryl group may explain the ETA dimerization reported by Schlievert and Gray (1989) or other polymorphic forms (Nauciel et al., 1969; Cunningham et al., 1976) and their conversion into a single form by reducing agents. ETA incubation for 2 h at 37°C with 2-mercaptoethanol (ME) followed by SDS–PAGE was shown to generate four bands of 5800, 7800, 16 000 and 20 000 Da and only one band of 8500 Da after 16-h incubation (Houston and Ferretti, 1981). The contention that this treatment led to total reduction of multimers to a 8500-Da monomeric form is not satisfactory and is not supported by ETA structure as presently known. No presence of subunits was found by Rasmussen and Wuepper (1981) upon ME treatment of the streptococcal proliferative factor (very likely identical to ETA by various criteria).

ETA molecules as well as the other serotypes may associate to streptococcal hyaluronic acid or to other proteins (see Wannamaker and Schlievert, 1988) which makes toxin purification sometimes difficult.

Structural and biochemical characteristics of ETB

An interesting feature of ETB is its identity or very close similarity to streptococcal proteinase as first reported by Gerlach et al. (1983). This enzyme, which has been studied for more than 40 years, is produced (see Elliott and Liu, 1970 for an overview) as a 44-kDa zymogen which is cleaved with reducing agents or proteolysis by trypsin into a 32-kDa active proteinase and a 12-kDa peptide. The proteinase precursor was later shown (Yonaha et al., 1982) to have a molecular weight of 36 708 and to undergo autocatalysis upon reduction to yield an active proteinase of 28 000 Da. Similarly, ETB was shown by SDS–PAGE (Gerlach et al., 1983) to consist of 30-kDa and 12-kDa components and to have the same amino acid composition as that of the proteinase purified in parallel from strains T19 and B220 according to the procedure of Elliott and Liu. Both ETB and proteinase cross-reacted by immunodiffusion tests. No ETB was produced when the bacteria were grown above pH 6.8 which avoids streptococcal proteinase production. The restriction of this production to the acidic pH range explains the discrepancies in the literature concerning ETB production.

The structural gene encoding ETB has been cloned in *E. coli* (Bohach *et al.*, 1988a) and its nucleotide sequence determined (Hauser and Schlievert, 1990). The toxin partially purified from this organism was mitogenic and immunologically related to ETB derived from *S. pyogenes*. The molecular weight reported was 29 300.

The structural relationship of ETB and streptococcal proteinase precursor was confirmed by the amino acid sequence of the toxin deduced from the nucleotide sequence of its gene (Hauser and Schlievert, 1990). ETB was shown to be initially translated as a 398-residue protein. Cleavage of a 27-amino acid signal peptide yields a mature 371-residue protein with a molecular weight of 40 314. Subsequent proteolysis results in the formation of multiple smaller intermediates and eventually a 253-residue breakdown product of 27 588 Da. The amino acid sequence of the proteinase reported by Tai *et al.* (1976) and Yonaha *et al.* (1982) and that inferred from the gene sequence clearly indicate that SEB and the proteinase are variants of the same proteins.

Structural characteristics of ETC

The nucleotide sequence of the gene encoding ETC determined by Goshorn *et al.* (1988) allowed the primary structure of this toxin to be deduced. The mature polypeptide contained 208 amino acids (calculated M_r 24 354). A signal peptide of 25 residues was reported. The sequence homology with ETA is discussed below.

Toxin stability

Purified ETA is relatively heat-stable. No loss of skin reactivity was observed after 1 h heating at 80°C (Nauciel *et al.*, 1968) and that of pyrogenicity after 30 min incubation at 65°C or after boiling for 2 min (Kim and Watson, 1970). Boiling at 100°C with SDS did not destroy toxin immunoprecipitation capacity with antiserum (Gerlach *et al.*, 1987). The biological activity was not affected by streptococcal proteinase, trypsin, pepsin or pronase E treatment (Gerlach *et al.*, 1981). In contrast, trypsin, pepsin and pronase E inactivated ETC. This toxin was not digested after 24 h incubation with papain and streptococcal proteinase (Ozegowski *et al.*, 1984).

Structural and antigenic relatedness between the three staphylococcal and streptococcal toxin groups

Structural homology

SEB, SEC1 and ETA show sequence homology particularly in the C-terminal portion (Weeks and Ferretti, 1986; Johnson *et al.*, 1986a; Bohach and Schlievert, 1987b). SEC1 and SEB shared 69% homology, ETA and SEB 52% and SEC1 and ETA 49% homology (Bohach *et al.*, 1988b). The homology of SEA with SEB, SEC1 and ETA was lower (Betley and Mekalanos, 1988). ETC sequence was more related to that of ETA and less to that of SEs. Amino acid alignments of ETC with ETA, SEB and SEC1 revealed only a few clusters of conservation, particularly in the carboxyl halves of the four proteins (Goshorn and Schlievert, 1988).

Table 7 shows a comparison of the percentage of homology of the nucleotide sequences of four SE genes and ETA. ETA and SEB genes shared 71% homology. No significant homology between TSST-1 and the other toxins was found with regard to amino acid and nucleotide sequences (Blomster-Hautamaa *et al.*, 1986a; Betley and Mekalanos, 1988). Monte-Carlo analysis of the mature amino acid sequences of the three toxin groups indicated that SEB, SEC1, SEC2 and ETA form a closely related cluster of toxins, that SEA, SED and SEE form another cluster and that TSST-1 and ETB

Table 7. Nucleotide sequence comparison between structural genes for staphylococcal enterotoxins (*se*) and streptococcal erythrogenic (pyrogenic) exotoxin A (*speA*)[a]

| Toxin gene | % Homology of | | | | |
	sea^+	seb^+	sec_1^+	see^+	$speA$
sea+	100	54	49	84	49
seb+		100	75	52	61
sec₁+			100	50	60
see+				100	51
speA					100

[a]From Couch *et al.* (1988). See Betley *et al.* (1990) for gene nomenclature.

are much less closely related to the other toxins. Computer alignments of the amino acid sequences of all these toxins (except ETB) were reported by Marrack and Kappler (1990) who showed that TSST-1 sequence can be aligned with the other toxin sequences albeit with some difficulty. Table 8 summarizes the data based on Monte-Carlo analyses of amino acid sequence similarities between ETA, ETB, SEA, SEB, SEC1 and TSST-1.

The conserved structures in the enterotoxins and the erythrogenic toxins suggest a common superfamily of toxin genes. Have these diverse genes evolved from a single progenitor or do they represent a family of genes structurally and functionally similar yet derived from distinct ancestral genes? As no obvious link between TSST-1 and the other toxins was found at the amino acid sequence level, the functional relatedness of the former to the others might be due to convergent evolution and/or to similar 3-dimensional configuration. The presence of most toxin genes of ETs and SEs on mobile elements (Table 4) may explain a possible dissemination between *S. aureus* and *S. pyogenes*. These Gram-positive cocci share a number of characteristics: both produce several analogous gene products and both are approximately 70% A+T rich.

Antigenic homology

The structural homology is more or less reflected at the antigenic level investigated by means of polyclonal and monoclonal antibodies. Immunological cross-reactivity between ETA, SEB and SEC1 was demonstrated by double immunodiffusion in agar, immunoblot and immunodot analyses (Hynes *et al.*, 1987). The cross-reactive epitopes are very likely located on the C-terminal regions. Bohach *et al.* (1988b) investigated the cross-reactivity of the three toxins by ELISA neutralization tests of toxin mitogenicity and immunization. Ten IgG_1 monoclonal antibodies (MAb) raised against SEC1 showed little or no cross-reactivity by ELISA analysis but many of these neutralized the mitogenic effect of SEB, ETA or both. Two anti-SEC1 MAb (IgM) and eight anti-ETA MAb (IgM) showed cross-reactivity in both ELISA and neutralization assays. No cross-reactivity was observed with ETC or TSST-1. Rabbits immunized against SEB, SEC1 or ETA were resistant (in terms of lethality) to challenge by the homologous toxins in the presence of 50% lethal dose of endotoxin. Immunity to ETA provided protection against SEB but not SEC1. The variations in cross-reactivity may be due to differences in 3-dimensional conformation.

Common biological properties

A striking feature of the toxins described in this chapter is the sharing of a number of major biological properties as listed in Table 9 and described in this section. These toxins appear therefore as a family of structurally (except TSST-1 ?) and functionally related proteins.

Table 8. Monte-Carlo analysis of the sequence similarities between mature amino acid sequences of TSST-1, erythrogenic toxins A, C and enterotoxins A, B, C1[a]

Probe	Monte-Carlo score for compared sequence					
	ETA	ETC	SEA	SEB	SEC1	TSST-1
ETA		11.0	19.4	55.8	34.6	3.72
ETC	10.2		6.49	5.31	6.32	1.79
SEA	12.6	5.82		20.1	10.1	3.63
SEB	31.7	4.40	14.8		54.8	4.18
SEC1	37.5	4.79	15.1	64.5		3.08
TSST-1	5.43	1.29	3.06	5.20	3.23	

[a]From Goshorn and Schlievert (1989). Score < 3.0: no significant homology; 3.00–6.0: possible homology; > 6.0: probable homology.

Table 9. Common biological properties of *S. aureus* enterotoxins, TSST-1 and *S. pyogenes* erythrogenic (pyrogenic) toxins[a]

1. Polyclonal mitogenic stimulation of T lymphocytes

2. Binding to class II MHC proteins and specificity for particular Vβs of T cell receptor ('superantigenic' properties)

3. Cytokine production of toxin-stimulated immunocytes

4. Induction of hyperthermia (pyrogenic effect)

5. Enhancement of host susceptibility to lethal shock by bacterial endotoxins (lipopolysaccharides)

6. Reticuloendothelial system blockade (impairment of clearance functions)

7. Alteration of one or several of the following immune functions: (a) suppression and/or enhancement of the synthesis of IgM or other isotypes, (b) suppression and/or enhancement of cell-mediated responses, (c) tolerance induction in rabbits.

[a]See text for references.

Polyclonal mitogenicity on T lymphocytes

Enterotoxins

Early *in vitro* studies in the 1970s showed that SEs are polyclonal mitogens on human and murine T lymphocytes (see Langford *et al.*, 1978; Vroegop and Buxser, 1989 for references). B lymphocytes are insensitive. These toxins, and particularly SEA, are the most potent mitogens known. They are mitogenic and elicit cytokine release at concentrations of $10^{-9}-10^{-12}$ M (see Carlsson *et al.*, 1988; Fischer *et al.*, 1989; Fleischer *et al.*, 1989) and even at lower concentrations ($10^{-13}-10^{-16}$ M) on human T cells (Johnson and Magazine, 1988; Pontzer *et al.*, 1989). That only few molecules per T cell are sufficient for stimulation indicates a highly selective cell–mitogen interaction.

A study of the mitogenic effect of SEA, SEB, SEC and TSST-1 by Uchimaya *et al.* (1989a) on human peripheral blood or cord blood cells and on murine spleen and thymus cells showed that the four toxins transformed T lymphocytes at

concentrations of 100 fg to 1 pg/ml. SEA was the strongest. In the murine system only SEA and TSST-1 were strong mitogens at concentrations of 1 pg/ml, whereas SEB and SEC1 appeared much weaker. A similar finding was reported by Fujikawa *et al.* (1989) for rabbit spleen cells. These results are in agreement with the data of Vroegop and Buxser (1989) who showed that SEA was able to elicit significant mitogenesis of murine spleen cells at 100-fold lower concentrations (10^{-4} μg/ml) than SEB (10^{-2} μg/ml). The study by these authors of SEA or SEB mitogenicity on the spleen cells of five different inbred strains of mice showed that the magnitude of the mitogenic response was highly strain dependent. The strain maximally responsive to SEA (SJL) was different from the strain maximally responsive to SEB (BALB/CJ). SEs were unable to induce mitogenesis of spleen cells from nude mice (see Vroegop and Buxser, 1989).

The *in vivo* effect of intravenously administered SEA on the peripheral blood lymphocytes (PBL) of monkey baboons reported by Zehavi-Willner *et al.* (1984) showed an initial lymphocytic leukopenia lasting *ca.* 24 h. The lymphocytes collected during this stage had normal or decreased [^{3}H]thymidine-incorporating activity. This activity increased 5- to 6-fold after 48 h and 7- to 8-fold after SEA administration and was concomitant with the conversion of lymphopaenia into lymphocytosis. The lymphocytes isolated 24 h after SEA challenge did not respond to the mitogenic action of the toxin *in vitro*, whereas those collected at later stages were fully activated by the toxin.

TSST-1

TSST-1 is also a potent mitogen for human, rabbit and mouse T lymphocytes (see Poindexter and Schlievert, 1985b; Blomster-Hautamaa and Schlievert, 1988a; Naidu *et al.*, 1989; Uchiyama *et al.*, 1989a). Calvano *et al.* (1984) showed that both CD4$^+$ (helper) and CD8$^+$ (cytotoxic/suppressor) human T cells responded to the toxin over a range of 1 ng to 5 μg of TSST-1/ml. Non-adherent accessory cells (monocytes/macrophages) even irradiated were essential for T cell stimulation. The mitogenic effect of the toxin was still present following heat-treatment of this protein at 100°C for 60 min. The induction of CD8$^+$ blasts may explain the defective humoral immune response associated with TSS. A high proliferation of murine cells with as low as 10^{-5} µg of toxin/2.5×10^5 cells was reported (Blomster-Hautamaa *et al.*, 1986b). The other recent studies on T cell mitogenicity elicited by SEs or TSST-1 will be mentioned later.

Erythrogenic toxins

The non-specific polyclonal mitogenicity of the extracellular proteins of group A streptococci are attributable to erythrogenic toxins A, B and C (see Alouf, 1980 and Wannamaker and Schlievert, 1988 for a review and historical background) and to a number of other more or less characterized proteins (see Seravalli and Taranta, 1974; Abe and Alouf, 1976; Cavaillon *et al.*, 1979; Cavaillon and Alouf, 1982; Gray, 1979; Quagliata *et al.*, 1982; Knöll *et al.*, 1983; Suzuki and Vogt, 1986). Only the mitogenicity of purified ETA, ETB and ETC will be considered here. It was first reported by Kim and Watson (1972) for human lymphocytes, Hřibalovà and Pospisil (1973) for rabbit lymphocytes and Nauciel (1973) for human, rabbit, guinea-pig and mouse lymphocytes. Detailed investigations were later reported by Abe and Alouf (1976) and Abe *et al.* (1980), Knöll *et al.* (1981, 1983), Barsumian *et al.* (1978b), Cavaillon *et al.* (1979), Gray (1979) and Regelmann *et al.* (1982). Only T lymphocytes are transformed. Most data indicated that ETs are very potent mitogens since as low as 10^{-4} µg/ml ($\sim 3 \times 10^{-12}$ M) of toxin and sometimes less was sufficient to elicit human lymphocyte proliferation. Mouse lymphocytes are much less sensitive. It was demonstrated (Cavaillon *et al.*, 1982; Regelmann *et al.*, 1982; Alouf *et al.*, 1986) that adherent cells (even irradiated) were requisite for T cell mitogenesis in contrast to the report of Schlievert *et al.* (1979d). The role of monocytes/macrophages in this process was clearly established in the recent works on the mechanism of lymphocyte proliferation induced by SEs, TSST-1 and ETs.

According to Schlievert *et al.* (1980), cell surface gangliosides or sialic acid at the cell surface may act as binding sites for ETs on the basis of the partial inhibition of ET-induced mitogenicity by these components or by α-methyl-D-mannopyranoside or galactose. This contention was not supported by the ETA–gold binding studies of Scriba *et al.* (1987) and Wagner *et al.* (1988). Lymphocyte treatment with neuraminidase increased toxin binding and only slight inhibition of binding was observed in the presence of gangliosides.

ETA was shown by Alouf *et al.* (1986) to stimulate human CD4$^+$ T lymphocytes, whereas CD8$^+$ T lymphocytes were almost insensitive. This finding was correlated by the study of ETA binding by human peripheral blood lymphocytes reported by Wagner *et al.* (1988) who found by light and electron microscopy investigations using ETA–gold conjugate, mouse OKT4 and OKT8 monoclonal antibodies and mouse-IgG–gold that ETA binding was confined to CD4$^+$ (helper) T cells. In contrast to these findings Fleischer *et al.* (1991) reported that both human CD4$^+$ and CD8$^+$ T cell clones were transformed by ETA, as also reported by Regelmann *et al.* (1982) and Knöll *et al.* (1988b). These divergent results remain to be explained.

In vitro infection of human CD4$^+$ cells with HIV-1 in the presence of ETA as mitogen led to 6- to 10-fold higher yield of virus as compared to similar lymphocyte stimulation with haemagglutinin (PHA) (Alouf *et al.*, 1986). This finding may be of clinical significance for subjects contaminated by HIV-1 who become infected with mitogen-producing streptococci.

'Superantigenic' properties: binding to MHC class II proteins of adherent cells and to Vβ domain of T cell receptor

Unlike many T cell polyclonal mitogens (PHA, ConA, etc.), which all bind directly to T lymphocytes, SEs, TSST-1 and ETs do not bind directly to these cells (see Fischer *et al.*, 1989). T cell activation of and subsequent lymphokine release required prior toxin binding to accessory cells (macrophages/monocytes, B cells or cell lines) expressing major histocompatibility complex (MHC) class II (HLA/I-A, I-E) molecules on their surface (see Carlsson *et al.*, 1988; Scholl *et al.*, 1989a,b; Fischer *et al.*, 1989; Russell *et al.*, 1990). The requirement for toxin binding by class II molecules is consistent with the observations that neither interleukin 1 (IL1), nor interleukin 2 (IL2) could substitute for this requirement (Chatila *et al.*, 1988; Carlsson *et al.*, 1988).

Toxin–accessory cell interaction

The quantitative aspects of the binding of SEA by murine splenic lymphocytes (Buxser *et al.*, 1981) and human splenic lymphocytes (Ezpechuk *et al.*, 1983) as well as that of TSST-1 by human peripheral blood mononuclear cells (PMBC) (Poindexter and Schlievert, 1987) have been reported. In a similar study on these cells based on competition experiments (Lee *et al.*, 1990), both SEA and TSST-1 (but not the other enterotoxins) bound to the same receptors. The binding epitopes appeared to be overlapping or separate. That SEA compete for TSST-1 receptors may explain the similarity of their pathophysiological effects and of the possible role of SEA in TSS (McCollister *et al.*, 1990).

The toxins bind to AC without internalization and processing, in contrast to antigens in T cell-dependent immune response. Toxin binding by AC class II molecules occurs regardless of allelic form of these molecules since allogenic or even xenogenic class II molecules can reconstitute the T cell response to the toxins (Fleischer *et al.*, 1989; Mollick *et al.*, 1989).

It might be not coincidental that a sequence similarity has been found between a stretch of sequence in the second half of SEs, TSST-1, ETA and ETC molecules and a sequence at the COOH-terminal end of human and mouse invariant chain, a polypeptide associated with nascent MHC class II molecule (Marrack and Kappler, 1990).

That the presence of class II molecules on AC is critical for T cell transformation by the AC-bound toxin is illustrated by transfection: a non-stimulatory cell was rendered stimulatory for SE- or TSST-1-induced T cell activation if transfected with MHC class II gene (Fleischer *et al.*, 1989; Mollick *et al.*, 1989; Scholl *et al.*, 1989b; Uchiyama *et al.*, 1989b,c). In addition, anti HLA-DR (but not anti HLA-DQ) antibodies were shown to inhibit TSST-1 binding (Scholl *et al.*, 1989b; Uchiyama *et al.*, 1989b) and SEA binding (Fischer *et al.*, 1989) on AC. By using chimaeric α- and β-chains of DR and DP expressed at the surface of the transfected murine fibroblasts, Karp *et al.* (1990) demonstrated the high-affinity binding of TSST-1 by the non-polymorphic α-1 domain of the HLA-DR molecule (but not DP molecule). They also found that toxin binding did not prevent subsequent binding of a DR-restricted antigenic peptide.

According to Fleischer (1989), SEs, TSST-1 and ETA can activate T cells in the absence of class II molecules if the toxins are co-crosslinked on beads together with anti-CD8 or anti-CD2 antibodies or if they are immobilized on plastic or if phorbol esters are added.

The interaction of TSST-1 with human epithelial cells has been investigated by Kushnaryov *et al.* (1984) who reported toxin internalization by receptor-mediated endocytosis. It remains to be explained why this process does not take place in the case of accessory cells.

Vβ chain–toxin interaction

Another important feature in cell–toxin interaction is that T cell activation by the intact toxin bound to AC is clonally specific and therefore that the toxins are not indiscriminate polyclonal mitogens, as is the case for lectins which stimulate every T cell. Only certain T lymphocyte subpopulations bearing particular Vβ structures on their T cell antigen receptor (TCR) are stimulated (see Marrack and Kappler, 1990; Misfeldt, 1990 for a review). This property is characteristic of the group of molecules designated as 'superantigens' (Fleischer *et al.*, 1989; Janeway *et al.*, 1989; White *et al.*, 1989; Matis, 1990). To this group belong the enigmatic murine Mls (minor lymphocyte stimulating) self antigens, *Mycoplasma arthriditis* soluble antigen (MAS) and possibly other mitogenic protein toxins in addition to the toxins studied here.

TCR is a heterodimer made up of α- and β-chains or γ- and δ-chains. In the immune response, the specificity of TCRs for the antigen presented in the context of MHC class I or class II proteins is determined by all of the variable elements of constitutive chains. For the α/β receptor the five germline-encoded variable elements (Vα, Jα, Vβ, Dβ and Jβ) as well as non-germline-encoded amino acids contribute to the receptor combining site (see Matis, 1990). As first shown for SEA and SEB by Fleischer and Schrezenmeier (1988) and thereafter by many authors (Table 10), the toxins stimulated human and murine T cells in a Vβ-specific fashion (Table 11).

In the case of murine T cells, SEA activates the cells bearing Vβ1, 3 and 11 but not those bearing the Vβ8 family. In contrast, SED strongly stimulated Vβ8 family in addition to those bearing Vβ3, Vβ7 and Vβ11. Cells expressing Vβ3 were

Table 10. T lymphocyte activation by 'superantigenic' staphylococcal enterotoxins, TSST-1 and streptococcal erythrogenic toxin A

Toxin	T cells	References
SEA	Human	1–4
SEB	Mouse	5, 6
SEA, SEB	Human	8, 9, 10
SEB, SEC1, SEE	Human	11
SEA, SEB, SEC1, SEC2, SEC3, SED, SEE	Human Mouse	12, 13 14
TSST-1	Human Mouse	7, 11, 12, 13, 15, 16 14, 17, 18
ETA	Human	13

(1) Carlsson *et al.* (1988); (2) Fischer *et al.* (1989); (3) Mollick *et al.* (1989); (4) Lando *et al.* (1990); (5) White *et al.* (1989); (6) Marrack *et al.* (1990); (7) Scholl *et al.* (1989a); (8) Fleischer and Schrezenmeier (1988); (9) Fleischer *et al.* (1989); (10) Fraser (1989); (11) Choi *et al.* (1989); (12) Kappler *et al.* (1989); (13) Fleischer *et al.* (1991); (14) Callahan *et al.* (1990); (15) Uchiyama *et al.* (1989b); (16) Scholl *et al.* (1989b); (17) Uchiyama *et al.* (1989c); (18) Norton *et al.* (1990).

not stimulated by SEE, whereas all other SE serotypes and TSST-1 were active. The most closely related toxins SEC1, SEC2 and SEC3 had similar but not identical patterns of Vβ specificity. Each of these SEC subtypes stimulated to some degree cells bearing Vβ3 and Vβ8.2 but SEC2 also stimulated Vβ10$^+$ cells whereas SEC1 and SEC3 did not. SEC1 stimulated Vβ3$^+$ cells to a much greater extent than did either SEC2 or SEC3, reflecting affinity differences. TSST-1 stimulated Vβ15$^+$ T cells in contrast to SEs, whereas the Vβ8 family, which is a common target of SEB, SEC subtypes and SED, were not stimulated by TSST-1. Interestingly, murine T cells bearing certain Vβ families such as Vβ2, 6, 9, 12, 14 and 16 were not induced to proliferate by any of these toxins (Callahan *et al.*, 1990).

The similarities between mice and humans in the T cell response to the toxins are striking. In both cases T cells bearing particular Vβs dominate the response to each toxin (Choi *et al.*, 1989; Callahan *et al.*, 1990). For example, cells bearing the human Vβ13.2 element responded to SEC2 whereas those bearing Vβ13.1 did not (Choi *et al.*, 1990). Sequence analysis of murine and human TCR Vβ genes and amino acid sequences showed that there are some homologies between Vβ

Table 11. T cell Vβ specificities of staphylococcal enterotoxins, TSST-1 and streptococcal erythrogenic toxin A

Toxin	Vβ specificity	
	Murine Vβ chain	Human Vβ chain
SEA	1, 3, 10, 11, 17	?
SEB	3, 7, 8.1, 8.2, 8.3, 11, 17	3, 12, 14, 15, 17, 20
SEC1	3, 8.2, 8.3, 11, 17	12
SEC2	3, 8.2, 10, 17	12, 13.2, 14, 15, 17, 20
SEC3	3, 7, 8.1, 8.2	5, 12
SED	3, 7, 8.1, 8.2, 8.3, 11, 17	5, 12
SEE	11, 15, 17	5.1, 6.1, 6.2, 6.3, 8, 18
TSST-1	3, 10, 15, 17	2
ETA	?	8

Data from Choi *et al.* (1989); Callahan *et al.* (1990); Marrack and Kappler (1990); Marrack *et al.* (1990); Fleischer *et al.* (1991); Choi *et al.* (1990); Misfeldt (1990).

regions. TSST-1 stimulated human T cells bearing Vβ2 and mouse T cells bearing the homologous Vβ15. Similarly SEB and SEC1 but not TSST-1 or SEE stimulated the human Vβ12, 14, 15 and 17 T cells and the homologous Vβ8 murine T cells. SEE stimulated the human Vβ8, 6.1 and 18 and the homologous murine Vβ11. However, similar response patterns by T cells bearing homologous Vβs were not always observed, as found for the murine Vβ3 T cells which responded to SEA, SEB, SECs, SED and TSST-1 whereas the closest human Vβ10 analogue did not (Choi et al., 1989; Callahan et al., 1990).

Fleischer et al. (1991) investigated the stimulation of human T lymphocytes with SEs, TSST-1 and ETA. They showed that at least 50% of these cells responded to each of these toxins, whereas less than 5% responded to MAS. Preferential but not exclusive stimulation of T cells carrying certain Vβs was shown. Cloned T cells expressing Vβ5 or Vβ8 were tested with the different toxins. A depletion of Vβ5$^+$ was shown after stimulation of SEB, whereas ETA remarkably stimulated Vβ8$^+$ T cells.

The studies mentioned here have focused on T cells expressing αβ TCR. However, γδ-expressing T cells may also be clonally expanded in an MHC class II-dependent manner, as shown for a CD4$^-$ CD8$^+$ TCR$^+$ cell clone stimulated with SEA (Fleischer and Schrezenmeier, 1988; Fleischer, 1989). Recently, SEA was shown to be specifically recognized by human T cells bearing γδ receptors with the Vγ9 region (Rust et al., 1990). The response was a specific dose-dependent killing of enterotoxin-coated T cells but not a proliferative process as for α$^+$β$^+$ T cells. This cytotoxic response required the expression of HLA class II molecules by the target cell.

The superantigenic toxins appear, therefore, as functionally bivalent mitogenic proteins which stimulate T cells through the formation of a trimolecular complex formed by the intact toxin molecule bound to a non-polymorphic region of MHC class II molecules on AC and to TCR (see Fleischer et al., 1989; Karp et al., 1990; Lando et al., 1990). Toxin affinity for both structures is probably related to the intact protein structure of the superantigen, given that proteolytic treatment (Marrack and Kappler, 1990) suppresses mitogenic activity (Fraser, 1989). It seems likely that when

bound to class II molecule the toxin becomes conformationally modified, rendering it capable of interacting with and activating T cells by itself or as a complex together with class II molecule (Fischer et al., 1989). Physical contacts between the monomorphic DR-α chain and the Vβ portion of the TCR in toxin interaction is strongly suggested by the binding of TSST-1 in the chimaeric system described by Karp et al. (1990). A structural model for the amino acids of Vβ13.1 and Vβ13.2 regions reacting with SEC2 has been developed by Choi et al. (1990).

Role of superantigenic toxins in pathogenesis

According to Kappler et al. (1989) the ability of these superantigenic toxins to stimulate populations of T cells bearing particular Vβs may be related to the differential resistance of different individuals to the effects of these toxins and also to the ability of microbial attack to induce immune consequences such as autoimmunity in certain individuals. Interestingly, mice can be made tolerant to SEB by neonatal injection of the toxin. This treatment deletes all SEB-reactive T cells expressing the appropriate Vβ specificites (White et al., 1989). Marrack et al. (1990) reported that the pathological effects of SEB in susceptible mice reflected by weight loss, thymus depletion and immunosuppression may be the consequence of T cell activation, since animals containing few SEB-reactive T cells because of genetic defects or immunosuppressed by cyclosporin lose little or no weight after challenge with SEB. By their stimulation of T cells the toxins may alter the host's immune system, possibly creating some sort of advantage for the pathogen. The finding that the murine T cell response to the toxins produced by the human pathogenic staphylococci and streptococci was much weaker suggests an evolutionary adaptation of the immune system of the host (Fleischer et al., 1990; Marrack et al., 1990).

B cell transformation

Under certain conditions, TSST-1 was shown to elicit human B cell proliferation (Mourad et al., 1989). The toxin bound with high affinity and with saturation kinetics to MHC class II antigens on purified tonsillar B lymphocytes. No proliferation of the cells or Ig production was observed unless irradiated T cells are present. In this case, the resting B cells were transformed and differentiated into Ig-secreting lymphocytes. This process was shown to proceed via MHC-unrestricted cognate T/B cell interaction involving the TCR/CD3 complex and the MHC class II antigen/toxin complex. TSST-1-dependent B cell proliferation increased as a function of the load of irradiated cells added. In contrast, Ig production was dependent on critical T cell/B cell ratio and declined for higher T cell loads. This finding may explain the inhibition of Ig synthesis by TSST-1 and very likely by SEs and ETs.

Toxin-induced cytokine production

The clinical disorders including death and the *in vivo* pathogenic effects elicited by SEs, TSST-1 and ETs are necessarily the consequence of either direct damaging effects on host tissues or effects of mediators released by the host in response to the toxins excreted by the bacteria infecting the host. Several lines of evidence suggest that these mediators play a major role in the pathogenesis of the relevant diseases without excluding a synergistic participation of other factors and mediators, particularly endotoxins, in the overall pathological process. Among the mediators proven or postulated to be involved in this process are the cytokines produced by both AC and T lymphocytes following toxin binding and activation of these cells. A variety of cytokines (Table 12) have been shown to be released, sometimes in massive amounts upon *in vitro* challenge of murine and human monocytes/macrophages, T lymphocytes and many derived cell lines by SEs, TSST-1 and ETA.

Staphylococcal enterotoxins

Interferon-γ (IFN-γ) was the first cytokine shown to be released by human lymphocytes stimulated with SEA, as reported by Langford *et al.* (1978) and thereafter by many authors (see Table 12) upon challenge of human or murine T cells with SEs provided accessory cells are present. Interesting data on the induction of IFN-γ by SEB-stimulated human peripheral blood mononuclear cells were provided by Lee *et al.* (1990). By using an immunogold–silver method to label cell surface antigens, it was possible to establish that

Table 12. Cytokine production by human or murine immunocytes stimulated by enterotoxins, TSST-1 and erythrogenic toxin A

Cytokines	Toxins	References
Interferon-γ	Enterotoxins	1–11
	TSST-1	10, 12
	ETA	13, 14
Interleukin 1	TSST-1	12, 15–19
	Enterotoxins	5, 6, 8, 20–23
Interleukin 2	TSST-1	24–28
	ETA	29
Interleukin 3	TSST-1	19
	ETA	30
Interleukin 6	ETA	30
Colony-stimulating factors	TSST-1	19
Tumour necrosis factor-α	Enterotoxins	11, 31
	TSST-1	12, 32, 33
Tumour necrosis factor-β (lymphotoxin)	Enterotoxins	11
	TSST-1	12

(1) Langford *et al.* (1978); (2) Georgiades and Johnson (1981a,b); (3) Arbeit *et al.* (1982); (4) Torres *et al.* (1982); (5) Clark *et al.* (1984); (6) Carlsson and Sjögren (1985); (7) Carlsson *et al.* (1988); (8) Uchiyama *et al.* (1989a); (9) Pontzer *et al.* (1989); (10) Micusan *et al.* (1989); (11) Fischer *et al.* (1990); (12) Jupin *et al.* (1988); (13) Cavaillon *et al.* (1982); (14) Tonew *et al.* (1982); (15) Ikejima *et al.* (1984); (16) Hirose *et al.* (1985); (17) Parsonnet *et al.* (1985); (18) Beezhold *et al.* (1989); (19) Galelli *et al.* (1989); (20) White *et al.* (1989); (21) Fischer *et al.* (1989); (22) Callahan *et al.* (1990); (23) Choi *et al.* (1990); (24) Micusan *et al.* (1986); (25) Uchiyama *et al.* (1986); (26) Uchiyama *et al.* (1989b); (27) Chatila *et al.* (1988); (28) Norton *et al.* (1990); (29) Alouf *et al.* (manuscript in preparation); (30) Knöll *et al.* (XIth Lancefield Symposium Streptococci, Siena, September 1990); (31) Fast *et al.* (1988); (32) Parsonnet and Gillis (1988); (33) De Azavedo *et al.* (1988b).

the majority of IFN-γ-producing cells in response to SEB displayed the CD3, CD4, CD25 and OKT11 markers and the pan T cell surface antigen T11 (CD2).

SEs are also potent inducers of interleukin2 (IL2) (Table 12) and IL2 receptor (Carlsson and Sjögren, 1985). The mitogenic activity has been shown to be dependent on IL2 synthesis (Johnson and Magazine, 1988). The Jurkat Vβ8⁺ cell line produced very high amounts of IL2 upon stimulation by SED, SEE and ETA, in contrast to other SEs and TSST-1 which were almost ineffective (Fleischer et al., 1990). IFN-γ produced by SEA-stimulated murine spleen cells was shown to be regulated by IL2 (Torres et al., 1982). According to Garcia-Peñarrubia et al. (1989), IFN-γ and IL2 are thought to be involved in the selective proliferation of natural killer cells under the effects of SEB.

The production of tumor necrosis factor-α (TNF-α), also known as cachetin, with human monocytes stimulated by SEB and SEC1 has been reported (Fast et al., 1989). A detailed study of TNF-α and TNF-β (lymphotoxin) production and kinetics by human monocytes and T lymphocytes respectively, was undertaken by Fischer et al. (1990). These cytokines were induced by less than 1 pg/ml of SEA and required the presence of both monocytes and T cells. The CD4⁺ 45R⁻ memory Th cells but not CD4⁺ 45R⁺ and CD8⁺ cells supported TNF-α production in monocytes. A four-fold higher frequency of TNF-β-producing cells was demonstrated among CD4⁺ vs. CD8⁺ cells. The CD4⁺ 45R⁻ T cell subset was an efficient producer of both TNF-β and IFN-γ whereas the CD4⁺ 45R⁺ T cell subset produced significant amounts of TNF-β but only marginal amounts of IFN-γ. These findings suggest that toxin-stimulated cell subsets use different regulation pathways for INF-γ. Purified monocytes stimulated with SEA in the absence of T cells did not produce TNF-α. The addition of several cytokines such as IL2, IL4, IL6 and INF-γ or a combination of these failed to substitute for T cells, indicating that a signal from these cells is required for the monocytes.

TSST-1

Massive release of IL1 by human blood monocytes challenged with TSST-1 preparations was first reported by Ikejima et al. (1984) and confirmed with highly purified TSST-1 by Parsonnet et al. (1985) and Jupin et al. (1988). A similar finding was reported by Galelli et al. (1989) and Beezhold et al. (1989) on murine monocytes. TSST-1 appeared as a more potent inducer of IL1 than endotoxin, sometimes by one or two orders of magnitude. The exposure of macrophages to both TSST-1 and LPS resulted in a synergistic induction of IL-1 (Beezhold et al., 1989).

Large amounts of IL2 were produced and murine T cells challenged with TSST-1 (Micusan et al., 1986; Uchiyama et al., 1986, 1989a,b; Jupin et al., 1988; Chatila et al., 1988). As low as 0.1 ng of toxin induced significant release of IL2. The expression of IL2 receptor after stimulation with TSST-1 was shown by Chatila et al. (1988) and Micusan et al. (1989). The toxin induced IL2 receptor and IL2 synthesis in a strictly monocyte-dependent manner. Neither IL1 nor IL2 could substitute for the monocyte requirement (Chatila et al., 1988).

The production of INF-γ by TSST-1-stimulated human blood lymphocytes was shown by Jupin et al. (1988) and Micusan et al. (1989). The levels of INF-γ were similar to those induced by SEA. The release of high amounts of TNF-α and TNF-β by TSST-1-stimulated human peripheral blood monocytes and lymphocytes respectively was shown for the first time by Jupin et al. (1988), who also found that lymphocyte costimulation with phorbol myristate acetate strongly potentiated the release of INF-γ, IL2 and TNF-β. As low as 0.1 ng of TSST-1 triggered the release of TNF-α by the monocytes, indicating that this toxin is more potent than endotoxin. The release of this cytokine was also shown by Parsonnet and Gillis (1988) and Fast et al. (1988, 1989). TNF-α release by rabbit monocytes was reported by De Azavedo et al. (1988b). High levels of TNF in rabbit sera were detected by Ikejima et al. (1988) following TSST-1 injection.

Galelli et al. (1989) reported for the first time that TSST-1 had a potent stimulatory effect on the induction in vitro of haemopoietic growth factors by murine spleen cell cultures challenged with nanogram amounts of TSST-1. The factors released were colony-stimulating factors (CSF) identified as IL3 and G-CSF, M-CSF and GM-CSF, which allowed the proliferation and differentiation of granulocyte–macrophage progenitor cells of mouse bone marrow. It is very likely that production of these cytokines may be in part mediated by IL1 which is known to stimulate CSF production in various experimental systems. The *in vivo* induction of CSF was also demonstrated in mouse serum within 1 h after injection of nanogram amounts of toxin (Galelli *et al.*, 1989).

Erythrogenic toxin A

The stimulation of murine spleen cells by ETA and other extracellular mitogens released in parallel with this toxin was shown (Cavaillon *et al.*, 1982) to elicit the production of IFN-γ and two IFN-dependent enzymes 2-5 A synthetase and protein kinase. A similar finding was reported by Tonew *et al.* (1982) for ETA and ETB. Large amounts of TNF-α were also released (Fast *et al.*, 1988). ETA elicited the release of IL2 and the expression of IL2 receptor by human T lymphocytes (Korinek, Alouf and Gluckman, manuscript in preparation). IL3 and IL6 were also shown to be elicited (Knöll, Cavaillon, Alouf and Köhler, Abstract, XIth Lancefield Symposium, Siena, 1990).

Pathological significance of cytokine release

It is tempting to consider that the induction by SEs, TSST-1 and ETA of the various monokines and lymphokines such as IL1, IL2, IL3, IL6, TNF and IFN-γ over the physiologically required amounts and often massively might be responsible for the clinical manifestations associated with TSS and TSS-like symptoms and other pathological disorders (see Ikejima *et al.*, 1984; Jupin *et al.*, 1988; Uchiyama *et al.*, 1989a,b; Galelli *et al.*, 1989; Norton *et al.*, 1990; Marrack *et al.*, 1990).

All the mediators mentioned above have pleiotropic biological activities on organ and cell systems leading to the various clinical disorders observed (see Parsonnet, 1989 for an overview). The endogenous pyrogens IL1 and TNF may explain the high fever and some of the multisystem disorders of TSS. IFN-γ has a wide array of immunoregulatory effects, including modulation of the production and activities of IL1 and TNF-α. It is known that individuals given large doses of IL2 manifest serious disorders, some of which are similar to TSS, and combined administration of TNF-α and IFN-γ show toxic manifestations

similar to those observed upon administration of IL2 (Uchiyama *et al.*, 1989b,c). *In vivo* experiments in rabbits showed that the injection of TNF-α and IL1 induced a sharp fall of mean arterial pressure in the rabbit and shock-like state followed by death for doses of *ca.* 5 μg/kg (Ikejima *et al.*, 1988). The *in vivo* CSF induction by TSST-1 in mice allow us to speculate (Galelli *et al.*, 1989) that the serious local inflammation observed in TSS might be related to the induction of these factors in humans.

Besides the cytokine-mediated effects, direct interactions of TSST-1 or ETs with renal tubular cells or vascular endothelial cells are likely to be involved in the pathological effects of these toxins (Keane *et al.*, 1986; Stevens *et al.*, 1989). SEB-induced leukotriene production (Scheuber *et al.*, 1987) may also contribute to those effects.

Pyrogenicity

This property is a prominent feature of SEs, TSST-1 and ETs as shown by the rapid increase of body temperature in experimental animals challenged with these toxins. The rapid onset of high fever in TSS and toxic-shock-like syndrome or in other clinical manifestations of acute staphylococcal and streptococcal diseases is very likely due (at least partially) to the pyrogenicity of the *in vivo* released toxins. It is well-documented that fever is mediated by endogenous pyrogenic proteins produced by toxin-activated macrophages, monocytes and other phagocytic cells acting on central nervous system temperature-regulating centres (particularly the hypothalamus). IL1 is identical with the factor(s) originally described as endogenous (or leukocytic) pyrogen(s). Later it became evident that in addition to IL1, TNF-α and IL6 are also potent endogenous pyrogens (see Bendtzen, 1988).

The pyrogenicity of SEs administered orally or by i.v. or intracerebroventricular routes has been investigated in rabbits, cats, monkeys and goats (see Clark and Page, 1968; Brunson and Watson, 1974; Van Miert *et al.*, 1984). In most reports the i.v. injection triggered biphasic febrile response, peaking usually 1 h and 3 h after toxin challenge. Maximum peak fever in the rabbit after SEA injection was obtained with *ca.* 1 μg/kg (i.v. route). A monophasic pyrogenic effect was reported by Bohach and Schlievert (1987a) with recombinant SEC1 injected (i.v. route) in rabbits. Temperature rise was 1.25°C over a 4-h period. Repeated i.v. injections of SEA or SEB in cats (Clark and Page, 1968) and rabbits (Brunson and Watson, 1974) produced a pyrogenic 'tolerance' effect (attenuation of febrile response).

TSST-1 pyrogenicity has been reviewed by Blomster-Hautamaa and Schlievert (1988a). Toxin injection (i.v.) in rabbit was shown by Schlievert et al. (1981) to induce a gradual linear fever response, peaking at 4 h. The minimum pyrogenic dose (MPD-4)/kg, defined as the dose required to produce an average fever response of 0.5°C in rabbit after 4 h, was 0.15 µg. An increase of 1°C was elicited by 7.5 µg/kg regardless of the route of administration (i.v., i.d. or s.c.) suggesting that TSST-1 is highly mobile. Fever induction was directly proportional to the logarithm of toxin dose. A biphasic fever response was reported by Igarashi et al. (1984) with purified TSST-1 preparation. The MPD was 0.018 µg/kg. Similarly a biphasic fever was reported by Ikejima et al. (1984) in rabbit injected intravenously with the culture filtrate derived from TSS-associated strains. The temperature was elevated by 0.5–0.8°C at 1.5 h after injection then fell before increasing to a major peak (1.2–1.7°C) around 4–5 h post-injection. The plasma collected at 5 h injected into a second set of rabbits produced a brief monophasic fever typical of endogenous pyrogens. Hyperthermia was investigated by Bulanda et al. (1989) in conventional and germ-free piglets challenged with TSST-1 (s.c. route). A 2°C increase of body temperature in 8 h (for 10 µg of toxin/kg) was observed in conventional animals, whereas the reaction of germ-free animals was less pronounced. No temperature changes in adults were observed for as high as 100 µg of toxin.

Daily injections of the culture filtrates induced pyrogenic tolerance in rabbits which, however, remained responsive to human leukocyte pyrogen (IL1) as shown by Ikejima et al. (1984). They concluded that in TSS, the sudden fever and probably other components of the acute-phase response, may be attributed to a massive release of IL1. However, one should not rule out that fever during the course of staphylococcal infection may be initiated by a variety of stimuli other than toxins (e.g. bacterial cell wall peptidoglycans and various molecules of the inflammatory cascade). Schlievert and co-workers attributed the fever elicited by TSST-1 to direct interaction with the preoptic area of the anterior hypothalamus stimulating increased production of prostaglandin E rather than to an indirect effect through endogenous pyrogen release, as also postulated for ET action (see below). A detailed overview of the various microbial products and endogenous mediators which may be involved in the pathogenesis of TSS and the various physiological disturbances including hyperthermia observed in vivo has been reported by Parsonnet (1989).

Fever induction by ETs has been widely investigated (see Alouf, 1980; Murai et al., 1987; Wannamaker and Schlievert, 1988). The i.v. injection of the toxins in rabbits was characterized in most reports by latency of 30–60 min and then by a gradual monophasic increase of the temperature up to 4–5 h (Kim and Watson, 1970; Schlievert and Watson, 1978; Hříbalová et al., 1980; Murai et al., 1987). The pyrogenic effect was directly proportional to the logarithm of toxin dose. A biphasic response was also reported by Brunson and Watson (1974) for ETA, ETB and ETC as previously shown by Schuh et al. (1970) with culture filtrates from various toxin-producing strains. The minimal pyrogenic dose of ETA or ETC determined either 3 or 4 h after i.v. injection ranged from 0.017 to 0.1 µg/kg except for the toxin used by Murai et al. (1987) who found a value of 1 µg/kg. ETB pyrogenicity was weaker. A study of ETC pyrogenicity in mice was investigated by Schlievert and Watson (1978), who also demonstrated the high diffusibility of this toxin since similar response peaking at 4 h post-injection was observed whatever the injection route (i.v., i.m., i.d. or s.c.).

Similarly to SEs and TSST-1, repeated injections of culture filtrates (Schuh et al., 1970) or purified toxins (Hříbalová et al., 1980; Wannamaker and Schlievert, 1988) resulted in a pyrogenic tolerance specific to the serotype of the injected toxin. A normal pyrogenic effect was observed in tolerant animals challenged with the other serotypes, enterotoxins or endotoxins, demonstrating immunological specificity in response to the pyrogen (Brunson and Watson, 1974; Hříbalová et al., 1980). The finding of serotype D by Schuh et al. (1970) was based on this property.

The mechanism(s) underlying the pyrogenic effects of ETs is controversial. According to

Hřibalová *et al.* (1979), these effects are mediated by leukocyte endogenous pyrogen on the basis of pyrogen release *in vitro* by toxin-stimulated macrophages and of the pyrogenic response to toxin in rabbit previously administered antilymphocytic serum. An alternative direct mechanism was proposed by Schlievert and Watson (1978) in their study of the pyrogenicity in rabbit of ETC (SPE C according to their nomenclature). They reported that the toxin crosses the blood–brain barrier to elicit fever by direct stimulation of the hypothalamus centre rather than indirectly by endogenous pyrogen. Whether ETA and ETB act similarly has not been reported to our knowledge. It was recently shown in the same laboratory (Fast *et al.*, 1989), that ETA (SPE A) incubated *in vitro* with human PBMC released TNF. The release of both TNF and IL6 by ETA was reported by Knöll *et al.* (see section VIII.C.3). These two cytokines in addition to IL1 are endogenous pyrogens.

Enhancement of host susceptibility to lethal shock by endotoxins

The susceptibility of rabbits and other animals to the lethal shock induced by Gram-negative bacterial endotoxins (lipopolysaccharides) was shown to be greatly potentiated after previous injection of staphylococcal enterotoxins (Sugiyama *et al.*, 1964; Bohach *et al.*, 1988b), TSST-1 (Schlievert, 1982; Igarashi *et al.*, 1984, 1989; De Azavedo and Arbuthnott, 1984; Reeves *et al.*, 1986) and erythrogenic toxins (Kim and Watson, 1970; Schlievert and Watson, 1978; Murai *et al.*, 1987; Bohach *et al.*, 1988b). The animals (rabbits in most cases) receive an i.v. injection of appropriate doses of toxin (e.g. 0.1–1 µg/kg) followed 3 or 4 h later by an i.v. injection of various doses of endotoxin (generally from 1/50 to 1/500 of LD_{50} for rabbits which is between 500 and 1000 µg/kg). Deaths are recorded over 48 h. A detailed technical protocol has been described by Bohach and Schlievert (1988).

ETA greatly enhanced the susceptibility to endotoxin shock in rabbits, monkeys and mice by as much as 100 000-fold (in American Dutch belted rabbits) as shown by Kim and Watson (1970). Myocardial and liver damage were observed in surviving animals. A similar effect was observed with ETC by Schlievert and Watson (1978), who showed that the ability to produce fever and to enhance lethal endotoxin shock were separate functions of the toxin. The mechanism of potentiation effect was investigated by Murai *et al.* (1987) who observed severe pathophysiological changes in Japanese white rabbits. These changes included transient hyperglycaemia followed by profound hypoglycaemia, elevation of the blood lipoperoxide level and an acute increase in plasma β-glucuronidase activity suggesting a general potentiation of physiologic failures. Recently, Murai *et al.* (1990) showed that the macrophages from toxin-treated rabbits exhibited hyperreactivity to endotoxin as assessed by their increased consumption of glucose. This enhancing effect was also reflected by potentiation of the febrile response to endotoxin in rabbits pretreated with ET suggesting an increased production of endogenous pyrogens. It is likely that other cytokines are massively released which may contribute to the lethal enhancing effect.

A great deal of work has been done concerning the enhancing effects of TSST-1. A 50 000-fold enhanced susceptibility to endotoxin in rabbit was reported by Schlievert (1982). The TSST-1 preparation used by Igarashi *et al.* (1984) was more potent, suggesting a higher state of toxin purification as also considered by Reiser *et al.* (1983). Rabbits preimmunized against endotoxins are protected. Specific pathogen-free and conventional New Zealand rabbits were more susceptible than Dutch belted rabbits to the lethal effect of TSST-1 itself as well as to toxin-enhanced endotoxic shock (De Azavedo and Arbuthnott, 1984). The reason for this difference in susceptibility was not clear. Reeves *et al.* (1986) observed that enhancement did not occur when both TSST-1 and endotoxin are injected subcutaneously; however, lethal shock did occur when endotoxin was injected intravenously after s.c. or i.v. injection of TSST-1. The death from TSST-1-potentiated endotoxin shock is generally preceded by a variety of pathological effects reported by Reeves *et al.* (1986). Blomster-Hautamaa and Schlievert (1988a), Bulanda *et al.* (1989) and Keane *et al.* (1986) showed that the *in vitro* necrosis of renal tubular cells elicited by endotoxin was augmented at least 1000-fold by pretreatment of these cells with TSST-1. In the latter case, less than 0.1 ng of endotoxin is required for cell damage. The oxidative metabolism of arachidonic acid and generation of reactive hydroxyl radicals (·OH) appeared to mediate cell injury. The enhancement of endotoxin lethal shock by TSST-1 has been also observed in chick embryos (De Azavedo *et al.*, 1988a).

The mode of action of TSST-1 and endotoxin alone or in combination in the system was investigated *in vitro* by Drumm *et al.* (1989) on primary chick embryo cells by means of light and scanning electron microscopy. The main damaging effect was the synergistic contribution of LPS to cause detachment and clumping of cell monolayers. Whether endotoxin could augment TSST-1-induced production of IL1 by murine macrophages was investigated by Beezhold *et al.* (1989). The macrophages from C57Bl/6 mice or the endotoxin-resistant C3H/HeJ mice were challenged with TSST-1 alone or associated with endotoxin. The costimulation of C57Bl/6 macrophages with both toxins resulted in a synergistic induction of IL1.

The systemic pathology produced by TSST-1 alone or in combination with exotoxin in various animal models (Van Miert *et al.*, 1984; De Azavedo and Arbuthnott, 1984; Reeves *et al.*, 1986; Stone and Schlievert, 1987; Bulanda *et al.*, 1989; Melish *et al.*, 1989) or with TSS-associated *S. aureus* strains implanted *in vivo* in artificial infection chambers (see De Azavedo, 1989; Melish *et al.*, 1989, for references) was strikingly similar in many (but not all) respects to that seen in humans who died of TSS. The data reported by certain authors led to the hypothesis, as also suggested by Schlievert (1983), that TSST-1 exerts its host sensitization by endogenous endotoxins deriving from mucosal colonization with opportunistic Gram-negative bacteria. Bacterial growth was postulated to be due to the suppression of host immune response by TSST-1 and/or to the blockade of reticuloendothelial system activity and thereby the inhibition of endogenous endotoxin clearance. These contentions appeared to be supported by (i) the finding that polymyxin B which binds and inactivates endotoxins neutralized the effects of TSST-1 (De Azavedo and Arbuthnott, 1984; Stone and Schlievert, 1987), (ii) the detection of endotoxin in acute-phase sera but not in convalescent-phase sera (Stone and Schlievert, 1987), (iii) the lower susceptibility of germ-free piglets than corresponding conventional animals to the pathogenic effects of TSST-1 (Bulanda *et al.*, 1989). However, other observations suggest that TSST-1 does not act

synergistically with endotoxins. Melish *et al.* (1989) reported that TSST-1 toxicity was not reduced in the presence of polymyxin B or when anti-LPS antibodies are administered concurrently. In fact, in animals given polymyxin B renal failure was accelerated. In a rabbit model of TSS using a constant s.c. infusion of TSST-1, endotoxin did not appear to be a requisite for TSS (Parsonnet *et al.*, 1989). Fukijawa *et al.* (1986) reported that no measurable amount of endotoxin was detected in the blood of rabbits challenged with sublethal amounts of TSST-1 up to 100 μg/kg, indicating that TSST-1 does not lead bacterial endotoxin from other body sites into the blood and that the fever elicited by TSST-1 might not originate from endotoxin in the blood.

On the other hand, the effects of various drugs on the pyrogenicity of TSST-1 and its capacity to enhance lethal endotoxic shock were investigated by Igarashi *et al.* (1989). Antipyretic agents (aspirin, aminopyrine, indomethacine) as well as cyclosporin decreased or suppressed the febrile response of TSST-1 but did not inhibit the enhancement of lethality by endotoxin. Methylprednisolone did not decrease the fever but inhibited the lethal effect of the various agents reported to suppress endotoxic shock experimentally; only chlorpromazine decreased the fever and none inhibited the lethal effects.

Reticuloendothelial system (RES) blockade

Erythrogenic toxins depressed significantly the clearance function of the RES in rabbits and mice as measured by the reduction of the rate of clearance of colloidal carbon (Hanna and Watson, 1965; Hřibalová, 1979) or ^{51}Cr-labelled sheep red blood cells (Cunningham and Watson, 1978a) from blood after i.v. injection of the toxins. In addition, toxin treatment failed to clear endotoxin from the circulation in parallel with the inhibition of RNA synthesis in Kupffer cells (see Wannamaker and Schlievert, 1988). A similar inhibition of colloidal carbon clearance in rabbits was observed (Schlievert, 1983) when both TSST-1 and endotoxin were injected simultaneously, whereas TSST-1 alone failed to induce measurable RES blockade. Fukijawa *et al.* (1986) showed that TSST-1 inhibited the clearance of endotoxin in the blood.

This clearance is thought to be mainly done by the RES.

Alteration and modulation of humoral and cell-mediated immune functions

SEs, TSST-1 and ETs are potent immunosuppressors of the antibody response toward T-dependent antigens. The classical approach used is the inhibition of the *in vitro* synthesis of IgM antibodies to sheep erythrocytes in mouse or rabbit splenocyte as measured in a direct Jerne plaque-forming cells (PFC) assay. PFC inhibition by SEB was observed for as low as 10 ng/ml and was complete for 12.5 μg/ml (Smith and Johnson, 1975). A detailed study of the immunosuppressive activity of this toxin undertaken in a Mishell-Dutton mouse splenocyte culture system (Donnelly and Rogers, 1982; Holly *et al.*, 1988) showed the generation of suppressor cells bearing the CD8 phenotype. These cells inhibited primary and secondary responses *in vitro*. Taub *et al.* (1989) established that the activity of the SEB-induced suppressor cell population is mediated by a soluble 26-kDa I-J restricted suppressor factor. In contrast, the suppressive activity was not restricted at the IgH locus, as are certain suppressor factors.

IgM synthesis was shown to be inhibited by very low concentrations of TSST-1 (Schlievert, 1983). As low as 0.1 ng per 10^7 mouse splenocytes inhibited this synthesis by 64% (Blomster-Hauta-maa *et al.*, 1986b). The toxin also suppressed all Ig classes production by human B lymphocytes stimulated with pokeweed mitogen (Poindexter and Schlievert, 1986). This *in vitro* suppression of both human and murine immunoglobulins is not due to direct effect of the toxin on B lymphocytes but occurs through the release of a soluble suppressor by T cells. This finding is consistent with the impairment of immune functions in TSS patients as reflected by the low antibody titre against the toxin in these patients (Crass and Bergdoll, 1986).

In a series of investigations by Hanna and his co-workers (see Hanna *et al.*, 1980; Wannamaker and Schlievert, 1988, for references; Cunningham and Watson, 1987b; Cavaillon *et al.*, 1983), the erythrogenic toxins were shown to suppress the *in vitro* IgM response against sheep erythrocytes in rabbits and mice. The suppression of PFC response

occurred at about 4 days. It was shown to result from an activation of T suppressor cells by the toxins and not from inactivation of T helper cells (Schlievert, 1980a). The immunosuppression was followed by a late delayed PFC burst of IgG synthesis at 10–12 days. This immuno-enhancement of antibody response to the erythrocytes (called deregulation by Hanna *et al.*, 1980) was shown to be a T cell-dependent property of ETs (Cunningham and Watson, 1978b) which probably results from a loss of T suppressor functions (Misfeldt and Hanna, 1981). Gray *et al.* (1988) reported the enhancement by ETA (blastogen A) of the cytotoxic activity of peripheral blood cells of rheumatic heart disease patients against tumour cells (leukaemia cell line K562).

The *in vivo* suppression of antibody response to sheep erythrocytes by ETs was established by Hanna and Watson (1968) in the rabbit and by Cunningham and Watson (1978a) in the mouse. Similarly, *in vivo* humoral and/or cellular immune responses were suppressed by SEs (Smith and Johnson, 1975; Pinto *et al.*, 1978; Zehavi-Willner *et al.*, 1984; Kawaguchi-Nagata *et al.*, 1985; Marrack *et al.*, 1990).

The modulation of various cellular immune functions in human and murine systems by SEs and TSST-1 have been reported. As with other polyclonal T-cell mitogens, SEB (Zehavi-Willner and Berke, 1986) and SEA (Fleischer and Schrezenmeier, 1988) were shown to induce non-specific, MHC class II-dependent T lymphocyte cytotoxicity. This finding was extended later to other SEs, TSST-1 and ETA (Fleischer *et al.*, 1991). SEA was also shown to increase human natural killer cells and induce suppressor cells (Platsoucas *et al.*, 1986; Garcia-Peñarrubia *et al.*, 1989). The suppression of the cytolytic T cell activity by SEB-induced suppressor cells and the role of IL2 in this process was investigated by Lin *et al.* (1986). Toxin-mediated suppression of allograft rejection was reported by Pinto *et al.* (1978).

Conclusion and trends for future research

The length of this chapter reflects the considerable progress made within the past 5 years and the attempt to bring together the various specific and common features of staphylococcal enterotoxins, toxic-shock syndrome toxin-1 and streptococcal erythrogenic (pyrogenic) toxins.

In the general surveys or books previously published, these toxins were reviewed separately. We thought that it would be more useful for the reader to have an integrated presentation of the bacteriological, structural, genetic, biological and pathophysiological aspects of this family of bacterial toxins. Their similarities are striking in many respects and such an integrated survey allows a better understanding of their properties and modes of action.

Within less than a decade and in certain instances within the last two years, the telling achievements recorded may be summarized as follows:

(1) The structural genes of all enterotoxins so far known, TSST-1 and erythrogenic toxins A, B and C have been cloned and their nucleotide sequences determined. Consequently the primary structure of all toxins is presently known.
(2) The discovery of TSST-1 and the intensive investigations on the multifaceted aspects of this toxin and TSS.
(3) The explosive new field in immunological research opened by the demonstration of the superantigenic properties of the three toxin groups and their interaction with the Vβ (and Vγ) region of T cell receptors.
(4) The demonstration of the induction of many components of the cytokine network by the toxins. This finding is a critical and major step for the understanding of the molecular mechanisms of action which are still far from clear.

The next important phase of research must involve the determination or the prediction of the tertiary structure of the toxins and how this relates to biological activity. In this respect, the determination of the molecular structure of the toxin molecule on the one hand and that of the Vβ or Vγ regions as well as that of class II MHC molecules on the other hand, will certainly be one of the major areas of research during the next years.

From the epidemiological point of view, the recent (re)appearance in the 1980s of the streptococ-

cal toxic-shock-like syndrome raises the question of the surge of new virulence of group A streptococci and thereby that of an 'old' (erythrogenic) toxin, a mutated version of it or new toxin(s) awaiting discovery.

We hope that we were able in this chapter to convey to the reader a picture of the present state of the art of this fascinating family of staphylococcal and streptococcal shock-inducing toxins.

Acknowledgements

The authors would like to thank Miss Luce Cayrol for her excellent and invaluable secretarial assistance in the preparation of this manuscript.

References

Abe, Y. and Alouf, J.E. (1976). Partial purification of exocellular lymphocyte mitogen from *Streptococcus pyogenes*. *Jap. J. Exp. Med.* **46**, 363–9.

Abe, Y., Alouf, J.E., Kurihara, T. and Kawashima, H. (1980). Species-dependent response to streptococcal lymphocyte mitogens in rabbits, guinea pig and mice. *Infect. Immun.* **29**, 814–18.

Abe, Y., Nakano, S., Nakahara, T., Kamezawa, Y., Kato, I., Ushijima, H., Yoshino, K., Ito, S., Noma, S., Okitsu, S. and Tajima, M. (1990). Detection of serum antibody by the antimitogen assay against streptococcal erythrogenic toxins. Age distribution in children and the relation to Kawasaki disease. *Pediat. Res.* **27**, 11–15.

Aliu, B. and Bergdoll, M.S. (1989). Characterization of staphylococci from patients with toxic shock syndrome. *J. Clin. Microbiol.* **26**, 2427–8.

Alouf, J.E. (1980). Streptococcal toxins (streptolysin O, streptolysin S, erythrogenic toxin). *Pharmacol. Ther.* **11**, 661–718.

Alouf, J.E. (1990). Immunocytotropic bacterial protein toxins. In: *Bacterial Protein Toxins* (eds R. Rappuoli, J.E. Alouf, P. Falmagne, F. Fehrenbach and J.H. Freer), Zbl. Bakt. Suppl. 19, pp. 453–60. Gustav Fischer Verlag, Stuttgart.

Alouf, J.E., Geoffroy, C., Klatzmann, D., Gluckman, J.C., Gruest, J. and Montagnier, L. (1986). High production of the AIDS virus (LAV) by human T lymphocytes. *J. Clin. Microbiol.* **24**, 639–41.

Arbeit, R.D., Leary, P.L. and Levin, M.L. (1982). Gamma interferon production by combinations of human peripheral blood lymphocytes, monocytes and culture macrophages. *Infect. Immun.* **35**, 383–90.

Arbuthnott, J.P. (1988). Toxic shock syndrome: a multisystem conundrum. *Microbiol. Sci.* **5**, 13–16.

Barbour, A.G. (1981). Vaginal isolates of *Staphylococcus*

aureus associated with toxic-shock syndrome. *Infect. Immun.* **33**, 442–9.

Barsumian, E.L., Cunningham, C.M., Schlievert, P.M. and Watson, D.W. (1978a). Heterogeneity of group A streptococcal pyrogenic exotoxin type B. *Infect. Immun.* **20**, 512–18.

Barsumian, E.L., Schlievert, P.M. and Watson, D.W. (1978b). Non-specific and specific immunological mitogenicity by group A streptococcal pyrogenic exotoxins. *Infect. Immun.* **22**, 681–8.

Bayles, K.W. and Iandolo, J.J. (1989). Genetic and molecular analyses of the gene encoding staphylococcal enterotoxin D. *J. Bacteriol.* **171**, 4799–806.

Beezhold, D.H., Best, G.K., Bonventre, P.F. and Thompson, M. (1989). Endotoxin enhancement of toxic shock syndrome toxin 1-induced secretion of interleukin 1 by murine macrophages. *Rev. Infect. Dis.* **11** (Suppl. 1), S289–93.

Bendtzen, K. (1988). Interleukin 1, interleukin 6 and tumor necrosis factor in infection, inflammation and immunity. *Immunol. Lett.* **19**, 183–92.

Bennett, R.W. (1986). Detection and quantitation of Gram-positive non spore forming pathogens and their toxins. In: *Foodborne Microorganisms and their Toxins: Developing Methodology* (eds M.D. Pierson and N.J. Stern), pp. 345–92. Marcel Dekker, New York.

Bergdoll, M.S. (1979). Staphylococcal intoxications. In: *Fooodborne Infections and Intoxications* (eds H. Riemann and F.L. Bryan), pp. 443–93. Academic Press, New York.

Bergdoll, M.S. (1983). Enterotoxins. In: *Staphylococci and Staphylococcal Infections*, Vol. 2 (eds C.S.F. Easmon and C. Adlam), pp. 559–98. Academic Press, New York.

Bergdoll, M.S. (1986). The role of the staphylococcal enterotoxins in staphylococcal disease. In: *Bacterial Protein Toxins* (eds P. Falmagne, J.E. Alouf, F.J. Fehrenbach, J. Jeljaszewicz and M. Thelestam), Zbl. Bakt. Mikrobiol. Hyg. I, Suppl. 15, ppp. 321–6. Gustav Fischer Verlag, Stuttgart.

Bergdoll, M.S. (1988). Monkey feeding test for staphylococcal enterotoxins. In: *Methods in Enzymology*, Vol. 165, *Microbial Toxins: Tools in Enzymology* (ed. S. Harshman), pp. 324–33. Academic Press, San Diego.

Bergdoll, M.S. and Schlievert, P.M. (1984). Toxic shock syndrome toxin. *Lancet* 2, 691.

Bergdoll, M.S., Crass, B.A., Reiser, B.F., Robbins, R.N. and Davis, J.P. (1981). A new staphylococcal enterotoxin, enterotoxin F, associated with toxic shock syndrome *Staphylococcus aureus* isolates. *Lancet* 1, 1017–21.

Betley, M.J. and Mekalanos, J.J. (1988). Nucleotide sequence of type A staphylococcal enterotoxin gene. *J. Bacteriol.* **170**, 34–41.

Betley, M.J., Schlievert, P.M., Bergdoll, M.S., Bohach, G.A., Iandolo, J.J., Khan, S.A., Pattee, P.A. and Reiser, R.R. (1990). Staphylococcal gene nomenclature. *ASM News* **50**, 182.

Blomster-Hautamaa, D.A. and Schlievert, P.M. (1988a). Non enterotoxic staphylococcal toxins. In: *Bacterial Toxins, Handbook of Natural Toxins*, Vol. 4 (eds M.C. Hardegree and A.T. Tu), pp. 297–330. Marcel Dekker, New York.

Blomster-Hautamaa, D.A. and Schlievert, P.M. (1988b). Preparation of toxin shock syndrome toxin-1. In: *Methods in Enzymology*, Vol. 165, *Microbial Toxins: Tools in Enzymology* (ed. S. Harshman), pp. 37–43. Academic Press, San Diego.

Blomster-Hautamaa, D.A., Kreiswirth, B.N., Kornblum, J.S., Novick, R.P. and Schlievert, P.M. (1986a). The nucleotide and partial amino acid sequence of toxic shock syndrome toxin-1. *J. Biol. Chem.* **261**, 15783–6.

Blomster-Hautamaa, D.A., Novick, R.P. and Schlievert, P.M. (1986b). Localization of biologic functions of toxic shock syndrome toxin-1 by use of monoclonal antibodies and cyanogen bromide-generated toxin fragments. *J. Immunol.* **137**, 3572–6.

Blomster-Hautamaa, D.A., Kreiswirth, B.N., Novick, R.P. and Schlievert, P.M. (1986c). Resolution of highly purified toxic shock syndrome toxin 1 into two distinct proteins by isoelectric focusing. *Biochemistry* **25**, 54–9.

Bloomster, T.G. and Watson, D.W. (1982). Recent trends in streptococcal pyrogenic exotoxin production by group A streptococci. In: *Basic Concepts of Streptococci and Streptococcal Diseases* (eds S.E. Holm and P. Christensen), pp. 139–41. Reedbooks, Chertesey, Surrey.

Bohach, G.A. and Schlievert, P.M. (1987a). Expression of staphylococcal enterotoxins C1 in *Escherichia coli*. *Infect. Immun.* **55**, 428–32.

Bohach, G.A. and Schlievert, P.M. (1987b). Nucleotide sequence of the staphylococcal enterotoxin C1 gene and relatedness to other pyrogenic toxins. *Molec. Gen. Genet.* **209**, 15–20.

Bohach, G.A. and Schlievert, P.M. (1988). Detection of endotoxin by enhancement with toxic shock syndrome toxin-1. In: *Methods in Enzymology*, Vol. 165, *Microbial Toxins: Tools in Enzymology* (ed. S. Harshman), pp. 302–6. Academic Press, San Diego.

Bohach, G.A. and Schlievert, P.M. (1989). Conservation of the biologically active portions of staphylococcal enterotoxins C1 and C2. *Infect. Immun.* **57**, 2249–52.

Bohach, G.A., Hauser, A.R. and Schlievert, P.M. (1988a). Cloning of the gene, *speB*, for streptococcal pyrogenic exotoxin type B in *Escherichia coli*. *Infect. Immun.* **56**, 1665–7.

Bohach, G.A., Hovde, C.J., Handley, J.P. and Schlievert, P.M. (1988b). Cross-neutralization of staphylococcal and streptococcal pyrogenic toxins by monoclonal and polyclonal antibodies. *Infect. Immun.* **56**, 400–4.

Bohach, G.A., Handley, J.P. and Schlievert, P.M. (1989). Biological and immunological properties of the carboxyl terminus of staphylococcal enterotoxin C1. *Infect. Immun.* **57**, 23–8.

Bonventre, P.F., Weckbach, L., Staneck, J., Schlievert, P.M. and Thompson, M. (1983). Production of staphylococcal enterotoxin F and pyrogenic exotoxin

C by *Staphylococcus aureus* isolates from toxic shock syndrome-associated sources. *Infect. Immun.* **40**, 1023–9.

Bonventre, P.F., Thompson, M.R., Adinolfi, L.E., Gillis, Z.A. and Parsonnet, J. (1988) Neutralization of toxic shock syndrome toxin-1 by monoclonal antibodies *in vitro* and *in vivo*. *Infect. Immun.* **56**, 135–41.

Brunson, K.W. and Watson, D.W. (1974). Pyrogenic specificity of streptococcal exotoxins, staphylococcal enterotoxin and gram-negative endotoxins. *Infect. Immun.* **10**, 347–51.

Bulanda, M., Zaleska, M., Mandel, L., Talafantova, M., Travnicek, J., Kunstmann, G., Mauff, G., Pulverer, G. and Heczko, P.B. (1989). Toxicity of staphylococcal toxic shock syndrome toxin 1 for germ-free and conventional piglets. *Rev. Infect. Dis.* **11** (Suppl. 1), S248–53.

Buxser, S., Bonventre, P.F. and Archer, D.L. (1981). Specific receptor binding of staphyloccal enterotoxins of murine splenic lymphocytes. *Infect. Immun.* **33**, 827–33.

Callahan, J.E., Herman, A., Kappler, J.W. and Marrack, P. (1990). Stimulation of B10-BRT cells with superantigenic staphylococcal toxins. *J. Immunol.* **144**, 2473–9.

Calvano, S.E., Quimby, F.W., Antonacci, A.C., Reiser, R.F., Bergdoll, M.S. and Dineen, P. (1984). Analysis of the mitogenic effects of toxic shock toxin on human peripheral blood mononuclear cells *in vitro*. *Clin. Immunol. Pathol.* **33**, 99–110.

Carlsson, R. and Sjögren, H.O. (1985). Kinetics of IL-2 and interferon-gamma production, expression of IL-2 receptors and cell proliferation in human mononuclear cells exposed to staphylococcal enterotoxin A. *Cell. Immunol.* **96**, 175–83.

Carlsson, R., Fischer, H. and Sjögren, H.O. (1988). Binding of staphylococcal enterotoxin A to accessory cells is a requirement for its ability to activate human T cells. *J. Immunol.* **140**, 2484–8.

Cavaillon, J.M. and Alouf, J.E. (1982). Mitogenicity of streptococcal extracellular products and antagonism with concanavalin A. *Immunol. Lett.* **5**, 317–22.

Cavaillon, J.M., Geoffroy, C. and Alouf, J.E. (1979). Purification of two extracellular streptococcal mitogens and their effect on human, rabbit and mouse lymphocytes. *J. Clin. Lab. Immunol.* **2**, 155–63.

Cavaillon, J.M., Rivière, Y., Svab, J., Montagnier, L. and Alouf, J.E. (1982). Induction of interferon by *Streptococcus pyogenes* extracellular products. *Immunol. Lett.* **5**, 323–6.

Cavaillon, J.M., Leclerc, C. and Alouf, J.E. (1983). Polyclonal antibody-forming cell activation and immunomodulation of the *in vitro* immune response induced by streptococcal extracellular products. *Cell. Immunol.* **76**, 200–6.

Chatila, T., Wood, N., Parsonnet, J. and Geha, R.S. (1988). Toxic shock syndrome toxin-1 induces inositol-phospholipid turnover, protein kinase C

translocation, and calcium mobilization in human T cells. *J. Immunol.* **140**, 1250–5.

Chesney, P.J., Bergdoll, M.S., Davis, J.P. and Vegeront, J.M. (1984). The disease spectrum, epidemiology, and etiology of toxic-shock syndrome. *Ann. Rev. Microbiol.* **38**, 315–38.

Choi, Y., Kotzin, B., Herron, L., Callahan, J., Marrack, P. and Kappler, J. (1989). Interaction of *Staphylococcus aureus* toxin 'superantigen' with human T cells. *Proc. Natl. Acad. Sci. USA* **86**, 8941–5.

Choi, Y.W., Herman, A., DiGiusto, D., Wade, T., Marrack, P. and Kappler, J. (1990). Residues of the variable region of the T-cell-receptor β-chain that interact with *S. aureus* toxin superantigens. *Nature* **346**, 471–3.

Clark, W.G. and Page, J.S. (1968). Pyrogenic responses to staphylococcal enterotoxins A and B in cats. *J. Bacteriol.* **96**, 1940–6.

Clark, B.R., Mills, B.J., Horikoshi, K., Shively, J.E. and Todd, C.W. (1984). Production of mitogen-free immune interferon and T-cell growth factor by human peripheral blood lymphocytes induced with biotin-labeled staphylococcal enterotoxin A. *J. Immunol. Meth.* **67**, 371–7.

Clyne, M., De Azavedo, J., Carlson, E. and Arbuthnott, J.P. (1988). Production of gamma-hemolysin and lack of production of alpha-hemolysin by *Staphylococcus aureus* strains associated with toxic-shock syndrome. *J. Clin. Microbiol.* **26**, 535–9.

Cohen, J.O. (1969). Effect of culture medium composition and pH on the production of M protein and proteinase by group A streptococci. *J. Bacteriol.* **99**, 737–44.

Cone, L.A., Woodard, D.R., Schlievert, P.M. and Tomory, G.S. (1987). Clinical and bacteriologic observations of a toxic shock-like syndrome due to *Streptococcus pyogenes*. *New Engl. J. Med.* **317**, 146–9.

Couch, J.L. and Betley, M.J. (1989). Nucleotide sequence of the type C3 staphylococcal enterotoxin gene suggests that intergenic recombination causes antigenic variation. *J. Bacteriol.* **171**, 4507–10.

Couch, J.L., Soltis, M.T. and Betley, M.J. (1988). Cloning and nucleotide sequence of the type E staphylococcal enterotoxin gene. *J. Bacteriol.* **170**, 2954–60.

Crass, B.A. and Bergdoll, M.S. (1986). Toxin involvement in toxic shock syndrome. *J. Infect. Dis.* **153**, 918–26.

Cremer, N. and Watson, D.W. (1960). Host–parasite factors in group A streptococcal infections. A comparative study of streptococcal pyrogenic toxins and Gram-negative bacterial endotoxin. *J. Exp. Med.* **112**, 1037–53.

Cunningham, C.M. and Watson, D.W. (1978a). Alteration of clearance function by group A streptococcal pyrogenic exotoxin and its relation to suppression of the antibody response. *Infect. Immun.* **19**, 51–7.

Cunningham, C.M. and Watson, D.W. (1978b). Suppression of antibody response by group A streptococcal pyrogenic exotoxin and characterization of the cells involved. *Infect. Immun.* **19**, 470–6.

Cunningham, C.M., Barsumian, E.L. and Watson, D.W. (1976). Further purification of group A streptococcal pyrogenic exotoxin and characterization of the purified toxin. *Infect. Immun.* **14**, 767–75.

Davis, J.P., Chesney, P.J., Wand, P. and La Venture, M. (1980). Investigation and laboratory team. Toxic-shock syndrome: epidemiological features, recurrence, risk factors and prevention. *New Engl. J. Med.* **303**, 1429–35.

De Azavedo, J.C.S. (1989). Animal models for toxic shock syndrome: overview. *Rev. Infect. Dis.* **11** (Suppl. 1), S205–9.

De Azavedo, J.C.S. and Arbuthnott, J.P. (1984). Toxicity of staphylococcal toxic shock syndrome toxin 1 in rabbits. *Infect. Immun.* **46**, 314–17.

De Azavedo, J.C.S., Lucken, R.N. and Arbuthnott, J.P. (1988a). Chick embryo assay for staphylococcal toxic shock syndrome toxin-1. In: *Methods in Enzymology*, Vol. 165, *Microbial Toxins: Tools in Enzymology* (ed. S. Harshman), pp. 344–7. Academic Press, San Diego.

De Azavedo, J.C.S., Drumm, A., Jupin, C., Parant, M., Alouf, J.E. and Arbuthnott, J.P. (1988b). Induction of tumour necrosis factor by staphylococcal toxic shock toxin 1. *FEMS Microbiol. Immunol.* **47**, 69–74.

Dick, G.F. and Boor, A.K. (1935). Scarlet fever toxin. I. A method of purification and concentration. *J. Infect. Dis.* **57**, 164–73.

Dick, G.F. and Dick, G.H. (1924). A skin test for susceptibility to scarlet fever. *J. Am. Med. Assoc.* **82**, 265–6.

Dickgiesser, N. (1988). Genetic studies on toxic shock syndrome toxin-1 (TSST-1) producing *Staphylococcus aureus* strains from Austria, the USA and West Germany. *Zbl. Bakt. Mikrobiol. Hyg. A-Med* **267**, 506–9.

Donnelly, R.P. and Rogers, T.J. (1982). Immunosuppression induced by staphylococcal enterotoxin B. *Cell. Immunol.* **72**, 166–77.

Drumm, A., De Azavedo, J.C.S. and Arbuthnott, J.P. (1989). Damaging effect of toxic shock syndrome toxin 1 on chick embryo cells in vitro. *Rev. Infect. Dis.* **11** (Suppl. 1), S275–80.

Edwin, C. (1989). Quantitative determination of staphylococcal enterotoxin A by an enzyme-linked immunosorbent assay using a combination of polyclonal and monoclonal antibodies and biotin-streptavidin interaction. *J. Clin. Microbiol.* **27**, 1496–501.

Edwin, C. and Kass, E.H. (1989). Identification of functional antigenic segments of toxic shock syndrome toxin 1 by differential immunoreactivity and by differential mitogenic responses of human peripheral blood mononuclear cells, using active toxin fragments. *Infect. Immun.* **57**, 2230–6.

Edwin, C., Tatini, S.R. and Maheswaran, S.K. (1986). Specificity and cross-reactivity of staphylococcal enterotoxin A monoclonal antibodies with enterotox-

ins B, C1, D and E. *Appl. Environ. Microbiol.* **52**, 1253–7.

Edwin, C., Parsonnet, J. and Kass, E.H. (1988). Structure-activity relationship of toxic-shock-syndrome toxin-1: derivation and characterization of immunologically and biologically active fragments. *J. Infect. Dis.* **158**, 1287–95.

Elliott, S.D. and Liu, T.Y. (1970). Streptococcal proteinase. *Methods Enzymol.* **19**, 252–61.

Ende, I.A., Terplan, G., Kickhofen, B. and Hammer, D. (1983). Chromatofocusing: a new method for purification of staphylococcal enterotoxins B and C1. *Appl. Environ. Microbiol.* **46**, 1323–30.

Eriksson, K.O., Naidu, A.S., Kibar, F., Wadström, T. and Hjerten, S. (1989). Surface hydrophobicity and electrophoretic mobilities of staphylococcal exotoxins with special reference to toxic shock syndrome toxin-1. *APMIS* **97**, 1081–7.

Ezepchuk, Y.V., Noskova, P., Aspetov, D., Novokhatsky, A.S. and Noskov, A.N. (1983). Binding of staphylococcal enterotoxin A (SEA) with human splenic lymphocytes. *J. Biochem.* **15**, 285–8.

Fast, D.J., Schlievert, P.M. and Nelson, R.D. (1988). Nonpurulent response to toxic-shock syndrome toxin-1 producing *Staphylococcus aureus*. *J. Immunol.* **140**, 949–53.

Fast, D.J., Schlievert, P.M. and Nelson, R.D. (1989). Toxic shock syndrome-associated staphylococcal and streptococcal pyrogenic toxins are potent inducers of tumor necrosis factor production. *Infect. Immun.* **57**, 291–4.

Feldman, C.A. (1962). Staphylococcal scarlet fever. *New Engl. J. Med.* **267**, 877–8.

Fey, H., Pfister, H. and Rüegg, O. (1984). Comparative evaluation of different enzyme-linked immunosorbent assay systems for the detection of staphylococcal enterotoxins A, B, C and D. *J. Clin. Microbiol.* **19**, 34–8.

Fischer, H., Dohlsten, M., Lindvall, M., Sjögren, H.O. and Carlsson, R. (1989). Binding of staphylococcal enterotoxin A to HLA-DR on B cell lines. *J. Immunol.* **142**, 3151–7.

Fischer, H., Dohlstein, M., Andersson, U., Hedlung, G., Ericsson, P., Hansson, J. and Sjögren, H.O. (1990). Production of TNF-α and TNF-β by staphylococcal enterotoxin A activated human T cells. *J. Immunol.* **144**, 4663–9.

Fleischer, B. (1989). Bacterial toxins as probes for the T cell antigen receptor. *Immunol. Today* **10**, 262–4.

Fleischer, B. and Schrezenmeier, H. (1988). T cell stimulation by staphylococcal enterotoxins. *J. Exp. Med.* **167**, 1697–707.

Fleischer, B., Schrezenmeier, H. and Conradt, P. (1989). T lymphocyte activation by staphylococcal enterotoxins: role of class II molecules and T cell surface structures. *Cell. Immunol.* **120**, 92–101.

Fleischer, B., Gerardy-Schahn, R., Carrel, S., Gerlach, D. and Köhler, W. (1991). An evolutionary conserved mechanism of T cell activation by microbial toxins. *J. Immunol.* **146**, 11–17.

Fraser, J.D. (1989). High-affinity binding of staphylococcal enterotoxins A and B to HLA-DR. *Nature* **339**, 221–3.

Freer, J.H. Arbuthnott, J.P. (1983). Toxins of *Staphylococcus aureus*. *Pharmacol. Ther.* **19**, 55–106.

Fukijawa, H., Igarashi, H., Usami, H., Tanaka, S. and Tamura, H. (1986). Clearance of enterotoxin from blood of rabbits injected with staphylococcal toxic shock syndrome toxin-1. *Infect. Immun.* **52**, 134–7.

Fukijawa, H., Takayama, H., Uchiyama, T. and Igarashi, H. (1989). Bindings of toxic shock syndrome toxin-1 and staphylococcal enterotoxins A, B, and C to rabbit spleen cells. *Microbiol. Immunol.* **33**, 381–90.

Galelli, A., Anderson, S., Charlot, B. and Alouf, J.E. (1989). Induction of murine hemopoietic growth factors by toxic shock syndrome toxin-1. *J. Immunol.* **142**, 2855–63.

Garbe, P.L., Arko, R.J., Reingold, A.L., Graves, L.M., Hayes, P.S., Hightower, A.W., Chandler, F.W. and Broome, C.V. (1985). *Staphylococcus aureus* isolates from patients with non-menstrual toxic shock syndrome. Evidence for additional toxins. *J. Am. Med. Assoc.* **253**, 2538–42.

Garcia-Peñarrubia, P., Lennon, M.P., Koster, F.T., Kelley, R.O. and Bankhurst, A.D. (1989). Selective proliferation of natural killer cells among monocyte-depleted peripheral blood mononuclear cells as a result of stimulation with staphylococcal enterotoxin B. *Infect. Immun.* **57**, 2057–65.

Gaskill, M.E. and Khan, S.A. (1988). Regulation of the enterotoxin B gene in *Staphylococcus aureus*. *J. Biol. Chem.* **263**, 6276–80.

Gaworzewska, E.T. and Hallas, G. (1989). Letter to the editor. *New Engl. J. Med.* **321**, 1546.

Geoffroy, C. and Alouf, J.E. (1988). Production, purification and assay of streptococcal erythrogenic toxin. In: *Methods in Enzymology*, Vol. 165, *Microbial Toxins: Tools in Enzymology* (ed. S. Harshman), pp. 64–7. Academic Press, San Diego.

Georgiades, J.A. and Johnson, H.M. (1981a). Partial purification and characterization of human immune interferon. *Methods Enzymol.* **78**, 536–9.

Georgiades, J.A. and Johnson, H.M. (1981b). Partial purification and characterization of mouse immune interferon. *Methods Enzymol.* **78**, 545–52.

Gerlach, D., Knöll, H. and Köhler, W. (1980). Isolierung und Characterisierung von erythrogenen Toxinen. I. Untersuchungen des von *Streptococcus pyogenes* Stamm NY-5 gebildeten Toxins A. *Zbl. Bakt. Hyg. I. Abt. Orig.A* **247**, 177–91.

Gerlach, D., Knöll, H., Köhler, W. and Ozegowski, J.H. (1981). Isolation and characterization of erythrogenic toxins of *Streptococcus pyogenes*. 3. Comparative studies of type A erythrogenic toxins. *Zbl. Bakt. Hyg. I. Abt. Orig. A* **250**, 277–86.

Gerlach, D., Knöll, H., Köhler, W., Ozegowski, J.H. and Hříbalová, V. (1983). Isolation and characterisation of erythrogenic toxins. V. Identity of erythrogenic toxin type B and streptococcal proteinase

precursor. *Zbl. Bakt. Hyg. I. Abt. Orig. A* **255**, 221–33.

Gerlach, D., Ozegowski, J.-H., Knöll, H. and Köhler, W. (1986). Isolierung und Characterisierung erythrogener Toxine. XVIII. Die Reinigung eines biologisch aktiven Proteins vom Molekulargewicht 10 000 (LMP-10K) aus Kulturfiltraten des *Streptococcus pyogenes* Stamm NY-5. Beziehung zum erythrogenen Toxin Typ A. *Zbl. Bakt. Hyg. A* **261**, 75–84.

Gerlach, D., Köhler, W., Knöll, H., Moravek, L., Weeks, C.R. and Ferretti, J.J. (1987). Purification and characterization of *Streptococcus pyogenes* erythrogenic toxin type A produced by a cloned gene in *Streptococcus sanguis*. *Zbl. Bakt. Hyg. A* **266**, 347–58.

Ginsburg, I. (1972). Mechanisms of cell and tissue injury induced by group A streptococci: Relation to poststreptococcal sequelae. *J. Infect. Dis.* **126**, 294–334, 419–50.

Goshorn, S.C. and Schlievert, P.M. (1988). Nucleotide sequence of streptococcal pyrogenic exotoxin type C. *Infect. Immun.* **56**, 2518–20.

Goshorn, S.C. and Schlievert, P.M. (1989). Bacteriophage association of streptococcal pyrogenic exotoxin type C. *J. Bacteriol.* **171**, 3068–73.

Goshorn, S.C., Bohach, G.A. and Schlievert, P.M. (1988). Cloning and characterization of the gene, *speC*, for pyrogenic exotoxin type C from *Streptococcus pyogenes*. *Molec. Gen. Genet.* **212**, 66–70.

Gray, E.D. (1979). Purification and properties of an extracellular blastogen produced by group A streptococci. *J. Exp. Med.* **149**, 1438–49.

Gray, E.D., Abdin, Z.H., El Kholy, A., Mansour, M., Miller, L.C., Zaher, S., Kamel, R. and Regelmann, W.E. (1988). Augmentation of cytotoxic activity by mitogens in rheumatic heart diseases. *J. Rheumatol.* **15**, 1672–6.

Grójec, P.L. and Jeljaszewicz, J. (1985). Staphylococcal leukocidin, Panton-Valentine type. *J. Toxicol.-Toxin Rev.* **4**, 133–89.

Hahn, H.A., Pickenhahn, P., Lenz, W. and Brandis, H. (1986). An avidin-biotin ELISA for the detection of staphylococcal enterotoxins A and B. *J. Immunol. Meth.* **92**, 25–59.

Hall, W.W., Blander, S., Kakani, R., Fischetti, V.A. and Zabriskie, J.B. (1989). Identification of a new toxin from a strain of *Staphylococcus aureus* isolated from a patient with non menstrual toxic shock syndrome. *Rev. Infect. Dis.* **11** (Suppl. 1), S130–5.

Hallas, G. (1985). The production of pyrogenic exotoxins by group A streptococci. *J. Hyg. Camb.* **95**, 47–57.

Hanna, E.E. and Watson, D.W. (1965). Host–parasite relationships among group A streptococci. III. Depression of reticuloendothelial function by streptococcal erythrogenic exotoxins. *J. Bacteriol.* **89**, 154–8.

Hanna, E.E. and Watson, D.W. (1968). Host–parasite relationship among group A streptococci. IV. Suppression of antibody response by streptococcal pyrogenic exotoxin. *J. Bacteriol.* **95**, 14–21.

Hanna, E.E., Hale, M.L. and Misfeldt, M.L. (1980).

Deregulation of mouse antibody-forming cells by streptococcal pyrogenic exotoxin (SPE). III. Modification of T-cell-dependent plaque-forming cell responses of mouse immunocytes is a common property of highly purified and crude preparation of SPE. *Cell. Immunol.* **56**, 247–57.

Hauser, A.R. and Schlievert, P.M. (1990). Nucleotide sequence of the streptococcal pyrogenic exotoxin type B gene and relationship between the toxin and the streptococcal proteinase precursor. *J. Bacteriol.* **172**, 4536–42.

Hirose, A., Ikejima, T. and Gill, D.M. (1985). Established macrophagelike cell lines synthesize interleukin-1 in response to toxic shock syndrome toxin. *Infect. Immun.* **50**, 765–70.

Hirsch, M.L. and Kass, E.H. (1986). An annotated bibliography of toxic shock syndrome. *Rev. Infect. Dis.* **8** (Suppl. 1), S1–108.

Ho, G., Campbell, W.H., Bergdoll, M.S. and Carlson, E. (1989). Production of toxic shock syndrome toxin variant by *Staphylococcus aureus* strains associated with sheep, goats and cows. *J. Clin. Microbiol.* **27**, 1946–8.

Holly, M., Lin, Y.S. and Rogers, T.C. (1988). Induction of suppressor cells by staphylococcal enterotoxin B: identification of a suppressor-effector cell circuit in the generation of suppressor-effector cell. *Immunology* **64**, 643–8.

Holm, S.E. and Falsen, B. (1967). An antigen-free medium for cultivation of beta-hemolytic streptococci. *Acta Pathol. Microbiol. Scand.* **69**, 264–76.

Holm, S.E., Ekedahl, C. and Bergholm, A.-M. (1978). Comparison of antibiotic assays using different experimental models and their possible clinical significance. *Scand. J. Infect. Dis. Suppl.* **14**, 214–20.

Hooker, S.B. and Follensby, E.M. (1934). Studies on scarlet fever. II. Different toxins produced by hemolytic streptococci of scarlatinal origin. *J. Immunol.* **27**, 177–93.

Hosotsubo, K.K., Hosotsubo, H., Nishijima, M.K., Nishimura, M., Taenaka, N. and Yoshiya, I. (1989). Rapid screening for *Staphylococcus aureus* infection by measuring enterotoxin B. *J. Clin. Microbiol.* **27**, 2794–8.

Houston, C.W. and Ferretti, J.J. (1981). Enzyme-linked immunosorbent assay for detection of type A streptococcal exotoxin. Kinetics and regulation during growth of *Streptococcus pyogenes*. *Infect. Immun.* **33**, 862–9.

Hovde, C.J., Hackett, S.P. and Bohach, G.A. (1990). Nucleotide sequence of the staphylococcal enterotoxin C3 gene. Sequence comparison of all three type C staphylococcal enterotoxins. *Molec. Gen. Genet.* **220**, 329–33.

Hříbalová, V. (1979). Effect of scarlet fever toxin on the phagocytic activity of the reticuloendothelial system. *Folia Microbiol. (Praha)* **24**, 415–27.

Hříbalová, V. and Pospíšil, M. (1973). Lymphocyte-stimulating activity of scarlet fever toxin. *Experientia* **29**, 704–5.

Hříbalová, V., Castrovan, A. and Pekarek, J. (1979). Influence of antilymphocyte and antipolymorphonuclear sera on the pyrogenic effect of scarlet fever toxin. *Folia Microbiol. (Praha)* **24**, 428–34.

Hříbalová, V., Knöll, H., Gerlach, D. and Köhler, W. (1980). Purification and characterization of erythrogenic toxins. II. Communication: *In vivo* biological activities of erythrogenic toxin produced by *Streptococcus pyogenes* strain NY-5. *Zbl. Bakt. Hyg. A* **248**, 314–22.

Huang, I.-Y., Hughes, J.L., Bergdoll, M.S. and Schantz, E.J. (1987). Complete amino acid sequence of staphylococcal enterotoxin A. *J. Biol. Chem.* **262**, 7006–13.

Humphreys, H., Keane, C.T., Hone, R., Pomeroy, H., Russel, R.J., Arbuthnott, J.P. and Coleman, D.C. (1989). Enterotoxin production by *Staphylococcus aureus* isolates from cases of septicaemia and from healthy carriers. *J. Med. Microbiol.* **28**, 163–72.

Hynes, W.L., Weeks, C.R., Iandolo, J.J. and Ferretti, J.J. (1987). Immunologic cross-reactivity of type A streptococcal exotoxin (erythrogenic toxin) and staphylococcal enterotoxins B and C1. *Infect. Immun.* **55**, 837–8.

Iandolo, J.J. (1989). Genetic analysis of extracellular toxins of *Staphylococcus aureus*. *Ann. Rev. Microbiol.* **43**, 375–402.

Iandolo, J.J. and Tweten, R.K. (1988). Purification of staphylococcal enterotoxins. In: *Methods in Enzymology*, Vol. 165, *Microbial Toxins: Tools in Enzymology* (ed. S. Harshman), pp. 43–52. Academic Press, San Diego.

Igarashi, H., Fukijawa, H., Usami, H., Kawabata, A. and Morita, T. (1984). Purification and characterization of *Staphylococcus aureus* FRI 1169 and 587 toxic shock syndrome exotoxins. *Infect. Immun.* **44**, 175–81.

Igarashi, H., Fujikawa, H. and Usami, H. (1989). Effect of drugs on the pyrogenicity of toxic shock syndrome toxin 1 and its capacity to enhance the susceptibility to the lethal effects of endotoxic shock in rabbits. *Rev. Infect. Dis.* **11** (Suppl. 1), S210–13.

Ikejima, T., Dinarello, C.A., Gill, D.M. and Wolff, S.M. (1984). Induction of human interleukin-1 by a product of *Staphylococcus aureus* associated with toxic shock syndrome. *J. Clin. Invest.* **73**, 1312–20.

Ikejima, T., Okusawa, S., van der Meer, J.W.M. and Dinarello, C.A. (1988). Induction by toxic-shock-syndrome toxin-1 of a circulating tumor necrosis factor-like substance in rabbits and of immunoreactive tumor necrosis factor and interleukin-1 from human mononuclear cells. *J. Infect. Dis.* **158**, 1017–25.

Immerman, R.P. and Greenman, R.L. (1987). Toxic shock syndrome associated pyomyositis caused by a strain of *Staphylococcus aureus* that does not produce toxic-shock syndrome toxin-1. *J. Infect. Dis.* **156**, 505–7.

International Symposium on Toxic Shock Syndrome (1989). Proceedings of Atlanta (Georgia) Symposium, 15–18 November 1987. *Rev. Infect. Dis.* **11** (Suppl. 1), S1–333.

James, J.F., Chu, M.C., Lee, L., Peck, S.A., McKissic, C., Sullivan, H., Frogner, K. and Melish, M. (1989). Effect of magnesium on *in vitro* production of toxic shock syndrome toxin 1. *Rev. Infect. Dis.* **11** (Suppl. 1), S157–66.

Janeway, C.A., Jr., Yagi, J., Conrad, P.J., Katz, M.E., Jones, B., Vroegop, S. and Buxser, S. (1989). T-cell responses to Mls and to bacterial proteins that mimic its behavior. *Immunol. Rev.* **107**, 61–88.

John, M.B. Jr. and Khan, S.A. (1988). Staphylococcal enterotoxin B is associated with a discrete genetic element. *J. Bacteriol.* **170**, 4033–9.

Johnson, H.M. and Magazine, H.L. (1988). Potent mitogenic activity of staphylococcal enterotoxin A requires induction of interleukin 2. *Int. Arch. Allergy Appl. Immunol.* **87**, 87–90.

Johnson, L.P. and Schlievert, P.M. (1984). Group A streptococcal phage T12 carries the structural gene for pyrogenic exotoxin type A. *Molec. Gen. Genet.* **194**, 52–6.

Johnson, L.P., L'Italien, J.J. and Schlievert, P.M. (1986a). Streptococcal pyrogenic exotoxin type A (scarlet fever toxin) is related to *Staphylococcus aureus* enterotoxin B. *Molec. Gen. Genet.* **203**, 354–6.

Johnson, L.P., Tornai, M.A. and Schlievert, P.M. (1986b). Bacteriophage involvement in group A streptococcal pyrogenic exotoxin A production. *J. Bacteriol.* **166**, 623–7.

Jones, C.L. and Khan, S.A. (1986). Nucleotide sequence of the enterotoxin B gene from *Staphylococcus aureus*. *J. Bacteriol.* **166**, 29–33.

Jupin, C., Anderson, S., Damais, C., Alouf, J.E. and Parant, M. (1988). Toxic shock syndrome toxin 1 as an inducer of human tumor necrosis factors and gamma interferon. *J. Exp. Med.* **167**, 752–61.

Kamezawa, Y.M. and Nakahara, T. (1989). Purification and characterization of streptococcal erythrogenic toxin type A produced by *Streptococcus pyogenes* strain NY-5 cultured in the synthetic medium NCTC-135. Comparison with the dialyzed medium (TP-medium)-derived toxin. *Microbiol. Immunol.* **33**, 183–94.

Kamezawa, Y.M., Nakahara, T., Abe, Y. and Kato, I. (1990). Increased vascular permeability, erythema, and leukocyte emigration induced in rabbit skin by streptococcal erythrogenic toxin type A. *FEMS Microbiol. Lett.* **68**, 159–62.

Kappler, J., Kotzin, B., Herron, L., Gelfand, E.W., Bigler, R.D., Boylston, A., Carrel, S., Posnett, D.N., Choi, Y. and Marrack, P. (1989). V beta-specific stimulation of human T cells by staphylococcal toxins. *Science* **244**, 811–13.

Karp, D.R., Teletski, C.L., Scholl, P., Geha, R. and Long, E.E. (1990). The $\alpha 1$ domain of the HLA-DR molecule is essential for high-affinity binding of the toxic shock syndrome toxin-1. *Nature* **346**, 474–6.

Kass, E.H. (1989). Magnesium and the pathogenesis of toxic shock syndrome. *Rev. Infect. Dis.* **11** (Suppl. 1), S167–73.

Kawaguchi-Nagata, K., Okamura, H., Shoji, K., Kanagawa, H., Semma, M. and Shinagawa, K. (1985).

Immunomodulating activities of staphylococcal enterotoxins. I. Effects on *in vivo* antibody responses and contact sensitivity reaction. *Microbiol. Immunol.* **29**, 183–93.

Keane, W.F., Gekker, G., Schlievert, P.M. and Peterson, P.K. (1986). Enhancement of endotoxin-induced isolated renal tubular cell injury by toxic-shock syndrome toxin-1. *Am. J. Pathol.* **122**, 169–76.

Kienle, E. and Buschmann, H.G. (1989). Specificity, cross-reactivity and competition profile of monoclonal antibodies to staphylococcal enterotoxins B and C1 detected by indirect enzyme-linked immunosorbent assays. *Med. Microbiol. Immunol.* **178**, 127–33.

Kim, Y.B. and Watson, D.W. (1970). A purified group A streptococcal pyrogenic exotoxin. Physiochemical and biological properties including the enhancement of susceptibility to endotoxin lethal shock. *J. Exp. Med.* **131**, 611–28.

Kim, Y.B. and Watson, D.W. (1972). Streptococcal exotoxins: biological and pathological properties. In: *Streptococci and Streptococcal Diseases* (eds. L.W. Wannamaker and J.H. Matsen), pp. 33–50. Academic Press, New York and London.

Kirkland, J.J., Ryan, C.A., Kohrman, K.A. and Danneman, P.J. (1989). Growth of *Staphylococcus aureus* and synthesis of toxic shock syndrome toxin 1 in different *in vitro* systems. *Rev. Infect. Dis.* **11** (Suppl. 1), S188–95.

Knöll, H., Hříbalová, V., Gerlach, D. and Köhler, W. (1981). Purification and characterization of erythrogenic toxins. IV. Commun. Mitogenic activity of erythrogenic toxin produced by *Streptococcus pyogenes* strain NY-5. *Zbl. Bakt. Hyg. A* **251**, 15–26.

Knöll, H., Holm, S.E., Gerlach, D. and Köhler, W. (1982). Tissue cages for study of experimental streptococcal infection in rabbits. I. Production of erythrogenic toxins *in vivo*. *Immunobiology* **162**, 128–40.

Knöll, H., Gerlach, D., Ozegowski, J.H., Hříbalová, V. and Köhler, W. (1983). Mitogenic activity of isoelectrically focused erythrogenic toxin preparations and culture supernatants of group A streptococci. *Zbl. Bakt. Hyg. A* **256**, 49–60.

Knöll, H., Holm, S.E., Gerlach, D., Kühnemund, O. and Köhler, W. (1985). Tissue cages for study of experimental streptococcal infection of rabbits. II. Humoral and cell mediated immune response to erythrogenic toxins. *Immunobiology* **169**, 116–27.

Knöll, H., Holm, S.E., Gerlach, D., Ozegowski, J.H. and Köhler, W. (1988a). Tissue cages for study of experimental streptococcal infection in rabbits. III. Influence of immunization with erythrogenic toxin A (ET A) and its toxoid on subsequent infection with an ET A producing strain. *Zbl. Bakt. Hyg. A* **269**, 366–76.

Knöll, H., Gerlach, D., Köhler, W., Holm, S.E. and Wagner, B. (1988b). Biological and biochemical activities of a toxoid of erythrogenic toxin type A. *Zbl. Bakt. Hyg. A* **269**, 468–78.

Knöll, H., Gerlach, D., Srámek, J., Reichardt, W. and

Köhler, W. (1990). Changes in erythrogenic pattern in group A: results from scarlet fever patients in correlation to the antibody response. In: *Bacterial Protein Toxins* (eds R. Rappuoli, J.E. Alouf, P. Falmagne, F. Fehrenbach and J.H. Freer), Zbl. Bakt. Suppl. 19, pp. 409–11. Gustav Fischer Verlag, Stuttgart.

Köhler, W. (1990). Streptococcal toxic syndrome. *Zbl. Bakt.* **272**, 257–64.

Köhler, W., Gerlach, D. and Knöll, H. (1987). Streptococcal outbreaks and erythrogenic toxin type A. *Zbl. Bakt. Hyg. A* **266**, 104–15.

Köhler, W., Knöll, H., Gerlach, D., Holm, S.E. and Alouf, J.E. (1990). Partial characterization of highly toxic group A streptococcus strains. In: *Bacterial Protein Toxins* (eds R. Rappuoli, J.E. Alouf, P. Falmagne, F. Fehrenbach and J.H. Freer), Zbl. Bakt. Suppl. 19, pp. 413–15. Gustav Fischer Verlag, Stuttgart.

Kokan-Moore, N.P. and Bergdoll, M.S. (1989a). Effect of chemical modification of histidine and tyrosine residues in toxic shock syndrome toxin 1 on the serologic and mitogenic activities of the toxin. *Infect. Immun.* **57**, 1901–5.

Kokan-Moore, N.P. and Bergdoll, M.S. (1989b). Determination of biologically active region in toxic shock syndrome toxin-1. *Rev. Infect. Dis.* **11** (Suppl. 1), S125–9.

Kreger, A.S., Cuppari, G. and Taranta, A. (1972). Isolation by electrofocusing of two lymphocyte mitogens produced by *Staphylococcus aureus. Infect. Immun.* **5**, 723–7.

Kreiswirth, B.N. (1989). Genetics and expression of toxic shock syndrome toxin 1: overview. *Rev. Infect. Dis.* **11** (Suppl. 1), S97–100.

Kreiswirth, B.N., Handley, J.P., Schlievert, P.M. and Novick, R.P. (1987). Cloning and expression of streptococcal pyrogenic exotoxin A and staphylococcal toxic shock syndrome in *Bacillus subtilis. Molec. Gen. Genet.* **208**, 84–7.

Krejci, L.E., Stock, A.A., Samigar, E.C. and Kraemer, E.O. (1942). Studies on the haemolytic streptococcus. V. The electrophoretic isolation of the erythrogenic toxin of scarlet fever and the determination of its chemical and physical properties. *J. Biol. Chem.* **142**, 785–802.

Kushnaryov, V.M., MacDonald, H.S., Reiser, R. and Bergdoll, M. (1984). Staphylococcal toxic shock toxin specifically binds to cultured human epithelial cells and is rapidly internalized. *Infect. Immun.* **45**, 566–71.

Kuntsmann, G., Schroder, E., Hasbach, H. and Pulverer, G. (1989). Immune response to toxic-shock-syndrome toxin-1 and to staphylococcal enterotoxin-A, enterotoxin-B and enterotoxin-C in *Staphylococcus aureus* infections. *Zbl. Bakt.* **271**, 486–92.

Lando, P.A., Dohlsten, M., Kalland, T., Sjögren, H.O. and Carlsson, R. (1990). The TCR-CD3 complex is required for activation of human lymphocytes with staphylococcal enterotoxin A. *Scand. J. Immunol.* **31**, 133–8.

Langford, M.P., Stanton, G.J. and Johnson, H.M. (1978). Biological effects of staphylococcal enterotoxin A on human peripheral lymphocytes. *Infect. Immun.* **22**, 62–8.

Lapeyre, C., Kaveri, S.V., Janin, F. and Strosberg, A.D. (1987). Production and characterization of monoclonal antibodies to staphylococcal enterotoxins: use in immunodetection and immunopurification. *Molec. Immunol.* **24**, 1243–54.

Lee, C.L.Y., Lee, S.H.S., Jay, F.J. and Rozee, K.R. (1990). Immunobiological study of interferon-gamma-producing cells after staphylococcal enterotoxin B stimulation. *Immunology* **70**, 94–9.

Lee, P.K. and Schlievert, P.M. (1989). Quantification and toxicity of group A streptococcal pyrogenic exotoxins in an animal model of toxic shock syndrome-like illness. *J. Clin. Microbiol.* **27**, 1890–2.

Lei, Z., Reiser, R.F. and Bergdoll, M.S. (1988). Chromatofocusing in the purification of staphylococcal enterotoxin D. *J. Clin. Microbiol.* **26**, 1236–7.

Lin, Y.-S., Patel, M.R., Linna, T.J. and Rogers, T.J. (1986). Suppression of cytolytic T-cell activity by staphylococcal enterotoxin B induced suppressor cells: role of interleukin 2. *Cell. Immunol.* **103**, 147–59.

Lin, Y.-S., Largen, M.T., Newcomb, J.R. and Rogers, T.J. (1988). Production and characterisation of monoclonal antibodies for staphylococcal enterotoxin B. *J. Med. Microbiol.* **27**, 263–70.

Mahmood, R. and Khan, S.A. (1990). Role of upstream sequences in the expression of the staphylococcal enterotoxin B gene. *J. Biol. Chem.* **265**, 4652–6.

Marrack, P. and Kappler, P. (1990). The staphylococcal enterotoxins and their relatives. *Science* **248**, 705–11.

Marrack, P., Blackman, M., Kushnir, E. and Kappler, J. (1990). The toxicity of staphylococcal enterotoxin B in mice is mediated by T cells. *J. Exp. Med.* **171**, 455–64.

Matis, L.A. (1990). The molecular basis of T-cell specificity. *Ann. Rev. Immunol.* **8**, 65–82.

Mauff, G., Röhrig, I., Ernzer, U., Lenz, W., Bergdoll, M. and Pulverer, G. (1983). Enterotoxinogenicity of *Staphylococcus aureus* strains from clinical isolates. *Eur. J. Clin. Microbiol.* **2**, 321–6.

McCloskey, R.V. (1973). Scarlet fever and necrotising fasciities caused by coagulase-positive hemolytic *Staphylococcus aureus* phage 85. *Ann. Intern. Med.* **78**, 85–7.

McCollister, B.D., Kreiswirth, B.N., Novick, R.P. and Schlievert, P.M. (1990). Production of toxic shock syndrome-like illness in rabbits by *Staphylococcus aureus* D 4508: association with enterotoxin A. *Infect. Immun.* **58**, 2067–70.

McMillian, R.A., Bloomster, T.A., Saeed, A.M., Henderson, K.L., Zinn, N.E., Abernathy, R., Watson, D.W. and Greenberg, R.N. (1987). Characterization of a fourth streptococcal pyrogenic exotoxin (SPE D). *FEMS Microbiol. Lett.* **44**, 317–22.

Melconian, A.K., Brun, Y. and Fleurette, J. (1983a). Enterotoxin production, phage typing and serotyping of *Staphylococcus aureus* strains isolated from clinical materials and food. *J. Hyg. Camb.* **91**, 235–42.

Melconian, A.K., Flandrois, J.P. and Fleurette, J. (1983b). Production and purification of *Staphylococcus aureus* toxic shock toxin. *Curr. Microbiol.* **9**, 325–8.

Melish, M.E., Murata, S., Fukunga, C., Frogner, K., Hirata, S. and Wong, C. (1989). Endotoxin is not an essential mediator in toxic shock syndrome. *Rev. Infect. Dis.* **11** (Suppl. 1), S219–28.

Meyer, R.F., Miller, L., Bennett, R.W. and Macmillan, J.D. (1984). Development of a monoclonal antibody capable of interacting with five serotypes of *Staphylococcus aureus* enterotoxin. *Appl. Environ. Microbiol.* **47**, 283–7.

Micusan, V.V., Mercier, G., Bhatti, A.R., Reiser, R.F., Bergdoll, M.S. and Oth, D. (1986). Production of human and murine interleukin-2 by toxic shock syndrome toxin-1. *Immunology* **58**, 203–8.

Micusan, V.V., Desrosiers, M., Gosselin, J., Mercier, G., Oth, D., Bhatti, A.R., Heremans, H. and Billian, A. (1989). Stimulation of T cells and induction of interferon by toxic shock syndrome toxin 1. *Rev. Infect. Dis.* **11** (Suppl. 1), S305–12.

Mills, J.T., Dodel, A. and Kass, E.H. (1986). Regulation of staphylococcal toxic shock syndrome toxin-1 and total exoprotein production by magnesium ion. *Infect. Immun.* **53**, 663–70.

Misfeldt, M.L. (1990). Microbial 'superantigens'. *Infect. Immun.* **58**, 2409–13.

Misfeldt, M.L. and Hanna, E.E. (1981). Deregulation of mouse antibody-forming cells by streptococcal pyrogenic exotoxin. IV. Fractionation of a T-cell subpopulation which generates SPE-induced deregulation of anti-TNF PFC responses. *Cell. Immunol.* **57**, 20–7.

Mochmann, H., Akatov, A.K., Khatenever, M.L., Richter, U., Kuschko, I.W. and Karsch, W. (1981). Studies on enterotoxin production by strains of staphylococcus of different origin obtained from USSR. *Zbl. Bakt. Mikrobiol. Hyg. A* (Suppl. 10). 377–80.

Mollick J.A., Cook, R.G. and Rich, R.R. (1989). Class II MHC molecules are specific receptors for staphylococcus enterotoxin A. *Science* **244**, 817–20.

Mourad, W., Scholl, P., Diaz, A., Geha, R. and Chatila, T. (1989). The staphylococcal toxic shock syndrome toxin 1 triggers B cell proliferation via major histocompatibility complex-unrestricted cognate T/B cell interaction. *J. Exp. Med.* **170**, 2011–22.

Murai, T., Ogawa, Y., Kawasaki, H. and Kanoh, S. (1987). Physiology of the potentiation of lethal endotoxin shock by streptococcal pyrogenic exotoxin in rabbits. *Infect. Immun.* **55**, 2456–60.

Murai, T., Ogawa, Y. and Kawasaki, H. (1990). Macrophage hyperreactivity to endotoxin induced by streptococcal pyrogenic exotoxin in rabbits. *FEMS Microbiol. Lett.* **68**, 61–4.

Murphy, B.G., Kreiswirth, B.N., Novick, R.P. and Schlievert, P.M. (1988). Localization of biologically

important epitope on toxic-shock-syndrome toxin-1. *J. Infect. Dis.* **158**, 549–55.

Naidu, A.S., Eriksson, K.O., Hallberg, T., Lindberg, J., Liao, J.-L., Yao, K., Wadström, T. and Hjertén, S. (1989). Mitogenic properties of two distinct forms of toxic shock toxin-1 separated on hydroxyapatite by high-performance liquid chromatography. *APMIS* **97**, 1088–96.

Nauciel, C. (1973). Mitogenic activity of purified streptococcal erythrogenic toxin on lymphocytes. *Ann. Immunol. (Inst Pasteur)* **114**, 796–811.

Nauciel, C., Raynaud, M. and Bizzini, B. (1968). Purification et propriétés de la toxine érythrogène du streptocoque. *Ann. Inst. Pasteur* **114**, 796–811.

Nauciel, C., Blass, J., Mangalo, R. and Raynaud, M. (1969). Evidence for two molecular forms of streptococcal erythrogenic toxin. Conversion to a single form by 2-mercaptoethanol. *Eur. J. Biochem.* **11**, 160–4.

Nida, S.K. and Ferretti, J.J. (1982). Phage influence on the synthesis of extracellular toxins in group A streptococci. *Infect. Immun.* **36**, 746–50.

Norton, S.D., Schlievert, P.M., Novick, R.P. and Jenkins, M.K. (1990). Molecular requirements for T cell-activation by the staphylococcal toxic shock syndrome toxin-1. *J. Immunol.* **144**, 2089–95.

Ogburn, C.A., Harris, T.N. and Harris, S. (1958). Extracellular antigens in steady-state cultures of the hemolytic streptococcus. Production of proteinase at low pH. *J. Bacteriol.* **76**, 142–51.

Otero, A., Garcia, M.L., Garcia, M.C., Moreno, B. and Bergdoll, M.S. (1990). Production of staphylococcal enterotoxins C1 and C2 and thermonuclease throughout the growth cycle. *Appl. Environ. Microbiol.* **56**, 555–9.

Ozegowski, J.-H., Knöll, H., Gerlach, D. and Köhler, W. (1984). Isolierung und Characterizierung von erythrogenen Toxinen. VII. Bestimmung des von *Streptococcus pyogenes* gebildeten erythrogenen Toxins Typ C. *Zbl. Bakt. Hyg. A* **257**, 38–50.

Parker, M.T. (1984). Streptococcal diseases. In: *Topley and Wilson's Principles of Bacteriology, Virology and Immunity*, Vol. III, pp. 225–53, 7th edn. Arnold, London.

Parsonnet, J. (1989). Mediators in the pathogenesis of toxic shock syndrome: overview. *Rev. Infect. Dis.* **11** (Suppl. 1), S263–9.

Parsonnet, J. and Gillis, Z.A. (1988). Production of tumor-necrosis factor by human monocytes in response to toxic shock syndrome toxin-1. *J. Infect. Dis.* **158**, 1026–33.

Parsonnet, J., Hickman, R.K., Eardley, D.D. and Pier, G.B. (1985). Induction of human interleukin-1 by toxic-shock-syndrome toxin-1. *J. Infect. Dis.* **151**, 514–22.

Parsonnet, J., Harrison, A.E., Spencer, S.E., Reading, A., Parsonnet, K.C. and Kass, E.H. (1987). Non production of toxic shock syndrome toxin 1 by coagulase-negative staphylococci. *J. Clin. Microbiol.* **25**, 1370–2.

Parsonnet, J., Gillis, Z.A., Richter, A.G. and Pier, G.B. (1989). A rabbit model of toxic shock syndrome using a constant subcutaneous infusion of toxic shock syndrome toxin-1. *Infect. Immun.* 55, 1070–6.

Pinto, M., Torten, M. and Birnbaum, S.C. (1978). Suppression of the *in vivo* humoral and cellular immune response by staphylococcal enterotoxin B. *Transplantation* 25, 320–4.

Platsoucas, C.D., Oleszak, E.L. and Good, R.A. (1986). Immunomodulation of human leukocytes by staphylococcal enterotoxin A: augmentation of natural killer cells and induction of suppressor cells. *Cell. Immunol.* 97, 371–85.

Poindexter, N.J. and Schlievert, P.M. (1985a). The biochemical and immunological properties of toxic-shock syndrome toxin-1 (TSST-1) and association with TSS. *J. Toxicol.-Toxin Rev.* 4, 1–39.

Poindexter, N.J. and Schlievert, P.M. (1985b). Toxic-shock toxin syndrome toxin 1-induced proliferation of lymphocytes: comparison of the mitogenic response of human, murine and rabbit lymphocytes. *J. Infect. Dis.* 151, 65–72.

Poindexter, N.J. and Schlievert, P.M. (1986). Suppression of immunoglobulin-secreting cells from human peripheral blood by toxic-shock-syndrome toxin-1. *J. Infect. Dis.* 153, 772–9.

Poindexter, N.J. and Schlievert, P.M. (1987). Binding of toxic shock syndrome toxin-1 to human peripheral blood mononuclear cells. *J. Infect. Dis.* 156, 122–9.

Pontzer, C.H., Russell, J.K. and Johnson, H.M. (1989). Localization of an immune functional site on staphylococcal enterotoxin A using the synthetic peptide approach. *J. Immunol.* 143, 280–4.

Quagliata, F., Beckerdite-Quagliata, S., Alsobrooke, R.A., Russo, C., Indiveri, F., Pellegrino, M.A. and Ferrone, S. (1982). Functional and immunological characterization of the stimulation of human lymphocytes with mitogen from group A streptococci (SM). *Cell. Immunol.* 68, 146–54.

Quimby, F. and Nguyen, H.T. (1985). Animal studies of toxic shock syndrome. *Crit. Rev. Microbiol.* 12, 1–44.

Quinn, R.W. (1989). Comprehensive review of morbidity and mortality trends for rheumatic fever, streptococcal disease and scarlet fever: the decline of rheumatic fever. *Rev. Infect. Dis.* 11, 928–53.

Rasmussen, E.O. and Wuepper, K.D. (1981). Purification and characterization of streptococcal proliferative factor. *J. Invest. Dermatol.* 77, 246–9.

Reeves, M.W. (1989). Effect of trace metals on the synthesis of toxic shock syndrome toxin-1. *Rev. Infect. Dis.* 11 (Suppl. 1), S145–9.

Reeves, M.W., Arko, R.J., Chandler, F.W. and Bridges, N.B. (1986). Affinity purification of staphylococcal toxic shock toxin 1 and its pathologic effects in rabbits. *Infect. Immun.* 51, 431–9.

Reiser, R.F., Robbins, R.N., Khoe, G.P. and Bergdoll, M.S. (1983). Purification and some physicochemical properties of toxic-shock toxin. *Biochemistry* 22, 3907–12.

Regelmann, W.E., Gray, E.D. and Wannamaker, L.W. (1982). Characterization of the human cellular immune response to purified group A streptococcal blastogen A. *J. Immunol.* 128, 1631–6.

Reynolds, D., Tranter, H.S., Sage, R. and Hambleton, P. (1988). Novel method for purification of staphylococcal enterotoxin A. *Appl. Environ. Microbiol.* 54, 1761–5.

Rizkallah, M.F., Tolyamat, A., Martinez, J.S., Schlievert, P.M. and Ayoub, E.M. (1989). Toxic shock syndrome caused by a strain of *Staphylococcus aureus* that produces enterotoxin C but not toxic shock syndrome toxin-1. *Am. J. Dis. Child.* 143, 848–9.

Robern, H., Stavrick, S. and Dickie, N. (1975). The application of QAE Sephadex for the purification of two staphylococcal enterotoxins. I. Purification of enterotoxin C2. II. Purification of enterotoxin A. *Biochim. Biophys. Acta* 393, 148–58, 159–64.

Robbins, R.N., Kelly, B.J., Hehl, G.L. and Bergdoll, M.S. (1989). Effect of tampon wraps on production of toxic shock syndrome toxins 1. *Rev. Infect. Dis.* 11 (Suppl. 1), S197–202.

Rosten, P.M., Bartlett, K.H. and Chow, A.W. (1987). Detection and quantitation of toxic shock syndrome toxin-1 *in vitro* and *in vivo* by a non competitive enzyme-linked immunosorbent assay. *J. Clin. Microbiol.* 25, 327–32.

Rosten, P.M., Bartlett, K.H. and Chow, A.W. (1989). Purification and purity assessment of toxic shock syndrome toxin 1. *Rev. Infect. Dis.* 11 (Suppl. 1), S110–15.

Russell, J.K., Pontzer, C.H. and Johnson, H.M. (1990). The I A.[b] region (65–85) is a binding site for the superantigen, staphylococcal enterotoxin A. *Biochem. Biophys. Res. Commun.* 168, 696–701.

Rust, C.J.J., Verreck, F., Vietor, H. and Koning, F. (1990). Specific recognition of staphylococcal enterotoxin A by human T cell bearing receptors with the Vγ9 region. *Nature* 346, 572–4.

Scheuber, P.H., Denzlinger, C., Wilker, D., Beck, G., Keppler, D. and Hammer, D.-K. (1987). Staphylococcal enterotoxin B as a non immunological mast cell stimulus in primates: the role of endogenous cysteinyl leukotrienes. *Int. Arch. Allergy Appl. Immunol.* 82, 289–91.

Schlievert, P.M. (1980a). Activation of murine T-suppressor lymphocytes by group A streptococcal and staphylococcal pyrogenic exotoxins. *Infect. Immun.* 28, 876–80.

Schlievert, P.M. (1980b). Purification and characterization of staphylococcal pyrogenic exotoxin B. *Biochemistry* 19, 6204–8.

Schlievert, P.M. (1981). Scarlet fever: role of pyrogenic exotoxins. *Infect. Immun.* 31, 732–6.

Schlievert, P.M. (1982). Enhancement of host susceptibility to lethal endotoxin shock by staphylococcal pyrogenic exotoxin type C. *Infect. Immun.* 36, 123–8.

Schlievert, P.M. (1983). Alteration of immune function by staphylococcal pyrogenic exotoxin type C: possible role in toxic-shock syndrome. *J. Infect. Dis.* **147**, 391–8.

Schlievert, P.M. (1988). Immunochemical assays for toxic shock syndrome toxin-1. In: *Methods in Enzymology*, Vol. 165, *Microbial Toxins: Tools in Enzymology* (ed. S. Harshman), pp. 339–44. Academic Press, San Diego.

Schlievert, P.M. and Gray, E.D. (1989). Group A streptococcal pyrogenic exotoxin (scarlet fever toxin) type A and blastogen A are the same protein. *Infect. Immun.* **57**, 1865–7.

Schlievert, P.M. and Watson, D.W. (1978). Group A streptococcal pyrogenic exotoxin: pyrogenicity, alteration of blood-brain barrier and separation of sites for pyrogenicity and enhancement of lethal endotoxic shock. *Infect. Immun.* **21**, 753–63.

Schlievert, P.M., Bettin, K.M. and Watson, D.W. (1977). Purification and characterisation of group A streptococcal pyrogenic exotoxin type C. *Infect. Immun.* **16**, 673–9.

Schlievert, P.M., Schoettle, D.J. and Watson, D.W. (1979a). Purification and physicochemical and biological characterization of a staphylococcal pyrogenic exotoxin. *Infect. Immun.* **23**, 609–17.

Schlievert, P.M., Bettin, K.M. and Watson, D.W. (1979b). Reinterpretation of the Dick test: Role of group A streptococcal pyrogenic exotoxin. *Infect. Immun.* **26**, 467–72.

Schlievert, P.M., Bettin, K.M. and Watson, D.W. (1979c). Production of pyrogenic exotoxin by groups of streptococci, association with group A. *J. Infect. Dis.* **140**, 676–81.

Schlievert, P.M. Schoettle, D.J. and Watson, D.W. (1979d). Nonspecific T-lymphocyte mitogenesis by pyrogenic exotoxins from group A streptococci and *Staphylococcus aureus*. *Infect. Immun.* **25**, 1075–7.

Schlievert, P.M., Schoettle, D.J. and Watson, D.W. (1980). Ganglioside and monosaccharide inhibition of nonspecific lymphocyte mitogenicity by group A streptococcal pyrogenic exotoxin. *Infect. Immun.* **27**, 276–9.

Schlievert, P.M., Shands, K.N., Dan, B.B., Schmid, G.P. and Nishimura, R.D. (1981). Identification and characterization of an exotoxin from *Staphylococcus aureus* associated with toxic shock syndrome. *J. Infect. Dis.* **143**, 509–16.

Schmidt, J.J. and Spero, L. (1983). The complete amino acid and sequence of enterotoxin C1. *J. Biol. Chem.* **258**, 6300–6.

Scholl, P.R., Diez, A. and Geha, R.S. (1989a). Staphylococcal enterotoxin B and toxin-shock syndrome toxin-1 bind to distinct sites on HLA-DR and HLA-DQ molecules. *J. Immunol.* **143**, 2583–8.

Scholl, P., Diez, A., Mourad, W., Parsonnet, J., Geha, R.S. and Chatila, T. (1989b). Toxic shock syndrome toxin 1 binds to major histocompatibility complex class II molecules. *Proc. Natl. Acad. Sci. USA* **86**, 4210–14.

Schuh, V., Hřibalová, V. and Atkins, E. (1970). The pyrogenic effect of scarlet fever toxin. IV.

Pyrogenicity of strain C203U filtrate: comparison with some classic characteristics of the known types of scarlet fever toxin. *Yale J. Biol. Med.* **43**, 31–42.

Schwab, J.H., Watson, D.W. and Cromartie, W.J. (1953). Production of generalized Schwartzman reaction with group A streptococcal factors. *Proc. Soc. Exp. Biol. Med.* **82**, 754–61.

Scott, D.F., Kling, J.M. and Best, G.K. (1986). Immunological protection of rabbits infected with *Staphylococcus aureus* isolates from patients with toxic shock syndrome. *Infect. Immun.* **53**, 441–4.

Scriba, S., Wagner, B., Wagner, M., Gerlach, D. and Köhler, W. (1987). Investigations on the binding of erythrogenic toxin A of *Streptococcus pyogenes* on human peripheral blood lymphocytes. I. Light and electron microscopical demonstration of cell surface receptors using colloidal gold-labelled toxin. *Zbl. Bakt. Hyg. A* **266**, 478–90.

See, R.H. and Chow, A.W. (1989). Microbiology of toxic shock syndrome: overview. *Rev. Infect. Dis.* **11** (Suppl. 1), S55–60.

See, R.H., Krystal, G. and Chow, A.W. (1990). Binding competition of toxic shock syndrome toxin 1 and other staphylococcal exoproteins for receptors on human peripheral blood mononuclear cells. *Infect. Immun.* **58**, 2392–6.

Seravalli, E. and Taranta, A. (1974). Lymphocyte transformation and macrophage migration inhibition by electrofocused and gel-filtered fractions of group A streptococcal filtrate. *Cell. Immunol.* **14**, 366–75.

Shands, K.N., Schmid, G.P., Dan, B.B., Guidotti, R.J., Hargrett, N.T., Anderson, R.L., Hill, D.L., Broome, C.V., Band, J.D. and Fraser, D.W. (1980). Toxic shock syndrome in menstruating women: its association with tampon use and *Staphylococcus aureus* and the clinical features in 52 cases. *New Engl. J. Med.* **303**, 1436–42.

Singh, B.R. and Betley, M.J. (1989). Comparative structural analysis of staphylococcal enterotoxins A and E. *J. Biol. Chem.* **264**, 4404–11.

Singh, B.R., Kokan-Moore, N.P. and Bergdoll, M.R. (1988a). Molecular topography of toxic shock syndrome toxin 1 as revealed by spectroscopic studies. *Biochemistry* **27**, 8730–5.

Singh, B.R., Evenson, M.L. and Bergdoll, M.R. (1988b). Structural analysis of staphylococcal enterotoxins B and C1 using circular dichroism and fluorescence spectroscopy. *Biochemistry* **27**, 8735–41.

Smith, B.G. and Johnson, H.M. (1975). The effect of staphylococcal enterotoxins on the primary *in vitro* immune response. *J. Immunol.* **115**, 575–8.

Smith, J.L., Bencivengo, M.M., Buchanan, R.L. and Kunsch, C.A. (1986a). Enterotoxin A production in *Staphylococcus aureus*: inhibition by glucose. *Arch. Microbiol.* **144**, 131–6.

Smith, J.L., Bencivengo, M.M. and Kunsch, C.A. (1986b). Enterotoxin A synthesis in *Staphylococcus aureus*: inhibition by glycerol and maltose. *J. Gen. Microbiol.* **132**, 3375–80.

Soltis, M.T., Mekalanos, J.J. and Betley, M.J. (1990).

Identification of a bacteriophage containing a silent staphylococcal variant enterotoxin gene (sezA⁺). *Infect. Immun.* **58**, 1614–19.

Spero, L., Griffin, B.Y., Middlebrook, J.L. and Metzger, J.F. (1976). Effect of a single and double peptide bond scission by trypsin on the structure and activity of staphylococcal enterotoxin C. *J. Biol. Chem.* **251**, 5580–8.

Spero, L., Johson-Winegar, A. and Schmidt, J.J. (1988). Enterotoxins of staphylococci. In: *Bacterial Toxins, Handbook of Natural Toxins*, Vol. 4 (eds M.C. Hardegree and A.T. Tu), pp. 131–63. Marcel Dekker, New York.

Stevens, D.L., Tanner, M.H., Winship, J., Swarts, R., Ries, K.M., Schlievert, P.M. and Kaplan, E. (1989). Severe group A streptococcal infections associated with toxic shock like syndrome and scarlet fever toxin A. *New Engl. J. Med.* **321**, 1–7.

Stock, A.H. (1939). Studies of the hemolytic streptococcus. I. Isolation and concentration of erythrogenic toxin of the NY5-strain of hemolytic streptococcus. *J. Immunol.* **36**, 480–97.

Stock, A.H. and Lynn, R.J. (1961). Preparation and properties of partially purified erythrogenic toxin B of group A streptococci. *J. Immunol.* **86**, 561–6.

Stone, R.L. and Schlievert, P.M. (1987). Evidence for the involvement of endotoxin in toxic shock syndrome. *J. Infect. Dis.* **155**, 682–9.

Strickler, M.P., Neill, R.J., Stone, M.J., Hunt, R.E., Brinkley, W. and Gemski, P. (1989). Rapid purification of staphylococcal enterotoxin B by higher-pressure liquid chromatography. *J. Clin. Microbiol.* **27**, 1031–5.

Sugiyama, H., McKissic, E.M., Bergdoll, M.S. and Heller, B. (1964). Enhancement of bacterial endotoxin lethality by staphylococcal enterotoxin. *J. Infect. Dis.* **114**, 111–18.

Suzuki, M. and Vogt, A. (1986). Mitogenic activity of extracellular cationic products produced by group A streptococci; analysis of the lymphocyte response. *Clin. Exp. Immunol.* **66**, 132–8.

Tai, J.Y., Kortt, A.A., Liu, T.Y. and Elliott, S.D. (1976). Primary structure of streptococcal proteinase. III. Isolation of cyanogen bromide peptides: complete covalent structure of the polypeptide chain. *J. Biol. Chem.* **251**, 1955–9.

Taranta, A. (1974). Lymphocyte mitogens of staphylococcal origin. *Ann. N.Y. Acad. Sci.* **236**, 362–75.

Taub, D.D., Liu, Y.S., Hu, S.-C. and Rogers, T.C. (1989). Immunomodulatory activity of an I-J restricted suppressor factor. *J. Immunol.* **143**, 813–20.

Taylor, D. and Holland K.T. (1988). Effect of dilution rate and Mg²⁺ limitation on toxin shock syndrome toxin-1 production by *Staphylococcus aureus* grown in defined continuous culture. *J. Gen. Microbiol.* **134**, 719–23.

Thompson, N.E., Ketterhagen, M.J. and Bergdoll, M.S. (1984). Monoclonal antibodies to staphylococcal enterotoxins B and C: cross-reactivity and localisation of epitopes on tryptic fragments. *Infect. Immun.* **45**, 281–5.

Thompson, N.E., Razdan, M., Kuntsmann, G., Aschenbach, J.M., Evenson, M.L. and Bergdoll, M.S. (1986). Detection of staphylococcal enterotoxins by enzyme-linked immunosorbent assays: comparison of monoclonal and polyclonal antibody systems. *Appl. Environ. Microbiol.* **51**, 885–90.

Tierno, P.M. Jr. and Hanna, B. (1989). Ecology of toxic shock syndrome: amplification of toxic shock syndrome toxin 1 by materials of medical interest. *Rev. Infect. Dis.* **11** (Suppl. 1). S182–6.

Todd, J., Fishant, M., Kapral, F. and Welch, T. (1978). Toxic-shock syndrome associated with phage-group-1 staphylococci. *Lancet* **ii**, 1116–18.

Tomar, R.H. (1976). Delayed skin reactor from streptokinase-streptodornase. Stability studies and amino acid analysis. *Int. Arch. Allergy Appl. Immunol.* **50**, 220–4.

Tonew, E., Gerlach, D., Tonew, M. and Köhler, W. (1982). Induktion von Immuninterferon durch erythrogene Toxine A und B des *Streptococcus pyogenes*. *Zbl. Bakt. Hyg. Orig. A* **252**, 463–71.

Torres, B.A., Farrar, W.L. and Johnson, H.M. (1982). Interleukin 2 regulates immune interferon (INFγ) production by normal and suppressor cell cultures. *J. Immunol.* **128**, 2217–19.

Uchiyama, T., Kamagata, Y., Wakai, M., Reiser, R.F., Bergdoll, M.S. and Oth, D. (1986). Study of the biological activities of toxic shock syndrome toxin-1. Proliferative response and interleukin-2 production by T cells stimulated with the toxin. *Microbiol. Immunol.* **30**, 469–83.

Uchiyama, T., Kamagata, Y., Yan, X.-J., Kawachi, A., Fujikawa, H., Igarashi, H. and Okubo, M. (1989a). Relative strength of the mitogenic and interleukin-2-production-inducing activities of staphylococcal exotoxins presumed to be causative exotoxins of toxic shock syndrome: toxic shock syndrome toxin-1 and enterotoxins A, B and C to murine and human T cells. *Clin. Exp. Immunol.* **75**, 239–44.

Uchiyama, T., Imanishi, K., Saito, S., Araake, M., Yan, X.-J., Fujikawa, H., Igarashi, H., Kato, H., Obata, F., Kashiwagi, N. and Inoko, H. (1989b). Activation of human T cells by toxic shock syndrome toxin-1: the toxin binding structures expressed on human lymphoid cells acting as accessory cells are HLA class II molecules. *Eur. J. Immunol.* **19**, 1803–9.

Uchiyama, T., Takamuda, T., Imanishi, K., Araake, M., Saito, S., Yan, X.-J., Fujikawa, H., Igarashi, H. and Yamaura, N. (1989c). Activation of murine T cells by toxic shock syndrome toxin-1. The toxin-binding structures expressed on murine accessory cells are MHC class II molecules. *J. Immunol.* **143**, 3175–82.

Van Miert, S.J.P., van Duin, C.T.M. and Schotman, A.J.H. (1984). Comparative observations of fever and associated clinical, hematological and blood biochemical changes after intravenous administration of staphylococcal enterotoxins B and F (toxic-shock syndrome toxin-1 in goats). *Infect. Immun.* **46**, 354–60.

Vroegop, S.M. and Buxser, S.E. (1989). Cell surface molecules involved in early events in T-cell mitogenic stimulation by staphylococcal enterotoxins. *Infect. Immun.* **57**, 1816–24.

Wagner, B., Scriba, S., Wagner, M. and Köhler, W. (1988). Investigations on the binding of erythrogenic toxin A of *Streptococcus pyogenes* on human peripheral blood lymphocytes. II. Identification of toxin-binding lymphocytes and characterization of the receptor. *Zbl. Bakt. Hyg. A* **267**, 404–13.

Wannamaker, L.W. and Schlievert, P.M. (1988). Exotoxins of group A streptococci. In: *Bacterial Toxins, Handbook of Natural Toxins* (eds M.C. Hardegree and A.T. Tu), pp. 267–95. Marcel Dekker, New York.

Watson, D.W. (1960). Host–parasite factors in group A streptococcal infections. Pyrogenic and other effects of immunologic distinct exotoxins related to scarlet fever toxins. *J. Exp. Med.* **111**, 255–84.

Weckbach, L.S., Thompson, M.R., Staneck, J.L. and Bonventre, P.F. (1984). Rapid screening assay for toxic shock syndrome toxin production by *Staphylococcus aureus*. *J. Clin. Microbiol.* **20**, 18–22.

Weeks, C.R. and Ferretti, J.J. (1984). The gene for type A streptococcal exotoxin (erythrogenic toxin) is located in bacteriophage T12. *Infect. Immun.* **46**, 531–6.

Weeks, C.R. and Ferretti, J.J. (1986). Nucleotide sequence of the type A streptococcal exotoxin (erythrogenic toxin) gene from *Streptococcus pyogenes* bacteriophage T12. *Infect. Immun.* **52**, 144–50.

Wells, D.E., Reeves, M.W., McKinney, R.M., Graves, L.M., Olsvik, O., Bergan, T. and Feeley, J.C. (1987). Production and characterization of monoclonal antibodies to toxic shock syndrome toxin 1 and use of monoclonal antibody in a rapid, one-step enzyme-linked immunosorbent assay for detection of picogram quantitites of toxic shock syndrome toxin 1. *J. Clin. Microbiol.* **25**, 516–21.

White, J., Herman, A., Pullen, A.M., Kubo, R., Kappler, J.W. and Marrack, P. (1989). The V beta-specific superantigen staphylococcal enterotoxin B: stimulation of mature T cells and clonal deletion in neonatal mice. *Cell* **56**, 27–35.

Whiting, J.L., Rosten, P.M. and Chow, A.W. (1989). Determination by Western blot (immunoblot) of seroconversions to toxic shock syndrome (TSS) toxin 1 and enterotoxin A, B, C during infection with TSS and non-Tss associated *Staphylococcus aureus*. *Infect. Immun.* **57**, 231–4.

Williams, R.T., Wehr, C.T., Rogers, T.J. and Bennett, R. (1983). High-performance liquid chromatography of staphylococcal enterotoxin B. *J. Chromatogr.* **266**, 179–86.

Willoughby, R. and Greenberg, R.N. (1983). The toxic shock syndrome and streptococcal pyrogenic exotoxins. *Ann. Intern. Med.* **98**, 559.

Wong, A.V.L. and Bergdoll, M.S. (1990). Effect of environmental conditions on production of toxic shock syndrome toxin 1 *Staphylococcus aureus*. *Infect. Immun.* **58**, 1026–9.

Yonaha, K., Elliott, S.D. and Liu, T.Y. (1982). Primary structure of zymogen of streptococcal proteinase. *J. Protein Chem.* **1**, 317–34.

Yu, C.-E. and Ferretti, J.J. (1989). Molecular epidemiologic analysis of the type A streptococcal exotoxin (erythrogenic toxin) gene (*speA*) in clinical *Streptococcus pyogenes* strains. *Infect. Immun.* **57**, 3715–19.

Yu, C.-E. and Ferretti, J.J. (1991). Frequency of the erythrogenic toxin B and C genes (*speB* and *speC*) among clinical isolates of group A streptococci. *Infect. Immun.* **59**, 211–15.

Zehavi-Willner, T. and Berke, G. (1986). The mitogenic activity of staphylococcal enterotoxin B (SEB): a monovalent T cell mitogen that stimulates cytolytic T lymphocytes but cannot mediate their lytic action. *J. Immunol.* **137**, 2682–7.

Zehavi-Willner, T., Shenberg, E. and Barnea, A. (1984). *In vivo* effect of staphylococcal enterotoxin A on peripheral blood lymphocytes. *Infect. Immun.* **44**, 401–5.

18

The Family of Shiga and Shiga-like Toxins

David W.K. Acheson, Arthur Donohue-Rolfe and Gerald T. Keusch

Department of Medicine, Division of Geographic Medicine and Infectious Diseases, New England Medical Center and Tufts University School of Medicine, Boston, MA 02111, USA

Introduction

The first complete description of Shiga's bacillus (the prototype organism of the genus *Shigella* (now named *Shigella dysenteriae* type 1) was accomplished by the Japanese microbiologist Kiyoshi Shiga in 1898, following an extensive epidemic of lethal dysentery in Japan (Shiga, 1898). With the emphasis on toxins in pathogenesis of bacterial diseases at the end of the nineteenth century, a search for *Shigella* toxins was inevitable, and by 1900 Simon Flexner made the initial observations suggesting the existence of Shiga toxin, although much of the effect he observed was due to endotoxin (Flexner, 1900). In 1903, Neisser and Shiga reported on the lethal effects of bacterial extracts of Shiga's bacillus given to rabbits; however, the credit for the discovery of Shiga toxin is generally accorded to Conradi, who described many of its properties in the same year. Among the most prominent effects of crude toxin was its ability to cause limb paralysis followed by death when given parenterally to certain experimental animals. These neurological manifestations led to the designation of the factor as Shiga 'neurotoxin', a term which remained in use for nearly 70 years.

The first two decades of study of Shiga toxin were dominated more by the technical difficulties in separating protein toxins with biological activity from lipopolysaccharide endotoxin than by the task of determining relevance to pathogenesis of disease (Keusch *et al.*, 1986a). Definitive evidence that the protein 'neurotoxin' was separable from lipopo-lysaccharide was not obtained until 34 years after the original report of Shiga toxin, when chemical fractionation was employed (Boivin and Mesro-beanu, 1937). This has been confirmed many times since, using increasingly sophisticated methods for protein purification.

From the time of its description until today, the role of Shiga toxin in the pathogenesis of shigellosis has been the subject of debate (Keusch *et al.*, 1986a). For the first five decades, ability to interpret studies was clouded initially by the inevitable contamination of neurotoxin with large quantities of endotoxin, and later by the lack of model systems suitable to study pathogenesis of intestinal manifestations of the disease. Thus, most models in animals employed parenteral inoculation of what turned out later to be impure preparations of toxins, and focused on the paralysis caused by the protein Shiga 'neurotoxin'. It did not seem to matter that such systemic manifestations were not representative of events occurring in the intestine during acute shigellosis.

In 1953, van Heyningen and colleagues achieved a significant increase in purification and yield of Shiga toxin, although later studies were to demonstrate that this preparation had at least 20 protein bands and was no more than 5% toxin (Keusch and Jacewicz, 1975). This 'purified' toxin preparation was used to study pathogenesis of the neurotoxin effect in animals (Bridgewater *et al.*, 1955; Howard, 1955). These studies suggested that the underlying lesion caused by Shiga toxin was focal endothelial cell damage, resulting in focal bleeding in the spinal cord and secondary neurological manifestations. In 1960, Vicari *et al.* found that van Heyningen's preparation of Shiga toxin was cytotoxic to certain epithelial cells in culture; however, the possible significance of this was overlooked. At the end of the decade, an extensive review of the published literature concluded that Shiga toxin played no role in pathogenesis of shigellosis (Engley, 1952).

In 1969, an extensive outbreak of epidemic *S. dysenteriae* type 1 infection occurred in Central America and Mexico, with significant morbidity and mortality (Gangarosa *et al.*, 1970). This led to a renewed interest in Shiga toxin as a virulence factor of *Shigella*. By 1972 it was reported that cell-free supernatants of *S. dysenteriae* 1 cultures could reproduce the major features of human shigellosis in the rabbit ligated ileal loop model, including intestinal fluid secretion and inflammatory enteritis (Keusch *et al.*, 1972a,b). Since then, great progress has been made in the biochemistry, physiology and genetics of Shiga toxin. In recent years the toxin genes have been identified and sequenced, a mechanism for iron regulation of production has been defined, the purification of toxin and elucidation of its subunit structure has been accomplished, leading to the discovery of the biochemical mechanism of action of the A subunit and the recent crystallization of the B subunit, and the mammalian cell membrane receptor has been described.

In the past decade, similar protein toxins (called 'Shiga-like toxins') (O'Brien and Holmes, 1987) have been shown to be produced by a number of pathogenic *E. coli* strains lacking the ability to invade cells or produce the classical heat-labile or heat-stable *E. coli* toxins. These toxins are related to Shiga toxin and share the same binding and enzymatic specificities, and their association with certain diseases has provided new evidence for a role of cytotoxins in pathogenesis. This chapter will review these new data about the Shiga family of toxins, including Shiga toxin and the Shiga-like toxins (SLTs).

Shiga toxin

Toxin structure

In the past decade, new methods for purification of Shiga toxin and related SLTs have provided enough pure toxin for careful study of their physico-chemical properties. Purification of Shiga toxin remained a problem until the early 1980s when small amounts (a few micrograms) of highly purified toxin were obtained by Olsnes and Eiklid (1980) and O'Brien *et al.* (1980) by ion-exchange column or antibody affinity chromatography, respectively. The first successful large-scale purification was reported by Donohue-Rolfe *et al.* in 1984. Their method employed chromatofocusing as the principal technique and resulted in a yield of nearly 1 mg pure toxin from 3 litres of growth. The purified toxin was isoelectric at pH 7.2, and consisted of two peptide bands of approximately 7 and 32 kDa. Toxin production was enhanced by low iron concentration, and steadily diminished as ferric iron concentration rose above 0.15 mM. Although O'Brien *et al.* (1980) used chelex-treated medium to reduce its iron content, Donohue-Rolfe *et al.* (1984) found that the iron content of Syncase medium was sufficiently low to support maximum toxin production without the need for further manipulation. Recently, Donohue-Rolfe *et al.* (1989a) have reported a creative and more efficient method for purification that involves coupling a glycoprotein toxin receptor analogue present in hydatid cyst fluid, P1-blood group active glycoprotein (P1gp), to Sepharose 4B and using this as an affinity chromatography matrix. Toxin bound to immobilized P1gp and could be eluted with 4.5 M $MgCl_2$. Renatured toxin was fully active, and as much as 10 mg of pure toxin could be obtained from 20-litre fermenter cultures, and considerably more in the case of inducible phage-encoded SLTs. This method has proved useful in obtaining large amounts of purified 'Shiga-like' toxins produced by certain *E. coli* strains and other Gram-negative bacteria (see below).

Pure toxin separates into two peptide bands in SDS polyacrylamide gels under reducing conditions (Donohue-Rolfe *et al.*, 1984) (Fig. 1). The larger 32-kDa A subunit has been isolated and shown to mediate the inhibitory effect of Shiga toxin on

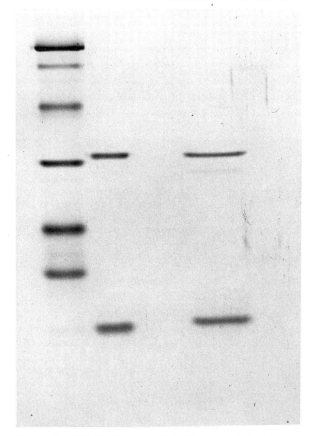

Figure 1. SDS–PAGE of purified Shiga toxin. The left lane is a set of molecular weight standards. The middle lane is Shiga toxin, demonstrating the larger A subunit and the smaller B subunit. The right lane shows the A1 subunit after nicking and reduction as a faint band under the A subunit.

protein synthesis in cell-free systems (Reisbig *et al.*, 1981). The smaller 7-kDa B subunit has also been separated and shown to bind to receptor-positive cells and to competitively inhibit both the binding and cytotoxicity of holotoxin (Donohue-Rolfe *et al.*, 1989b; Mobassaleh *et al.*, 1988). The B subunit is required for the action of toxin on intact cells by interacting with cell surface receptors which it recognizes (Donohue-Rolfe *et al.*, 1989b; Mobassaleh *et al.*, 1988). Cross-linking toxin with the heterobifunctional reagent dimethylpimelimidate results in the formation of a ladder of complexes of B subunits with one another, and with the A subunit (Donohue-Rolfe *et al.*, 1984). The largest cross-linked band present was consistent with a complex of 1A–5B. The same workers achieved even more convincing evidence of the 1A–5B

structure by first separating and isolating the A and B subunits, iodinating only the B subunit, and subjecting unlabelled A and iodinated B subunits to the cross-linking procedure (Donohue-Rolfe *et al.*, 1989b). Thus, all labelled cross-linked proteins were due to the presence of B subunits, and no confusion with single A or cross-linked A–A subunits was possible. Once again, a ladder of cross-linked peptides was obtained, up to a complex consistent with 1A–5B.

The B subunit has recently been crystallized (Hart *et al.*, 1991) using B subunit obtained from a hyperproducing *E. coli* containing a cloned segment of the identical SLT-I B subunit gene and purified by the P1gp affinity column method (Calderwood *et al.*, 1990). Cross-linking of isolated B subunits results in formation of dimers to pentamers (Fig. 2). The crystals obtained have been subjected to high-resolution X-ray analysis. Surprisingly, there was a four-fold symmetry axis within the crystal suggesting that the B subunits exist naturally as tetramers and not pentamers. It may be that the higher than 4B complexes seen with chemical cross-linking are due to polymerization of B tetramers. Definitive toxin stoichiometry will have to await X-ray diffraction of the crystallized holotoxin.

Mechanism of action

Reisbig *et al.* (1981) first reported that Shiga toxin could irreversibly inhibit protein synthesis by a highly specific action on the 60S mammalian ribosomal subunit. Brown *et al.* (1980) and others (O'Brien *et al.*, 1980; Olsnes *et al.*, 1981) found that the A subunit was activated by proteolysis and reduction, leading to the removal of a small A2 peptide from the biologically active A1 subunit. In these properties, Shiga toxin resembled the plant toxin, ricin. The possible structural relevance of this became apparent in 1987 when Calderwood *et al.* (1987) cloned and sequenced the virtually identical Shiga-like toxin-I and found significant homology between the A subunits of SLT-I and the previously sequenced ricin (7.9 standard deviations above the mean by the IFIND program for homology, and 9.3 standard deviations above the mean by the FASTP program). In the 73 amino acid segment from residues 138–210 of Shiga toxin there were 32% identical and 53%

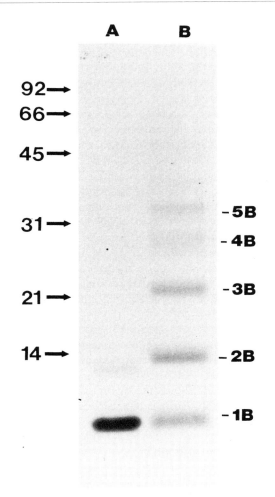

Figure 2. Cross-linking of purified B subunit obtained from an *E. coli* containing a cloned segment of the Shiga toxin gene which hyperproduces the B subunit monomer (lane A). This was cross-linked with dimethylpimelimidate and subjected to SDS–PAGE in a 15% gel and stained with Coomassie brilliant blue. The location of molecular weight markers is shown by the arrows on the left. (Reprinted with permission from Calderwood *et al.*, 1990, *Infect. Immun.* **58**, 2977–82.)

family of toxins. This striking homology between prokaryotic and eukaryotic toxins can be explained by convergent or divergent evolution, although given the exquisite specificity of the enzymatic activity, evolution from a common ancestor is the more likely explanation (Calderwood *et al.*, 1987). Molecules with high degrees of homology to ricin A chain have been identified in other plants, where they are postulated to serve autoregulatory functions. The function of these toxins in Shiga bacillus and in SLT-producing *E. coli* remains unclear, however.

The B subunit of Shiga toxin has been purified and sequenced (Seidah *et al.*, 1986). It is a 69 amino acid peptide with a molecular mass of 7691 Da, containing a single disulphide bridge but no potential glycosylation sites. By the Chou and Fasman calculations, the sequence predicts that 91% of all resides are associated with defined structural elements, including β-sheet (30%), α-helix (26%) and β-turns (26%). The segment between residues 36 and 52 is highly hydrophobic, whereas a surface-exposed, hydrophilic domain is found in the region from positions 10 to 20. Antibodies to a synthetic peptide spanning this hydrophilic domain are reported to neutralize cytotoxin activity in cell culture and neurotoxin activity in mice (Harari *et al.*, *1988*).

Toxin genetics

The structural genes for both the A and B subunits of Shiga toxin are chromosomally located in *S. dysei teriae* 1 (Strockbine *et al.*, 1988). These genes, designated *stxA* and *stxB*, are present on a single transcriptional unit, with the former preceding the latter. The genes are separated by 12 base pairs. Shiga toxin and SLT-I are highly conserved, and there were only three nucleotide differences in three codons of the A subunit of SLT-I, only one of which led to an amino acid change (threonine 45 for serine 45). Calculated molecular masses for the processed A and B subunits were 32 225 and 7691, respectively. Signal peptides of 22 and 20 residues were present for the A and B subunits, and the *stx* operon had putative promoter- and ribosomal-binding sites (Strockbine *et al.*, 1988).

The production of both Shiga toxin and SLT-I is regulated by the concentration of iron in the growth medium. This iron regulation involves the

chemically conserved residues. The same year, Endo and Tsurugi (1987) determined the enzymatic activity of ricin as an *N*-glycosidase that hydrolysed adenine 4324 of the 28S ribosomal RNA of the 60S ribosomal subunit. Shortly thereafter, Endo and colleagues (1988) found that Shiga toxin had the identical enzymatic specificity as ricin. The Shiga-like toxins also share this property, which is one of the criteria used to define the Shiga control gene, *fur*, whose protein product acts as a negative repressor for transcription. In the promoter regions of both Shiga toxin and SLT I operons there is a 21 base pair dyad repeat that appears to be a binding site for the Fur protein (Calderwood and Mekalonos, 1987; Strockbine *et al.*, 1988).

A single gene copy was present in *S. dysenteriae* 1. Using a probe specific for the A subunit of the virtually identical SLT-I, and DNA from the four *Shigella* species, strong homology was seen only in DNA from *S. dysenteriae* type 1 strains. Weak homology was detected with a strain of *S. flexneri*, and no homology was seen in *S. boydii* or *S. sonnei* (Strockbine *et al.*, 1988).

Toxin receptors

The search for Shiga toxin receptors on mamalian cells began in 1977, when Keusch *et al.* reported that toxin-sensitive cells in tissue culture removed toxin bioactivity from the medium, whereas toxin-resistant cells did not. These studies also suggested that the receptor was carbohydrate in nature, and that the toxin was a sugar-binding protein or lectin. Based on *in vitro* studies using isolated glycolipids, Lindberg and colleagues (1987) later reported that toxin bound to the P blood group active glycolipid, Gb_3, which consists of a trisaccharide, galactose-$\alpha 1 \rightarrow 4$-galactose-$\beta 1 \rightarrow 4$-glucose, linked to ceramide, and that Gb_3 inhibited biological activity of Shiga toxin in cell culture systems. Jacewicz *et al.* (1986) also extracted a toxin-binding constituent from toxin-sensitive HeLa cells and from a relevant intestinal target tissue, rabbit jejunal microvillus membranes (MVMs), already known to express a specific binding site for Shiga toxin (Fig. 3) (Fuchs *et al.*, 1986). The MVM binding site was shown to be Gb_3 by thin-layer chromatography (TLC) methods (Jacewicz *et al.*, 1986), and later confirmed by high-performance liquid chromatography (HPLC) of derivatized glycolipds, which also demonstrated it to be the hydroxylated fatty acid variety of Gb_3 (Mobassaleh *et al.*, 1989). In addition, HeLa cells contained a toxin binding glycolipid that migrated similarly to the disaccharide galabiosylceramide (Gal-$\alpha 1 \rightarrow 4$-Gal-$\beta 1$-ceramide) on TLC. Toxin also bound to the P1 blood group antigen present in human erythrocytes (Jacewicz *et al.*, 1986), a Gal-$\alpha 1 \rightarrow 4$-Gal terminal

pentasaccharide linked to ceramide. This finding has served as the basis for development of an affinity chromatography purification of Shiga toxin using a P1 antigen present on glycoproteins found in hydatid cyst fluid (Donohue-Rolfe *et al.*, 1989a), and a toxin ELISA using the hydatid cyst fluid as the coating for capture of antigen (Acheson *et al.*, 1990b).

The function of Gb_3 as a receptor has been shown in several ways. Most relevant to the intestinal effects of Shiga toxin is the work of Mobassaleh *et al.* (1988). They reported that infant rabbits are not susceptible to the fluid secretory effects of Shiga toxin before 16 days of age, and that the age-related sensitivity correlated with developmentally regulated Gb_3 levels in rabbit intestinal MVMs. Mobassaleh found very low levels of Gb_3 in MVM from young animals, and a rapid increase in Gb_3 content in animals from 16 to 24 days of age (Fig. 4), in parallel with increasing intestinal responses to toxin. More recently, these authors have presented preliminary evidence that this developmental regulation of Gb_3 content is due, at least in part, to age-related increases in the microsomal biosynthetic enzyme, UDP-galactose: lactosylceramide galactosyltransferase (Mobassaleh *et al.*, 1991).

Additional evidence suggesting that Gb_3 is the receptor mediating the fluid secretory response of rabbit intestine has been gathered by Kandel *et al.* (1989). These investigators found that Shiga toxin leads to a significant decrease in neutral sodium absorption in rabbit jejunum, with no alteration in substrate-coupled sodium absorption or active chloride secretion. This suggested that toxin acted on the absorptive villus cell and not on the secretory crypt cell. When villus and crypt cells were isolated from rabbit jejunum, it was found that the former cell type alone expressed Gb_3, bound toxin, and were susceptible to its effect on protein synthesis (Fig. 5). The interpretation of these data is that Shiga toxin targets the villus cell because it expresses the Gb_3 receptor, and that inhibition of villus cell protein synthesis results in diminished sodium absorptive capacity. In the presence of continued chloride secretion from the unaffected crypt cell, net salt and water accumulation occurs in the lumen of the gut.

A variety of studies in cell culture have provided evidence that Gb_3 is also critical for the protein

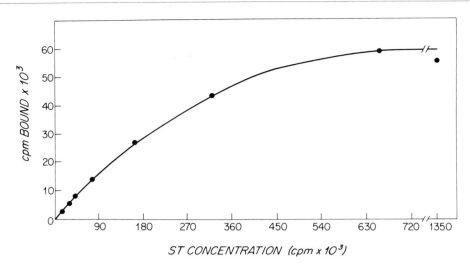

Figure 3. Increasing amounts of ^{125}I-labelled purified Shiga toxin were added to 0.75 mg of rabbit jejunal microvillus membrane (MVM) protein at 4°C. MVMs were washed on a 0.22-μm polyvinylidine fluoride (Durapore) filter (Millipore Corp., Bedford, MA, USA) and MVM-bound radioactivity was measured.

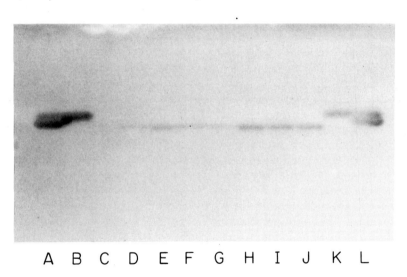

A   B   C   D   E   F   G   H   I   J   K   L

Figure 4. High-performance thin-layer chromatography of neutral glycolipids extracted from jejunal microvillus membranes (MVM) of rabbits at different ages. These have been overlain with [^{125}I]-Shiga toxin and autoradiography performed. Lanes A and L are a commercial Gb$_3$ standard and lanes B and K are purified human B erythrocyte non-hydroxylated Gb$_3$. Lanes C – J are MVM from rabbits of different ages: C, neonate; D, 16 days; E, 18 days; F, 20 days; G, 22 days; H, 24 days; I, 26 days; J, adult. (Reprinted with permission from Mobassaleh *et al.*, 1988, *J. Infect. Dis.* **157**, 1023–31.)

synthesis inhibitory effect of Shiga toxin. Sensitive cell lines, such as HeLa or Vero, express Gb$_3$ (Jacewicz *et al.*, 1986, 1989; Lingwood *et al.*, 1987; Waddell *et al.*, 1988), and refractory cells, such as CHO cells or selected Daudi cell lines, do not (Cohen *et al.*, 1987; Jacewicz *et al.*, 1989; Waddell

et al., 1990). In addition, very sensitive HeLa 229 cells have significantly more Gb$_3$ than the more resistant HeLa CCL-2 cell line (2.3 × 10^3 pmol/mg cell protein vs 0.4 × 10^3 pmol/mg cell protein) (Jacewicz *et al.*, 1989). Analysis of HeLa cell toxin receptors using a Scatchard model has demonstrated two distinct sites, one of high affinity/low capacity and the second of low affinity/high capacity. When such HeLa cells are cloned by limiting dilution methods in soft agar, clones can be obtained that vary in sensitivity to toxin by over 1 million fold (Jacewicz et al., 1989). Sensitivity is also directly related to Gb_3 content, although not in a linear fashion. Toxin response is also inhibited by pretreatment of cells with tunicamycin, which blocks synthesis of N-linked glycoproteins (Keusch et al., 1986b). This finding

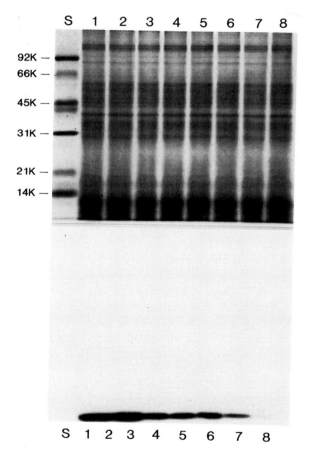

Figure 5. SDS–PAGE of proteins extracted from isolated epithelial cells from rabbit jejunum. A gradient of cells from villus (lanes 1–3) to crypt (lane 8) is obtained. The upper panel has been stained with Coomassie brilliant blue. The lower panel has been overlain with [^{125}I]-Shiga toxin and autoradiography performed. A single band at the dye front, identified as Gb_3, binds toxin in the villus cells. As the gradient approaches the crypt cell, this binding site disappears and is absent in the crypt cell fraction. (Reprinted with permission from Kandel et al., 1989, J. Clin. Invest. **84**, 1509–17.)

led to the suggestion that glycoprotein toxin receptors were present on HeLa cells. Since the primary effect of tunicamycin was to diminish the number of high-affinity binding sites (Jacewicz et al., 1989), this was presumed to be the functional receptor. However, in parallel with the decrease in receptor number, there was a decrease in amount of Gb_3 present as well (Jacewicz et al., 1989), suggesting that the high-affinity site was glycolipid, and not glycoprotein, in nature. If true, it is not clear how tunicamycin might affect glycolipid synthesis. One interpretation of this rather surprising finding is that the effect of tunicamycin on toxin sensitivity is indirect, possibly via a tunicamycin-sensitive glycoprotein involved in regulating Gb_3 levels in HeLa cells.

Attempts have also been made to increase toxin sensitivity by incorporating Gb_3 into plasma membranes. In one study, Waddell et al. (1990) incubated selected toxin-resistant mutant Daudi cells with glycolipids incorporated into phosphatidylethanolamine:phosphatidylserine liposomes. It was already known that Gb_3 present in liposomes bound toxin and competitively inhibited toxin binding in cell culture (Jacewicz et al., 1986). In Waddell's experiments, the exogenously added Gb_3 became cell-associated, toxin-binding sites appeared, and cytotoxicity was detected. Our own laboratory has employed a different strategy to increase the cellular content of Gb_3 by incubating cells with a competitive inhibitor of α-galactosidase A, 1,5-dideoxy-1,5-imino-D-galactitol (DIG). Selected toxin-resistant HeLa cell clones treated with DIG significantly increase Gb_3 content, toxin binding, and sensitivity to the biological effects of Shiga toxin (M. Jacewicz, P. Daniel, S.K. Gross, R.K. McCluer, and G.T. Keusch, unpublished).

Shiga-like toxins

History

In the latter half of the 1970s, Konowalchuk et al. (1977) reported that culture filtrates of several different strains of *Escherichia coli* contained a cytotoxin activity capable of killing Vero cells. This cytotoxicity was heat labile, and not neutralized by antisera to the classical cholera-like *E. coli* heat-labile enterotoxin. Konowalchuk subsequently investigated 136 *E. coli* strains and found only 10 to be capable of producing this Vero cell cytotoxin. Eight of the strains were associated with diarrhoea (seven in human infants, and one in a weanling pig) and two were isolated from cheese. It soon became apparent that the cytotoxicity was not due to a single protein. Investigation of one strain (*E. coli* 026 H30) revealed two variant toxins, with a pI of 7.2 and 6.8 (Konowalchuk *et al.*, 1978). Following these observations, Wade *et al.* (1979) in England noted the presence of cytotoxin-producing *E. coli* 026 strains in association with bloody diarrhoea. This finding was supported by reports from Scotland *et al.* (1979) in the UK and Wilson and Bettleheim (1980) in New Zealand. Several excellent reviews about SLTs have been published recently (O'Brien and Holmes, 1987; Karmali, 1989; Brunton, 1990), and the reader is referred to these publications for further historical information.

Nomenclature for this group of cytotoxins remains confusing. While they were initially called Vero toxins, there are several compelling and historical reasons to classify them as Shiga-like toxins (SLTs). First of all, by 1982 O'Brien *et al.* reported that certain of the *E. coli* cytotoxins capable of killing HeLa cells were neutralizable by polyclonal antisera raised against Shiga toxin. These subsequently were defined as SLT-I (or VT-I), and were shown to differ from Shiga toxin by just one amino acid substitution in the A subunit. Secondly, extracts of *E. coli* containing the cytotoxin activity resulted in fluid secretion in rabbit ileal loops (enterotoxicity), caused paralysis and death in mice (neurotoxicity), and inhibited protein synthesis in HeLa cells (cytotoxicity), in common with Shiga toxin. Thirdly, these toxins have recently been shown to share the same binding site and also to possess the same highly specific enzymatic activity as Shiga toxin. Finally, the *E. coli* cytotoxins are neither made by Vero cells nor are they specific for Vero cells. It would be far better and less confusing if a uniform nomenclature were adopted.

Purification and characterization

In 1983, the first reports appeared incriminating cytotoxin-producing *E. coli* of serotype 0l57:H7 in outbreaks of haemorrhagic colitis and haemolytic uraemic syndrome (HUS) (Riley *et al.*, 1983;

Johnson *et al.*, 1983; Karmali *et al.*, 1983). At the same time, Williams-Smith *et al.* (1983) established that the genes encoding cytotoxin in *E. coli* 026:H19 were phage-mediated. They found two distinct phage types, which were designated H19A and H19B. O'Brien *et al.* (1984a) then reported that *E. coli* K12 could acquire the ability to produce SLT after lysogenization by either of two different bacteriophages from *E. coli* 0157:H7, strain 933, designated 933J and 933W. Phage 933J was closely related to H19B, and produced a toxin distinct from the cytotoxin encoded by 933W, although both were toxic to the same cell lines, caused paralysis and death in mice, and a positive ileal loop response in rabbits (Strockbine *et al.*, 1986). The cytotoxin associated with 933J was predominantly cell-associated, and because it could be neutralized with antiserum raised against Shiga toxin from *S. dysenteriae* type 1, it was termed Shiga-like toxin-I or SLT-I. In contrast, the cytotoxin from 933W was less cell-associated than SLT-I, was not neutralized by polyclonal antiserum to Shiga toxin, and was designated SLT-II (Strockbine *et al.*, 1988).

A third major group of SLTs has since been isolated from *E. coli* strains associated with oedema disease of swine (Marques *et al.*, 1987), although more recently similar toxins have been associated with human disease as well (Oku *et al.*, 1989; Gannon *et al.*, 1990). This group of SLTs share the same biological properties as SLT-I and -II, are much more cytotoxic to Vero cells than HeLa cells, and are neutralizable with antiserum to SLT-II but not SLT-I. For these reasons, toxins falling into this group have been called SLT-II variants (SLT−IIv), or occasionally SLT-IIe, for oedema disease-associated toxins.

The close relationship of SLT-I to Shiga toxin has been known since 1983. O'Brien and LaVeck (1983) purified SLT-I from *E. coli* H30 using a multistep procedure and found the mobilities of the two subunits to be identical with Shiga toxin from *S. dysenteriae* type I on SDS–PAGE, with an A subunit in the 31.5 kDa range and a B subunit in the 4–15 kDa range. the pI of the two toxins was also similar (7.03 ± 0.02).

In contrast, Yutsudo *et al.* (1987) purified a protein from *E. coli* 0157:H7, also using a multistep procedure, and found it to consist of A and B subunits somewhat larger than Shiga toxin (35 kDa and 10.7 kDa, respectively). This protein was isoelectric at pH 4.1 and was cytotoxic to Vero cells at a similar concentration as Shiga toxin. However, as it was not neutralized by antiserum to Shiga toxin, it fulfilled the criteria for SLT-II. A more complex purification scheme utilizing monoclonal antibody affinity chromatography has also been described for SLT-II (Downes et al., 1988). Toxin purified by this method consisted of an A subunit (32 kDa) and multiple copies of a B subunit (10.2 kDa), with an acidic pI (5.2) compared to SLT-I. The possibility that SLT-II might be a family of several related proteins was raised when Head et al. (1988) purified a cytotoxin from E. coli strain 32511, which behaved immunologically like SLT-II but had a pI of 6.5.

Head et al. (1988) also demonstrated an increase in toxin yield following the addition of mitomycin C to the culture medium. This induction of toxin expression by mitomycin C is a real phenomenon (Table 1), and has been utilized in conjunction with P1 glycoprotein affinity chromatography to purify large amounts of SLT-I and -II from E. coli C600(933J/W) (Fig. 6) (Acheson et al., 1989; Donohue-Rolfe et al., 1989). Like the SLT-II described by Downes et al. (1988), the 933W toxin had a pI of 5.2, and was composed of one A and multiple B subunits. Preliminary cross-linking data, using dimethylpimelimidate, suggest that the stoichiometry of SLT-II holotoxin is one A subunit and five B subunits (Acheson et al., 1989), and behaves similar to Shiga toxin under the same conditions (Donohue-Rolfe et al., 1984).

This same method has also been used to purify SLT-I B subunit from an E. coli containing a plasmid with a B subunit gene insert of SLT-I

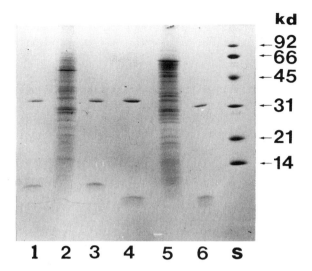

Figure 6. Purification of SLTs by P1 glycoprotein receptor analogue affinity chromatography. Toxins were eluted from the column in 4.5 M MgCl$_2$, and subjected to SDS–PAGE. Gels were stained with Coomassie brilliant blue. Lane 1, SLT-IIv (human); lane 2, crude cell lysate of SLT-IIv (human); lane 3, SLT-II; lane 4, SLT-IIv (porcine); lane 5, crude cell lysate of SLT-IIv (porcine); lane 6, SLT-I; lane S, molecular weight standards.

under the control of the trc promoter. When cultures are induced by addition of isopropyl-β-D-thiogalactopyranoside, the B subunit is secreted as a multimer in the absence of an A subunit. It is functionally identical to SLT-I (and Shiga toxin) B subunit from parent strains (Calderwood et al., 1990). Milligram amounts have been purified by binding to, and elution from, P1gp affinity columns as reported for the holotoxins (Donohue-Rolfe et al., 1989). This preparation has been used for

Table I. Effect of mitomycin C on SLT-I and -II production in recA$^+$ and recA$^-$ E. coli

Strain	Toxin	recA background	Toxin yield (ng/ml)	
			—	mitomycin C +
C600 (933W)	SLT-II	+	52	8887
HB101 (933W)	SLT-II	−	<2	<2
C600 (H19B)	SLT-I	+	1783	4135
HB101 (H19B)	SLT-I	−	1905	636

crystallization and structural analysis by X-ray crystallography as noted above (Hart *et al.*, 1991).

A unique observation by Padhye *et al.* (1986), as yet to be confirmed, suggests that additional *E. coli* cytotoxins exist. These workers isolated a cytotoxin from *E. coli* 0157:H7 strain 932 which was not neutralizable with antisera to Shiga toxin, had a pI of 5.2, and thus resembled SLT-II. However, it appeared to lack A and B subunits and migrated instead as a single 64-kDa band when subjected to SDS–PAGE. Studies of its biological effects in mice suggested that this toxin may be responsible for colonic mucosal haemorrhage and lymphoid tissue lesions (Padhye *et al.*, 1987).

MacLeod and Gyles (1989) have attempted to optimize conditions for SLT-IIv production from an oedema disease strain, *E. coli* 412, and a strain containing the cloned genes for SLT-IIv, *E. coli* TB1[pCG6]. This toxin was predominantly cell-associated, and maximum yield was obtained with use of syncase broth, incubation at 37°C for 24 h with shaking, and initial adjustment of the medium pH to 6.5. Strain TB1pCG6, containing the cloned SLT-IIv gene, released toxin mainly into the culture medium when grown in the presence of mitomycin C. This is interesting, since mitomycin C is believed to increase toxin by inducing the phage expressing SLT genes, and TB1pCG6 does not possess a phage toxin gene. The mechanism for toxin induction by mitomycin C in this strain remains uncertain.

Other workers have identified SLT-IIvs (defined by greater cytotoxic activity against Vero compared with HeLa cells) from humans. Oku *et al.* (1989) purified toxin from a strain isolated from a patient with HUS. This consisted of A and B subunits and had a pI of 6.1. Gannon *et al.* (1990) also described a variant of SLT-IIv, from a human diarrhoeal isolate of *E. coli* H.1.8 (0128:B12). This toxin, while more cytotoxic to Vero than HeLa cells, was more heat stable than the SLT-IIv from *E. coli* strain 412 studied by MacLeod and Gyles (1989).

Recent data in our laboratory demonstrate that the mechanism whereby mitomycin C causes an increase in SLT-I and -II production is via induction of bacteriophage, resulting in an increase in the copy number of toxin genes. Thus, we have found under a variety of conditions, that whenever mitomycin C induces toxin there is a concomitant increase in the number of phage particles in the culture. Another known inducer of phage replication, ultraviolet irradiation, also increases both phage and toxin production in irradiated cultures of C600(933W). In contrast, mitomycin C has no effect on toxin production from *S. dysenteriae* type I, strain 60R, in which the toxin gene is not phage-associated (Acheson *et al.*, 1990a). Additional proof that mitomycin C induces toxin via its effect on phage replication has been obtained by introducing the transforming phage in *E. coli* strains with a recA⁻ background. Phage induction is known to depend on the RecA gene product, which also functions in DNA repair responses. DNA damage, whether due to UV energy or alkylating agents such as mitomycin C, activates the RecA protease, which cleaves the LexA gene product. LexA is a repressor of several unlinked genes that encode DNA repair proteins; when LexA is cleaved, these so-called damage-inducible (din) genes are de-repressed. RecA protease also cleaves phage repressor proteins, thus leading to the transition from the lysogenic to the lytic phase. If the same control mechanism is applicable to SLT-transforming phage, mitomycin C should be an ineffective inducer if the RecA system is knocked out. This is precisely what has been found in a recA⁻ *E. coli* containing the SLT-transforming phage (Table 1) (A. Donohue-Rolfe, I. Muehldorfer and G.T. Keusch, unpublished).

Mechanism of action

Igarashi *et al.* (1987) found that SLT-I from *E. coli* 0157:H7 inactivated the 60S ribosomal subunits of rabbit reticulocytes and blocked elongation-factor-1-dependent binding of aminoacyl-tRNA to ribosomes. This effect is due to the toxin A subunit, an enzyme catalysing the cleavage of the *N*-glycosidic bond in adenosine 4324 of the 28S rRNA of the 60S ribosomal subunit (Endo *et al.*, 1988). This enzymatic specificity is shared with Shiga toxin and the toxic plant lectin, ricin (Endo and Tsurugi, 1987). When Hovde *et al.* (1988) substituted glutamic acid 167 in SLT-I with aspartic acid 167, a conservative substitution, there was still a marked effect on the functional activity of the A subunit. This finding is consistent with the report of Jackson *et al.* (1990a) who used site-directed mutagenesis of the A subunit of SLT-II to replace glutamic 166 with aspartic acid. Again, the substitution resulted in a 100-fold reduction in the toxin's ability to inhibit protein synthesis.

The A subunits of SLT-I and -II are composed of two peptides, A1 and A2, which are derived by proteolytic cleavage (nicking) and reduction of the A subunit, as with Shiga toxin (Brown et al., 1980; Olsnes et al., 1981). Nicking the SLT-I A subunit has no effect on toxin heat stability or binding to, and cytotoxic activity on, Vero cells (Kongmuang et al., 1988).

Genetics

The structural genes for SLT-I and -II are encoded on the bacteriophage. In contrast, the genes for SLT-IIv are not (Marques et al., 1987). Jackson et al. (1987a) cloned two SLT-I genes from 933J bacteriophage, which were designated slt-IA and slt-IB. These genes were orientated on a single transcriptional unit with the A subunit gene preceding the B subunit gene. Putative ribosome binding sites were found 5′ to both structural genes. Translation of the SLT-I nucleotide sequence revealed that both A and B subunits were synthesized with signal peptides. The calculated molecular weight of the A subunit was 32 211 Da and the B subunit 7 690 Da. When compared with Shiga toxin three nucleotide changes in three separate codons were detected in the A subunit. This resulted in only one amino acid alteration in SLT-I (threonine 45 to serine 45). The amino acid sequences of the B subunits were identical. Signal peptides, iron regulation operator sequences and ribosome-binding sites were also identical between Shiga toxin and SLT-I (Strockbine et al., 1988).

The genes for SLT-II were cloned from E. coli 933W and compared with SLT-I by Newland et al. (1987). The genes for SLT-II were organized in a similar way to the SLT-I gene, with the coding sequences for the A subunit adjacent to that of the B subunit. Comparison of the nucleotide sequence of the A and B subunits of SLT-I with SLT-II showed 57% and 60% homology respectively, with 55% and 57% amino acid homology (Jackson et al., 1987b). In view of this degree of homology it is surprising that these two molecules are so consistently reported to be immunologically distinct when reacted with polyclonal antisera. Donohue-Rolfe et al. (1989a), however, succeeded in produc-

ing and isolating monoclonal antibodies to SLT-II which are cross-reactive with Shiga toxin (and SLT-I). One of the four monoclonal antibodies obtained was directed against the toxin B subunits and was capable of neutralizing both toxins in cell culture cytotoxicity assays.

The SLT-IIv of porcine origin has been cloned, sequenced and compared with SLT-II. There is 94% nucleotide sequence homology between the A subunits (93% deduced amino acid sequence homology) and 79% nucleotide sequence homology between the B subunits (84% deduced amino acid sequence homology). The degree of homology between SLT-IIv and SLT-I was 55–60%, which is similar to the degree of homology between SLT-I and SLT-II (Weinstein et al., 1988b). Comparison of the nucleotide sequence between one SLT-IIvp and one SLT-IIvh toxin, revealed 98% sequence homology in the B subunit gene and 70.6% homology in the A subunit gene (Gannon et al., 1990).

Toxin receptors

One of the main differentiating features of SLT-II and -IIv is their differential cytotoxic activities on HeLa and Vero cells. The latter is substantially more active on Vero cells, reportedly due to differences in receptor specificity of the two toxins mediated by important differences in the B subunits (Weinstein et al., 1989). DeGrandis et al. (1989) showed that SLT-IIv from a pig isolate (SLT-vp) bound less well to Gb_3 separated on TLC plates compared with SLT-I or -II. Instead, SLT-IIvp bound preferentially to globotetraosylceramide (Gb_4), another neutral glycolipid which has an internal Gal-α1→4-Gal disaccharide. Samuel et al. (1990) compared binding specificity of SLT-IIvh from a human isolate and SLT-IIvp with SLT-II. By preparing E. coli strains carrying various fusions between SLT-II and -IIv genes for the A and the B subunits, hybrid SLT-II/IIv toxins were obtained. These studies confirmed that the B subunits conferred cell specificity for the cytotoxic effects. Unlike SLT-II and SLT-IIvh, both of which bound to terminal Galα1-4 Gal moieties in galabiosylceramide and Gb_3, SLT-IIvp preferentially bound to internal Galα1-4 Gal sequences in globotetraosylceramide and galactosylglobotetraosylceramide. These specificities were important in *vivo* as well, as there was a 400-fold difference in mouse lethality between SLT-II and SLT-IIvp.

Detection of Shiga toxin and SLTs

Several methods have been reported for the detection of Shiga toxin and SLT-I and -II. A sensitive and specific ELISA for these toxins has been described using plates coated with P1 glycoprotein-containing hydatid cyst fluid to capture toxin antigen. Bound antigen could then be detected by an appropriate polyclonal rabbit antiserum (Acheson *et al.*, 1990b). An ELISA using purified Gb$_3$ as the capture molecule has also been reported for detection of SLT-I (Ashkenazi and Cleary, 1989, 1990). Other ELISAs have been described by Downes *et al.* (1989) and Basta *et al.* (1989).

Gene probes for SLT-I and -II

With the nucleotide sequence of SLT-I and -II available, several groups have investigated the use of oligonucleotide probes for the identification of SLT-producing *E. coli* (Karch and Meyer, 1989a; Brown *et al.*, 1989; Meyer *et al.*, 1989). The polymerase chain reaction (PCR) has also been used for the identification of SLT genes (Karch and Meyer, 1989b; Pollard *et al.*, 1990; Bettelheim *et al.*, 1990). PCR has been used for the detection of SLT-producing *E. coli* in food and faecal specimens. This technique may prove to be useful for epidemiological studies and in disease prevention when used to detect contaminated food (Samadpour *et al.*, 1990). Probes have already been used to study the epidemiology of toxin-producing strains. For example, Suthienkul *et al.* (1990) have found that cattle in Thailand are commonly infected with SLT-producing *E. coli* and thus may be an important source of human infection. Similar observations in cattle have recently been published from Sheffield, England (Chapman *et al.*, 1989) and Hamburg, Germany (Montenegro *et al.*, 1990). Smith *et al.* (1988) reported that SLT-I and -II-producing strains of *E. coli* were present in both pigs and cattle with enteric disease.

Clinical associations of Shiga and Shiga-like toxins

The evidence implicating Shiga toxin in pathogenesis of shigellosis is not conclusive, and is based on animal models and *in vitro* studies in cell culture. None of the animal models used, including oral infection of primates, truly mimics human infection. Although Shiga toxin causes fluid secretion in rabbits (Keusch *et al.*, 1972a), results in inflammatory enteritis in the rabbit model (Keusch *et al.*, 1972b) and is cytotoxic to human colonic epithelial cells (Moyer *et al.*, 1987), and thus can cause manifestations of clinical shigellosis, the interpretation is complicated because *Shigella* are invasive and multiply within epithelial cells (Sansonetti, 1991). Therefore, at least some of the pathophysiology may be due to invasion *per se*, and the inflammatory response it induces. Fontaine *et al.* (1988) created a Shiga toxin deletion mutant strain of *S. dysenteriae* 1, and have compared its clinical effects in the monkey with the wild-type strain. The tox$^-$ mutant caused disease, but it was much less haemorrhagic, suggesting that the toxin plays a role in the inflammatory dysenteric phase of shigellosis but is not a necessary factor for the initiation of clinical disease. The potential role of Shiga toxin in HUS will be discussed below.

Evidence that the Shiga-like toxins are involved in disease pathogenesis is also predominantly circumstantial. There is, however, strong epidemiological evidence for association of SLT-I and/or -II-producing *E. coli* strains with outbreaks of haemorrhagic colitis, haemolytic uremic syndrome (HUS) and thrombotic thrombocytopaenic purpura (TTP) (Riley *et al.*, 1983; Bopp *et al.*, 1987; Carter *et al.*, 1987; Johnson *et al.*, 1983; Remis *et al.*, 1984). These organisms have also been associated with pseudomembranous colitis, acute exacerbations of ulcerative colitis and enterocolitis (Hunt *et al.*, 1989; Ljungh *et al.*, 1988; Von Wulfen *et al.*, 1989), although a causal relationship is not established. During a two-year prospective study between 1984 and 1988 in Alberta, Canada, SLT-producing *E. coli* were among the most common causes of bacterial diarrhoea, and although due predominantly to 0157:H7, other *E. coli* serotypes were also incriminated (Pai *et al.*, 1986b). Despite the large number of *E. coli* serotypes able to produce SLT-I and/or -II (Karmali, 1989), in a recent study by Smith *et al.* (1990) in which 449 strains of enteropathogenic *E. coli* were examined for SLT-I or -II, only 16 were found to be toxin-positive, and 15 of those were of serogroup 016 or 0128.

Shiga-like toxins may also be virulence factors in organisms other than *E. coli*. O'Brien *et al.* (1984b) have shown that human isolates of *Vibrio parahemolyticus* are able to produce cytotoxins which are neutralized by antiserum to Shiga toxin, but not by antiserum to cholera toxin. Moore *et al.* (1988) reported that certain *Campylobacter jejuni* and *C. coli* strains elaborated a low-level cytotoxin which was neutralizable with antisera to Shiga toxin. Despite this immunological cross-reactivity there was no significant nucleotide sequence homology between the SLT-I genes from *E. coli* phage 933J and the 'SLT genes' of *Campylobacter*.

Unfortunately, there is no totally satisfactory animal model of haemorrhagic colitis or HUS available, although many attempts to develop one have been made. Keenan *et al.* (1986) compared the morphological effects of Shiga toxin and SLT from *E. coli* 0157:H7 on rabbit intestine and found the effects of the two toxins to be very similar. Both toxins caused a dose-dependent effect, with villus blunting and a decrease in the villus/crypt ratio. Barrett *et al.* (1989a) have induced diarrhoea and haemorrhagic gut lesions by placing an osmotic pump that allowed continuous infusion of SLT-II into the peritoneal space of rabbits, although none of the animals developed renal failure. Further experiments showed that addition of lipopolysaccharide (LPS) to SLT-II had a significant synergistic effect on the lethal activity of SLT-II; however, pretreatment of rabbits with LPS protected them from the lethal effect of SLT-II challenge (Barrett *et al.*, 1989b). The authors concluded that LPS acts in a macrophage-mediated manner to enhance SLT-II-induced injury.

Gnotobiotic piglets infected with SLT-II-producing *E. coli* 0157:H7 develop diarrhoea and brain lesions similar to those found in oedema disease of swine. This phenomenon appears to be a direct consequence of the toxin, since toxin-negative strains fail to cause the cerebral lesions (Francis *et al.*, 1989).

Pai *et al.* (1986a) infected rabbits with live *E. coli* 0157:H7 and caused a bloody diarrhoea. They also found that the presence of free toxin in the mid-colon correlated with diarrhoea. Organisms were not invasive and challenge with toxin alone led to similar manifestations as in animals infected with whole living organisms. This is presumably due an effect on the colon, since the jejunum of rabbits of this age does not express Gb_3 and is not affected by Shiga toxin (Mobassaleh *et al.*, 1988). It is also possible that the effect observed by Pai *et al.* was mediated by the SLT-II produced by the 0157:H7 challenge organism. Wadolkowski *et al.* (1990) have recently described a streptomycin-treated mouse model susceptible to infection with *E. coli* 0157:H7. Infected mice became colonized and developed acute necrosis of the renal proximal convoluted tubules. This was specifically associated with production of SLT-II, although there was little pathological similarity between the tubular lesion found in the mice and the glomerular lesions in humans with HUS.

Shiga and Shiga-like toxins and haemolytic uraemic syndrome

Karmali *et al.* (1985) first noted a link between SLT-producing *E. coli* and HUS, a finding which has subsequently been corroborated by many investigators (Karmali, 1989). Recent studies from Britain and the United States, in which the association of HUS and SLT-producing *E. coli* was examined, have shown that the majority of *E. coli* isolates produced SLT-II either alone or together with SLT-I. In England, the 0157:H7 serotype was most common, although 10 other SLT-producing serotypes were also identified (Milford and Taylor, 1990; Milford *et al.*, 1990). The same group also reported the isolation of SLT-producing *E. coli* from three control groups without HUS. The percentage of patients in whom SLT-producing *E. coli* were found was as follows: 8% of patients with bloody diarrhoea, 6% of patients with non-bloody diarrhoea and 4% of healthy controls (Kleanthous *et al.*, 1990). In the states of Minnesota and Washington in the US, SLT-II-producing *E. coli* 0157:H7 are the most common isolates from HUS patients (Ostroff *et al.*, 1989; Martin *et al.*, 1990). In contrast, in Buenos Aires, Argentina, which probably has the highest reported incidence of HUS in the world, non-0157:H7 SLT-producing *E. coli* are the principal causative organisms (Lopez *et al.*, 1989). This last study, and the data of Milford *et al.* (1989), underscore the importance of looking for SLT-producing strains in HUS patients and using a technique which is not specific for serotype 0157:H7.

An intriguing situation which adds to the epidemiological base suggesting the possible role of SLTs in HUS is that of cancer-associated HUS. This entity has been linked to the use of the drug mitomycin C as a chemotherapeutic agent (Lesesne *et al.*, 1989). Since mitomycin C dramatically increases the levels of SLT-I and -II produced *in vitro* (Acheson *et al.*, 1989), it is possible that therapeutic levels of mitomycin C during cancer treatment may also exert a similar inductive effect on SLT-producing *E. coli* that may be present in the gut flora (Acheson and Donohue-Rolfe, 1989). If so, then *in situ* induction of SLT toxin production may initiate or contribute to development of HUS, whereas in the absence of the inducing agent these organisms do not produce sufficient toxin to be pathogenic.

The mechanism by which SLT may be involved in the pathogenesis of HUS is at present unknown, although accumulating evidence suggests that these toxins may act primarily on endothelial cells (Obrig *et al.*, 1988; Kavi and Wise, 1989; Cleary and Lopez, 1989; Milford and Taylor, 1990). One study reports that SLT decreases prostacyclin synthesis by endothelial cells (Karch *et al.*, 1988).

Summary

It is now apparent that Shiga toxin is the prototype of a much larger family of related toxins, which are best called Shiga-like toxins because of their structural, biochemical and physiological homology with the previously described Shiga toxin. SLT-I and -II, as initially described, are only two members of a much larger family of Shiga-like toxins, which have been implicated in both human and porcine disease. These toxins are specific inhibitors of protein synthesis, acting enzymatically to inactivate the 60S ribosomal subunit. They all bind to globo series neutral glycolipids, which are present in the plasma membrane of some, but not all cells. The distribution of receptor glycolipids determines cell susceptibility to these toxins. Shiga toxin is specifically produced by *Shigella dysenteriae* type 1. SLTs are clearly produced by *E. coli* present in various animal reservoirs all over the world, and are being detected in an ever-increasing number of *E. coli* serotypes and in other bacterial species as well. Although their precise role in disease pathogenesis has yet to be elucidated there is accumulating evidence that these toxins are important in the pathgenesis of both diarrhoeal disease and haemolytic uraemic syndrome.

Acknowledgements

D.W.K. Acheson gratefully acknowledges the financial support of the Wellcome Trust, London, UK. This work was supported by Grants AI-16242 and AI-20235 from the National Institute of Allergy and Infectious Disease and Grant DK-34928 from the National Institute of Diabetes and Digestive and Kidney Diseases, National Institutes of Health, Bethesda, MD, USA, and by a Health Sciences for the Tropics Partnership in Research and Training Grant from the Rockefeller Foundation, New York, USA.

References

Acheson, D.W.K. and Donohue-Rolfe, A. (1989). Cancer associated hemolytic uremic syndrome. A possible role of mitomycin C in relation to Shiga-like toxins. *J. Clin. Oncol.* 7, 1943.

Acheson, D.W.K., Kane, A.V., Keusch, G.T. and Donohue-Rolfe, A. (1989). High yield purification and subunit characterization of Shiga-like toxin II. *Abstracts of the Interscience Conference on Antimicrobial Agents and Chemotherapy*, American Society for Microbiology, Washington, DC, Abstract 901, p. 242.

Acheson, D.W.K., Muhldorfer, I., Kane, A., Keusch, G.T. and Donohue-Rolfe, A. (1990a). Bacteriophage induction as a cause of increased Shiga-like toxin synthesis in *E. coli. Abstr. Am. Soc. Microbiol.*

Acheson, D.W.K., Keusch, G.T., Lightowers, M. and Donohue-Rolfe, A. (1990b). Enzyme linked immunosorbent assay for Shiga toxin and Shiga-like toxin II using P1 glycoprotein from hydatid cysts. *J. Infect. Dis.* 161, 134–7.

Ashkenazi, S. and Cleary, T.G. (1989). Rapid method to detect Shiga toxin and Shiga-like toxin I based on binding to globotriaosylceramide (GB3), their natural receptor. *J. Clin. Microbiol.* 27, 1145–50.

Ashkenazi, S. and Cleary, T.G. (1990). A method for

detecting Shiga toxin and Shiga-like toxin I in pure and mixed culture. *J. Med. Microbiol.* **32**, 255–61.

Barrett, T.J., Potter, M.E. and Wachsmuth, I.K. (1989a). Continuous peritoneal infusion of Shiga-like toxin II (SLT II) as a model for SLT II-induced diseases. *J. Infect. Dis.* **159**, 774–7.

Barrett, T.J., Potter, M.E. and Wachsmuth, I.K. (1989b). Bacterial endotoxin both enhances and inhibits the toxicity of Shiga-like toxin II in rabbits and mice. *Infect. Immun.* **57**, 3434–7.

Basta, M., Karmali, M. and Lingwood, C. (1989). Sensitive receptor-specified enzyme-linked immunosorbent assay for *Escherichia coli* verocytotoxin. *J. Clin. Microbiol.* **27**, 1617–22.

Bettelheim, K.A., Brown, J.E., Lolekha, S. and Echeverria, P. (1990). Serotypes of *Escherichia coli* that hybridized with DNA probes for genes encoding Shiga-like toxin I, Shiga-like toxin II, and serogroup 0157 enterohemorrhagic *E. coli* fimbriae isolated from adults with diarrhoea in Thailand. *J. Clin. Microbiol.* **28**, 293–5.

Boivin, A. and Mesrobeanu, L. (1937). Recherches sur les toxines des bacilles dysentériques. Sur l'identité entre la toxine thermolabile de Shiga et l'exotoxine présénte dans les filtratés des cultures sur bouillon de la même bactérie. *C. r. Soc. Biol.* **126**, 323–5.

Bopp, C.A., Greene, K.D., Downes, F., Sowers, E.G., Wells, J.G. and Wachsmuth, I.K. (1987). Unusual verotoxin-producing *Escherichia coli* associated with hemorrhagic colitis. *J. Clin. Microbiol.* **25**, 1485–9.

Bridgewater, F.A.J., Morgan, R.S., Rowson, K.E.K. and Payling-Wright, G. (1955). The neurotoxin of *Shigella shigae*. Morphological and functional lesions produced in the central nervous system of rabbits. *Br. J. Exp. Pathol.* **36**, 447–53.

Brown, J.E., Ussery, M.A., Leppla, S.H. and Rothman, S.W. (1980). Inhibition of protein synthesis by Shiga toxin. Activation of the toxin and inhibition of peptide elongation. *FEBS Lett.* **117**, 84–8.

Brown, J.E., Sethabutr, O., Jackson, M.P., Lolekha, S. and Echeverria, P. (1989). Hybridization of *Escherichia coli* producing Shiga-like toxin I, Shiga-like toxin II, and a variant of Shiga-like toxin II with synthetic oligonucleotide probes. *Infect. Immun.* **57**, 2811–14.

Brunton, J.L. (1990). The Shiga family: molecular nature and possible role in disease. In: *Molecular Basis of Bacterial Pathogenesis* (eds B.H., Iglewski and Y.L., Clark), pp. 377–98. Academic Press, New York.

Calderwood, S.B. and Mekalanos, J.J. (1987). Iron regulation of Shiga-like toxin expression in *Escherichia coli* is mediated by the *fur* locus. *J. Bacteriol.* **169**, 4759–64.

Calderwood, S.B., Auclair, F., Donohue-Rolfe, A., Keusch, G.T. and Mekalanos, J.J. (1987). Nucleotide sequence of the Shiga-like toxin genes of *Escherichia coli*. *Proc. Natl. Acad. Sci. USA* **84**, 4364–8.

Calderwood, S.B., Acheson, D.W.K., Goldberg, M.B., Boyko, S.A. and Donohue-Rolfe, A. (1990). A system for production and rapid purification of large amounts of Shiga toxin/Shiga-like toxin I B subunit. *Infect. Immun.* **58**, 2977–82.

Carter, A.O., Borczyk, A.A., Carlson, J.A.K., Harvey, B., Hockin, J.C., Karmali, M.A., Krishnan, C., Kron, D.A. and Lior, H. (1987). A severe outbreak of *Escherichia coli* 0157:H7-associated hemorrhagic colitis in a nursing home. *New Engl. J. Med.* **317**, 1496–1500.

Chapman, P.A., Wright, D.J. and Norman, P. (1989). Verotoxin-producing *Escherichia coli* infections in Sheffield: cattle as a possible source. *Epidem. Inf.* **102**, 439–45.

Cleary, T.G. and Lopez, E.L. (1989). The Shiga-like toxin-producing *Escherichia coli* and hemolytic uremic syndrome. *Pediatr. Infect. Dis. J.* **8**, 720–4.

Cohen, A., Hannigan, G.E., Williams, B.R.G. and Lingwood, C. A. (1987). Roles of globotriosyl- and galabiosylceramide in Verotoxin binding and high affinity interferon receptor. *J. Biol. Chem.* **262**, 17088–91.

Conradi, H. (1903). Über lösliche, durch aseptische Autolyse erhatlene Giftstoffe von Ruhr- und typhus-bazillen. *Dt. med. Wschr.* **20**, 26–8.

DeGrandis, S., Law, H., Brunton, J., Gyles, C. and Lingwood, C.A. (1989). Globotetraosylceramide is recognized by the pig edema disease toxin. *J. Biol. Chem.* **264**, 12520–5.

Donohue-Rolfe, A., Keusch, G.T., Edson, C., Thorley-Lawson, D. and Jacewicz, M. (1984). Pathogenesis of *Shigella* diarrhea. IX. Simplified high yield purification of Shigella toxin and characterization of subunit composition and function by the use of subunit-specific monoclonal and polyclonal antibodies. *J. Exp. Med.* **160**, 1767–81.

Donohue-Rolfe, A., Acheson, D.W.K., Kane, A.V. and Keusch, G.T. (1989a). Purification of Shiga toxin and Shiga-like toxins I and II by receptor analogue affinity chromatography with immunobilized P1 glycoprotein and the production of cross-reactive monoclonal antibodies. *Infect. Immun.* **57**, 3888–93.

Donohue-Rolfe, A., Jacewicz, M. and Keusch, G.T. (1989b). Isolation and characterization of functional Shiga toxin subunits and renatured holotoxin. *Molec. Microbiol.* **3**, 1231–6.

Downes, F., Barrett, T.J., Green, J.H., Aloisio, C.H., Spika, J.S., Strockbine, N.A. and Wachsmuth, I.K. (1988). Affinity purification and characterization of Shiga-like toxin II and production of toxin-specific monoclonal antibodies. *Infect. Immun.* **56**, 1926–33.

Downes, F., Green, J.H., Greene, K., Strockbine, N., Wells, J.G. and Wachsmuth, I.K. (1989). Development and evaluation of enzyme-linked immunosorbent assays for detection of Shiga-like toxin I and Shiga-like toxin II. *J. Clin. Microbiol.* **27**, 1292–7.

Endo, Y. and Tsurugi, K. (1987). RNA N-glycosidase activity of ricin A-chain. Mechanism of action of the toxic lectin ricin on eukaryotic ribosomes. *J. Biol. Chem.* **262**, 8128–230.

Endo, Y., Tsurugi, K., Yutsudo, T., Takeda, Y.,

Ogasawara, T. and Igarashi, E. (1988). Site of action of a Vero toxin (VT2) from *Escherichia coli* 0157:H7 and of Shiga toxin on eukaryotic ribosomes. *Eur. J. Biochem.* **171**, 45–50.

Engley, F.B. Jr. (1952). The neurotoxin of *Shigella dysenteriae* (SHIGA). *Bact. Rev.* **16**, 407–16.

Flexner, S. (1900). On the etiology of tropical dysentery. *Bull. Johns Hopkins Hospital* **11**, 231–42.

Fontaine, A., Arondel, J. and Sansonetti, P.J. (1988). Role of Shiga toxin in pathogenesis of bacillary dysentery studied by using a tox- mutant of *Shigella dysenteriae* 1. *Infect. Immun.* **56**, 3099–109.

Francis, D.H., Moxley, R.A. and Andracs, C.Y. (1989). Edema disease-like brain lesions in gnotobiotic piglets infected with *Escherichia coli* serotype 0157:H7. *Infect. Immun.* **57**, 1339–42.

Fuchs, G., Mobassaleh, M., Donohue-Rolfe, A., Montgomery, R.K., Grand, R.J. and Keusch, G.T. (1986). Pathogenesis of *Shigella* diarrhea: rabbit intestinal cell microvillus membrane binding site for *Shigella* toxin. *Infect. Immun.* **53**, 372–7.

Gangarosa, E.J., Perera, D.R., Mata, L.J., Mendizábal-Morris, C. Guzmán, G. and Reller, L.B. (1970). Epidemic Shiga bacillus dysentery in Central America. II. Epidemiologic studies in 1969. *J. Infect. Dis.* **122**, 181–90.

Gannon, V.P., Teerling, C., Masri, S.A. and Gyles, C.L. (1990). Molecular cloning and nucleotide sequence of another variant of the *Escherichia coli* Shiga-like toxin II family. *J. Gen. Microbiol.* **136**, 1125–35.

Harari, I., Donohue-Rolfe, A., Keusch, G. and Arnon, R. (1988). Synthetic peptides of Shiga toxin B subunit induce antibodies which neutralize its biological activity. *Infect. Immun.* **56**, 1618–24.

Hart, P.J., Monzingo, A.F., Donohue-Rolfe, A., Keusch, G.T., Calderwood, S.B. and Robertus, J.D. (1991). Crystallization of the B chain of Shiga-like toxin I from *Escherichia coli*. *J. Mol. Biol.* **218**, 691–4.

Head, S.C., Petric, M., Richardson, S., Roscoe, M. and Karmali, M.A. (1988). Purification and characterization of verocytotoxin 2. *FEMS Microbiol. Lett.* **51**, 211–16.

Hovde, C.J., Calderwood, S.B., Mekalanos, J.J. and Collier, R.J. (1988). Evidence that glutamic acid 167 is an active-site residue of Shiga-like toxin I. *Proc. Natl. Acad. Sci. USA* **85**, 2568–72.

Howard, J.G. (1955). Observations on the intoxication produced in mice and rabbits by the neurotoxin of *Shigella shigae*. *Br. J. Exp. Pathol.* **36**, 439–46.

Hunt, C.M., Harvey, J.A., Youngs, E.R., Irwin, S.T. and Reid, T.M. (1989). Clinical and pathological variability of infection by enterohaemorrhagic (Verocytotoxin producing) *Escherichia coli*. *J. Clin. Pathol.* **42**, 847–52.

Igarashi, K., Ogasswara, T., Ito, K., Yutsudo, T. and Takada, Y. (1987). Inhibition of elongation factor 1-dependent aminoacyl-tRNA binding to ribosomes by Shiga-like toxin I (VTI) from *Escherichia coli* 0157:H7 and by Shiga toxin. *FEMS Microbiol. Lett.* **44**, 91–4.

Jacewicz, M., Clausen, H., Nudelman, E., Donohue-Rolfe, A. and Keusch, G.T. (1986). Pathogenesis of *Shigella* diarrhea. XI. Isolation of a shigella toxin-binding glycolipid from rabbit jejunum and HeLa cells and its identification as globotriaosylceramide. *J. Exp. Med.* **163**, 1391–404.

Jacewicz, M., Feldman, H.A., Donohue-Rolfe, A., Balasubramanian, K.A. and Keusch, G.T. (1989). Pathogenesis of *Shigella* diarrhea. XIV. Analysis of Shiga toxin receptors on cloned HeLa cells. *J. Infect. Dis.* **159**, 881–9.

Jackson, M.P., Newland, J.W., Holmes, R.K. and O'Brien, A.D. (1987a). Nucleotide sequence analysis of the structural genes for Shiga-like toxin I encoded by bacteriophage 933J from *Escherichia coli*. *Microb. Pathogen.* **2**, 147–53.

Jackson, M.P., Neill, R.J., O'Brien, A.D., Holmes, R.K. and Newland, J.W. (1987b). Nucleotide sequence analysis and comparison of the structural genes for Shiga-like toxin I and Shiga-like toxin II encoded by bacteriophages from *Escherichia coli* 933. *FEMS Microbiol. Lett.* **44**, 109–14.

Jackson, M.P., Deresiewicz, R.L. and Calderwood, S.B. (1990a). Mutational analysis of the Shiga toxin and Shiga-like toxin II enzymatic subunits. *J. Bacteriol.* **172**, 3346–50.

Jackson, M.P., Wadolkowski, E.A., Weinstein, D.L., Holmes, R.K. and O'Brien, A.D. (1990b). Functional analysis of the Shiga toxin and Shiga-like toxin type II variant binding subunits by using site-directed mutagenesis. *J. Bacteriol.* **172**, 653–8.

Johnson, W.M., Lior, H. and Beznason, G.S. (1983). Cytotoxic *Escherichia coli* 0157:H7 associated with haemorrhagic colitis in Canada. *Lancet* **i**, 76.

Kandel, G., Donohue-Rolfe, A., Donowitz, M. and Keusch, G.T. (1989). Pathogenesis of Shigella diarrhea XVI. Selective targetting of Shiga toxin to villus cells of rabbit jejunum explains the effect of the toxin on intestinal electrolyte transport. *J. Clin. Invest.* **84**, 1509–17.

Karch, H. and Meyer, T. (1989a). Evaluation of oligonucleotide probes for identification of Shiga-like-toxin-producing *Escherichia coli*. *J. Clin. Microbiol.* **27**, 1180–6.

Karch, H. and Meyer, T. (1989b). Single primer pair for amplifying segments of distinct Shiga-like-toxin genes by polymerase chain reaction. *J. Clin. Microbiol.* **27**, 2751–7.

Karch, H., Bitean, M., Pietsch, R., Stenger, K., von Wulffen, H., Heesemann, J. and Dusing, R. (1988). Purified verotoxins of *Escherichia coli* 0157:H7 decrease prostacyclin synthesis by endothelial cells. *Microb. Pathogen.* **5**, 215–221.

Karmali, M.A. (1989). Infection by verocytotoxin-producing *Escherichia coli*. *Clin. Microbiol. Rev.* **2**, 15–38.

Karmali, M.A., Steele, B.T., Petric, M. and Lim, C. (1983). Sporadic cases of hemolytic-uraemic syndrome associated with faecal cytotoxin and cytotoxin-producing *Escherichia coli* in stools. *Lancet* **ii**, 619–20.

Karmali, M.A., Petric, M., Lim, C., Fleming, P.C., Arbus, G.A. and Lior, H. (1985). The association between idiopathic hemolytic uremic syndrome and infection by verotoxin producing *Escherichia coli*. *J. Infect. Dis.* **151**, 775–82.

Kavi, J. and Wise, R. (1989). Causes of the hemolytic uraemic syndrome. *Br. Med. J.* **298**, 65–6.

Keenan, K.P., Sharpnack, D.D., Collins, H., Formal, S.B. and O'Brien, A.D. (1986). Morphologic evaluation of the effects of Shiga toxin and *Escherichia coli* Shiga-like toxin on the rabbit intestine. *Am. J. Pathol.* **125**, 69–80.

Keusch, G.T. and Jacewicz, M. (1975). Pathogenesis of *Shigella* diarrhea. V. Relationship of Shiga enterotoxin, neurotoxin, and cytotoxin. *J. Infect. Dis.* **131**, S33–S39.

Keusch, G.T. and Jacewicz, M. (1977). Pathogenesis of *Shigella* diarrhea. VII. Evidence for a cell membrane toxin receptor involving β1→4 linked *N*-acetyl-D-glucosamine oligomers. *J. Exp. Med.* **146**, 535–46.

Keusch, G.T., Donohue-Rolfe, A. and Jacewicz, M. (1986a). *Shigella* toxin(s): Description and role in diarrhea and dysentery. In: *Pharmacology of Bacterial Toxins* (eds F. Dorner and J. Drews), pp. 235–70. Pergamon Press, Oxford.

Keusch, G.T., Jacewicz, M., and Donohue-Rolfe, A. (1986b). Pathogenesis of *Shigella* diarrhea: evidence for N-linked glycoprotein *Shigella* receptors and receptor modulation by β galactosidase. *J. Infect. Dis.* **153**, 238–48.

Keusch, G.T., Grady, G.F., Mata, L.J. and McIver, J. (1972a). The pathogenesis of shigella diarrhea. I. Enterotoxin production by *Shigella dysenteriae* 1. *J. Clin. Invest.* **51**, 1212–18.

Keusch, G.T., Grady, G.F., Takeuchi, A. and Sprinz, H. (1972b). The pathogenesis of *Shigella* diarrhea. II Enterotoxin induced acute enteritis in the rabbit ileum. *J. Infect. Dis.* **126**, 92–5.

Kleanthous, H., Smith, H.R., Scotland, S.M., Gross, R.J., Rowe, B., Taylor, C.M. and Milford, D.V. (1990). Hemolytic uraemic syndromes in the British Isles, 1985–8; association with verocytotoxin producing *Escherichia coli*. Part 2: microbiological aspects. *Arch. Dis. Child.* **65**, 722–7.

Kongmuang, U., Honda, T. and Miwatani, T. (1988). Effect of nicking Shiga-like toxin I of enterohaemorrhagic *Escherichia coli*. *FEMS Microbiol. Lett.* **56**, 105–8.

Konowalchuk, J., Speirs, J.I. and Stavric, S. (1977). Vero response to a cytotoxin of *Escherichia coli*. *Infect. Immun.* **18**, 775–9.

Konowalchuk, J., Dickie, N., Stavrie, S. and Speirs, J.I. (1978). Properties of an *Escherichia coli* cytotoxin. *Infect. Immun.* **20**, 575–7.

Lesesne, J.B., Rothschild, N., Erickson, B., Korec, S., Sisk, R., Keller, J., Arbus, M., Woolley, P.V., Chiazze, L., Schein, P.S. and Neefe, J.R. (1989). Cancer-associated uremic syndrome: analysis of 85 cases from a national registry. *J. Clin. Oncol.* **7**, 781–9.

Lindberg, A.A., Brown, J.E., Stromberg, N., Westling-Ryd, M., Schultz, J.E. and Karlson, K. (1987). Identification of the carbohydrate receptors for Shiga toxin produced by *Shigella dysenteriae* type 1. *J. Biol. Chem.* **262**, 1779–85.

Lingwood, C.A., Law, H., Richardson, S., Petric, M., Bruton, J.L., DeGrandis, S. and Karmali, M. (1987). Glycolipid binding of purified and recombinant *Escherichia coli* produced verotoxin *in vitro*. *J. Biol. Chem* **262**, 8834–9.

Ljungh, A., Erickson, M., Erickson, O., Henter, J.I. and Wadstrom, T. (1988). Shiga-like toxin production and connective tissue protein binding of *Escherichia coli* isolated from a patient with ulcerative colitis. *Scand. J. Infect. Dis.* **20**, 443–6.

Lopez, E.L., Diaz, M., Grinstein, S., Devoto, S., Mendilaharzu, F., Murray, B.E., Ashenazi, S., Rubeglio, E., Woloj, M., Vasquez, M., Turco, M., Pickering, L.K. and Cleary, T.G. (1989). Hemolytic uremic syndrome and diarrhea in Argentine children: The role of Shiga-like toxins. *J. Infect. Dis.* **160**, 469–75.

MacLeod, D.L. and Gyles, C.L. (1989). Effects of culture conditions on yield of Shiga-like toxin-IIv from *Escherichia coli*. *Can. J. Microbiol.* **35**, 623–9.

Marques, L.R.M., Peiris, J.S.M., Cryz, S.J. and O'Brien, A.D. (1987). *Escherichia coli* strains isolated from pigs with edema disease produce a variant of Shiga-like toxin II. *FEMS Microbiol. Lett.* **44**, 33–8.

Martin, D.L., MacDonald, K.L., White, K.E., Soler, J.T. and Osterholm, M.T. (1990). The epidemiology and clinical aspects of the hemolytic uremic syndrome in Minnesota. *New Engl. J. Med.* **323**, 1161–7.

Meyer, T., Bitean, M., Sandkamp, O. and Karob, H. (1989). Synthetic oligo-deoxyribonucleotide probes to detect verocytotoxin-producing *Escherichia coli* in diseased pigs. *FEMS Microbiol. Lett.* **57**, 247–52.

Milford, D.V. and Taylor, C.M. (1990). New insights into the hemolytic uraemic syndromes. *Arch. Dis. Child.* **65**, 713–15.

Milford, D.V., Taylor, C.M., Guttridge, B., Hall, S.M., Rowe, B. and Kleanthous, H. (1990). Hemolytic uraemic syndrome in the British Isles 1985–8; association with verocytotoxin producing *Escherichia coli*. Part 1: clinical and epidemiological aspects. *Arch. Dis. Child.* **65**, 716–21.

Mobassaleh, M., Donohue-Rolfe, A., Jacewicz, M., Grand, R.J. and Keusch, G.T. (1988). Pathogenesis of Shigella diarrhea: Evidence for a developmentally regulated glycolipid receptor for Shigella toxin involved in the fluid secretory response of rabbit small intestine. *J. Infect. Dis.* **157**, 1023–31.

Mobassaleh, M., Gross, S.K., McCluer, R.H., Donohue-Rolfe, A. and Keusch, G.T. (1989). Quantitation of the rabbit intestinal glycolipid receptor for Shiga toxin. Further evidence for the developmental regulation of globotriaosylceramide in microvillus membranes. *Gastroenterology* **97**, 384–91.

Mobassaleh, M., Gross, S.K., McCluer, R.H. and Keusch, G.T. (1991). Distribution, subcellular localization and elucidation of the developmental regulation of Shiga toxin receptor synthesis in rabbit small intestine. *Abstracts, Annual Meeting, American Gastroenterological Association*. AGA, Washington, DC.

Montenegro, M.A., Bulte, M., Trumpf, T., Aleksic, S., Reuter, G., Bulling, E. and Helmuth, R. (1990). Detection and characterization of fecal verotoxin-producing *Escherichia coli* from healthy cattle. *J. Clin. Microbiol.* **28**, 1417–21.

Moore, M.A., Blaser, M.J., Perez-Perez, G.I. and O'Brien, A.D. (1988). Production of a Shiga-like cytotoxin by *Campylobacter*. *Microb. Pathogen.* **4**, 455–62.

Moyer, M.P., Dixon, P.S., Rothman, S.W. and Brown, J.E. (1987). Cytotoxicity of Shiga toxin for primary cultures of human colonic and ileal epithelial cells. *Infect. Immun.* **55**, 1533–5.

Neisser, M. and Shiga, K. (1903). Ueber freie receptoren von typhus- und dysenterie-bazillen und über das dysenterie-toxin. *Dt. med. Wschr.* **29**, 61–2.

Newland, J.W., Strockbine, N.A. and Neill, R.J. (1987). Cloning of genes for production of *Escherichia coli* Shiga-like toxin type II. *Infect. Immun.* **5**, 2675–80.

O'Brien, A.D. and Holmes, R.K. (1987). Shiga and Shiga-like toxins. *Microbiol. Rev.* **51**, 206–20.

O'Brien, A.D. and LaVeck, G.D. (1983). Purification and characterization of Shigella dysenteriae I-like toxin produced by *Escherichia coli*. *Infect. Immun.* **40**, 675–83.

O'Brien, A.D., LaVeck, G.D., Griffin, D.E. and Thompson, M.R. (1980). Characterization of *Shigella dysenteriae* 1 (Shiga) toxin purified by anti-Shiga toxin affinity chromatography. *Infect. Immun.* **30**, 170–9.

O'Brien, A.D., LaVeck, G.D., Thompson, M.R. and Formal, S.B. (1982). Production of *Shigella dysenteriae* type I-like cytotoxin by *Escherichia coli*. *J. Infect. Dis.* **146**, 763–89.

O'Brien, A.D., Newland, J.W., Miller, S.F., Holmes, R.K., Williams-Smith, H. and Formal, S.B. (1984a). Shiga-like toxin-converting phages from *Escherichia coli* strains that cause hemorrhagic colitis or infantile diarrhea. *Science* **226**, 694–6.

O'Brien, A.D., Chen, M.E. and Holmes, R.K. (1984b). Environmental and human isolates of *Vibrio cholerae* and *Vibrio parhaemolyticus* produce *Shigella dysenteriae* I (Shiga)-like cytotoxin. *Lancet* i, 77–8.

Obrig, T.G., Del Vecchio, P.J., Brown, J.E., Moran, T.P., Rowland, B.M., Judge, T.K. and Rothman, S.W. (1988). Direct cytotoxic action of Shiga toxin on human vascular endothelial cells. *Infect. Immun.* **56**, 2373–8.

Oku, Y., Yutsudo, T., Hirayama, T., O'Brien, A.D. and Takeda, Y. (1989). Purification and some properties of a vero toxin from a human strain of *Escherichia coli* that is immunologically related to Shiga-like toxin II (VT2). *Microb. Pathogen.* **6**, 113–22.

Olsnes, S. and Eiklid, K. (1980). Isolation and charac-

terization of *Shigella shigae* cytotoxin. *J. Biol. Chem* **255**, 284–9.

Olsnes, S., Reisbig, R. and Eiklid, K. (1981). Subunit structure of Shigella cytotoxin. *J. Biol. Chem.* **256**, 8732–8.

Ostroff, S.M., Tarr, P.I., Neill, M.A., Lewis, J.H., Hargrett-Bean, N. and Kobayashi, J.M. (1989). Toxin genotypes and plasmid profiles as determinants of systemic sequelae in *Escherichia coli* 0157:H7 infections. *J. Infect. Dis.* **160**, 994–8.

Padhye, V., Kittell, F.B. and Doyle, M.P. (1986). Purification and physico-chemical properties of a unique vero cell cytotoxin from *Escherichia coli* 0157:H7. *Biochem. Biophys. Res. Commun.* **139**, 424–30.

Padhye, V.V., Beery, J.T., Kittell, F.B. and Doyle, M.P. (1987). Colonic hemorrhage produced in mice by a unique vero cell cytotoxin from an *Escherichia coli* strain that causes hemorrhagic colitis. *J. Infect. Dis.* **155**, 1249–53.

Pai, C.H., Kelly, J.K. and Meyers, G.I. (1986a). Experimental infection of infant rabbits with verotoxin-producing *Escherichia coli*. *Infect. Immun.* **51**, 16–23.

Pai, C.H., Ahmed, N., Lior, H., Johnson, W.M., Sims, H.V. and Woods, D.E. (1986b). Epidemiology of sporadic diarrhea due to verocytotoxin-producing *Escherichia coli*: a two year prospective study. *J. Infect. Dis.* **157**, 1054–7.

Pollard, D.R., Johnson, W.M., Lior, H., Tyler, D. and Rozee, K.R. (1990). Rapid and specific detection of verotoxin genes in *Escherichia coli* by the polymerase chain reaction. *J. Clin. Microbiol.* **28**, 540–5.

Reisbig, R., Olsnes, S. and Eiklid, K. (1981). The cytotoxic activity of *Shigella* toxin. Evidence for catalytic inactivation of the 60S ribosomal subunit. *J. Biol. Chem.* **256**, 8739–44.

Remis, R.S., MacDonald, K.L., Riley, L.W., Phur, N.D., Wells, J.G., David, B.R., Blake, P.A. and Cohen, M.L. (1984). Sporadic cases of hemorrhagic colitis associated with *Escherichia coli* 0157:H7. *Ann. Intern. Med.* **101**, 624–6.

Riley, L.W., Temis, R.S., Helgerson, S.D., McGee, H.B., Wells, J.G., Davis, B.R., Hebert, R.J., Olcott, E.S., Johnson, L.M., Hargrett, N.T., Blake, P.A. and Cohen, M.L. (1983). Hemorrhagic colitis associated with a rare *Escherichia coli* serotype. *New Engl. J. Med.* **308**, 681–5.

Samadpour, M., Liston, J., Ongerth, J.E. and Terr, P.I. (1990). Evaluation of DNA probes for detection of Shiga-like-toxin-producing *Escherichia coli* in food and calf fecal samples. *Appl. Environ. Microbiol.* **36**, 1212–15.

Samuel, J.E., Perera, L.P., Ward, S., O'Brien, A.D., Ginsburg, V. and Krivan, H.C. (1990). Comparison of the glycolipid receptor specificities of Shiga-like toxin type II and Shiga-like toxin type II variants. *Infect. Immun.* **58**, 611–18.

Sansonetti, P.J. (1991). Genetic and molecular basis of

epithelial cell invasion by *Shigella* species. *Rev. Infect. Dis.* 13 (Suppl. 3), **13**, S285–S292.

Scotland, S.M., Day, N.P. and Rowe, B. (1979). Production by strains of *Escherichia coli* of a cytotoxin (VT) affecting Vero cells. *Soc. Gen. Microbiol. Quart.* **6**, 156–7.

Seidah, N.G., Donohue-Rolfe, A., Lazure, C., Auclair, F., Keusch, G.T. and Chrétien, M. (1986). Complete amino acid sequence of *Shigella* toxin B-chain. A novel polypeptide containing 69 amino acids and one disulfide bridge. *J. Biol. Chem.* **261**, 13928–31.

Shiga, K. (1898). Ueber den Dysenteriebacillus (*Bacillus dysenteriae*). *Zbl. Bakt. Parasit. Abt. 1, Orig.* **24**, 817–24.

Smith, H.R., Scotland, S.M., Willshaw, G.A., Wray, C., McLaren, I.M., Cheasty, T. and Rowe, B. (1988). Verocytotoxin production and presence of VT genes in *Escherichia coli* strains of animal origin. *J. Gen. Microbiol.* **134**, 829–34.

Smith, H.R., Scotland, S.M., Stokes, N. and Rowe, B. (1990). Examination of strains belonging to enteropathogenic *Escherichia coli* serogroups for genes encoding EPEC adherence factor and Vero cytotoxins. *J. Med. Microbiol.* **31**, 235–40.

Strockbine, N.A., Marques, L.R.M., Newland, J.W., Williams-Smith, H., Holmes, R.K. and O'Brien, A.D. (1986). Two toxin-converting phages from *Escherichia coli* 0157:H7 strain 933 encode antigenically distinct toxins with similar biologic activities. *Infect. Immun.* **53**, 135–40.

Strockbine, N.A., Jackson, M.P., Sung, L.M., Holmes, R.K. and O'Brien, A.D. (1988). Cloning and sequencing of the genes for Shiga toxin from *Shigella dysenteriae* type I. *J. Bacteriol.* **170**, 1116–22.

Suthienkul, O., Brown, J.E., Seriwatana, J., Tienthongdee, S., Sastravaha, S. and Echeverria, P. (1990). Shiga-like-toxin-producing *Escherichia coli* in retail meats and cattle in Thailand. *Appl. Environ. Microbiol.* **56**, 1135–9.

van Heyningen, W.E. and Gladstone, G.P. (1953). The neurotoxin of *Shigella dysenteriae* 1. Production, purification and properties of the toxin. *Br. J. Exp. Pathol.* **34**, 202–16.

Vicari, G., Olitzki, A.L. and Olitzki, Z. (1960). The action of the thermolabile toxin of *Shigella dysenteriae* on cells cultivated *in vitro*. *Br. J. Exp. Pathol.* **41**, 179–89.

Von Wulfen, H., Russmann, H., Karch, H., Meyer, T., Bitzan, M., Kohrt, T.C. and Aleksic, S. (1989). Verocytotoxin-producing *Escherichia coli* 02:H5 isolated from patients with ulcerative colitis. *Lancet* **i**, 1449–50.

Waddell, T., Head, S., Petric, M., Cohen, A. and Lingwood, C. (1988). Globotriaosylceramide is specifically recognized by the *Escherichia coli* verocytotoxin 2. *Biochem. Biophys. Res. Commun.* **152**, 674–9.

Waddell, T., Cohen, A. and Lingwood, C.A. (1990). Induction of verotoxin sensitivity in receptor-deficient cell lines using the receptor glycolipid globotriaosylceramide. *Proc. Natl. Acad. Sci. USA* **87**, 7898–901.

Wade, W.G., Thom, B.T. and Evens, N. (1979). Cytotoxic enteropathogenic *Escherichia coli*. *Lancet* **ii**, 1235–6.

Wadolkowski, E.A., Burris, J.A. and O'Brien, A.D. (1990). Mouse model for colonization and disease caused by enterohemorrhagic *Escherichia coli*. *Infect. Immun.* **58**, 2438–45.

Weinstein, D.L., Holmes, R.K. and O'Brien, A.D. (1988a). Effects of iron and temperature on Shiga-like toxin I production by *Escherichia coli Infect. Immun.* **56**, 106–11.

Weinstein, D.L., Jackson, M.P., Samuel, J.E., Holmes, R.K. and O'Brien, A.D. (1988b). Cloning and sequencing of a Shiga-like toxin type II variant from an *Escherichia coli* strain responsible for edema disease of swine. *J. Bacteriol.* **170**, 4223–30.

Weinstein, D.L., Jackson, M.P., Perera, L.P., Holmes, R.K. and O'Brien, A.D. (1989). In vivo formation of hybrid toxins comprising Shiga toxin and the Shiga-like toxins and a role of the B subunit in localization and cytotoxic activity. *Infect. Immun.* **57**, 3743–50.

Williams-Smith, H., Green, P. and Parsell, Z. (1983). Vero cell toxins in *Escherichia coli* and related bacteria transfer by phage and conjugation and toxic action in laboratory animals, chickens and pigs. *J. Gen. Microbiol.* **129**, 3121–37.

Wilson, M.W. and Bettelheim, K.A. (1980). Cytotoxic *Escherichia coli* serotypes. *Lancet* **ii**, 201.

Yutsudo, T., Nakabayashi, N., Hirayama, T. and Takeda, Y. (1987). Purification and some properties of a Vero toxin from *Escherichia coli* 0157:H7 that is immunologically unrelated to Shiga-toxin. *Microb. Pathogen.* **3**, 21–30.

19

On the Characteristics of Carbohydrate Receptors for Bacterial Toxins: Aspects of the Analysis of Binding Epitopes

Karl-Anders Karlsson, Jonas Ångström and Susann Teneberg

Glycobiology, Department of Medical Biochemistry, University of Göteborg, PO Box 33031, S-400 33 Göteborg, Sweden

Introduction

The majority of receptors for bacterial toxins are carbohydrates (Eidels *et al.*, 1983; Middlebrook and Dorland, 1984; Jackson, 1990). Most of these contain sialic acid, which is an abundant monosaccharide on animal cell surfaces as part of various glycoconjugates (see examples in Table 1). One question that has been asked is whether toxin proteins that bind to structurally related gangliosides (sialic acid-containing glycolipids) bear similarities in their binding sites and thus are related in evolutionary terms. At present this is difficult to determine since no toxin has yet been defined in three dimensions, including localization of the receptor-binding site, either by X-ray crystallography or by NMR. However, recent progress in the characterization of carbohydrate receptors for microbes (Karlsson, 1989) allows an approximation of the binding epitope on a carbohydrate receptor, which invites comparative studies, but also deepens our understanding of microbe–host interactions in general. The purpose of the present contribution is to describe briefly what we know about the character of carbohydrate receptors and some consequences of this knowledge for current investigations.

Common for microbes appears to be the property of recognizing internally placed sequences of a receptor saccharide (Karlsson, 1989). This should have a biological meaning since it differs from most anti-carbohydrate antibodies (see Hakomori, 1984, and Thurin, 1988 for anti-glycolipid specificities), which bind to terminal parts. One advantage with the internal binding may be a more efficient shift of receptor specificity (shift of target cell, see example below). An interesting experimental consequence of this internal binding is the dissection of binding epitopes that we have devised (Karlsson, 1986, 1989; Karlsson *et al.*, 1986). We will illustrate this below. Two main aspects will be discussed by use of selected examples from the toxin field. One concerns the discovered families of receptor specificities, based on a common minimal receptor sequence but with a slight variation in binding depending on various neighbouring groups added to this sequence. The example is the Galα4Gal specificity and shiga-like toxins in comparison with P-fimbriated *E. coli*. In the second illustration we will discuss the binding epitope on ganglioside GM1 for cholera toxin, and also propose, based on this and a comparison with binding data for tetanus toxin, a probable evolutionary relation between these two toxins.

Experimental

Binding preferences of bacteria and toxins for glycolipid isoreceptors (various sequences with the

Table I. Selected examples of proposed carbohydrate receptors for bacterial toxins[a]

Cholera toxin

GalNAcβ4(NeuAcα3)Galβ4GlcβCer
Galβ3GalNAcβ4(NeuAcα3)Galβ4GlcβCer
GalNAcβ4Galβ3GalNAcβ4(NeuAcα3)Galβ4GlcβCer

E. coli heat-labile toxins[b]

Galβ3GalNAcβ4(NeuAcα3)Galβ4GlcβCer

Tetanus toxin[c]

NeuAcα3Galβ3GalNAcβ4(NeuAcα8NeuAcα3)Galβ4GlcβCer
NeuAcα8NeuAcα3Galβ3GalNAcβ4(NeuAcα8NeuAcα3)Galβ4GlcβCer

Botulinum toxins A, B, C, E and F[d]

NeuAcα3Galβ3GalNAcβ4(NeuAcα8NeuAcα3)Galβ4GlcβCer
NeuAcα8NeuAcα3Galβ3GalNAcβ4(NeuAcα8NeuAcα3)Galβ4GlcβCer

Delta toxin of Cl. perfringens[e]

GalNAcβ4(NeuAcα3)Galβ4GlcβCer
Galβ3GalNAcβ4(NeuAcα3)Galβ4GlcβCer
NeuAcα3Galβ3GalNAcβ4(NeuAcα3)Galβ4GlcβCer

Toxin A of Cl. difficile[f]

Galα3Galβ4GlcNAcβ3Galβ4GlcβCer
GalNAcβ3Galα3Galβ4GlcNAcβ3Galβ4GlcβCer
(not active)

Shiga toxin and shiga-like toxins[g]

Galα4GalβCer
Galα4Galβ4GlcβCer
Galα3Galα4Gal4βGlcβCer
GalNAcβ3Galα4Galβ4GlcβCer

Several isoreceptors may be given, with the probable common and internally placed epitope underlined. The condensed representation of sugar chains follows recent recommendations (Nomenclature, 1988).
[a]For general reviews on bacterial toxins and their receptors, see Eidels et al. (1983); Middlebrook and Dorland (1984); Jackson (1990).
[b]Although the same three sugars may be involved in the epitope for these toxins as for the cholera toxin, their binding epitopes should differ, probably with a different balance in the contribution from the three sugars.
[c]The epitope sequence underlined is only tentative (see also text and comparison with cholera toxin).
[d]The epitope sequence underlined is only tentative. The separate toxins probably differ slightly in their receptor-binding sites, preferring the listed isoreceptors somewhat different. Of interest is the finding of additional binding sites for free fatty acids (Takamizawa et al., 1986; Kamata et al., 1986; Kozaki et al., 1987), and also for GalβCer (toxin B), which was located on a separate chymotryptic fragment compared to the ganglioside recognition (Kozaki et al., 1987). A bilayer-close binding to the monoglycosylceramide may contribute to toxin penetration. One may note that the toxin-producing bacterial cells show a binding specificity for lactosylceramide, which also creates proximity to the membrane bilayer (Karlsson, 1989).
[e]According to available inhibition data (Jolivet-Reynaud and Alouf, 1983), the first sequence is the most active, indicating that the recognition epitope should differ both from that of cholera toxin and from those of the heat-labile toxins (see above).
[f]There is no evidence so far for an internal recognition of this toxin (Clark et al., 1987; S. Teneberg, C. Torres, I. Lönnroth, K.-A. Karlsson, unpublished results).
[g]The shiga toxin (Brown et al., 1983; Lindberg et al., 1987) and shiga-like toxin (unpublished) that we have compared appear to have very similar specificities. However, several recombinant and mutated shiga-like toxins show separate dependencies on neighbouring groups to the apparently required Gal–Gal sequence (Jackson, 1990; Samuel et al., 1990; O'Brien and Holmes, 1987). See also discussion of Table 2.

minimum receptor sequence in common) were obtained by binding to glycolipids separated on thin-layer plates or coated in microtitre wells (Magnani et al., 1980; Brown et al., 1983; Hansson et al., 1983, 1985; Karlsson and Strömberg, 1987). The computer-based calculation of conformations and molecular modelling was done in principle as described in the original publications using the HSEA (Thøgersen et al., 1982; Sabesan et al., 1984) and GESA (Paulsen et al., 1984) programs. The approach of epitope dissection on carbohydrate receptors for microbes has been described (Karlsson, 1986, 1989), but see also below.

Discussion

Table 1 shows receptor sequences for some bacterial toxins. This is not a complete summary of known specificities. The minimal sequence proposed to be recognized is underlined. As shown, all toxins but one are internal binders and many of them require sialic acid. Historically, some of the early data on toxin binding to gangliosides were considered contradictory, since traditional thinking based on the experience from antibodies excluded an internal binding. Considering some of the isoreceptors of Table 1, they are non-related structurally in their terminals although they are positive binders. Also, although the actual internal sequence may be present, the isoreceptor in question may be a negative binder (not shown), since neighbouring groups may for sterical reasons block out the binding. Therefore the binding data from studies of microbes require a careful interpretation.

The growing family of variant Galα4Gal binders

In Table 2 we have gathered several *E. coli* and toxin variants, which all seem to have in common this disaccharide (or part of it). As can be deduced from the table all eight ligands showed different binding patterns to the listed isoreceptors. This should mean that the proteins recognizing the receptors differ slightly in/at their binding sites, possibly with only one or a few amino acid substitutions. Thus a minor difference (limited mutation) may shift the binding pattern. In the case of the PapG and PrsG bacterial adhesins this change is sufficient to cause a shift of host animal from human to dog in the case of urinary tract infection (Strömberg et al., 1990). If we consider the isoreceptors of Table 2, globoside (third from top) is the dominating isoreceptor of human host cells, while the Forssman glycolipid (fourth from top) dominates in the dog. Also noteworthy was the finding (Strömberg et al., 1990) that although these two isoreceptors were not clearly discriminated using isolated glycolipids in the thin-layer overlay assay, they were apparently selectively recognized when placed in the intact cell membrane. The reason for this sophisticated effect is probably a slight conformational change induced on the saccharide chain when the glycolipid is inserted in the membrane, due to the bending back towards the membrane surface of the GalNAc-α3GalNAc extension. The reader may get an intuitive impression of this when looking at the model shown in Fig. 1.

Considering Pap-2 and the pig oedema toxin, the requirement for the Galα4Gal disaccharide is not absolute, since the adhesin was inhibited by a GalNAcβ3Gal disaccharide (Karr et al., 1989) and the toxin was found to bind also to the Galα3Gal isomer (DeGrandis et al., 1989). Shiga toxin (Brown et al., 1983; Lindberg et al., 1987) and the PapG and PrsG adhesins (Strömberg et al., 1990) did, however, bind only to the position-4 isomer. Concerning the other three shiga-like toxins (Samuel et al., 1990), the first two tolerated one and two monosaccharides, respectively, as extensions to the underlined disaccharide, but the last toxin bound only to a terminally placed epitope.

A molecular model of the computer-calculated conformation of globoside may reflect the receptor surface to which these proteins associate (Fig. 1). The Galα4Gal disaccharide produces an almost 90° bend of the chain and it is very likely that this convex surface represents the binding site. In the model, this side is exposed to the viewer with the ring hydrogens H-1 and H-2 of Galα, and H-1, H-3, H-4 and H-5 of Galβ indicated with corresponding numbers. These hydrogens make up a non-polar surface on the convex side of the disaccharide. On this side the methyl group (carbon coloured black) of the acetamido substituent of GalNAc is extending for a possible interaction with the protein ligand.

Table 2. Variety of binding patterns of *E. coli* and shiga-like toxins that recognize Galα4Gal-containing isoreceptors

Glycolipid structure	Bacteria			Toxins				
	PapG	PrsG	Pap-2	Shiga toxin	Pig oedema toxin	SLT	SLT vp	SLT vh
Galα4GalβCer				+		+	–	+
Galα4Galβ4GlcβCer	+	–		+	+	+	+	+
GalNAcβ3Galα4Galβ4GlcβCer	+	(+)	+	(+)	+	(+)	+	–
GalNAcα3GalNAcβ3Galα4Galβ4GlcβCer	+	+		–	(+)	–	(+)	–
Galβ3GalNAcβ3Galα4Galβ4GlcβCer	+	+		–		–	+	–
GalNAcβ3Galα3Galβ4GlcβCer	–	–	+	–	+			
Galα3Galα4Galβ4GlcβCer				+				
Reference	1	1	2	3,4	5	6	6	6

Binding data were obtained by overlay of ligand on thin-layer plates with separated glycolipids (except Pap-2, which was shown to be inhibited in its binding by GalNAcβ3Gal). PapG, PrsG and Pap-2 were cloned adhesions from *E. coli*. Only a limited number of isoreceptors have been included to illustrate the differences in binding. A positive sign, +, means a clear positive binding; a positive sign within parentheses, (+), means a weak binding; and a negative sign, –, means no binding.
(1) Strömberg *et al.* (1990); (2) Karr *et al.* (1989); (3) Brown *et al.* (1983); (4) Lindberg *et al.* (1987); (5) DeGrandis *et al.* (1989); (6) Samuel *et al.* (1990).

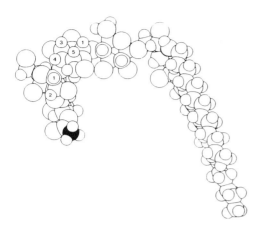

GalNAcβ3Galα4Galβ4GlcβCer

Figure 1. Computer-calculated conformation of globoside, GalNAcβ3Galα4Galβ4GlcβCer, projected to visualize the probable binding side for the eight variant microbial ligands of Table 2. The ring hydrogens of Galα, H-1, and H-2, and of Galβ, H-1, H-3, H-4 and H-5, have been numbered, and the methyl carbon of the acetamido group of GalNAc is shown in black. The indicated ring hydrogens form a continuous non-polar surface over Gal–Gal, which is bent about 90° to give a convex accessible epitope. Most ligands in Table 2 seem to require Gal–Gal; two of them, however, are satisfied with Galα. They have various dependencies for GalNAc: for some ligands the binding is improved, for others inhibited and in one case completely blocked. In several cases the indicated N-acetyl group is probably involved. This is an example of a family of closely related receptor specificities, where the complementary binding sites of the proteins should be structurally and evolutionary related.

For PapG (Table 2) this GalNAc slightly improves the binding compared with a terminal Galα4Gal (Bock *et al.*, 1985), which could imply an interaction with the acetamido group. Addition of another GalNAc or Gal does not change the binding. For PrsG, however, a terminal Galα4Gal is not sufficient, and binding appears only after addition of GalNAc. It further improves on extension with another GalNAcα or with Galβ. As these terminal sugars have quite different stereochemical presentations, they probably do not interact directly with the protein ligand but, rather, induce an optimal conformation of the minimal receptor sequence. In the case of shiga toxin, addition of GalNAc diminishes the binding whereas Galα has no effect. The oedema toxin has a more limited epitope since 4Gal may be replaced with a 3Gal and two sugars added block the binding. On the model this means that the epitope has shifted downwards and is limited to GalNAc–Gal. A very similar situation may exist for the Pap-2 adhesin, whose binding may be inhibited with this disaccharide. For the shiga-like toxin SLT-vh, finally, addition of GalNAc (the terminal sugar of the model in Fig. 1) completely blocks the binding, meaning that the epitope in this case is restricted to Galα4Gal, and that GalNAc probably is producing a sterical hindrance for access to the epitope by the protein ligand.

The impression given from analysis of these eight closely related variants of receptor binders is that the internal binding is absolutely necessary to allow these shifts (Karlsson, 1989). Similar epitope changes over a terminally located Gal–Gal surface would have limited effects, since the different epitopes would be accessible without discriminating hindrances from added neighbouring groups. Although the conclusion thus is that these mutational shifts may efficiently produce a number of biological variants of ligands with separate tropisms, the possibility still exists that the synthesis chemist may design a soluble receptor analogue that is capable of blocking all variants. In the present case such an analogue should be based on three sugars, the convex surface of the Gal–Gal sequence (Fig. 1) and a modified GalNAc to allow interaction also with SLT-vh (Table 2).

The binding epitope for cholera toxin on the GM1 ganglioside

Cholera toxin is one of the most extensively studied toxins (Eidels *et al.*, 1983; Middlebrook and Dorland, 1984). Although the target cell in cholera disease, the enterocyte of human small intestine appears to carry only GM1 as receptor, as shown by thin-layer chromatography overlay on extracted gangliosides (unpublished results), also this toxin has been found by us to recognize an internal sequence as revealed by binding to isoreceptors of other origins (extensions to GM1). As shown in Table 3, the classical GM1 receptor (no. 4) can be extended on the sialic acid (no. 11) or the terminal Gal (nos 12–15) without complete loss of activity. However, two sialic acids on the terminal Gal (no. 16) block binding. The ganglioside without the terminal Gal (no. 3) is also active, as is the extended sequence (Nakamura *et al.*, 1987) with an internal GalNAc (no.7) instead of a Glc as in GM1 (no. 4). This means that the minimally required

Table 3. Selected binding data for cholera toxin and GM1 isoreceptors (natural and modified sequences)

1	NeuAcα3Galβ4GlcβCer	−
2	GalβGalNAcβ4Galβ4GlcβCer	−
3	<u>GalNAcβ4</u>(NeuAcα3)Galβ4GlcβCer	+
4	Galβ3<u>GalNAcβ4</u>(NeuAcα3)Galβ4GlcβCer	+ + +
5	Galβ3<u>GalNAcβ4</u>(NeuGcα3)Galβ4GlcβCer[a,b]	+ + +
6	6-Ox-Galβ3<u>GalNAcβ4</u>(NeuAcα3)Galβ4GlcβCer[a]	+ + +
7	Galβ3<u>GalNAcβ4</u>(NeuGcα3)Galβ3GalNAcβ4Galβ4GlcβCer[b]	+ + +
8	Galβ3<u>GalNAcβ4</u>(R-NeuAcα3)Galβ4GlcβCer[c] R = Lucifer Yellow CH, Rhodamine, or DNP	+
9	Galβ3<u>GalNAcβ4</u>(Red-NeuAcα3)Galβ4GlcβCer[a]	−
10	Galβ3<u>GalNβ4</u>(Neuα3)Galβ4GlcβCer[d]	+
11	Galβ3<u>GalNAcβ4</u>(NeuAcα8NeuAcα3)Galβ4GlcβCer	+
12	NeuAcα3Galβ3<u>GalNAcβ4</u>(NeuAcα3)Galβ4GlcβCer	+
13	GalNAcβ4Galβ3<u>GalNAcβ4</u>(NeuAcα3)Galβ4GlcβCer	+
14	Fucα2Galβ3<u>GalNAcβ4</u>(NeuAcα3)Galβ4GlcβCer[b]	+
15	Galα3(Fucα2)Galβ3<u>GalNAcβ4</u>(NeuAcα3)Galβ4GlcβCer	+ +
16	NeuAcα8NeuAcα3Galβ3<u>GalNAcβ4</u>(NeuAcα3)Galβ4GlcβCer	−

The binding activity was in most cases assayed by binding of labelled toxin to glycolipids separated on thin-layer plates or coated in microtitre wells. Only rough relative binding preferences are given.
[a]Result reported elsewhere (Fishman et al., 1980).
[b]This sequence was recently reported and tested for binding (Nakamura et al., 1987).
[c]These derivatives were reported elsewhere (Spiegel, 1985).
[d]This derivative was reported elsewhere (Schengrund and Ringler, 1989).

sequence is the one underlined in several isoreceptors in Table 3. This is in part supported by substance no. 6, where the terminal Gal of GM1 has been oxidized by galactose oxidase, which does not affect binding (Fishman et al., 1980).

As discussed earlier (Karlsson, 1986, 1989; Karlsson et al., 1986) it appeared to us that this internal binding should allow an approximation of the surface produced by this minimal sequence that is recognized by the ligand, provided that the conformation is stable and does not change upon binding. We have named this approach epitope dissection. The obvious principle is that binding cannot occur on the side where sterical protrusions exist, but rather on a side with easy access to the binding epitope. It has been shown that the calculated conformation of GM1 is in perfect agreement with the conformation of the released

pentasaccharide in solution as analysed by NMR (Sabesan et al., 1984). And as oligosaccharides in general, especially branched ones, are rather rigid structures, molecular modelling should reveal some data of interest.

When considering some derivatives of GM1 added to Table 3, the N-glycolyl form (Fishman et al., 1980) of GM1 (no. 5) is as active as the N-acetyl form existing in human intestine. This part is not expected to interact with the toxin since these acyl groups have very different properties. The methyl group of the N-acetyl group is indicated in black in the GM1 model of Fig. 2 and is located in the downward region of the projection shown. In contrast, if the carboxyl group (the two oxygens are indicated in black in the model of Fig. 2) is reduced to the corresponding alcohol (Fishman et al., 1980), then the binding activity is lost

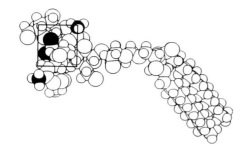

Galβ3GalNAcβ4(NeuAcα3)Galβ4GlcβCer

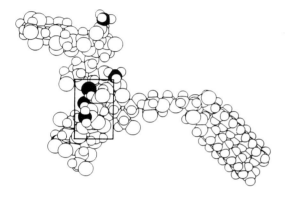

NeuAcα8NeuAcα3Galβ3GalNAcβ4(NeuAcα8NeuAcα3)Galβ4GlcβCer

Figure 2. Computer-calculated conformations of the receptor gangliosides for cholera (top) and tetanus toxins. The sequences under the models have been partially underlined to indicate the probable minimal, internally recognized sequences. The rectangles drawn on the models define the proposed epitopes, which are highly overlapping and should correspond to structurally and evolutionary related binding sites of the two toxins. In the GM1 model (top) the methyl carbons of GalNAc (top), NeuAc (bottom), and the two carboxyl oxygens are shown in black. In the lower model the same atoms are shown in black, and also the visible methyl carbons of the two added NeuAc moieties.

completely (no. 9 in Table 3). It has also been claimed (Schengrund and Ringler, 1989) that GM1 without *N*-acetyl groups is active (no. 10). Furthermore, the tail of sialic acid can be oxidized with periodate and substituted with rather bulky groups (no. 8 of Table 3) with retention of some activity (Spiegel, 1985). When comparing these binding data with the calculated conformation of GM1 (Fig. 2), the glycerol tail of the sialic acid, which is not essential for binding, is found to be on the opposite side of the essential carboxyl group.

If neither the *N*-acetyl group of the sialic acid nor of GalNAc is required, this should mean that the minimum epitope may be located within the rectangle drawn on the projection in Fig. 2, which coincides with the narrow interface of the three minimally required sugars at the branch of GM1, GalNAc(NeuAc)Gal. GM1 (no. 4 in Table 3) is a much better binder than GM2 (no. 3). This is probably due to an interaction of the toxin with the non-polar side of the added Gal, which can be seen with the four ring hydrogens, H-1, H-3, H-4 and H-5, on top to the left of the model of Fig. 2. This Gal may be substituted with various sugars (nos 12–15 in Table 3) on the back side of the projected model without complete loss of activity. Also, a second sialic acid linked to the tail of the first (no. 11), though found on the opposite side of the epitope, reduces the affinity to the same level as the previously mentioned substitutions.

It is therefore very likely that the cholera toxin interaction with its receptor, the ganglioside GM1, is limited to a minor part of the total receptor saccharide. This explains why several sugars may be added (on the terminal Gal or on NeuAc) without completely blocking the toxin access to the epitope. From a technical point of view this means that chemical synthesis of receptor analogues for various drug or affinity purposes may start with computer design from the epitope area defined in Fig. 2.

Relation in binding epitope between receptors for cholera and tetanus toxins

Cholera toxin produced by *Vibrio cholerae* and acting in the small intestine, and tetanus toxin produced by *Clostridium tetani* and acting on nerve cells, both have gangliosides as receptors (Eidels *et al.*, 1983; Middlebrook and Dorland, 1984). The biological receptor for cholera toxin is the GM1 ganglioside and one of the tetanus receptors is the ganglioside GQ1b. Their sequences and probable minimal binding sequences (underlined) are shown below the molecular models in Fig. 2.

There is no cross-reaction between the toxins and these two receptors. Thus, cholera toxin does not bind to GQ1b and tetanus toxin does not bind to GM1 when using the thin-layer chromatography overlay technique. However, as shown by the partially underlined sequences and by the rectangles in Fig. 2, the proposed internal binding epitopes are highly overlapping (unpublished results), although the terminal sequences are quite different. Our proposal is, therefore, that the complementary binding sites on the toxin proteins are closely related structurally and evolutionarily, although they are produced by different bacteria and act in different places. The details of this discussion will be reported elsewhere (manuscript in preparation).

This relation between epitopes, although not immediately obvious, is analogous to the eight members of the family of Gal–Gal binders discussed above. This type of variance may be quite common among receptor specificities for microbes. Another example is the great number of bacteria that recognize lactosylceramide, where there are several variant binders (Karlsson, 1989).

General conclusions

The recognition of internal sequences of carbohydrate receptors as a general property of microbial ligands has now been known for more than five years (Brown, et al., 1983; Hansson, et al., 1983; Bock et al., 1985; Karlsson, 1986, 1989; Karlsson et al., 1986; Lindberg et al., 1987). The biological meaning of this may be an economical shift of receptor specificity, which would not have been possible with terminal epitopes (Karlsson, 1989). The technical consequences of the internal binding for characterization of binding epitopes (Karlsson, 1986, 1989) now opens up interesting possibilities for the design of synthetic receptor analogues for various purposes. In theory, one product may be efficient within a whole family of receptor specificities. We have illustrated elsewhere the epitope dissection for several bacterial specificities based on computer modelling (Karlsson, 1988).

Glycolipid-based specificities have often been selected, especially for toxins (Eidels et al., 1983; Middlebrook and Dorland, 1984; Karlsson, 1989; Jackson, 1990). Glycolipids are strictly membrane-bound and do not usually appear in secretions, in contrast to glycoproteins. Glycolipids may also offer membrane proximity for ligand penetration. A recent paper (Pacuszka and Fishman, 1990) demonstrated the requirement of proximity for cholera toxin. The oligosaccharide was released

from the GM1 ganglioside and coupled directly to surface membrane proteins. Toxin-insensitive cells treated in this way readily bound the toxin but there was no biological effect compared with cells coated with the GM1 ganglioside. Thus the early model has been validated, where the membrane-close GM1 binding by the B subunits makes the penetration of the A subunit possible (Eidels et al., 1983; Middlebrook and Dorland, 1984).

Acknowledgements

Our own work described here was supported by grants from the Swedish Medical Research Council (no. 3967) and by Symbicom Ltd.

References

Bock, K., Breimer, M.E., Brignole, A., Hansson, G.C., Karlsson, K.-A., Larson, G., Leffler, H., Samuelsson, B.E., Strömberg, N., Svanborg Éden, C. and Thurin, J. (1985). Specificity of binding of a strain of uropathogenic Escherichia coli to Galα4Gal-containing glycosphingolipids. J. Biol. Chem. **260**, 8545–51.

Brown, J.E., Karlsson, K.-A., Lindberg, A.A., Strömberg, N. and Thurin, J. (1983). Identification of the receptor glycolipid for the toxin of Shigella dysenteriae. In: Glycoconjugates (eds M.A. Chester, D. Heinegård, A. Lundblad and S. Svensson), pp. 678–9. Rahms in Lund, Sweden.

Clark, G.F., Krivan, H.C., Wilkins, T.D. and Smith, D.F. (1987). Toxin A from Clostridium difficile binds to rabbit erythrocyte glycolipids with terminal Galα3Galβ4GlcNAc sequences. Arch. Biochem. Biophys. **257**, 217–29.

DeGrandis, S., Law, H., Brunton, J., Gyles, C. and Lingwood, C.A. (1989). Globotetraosylceramide is recognized by the pig edema disease toxin. J. Biol. Chem. **264**, 12520–5.

Eidels, L., Proia, R.L. and Hart, D.A. (1983). Membrane receptors for bacterial toxins. Microbiol. Rev. **47**, 596–620.

Fishman, P.H., Pacuszka, T., Holm, B. and Moss, J. (1980). Modification of ganglioside GM1. Effect of lipid moiety on choleragen action. J. Biol. Chem. **255**, 7657–64.

Hakomori, S.-i. (1984). Tumor-associated carbohydrate antigens. Ann. Rev. Immunol. **2**, 103–26.

Hansson, G.C., Karlsson, K.-A., Larson, G., Lindberg, A.A., Strömberg, N. and Thurin, J. (1983). Lactosylceramide is the probable adhesion site for major indigenous bacteria of the gastrointestinal tract. In: Glycoconjugates (eds M.A. Chester, D. Heinegård,

A. Lundblad and S. Svensson), pp. 631–2. Rahms in Lund, Sweden.

Hansson, G.C., Karlsson, K.-A., Larson, G., Strömberg, N. and Thurin, J. (1985). Carbohydrate-specific adhesion of bacteria to thin-layer chromatograms: a rationalized approach to the study of host cell glycolipid receptors. Analyt. Biochem. 146, 158–63.

Jackson, M.P. (1990). Structure-function analyses of Shiga toxin and the Shiga-like toxins. Microb. Pathogen. 8, 235–42.

Jolivet-Reynaud, C. and Alouf, J.E. (1983). Binding of Clostridium perfringens 125-I-labeled delta-toxin to erythrocytes. J. Biol. Chem. 258, 1871–7.

Kamata, Y., Kozaki, S., Sakaguchi, G., Iwamori, M. and Nagai, Y. (1986). Evidence for direct binding of Clostridium botulinum type E derivative toxin and its fragments to gangliosides and free fatty acids. Biochem. Biophys. Res. Commun. 140, 1015–19.

Karlsson, K.-A. (1986). Animal glycolipids as attachment sites for microbes. Chem. Phys. Lipids 42, 153–72.

Karlsson, K.-A. (1988). Current experience from the interaction of bacteria with glycosphingolipids. In: Molecular Mechanisms of Microbial Adhesion (eds L. Switalski, M. Höök and E. Beachey), pp. 77–96. Springer Verlag, New York.

Karlsson, K.-A. (1989). Animal glycosphingolipids as membrane attachment sites for bacteria. Ann. Rev. Biochem. 58, 309–50.

Karlsson, K.-A. and Strömberg, N. (1987). Overlay and solid-phase analysis of glycolipid receptors for bacteria and viruses. Methods Enzymol. 138, 220–32.

Karlsson, K.-A., Bock, K., Strömberg, N. and Teneberg, S. (1986). Fine dissection of binding epitopes on carbohydrate receptors for microbial ligands. In: Protein–Carbohydrate Interactions in Biological Systems (ed. D.L. Lark), pp. 207–13. Academic Press, London.

Karr, J.F., Nowicki, B., Truong, L.D., Hull, R.A. and Hull, S.I. (1989). Purified fimbriae from two cloned gene clusters of a single pyelonephritogenic strain adhere to unique structures in the human kidney. Infect. Immun. 57, 3594–600.

Kozaki, S., Ogasawara, J., Shimote, Y., Kamata, Y. and Sakaguchi, G. (1987). Antigenic structure of Clostridium botulinum type B neurotoxin and its interaction with gangliosides, cerebroside, and free fatty acids. Infect. Immun. 55, 3051–6.

Lindberg, A.A., Brown, J.E., Strömberg, N., Westling-Ryd, M., Schultz, J.E. and Karlsson, K.-A. (1987). Identification of the carbohydrate receptor for Shiga toxin produced by Shigella dysenteriae type 1. J. Biol. Chem. 262, 1779–85.

Magnani, J.F., Smith, D.F. and Ginsburg, V. (1980). Detection of gangliosides that bind cholera toxin: direct binding of 125-I-labeled toxin to thin-layer chromatograms. Analyt. Biochem. 109, 399–402.

Middlebrook, J.L. and Dorland, R.B. (1984). Bacterial

toxins: cellular mechanisms of action. Microbiol. Rev. 48, 199–221.

Nakamura, K., Suzuki, M., Inagaki, F., Yamakawa, T. and Suzuki, A. (1987). A new ganglioside showing choleragen action in mouse spleen. J. Biochem. 101, 825–35.

Nomenclature of glycoproteins (1988). J. Biol. Chem. 262, 13–18.

O'Brien, A.D. and Holmes, R.K. (1987). Shiga and Shiga-like toxins. Microbiol. Rev. 51, 206–20.

Pacuszka, T. and Fishman, P. (1990). Generation of cell-surface neoglycoproteins. GM1-neoglycoproteins are non-functional receptors for cholera toxin. J. Biol. Chem. 265, 7673–8.

Paulsen, H., Peters, T., Sinnwell, V., Lebuhn, R. and Meyer, B. (1984). Conformational analysis. 24. Determination of the conformations of trisaccharide and tetrasaccharide sequences of N-glycoproteins. The problem of the (1-6)-glycosidic bond. Liebigs Ann. Chem. 951–76.

Sabesan, S., Bock, K. and Lemieux, R.U. (1984). The conformational properties of the gangliosides GM2 and GM1 based on 1H and 13C NMR studies. Can. J. Chem. 62, 1034–45.

Samuel, J.E., Perera, L.P., Ward, S., O'Brien, A.D., Ginsburg, V. and Krivan, H.C. (1990). Comparison of the glycolipid receptor specificities of Shiga-like toxin type II and Shiga-like toxin type II variants. Infect. Immun. 58, 611–18.

Schengrund, C.-L. and Ringler, N.J. (1989). Binding of Vibrio cholera toxin and the heat-labile enterotoxin of Escherichia coli to GM1, derivatives of GM1, and non-lipid oligosaccharide polyvalent ligands. J. Biol. Chem. 264, 13233–7.

Spiegel, S. (1985). Fluorescent derivatives of ganglioside GM1 function as receptors for cholera toxin. Biochemistry 24, 5947–52.

Strömberg, N., Marklund, B.-I., Lund, B., Ilver, D., Hamers, A., Gaastra, W., Karlsson, K.-A. and Normark, S. (1990). Host-specificity of uropathogenic Escherichia coli depends on differences in binding specificity to Galα4Gal-containing isoreceptors. EMBO J. 9, 2001–10.

Takamizawa, K., Iwamori, M., Kozaki, S., Sakaguchi, G., Tanaka, R., Takayama, H. and Nagai, Y. (1986). TLC immunostaining characterization of Clostridium botulinum type A neurotoxin binding to gangliosides and free fatty acids. FEBS Lett. 201, 229–32.

Thøgersen, H., Lemieux, R.U., Bock, K. and Meyer, B. (1982). Further justification of the exoanomeric effect. Conformational analysis based on NMR of oligosaccharides. Can. J. Chem. 60, 44–57.

Thurin, J. (1988). Binding sites of monoclonal anti-carbohydrate antibodies. In: Carbohydrate–Protein Interaction, Current Topics in Microbiology and Immunology (eds A.E. Clark and I.A. Wilson), Vol. 139, pp. 59–79. Springer Verlag, Berlin.

20

Genetic Analysis of the *In Vivo* Role of Bacterial Toxins

Timothy J. Foster

Microbiology Department, Moyne Institute, Trinity College, Dublin 2, Ireland

Introduction

A particularly valuable approach for analysing the role of toxins in the pathogenesis of bacterial diseases is to isolate mutants deficient in toxin production (Tox⁻) and to compare their behaviour with toxigenic (Tox⁺) wild-type strains in experimental infections (Sparling, 1983). Tox⁻ mutants may also be of value as potential live vaccines (Mekalanos *et al.*, 1983; Kaper *et al.*, 1984a,b). However, many early studies with Tox⁻ mutants were flawed because of the possibility that the strains had more than one mutation. It is important to be certain that the mutants are only defective in the gene of interest.

The classical method for generating mutations is to treat a population of bacterial cells with a chemical mutagen or with ultraviolet light. Optimum mutagenic activity usually causes substantial killing of bacteria. Defects in genes other than the gene of interest, including ones which contribute to virulence, will almost certainly be caused. In addition, one has to distinguish between mutations affecting structural genes and those that affect regulatory loci. These problems cloud interpretation of early attempts to dissect pathogenic mechanisms using genetics.

Site-specific mutations

Modern molecular genetic techniques, viz. transposon mutagenesis and allele replacement, have allowed single-site Tox⁻ mutants to be isolated in many pathogenic bacteria. This review discusses methods for isolating site-specific mutations and shows how Tox⁻ mutants have contributed to understanding of the role of toxins in pathogenesis of bacterial diseases.

Transposons

The major advantages of using a transposon to isolate site-specific mutations are as follows (reviewed by de Bruijn and Lupski, 1984; Foster, 1984; Berg *et al.*, 1989): (i) the technique is applicable to many different genera of bacteria; (ii) high-frequency mutagenesis occurs with low rates of multiple mutations; (iii) controls can be easily performed to verify that a single copy of the transposon is present in the genome (see below); (iv) it is not necessary to have cloned the toxin gene; (v) mutations are easy to isolate and verify; (vi) a drug resistance marker is associated with the mutational insertion; and (vii) transposon insertion mutations can be used to clone the inactivated gene.

The applicability of any particular transposon to mutagenesis of pathogenic bacteria depends on the transposition functions and drug resistance markers. Failure of either to be expressed in the host to be mutagenized will prevent its use. For example Tn*10* has a relatively narrow host range and is mainly confined to enteric bacteria related to *E. coli*, whereas Tn5 has a broader activity and has been used for mutagenesis in many different genera of Gram-negative bacteria. A detailed discussion of the problems associated with Tn5 mutagenesis in *Rhizobium* (Simon *et al.*, 1986) will be relevant to the use of the transposon in Gram-negative pathogens. The conjugative transposons of streptococci appear to have a relatively broad host range among Gram-positive bacteria (Clewell and Gawron-Burke, 1986).

Delivery systems

In order to isolate transposon insertion mutations a mechanism for eliminating the donor replicon and for selecting for transposition into the chromosome is required. Several transposon delivery systems have been described for use in Gram-positive and Gram-negative bacteria.

ts Plasmid vectors. Plasmids with temperature-sensitive replication systems can be used as vectors for selecting for transposition into the chromosome. The *ts* plasmid carrying the transposon is transferred into the Tox$^+$ strain at a permissive temperature. Selection at a restrictive temperature for the drug resistance marker encoded by the transposon results in a reduced efficiency of plating. The majority of survivors have a copy of the transposon in the chromosome. This can be verified by scoring for loss of the drug resistance determinants expressed by the vector and by Southern hybridization (see below). Transposon delivery systems using *ts* plasmid vectors are available for both Gram-positive and Gram-negative bacteria (reviewed by Foster, 1984; Berg *et al.*, 1989).

Conjugative transposons. The conjugative streptococcal transposons Tn916 and Tn1545 can promote transposition in several different species of Gram-positive bacteria and have been used to study toxins of group A and B streptococci, *S. faecalis* and *Listeria monocytogenes* (Gaillard *et al.*, 1986; Weiser and Rubens, 1987; Kathariou *et al.*, 1987). The transposon-carrying donor strain is mated with the toxigenic recipient on the surface of a filter. The transposon promotes conjugation with the recipient. Transconjugants are obtained by selecting for expression of a drug resistance marker present on the transposon. Each colony represents an independent transfer/transposition event.

Broad host range suicide plasmids in Gram-negative bacteria. Early suicide vectors incorporated bacteriophage Mu into a broad host range conjugative plasmid (Beringer *et al.*, 1978). These plasmids cannot survive in certain Gram-negative bacteria. Thus selection for the drug resistance marker associated with the transposon often results in derivatives where the transposon has moved into the chromosome. One such plasmid, pJB4JI (Beringer *et al.*, 1978), was used to isolate Tn5 mutations affecting protease production in *Aeromonas hydrophila* (Leung and Stevenson, 1988).

Another type of suicide vector allows selection for Tn5 transposition in Gram-negative bacteria in which plasmid ColE1 cannot replicate (reviewed by Simon *et al.*, 1986). The suicide vector plasmid carries the transposon, the *mob* site of the RP4 *tra* operon and the replication system of ColE1, which only functions in bacteria closely related to *E. coli*. The donor *E. coli* strain carries a second element (a non-transferable or integrated plasmid) which expresses the broad host range transfer functions. Thus mating with a pathogenic recipient results in transfer of the vector plasmid into the cytoplasm of the cell, where it fails to replicate. Bacteria expressing the drug resistance specified by the transposon should have a copy of the element in the chromosome.

The suicide vector mentioned above cannot be used in Gram-negative bacteria in which ColE1 can replicate. Another suicide vector gets round this by using the replication system of plasmid R6K (Miller and Mekalanos, 1988). The vector plasmid carries the replication origin of plasmid R6K but lacks the *pir* gene which codes for a protein required to initiate replication. This is supplied in the donor strain by a lysogenic λ*pir* phage. The host also expresses the RP4 *tra* functions which promote transfer of the Mob$^+$ vector by conjugation. This vector has been exploited for mutagenesis with the Tn5-derived transposon Tn*phoA* (Manoil and Beckwith, 1985; Taylor *et al.*, 1987) which allows detection of insertions in genes coding for secreted proteins. Tn*phoA* is particularly valuable for identifying novel cell surface-associated or extracellular virulence functions in Gram-negative pathogens.

Controls

Most transposon mutations can revert to wild-type by precise excision (Berg *et al.*, 1989). If the reversion frequency is sufficiently high it will be possible to isolate revertants and demonstrate that virulence has been restored. This shows that secondary mutations were not responsible for loss of virulence. It might be difficult to isolate revertants if the frequency is low. This would preclude direct screening of individual colonies. However, reversion to an easily scorable phenotype (e.g. haemolysis) should be detectable on a semi-confluent plate. Alternatively colony immunoblotting could be used. Another method for attributing the observed loss of virulence to the missing toxin is to transfer the cloned wild-type *tox* gene on a plasmid into the mutant strains.

When bacteria have been growing *in vivo* it is important to check that the cells isolated from the infected animal still carry the mutation. This is particularly important if no difference in virulence is found between mutants and wild-type. Wild-type revertants might have a growth advantage *in vivo*. This seems to be the case with *Listeria monocytogenes* haemolysin mutants which are out-grown by wild-type revertants when large inocula are used (Kathariou *et al.*, 1987). The haemolysin has a major role in preventing bacteria being killed following phagocytosis.

It is important to perform Southern blot hybridization experiments to show that only one copy of the transposon is present in the chromosome of the mutant and that the defect is located in the structural gene for the toxin. Genomic DNA is cut with a restriction enzyme which does not cleave the transposon. The fractionated DNA is probed with labelled transposon DNA. If more than one band hybridizes it is likely that additional copies of the transposon are present in the chromosome of the mutant strain. A *tox* gene used to probe wild-type and mutant genomic DNA will show that the *tox* structural gene is disrupted by the transposon.

Advantages and disadvantages

Transposon mutations are usually quite easy to isolate given efficient selection for transposition, random insertion of the transposon into target DNA sequences and a good screening system for the mutant phenotype. It is a question of patiently screening several thousand colonies either directly on the plates selective for transposition or after replica plating. The frequency at which a mutation in an individual gene is isolated will depend on the size of the target, the size of the chromosome and random insertion of the transposon. If the size of the bacterial chromosome is 2×10^6 bp then an insertion in a gene of 1000 bp should occur once in every 2000 independent transpositions.

Problems encountered with the use of transposon mutations in virulence studies include (i) reversion of the mutation and (ii) polarity. Many transposon insertion mutations can revert to wild-type by precise excision. If revertants are present in the inoculum used in virulence studies this might alter the outcome of the infection. When 10^6 *Listeria monocytogenes* haemolysin-defective mutants were inoculated into mice the majority of bacteria recovered from spleens were Hly$^+$ revertants (Kathariou *et al.*, 1987). These were not recovered when 10^3 bacteria were adminstered because the reversion frequency was lower than 1×10^{-3}. Also strains of *Vibrio cholerae* E1 Tor which produce high levels of cholera enterotoxin seem to have a growth advantage over hypotoxigenic strains in the small intestine. This was suggested because bacteria isolated from infected patients often carried tandemly duplicated copies of the *ctx* gene and this correlated with higher levels of expression of toxin, presumably due to a gene-dosage effect (Mekalanos, 1983). In addition Tox$^+$ revertants of a hypotoxigenic strain were selected during *in vivo* growth (Mekalanos, 1983).

Such potentially unstable mutations could not be used in live vaccines. However, Tn*10* insertions can undergo a genetic rearrangement which prevents subsequent reversion by causing a deletion adjacent to the site of the transposon (Foster *et al.*, 1981). The tetracycline resistance marker of Tn*10* is lost and a deletion of DNA on one side of the insertion occurs. This phenomenon has been exploited in the generation of stable aromatic-dependent mutants of *Salmonella* species which have potential as live vaccines (Dougan *et al.*, 1987; Edwards and Stocker, 1988).

Polarity occurs if the transposon has inserted in a gene located promoter-proximal (5′) to other genes in the same transcription unit (reviewed by Foster, 1984; Berg *et al.*, 1989). If the distal gene codes for another virulence factor or for a function that affects growth or survival *in vivo* then changes in the virulence phenotype will not be exclusively due to loss of the toxin. If the downstream gene codes for a component of the bipartite toxin then polarity is not a problem. The only way to circumvent caveats being raised about the value of site-specific mutations in virulence studies is to isolate in-frame deletions by gene replacement (see below) and to show by DNA sequencing and transcription studies that the transcription unit is monocistronic.

Replacement mutagenesis

The technique of allele-replacement mutagenesis was first described by Ruvkun and Ausubel (1981) for directed mutagenesis of chromosomal genes in *Rhizobium*. They called the technique site-directed mutagenesis but this term has been replaced by allele-replacement or gene-replacement to avoid confusion of mutagenesis with oligonucleotides.

The gene to be mutated must first have been cloned. It is preferable, but not essential that the DNA sequence be known. A mutation is introduced into the coding sequence by *in vitro* recombinant DNA techniques or by transposon mutagenesis. In the former case this involves cloning a fragment of DNA capable of expressing a selectable marker (usually antibiotic resistance) into a restriction site located in the coding sequence (Fig. 1). The mutation may be a simple insertion in the restriction site or a substitution mutation where the inserted fragment replaces sequences located in the gene. The latter mutation is preferable because it creates a deletion. The second approach and the one used by Ruvkun and Ausubel (1981) in their pioneering experiments was to insert a transposon into the cloned plasmid-located gene in *E. coli*.

The next stage is to introduce the mutated gene into the wild-type host and to allow recombination to occur between the wild-type locus in the chromosome and the mutated plasmid locus. The wild-type sequences are replaced by the mutated ones (Fig. 1).

The frequency of double recombination compared to plasmid integration will depend on the length and degree of homology in sequences flanking the mutation, the length of the insertion,

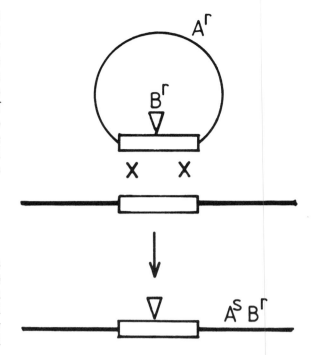

Figure 1. Allele-replacement mutagenesis. The open box represents the toxin gene to be inactivated. The circle shows that plasmid vector carrying a toxin gene inactivated by insertion of a selectable marker B^r. The marker A^r is associated with the plasmid. Recombination occurs on both sides of the insertion resulting in the mutation being introduced into the chromosomal locus. Recombinants are recognized after elimination of the plasmid by the A^s B^r phenotype.

and in the case of substitutions, the length of the deleted DNA.

Controls must be performed to ensure that *bona fide* recombination has occurred, and in the case of transposon insertions, to check that the transposon has not moved by transposition into another gene in the chromosome.

Selective methods

Recombination between plasmid and chromosome does not occur in the majority of cells carrying the vector plasmid. It is thus necessary to eliminate the plasmid from the population in order to expose the rare recombinants where the mutant allele is present in the chromosome. Two of the procedures are the same as those used to select transposon insertions, i.e. a *ts* plasmid replicon and a suicide broad host range conjugative vector.

ts vectors. The *ts* vector plasmid carrying the mutated gene is transferred into the host to be mutagenized. Growth of bacteria at a restrictive temperature in liquid medium in the absence of antibiotic selection causes plasmid-free cells to emerge. The culture is then plated on media containing an antibiotic selective for the marker associated with the mutation and colonies are screened for recombinants.

Conjugative suicide vectors. Two conjugative suicide vectors similar to those described for transposon mutagenesis have been developed for replacement mutagenesis in Gram-negative bacteria. One is based on the pSUP series of pBR325*mob* vectors (Simon *et al.*, 1983) and selects recombinants in bacteria which cannot support the replication of ColE1 (e.g. *Aeromonas hydrophila*; Chakraborty *et al.*, 1987). The other employs a vector such as pJM703.1 (Miller and Mekalanos, 1988) which requires the R6K *pir* product to replicate. This can be exploited for mutagenesis in bacteria which support the replication of ColE1-based vectors (e.g. *Shigella dysenteriae*; Fontaine *et al.*, 1988). The suicide plasmid (carrying the *in vitro*-constructed mutation) is transferred from *E. coli* to the host to be mutated. This usually occurs at a sufficiently high frequency for recombinants to be selected directly. The majority of transconjugants carry a copy of the plasmid integrated in the chromosome by a single cross-over (Fig. 2). Allele-replacement mutations occur by double recombination (Fig. 1) which is a rarer event. The double recombinants can be detected by replica-plating to detect loss of antibiotic resistance encoded by the plasmid backbone.

Plasmid incompatibility. Plasmid incompatibility has been used to eliminate vector plasmids, for example in the isolation of *Vibrio cholerae* mutations defective in cholera toxin (Mekalanos *et al.*, 1983; Kaper *et al.*, 1984a, b) and *S. aureus* mutants lacking α-toxin and protein A (O'Reilly *et al.*, 1986; Patel *et al.*, 1987). A strain carrying both the vector plasmid with the mutated toxin gene and an incompatible plasmid is constructed by appropriate mating and antibiotic selection. The bacteria are then grown in liquid medium selecting for the incompatible plasmid. Growth must be continued for a number of generations until the vector plasmid has either been eliminated totally

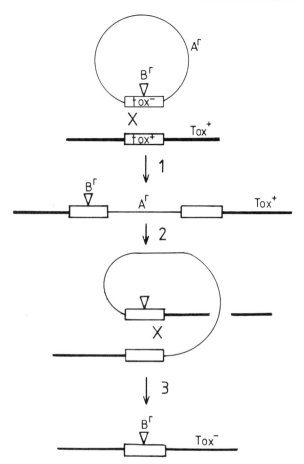

Figure 2. Allele-replacement in two steps. The open box represents the toxin gene being inactivated. The circle is the plasmid vector carrying the toxin gene inactivated by insertion of a selectable marker (B^r). (1) Integration of the plasmid into the chromosomal toxin locus by a single recombination event to give an A^r B^r Tox$^+$ derivative. (2) Loss of the plasmid by a second recombination event occurring on the other side of the inserted marker. (3) Recovery of the mutant by identification of B^r A^s Tox$^-$ derivatives.

from the population or from a sufficient proportion of it to enable recombinants to be detected. It is important that the dilution factor is less than the frequency of recombination, otherwise mutants will be lost at each dilution step. With *S. aureus* organisms are grown for 100–200 generations by performing successive 100-fold dilutions in fresh drug-containing broth.

Direct selection vectors. Vectors have been constructed which allow direct selection for a recombinational replacement mutation without having to screen large numbers of colonies or to grow extensively in broth. Most importantly drug-sensitive deletion mutations can be recombined into the chromosome without difficulty despite the absence of a selective marker for the mutation (Fig. 3). The RTP (return to pertussis) vector (Stibitz *et al.*, 1986) is a variation on the broad host range suicide plasmid theme described above. In addition to the *mob* site and narrow host range replicon the vector carries a dominant marker, a wild-type *rpsL* gene cloned from *E. coli*. The *rpsL*

gene codes for a ribosomal protein which is involved in binding of streptomycin. Str^r mutants have alterations in the *rpsL* gene. Of crucial importance for the use of the RTP vector is the fact that the sensitive allele of *rpsL* is dominant to the resistant one. Thus cells which carry both a sensitive and resistant copy are phenotypically sensitive to the drug. Resistance can only be expressed if the dominant wild-type allele is lost. This is exploited in RTP vectors because the dominant Str^s marker provides positive selection for loss of the integrated RTP plasmid. Initially a conventional drug-resistant replacement or transposon insertion mutation is selected. Then the RTP vector carrying the deleted *tox* gene is introduced into the marked Tox$^-$ Str^r strain (Fig. 3). The RTP vector integrates into the chromosome by a single cross-over between homologous sequences (Campbell-type insertion; Fig. 3). These cells are Str^s because of the dominant effect of the wild-type allele of *rpsL*. Then Str^r cells are selected. The cells express the recessive Str^r markers when the integrated plasmid is lost by excision (Fig. 3). Some recombination events occur on the other side of the recombination site used for integrating the plasmid, thus introducing the deletion mutation into the chromosome. The RTP vector will presumably function in any host in which the *E. coli rpsL* gene is expressed and the defective S12 protein can be incorporated into ribosomes.

Isolation of deletion mutations lacking a drug resistance marker: colony hybridization. Strains to be used as live vaccines must not have drug resistance markers associated with mutations. Techniques have been devised to isolate deletion mutations lacking antibiotic resistance markers.

If a positive selection vector such as RTP is not available, more laborious methods for detecting deletion mutations lacking a selective drug resistance marker must be employed. First a deletion/substitution mutation with a selectable drug resistance marker is constructed by one of the methods described above. Next a suicide vector is constructed which has a deletion in *tox* DNA sequences in a different part of the gene to the first mutation (Fig. 4). Recombinants are detected by colony hybridization using a probe which recognizes DNA sequences missing in the original substitution mutation but which are present in the second

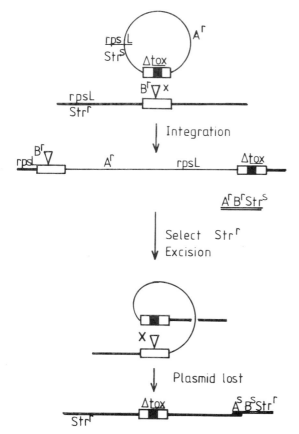

Figure 3. Isolation of drug-sensitive deletion mutants. The box represents the toxin gene to be inactivated. The return to pertussis (RTP) vector is shown by the circle. It carries the *tox* gene with a deletion mutation (Δ*tox*), a wild type copy of the *rpsL* gene and a selectable marker A^r. Integration of the suicide plasmid into the chromosome occurs via a single cross-over. The streptomycin-resistant cells become Str^s due to the dominance of the plasmid-located wild-type *rpsL* gene. Loss of the plasmid is selected by plating for Str^r derivatives. If the cross-over occurs on the other side of the Δ*tox* mutation compared to the integration event the mutation remains in the chromosome.

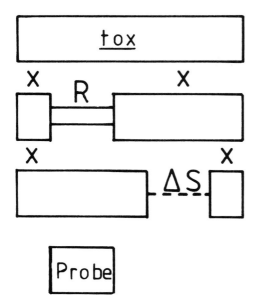

Figure 4. Blotting technique for identifying a chromosomal Δ*tox* mutation. Allele-replacement mutagenesis must be performed in two stages. The first is to introduce a substitution mutation into the chromosomal *tox* gene with sequences coding for a selectable marker R. A second round of replacement mutagenesis occurs with a plasmid carrying a Δ*tox* mutation. The deletion spans a region of the gene retained in the first mutation. Recombinants can be detected by colony hybridization using a probe from the region indicated.

construction. Deletion mutations in the enterotoxin genes of *Vibrio cholerae* have been isolated by this method (Mekalonos *et al.*, 1983).

Isolation of deletion mutations lacking a drug resistance marker: penicillin screening to enrich recombinants. The efficiency of detection of drug-sensitive mutants can be enhanced if the proportion of recombinants in the population can be enriched. A drug-resistant insertion mutation is constructed in the gene of interest using a marker which confers resistance to a bacteristatic drug and a plasmid carrying the *tox* deletion mutation is then introduced. Drug-sensitive cells are enriched by growing in the presence of the drug to which the insertion mutation confers resistance along with a bactericidal concentration of a β-lactam antibiotic. The bacteristatic antibiotic is added to an exponentially growing culture prior to the β-lactam. The drug-sensitive cells are prevented from growing by the bacteristatic drug and are thus not susceptible

to lysis by the β-lactam. Suitable markers for this procedure are tetracycline, chloramphenicol and erythromycin resistance. Aminoglycoside resistance determinants are not suitable. Deletions in the cholera enterotoxin genes have been selected by this method (Kaper *et al.*, 1984b).

Advantages and disadvantages of replacement mutations

The major advantage of replacement mutagenesis is that a precisely constructed mutation can be placed in the chromosome. The potential difficulties with polarity that could occur with insertion mutations can be overcome by constructing in-frame deletions. Intragenic deletions which result in a shift in the reading frame may also cause polar effects, although this is likely to be much weaker than polarity caused by insertions. Also the mutations are stable and drug-sensitive deletions are suitable for use as live vaccine strains.

Controls

Many of the controls described above for transposon mutants should be performed with replacement mutants. However, it is unlikely that wild-type revertants will be encountered with *in vitro* constructed mutations, particularly those involving loss of DNA sequences. Thus attributing the altered virulence phenotype to the missing toxin must be performed by restoring the Tox⁺ phenotype on a plasmid carrying the wild-type allele.

Integrating plasmid

A relatively simple method for directly inactivating a gene involves plasmid integration. The gene of interest must be cloned and mapped to allow identification of an intragenic restriction fragment. This fragment is incorporated into a suicide vector which is transferred into the host to be mutated and restrictive conditions applied. The plasmid can survive only by integration between homologous sequences on the plasmid and chromosome (Campbell-type insertion; Fig. 5). The integrated copy of the plasmid is flanked by two defective copies of the *tox* gene, thus creating a mutation. If the gene fragment in the integrating plasmid has intact 3′ or 5′ termini then one copy of the duplicated gene will still be wild-type and thus

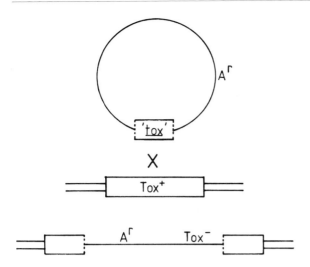

Figure 5. Gene inactivation by plasmid integration. A suicide plasmid carrying an intragenic fragment ('*tox*') from the *tox* gene to be inactivated integrates into the chromosome by a single cross-over event resulting in disruption of the chromosomal *tox* gene. The plasmid expresses a selectable marker A^r.

integration of the plasmid will not cause a mutant phenotype.

Plasmid insertion mutations are potentially unstable because the large, directly duplicated *tox* gene fragments provide substrates for homologous recombination and excision of the plasmid. However, for preliminary experiments such mutations could be very useful.

Negative bacteriophage conversion

The attachment sites for certain lysogenic bacteriophages are contained within genes coding for lipase and β-toxin of *Staphylococcus aureus*. Phage L54a lysogenizes *S. aureus* by integrating in the lipase gene (Lee and Iandolo, 1985), while several types of phage can integrate in the β-toxin gene (Coleman *et al.*, 1986). Lysogens no longer express the relevant extracellular protein and can thus be regarded as site-specific insertion mutations. The insertions are potentially unstable (because phage can excise) and could be polar. In contrast to plasmid integration, lysogenization is promoted by a site-specific phage-encoded recombinase.

Transfer of toxin genes into naturally occurring non-toxigenic strains

Another genetic approach which allows the *in vivo* role of a toxin to be studied is to introduce the *tox* gene into a naturally occurring non-toxigenic strain. This can be achieved by natural gene transfer processes (e.g. generalized transduction, lysogenization with a positive converting phage: specialized transduction or Hfr-promoted conjugation) or by recombinant DNA methods. The Tox$^+$ construct is then compared with the Tox$^-$ strain in experimental infections.

Several bacterial toxins are encoded by genes located on lysogenic bacteriophages or plasmids. Diphtheria toxin of *Corynebacterium diphtheriae* (Groman and Eaton, 1955), enterotoxin A of *Staphylococcus aureus* (Betley and Mekalanos, 1985), shiga-like toxin (Vero-toxin) of entero-invasive *Escherichia coli* (O'Brien *et al.*, 1984) and the pyrogenic exotoxin of *Streptococcus pyogenes* (Zabriskie, 1964) are bacteriophage-controlled. Enterotoxin D (Bayles and Iandolo, 1989) and epidermolytic toxin B (O'Toole and Foster, 1986) of *S. aureus*, haemolysin (Smith and Halls, 1967) and heat-labile and heat-stable enterotoxins of enterotoxigenic *E. coli* (reviewed by Elwell and Shipley, 1980; Betley *et al.*, 1986; Finn *et al.*, 1984) and the lethal toxin of *Bacillus anthracis* (Mikesell *et al.*, 1983) are plasmid-encoded. Also chimaeric plasmids carrying cloned *tox* genes could be transferred into avirulent non-toxigenic strains as was done in the analysis of TSST-1 of *Staphylococcus aureus* (de Azevedo *et al.*, 1985).

Controls must be performed to demonstrate that virulence is due to the toxin expressed by the accessory genetic element. The best control is an isogenic plasmid with a deletion or insertion mutation in the *tox* structural gene which is also transferred into the naturally occurring Tox$^-$ host. This approach has been employed to study the *in vivo* role of TSST-1 of *S. aureus* and the haemolysin of *Escherichia coli*.

Genetic manipulation of toxin genes in Gram-positive pathogens

Streptococci and *Listeria monocytogenes*

Mutants of streptococci and *L. monocytogenes* defective in haemolysin expression have been used in infection models to demonstrate involvement of the missing toxins in virulence. The conjugative streptococcal transposons Tn*916* and Tn*1545* were used to generate mutants defective in the haemolysin (listeriolysin) of *L. monocytogenes*, the β-haemolysin of group B streptococci and the haemolysin of *Streptococcus faecalis*. A pneumolysin-deficient mutant of *S. pneumoniae* was isolated by directed plasmid insertion.

The H1y$^-$ mutants of *S. faecalis* had a reduced LD$_{50}$ when injected intraperitoneally into mice (Ike *et al.*, 1984). This model was considered to be relevant to human peritoneal sepsis which can be caused by this organism. Thus the haemolysin may be a major virulence factor in peritonitis.

In contrast, the β-haemolysin-defective mutants of a group B streptococcus had the same LD$_{50}$ as the wild-type when injected subcutaneously in rats (Weiser and Rubens, 1987). However, the authors pointed out that systemic infection was being tested by their model. Perhaps the β-haemolysin contributes to resistance to local defences such as might occur at mucosal surfaces.

Studies with the haemolysin-defective mutants of *L. monocytogenes* have clearly established that this toxin is a major virulence factor in systemic disease (Gaillard *et al.*, 1986; Kathariou *et al.*, 1987). Hly$^-$ mutants had a lower LD$_{50}$ after i.v. inoculation into mice. In infections with wild-type bacteria abscesses were formed in the liver and spleen and there was a rapid increase in bacterial counts recovered from these organs (Kathariou *et al.*, 1987). In contrast, bacterial numbers declined after infection with a low dose of an Hly$^-$ mutant. However, if a large inoculum was used the number of viable bacteria recovered from spleens did not decline but the majority of organisms were Hly$^+$ revertants. This shows that haemolytic bacteria have a growth advantage *in vivo*, probably because they can survive intracellularly following phagocytosis. Recent studies have shown that *L. monocytogenes* is rapidly endocytosed by non-professional phagocytes and that a novel extracellular protein is required to stimulate this process (Kuhn and Goebel, 1989). Haemolysin is not involved in induced phagocytosis but is required for intracellular survival and seems to disrupt the phagosomal membrane, allowing bacteria to escape into the cytoplasm and evade lysosomal killing mechanisms (Kuhn *et al.*, 1988).

An important role for the thiol-activated toxin pneumolysin in the virulence of *Streptococcus pneumoniae* was suggested by mouse infection experiments with a mutant derived by directed plasmid insertion (Berry *et al.*, 1989). The LD$_{50}$ of the Tox$^-$ mutant was 10–100-fold higher than the wild-type. Also the mutants were cleared more rapidly from blood, suggesting that the toxin interferes with phagocytic clearance *in vivo*, as it has been shown to do *in vitro* (Nandoskar *et al.*, 1986). Virulence was restored in a wild-type derivative isolated by back transformation (Berry *et al.*, 1989). It is interesting to note that the loss of pneumolysin by *S. pneumoniae* had a less severe effect on virulence for mice compared with a listeriolysin-deficient mutant of *L. monocytogenes* (Gaillard *et al.*, 1986).

Staphylococcus aureus

S. aureus is a complex organism which can express many potential virulence factors, including five cytolytic toxins as well as enterotoxins, toxic-shock syndrome toxin and epidermolytic toxins. Mutants of *S. aureus* defective in α-toxin and β-toxin have been isolated by replacement mutagenesis and bacteriophage conversion, respectively (O'Reilly *et al.*, 1986; Patel *et al.*, 1987; Bramley *et al.*, 1989). This has permitted investigation of the *in vivo* role of these toxins in mouse infection models.

Mutants lacking α-toxin had a much reduced ability to kill mice after intraperitoneal or intra-mammary (i.m.) inoculation (Patel *et al.*, 1987), confirming that α-toxin is the major lethal toxin of *S. aureus*. Also, when injected subcutaneously (s.c.) in mice, smaller abscesses were formed by α-toxin-deficient mutants than by the wild-type, suggesting that the toxin has an important role in combating phagocytic cells and causing necrosis of surrounding host tissues (O'Reilly *et al.*, 1986; Patel *et al.*, 1987). When injected into the mouse mammary gland histopathological examination showed that bacteria quickly became associated with macrophages, but that these failed to eliminate the bacteria (Bramley et al., 1989). Indeed, bacteria appeared to proliferate inside macrophages and eventually to lyse. After 24 h both in the s.c. and the i.m. models macrophages were absent from the area of bacterial growth. This may have been due in part to the anti-chemotactic effect of α-toxin.

It has been proposed from studies with cultured endothelial cells that α-toxin plays an important role in the pathogenesis of endocarditis (Vann and Proctor, 1988). Bacteria adhere to the surface of the endothelial cells and become phagocytosed (Lowy et al., 1988). The endothelial cells were killed when they were infected by an α-toxin-producing strain but they survived uptake of an α^- mutant. By extrapolating these events to the in vivo situation, a model for invasive endocarditis caused by S. aureus has been proposed. The first stage in formation of a vegetative growth in the endocardium is infection and lysis of endothelial cells by α-toxin-producing bacteria. A fibrin clot forms in the area of tissue damage and a focus of bacterial growth occurs. It will be interesting to determine the ability of the α^- mutant to cause endocarditis in animals.

Prior to the study by Bramley et al. (1989) with mutants there was no strong evidence that β-toxin had a role in staphylococcal pathogenesis. However, it appears that β-toxin plays a role in promoting bacterial growth in the infected mouse mammary gland. An $\alpha^- \beta^-$ double mutant grew poorly in the mammary gland while an $\alpha^- \beta^+$ strain was recovered in much higher numbers. It was interesting to note that both the $\alpha^- \beta^+$ and the $\alpha^+ \beta^-$ mutants actually grew to higher levels than the $\alpha^+ \beta^+$ wild-type strain. This is possibly due to the antagonism that has been found to occur between the toxins in vitro when cytolytic activity was measured (Elek and Levy, 1954). This antagonism did not extend to lethal effects because $\alpha^+ \beta^-$ strains caused similar numbers of deaths to the wild-type strain. Histopathological examination suggested that, in contrast to α-toxin, β-toxin actually promoted a neutrophil influx. However, pathological changes were attributed to β-toxin, including some neutrophil killing and some damage to secretory epithelial cells.

It will be interesting to extend these infection studies to the lactating dairy cow to determine if β-toxin is important in bovine mastitis. Epidemiological evidence of bovine mastitis caused by human strains of S. aureus suggests that β-toxin may well be a virulence factor and that selection may occur in vivo for β-toxigenic strains (Hummel et al., 1978). Occasionally mastitis was shown to be caused by infection of cows with human commensal strains which were predominantly non-β-toxigenic staphylokinase producers, probably due to lysogenization with a double converting phage. The bovine isolates were identical in all respects to the human strains apart from expression of β-toxin and loss of staphylokinase production, presumably due to selection of variants cured of the lysogenic phage.

Genetic manipulation has contributed to the understanding that toxic-shock syndrome toxin-1 is responsible for the symptoms of toxic shock seen in a rabbit uterine model (de Azavedo et al., 1985). The cloned tss gene was introduced into a naturally occurring TSST-1$^-$ isolate on a plasmid. The bacteria were placed in a diffusion chamber and surgically implanted into the uterus of rabbits. The majority of animals infected with the TSST-1$^+$ organisms died, while those infected with bacteria carrying the plasmid with a deletion in the tss gene survived. The tss gene in TSS-associated clinical isolates has recently been inactivated by gene replacement (B.N. Kreiswirth, R.P. Novick, R. Sloane, J.C.S. de Azevedo, P.J. Hartigan, J.P. Arbuthnott and T.J. Foster, unpublished results). In vivo tests with these mutants confirmed the findings described above.

Genetic manipulation of toxin genes of Gram-negative pathogens

Aeromonas hydrophila

Aeromonas hydrophila causes infections in mammals and in fish. Genetic studies have demonstrated that the haemolysin aerolysin is a major factor in systemic infections in mice, and by extrapolation in human infections (Chakraborty et al., 1987). An aerolysin-deficient mutant with a substitution mutation in the structural gene was isolated by replacement mutagenesis using a broad host range conjugative suicide vector. The mutant had a higher LD_{50} in intraperitoneal and subcutaneous infections. Fewer bacteria were isolated from livers and spleens of animals infected with mutants. The dermonecrosis formed after subcutaneous injection with the mutant was less severe and healed more quickly.

Analysis of mutants has shown that protease is a major factor in the ability of *A. hydrophila* to cause haemorrhagic septicaemia in fish (Leung and Stevenson, 1988). This is consistent with the finding that purified protease from *Aeromonas salmonicida* causes tissue damage resembling the furuncular lesion caused by bacterial infection when injected intramuscularly into fish (Fyfe *et al.*, 1986). Protease-deficient mutants of *A. hydrophila* were isolated by transposon Tn5 mutagenesis using a broad host range suicide vector pJB4JI. The mutants had a higher LD_{50} after intramuscular injection. No liquefaction occurred at the site of injection and bacterial numbers were reduced over a period of 7 days. Mutant bacteria could not be recovered from kidneys, showing that systemic spread did not occur. Protease-producing bacteria survived better in fish serum *in vitro* because the protease supports bacterial growth by providing nutrients and lessens serum killing. Thus wild-type bacteria can survive better *in vivo* and can spread more rapidly.

The role of haemolysin in extra-intestinal infections caused by Escherichia coli

Some strains of *E. coli* cause extra-intestinal infections in man. Many pathogenic strains produce an extracellular haemolysin which may be involved in pathogenesis.

Gene cloning and transposon mutagenesis have clearly shown that the haemolysin is an important virulence factor in extra-intestinal infection models in rats (Welch *et al.*, 1981; Welch and Falkow, 1984; Marré *et al.*, 1986). However, pathogenesis is multifactorial and factors other than haemolysin are required for full virulence. The peritonitis model mimics the human disease with inflammation and multiple abscess formation. The cloned *hly* determinant was introduced into a non-haemolytic faecal strain of *E. coli*. Haemolysin was shown to be a major cause of symptoms because bacteria harbouring a plasmid with a transposon Tn*1* insertion in the *hly* gene were less virulent (Welch *et al.*, 1981). Also, more deaths occurred in rats

infected with strains which expressed high levels of haemolysin *in vitro* (Welch and Falkow, 1984).

Marré *et al.* (1986) isolated a spontaneous mutant of a pyelonephritic strain of *E. coli* which failed to express adhesins, serum resistance and haemolysin. The mutant had reduced virulence in the rat pyelonephritis model because lower numbers of bacteria were recovered from infected kidneys. The missing factors were returned to the mutant strain in a step-wise fashion using recombinant plasmids. The presence of haemolysin alone caused a small increase in virulence. An engineered strain expressing the three traits had a similar level of virulence to the wild-type, clearly showing that the pathogenesis of pyelonephritis is multifactorial.

The role of shiga toxin in the pathogenesis of dysentery

Site-specific mutants defective in shiga toxin have proved invaluable in elucidating the function of the toxin in the pathogenesis of dysentery.

The shiga toxin gene of *Shigella dysenteriae* type I was inactivated by replacement mutagenesis using a conjugative suicide vector (Fontaine *et al.*, 1988). *In vitro* experiments with cultured cells showed that the toxin was not involved in promoting invasion and intracellular growth of bacteria or in the killing of infected cells. This is consistent with results from a detailed analysis of dysentery caused by intragastric inoculation of bacteria in macaque monkeys. Both Tox$^+$ and Tox$^-$ bacteria caused lethal dysentery. The amount of pus and mucus in the stools was indistinguishable and the number and severity of abscesses in the gastric mucosa were similar. The major difference was the absence of blood in the stools of animals infected with the Tox$^-$ mutant and the lack of haemorrhage in the colon. Histopathology suggested that toxin was responsible for destroying blood capillaries in connective tissue and for an influx of inflammatory cells into the lumen of the intestine. Thus shiga toxin is not a diarrhoeaogenic toxin acting on intestinal epithelia but is necrotic and haemorrhagic towards the vascular system in the connective tissue of the colon.

The role of toxins in pathogenesis of Bordetella pertussis infections

B. pertussis expresses several toxins which are likely to be of importance in human disease (Weiss and

Hewlett, 1986). These include an extracellular adenylate cyclase, the pertussis toxin and a haemolysin. Pertussis toxin is an ADP-ribosyltransferase which modifies a GTP-binding protein inside susceptible cells (Weiss and Hewlett, 1986). The notion that the toxin is a major virulence factor was studied using Tox$^-$ Tn5 mutants and a deletion mutant (Weiss et al., 1984; Monack et al., 1989) and by transferring the pertussis toxin genes on a recombinant plasmid into the normally non-toxigenic but closely related organism B. parapertussis (Monack et al., 1989). Expression of pertussis toxin was clearly associated with the leukocytotoxic and reactogenic activities of B. pertussis (Monack et al., 1989).

The B. pertussis extracellular adenylate cyclase has the unusual property of being activated by the eukaryotic regulatory protein calmodulin (Glaser et al., 1988). It appears that cyclase enters eukaryotic cells and elicits unregulated synthesis of cAMP. Recently genetic analysis suggested that haemolysin and adenylate cyclase activities are specified by the same protein (Glaser et al., 1988). The failure of Weiss et al. (1983) to obtain a Hly$^+$ Adc$^-$ mutant could be explained if the cyclase activity was located in an N-terminal domain of the protein and the haemolysin in the C-terminus. The Tn5 mutant which caused the Adc$^+$ Hly$^-$ phenotype presumably expressed a truncated protein lacking the haemolysin domain.

A Tn5 mutant defective only in haemolysin showed a moderate increase (100-fold) in LD$_{50}$ (Weiss et al., 1984). A double mutant defective in both adenylate cyclase and haemolysin had a >10^4-fold increase in LD$_{50}$ compared to the wild-type. This suggests that both activities are important in virulence.

Vibrio cholerae mutants defective in enterotoxin

Initially the broad host range plasmid incompatibility method first reported by Ruvkun and Ausubel (1981) was used to isolate replacement mutations defective in cholera enterotoxin (Ctx). This can be now achieved much more readily with suicide vectors.

There is no doubt that the toxin is the major cause of diarrhoea in infected patients (Finkelstein, 1988). One reason for isolating Ctx$^-$ mutants was to evaluate the possibility of using such strains as live vaccines (Mekalanos et al., 1983; Kaper et al.,

1984a,b). For approval by regulatory authorities this required isolating deletion mutants which did not express antibiotic resistance. Two rounds of replacement mutagenesis were necessary to achieve this. Initially drug-resistant substitution mutations were isolated which were then replaced by a ctx deletion.

An additional complication was the need to isolate Ctx$^-$ mutants of the Classical biotype of V. cholerae which carries two unlinked copies of the ctx genes (Pearson and Mekalanos, 1982). (Early Ctx$^-$ derivatives of V. cholerae Classical strains were almost certainly defective in the toxR regulatory gene; Mekalanos, 1985.) Recombination had to occur twice in the same cell to inactivate both copies of ctx. One type of ctx mutation was an in-frame deletion in the ctxA gene which removed the enzymatically active A1 domain of the toxin but allowed expression of a truncated peptide expressing the A2 domain and subunit B. It was hoped that immunization with this mutant would provide antitoxic immunity without symptoms of enterotoxic activity. Another mutant had a deletion which caused an A$^-$B$^-$ phenotype.

Human volunteer trials performed with genetically engineered strains showed that immunity to challenge with a wild-type Ctx$^+$ organism was conferred by exposure to both A$^-$B$^-$ and A$^-$B$^+$ mutants (Levine et al., 1988). The latter induced both antitoxic and bactericidal immunity while the former only induced antibacterial immunity. Thus protection can be achieved without antitoxic immunity, presumably involving antibodies which prevent colonization or which kill attached bacteria before they have a chance to proliferate and synthesize enterotoxin. However, immunization produced a mild diarrhoea in most vaccinees. In order to identify factors responsible for residual diarrhoea the haemolysin was inactivated by replacement mutagenesis in a Ctx$^-$ vaccine strain, but this did not reduce reactogenicity (Levine et al., 1988). It is suspected that the shiga-like toxin (SLT) produced by some strains of V. cholerae is responsible. A naturally occurring Slt$^-$ strain in which a Ctx$^-$ mutation was constructed caused a much less severe reaction. Construction of a double Ctx$^-$ Slt$^-$ mutant would be very useful for sorting out this problem.

References

Bayles, K.W. and Iandolo, J.J. (1989). Genetic and molecular analyses of the gene encoding staphylococcal enterotoxin D. *J. Bacteriol.* **171**, 4799–806.

Berg, C.M., Berg, D.E. and Groisman, E.A. (1989). Transposable elements and the genetic engineering of bacteria. In: *Mobile DNA* (eds D.E. Berg and M.M. Howe), pp. 879–925. American Society for Microbiology, Washington, DC.

Beringer, J.E., Beynon, J.L. Buchanan-Wollaston, A.V. and Johnston, A.W.B. (1978). Transfer of the drug resistance transposon Tn5 to *Rhizobium*. *Nature (Lond.)* **276**, 633–4.

Berry, A.M., Yother, J., Briles, D.E., Hansman, D. and Paton, J.C. (1989). Reduced virulence of a defined pneumolysin-negative mutant of *Streptococcus pneumoniae*. *Infect. Immun.* **57**, 2037–42.

Betley, M.J. and Mekalanos, J.J. (1985). Staphylococcal enterotoxin A is encoded by a phage. *Science* **229**, 185–7.

Betley, M.J., Miller, V.J. and Mekalanos, J.J. (1986). Genetics of bacterial enterotoxins. *Ann. Rev. Microbiol.* **40**, 577–605.

Bramley, A.J., Patel, A.H., O'Reilly, M., Foster, R. and Foster, T.J. (1989). Roles of alpha-toxin and beta-toxin in virulence of *Staphylococcus aureus* for the mouse mammary gland. *Infect. Immun.* **57**, 2489–94.

Chakraborty, T., Huhle, B., Hof, H., Bergbauer, H. and Goebel, W. (1987). Marker exchange mutagenesis of the aerolysin determinant in *Aeromonas hydrophila* demonstrates the role of aerolysin in *A. hydrophila*-associated systemic infections. *Infect. Immun.* **55**, 2274–80.

Clewell, D.B. and Gawron-Burke, C. (1986). Conjugative transposons and the dissemination of antibiotic resistance in streptococci. *Ann. Rev. Microbiol*, **40**, 635–59.

Coleman, D.C., Arbuthnott, J.P., Pomeroy, H.M. and Birkbeck, T.H. (1986). Cloning and expression in *Escherichia coli* and *Staphylococcus aureus* of the beta-lysin determinant from *Staphylococcus aureus*: evidence that bacteriophage conversion of beta-lysin activity is caused by insertional inactivation of the beta-lysin determinant. *Microb. Pathogen.* **1**, 549–64.

de Azevedo, J.C.S., Foster, T.J., Hartigan, P.J., Arbuthnott, J.P., O'Reilly, M., Kreiswirth, B.N. and Novick, R.P. (1985). Expression of the cloned toxic shock syndrome toxin 1 gene (*tst*) in vivo with a rabbit uterine model. *Infect. Immun.* **50**, 304–9.

de Bruijn, F.J. and Lupski, J.R. (1984). The use of transposon Tn5 mutagenesis in the rapid generation of correlated physical and genetic maps of DNA segments cloned into multicopy plasmids – a review. *Gene* **27**, 131–49.

Dougan, G., Maskell, D., Pickard, D. and Hormaeche, C. (1987). Isolation of stable *aroA* mutants of *Salmonella typhi* Ty2: properties and preliminary characterization in mice. *Molec. Gen. Genet.* **207**, 402–5.

Edwards, M.F. and Stocker, B.A.D. (1988). Construction of ΔaroA his Δpur strains of *Salmonella typhi*. *J. Bacteriol.* **170**, 3991–5.

Elek, S.D. and Levy, E. (1954). The nature of discrepancies between haemolysins in culture filtrates and plate haemolysin patterns of staphylococci. *J. Pathol. Bacteriol.* **68**, 31–40.

Elwell, L.P. and Shipley, P.L. (1980). Plasmid-mediated factors associated with virulence of bacteria to animals. *Ann. Rev. Microbiol.* **34**, 465–96.

Finkelstein, R.A. (1988). Cholera, the cholera enterotoxins, and the cholera enterotoxin-related enterotoxin family. In: *Immunochemical and Molecular Genetic Analysis of Bacterial Pathogens*, FEMS Symposium No. 40 (eds P. Owen and T.J. Foster), pp. 85–102. Elsevier, Amsterdam.

Finn, C.W., Silver, R.P., Habig, W.H. and Hardegree, M.C. (1984). The structural gene for tetanus toxin is on a plasmid. *Science* **224**, 881–4.

Fontaine, A., Arondel, J. and Sansonetti, P.J. (1988). Role of shiga toxin in the pathogenesis of bacillary dysentery, studied by using a Tox⁻ mutant of *Shigella dysenteriae* 1. *Infect. Immun.* **56**, 3099–109.

Foster, T.J. (1984). Analysis of plasmids with transposons. In: *Methods in Microbiology*, Vol. 17 (eds P.M. Bennett and J. Grinsted), pp. 197–226. Academic Press, London.

Foster, T.J., Lundblad, V., Hanley-Way, S. and Kleckner, N. (1981). Three Tn10 associated excision events: relationship to transposition and role of direct and inverted repeats. *Cell* **23**, 215–27.

Fyfe, L., Finley, A., Coleman, G. and Munro, A.L.S. (1986). A study of the pathological effect of isolated *Aeromonas salmonicida* extracellular protease on Atlantic salmon, *Salmo salar* L. *J. Fish Dis.* **9**, 403–9.

Gaillard, J.L., Berche, P. and Sansonetti, P. (1986). Transposon mutagenesis as a tool to study the role of hemolysin in the virulence of *Listeria monocytogenes*. *Infect. Immun.* **52**, 50–5.

Glaser, P., Ladant, D., Sezer, O., Pichot, F., Ullmann, A. and Danchin, A. (1988). The calmodulin-sensitive adenylate cyclase of *Bordetella pertussis*: cloning and expression in *Escherichia coli*. *Molec. Microbiol.* **2**, 19–30.

Groman, N.B. and Eaton, M. (1955). Genetic factors in *Corynebacterium diphtheriae* conversion. *J. Bacteriol.* **70**, 637–40.

Hummel, R., Witte, W. and Kemmer, G. (1978). Zur frage der wechselseitigen übertragung von *Staphylococcus aureus* zwischen mensch und rind und der milieuadaptation der hamolysin- und fibrinolysinbildung. *Arch. Exp. Vet. Med. Leipzig* **32**, 287–98.

Ike, Y., Hashimoto, H. and Clewell, D.B. (1984). Hemolysin of *Streptococcus faecalis* subspecies *zymogenes* contributes to virulence in mice. *Infect. Immun.* **45**, 528–30.

Kaper, J.B., Lockman, H., Baldini, M.M. and Levine, M.M. (1984a). Recombinant nontoxigenic *Vibrio cholerae* strains as attenuated cholera vaccine candidates. *Nature* **308**, 655–8.

Kaper, J.B., Lockman, H., Baldini, M.M. and Levine, M.M. (1984b). A recombinant live oral cholera vaccine. *Biotechnology* **1**, 345–9.

Kathariou, S., Metz, P., Hof, H. and Goebel, W. (1987). Tn*916*-induced mutations in the hemolysin determinant affecting virulence of *Listeria monocytogenes*. *J. Bacteriol.* **169**, 1291–7.

Kuhn, M. and Goebel, W. (1989). Identification of an extracellular protein of *Listeria monocytogenes* possibly involved in intracellular uptake by mammalian cells. *Infect. Immun.* **57**, 55–61.

Kuhn, M., Kathariou, S. and Goebel, W. (1988). Hemolysin supports survival but not entry of the intracellular bacterium *Listeria monocytogenes*. *Infect. Immun.* **56**, 79–82.

Laird, W.J., Aaronson, W., Silver, R.P., Habig, W.H. and Hardegree, W.C. (1980). Plasmid-associated toxigenicity in *Clostridium tetani*. *J. Infect. Dis.* **142**, 623.

Lee, C.Y. and Iandolo, J.J. (1985). Mechanism of bacteriophage conversion of lipase activity in *Staphylococcus aureus*. *J. Bacteriol.* **164**, 288–93.

Leung, K.Y. and Stevenson, R.M.W. (1988). Tn*5*-induced protease-deficient strains of *Aeromonas hydrophila* with reduced virulence for fish. *Infect. Immun.* **56**, 2639–44.

Levine, M.M., Kaper, J.B., Herrington, D., Losonsky, G., Morris, J.G., Clements, M.L., Black, R.E., Tall, B. and Hall, R. (1988). Volunteer studies of deletion mutants of *Vibrio cholerae* 01 prepared by recombinant techniques. *Infect. Immun.* **56**, 161–7.

Lowy, F.D., Fant, J., Higgins, L.L., Ogawa, S.K. and Hatcher, V.B. (1988). *Staphylococcus aureus* – human endothelial cell interactions. *J. Ultrastruct. Molec. Struc. Res.* **98**, 137–46.

Manoil, C. and Beckwith, J. (1985). Tn*phoA*: a transposon probe for protein export signals. *Proc. Natl. Acad. Sci. USA* **82**, 8129–33.

Marré, R., Hacker, J., Henkel, W. and Goebel, W. (1986). Contribution of cloned virulence factors from uropathogenic *Escherichia coli* strains to nephropathogenicity in an experimental rat pyelonephritis model. *Infect. Immun.* **54**, 761–7.

Mekalanos, J.J. (1983). Duplication and amplification of toxin genes in *Vibrio cholerae*. *Cell* **35**, 253–63.

Mekalanos, J.J. (1985). Cholera toxin: genetic analysis, regulation, and role in pathogenesis. *Curr. Topics Microbiol. Immunol.* **118**, 97–118.

Mekalanos, J.J., Swartz, D.J., Pearson, G.N.D., Harford, N., Groyne, F. and de Wilde, M. (1983). Cholera toxin genes: nucleotide sequence, deletion analysis and vaccine development. *Nature* **306**, 551–7.

Mikesell, P., Ivins, B.E., Ristroph, J.D. and Dreier, T.M. (1983). Evidence for plasmid-mediated toxin production in *Bacillus anthracis*. *Infect. Immun.* **39**, 371–6.

Miller, V.L. and Mekalanos, J.J. (1988). A novel suicide vector and its use in construction of insertion

mutations: osmoregulation of outer membrane proteins and virulence determinants in *Vibrio cholerae* requires *toxR*. *J. Bacteriol.* **170**, 2575–83.

Monack, D., Munoz, J.J., Peacock, M.G., Black, W.J. and Falkow, S. (1989). Expression of pertussis toxin correlates with pathogenesis in *Bordetella* species. *J. Infect. Dis.* **159**, 205–10.

Nandoskar, M., Ferrante, A., Bates, E.J., Hurst, N. and Paton, J.C. (1986). Inhibition of human monocyte respiratory burst, degranulation, phospholipid methylation and bactericidal activity by pneumolysin. *Immunology* **59**, 515–20.

O'Brien, A.D., Newland, J.W., Holmes, R.K., Smith, H.W. and Formal, S.B. (1984). Shiga-like toxin-converting phages from *Escherichia coli* strains that cause hemorrhagic colitis or infantile diarrhea. *Science* **226**, 294–6.

O'Reilly, M., de Azevedo, J.C.S., Kennedy, S. and Foster, T.J. (1986). Inactivation of the alpha-haemolysin gene of *Staphylococcus aureus* 8325-4 by site-directed mutagenesis and studies on expression of its haemolysins. *Microb. Pathogen.* **1**, 125–38.

O'Toole, P.W. and Foster, T.J. (1986). Epidermolytic toxin serotype B of *Staphylococcus aureus* is plasmid-encoded. *FEMS Microbiol. Lett.* **36**, 311–14.

Patel, A.H., Nowlan, P., Weavers, E.D. and Foster, T.J. (1987). Virulence of protein A-deficient mutants of *Staphylococcus aureus* isolated by allele-replacement. *Infect. Immun.* **55**, 3103–10.

Pearson, G.N.D. and Mekalanos, J.J. (1982). Molecular cloning of the *Vibrio cholerae* enterotoxin genes in *Escherichia coli*. *Proc. Natl. Acad. Sci. USA* **79**, 2976–80.

Peterson, K.M. and Mekalanos, J.J. (1988). Characterization of the *Vibrio cholerae* ToxR regulon: identification of novel genes involved in intestinal colonization. *Infect. Immun.* **56**, 2822–9.

Ruvkun, G.B. and Ausubel, F.M. (1981). A general method for site-directed mutagenesis in prokaryotes. *Nature* **289**, 85–9.

Simon, R., Priefer, U. and Puhler, A. (1983). A broad host range mobilization system for *in vivo* genetic engineering: transposon mutagenesis in Gram negative bacteria. *Biotechnology* **1**, 784–91.

Simon, R., O'Connell, M., Labes, M. and Puhler, A. (1986). Plasmid vectors for the genetic analysis and manipulation of *Rhizobia* and other Gram-negative bacteria. In: *Methods in Enzymology*, Vol. 118 (eds A. Weissbach and H. Weissbach), pp. 640–59. Academic Press, London and New York.

Smith, H.W. and Halls, S. (1967). The transmissible nature of the genetic factor in *Escherichia coli* that controls haemolysin production. *J. Gen. Microbiol.* **47**, 153–61.

Sparling, P.F. (1983). Applications of genetics to studies of bacterial virulence. *Phil. Trans. R. Soc. Lond.* **B303**, 199–207.

Stibitz, S., Black, W. and Falkow, S. (1986). The construction of a cloning vector designed for gene replacement in *Bordetella pertussis*. *Gene* **50**, 133–40.

Taylor, R.K., Miller, V.L., Furlong, D.B. and Mekalanos, J.J. (1987). Use of *phoA* gene fusions to identify a pilus colonization factor coordinately regulated with cholera toxin. *Proc. Natl. Acad. Sci. USA* **84**, 2833–7.

Vann, J.M. and Proctor, R.A. (1988). Cytotoxic effects of ingested *Staphylococcus aureus* on bovine endothelial cells: role of *S. aureus* α-hemolysin. *Microb. Pathogen.* **4**, 443–53.

Weiser, J.N. and Rubens, C.E. (1987). Transposon mutagenesis of group B streptococcus beta-hemolysin biosynthesis. *Infect. Immun.* **55**, 2314–16.

Weiss, A.A. and Hewlett, E.L. (1986). Virulence factors of *Bordetella pertussis*. *Ann. Rev. Microbiol.* **40**, 661–86.

Weiss, A.A., Hewlett, E.L., Myers, G.A. and Falkow, S. (1983). Tn5-induced mutations affecting virulence factors of *Bordetella pertussis*. *Infect. Immun.* **42**, 33–41.

Weiss, A.A., Hewlett, E.L., Myers, G.A. and Falkow, S. (1984). Pertussis toxin and extracytoplasmic adenylate cyclase as virulence factors of *Bordetella pertussis*. *J. Infect. Dis.* **150**, 219–22.

Welch, R.A. and Falkow, S. (1984). Characterization of *Escherichia coli* hemolysins conferring quantitative differences in virulence. *Infect. Immun.* **43**, 156–60.

Welch, R.A., Dellinger, E.P., Minshew, B. and Falkow, S. (1981). Haemolysin contributes to virulence of extraintestinal *E. coli* infections. *Nature* **294**, 665–7.

Zabriskie, J.B. (1964). The role of temperate bacteriophage in the production of erythrogenic toxin by group A streptococci. *J. Exp. Med.* **119**, 761–79.

21

Effects of Bacterial Toxins on Activity and Release of Immunomediators

W. König, J. Scheffer, J. Knöller, W. Schönfeld, J. Brom and M. Köller

RUHR-Universität Bochum, Medizinische Mikrobiologie und Immunologie, Arbeitsgruppe für Infektabwehr, Universitätsstraße 150, Postfach 150, 4630 Bochum 1, Germany

Role of effector cells (granulocytes, neutrophils, eosinophils, basophils, monocytes) in microbial infection

The emigration of inflammatory cells from the blood into the tissues represents one of the most important components of the inflammatory response (Bainton, 1988). In this regard a variety of mediators (cytokines: e.g. GM-CSF, IL1, IL3, IL5, IL8) as well as factors derived after cell activation (preformed or newly generated) are involved in the regulation of granulocyte recruitment. Microenvironmental factors (cell, tissue, microbial products) and mediators of inflammation may then determine the composition of the cellular inflammatory response (Fantone and Ward, 1982; König et al., 1988). The phagocytes not only respond to an external signal by directed chemoattraction but in addition are activated to generate bactericidal components (e.g. oxygen radicals) and also a variety of intracellular constituents (Becker, 1988; Gemmell, 1988; Rossi et al., 1983, 1989). It is evident by now that products of the inflammatory cascade modulate the lymphocyte response, thus creating a link between acute and chronic inflammatory events. The composition of the various mediators as well as the presence of the cell adhesion molecules may determine to what extent neutrophils, eosinophils and basophils penetrate into the inflamed tissue (Wawryk et al., 1989). Evidently, the inflammatory response is beneficial if the responsible stimulus can be eliminated or

may be deleterious to the host if the formation is excessive or the inactivation of the mediator inadequate. In allergic and inflammatory reactions attention has been focused on the composition and the contribution of cells at the various sites in recent studies.

It is well established that the various effector cells of the inflammatory response exert distinct functions. In this regard mast cells with a multitude of preformed and newly generated mediators are involved in acute as well as chronic inflammatory reactions (Ishizaka et al., 1987). Activation of mast cells via immunological and non-immunological mechanisms leads to the formation of lipid mediators (hydroxyeicosatetraenoic acids (HETEs), leukotrienes, platelet-activating factor), granular constituents (histamine, heparin, neutral proteases) as well as cytokines (GM-CSF, IL3, IL4, IL6) (Robinson and Holgate, 1985; Plaut et al., 1989; Wodnar-Filipowicz et al., 1989). The mast cells via their activation products not only determine the acute inflammatory response but also exert an important role in the influx of additional inflammatory cells. In this regard the dual role of eosinophils in inflammation has been recently elaborated (Tonnel et al., 1989).

Evidence has been provided that activated mast cells via preformed and newly generated mediators attract eosinophils to the inflammatory focus. Eosinophils are involved in protection, both by their downregulatory effects of immediate type hypersensitivity reactions and by their cytotoxic activity against parasitic larvae. They contribute to tissue damage by the release of cationic proteins which exert cytocidal functions. An eosinophilia is induced by IL5; IL3 alters the density of eosinophils which leads to a change in the secretory pattern for lipid mediators. Mediators with eosinophil chemotactic properties are LTB_4 for guinea-pig eosinophils and the platelet-activating factor (PAF) for human eosinophils (König et al., 1976; Wardlaw et al., 1986). Inflammatory mediators may also recruit basophils, which release histamine, leukotrienes and other cellular constituents and also modulate the releasability of these cells. Neutrophil leukocytes are chiefly involved in host defence against microorganisms. They are activated by chemotactic agonists generated upon infection. Subsequently, the neutrophils migrate out of the microvessels into the infected tissues. Infectious microbes are usually phagocytosed and killed. Upon chemotactic and phagocytotic activation the neutrophils generate superoxide and H_2O_2, a variety of bioactive lipids, release enzymes and other storage proteins (Ward et al., 1983; Baggiolini and Dewald, 1984). Some of these products are required for the killing; several of them may, however, induce inflammation and tissue damage as a consequence of neutrophil activation.

The inflammatory scenario is now envisaged as a result of a multitude of mediators among which cytokines and low molecular weight mediators act together. Cytokines (IL3, GM-CSF, IL8) may prime the target cells or contribute to a differential release of low molecular weight mediators (Atkinson et al., 1988a,b). In this regard microbial exotoxins as released effector molecules or expressed with additional microbial components, e.g. adhesins or haemagglutinins, are powerful inducers of the inflammatory response (König et al., 1985). In contrast to the cytotoxic functions, their role in the activation of immunocompetent cells as well as inflammatory effector cells is of utmost importance under physiological and pathophysiological conditions.

Important to the modulation of inflammation are several classes of lipid autacoids which have the collective ability to orchestrate virtually every aspect of the inflammatory process by primary, secondary and/or synergistic actions (Lewis and Austen, 1984; Parker, 1987; Salmon and Higgs, 1987; Bach, 1988). These lipid factors are not stored as preformed mediators (such as histamine,

serotonin, lysosomal enzymes) but are rapidly generated after cell stimulation or perturbation. Recent studies have suggested that these lipid autacoids are indispensable to the regulation of inflammation, and their potential contribution to tissue injury and diseases is currently being intensively investigated.

Little information exists up to now to what extent microorganisms (bacteria, viruses) and their products (exo-, endotoxins) trigger the cells (granulocytes, monocytes, mast cells, platelets) for the release of the newly generated mediators, e.g. the lipoxygenase factors and the platelet-activating factor (PAF) (Bremm et al., 1988a).

The present study combines our data with published results on the role of (1) bacterial toxins in mediator generation; (2) microbial pathogenicity factors (adhesins, haemolysins) in the onset and perpetuation of the inflammatory response; and (3) toxins in the modulation of the signal transduction cascade of inflammatory effector cells resulting in an impaired function.

Biochemistry of cell activation

Signal transduction

Leukocyte responses to chemotactic and other phlogistic stimuli are vital for host defence and thus substantial interest has been focused on defining the mechanisms of signal transduction in these cells. Two events are believed to be essential for eliciting a response after cell activation: a transient enhancement of $[Ca^{2+}]_i$ and activation of protein kinase C. These appear to be involved since phorbol esters and permeant diacylglycerols elicit the respiratory burst (Sha'áfi and Molski, 1988; Rossi et al., 1989). Receptor-mediated neutrophil responses are prevented by pretreatment of the cells with B. pertussis toxin, showing that functional GTP-binding proteins are necessary (Volpi et al., 1985). Studies on the activation of the respiratory burst have shown in addition that continuous receptor occupancy by an agonist is a prerequisite to maintaining NADPH-oxidase activity. The minimum sequence of neutrophil activation via stimulus-dependent signal transduction is the following (Fig. 1): the interaction between the ligand–receptor complex and a GTP-

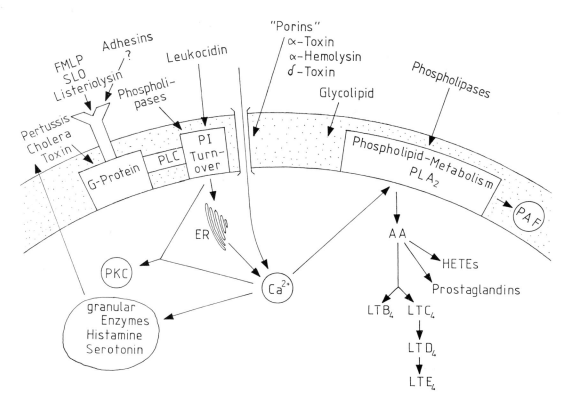

Figure I. Interaction of bacterial toxins with effector cells. Bacterial toxins activate the cells via (1) ligand–receptor interaction (fMLP, streptolysin O, listeriolysin), (2) formation of porins which allow the influx of divalent cations (α-haemolysin from *E. coli*, α-toxin and δ-toxin from *S. aureus*), (3) interaction with membranous phospholipids (glycolipid from *P. aeruginosa*), and (4) their intrinsic phospholipase activity (e.g. phospholipase C from *P. aeruginosa*). The external signal is processed intracellularly by (1) phosphatidylinositol turnover leading to inositol triphosphate (IP_3), (2) Ca^{2+} influx from external as well as internal stores, (3) activation of protein kinase C. Subsequently, kinases phosphorylate their specific target enzymes thus triggering the specific effector function of the cell (e.g. oxygen radical production, lipid mediator generation, enzyme release).

binding protein induces a consequent activation of a phosphatidylinositol-specific phospholipase C. This enzyme delivers inositoltriphosphate (IP_3) into the cytosol and diacylglycerol; the latter remains in the membrane. IP_3 liberates calcium from intracellular stores, raising $[Ca^{2+}]_i$ (Berridge and Irvine, 1984). Diacylglycerol and Ca^{2+} induce the translocation of protein kinase C to the plasma membrane and its subsequent activation. The rise in cytosolic calcium apparently activates phospholipase A_2 which hydrolyses phospholipid or phosphatide derived from diacylglycerol to release arachidonic acid.

Formation of lipid mediators

Mechanisms of arachidonic acid release by phospholipases

Stimulation of neutrophils with a variety of immunological and non-immunological compounds results in the release of arachidonic acid from membrane-bound phospholipids. The mechanism of arachidonic acid release as well as the characterization of the lipid pool from which it originates has been examined in several cell types (Chilton and Connel, 1988). The following pathways for the release of arachidonic acid have been proposed (Fig. 2):

(a) a phospholipase A$_2$-mediated release from phosphatidylcholine and phosphatidylethanolamine as substrates;

(b) a sequential action of phosphatidylinositol-specific phospholipase C and diacylglycerol lipase producing free arachidonic acid and monoacylglycerol;

(c) a PI-specific phospholipase C and diacylglycerol kinase, producing phosphatidic acid; a subsequent action via a phospholipase A$_2$ on phosphatidic acid produces free arachidonic acid.

Many investigations in neutrophils have indicated that phospholipase A$_2$ is the most active enzyme in these cells (Bauldry *et al.*, 1988).

Free arachidonic acid changes membrane fluidity, induces morphological changes in intact cells, activates the NADPH-oxidase and finally serves as substrate for the formation of the bioactive lipid mediators that can modulate cellular responses. It has been suggested that the cellular levels of free arachidonic acid may be controlled by two enzymes: the arachidonyl-CoA synthase and the acyl-CoA lysophosphatide acyltransferase. In many cells the activities of these enzymes exceed the activity of phospholipase A$_2$.

The reported pathways of arachidonic acid release are all stimulus-specific. The release of arachidonic acid via the action of phospholipase C in conjunction with a diglyceride kinase or by a combination of PLC and a diglyceride lipase occurs with fMLP-stimulated neutrophils. In contrast, the Ca-ionophore A23187 exclusively enhanced PLA$_2$-activity.

Opsonized zymosan and immune complexes induce stimulus-specific alterations of the lipid metabolism that are different from those induced by the Ca-ionophore A23187 (Godfrey *et al.*, 1987).

Conversion of arachidonic acid by neutrophils (Borgeat and Samuelsson, 1979) primarily results in the generation of 5-lipoxygenase transformation products (Fig. 3). Key products of this pathway are 5-HETE and leukotriene B$_4$ (LTB$_4$). These lipid metabolites are potent chemotactic agents for polymorphonuclear leukocytes and are also important mediators in the regulation of enzyme functions and the activation of PMNs. A direct biochemical relationship in which platelet-activating factor (1-*O*-alkyl-2-acetyl-*sn*-glycero-3-phosphocholine) shares a common biochemical pathway with leukotriene B$_4$ in neutrophils has been suggested. Choline-containing phospholipids as well as 1-*O*-alkyl-2-arachidonyl-GPC may serve as a substrate for PLA$_2$ and thus are precursors for LTB$_4$ and the platelet-activating factor. The released arachidonic acid is subsequently converted to LTB$_4$ while lysophosphatidylcholine or 1-*O*-alkyl-2-lyso-GPC are converted by acetylation to the platelet-activating factor (PAF).

Characterization of lipoxygenases

The generation of arachidonic acid from membrane phospholipids (e.g. phosphatidylethanolamine, -serine, diacylglycerol) is closely related to the subsequent formation of the biologically active lipid mediators, the lipoxygenase and cyclooxygenase products (Needleman *et al.*, 1986; Parker, 1987). The enzymatic subsets of the particular cell determine which classes of molecules are formed, e.g. prostaglandins, thromboxanes or leukotrienes, HETEs and lipoxins; moreover, defined cell types (e.g. monocytes and macrophages) possess both enzymatic pathways. In human PMNs the arachidonic acid is predominantly converted into lipoxygenase products.

The lipoxygenases (LO) belong to a family of enzymes which catalyse the insertion of a single oxygen molecule into unsaturated fatty acids, e.g. arachidonic acid at defined positions. The stereospecific action of these enzymes leads to the corresponding hydroperoxy-derivative of the fatty acid. Therefore, arachidonic acid can be converted principally into three HPETEs, the 5-, 12- and 15-HPETE, according to the stereo-specific orientation of its conjugated double-bond systems. All three HPETEs are formed by various mammalian cells (Spector *et al.*, 1988). The unstable HPETE is either reduced spontaneously or by enzymatic action of a peroxidase into the corresponding HETE. There is some evidence that lipoxygenases possess peroxidase activity (Siegel *et al.*, 1980). HPETEs and HETEs may be further oxygenated via a subsequent second (di-HETEs) or third (tri-HETEs) lipoxygenase attack.

The 5-lipoxygenase plays a central role in the inflammatory process, since it catalyses the key reactions of leukotriene formation leading to 5-HPETE and LTA$_4$, the precursor of LTC$_4$, LTD$_4$, LTE$_4$ and LTB$_4$. The enzyme is predominantly

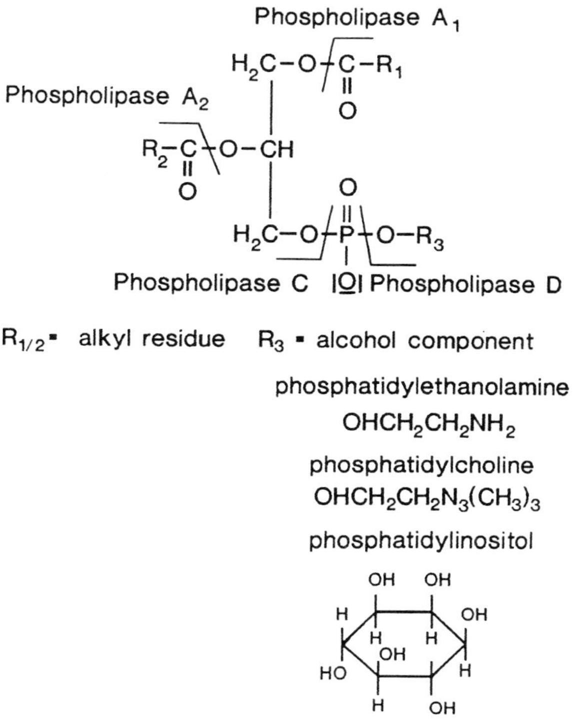

Phospholipase A$_1$

Phospholipase A$_2$

$$H_2C-O\!\!-\!\!C-R_1$$
$$\quad\quad\quad O$$

$$R_2-C-O-CH$$
$$\quad O$$

$$H_2C-O-P-O-R_3$$

Phospholipase C |O| Phospholipase D

R$_{1/2}$ = alkyl residue   R$_3$ = alcohol component

phosphatidylethanolamine

OHCH$_2$CH$_2$NH$_2$

phosphatidylcholine

OHCH$_2$CH$_2$N$_3$(CH$_3$)$_3$

phosphatidylinositol

Figure 2. Sites of action for different phospholipases.

present in inflammatory effector cells such as monocytes/macrophages and mast cells as well as eosinophil, neutrophil and basophil leukocytes (Borgeat and Samuelsson, 1979; Borgeat *et al.*, 1984; Rouzer *et al.*, 1982).

Human platelets are the predominant source of 12-lipoxygenase activity (Hamberg, 1976); they produce high amounts of 12-S-HETE upon stimulation with arachidonic acid, the Ca-ionophore and various microbial stimuli (glycolipid from *P. aeruginosa*, various haemolysins). The activation of the platelet-derived 12-LO occurs simultaneously with the cyclooxygenase pathway (leading to the production of TXB$_2$) suggesting a common control mechanism for both enzymes. Stimulation of peripheral monocytes and macrophages also leads to the generation of substantial amounts of 12-HETE, as is the case in endothelial and smooth muscle cells (Kühn *et al.*, 1985).

The sequence of the 15-lipoxygenase was recently determined in human eosinophils and reticulocytes (Sigal *et al.*, 1988a,b). This enzyme is present in epithelial cells, eosinophils, neutrophils and

platelets. The subsequent action of the 15-LO and the 5-LO leads to the formation of the lipoxins (Serhan *et al.*, 1984). The precise role of both lipoxins (which undergo rapid isomerization) is not yet clear since their biological effects are quite low as compared to other lipoxygenase products. The incubation of a mixed granulocyte–platelet suspension with the Ca-ionophore A23187 resulted in the formation of substantial amounts of lipoxins (Edenius *et al.*, 1988). Whether these mechanisms contribute to pathophysiological processes has to be clarified.

In various systems the HPETEs and HETEs generated by the different lipoxygenases exert potent biological functions, e.g. 5- and 12-HETE are chemotactic for neutrophils, 5-HPETE increases the Ca-ionophore and anti-IgE induced histamine release from basophils, 15-HETE released from epithelial cells leads to LTC$_4$ release from dog mastocytoma cells. The origin and biological roles of the different HETEs have been reviewed recently (Spector *et al.*, 1988).

Synthesis and metabolism of leukotrienes

The major known sources of leukotrienes *in vivo* are mast cells, basophils, eosinophils, neutrophils, monocytes and macrophages. Neutrophils generate LTB$_4$ which appears in the medium and is rapidly metabolized to 20-OH- and 20-COOH-LTB$_4$ derivatives, both of which are significantly less active in aggregation, chemotaxis and granule enzyme release than LTB$_4$ itself. LTC$_4$ production occurs primarily in monocytes, mast cells, eosinophils and basophils. In addition to the enzymatic hydrolysis of leukotriene A$_4$ into the chemotactic and chemokinetically active dihydroxyleukotriene B$_4$, granulocytes are able to conjugate LTA$_4$ with reduced glutathione to produce LTC$_4$. LTC$_4$ can undergo conversion into LTD$_4$ upon enzymatic hydrolysis of its peptide sidechain by a gamma-glutamyltranspeptidase (γ-GT). Further metabolism of LTD$_4$ via a dipeptidase results in the loss of the glycine molecule from LTD$_4$ and leads to the generation of LTE$_4$. The gamma-glutamyltranspeptidase activity is present in the microsomal fraction, whereas the dipeptidase activity correlates with the elution pattern of specific granules and the microsomal fraction. Upon challenge of human PMNs the enzymes are released from the cells into the cell supernatants (Raulf *et al.*, 1987).

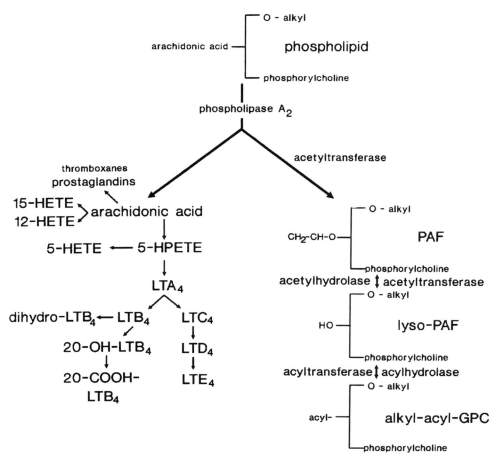

Figure 3. Generation of PAF and lipoxygenase products. Upon stimulation PAF is generated from phospholipids and arachidonic acid is released. In human granulocytes arachidonic acid is mainly converted via the 5-lipoxygenase pathway, leading to the formation of leukotrienes and HETEs.

There is increasing evidence for close collaboration between platelets, neutrophils, monocytes, lymphocytes and endothelial cells during the early phases of inflammatory and immunologic responses (Fitzpatrick *et al.*, 1984; Feinmark and Cannon, 1986; Maclouf and Murphy, 1988). LTA_4 generated from neutrophils can be transformed by lymphocytes, platelets and endothelial cells into LTB_4 and LTC_4. Through their effects on the production and action of other mediators, leukotrienes play an important role in cellular interactions (Table 1).

The LTB_4 receptor

Distinct receptor classes for chemotactic peptides or complement fragments were described on human neutrophils. The leukocyte LTB_4 receptor is primarily localized in the plasma membrane (Goldman *et al.*, 1985; Brom and König, 1989). The receptor classes transmit different biological responses of the cell. Occupation of the high-affinity receptor resulted in the adherence and chemotactic migration of PMNs; binding to the low-affinity receptor induced degranulation (Goldman and Goetzl, 1984). Interaction of the ligand with both receptor subpopulations led to an increase of intracellular calcium levels and transport exogenous LTB_4 to the metabolizing enzymes (Goldman *et al.*, 1985).

The LTB_4 receptor as well as the oligopeptide receptor are coupled to guanine nucleotide-binding proteins which control the affinity state of the receptor (Snyderman and Uhing, 1988). Appar-

Table I. Biological properties of PAF and lipoxygenase products

PAF	Chemotaxis of neutrophils and eosinophils Activation of neutrophils Enzyme release Bronchoconstriction Increase in vasopermeability Proliferation of $CD8^+$ cells Induction of hyperreactivity Influence on lymphocyte function Cytokine induction Platelet aggregation
LTB_4	Chemotaxis of neutrophils Activation of granulocytes Proliferation of $CD8^+$ cells IFN-γ synthesis and release Influence on lymphocyte functions
LTC_4, LTD_4, LTE_4	Bronchoconstriction Increase in vasopermeability Influence on lymphocyte functions Mucus production

ently, the high-affinity state is represented by the contact of the receptor to the G protein free of either GTP or GDP. The amount of chemoattractant receptor is not static on the cell surface. High-affinity receptors are converted to the low-affinity state after addition of GDP, GTP and non-hydrolysable analogues or by pertussis toxin (Goldman *et al.*, 1985). Sodium fluoride and phorbol myristate acetate resulted in a profound loss of high-affinity as well as low-affinity receptors, indicating a regulatory role of protein kinase C for LTB_4 receptor expression (Brom *et al.*, 1988b; O'Flaherty *et al.*, 1986).

Role of PAF in inflammation

PAF is formed *de novo* following immunological and non-immunological stimuli from a number of inflammatory cells including alveolar macrophages, eosinophils, neutrophils and platelets. The biosynthetic pathway involves activation of a calcium-dependent phospholipase A_2 which cleaves mem-

brane-bound ether-linked phospholipids, resulting in the formation of lyso-PAF, the precursor for PAF. Lyso-PAF has similar physico-chemical properties to PAF, although it is devoid of any biological activity. Lyso-PAF may then be acted on by an acetyltransferase which acetylates lyso-PAF at the *sn*-2 position of the molecule. PAF is rapidly degraded *in vivo* by a cytosolic acetylhydrolase enzyme resulting in the formation of lyso-PAF (Pinckard *et al.*, 1988). Lyso-PAF may then be taken up by the membrane and reincorporated into membrane lipids. It is of interest that certain cells, most notably neutrophils, monocytes and lung mast cells, can synthesize PAF, but much of the synthesized PAF is retained within the cell and not released into the extracellular milieu. PAF has been implicated as an important mediator in acute and chronic inflammatory processes (Braquet and Rola-Pleszczynski, 1987).

Induction of mediators (leukotrienes, PAF) by toxins

Bacteria release a variety of factors and proteolytic enzymes such as the phospholipase A_2 and C which may lead directly or indirectly to arachidonic acid release from phospholipids. An important group of bacterial pathogenicity factors are the haemolysins which receptor-dependently or -independently interact with target cells (Bremm *et al.*, 1983; König, W. *et al.*, 1986). To this group belong, for example, the α-toxin from *S. aureus*, the thiol-activatable toxins (e.g. streptolysin O, alveolysin, listeriolysin) as well as the haemolysins (heat-stable glycolipid, heat-labile phospholipase C) from *P. aeruginosa* and the α-haemolysin from *E. coli*. In addition to these haemolysins which act directly there are factors which, on interaction with additional components (phospholipase C, sphingomyelinase C), lead to haemolytic activity. To these belong the CAMP factor of β-haemolytic streptococci group B (*Str. agalactiae*) (Sterzik *et al.*, 1986). In the past the haemolysins were mainly considered with regard to their toxic activities for erythrocytes, but our recent data suggest that they are important inducers of inflammatory mediators. The amount of lipid mediator generated is strictly dependent on the toxin structure as well as its capability to induce and inactivate leukotrienes.

As an example, leukotriene formation by receptor-dependent mechanisms induced by thiol-activatable toxins is described in Table 2. Incubation of human granulocytes with these toxins (e.g. streptolysin O, alveolysin) led to a time-dependent generation of 5-HETE and leukotrienes (Bremm *et al.*, 1984, 1985). LTC_4 was the predominant product in contrast to cells stimulated with the calcium-ionophore A23187 or the α-haemolysin of *E. coli* (Scheffer *et al.*, 1985). At a concentration of 0.1 HU of alveolysin 5.1 ± 3.0 ng of LTC_4 was released which increased to 25.1 ± 5.6 ng at a concentration of 10 HU. At higher concentrations a decreased formation was observed. With SLO and θ-toxin a similar pattern became apparent. The amount of LTC_4 released proved to be optimal at concentrations of 10 HU for the various toxins and decreased with higher concentrations. Maximal LTB_4 activity was obtained with 100 HU of either toxin. At the optimal stimulatory concentrations LTB_4 levels were approximately three- to five-fold less than LTC_4 levels. The release of leukotriene-inducing (5-LO) and -metabolizing activities (γ-glutamyltranspeptidase, dipeptidase) was studied by incubating the supernatant of toxin-stimulated PMNs with [^{14}C]-arachidonic acid. With increasing concentrations of the toxins (1–100 HU) an increased formation of 5-HETE and 5-HPETE occurred. Caffeic acid as well as esculetin, an inhibitor of the 5-lipoxygenase pathway suppressed the formation of 5-HPETE by more than 50% (Bremm *et al.*, 1985). In kinetic experiments it became apparent that the pretreated neutrophils released an activity that metabolized arachidonic acid after only 1 min of incubation to a high degree. Optimal formation of 5-HETE and 5-HPETE was obtained with the supernatants from cells after 15–20 min of stimulation. Toxin-pretreated granulocytes also released γ-glutamyl-transpeptidase and dipeptidase activities. The release of leukotrienes occurred under non-cytotoxic concentrations as assessed by the determination of lactate dehydrogenase in the stimulated cell supernatants. With higher concentrations of the toxins (above 100 HU) even a decrease in leukotriene release was observed.

Our observations raised the question whether toxins at non- or sublytic concentrations lead to cell activation before the process of irreversible cell death. Indeed, Hirayama and Kato (1984) recently analysed the cytotoxic action of leukocidin from *P. aeruginosa*. The destruction of leukocytes by the toxin was reduced in the absence of Ca^{2+} and stimulated by the addition of calcium-

Table 2. Release of LTC_4 and LTB_4 from human granulocytes

		LTC_4	LTB_4
Alveolysin (*Bacillus alvei*)	10 HU/ml	25.1 ± 15.4	4.2 ± 0.8
	1	11.9 ± 3.4	3.8 ± 0.6
	0.1	5.1 ± 3.0	0.7 ± 0.3
Streptolysin O	10	16.8 ± 1.5	2.2 ± 1.9
	1	11.7 ± 1.0	1.8 ± 1.5
	0.1	3.2 ± 2.2	1.8 ± 1.2
θ-Toxin (*Clostridium perfringens*)	10	22.0 ± 5.1	4.2 ± 1.1
	1	12.1 ± 3.1	3.6 ± 1.1
	0.1	10.2 ± 2.0	1.8 ± 0.4
E. coli α-haemolysin	0.25	5.0 ± 1.8	36.2 ± 7.4
Phospholipase C (heat-labile toxin of *P. aeruginosa*)		1.6 ± 0.3	4.1 ± 0.8
Glycolipid (heat-stable toxin of *P. aeruginosa*)		0	0

Human granulocytes were isolated from peripheral blood and incubated (1×10^7 cells/500 μl) in the presence of Ca^{2+}/Mg^{2+} (1/0.5 mM) with the various toxins. Alveolysin, streptolysin O and θ-toxin were activated with 20 mM cysteine.
After incubation (15 min) the cell supernatant was analysed by specific radioimmunoassays and HPLC.

ionophore A23187. Their studies indicated that the initial action of the toxin was to stimulate phosphatidic acid production, causing a rapid metabolic change of phosphatidylinositol correlating with the activities of phosphatidylinositol-specific phospholipase C and 1,2-diacylglycerolkinase. Leukotriene release by the toxins is strictly calcium dependent. Alouf et al. (1984) described the interaction of bacterial SH-activated cytolytic toxins with monomolecular films of phosphatidylcholine and various steroids as a model for toxin-induced membrane destruction. The overall potency of the four toxins tested was SLO > alveolysin = perfringolysin O > pneumolysin (Bremm et al., 1985). When alveolysin, SLO and θ-toxin were analysed at various concentrations it appeared that θ-toxin at 0.1 HU is more active than SLO and alveolysin. Our experiments, however, do stress the similarities of the toxins as to their capacity to release leukotrienes. The biochemical processes occur at early times of toxin cell interaction and mostly show a maximum of leukotriene release after 5–15 min of incubation. As was indicated thiol-activatable toxins release more LTC_4 than LTB_4, whereas after ionophore stimulation and during phagocytosis LTB_4 was the predominant product. Toxin-pretreated cells showed an enhanced LTB_4 metabolizing capacity. Thus, the capability of cells to metabolize LTB_4 into the less chemotactic or non-chemotactic products $20\text{-}OH\text{-}LTB_4$ and $20\text{-}COOH\text{-}LTB_4$ markedly increased. The increase in hydroxylase activity may account for the rather low amounts of LTB_4 and could provide an efficient mechanism to counteract inflammatory processes initiated by an excess of LTB_4. Exposure of the cells to thiol-activatable toxins leads to a rapid impairment of the ability of the neutrophils to respond to a subsequent stimulus (Bremm et al., 1987). As early as 5 min after exposure the cells were significantly less sensitive towards stimulation with either the Ca-ionophore or opsonized zymosan as compared to non-pretreated cells. The decrease in LTC_4 release was more pronounced compared to LTB_4 generation. These mechanisms may account for the fact that under toxin exposure cells will lose their capacity to produce chemotactically active LTB_4 and spasmogenically active LTC_4. In heavily burned patients and after bacterial infection an enhanced LTB_4-hydroxylase activity as well as

the reduced ability to produce leukotrienes was observed (Brom et al., 1988a; Köller et al., 1988).

In addition to leukotrienes the toxins also induced the formation of PAF from lyso-PAF after incubation with PMNs (Bremm et al., 1988a). Toxin-pretreated cells inhibited the metabolism of PAF into lyso-PAF as was shown with streptolysin O as well as δ-toxin of S. aureus (Bremm et al., 1988a). The results thus suggest that toxin-activated granulocytes acquire a functional loss of mediator release and metabolism (Bremm et al., 1987). One may therefore suggest that the phagocytes are able to generate lipid mediators on interaction with toxins; however, the phagocytes lose the capability to counterbalance the inflammatory response. Mediator formation is suppressed by inhibitors of calmodulin and protein kinase C. The results therefore suggest the following sequence of events: toxins bind to granulocytes via receptor-dependent or -independent processes and activate the enzymatic cascade involved in PAF generation and metabolism such as the acetylhydrolase and the acetyltransferase. The ongoing interaction may then change the pattern of PAF metabolism. This could lead to an enhanced inflammatory response and over time to a deactivation (exhaustion) of the granulocyte which is then unable to fulfil its vital function during host defence.

Effect of receptor-independent (unknown) toxins on human cells

Each stage of the phagocyte cell function (e.g. chemotaxis, ingestion, bacterial killing and release of inflammatory mediators) can be influenced by a number of bacterial toxins. Clinical isolates of S. aureus secrete a number of protein exotoxins including α-, β-, γ- and δ-toxins (Freer and Arbuthnott, 1983). These toxins, also named haemolysins, have been involved in a variety of disease processes, e.g. local and disseminated infections. S. aureus δ-toxin is a 26 amino acid polypeptide containing 14 hydrophobic residues and a high percentage of non-ionizable sidechain amino acids (Fitton et al., 1980; Alouf et al., 1988, 1989). It is differentiated from the other staphylococcal haemolysins by its ability to lyse horse and human erythrocytes; in addition, δ-toxin activity can be inhibited by phospholipids and sera. δ-Toxin also has many other effects on various cell systems such as activation of membrane phospholipase A_2, stimulation of prostaglandin synthesis and inhibition of the epidermal growth factor binding to cell surfaces. δ-Toxin possesses the structural characteristics of a typical surface active protein and can readily insert itself into hydrophobic membrane structures (Bhakoo et al., 1982; Mellor et al., 1988). Its mode of action at the cellular level has often been compared with that of melittin which shows many physico-chemically and biologically identical properties with δ-toxin (Bernheimer and Rudy, 1986).

We compared the effects of δ-toxin with the structurally related bee venom toxin melittin in granulocyte functions. δ-Toxin and melittin lead to a rapid Ca^{2+} influx as determined by fluorescence detection (Raulf et al., 1990).

Human polymorphonuclear granulocytes (1×10^7) stimulated with melittin (10 µg) in the presence of (Ca^{2+}/Mg^{2+} 1/0.5 mM) over various times generated different amounts of leukotrienes as detected by RP-HPLC. Leukotriene generation occurred in a time-dependent fashion. LTB_4 was generated and the amounts of LTB_4 reached a maximum after 5 min of incubation. At later times the amount of 20-OH- and 20-COOH-LTB_4 increased. In contrast to melittin the δ-toxin at a wide concentration range (10 ng–10 µg) was not able to induce leukotriene generation from PMNs. When human cells were preincubated with δ-toxin the subsequent response obtained from PMNs or lymphocytes, monocytes and basophils (LMB-fraction) was clearly dependent on the stimulus used for the subsequent cellular activation. In this regard PMNs pretreated with δ-toxin and subsequently stimulated with the calcium-iono-phore A23187 showed a significantly enhanced generation of LTB_4 and ω-oxidated LTB_4 metab-olites. Conversely, with opsonized zymosan as secondary stimulus, toxin-pretreated PMNs showed a decreased generation of 20-COOH-, 20-OH-LTB_4 and LTB_4.

In contrast to human neutrophils different results were obtained with human LMBs. Prestimulation of the cells with δ-toxin decreased the amount of subsequent LTB_4 generation induced by the Ca-ionophore and opsonized zymosan. The results therefore suggest that the toxin-induced leukotriene pattern is cell dependent. This may imply that monocytes respond to a toxin via a different signal transduction pathway as neutrophils. One may also argue that the monocytes after toxin interaction show an impaired release of low molecular weight mediators. Whether this is due to the number of generated cytokines which then favour a cellular immune response has to be clarified.

Studies on the metabolism of exogenously added [^{3}H]-LTB_4 revealed a decreased formation of ω-oxidation products, e.g. 20-OH- and 20-COOH-LTB_4, after incubation with δ-toxin; similar effects were demonstrated after prestimulation of PMNs with PMA, an activator of protein kinase C (Brom et al., 1988b). In contrast to the results obtained with PMNs, when preincubated with PMA, the δ-toxin did not lead to a significant down-regulation of the LTB_4-binding sites. Thus, it appears likely that preincubation of PMNs with δ-toxin affects leukotriene induction by either enhancing LTB_4 generation and/or inhibiting LTB_4 metabolism. The resulting effects were dependent on the subsequent stimulus used for leukotriene forma-tion.

Tomita et al. (1984), who compared oxygen production from the cells induced by δ-toxin, suggested a different transductional pathway for δ-toxin as compared to stimuli such as melittin. They suggested that the surface active properties of δ-toxin affect the phagocyte membrane and induce a calcium flux from the medium. In contrast to δ-toxin which was not able to induce leukotriene metabolism but could modulate it, several toxins including the α-toxin of S. aureus and the thiol-activated toxins (streptolysin O, alveolysin, and Clostridium perfringens θ-toxin) generate leukotri-enes from human neutrophils and mononuclear cells (Bremm et al., 1985). These toxins interact by defined membrane biochemical mechanisms and induce leukotrienes with chemotactic and spasmogenic properties. Whether different mem-brane biochemical events (pore formation, phos-phoinositol turnover, protein kinase C) are the prerequisite for leukotriene induction has to be elucidated.

Although the bee venom melittin and the δ-toxin are both amphiphilic polypeptides with strong membrane surface activities, differences do, however, exist. Melittin is predominantly hydrophobic unlike δ-toxin. These structural dif-ferences and the resulting properties of δ-toxin and melittin may account for their different effects.

Recently, we studied the modulation of platelet-activating factor (PAF) and of lyso-PAF metabolism (Kasimir *et al.*, 1990). Preincubation of neutrophils with either δ-toxin or melittin resulted in the formation of PAF from lyso-PAF. After 5 min of incubation exogenously added lyso-PAF was converted to PAF (80 ± 3% of total radioactivity) and alkylacyl-GPC (20 ± 3%) (Fig. 4). The newly generated PAF was rapidly metabolized to lyso-PAF and alkylacyl-GPC during the subsequent incubation period of 60 min. Further studies indicated that the metabolism of PAF into lyso-PAF and alkylacyl-GPC is inhibited in the presence of δ-toxin. Melittin had no significant effect on the metabolism of PAF.

Although PAF exerts potent immunomodulatory and proinflammatory functions *in vitro* the conditions which lead to PAF release are not clear. Several reports indicate that the phospholipid is released into the fluid phase, whereas others observe that PAF is largely retained by stimulated PMNs. Our data suggest that most PAF from toxin-treated neutrophils remains cell-associated and is not secreted with the medium. The results suggest that PAF may exert an intracellular role with regard to PMN function. Its possible intercellular effects apparently do not require the release of PAF into the fluid phase (Braquet and Rola-Pleszczynski, 1987). One possible mechanism for the effects of δ-toxin in clinical infections is related to its influence on the generation and metabolism of PAF. Clearly PAF possesses complex activities in different immunological events.

Interaction of bacterial adhesins, surface components and haemolysins with inflammatory effector cells

Comparison of P. aeruginosa haemolysins (heat-stable, heat-labile)

Pseudomonas aeruginosa is frequently involved in nosocomial infections. Patients suffering from

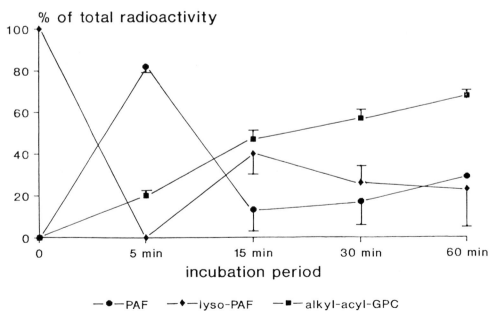

Figure 4. Effect of δ-toxin from *S. aureus* on the generation of PAF from lyso-PAF. [³H]-lyso-PAF was added to human PMNs (1 × 10⁷ cells, 2 min, 4°C) and the cells were subsequently treated with δ-toxin (0.15 μg/ml). The lipids were extracted and analysed by radioactive thin-layer scan. In the presence of δ-toxin a marked formation of PAF from lyso-PAF is observed, whereas in the appropriate buffer control lyso-PAF is metabolized exclusively into alkylacyl-GPC and formation of PAF is absent.

tumours, extensive burns and under immunosuppressive therapy are prone to *P. aeruginosa* infections (Morrison and Wenzel, 1984). The concomitant septicaemia results in a high lethality rate. Furthermore, mucoid variants of *P. aeruginosa* are isolated regularly from sputa of patients suffering from cystic fibrosis with a chronic infection of the lung. Recent research has focused on the pathogenicity of *P. aeruginosa* infections. A variety of virulence factors have been described, but the role of the individual factors is still unclear. Among these are cell-associated factors (e.g. lipopolysaccharide, pili, outer membrane proteins and mucus) and extracellular toxins such as exotoxin A, proteases, haemolysins, exotoxin S, lipase, lactoferrin and cytotoxin. *P. aeruginosa* produces two different haemolysins and a heat-labile protein identified as a 78-kDa protein with phospholipase C activity (Berka and Vasil, 1982). The heat-stable haemolysin consists of a group of glycolipids with detergent-like actions (Johnson and Boese-Marrazzo, 1980) which inhibit ciliar movement and enhance the enzymatic activity of PLC. PLC is produced in high concentrations from strains isolated from urinary tract infections and blood culture in septicaemia. Furthermore, PLC affects pulmonary surfactant, causes immunological reactions when injected locally into mice and interacts with several cell types.

We investigated clinical isolates of *P. aeruginosa* from severely burned patients and studied their capacity to induce inflammatory mediator release from rat mast cells or from human granulocytes (Bergmann *et al.*, 1989). The bacterial strains were characterized according to their cell-associated haemolysin activity as well as their secreted haemolysin and phospholipase C activities. Our data suggested that heat-stable and heat-labile toxins induce histamine and leukotriene release from rat mast cells and human granulocytes. *P. aeruginosa* strains expressing heat-labile haemolysin and phospholipase C induced histamine release from rat mast cells and leukotriene formation from human granulocytes; bacterial strains expressing heat-stable haemolysin were potent releasers of histamine but did not lead to leukotriene formation (Table 3). The mediator-inducing capacity was dependent on the growth characteristics.

The heat-labile haemolysin of *P. aeruginosa* obtained from the late logarithmic growth phase is cell-associated and can be detected within the culture supernatant during the stationary growth phase. Clinical isolates of *P. aeruginosa* expressed cell-bound PLC activity. There may be two different modes of PLC expression: cell-bound activity with or without subsequent secretion or non-cellbound and only extracellular PLC activity. All strains expressing PLC activity induced mediator release from mast cells and PMNs. The histamine-inducing capacity was dependent on the bacterial concentration. Maximal PLC activity was obtained with the 20-h culture. Heating of the bacterial supernatants induced in several cases a reduction in histamine release, indicating that for the majority of the strains at least two histamine-releasing principles are apparent – those which are heat-stable and those which are heat-labile.

In the past it has been shown that bacteria and various bacterial exotoxins activated granulocytes with the subsequent production of leukotrienes (König, W. *et al.*, 1986). Receptor-mediated as well as non-receptor-linked activation leads to the influx of Ca^{2+}, the activation of cellular PLC and the generation of arachidonic acid transformation products. The pattern of leukotriene release induced by washed *P. aeruginosa* bacteria or culture supernatants is similar to that obtained with the *E. coli* α-haemolysin (König, B. *et al.*, 1986; Scheffer *et al.*, 1985). More LTB_4 than LTC_4 was generated. An increase in *Pseudomonas* bacteria up to 5×10^8 led to an enhanced generation of ω-oxidation products which exceeded the amounts of LTB_4 (Bergmann *et al.*, 1989). Among the various clinical isolates were strains which expressed heat-stable haemolysin activity which has been identified as glycolipid. These supernatants and the purified glycolipid are potent activators of mast cells. At higher concentrations the glycolipid is toxic for the cells. The cytotoxic effect of the glycolipid may explain the failure to generate leukotrienes from granulocytes in the presence of the culture supernatants containing both haemolysins.

Berk *et al.* (1987) isolated two different haemolytic and enzymatically active proteins. These enzymes induce in 24 h a wheal and flare reaction which lasts for 7 days. Increasing amounts initiate a necrotic lesion after 2–4 days and lead to abscesses. These findings might be caused by the release of inflammatory mediators. LTB_4 is a chemoattractant that supports the accumulation of

Table 3. Induction of mediator release from rat peritoneal mast cells (RPMC) and human PMNs by bacteria and bacterial culture supernatants of *P. aeruginosa*

Strain		Histamine from RPMC (%)			Leukotrienes from PMN (ng)		
			Heat-stable[a]	LTB$_4$	20-OH-LTB$_4$	20-COOH-LTB$_4$	LTC$_4$
B21	bact.	14 ± 1	ND	5.5 ± 0.8	9.7 ± 2.1	3.6 ± 1.5	1.4 ± 0.6
	cult.	71 ± 6	16 ± 2	0	0	0	0
B10	bact.	35 ± 1	ND	4.6 ± 1.0	14.8 ± 1.0	5.0 ± 0.9	1.5 ± 0.6
	cult.	63 ± 7	20 ± 1	4.5 ± 1.0	0	0	2.2 ± 1.0
582f	bact.	29 ± 2	ND	5.3 ± 0.5	10.7 ± 1.6	4.1 ± 0.6	3.2 ± 1.1
	cult.	46 ± 2	0	4.1 ± 0.8	0	0	1.6 ± 0.3
6816A	bact.	0	ND	0	0	0	0
	cult.	96 ± 3	97 ± 1	0	0	0	0
B18	bact.	19 ± 2	ND	6.7 ± 0.6	17.6 ± 9.1	6.4 ± 1.4	2.2 ± 0.8
	cult.	83 ± 2	76 ± 4	0	0	0	0
B1	bact.	0	ND	0	0	0	0
	cult.	67 ± 6	23 ± 1	0	0	0	0
B2	bact.	0	ND	0	0	0	0
	cult.	54 ± 5	18 ± 3	0	0	0	0
B5	bact.	0	ND	0	0	0	0
	cult.	89 ± 3	93 ± 5	0	0	0	0
B22	bact.	0	ND	0	0	0	0
	cult.	77 ± 6	51 ± 3	0	0	0	0

cult. = culture supernatant; bact. = bacteria.
[a]After heat treatment of the culture supernatant.
ND = not determined.

neutrophil granulocytes. LTC$_4$ is able to enhance vascular permeability leading to hyperaemia and oedema. It was recently shown that perfusion of pig lungs with *P. aeruginosa* induces the generation of cyclooxygenase products (Chin Lee *et al.*, 1986). Furthermore, *P. aeruginosa* induced an accumulation of granulocytes in the lung when aerolized to C$_5$-deficient mice (Sordelli *et al.*, 1985).

Patients suffering from cystic fibrosis complicated by recurrent *P. aeruginosa* infections showed enhanced levels of leukotrienes in the sputa (O'Driscoll *et al.*, 1985); the accumulation of granulocytes in the lungs of these patients is a typical finding in the course of *P. aeruginosa* infection. In extension to these studies we determined the histamine release from basophils in patients suffering from cystic fibrosis and compared the data with an age-matched group of healthy donors (Schönfeld *et al.*, 1989). No significant differences in the basophil counts were determined between the CF and the control groups. However, the absolute histamine content per basophil was elevated in the CF group (2.6 ± 0.4 versus 1.4 ± 0.2 pg histamine/basophil). Stimulation of basophils with the Ca-ionophore (7.5 μM) and anti-IgE led to a significantly higher release of histamine per basophil in CF patients as compared to healthy donors. These data indicate that basophils in CF patients may have a greater potential to release mediators. Within the 14-day period of intravenous antibiotic treatment of the pulmonary infection (in 14 out of 15 cases *P. aeruginosa* was isolated from sputa samples) the histamine release per basophil (total histamine content) decreased to normal levels. This decline was accompanied by an improvement of the clinical condition of the patients ($n=8$) and reduction of *P. aeruginosa* isolates in sputa. In contrast, in three patients with sustained *P. aeruginosa* colonization of the upper airways and impaired lung function, histamine levels remained elevated. These results suggest that the histamine content of basophils as well as the release of histamine is increased in patients with cystic fibrosis and correlates with the clinical signs of the chronic infection.

The role of adhesins and α-haemolysin in E. coli

Escherichia coli strains cause infections of the gut (intestinal infections), the urinary tract (urinary tract infections, UTI), sepsis and are also the causative agents of newborn meningitis (NBM) (Cavalieri *et al.*, 1984). The infections which appear outside of the intestine have been termed extra-intestinal infections. Certain types of the O-antigen (O1, O6, O18, O75, O83) and capsule (K1, K5, K12, K15) are strongly associated with such infections, as are special fimbrial adhesins, haemolysins (Hly) and iron-binding chelators, such as aerobactin which takes up iron from the surrounding medium. About 50% of uropathogenic *E. coli* are haemolytic (Hughes *et al.*, 1982) and it has been shown that *E. coli* haemolysin represents a protein which may act as a cytolysin for normal tissue cells and for cells of the immune system (Gadeberg and Blom, 1986). Adherence to host cells represents the initial step in the course of bacterial infection. Adherence is mediated by fimbriae and agglutination. The adherence may or may not be inhibited by α-mannosides and is then termed mannose-sensitive (MSH) or mannose-resistant haemagglutination (MRH) (Blumenstock and Jann, 1982). The fimbriae can be subdivided according to their receptor requirements for adhesion. MS fimbriae recognize manno-oligosaccharide-containing glycoproteins. MR fimbriae can be subdivided according to their different receptor specificities – the P fimbriae of urinary tract infective *E. coli* strains recognize α-galactosyl-1,4-β-galactose as receptor (Korhonen *et al.*, 1986). MRH factors which recognize a sialic acid-containing receptor have been termed S fimbriae (Parkinen *et al.*, 1986). Strains with this binding specificity

occur more frequently in cases of newborn meningitis and sepsis.

We studied the role of various clinical *E. coli* strains that expressed different adhesins and/or generated haemolysin with regard to the induction of inflammatory mediators, e.g. histamine release from rat mast cells, the chemiluminescence response and the release of lipoxygenase transformation products from human polymorphonuclear neutrophils (Scheffer *et al.*, 1986). The degree of haemagglutination did not parallel the induction of the chemiluminescence response. When the formation of 5-HETE as compared to LTB$_4$ was studied it became apparent that haemolysin-negative bacteria with different adhesins induced more 5-hydroxyeicosatetraenoic (5-HETE) as compared to haemolysin-positive bacteria, which generated more leukotriene B$_4$. Among the leukotrienes more leukotriene B$_4$ as compared to leukotriene C$_4$ was released from peripheral leukocytes. Studies with rat mast cells showed that histamine release was dependent on the haemolysin activity expressed by washed bacteria or present within the bacterial culture supernatant (Scheffer *et al.*, 1985). Histamine release was markedly diminished when the haemolysin activity decayed.

Several haemolysin-negative bacteria with defined adhesins also released histamine in small amounts, suggesting that in addition to haemolysin other factors may contribute to mediator release. Histamine release from human basophils stimulated by the haemolysin-positive culture supernatants followed the same pattern as was observed for the rat mast cells (Groβ-Weege *et al.*, 1988; Thomas *et al.*, 1987). However, when the basophils were stimulated with the haemolysin-positive bacteria no linear correlation was observed between histamine release and haemolysin activity. Although the percentage of haemolysin activity as expressed by the lysis of sheep erythrocytes declined to zero, a significant release of histamine from human LMBs was still apparent. These results, however, were only obtained with washed haemolysin-positive bacteria, while haemolysin-negative bacteria induced no histamine release. It therefore appears that the basophil may either interact with cell surface components in the presence of haemolysin or that bacteria induced an intercellular activation which affects the capacity of histamine release from human basophils. In fact, the basophil-enriched fraction (LMB) also contains large numbers of monocytes and lymphocytes. One can speculate that the haemolysin-producing bacteria activate monocytes which release histamine-inducing factors, e.g. interleukin 1 (Subramanian and Bray, 1987) and histamine-releasing factors. A non-immunological induced activation was proposed to be mediated by lectins, e.g. carbohydrates of the bacterial outer membrane interact with lectins on the basophil cell membrane leading to histamine release (Norn *et al.*, 1987).

However, in contrast to the results of Norn *et al.* (1985) only haemolysin-producing *E. coli* strains induced a strong histamine release. The question therefore arises as to the cellular mechanisms involved in inflammation after interaction with the various bacteria. One may argue that in certain disease processes the primary target is the epithelial cell as compared to neutrophils and mast cells. Upon stimulation epithelial cells generate free arachidonic acid and peroxides, which then attract neutrophils and activate mast cells or macrophages (Chauncey *et al.*, 1988). Furthermore, free arachidonic acid can be metabolized to leukotrienes by cells (phagocytes, mast cells) that contain the appropriate transforming enzymes. It thus appears that the interaction of bacteria with various cells via defined mechanisms determines the degree of the inflammatory event which, after microbial colonization, leads to bacterial infection.

In order to elucidate the effect of defined pathogenicity factors, isogenic wild-type clones were constructed which differed in the expression of one single pathogenicity factor (haemolysin or adhesin) (Hacker *et al.*, 1983). In subsequent studies we investigated the role of bacterial mannose-resistant fimbriation of S fimbriae (Fim), mannose-resistant haemagglutination (S-MRH) and haemolysin (Hly) production by an *E. coli* parent and genetically cloned strains with regard to (i) their effect on histamine release from rat mast cells and (ii) generation of the chemiluminescence response, leukotriene and enzyme release from human polymorphonuclear granulocytes. Washed bacteria *E. coli* 764 Hly$^+$, *E. coli* 21085 Hly$^+$, *E. coli* 536 (Hly$^+$, Fim$^+$, MRH$^+$) as well as their culture supernatant were analysed at various times during their growth cycle. (König *et al.*, 1985; König, B. *et al.*, 1986). No differences exist between parent and cloned or mutant strains with

respect to their outer membrane proteins and lipopolysaccharide pattern. Washed bacteria (*E. coli* 764 and 21085 pANN202-312) which produced haemolysin, unlike Hly$^-$ strains induced high levels of histamine release from rat mast cells and led to a significant chemiluminescence response, enzyme and leukotriene release from human polymorphonuclear granulocytes. Bacterial culture supernatants from Hly$^+$-producing and -secreting strains showed similar results with the exception of *E. coli* 21085 pANN202-312 which is a haemolysin-producing but not a secretory strain. The presence of Fim and S-MRH potentiates mediator release (König *et al.*, 1989). The simultaneous presence of MRH and Fim (*E. coli* 536/21 pANN801-4) increased mediator released compared with MRH$^+$ Fim$^-$ strains (*E. coli* 536/21 pANN801) (Fig. 5). *E. coli* 536/21 (MSH$^-$, MRH$^-$, Fim$^-$, Hly$^-$) did not induce mediator release. Thus Hly$^+$ *E. coli* strains are potent inducers for the release of inflammatory mediators which occurs under non-cytotoxic concentrations.

The highly chemotactic LTB$_4$ is generated from neutrophils, thus providing a further influx of granulocytes. The concomitant release of LTC$_4$ may potentiate an inflammatory event such as the increase in permeability. Histamine and the lysosomal enzymes certainly contribute to inflammation and tissue destruction. These events are obviously more pronounced when bacteria possess MR adhesins and express S fimbriae. Thus, although haemolysin by itself already represents a potent pathogenicity factor, additional properties of the bacteria such as adherence may substantially potentiate the effect of haemolysin (König, B. *et al.*, 1986). The fact that the decay rate of cell-bound haemolysin is much slower than that of the secreted molecule provides the bacteria with the property of being able to induce an inflammatory response, thereby facilitating bacterial invasion.

Indeed adhesion by itself is already a trigger for the release of inflammatory mediators. In this regard *E. coli* strains with mannose-sensitive adhesions (MS-Fim) and haemagglutination properties (MSH), S-mannose-resistant adhesins (S-Fim) and S-haemagglutinating properties (S-MRH), P-mannose-resistant adhesins (P-Fim) and P-mannose-resistant haemagglutination (P-MRH) differ in the induction of cellular responses. The highest chemiluminescence response was obtained

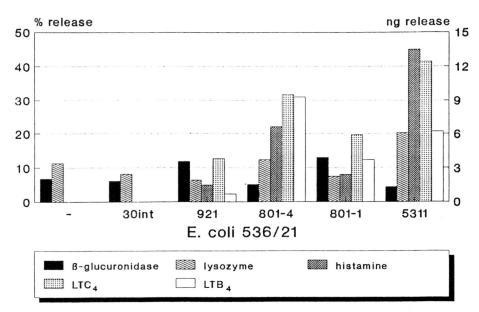

Figure 5. Leukotriene generation and enzyme release from human PMN and histamine release from rat peritoneal mast cells by adherent *E. coli* strains. The following isogenic strains were studied: *E. coli* 536/21 (Hly⁻, MRH⁻, Fim⁻). *E. coli* 536/21 pGB30int (Hly⁻, MSH⁺, MS-Fim⁺); *E. coli* 536/21 pANN921 (Hly⁻, P-MRH⁺, P-Fim⁺); *E. coli* pANN801-4 (Hly⁻, S-MRH⁺, S-Fim⁺); *E. coli* pANN801-1 (Hly⁻, S-MRH⁺, S-Fim⁻); *E. coli* 536/21 pANN5311 (Hly⁺, MRH⁻, Fim⁻). Human granulocytes (1×10^7) were incubated with the bacteria (5×10^8) for 10 min; the release of enzymes and leukotrienes into the cellular supernatant was determined. Histamine release from rat peritoneal mast cells (5×10^4 cells) was determined after 60 min of incubation with the various strains.

with the strains expressing MS-adhesins, MSH or S-MRH (König *et al.*, 1989). Similarly *E. coli* with P-adhesins and P-MRH induced a high chemiluminescence response but only slightly adhered to the surface of the PMN membrane. No leukotrienes were generated from granulocytes after interaction with strains without any adhesins, with MS-adhesins or MSH. The strain expressing P-adhesin and P-MRH induced only slight amounts of leukotrienes. The isogenic strain with S-adhesins and S-MRH was the most potent stimulus for leukotriene formation. The deletion in S-adhesins reduced the capacity of inflammatory mediator release. These results suggest that the chemiluminescence response, 5-lipoxygenase activation, leukotriene formation, histamine release and lysosomal enzyme release are dissociated events which are controlled by defined or less-defined receptor interactions; it appears that after cellular activation different and independent membrane biochemical processes are initiated.

The introduction of the haemolysin as a virulence factor into an adhesin-negative strain revealed

that haemolysin by itself induced an increase in adherence, chemiluminescence and led to a high leukotriene formation (Scheffer *et al.*, 1985). The fact that adhesins by themselves are able to trigger an inflammatory response may explain the pathophysiological prerequisites whereby bacteria may initiate defined disease processes such as pyelonephritis and meningitis; the concomitant release of haemolysin may then even potentiate the inflammatory response once the interaction of bacterial adhesin with tissue receptors has occurred.

Clearly, this interaction is specific for the granulocyte. When human LMBs are incubated with the genetically cloned *E. coli* expressing mannose-resistant haemagglutination, the subsequent formation of leukotrienes or histamine release was suppressed (Ventur *et al.*, 1990). An adhesin-specific as well as a common bacterial factor are apparently responsible for the deactivation of the monocytes and basophils to release low molecular weight mediators. In contrast, receptor-mediated events which include the recognition of Fc and complement receptors by the monocyte induce a low deactivation or even an enhanced generation of low molecular weight mediators. Apparently, this mechanism could play an important role during bacterial infection. The direct activation of granulocytes with MR-*E. coli* or haemolysin induces the release of mediators resulting in an inflammatory response which then provides the further influx of granulocytes. One may speculate that the monocyte deactivated for the release of low molecular weight mediators preferentially now operates as a phagocytosing and antigen-presenting cell. Obviously, the production of monokines, e.g. IL1 and TNF, is then stimulated. A humoral response is then facilitated. Interestingly, Fc-coated particles such as zymosan or bacteria activate the monocyte directly for the release of low molecular weight mediators. This may imply that once the antibody response has been established, the opsonized particle also induces leukotriene formation from monocytes and thus supports the influx of granulocytes. Antibody-coated Hly$^+$ bacteria showed decreased mediator induction as compared to Hly$^+$ bacteria without serum pretreatment. Phagocytosis thus appears to be facilitated and then leads to the elimination of the bacteria. These results suggest that, dependent from the stimulus, the monocytes and granulocytes act differently (Ventur *et al.*, 1990).

Comparison of cell-bound and secreted haemolysin from E. coli, Listeria *sp.*, Aeromonas sp. Serratia *sp.*

Haemolysins are a heterogeneous group of molecules. A variety of Gram-positive and Gram-negative bacteria are able to lyse erythrocytes. The potent role of haemolysins has been demonstrated in various animal models and cell cultures. In the past we compared four haemolysin-producing organisms (*Escherichia coli, Serratia marcescens, Aeromonas hydrophilia, Listeria monocytogenes*) for the release of inflammatory mediators (Scheffer *et al.*, 1988). Haemolysin-producing *E. coli* strains are often isolated from patients with extra-intestinal infections (Cavalieri *et al.*, 1984). *S. marcescens*, also a haemolysin-producing strain, is becoming recognized as an important opportunistic pathogen causing respiratory and urinary tract infections, bacteraemia, endocarditis, keratitis, arthritis and meningitis (Lyerly and Kreger, 1983). Clinical isolates of *S. marcescens* were previously shown to

produce various exoenzymes. A metalloprotease was suggested to induce pneumonia in laboratory animals, to enhance vascular permeability through the activation of a Hageman-dependent factor–Kallikrein pathway *in vitro* and to cause fibrinolysis (Kamata *et al.*, 1985). *A. hydrophila* has been associated with human diseases and food-borne infections (Chakraborty *et al.*, 1986). Many strains of this species elaborate various exotoxins, including two haemolysins, one of which is a cytotoxic haemolysin (aerolysin) (Asao *et al.*, 1984). In addition, *A. hydrophila* produces enterotoxin and proteases. Aerolysin has been isolated and purified to homogeneity (Asao *et al.*, 1986). Among the biological activities associated with the purified toxin are, for example, elicitation of vascular permeability at the site of injury, oedema and necrotic and lethal activities (Chakraborty *et al.*, 1986).

L. monocytogenes is a ubiquitous Gram-positive microorganism responsible for severe infections in humans and animals. Among various hypotheses, the secretion of a haemolysin (listeriolysin) has been proposed as an important mechanism promoting *L. monocytogenes* virulence (Kathariou *et al.*, 1987). This extracellular protein is a sulphydryl group-dependent cytotoxin which is antigenically related to streptolysin O (Parrisius *et al.*, 1986; Geoffroy *et al.*, 1987). *In vitro* it causes damage to the reticuloendothelial system and is lethal for experimental animals. Haemolysin-secreting as well as non-secreting *Listeria* are also a powerful tool to demonstrate the mechanisms of intracellular replication. It is currently suggested that bacteria internalized by professional and non-professional phagocytes are confined inside phagosomes. Under conditions of neutral pH and low Fe the bacteria secrete high amounts of exotoxins. Fusion of the phagosome with the lysosome allows the binding of the toxin molecule at low pH which then leads to the disruption of the intracellular membranes and toxin export into the cytoplasm of the macrophage (Berche *et al.*, 1989). The binding of listeriolysin to cholesterol may be a crucial event for its immunogenicity and potentiation of DTH. Listeriolysin-producing microorganisms also suppress the antigen presentation by macrophages.

For the studies the following strains were analysed: *E. coli* K12 pANN5211 (haemolysin-positive (Hly$^+$)); *E. coli* K12 (Hly$^-$); *E. coli* JM101 pUC19/1 (*S. marcescens* Hly$^+$); *E. coli* JM101 (*S. marcescens* Hly$^-$); *A. hydrophila* AB3 (Hly$^+$); *A. hydrophila* AB3 aer5 (Hly$^-$); *L. monocytogenes* WTD (Hly$^+$); *L. monocytogenes* M3D (Hly$^-$) (Fig. 6). Cloning and functional characterization of the chromosomal and plasmid-encoded haemolysin determinants were performed (Goebel and Hedgpeth, 1982; Berger *et al.*, 1982; Braun *et al.*, 1985; Chakraborty *et al.*, 1986; Kathariou *et al.*, 1987). The plasmid pANN5211 expresses the α-haemolysin determinant of a bacterial strain pathogenic for humans. *A. hydrophila* AB3 and *L. monocytogenes* WTD are haemolysin-producing wild-type strains. To study the expression of the *S. marcescens* haemolysin in a defined genetic background, the *E. coli* JM101 pUC19/1 clone was constructed (Braun *et al.*, 1985).

Our studies indicated that the expression of haemolysin activity is dependent on active cell growth and reached its optimum for the *E. coli* and the *S. marcescens* haemolysin at the midlogarithmic phase up to the onset of the lag phase (König *et al.*, 1987). The listeriolysin produced by *L. monocytogenes* and the aerolysin of *A. hydrophila* reached maximal activities in the stationary growth phase. The *E. coli* α-haemolysin, the *S. marcescens* haemolysin and the aerolysin are capable of inducing a time-dependent chemiluminescence response from human PMNs. The haemolysin-negative control strains expressed negligible activities. This pattern differed for the *L. monocytogenes* strains (Scheffer *et al.*, 1988). Both strains were highly active. It has been suggested that the release of superoxide dismutase may be responsible for the lack of increase of the chemiluminescence response. Welch *et al.* (1979) discussed, therefore, a possible role for superoxide dismutase in the pathogenesis of listeria infection.

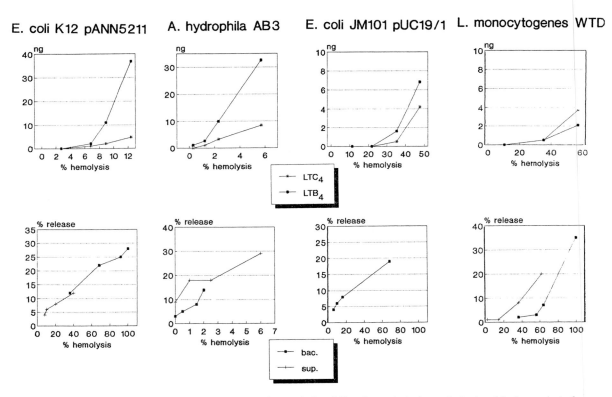

Figure 6. Leukotriene and histamine release by α-haemolysin of *E. coli*, aerolysin from *A. hydrophila*, haemolysin from *S. marcescens* and listeriolysin from *L. monocytogenes*. Human granulocytes were isolated from peripheral blood and incubated (1 × 10^7 cells/500 μl) in the presence of Ca^{2+}/Mg^{2+} (1/0.5 mM) with the individual toxins. Listeriolysin was activated with 20 mM cysteine. After incubation (10 min) the cell supernatant was analysed by radioimmunoassays for LTC$_4$ and LTB$_4$ and HPLC. Histamine release was studied with rat peritoneal mast cells (5 × 10^4 cells, 60 min incubation period) as target cells.

The degree of leukotriene release was dependent on the haemolysin type and the expression of haemolysin activity. As became apparent from [^{14}C]-arachidonic acid turnover studies, the *E. coli* haemolysin and the aerolysin induced the most pronounced release of LTB$_4$ and LTC$_4$ in a dose-dependent manner. LTB$_4$ levels exceeded those of LTC$_4$. In contrast, the *S. marcescens* haemolysin and the listeriolysin induced lower rates for leukotrienes. In general, LTC$_4$ levels were equal to LTB$_4$ levels in a dose-dependent manner. These data are in contrast to those obtained for the calcium-ionophore A23187 and during phagocytosis. The haemolysin-producing strains also triggered rat mast cells for histamine release. Unlike the results obtained with PMNs, all haemolysin-positive strains induced nearly the same histamine release in a dose-dependent manner. The listeriolysin-producing strain required a threshold to trigger mast cells. Bacteria expressing 60% haemolysin activity showed negligible release; beyond 60% haemolysis a sharp increase in histamine release occurred. The aerolysin-producing *Aeromonas* strain was the most potent stimulus. Strains expressing only 2% haemolysis already released 15% of the histamine from mast cells.

These studies clearly emphasize the differences of various haemolysins in inducing inflammatory mediator release from human cells. They also indicate that the various haemolysins are defined pathogenicity factors.

In vivo studies have demonstrated that mouse lethality increases when *E. coli* expresses haemolysin production (Hacker *et al.*, 1983). In addition, in the nephropathogenic model *E. coli* with mannose-resistant adhesins as well as haemolysin production expressed a higher degree of pathogenicity as compared to haemolysin-negative strains (Marré *et al.*, 1986). The role of thiol-activated toxins in mediator release has been shown. Thus, it appears likely that listeriolysin contributes to bacterial infection by its capacity to induce inflammatory mediator release. Aerolysin elicits in a direct manner vascular permeability and oedema formation. Several authors have described the enterotoxic activity of aerolysin without activation of adenylate cyclase (Gracey *et al.*, 1982; Hostacka *et al.*, 1982). Interestingly, arachidonic acid metabolites can stimulate the secretory process of isolated mucosa (Field *et al.*, 1984; Musch *et al.*, 1982).

The differences in the activity of the various haemolysins could be due to the fact that they trigger cells via distinct receptors or by pore formation (Benz *et al.*, 1989). It has been suggested that *E. coli* α-haemolysin and the α-toxin of *S. aureus* induce calcium influx in neutrophils and erythrocytes by formation of pores in the cytoplasm membrane (Bhakdi *et al.*, 1986; Jorgensen *et al.*, 1986). In addition, one must discuss different activation mechanisms of the various toxins. Transposon mutants of the *E. coli* α-haemolysin were also studied. A deletion of the amino acids isoleucine and threonine in positions 4 and 5 of the HlyA gene product markedly reduced the haemolysin activity in comparison to the unmodified protein. The induction of inflammatory mediators with these mutants showed that this deletion did not affect the amount of histamine release from mast cells. However, the formation of leukotrienes from human granulocytes as well as 12-HETE from platelets was markedly reduced.

Toxins, e.g. haemolysins, may also be readsorbed to a microbial matrix and thus trigger inflammatory mediator production. Using RBL cells as targets it became apparent that unlike in rat mast cells the α-haemolysin-containing bacterial supernatant required an additional bacterial factor to induce histamine release. In this regard, it is interesting that *E. coli* K12 in the presence of the haemolysin-containing supernatant or peptidoglycan provides the corresponding matrix for mediator induction (Fig. 7).

The phenomenon of erythrocyte lysis does not explain the pathophysiological events which occur after interaction of haemolysins with the various inflammatory cells (Table 4). It is apparent that phosphatidylinositol, cholesterol and ganglioside type III inhibit the lysis of *E. coli* α-haemolysin, while cholesterol inhibits the listeriolysin, and ganglioside type III affects the aerolysin-induced lysis. When the leukotriene-inducing capacity of the haemolysin-containing supernatants were analysed the following results were obtained: for the *E. coli* α-haemolysin a significant inhibition was obtained in the presence of cholesterol and ganglioside type III but not with phosphatidylinositol. No suppression was obtained for the aerolysin, while listeriolysin was inhibited with cholesterol. When histamine release from rat mast cells was studied, the following observations were obtained:

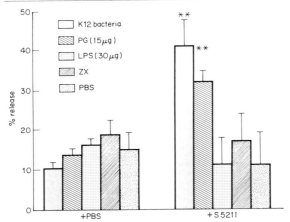

Figure 7. Modulation of histamine release from rat basophil leukaemia cells (RBL cells, 5×10^4 cells, 60 min, 37°C) by peptidoglycan (PG), lipopolysaccharide (LPS), opsonized zymosan (Zx) and *E. coli* K12 in the presence of the haemolysin-containing supernatant of *E. coli* pANN5211. PBS: buffer control. The data indicate that the haemolysin-containing supernatant needs a second signal (e.g. occupation of the glycan receptor by peptidoglycan or *E. coli* bacteria) to induce histamine release.

the *E. coli* α-haemolysin was inhibited by phosphatidylinositol, cholesterol and ganglioside type III, the listeriolysin only in the presence of cholesterol and the aerolysin only when ganglioside type III was present. These results clearly indicate the need to analyse the receptor structures for the various haemolysins on the effector cells as well as the

signal-transduction cascade for the toxin-induced inflammatory mediator release.

Modulation of the signal-transduction cascade by toxins – effect of toxins on inflammatory effector cells

Effect of toxins on the signal-transduction cascade

Major efforts are being directed to study the membrane biochemical events following toxin interaction with inflammatory and immunocompetent cells. In this regard defined toxins (a) are a powerful tool to modulate individual components of the signal-transduction cascade, and (b) induce immunological and inflammatory reactions; information about how the message is tranduced into the cell is of utmost importance.

The ability of leukocytes to participate in inflammatory responses is regulated by second messenger systems such as the calcium-mediated activation pathway and the cyclic AMP system. Activation of the leukocyte is initiated by the binding of chemoattractants to specific receptors on the cell surface. These events lead to a rapid formation of inositol phosphates, diacylglycerol and to a release of calcium from intracellular stores, suggesting that receptors for chemoattractants activate a phospholipase C.

Table 4. Inhibition of toxin-induced haemolysis and histamine release from rat peritoneal mast cells by phosphatidylinositol, cholesterol and ganglioside type III

	Inhibition of haemolysins ($n = 5$; $^*p < 0.001$)		
	α-Haemolysin *E. coli* K12 PANN 5211	Listeriolysin *L. monocytogenes* WTD	Aerolysin *A. hydrophila* AB3
Phosphatidylinositol (10 μg/ml)	59.7 ± 16.5*	22.7 ± 22.1	0.0 ± 10.1
Cholesterin (0.5 mg/ml)	84.1 ± 10.8*	75.7 ± 13.8*	13.3 ± 15.5
Ganglioside type III (1 mg/ml)	96.8 ± 3.7*	0.0 ± 20.3	40.9 ± 4.5*
	Inhibition of histamine release ($n = 3$)		
Phosphatidylinositol (10 μg/ml)	80.6 ± 22.6*	3.4 ± 27.5	1.2 ± 19.9
Cholesterin (0.5 mg/ml)	95.0 ± 1.5*	98.9 ± 0.9	7.8 ± 11.1
Ganglioside type III (1 mg/ml)	78.7 ± 22.2*	6.8 ± 23.0	66.9 ± 14.1*

Pertussis toxin (PT), an oligomeric AB-type toxin, can induce an ADP-ribosylation of the G protein α subunit of the adenylate cyclase complex in eukaryotic cells (Gilman, 1987). It has been suggested that a possible substrate of PT is G_i which may transduce inhibitory hormonal messages to adenylate cyclase (Katada and Ui, 1982). In addition, PT can have effects on cells by activation of second messenger pathways independent of its ability to ADP-ribosylate G proteins. For the mitogenic effects of PT on lymphocytes a specific receptor has been suggested (Rosoff et al., 1987). Activation of G proteins occurs when during ligand–receptor interactions the bound GDP is exchanged by GTP. The α subunit (Gα-GTP) dissociates from the β, γ chains (Gβ, γ) and modifies the activity of the effector enzymes. Non-hydrolysable nucleotide analogues of GTP such as GTPγS produce persistent activation of the α subunits (Harnett and Klaus, 1988).

The chemotactic peptide fMLP interacts with specific binding sites on PMNs and induces cellular events such as the 'respiratory burst' (Bokoch and Reed, 1979), cell adhesion (Patrone et al., 1980), chemotaxis (Lanni and Becker, 1983), release of arachidonic acid (Bokoch and Reed, 1980) and granular enzymes (Lanni and Becker, 1983), as well as the stimulation of the 5-lipoxygenase pathway (Jubiz et al., 1982). Recently, we provided evidence that incubation of PMNs with PT modulates the subsequent leukotriene generation induced by various stimuli (Hensler et al., 1989). A significant decrease in LTB_4 generation induced by fMLP was observed after pretreatment of the cells with PT. The effect of PT was time- and dose-dependent. In contrast, leukotriene generation induced by the Ca-ionophore A23187 was not affected. With opsonized zymosan as a secondary stimulus the inhibitory effects were much lower but significantly less than those observed when the cells were stimulated with fMLP. PT did not affect the leukotriene-metabolizing enzymes. It is well-established that the various stimuli interact with neutrophils via different membrane biochemical events. The Ca-ionophore induces Ca^{2+} influx, opsonized zymosan interacts with C3b and Fc receptors for binding and signal transduction. The chemotactic peptide fMLP interacts with

specific binding sites on cells. Thus, PT modulates leukotriene generation induced by stimuli which interact with specific receptors on the cell surface. The low changes in leukotriene generation observed in PT-treated cells stimulated by opsonized zymosan suggest that in this case PT-sensitive G proteins are only in part involved.

Becker et al. (1988) suggested that in the same cell, a receptor to one agonist interacts with a pertussis toxin-sensitive G protein, whereas a different receptor may interact with an insensitive G protein in order to activate a single effector enzyme, such as phospholipase C; which of the mechanisms is involved depends on the stimulus. PT treatment of cells also modulates histamine and LTC_4 release from human basophils depending on the stimulus. Warner et al. (1987) demonstrated the inhibition of the mediator release from PT-pretreated basophils after stimulation with fMLP and the anaphylatoxin C5a. The toxin did not affect the IgE-induced mediator release. To what extent the IgE-mediated signal proceeds via G proteins other than G_i or G_s has to be determined. Recently it was shown that PT can have effects on cells by activation of second messenger pathways independent of its ability to ADP-ribosylate G proteins (Rosoff et al., 1987). In this regard the mitogenic effect of PT holotoxin is mediated by the interaction of the B oligomer with CD3 (Gray et al., 1989).

Preincubation of PMNs with cholera toxin over a wide concentration range also suppressed the fMLP-induced leukotriene formation. These studies provided indirect evidence for the role of G proteins in inflammatory mediator formation. In this regard incubation of PMN with the non-hydrolysable GTP analogue, GTPγS, enhanced the LTB_4 generation after subsequent stimulation with fMLP in a time-dependent manner. Addition of GTPγS leads to a persistent activation of the α subunits of the G proteins which is followed by an increase in leukotriene formation. The mechanisms by which GTPγS acts in this system has still to be elucidated. GTPγS has been shown to enter permeabilized cells. Since a preincubation time of 1 h is required for an optimal effect, one may also suggest an uptake of GTPγS by the PMN. Addition of GTPγS did not alter the Ca-ionophore-induced stimulation. These results indicate that Ca-ionophore bypasses the activation of G proteins. A minute enhancement of leukotriene formation is observed in GTPγS-treated PMNs stimulated with opsonized zymosan. These results suggest that both PT-sensitive and -insensitive G proteins may be activated after stimulation with opsonized zymosan. It appears that they may be involved in the signal-transducing cascade to a different degree.

In recent studies it was established that fluoride ions (F^-) activate G proteins coupled to adenylate cyclase or to phospholipase C in the absence of receptor occupancy. Furthermore, human neutrophils can be stimulated by fluoride ions for superoxide anion production (Curnutte *et al.*, 1979), Ca mobilization (Strnad and Wong, 1985) and arachidonic acid release (Bokoch and Gilman, 1984). Incubation of human polymorphonuclear granulocytes, monocytes and platelets with sodium fluoride (NaF) results in a time- and dose-dependent generation of leukotrienes and 12-HETE (Brom *et al.*, 1989). This release was not influenced by pretreatment with pertussis or cholera toxin. It has been suggested that the fluoride anion complexes with aluminium contaminants to form AlF_4^- which then activates the G proteins by mimicking the γ-phosphate of GTP at the GTP-binding site. Inactivation of LTB_4 by the neutrophils via ω-oxidation into 20-OH- and 20-COOH-LTB_4 is inhibited by NaF (Brom *et al.*, 1991). In combination with other cell stimuli NaF showed modulatory effects such as an enhanced formation of leukotrienes when fMLP, opsonized zymosan, PMA and arachidonic acid were applied as stimuli. Prestimulation with NaF causes an increased binding of [^{3}H]-guanylylimidodiphosphate to isolated membrane preparations, indicating an enhanced exchange rate for GDP to GTP. These results suggested that a direct activation of GTP-binding proteins results in the generation of inflammatory mediators. Based on these observations studies were directed to elucidate the signal transduction of platelet activation after their stimulation with haemolysin-positive and -negative *E. coli* strains (König *et al.*, 1990).

Incubation of human platelets with the haemolysin-producing *E. coli* K12 pANN5211 induced the activation of protein kinase C, aggregation of platelets, calcium influx, low amounts of 12-HETE and release of serotonin from dense granules (König *et al.*, 1990). Non-haemolytic isogenic strains of *E. coli* 536/21 which differed only in their type of adhesins (MSH$^+$, MS-Fim$^{+/-}$, S-MRH$^+$, S-Fim$^+$, S-MRH$^+$, S-Fim$^-$, P-MRH$^+$, P-Fim$^+$) did not release serotonin or 12-HETE from human platelets. All haemolysin-negative bacteria except *E. coli* 536/21 without any adhesins were able to activate protein kinase C reversibly but did not induce calcium influx.

Activation of platelets with fluoride and activation of the GTP-binding protein was associated with protein kinase C activation, platelet aggregation, serotonin release and 12-HETE formation. In the presence of fluoride the cellular responses to *E. coli* K12 pANN5211 were not affected. The simultaneous stimulation of platelets with NaF and the non-haemolytic *E. coli* strains suppressed several of the NaF-induced platelet responses. Membrane preparations isolated from stimulated platelets with haemolysin-negative and haemolysin-positive *E. coli* showed an increased binding for guanylylimidodiphosphate. This increase was dependent on the haemolysin concentration. The data suggest that G regulatory proteins are involved in signal transduction by haemolysin-positive as well as by haemolysin-negative bacteria. NaF does not alter the response of platelets towards *E. coli* K12 pANN5211 with regard to platelet aggregation and serotonin release. In comparison to non-haemolytic *E. coli* strains, *E. coli* K12 pANN5211 does not inhibit 12-HETE release induced by NaF. Thus NaF and haemolysin do not counteract each other in the cellular transduction system. It has to be clarified whether non-haemolytic *E. coli* interact with the same or a different G protein as NaF which may inhibit several of the NaF-induced effects. Our data are consistent with a model in which platelet activation such as aggregation and dense granule release (serotonin) by haemolysin-positive *E. coli* requires the simultaneous formation of two signals, e.g. protein kinase C activation and calcium mobilization. Obviously the results are cell-dependent. When human granulocytes were studied the simultaneous activation of the cells with MR adhesins and Hly$^+$ *E. coli* enhanced LTB_4 formation. The results suggest that the *E. coli*-induced activation of granulocytes may include different G proteins.

Sepsis and septic shock

The diagnosis of septicaemia or sepsis is based upon the occurrence of bacteria (microbial products) present in the circulation or within tissue samples (biopsy) concomitant with the signs of septic illness which ultimately may turn into septic shock.

Animal studies have shown that during endotoxaemia 5-HETE is identified in the lung lymph of anaesthetized sheep (Ogletree et al., 1982). Cysteinyl-leukotrienes are also generated during the endotoxin shock, and the inactivation of leuktrienes is inhibited. Furthermore, cysteinyl-receptor antagonists prevent the experimental endotoxin shock (Hagmann et al., 1984, 1985).

Although the lipoxygenase factors exhibit potent physiologic and pathophysiologic activities, little work has been focused upon their role in severe traumatic injuries such as the burn syndrome. In animal models a rapid but reversible increase of N-acetyl-LTE$_4$ within the first 2 h was shown (Denzlinger et al., 1985); in addition, the involvement of cysteinyl-leukotrienes as mediators of the early burn oedema was suggested (Alexander et al., 1984).

We analysed the responsiveness of peripheral granulocytes from severely burned patients in longitudinal follow-up studies, since these patients are colonized by bacteria which quite often turn into microbial invasion (burn wound sepsis) (Winkler et al., 1987). The granulocytes from these patients showed a reduced capacity to generate LTB$_4$ after stimulation (Köller et al, 1989b) (Fig. 8). Granulocytes from patients who finally succumbed to burn wound sepsis showed a suppressed ability to generate leukotrienes as compared to granulocytes from survivors. The data are comparable with the observations of Braquet et al. (1985) who studied LTB$_4$ generation. A complete suppression of cellular reactions was also obtained for the basophil granulocyte. The basophils, although enhanced in number and intracellular histamine content, were not able to release histamine after challenge with anti-IgE or the calcium ionophore. In addition, we analysed the LTC$_4$ generation and the LTB$_4$ metabolism. We clearly demonstrated that LTC$_4$ generation correlated with the peripheral eosinophil cell count. The LTB$_4$ metabolism into ω-oxidated products was enhanced. In patients' PMNs an unknown LTB$_4$ metabolite was detected, which was not observed when PMNs from healthy donors were analysed (Brom et al., 1987). LTB$_4$ receptor expression was reduced on PMNs from severely burned patients (Brom et al., 1988a). Studies on the mechanism of these alterations indicated that the occurrence of immature PMNs (band cells) partly correlated with our observations (Köller et al., 1989a). When PAF was analysed it became apparent that the cells of the patients revealed an impaired metabolization of the mediator. The results suggest that these alterations in handling the inflammatory mediators may possibly contribute to the deleterious effects which result in burn sepsis.

Recent evidence suggests that granulocytes exposed to heat-treatment also show a reduced capacity to generate leukotrienes (Köller et al., 1989b). The previous exposure to heat (42°C) did not alter cellular viability. The results imply that various mechanisms, e.g. heat-shock proteins, priming and deactivation of granulocytes at various maturation stages, have to be considered. In fact, we provided evidence that pre-exposure of mature granulocytes to microbial exotoxins as well as endotoxins in addition to cellular activation deactivated the granulocyte for subsequent generation of LTB$_4$ (Bremm et al., 1987).

The decreased capacity to generate leukotrienes from PMNs preceded and/or accompanied the invasive bacterial infections, as was confirmed by quantitative bacterial counts within the respective biopsy specimen. The net result may be a suppressed chemotactic potential, which is followed by a reduced recruitment of phagocytes to sites of the burn wound and thus to an uncontrolled growth of bacteria. The continuous exposure of granulocytes towards bacterial toxins may further suppress the function of the phagocytes (Hartiala et al., 1985). Current research as to the role of growth factors, cytokines, as well as the signal-transduction cascade is necessary to elucidate the mechanisms which are required to activate and stabilize the granulocyte function.

Conclusion

Microbial toxins as soluble effectors or attached to the microbial surface are powerful tools to activate as well as counteract host-defence mechanisms.

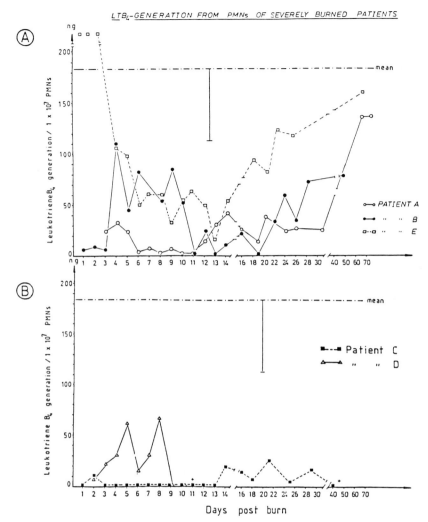

Figure 8. Concentration of LTB_4 in the supernatants of stimulated PMNs (1×10^7 cells stimulated with the Ca-ionophore A23187, 7.3 μM for 20 min) from severely burned patients. (A) Survivors ($n = 3$); (B) non-survivors. The broken line represents the mean value ± SD of healthy volunteers ($n = 8$).

While in the past major emphasis was directed towards their cytotoxic role it is increasingly evident that toxins at non-cytolytic concentrations modulate cellular functions. A variety of microbial exotoxins induce leukotriene or PAF formation which exert potent effects in acute and chronic inflammatory reactions. These mediators affect vasopermeability and also recruit additional inflammatory cells. Toxins also deactivate cells for their further functions. Toxin-pretreated cells may increase the metabolism of leukotriene B_4 into inactive products and/or also reduce chemotactic

LTB_4-receptor expression, thus leading to a deactivated granulocyte. Similar results were observed in septicaemia whereby granulocytes show a reduced leukotriene formation. In this regard mediators serve as local hormones of the microenvironment.

Toxins also modulate cellular functions by interacting with the signal-transduction cascade (e.g. protein kinase C, phosphoinositol turnover, G proteins). Thus, toxins are powerful probes in cell biology, membrane biochemistry and also interfere with cell-mediated immune processes, e.g. antigen presentation and cytokine formation, which also control the release of inflammatory mediators.

Acknowledgements

The authors gratefully appreciate the support of the Deutsche Forschungsgemeinschaft.

References

Alexander, F., Matthieson, M., Teoh, K.H.T., Huval, W.V., Lelcuk, S., Valeri, C.R., Shepro, D. and Hechtman, H.B. (1984). Arachidonic metabolites mediate early burn edema. *J. Trauma* **24**, 709–12.

Alouf, J.E., Geoffroy, C., Pattus, F. and Verger, R. (1984). Surface properties of bacterial sulfhydryl-activated cytolytic toxins. Interaction with monomolecular films of phosphatidylcholine and various sterols. *Eur. J. Biochem.* **141**, 205–10.

Alouf, J.E., Dufourq, J., Siffert, O. and Geoffroy, C. (1988). Comparative properties of natural and synthetic staphylococcal delta toxin and analogues, In: *Bacterial Protein Toxins* (eds F. Fehrenbach *et al.*), pp. 17–39. Gustav Fischer, Stuttgart and New York.

Alouf, J.E., Dufourcq, J., Siffert, O., Thiaudiere, E. and Geoffroy, C. (1989). Interaction of staphylococcal delta-toxin and synthetic analogues with erythrocytes and phospholipid vesicles. *Eur. J. Biochem.* **183**, 381–90.

Asao, T., Kinoshita, Y., Kozaki, S., Uemura, T. and Sakaguchi, G. (1984). Purification and some properties of *Aeromonas hydrophila* hemolysin. *Infect. Immun.* **46**, 122–7.

Asao, T., Kozaki, S., Kato, K., Kinoshita, Y., Otsu, K., Uemura, T. and Sakaguchi, G. (1986). Purification and characterization of an *Aeromonas hydrophila* hemolysin. *J. Clin. Microbiol.* **24**, 228–32.

Atkinson, Y.H., Marasco, W.A., Lopez, A.F. and Vadas, M.A. (1988a). Recombinant human tumor necrosis factor-α regulation of N-formyl-methionyl-leucylphenylalanine receptor affinity and function on human neutrophils. *J. Clin. Invest.* **81**, 759–65.

Atkinson, Y.H., Lopez, A.F., Marasco, W.A., Lucas, C.M., Wong, G.C., Burns, G.F. and Vadas, M.A. (1988b). Recombinant human granulocyte-macrophage colony stimulating factor (rH GM-CSF) regulates fMet-Leu-Phe receptors on human neutrophils. *Immunology* **64**, 519–25.

Bach, M.K. (1988). Lipid mediators of hypersensitivity. *Prog. Allergy* **44**, 10–98.

Baggiolini, M. and Dewald, B. (1984). Exocytosis by neutrophiles. *Contemp. Top. Immunobiol.* **14**, 221–46.

Bainton, D.F. (1988). Phagocytic cells: developmental biology of neutrophils and eosinophils. In: *Inflammation: Basic Principles and Clinical Correlates* (eds J.I. Gallin, I.M. Goldstein and R. Snyderman), pp. 265–81. Raven Press, New York.

Bauldry, S.A., Wyke, R.L. and Bass, D.A. (1988). Phospholipase A₂ activation in human neutrophils. *J. Biol. Chem.* **263**, 16787–95.

Becker, E.L. (1988). Membrane activation in immunologically relevant cells. In: *Progress in Allergy* (eds K. Ishizaka, P. Kallós, P.J. Lachmann and B.H. Waksman), pp. 11–16. S. Karger, Basel.

Benz, R., Schmid, A., Wagner, W. and Goebel, W. (1989). Pore formation by the *Escherichia coli* hemolysin: evidence for an association-dissociation equilibrium of the pore-forming aggregates. *Infect. Immun.* **57**, 887–95.

Berche, P., Gaillard, J.L. and Raveneau, J. (1989). Pathophysiology of *Listeria monocytogenes* infection. In: *Local Immmunity*, Vol. 5 (eds J.P. Revillard and N. Wierzbicki), pp. 23–41. Fondation Franco-Allemande.

Berger, H., Hacker, J., Juarez, A., Hughes, C. and Goebel, W. (1982). Cloning of the chromosomal determinants encoding hemolysin production and mannose-resistant hemagglutination in *Escherichia coli*. *J. Bacteriol.* **152**, 1241–7.

Bergmann, U., Scheffer, J., Schönfeld, W., Köller, M., Erbs, G., Müller, F.E. and König, W. (1989). Induction of inflammatory mediators (histamine, leukotrienes) from rat peritoneal mast cells and human granulocytes by *Pseudomonas aeruginosa* from burn patients. *Infect. Immun.* **57**, 2187–95.

Berk, R.S., Brown, D., Coutinho, I. and Meyers, D. (1987). In vivo studies with two phospholipase C fractions from *Pseudomonas aeruginosa*. *Infect. Immun.* **55**, 1728–30.

Berka, R.M. and Vasil, M.L. (1982). Phospholipase C (heat-labile hemolysin) of *Pseudomonas aeruginosa*: purification and preliminary characterization. *J. Bacteriol.* **152**, 239–45.

Bernheimer, A.W. and Rudy, B. (1986). Interactions between membranes and cytolytic peptides. *Biochem. Biophys. Acta* **864**, 123–41.

Berridge, J. and Irvine, R.F. (1984). Inositol triphosphate, a novel second messenger in cellular signal transduction. *Nature* **312**, 315–21.

Bhakdi, S., Mackman, N., Nicaud, J.M. and Holland, I.B. (1986). *Escherichia coli* hemolysin may damage target cell membranes by generating transmembrane pores. *Infect. Immun.* **52**, 63–9.

Bhakoo, M., Birbeck, T.H. and Freer, J.H. (1982). Interaction of *Staphylococcus aureus* delta-lysin with phospholipid monolayers. *Biochemistry* **21**, 6879–83.

Blumenstock, E. and Jann, K. (1982). Adhesion of piliated *Escherichia coli* strains to phagocytes: Differences between bacteria with mannose-sensitive pili and those with mannose-resistant pili. *Infect. Immun.* **35**, 264–9.

Bokoch, G.M. and Gilman, A.G. (1984). Inhibition of receptor-mediated release of arachidonic by pertussis toxin. *Cell* **39**, 301.

Bokoch, G.M. and Reed, P.W. (1979). Inhibition of the neutrophil oxidative response to a chemotactic peptide by inhibitors of arachidonic acid oxygenation. *Biochem. Biophys. Res. Comm.* **90**, 481–7.

Bokoch, G.M. and Reed, P.W. (1980). Stimulation of arachidonic metabolism in the polymorphonuclear leukocyte by an *N*-formylated peptide. *J. Biol. Chem.* **255**, 10223–6.

Borgeat, P. and Samuelsson, B. (1979). Metabolism of arachidonic acid in polymorphonuclear leukocytes structure analysis of novel hydroxylated compounds. *J. Biol. Chem.* **254**, 7865–9.

Borgeat, P., Fruteau de Laclos, B., Rabinovitch, H., Picard, S., Braquet, P., Hebert, J. and Laviolette, M. (1984). Generation and structures of the lipoxygenase products. Eosinophil-rich human polymorphonuclear leukocyte preparations characteristically release leukotriene C_4 on ionophore A 23187 challenge. *J. Allergy Clin. Immunol.* **74**, 310–15.

Braquet, M., Lavaud, P., Dormont, D., Garay, R., Ducousso, R., Guilbaud, J., Chignard, M., Borgeat, P. and Bracquet, P. (1985). Leukocyte functions in burn-injured patients. *Prostaglandins* **29**, 747–64.

Braquet, P. and Rola-Pleszczynski, M. (1987). Platelet-activating factor and cellular immune responses. *Immun. Today* **8**, 345–52.

Braun, V., Günther, H., Neuß, B. and Tautz, C. (1985). Hemolytic activity of *Serratia marcescens. Arch. Microbiol.* **141**, 371–6.

Bremm, K.D., Brom, H.J., König, W., Bohn, A., Theobald, K., Bhakdi, S., Lutz, F. and Fehrenbach, F.J. (1983). Bacteria and bacterial exotoxins induce leukotriene formation from human polymorphonuclear granulocytes. *Monographs in Allergy* **18**, 196–200.

Bremm, K.D., Brom, H.J., Alouf, J.E., König, W., Spur, B., Crea, A. and Peters, W. (1984). Generation of leukotrienes from human granulocytes by alveolysin from *Bacillus alvei. Infect. Immun.* **44**, 188–93.

Bremm, K.D., König, W., Pfeiffer, P., Rauschen, I., Theobald, K., Thelestam, M. and Alouf, J.E. (1985). Effect of thiol-activated toxins (streptolysin O, alveolysin, and theta toxin) on the generation of leukotrienes and leukotriene-inducing and -metabolizing enzymes from human polymorphonuclear granulocytes. *Infect. Immun.* **50**, 844–51.

Bremm, K.D., König, W., Thelestam, M. and Alouf, J.E. (1987). Modulation of granulocyte functions by bacterial exotoxin and endotoxins. *Immunology* **62**, 363–71.

Bremm, K.D., König, W., Brom, J., Thelestam, M., Fehrenbach, F.J. and Alouf, J.E. (1988a). Release of lipid mediators (leukotrienes, PAF) by bacterial toxins from human polymorphonuclear granulocytes. In: *Bacterial Protein Toxins* (eds Fehrenbach *et al.*), pp. 103–4. Gustav Fischer, Stuttgart.

Bremm, K.D., König, W., Hensler, T., Megret, F. and Alouf, J.E. (1988b). Modulation of mediator release (histamine, leukotrienes) from human leukocytes by *Bordetella pertussis* toxin. In: *Proceedings of the Conference organized by the Society of Microbiology and Epidemiology of the GDR* (eds S. Mebel, H. Stompe, M. Drescher and S. Rustenbach), pp. 208–16.

Brom, J. and König, W. (1989). Studies on the mechanisms of binding and metabolism of leukotriene B_4 by human neutrophils. *Immunology* **68**, 479–85.

Brom, C., Köller, M., Brom, J. and König, W. (1989). Effect of sodium fluoride on the generation of lipoxygenase products from human polymorphonuclear granulocytes, mononuclear cells and platelets: Indication for the involvement of G proteins. *Immunology* **68**, 240–6.

Brom, J., König, W., Köller, M., Gross-Weege, W., Erbs, G. and Müller, F. (1987). Metabolism of leukotriene B_4 by polymorphonuclear granulocytes of severely burned patients. *Prostaglandins, Leukotrienes Med.* **27**, 209–25.

Brom, J., Köller, M., Schönfeld, W., Knöller, J., Erbs, G., Muller, F.E. and König, W. (1988a). Decreased expression of leukotriene B_4 receptor sites on polymorphonuclear granulocytes of severely burned patients. *Prostaglandins, Leukotrienes Med.* **34**, 153–9.

Brom, J., Schönfeld, W. and König, W. (1988b). Metabolism of leukotriene B_4 by activated human polymorphonuclear granulocytes. *Immunology* **64**, 509–18.

Cavalieri, S.J., Bohach, G.A. and Snyder, I.S. (1984). *Escherichia coli* α-hemolysin: Characteristics and probable role in pathogenicity. *Microbiol. Rev.* **48**, 326–43.

Chakraborty, T., Huhle, B., Bergbauer, H. and Goebel, W. (1986). Cloning, expression, and mapping of the *Aeromonas hydrophila* aerolysin gene determinant in *Escherichia coli* K-12. *J. Bacteriol.* **167**, 368–74.

Chauncey, J.B., Simon, R.H. and Peters-Golden, M. (1988). Rat alveolar macrophages synthesize leukotriene B_4 and 12-hydroxyeicosatetraenoic from alveolar epithelial cell derived arachidonic acid. *Am. Rev. Respir. Dis.* **138**, 928–35.

Chilton, F.H. and Connel, T.R. (1988). 1-Ether linked phosphoglycerides: Major endogenous sources of arachidonate in human neutrophil. *J. Biol. Chem.* **263**, 5260–5.

Chin Lee, C., Sugerman, H.J., Tatum, J.L., Wriggt, J.P., Hirsch, P.D. and Hirsch, J.I. (1986). Effect of Ibuprofen on a pig *Pseudomonas* ARDS model. *J. Surg. Res.* **40**, 438–44.

Curnutte, J.T., Babior, B.M. and Karnovsky, M.L. (1979). Fluoride-mediated activation of the respiratory burst in human neutrophils: a reversible process. *J. Clin. Invest.* **63**, 637–41.

Denzlinger, C., Rapp, S., Hagmann, W. and Keppler, D. (1985). Leukotrienes as mediators in tissue trauma. *Science* **230**, 330.

Edenius, C., Haeggström, J. and Lindgren, J.A. (1988). Transcellular conversion of endogenous arachidonic acid to lipoxins in mixed human platelet-granulocyte suspensions. *Biochem. Biophys. Res. Commun.* **157**, 801–7.

Fantone, J.C. and Ward, P.A. (1982). Role of oxygen-derived free radicals and metabolites in leukocyte-dependent inflammatory reactions. *Am. J. Pathol.* **107**, 395–418.

Feinmark, S.J. and Cannon, P.J. (1986). Endothelial cell leukotriene C4 synthesis results from intercellular transfer of leukotriene A$_4$ synthesized by polymorphonuclear leukocytes. *J. Biol. Chem.* **261**, 16466–72.

Field, M., Musch, M.W., Miller, R.L. and Goetzl, E.J. (1984). Regulation of epithelial electrolyte transport by metabolites of arachidonic acid. *J. Allergy Clin. Immunol.* **74**, 382–5.

Fitton, J.E., Dell, A. and Shaw, W.V. (1980). The amino acid sequence of the delta haemolysin of *Staphylococcus aureus*. *FEBS Lett.* **115**, 209–12.

Fitzpatrick, F., Liggett, J., McGee, J., Bunting, S., Morton, D. and Samuelsson, B. (1984). Metabolism of leukotriene A$_4$ by human erythrocytes. A novel cellular source of leukotriene B4. *J. Biol. Chem.* **259**, 11403–7.

Freer, J.H. and Arbuthnott, J.P. (1983). Toxins of *Staphylococcus aureus*. *Pharmacol. Ther.* **19**, 55.

Gadeberg, O.V. and Blom, J. (1986). Morphological study of the in vitro cytotoxic effect of α-hemolytic *E. coli* bacteria and culture supernatants on human blood granulocytes and monocytes. *Acta. Pathol. Microbiol. Immunol. Scand.* Sect. B, **94**, 75–83.

Gemmell, C.G. (1988). Bacterial toxins and enzymes showing biological activity towards phagocytic cells: an overview. In: *Bacterial Protein Toxins* (eds F.J. Fehrenbach *et al.*), pp. 359–66. Gustav Fischer, Stuttgart and New York.

Geoffroy, C., Gaillard, J.L., Alouf, J.E. and Berche, P. (1987). Purification, characterization, and toxicity of the sulfhydryl-activated hemolysin listeriolysin O from *Listeria monocytogenes*. *Infect. Immun.* **55**, 1641–6.

Gilman, A.G. (1987). G-proteins: transducers of receptor-generated signals. *Ann. Rev. Biochem.* **56**, 615–49.

Godfrey, R.W., Manzi, R.M., Clark, M.A. and Hoffstein, S.T. (1987). Stimulus specific induction of phospholipid and arachidonic acid metabolism in human neutrophils. *J. Cell Biol.* **104**, 925–32.

Goebel, W. and Hedgpeth, J. (1982). Cloning and functional characterization of the plasmid-encoded hemolysin determinant of *Escherictia coli. J. Bacteriol.* **151**, 1290–8.

Goldman, D.W. and Goetzl, E.J. (1984). Heterogeneity of human polymorphonuclear leukocyte receptors for leukotriene B$_4$ identification of subset of high affinity receptors that transduce the chemotactic response. *J. Exp. Med.* **159**, 1027–41.

Goldman, D.W., Gifford, L.A., Olson, D.U. and Goetzl, E.J. (1985). Transduction by leukotriene B$_4$ receptors of increases in cytosolic calcium in human polymorphonuclear leukocytes. *J. Immunol.* **135**, 525–30.

Gracey, M., Burke, V. and Robinson, J. (1982). Aeromonas-associated gastroenteritidis. *Lancet* ii, 1304–6.

Gray, L.S., Huber, K.S., Gray, M.C., Hewlett, E.L. and Engelhard, V.H. (1989). Pertussis toxin effects on the T lymphocytes are mediated through CD3 and not by pertussis toxin catalyzed modification of a G protein. *J. Immunol.* **142**, 1631–8.

Groß-Weege, W., König, W., Scheffer, J. and Nimmich, W. (1988). Induction of histamine release from rat mast cells and human basophilic granulocytes by clinical *E. coli* isolates. Relation to hemolysin-producing and adhesion-expression. *J. Clin. Microbiol.* **26**, 1831–7.

Hacker, J., Hughes, C., Hof, H. and Goebel, W. (1983). Cloned hemolysin genes from *Escherichia coli* that cause urinary tract infection determine different levels of toxicity in mice. *Infect. Immun.* **42**, 57–63.

Hagmann, W., Denzlinger, C. and Keppler, D. (1984). Role of peptide leukotrienes and their hepatobiliary elimination in endotoxin action. *Circulat. Shock* **14**, 223–35.

Hagmann, W., Denzlinger, C. and Keppler, D. (1985). Production of peptido leukotrienes in endotoxin shock. *FEBS Lett.* **180**, 309–13.

Hamberg, M. (1976). On the formation of thromboxane B2 and 12-L-hydrox-5,8,10,14 eicosatetraenoic acid in tissues from the guinea pig. *Biochim. Biophys. Acta* **431**, 651–4.

Harnett, M.M. and Klaus, G.G.B. (1988). G protein regulation of receptor signalling. *Immunol. Today* **9**, 315–20.

Hartiala, K.T., Langlois, L., Goldstein, I.M. and Rosenbaum, J.T. (1985). Endotoxin induced selective dysfunction of rabbit polymorphonuclear leukocytes in response to endogenous chemotactic factors. *Infect. Immun.* **50**, 527–33.

Hensler, T., Raulf, M., Megret, F., Alouf, J.E. and König, W. (1989). Modulation of leukotriene generation by pertussis toxin. *Infect. Immun.* **57**, 3165–71.

Hirayama, T. and Kato, I. (1984). Mode of cytotoxic action of pseudomonal leukocidin on phosphatidylinositol metabolism and activation of lysosomal enzyme in rabbit leukocytes. *Infect. Immun.* **43**, 21–27.

Hostacka, A., Ciznar, I., Korych, B. and Karolcek, J. (1982). Toxic factors of *Aeromonas hydrophila* and *Pleisiomonas shigelloides*. *Zbl. Bakt. Hyg. A* **252**, 525–34.

Hughes, C., Müller, D., Hacker, J. and Goebel, W. (1982). Genetics and pathogenic role of *Escherichia coli* haemolysin. *Toxicon* **20**, 247–52.

Ishizaka, T., White, J.R. and Saito, H. (1987). Activation of basophils and mast cells for mediator release. *Int. Archs. Allergy Appl. Immunol.* **82**, 327–32.

Johnson, M.K. and Boese-Marrazzo, D. (1980). Production and properties of heat-stable extracellular hemolysin from *Pseudomonas aeruginosa*. *Infect. Immun.* **29**, 1028–33.

Jorgensen, S.E., Mulcahy, P.F. and Louis, C.F. (1986). Effects of *Escherichia coli* hemolysin on permeability of erythrocyte membranes to calcium. *Toxicon* **24**, 559–66.

Jubiz, W., Radmark, O., Malmsten, C., Hanson, G., Lindgren, J.A., Palmblad, J., Uden, A.M. and

Samuelsson, B. (1982). A novel leukotriene produced by stimulation of leukocytes with formyl-methionyl-leucyl phenylalanine. *J. Biol. Chem.* **257**, 6106–10.

Kamata, R., Yamamoto, T., Matsumoto, K. and Maeda, H. (1985). A serratial protease causes vascular permeability reaction by activation of the hageman factor-dependent pathway in guinea pigs. *Infect. Immun.* **48**, 747–53.

Kasimir, S., Schönfeld, W., Alouf, J.E. and König, W., (1990). Effect of staphylococcus aureus delta-toxin on human granulocyte functions and platelet-activating-factor metabolism. *Infect. Immun.* **58**, 1653–9.

Katada, T. and Ui, T.M. (1982). ADP ribosylation of the specific membrane protein of C6 cells by islet-activating protein associated with modification of adenylate cyclase activity. *J. Biol. Chem.* **257**, 7210–16.

Kathariou, S., Metz, P., Hof, H. and Goebel, W. (1987). Tn916-induced mutations in the hemolysin determinant affecting virulence of *Listeria monocytogenes*. *J. Bacteriol.* **169**, 1291–7.

Köller, M., König, W., Brom, J., Raulf, M., Groß-Weege, W., Erbs, G. and Müller, F.E. (1988). Generation of leukotrienes from polymorphonuclear granulocytes of severely burned patients. *J. Trauma* **28**, 733–40.

Köller, M., König, W., Brom, J., Erbs, G. and Müller, F.E. (1989a). Studies on the mechanism of granulocyte dysfunctions in severely burned patients. *J. Trauma* **29**, 435–44.

Köller, M., Brom, C., Brom, J. and König, W. (1989b). Heat shock induces alterations of the lipoxygenase pathway in human polymorphonuclear granulocytes. *Prostaglandins, Leukotrienes and Essential Fatty Acids* **38**, 99–106.

König, B., König, W., Scheffer, J., Hacker, J. and Goebel, W. (1986). Role of *Escherichia coli* alpha-hemolysin and bacterial adherence in infection: Requirement for release of inflammatory mediators from granulocytes and mast cells. *Infect. Immun.* **54**, 886–92.

König, B., Schönfeld, W., Scheffer, J. and König, W. (1990). Signal transduction in human platelets and inflammatory mediator release induced by genetically cloned hemolysin-positive and negative *Escherichia coli* strains. *Infect. Immun.* **58**, 1591–9.

König, W., Czarnetzki, B.M. and Lichtenstein, L.M. (1976). Eosinophil chemotactic factor (ECF). II. Release during phagocytosis of human polymorphonuclear leukocytes. *J. Immunol.* **117**, 235–45.

König, W., Scheffer, J., Bremm, K.D., Hacker, J. and Goebel, W. (1985). Role of bacterial adherence and toxin production from *Escherichia coli* on leukotriene generation from human polymorphonuclear granulocytes. *Int. Archs Allergy Appl. Immun.* **77**, 118–20.

König, W., Bremm, K.D., Scheffer, J., Kritschker, B., Schonfeld, W. and Theobald, K. (1986). Generation of mediators of inflammation (histamine, leukotrienes) from various cells by bacterial products,

In: *Bacterial Protein Toxins* (eds P. Falmagne, J.E. Alouf, F.J. Fehrenbach, J. Jeljaszewicz and M. Thelestam), pp. 131–9. Gustav Fischer Verlag, Stuttgart and New York.

König, W., Faltin, Y., Scheffer, J., Schöffler, H. and Braun, V. (1987). Role of cell-bound hemolysins as a pathogenicity factor for Serratia infections. *Infect. Immun.* **55**, 2554–61.

König, W., Knoller, J. and Schonfeld, W. (1988). Mediators and cellular interactions in the lung. In: *Local Immunity 'The Alveolar Macrophage'* (ed. C. Sorg), pp. 119–37. Regensberg & Biermannn, Stuttgart and New York.

König, W., König, B., Scheffer, J., Hacker, J. and Goebel, W. (1989). Role of cloned virulence factors (mannose resistant haemagglutination, mannose resistant adhesions) from uropathogenic *Escherichia coli* strains in the release of inflammatory mediators from neutrophils and mast cells. *Immunology* **67**, 401–7.

Korhonen, T.K., Virkola, R. and Holthöfer, H. (1986). Localization of binding sites for purified *Escherichia coli* P fimbriae in the human kidney. *Infect. Immun.* **54**, 328–32.

Kühn, H., Pönicke, K., Halle, W., Wiesner, R., Schewe, T. and Förster, W. (1985). Metabolism of [1-^{14}C] arachidonic acid by cultured calf aortic endothelial cells: Evidence for the presence of a lipoxygenase pathway. *Prostaglandins Leukotrienes Med.* **17**, 291–303.

Lanni, C. and Becker, E.L. (1983). Release of phospholipase A_2 activity from rabbit peritoneal neutrophils by f-Met-Leu-Phe. *Am. J. Pathol.* **113**, 90–4.

Lewis, R.A. and Austen, F. (1984). The biologically active leukotrienes biosynthesis, metabolism, receptors, functions and pharmacology. *J. Clin. Invest.* **73**, 889–97.

Lyerly, D.M. and Kreger, A.S. (1983). Importance of *Serratia* proteases in the pathogenesis of experimental *Serratia marcescens* pneumonia. *Infect. Immun.* **40**, 113–19.

Maclouf, J.A. and Murphy, R.C. (1988). Transcellular metabolism of neutrophil-derived leukotriene A_4 by human platelets. A potential source of leukotriene C4. *J. Biol. Chem.* **263**, 174–81.

Marré, R., Hacker, J., Henkel, W. and Goebel, W. (1986). Contribution of cloned virulence factors from uropathogenic *Escherichia coli* strains to nephropathogenicity in an experimental rat pyelonephritis model. *Infect. Immun.* **54**, 761–7.

Mellor, I.R., Thomas, D.H. and Sansom, M.S.P. (1988). Properties of ion channels formed by *Staphylococcus aureus* delta toxin. *Biochim. Biophys. Acta* **942**, 280–94.

Morrison, A.J. and Wenzel, R.P. (1984). Epidemiology of infections due to *Pseudomonas aeruginosa*. *Rev. Infect. Dis.* **6**(S3), 627–42.

Musch, M.W., Miller, R.J., Field, M. and Siegel, M.I. (1982). Stimulation of colonic secretion by lipoxygenase metabolites of arachidonic acid. *Science* **217**, 1255–6.

Needleman, P., Turk, J., Jakschik, B.A., Morrison, A.R. and Levkowith, J.B. (1986). Arachidonic acid metabolism. *Ann. Rev. Biochem.* **55**, 69–102.

Norn, S., Stahl, P., Jensen, C., Espersen, F. and Jarlov, J.O. (1985). Bacterial-induced histamine release. Examination of the bacterial cell wall components peptidoglycan, teichonic acid and protein A. *Agents and Actions* **16**, 273–6.

Norn, S., Stahl Skov, P., Jensen, C., Jarlov, J.O. and Espersen, F. (1987). Histamine release induced by bacteria. A new mechanism in asthma? *Agents and Actions* **20**, 29–34.

O'Driscoll, B.R.C., Cromwell, O. and Kay, B. (1985). Sputum leukotrienes in obstructive airway diseases. *Clin. Exp. Immunol.* **55**, 397–404.

O'Flaherty, J.T., Redman, J.F. and Jacobson, D.P. (1986). Protein kinase C regulates leukotriene B_4 receptors in human neutrophils. *FEBS Lett.* **206**, 279–82.

Ogletree, M.L., Oates, J.A., Brigham, K.L. and Hubbard, W.C. (1982). Evidence for pulmonary release of 5-hydroxyeicosatetraenoic acid (5-HETE) during endotoxemia in unanaesthetized sheep. *Prostaglandins* **23**, 459–68.

Parker, C.W. (1987). Lipid mediators produced through the lipoxygenase pathway. *Ann. Rev. Immunol.* **5**, 63–84.

Parkinen, J., Rogers, G.N., Korhonen, T., Dahr, W. and Finne, J. (1986). Identification of the O-linked sialyloligosaccharides of glycophorin A as the erythrocyte receptors for S-fimbriated *Escherichia coli*. *Infect. Immun.* **54**, 37–42.

Parrisius, J., Bhakdi, S., Roth, M., Tranum-Jensen, J., Goebel, W. and Seelinger, H.P.R. (1986). Production of listeriolysin by beta-hemolytic strains of *Listeria monocytogenes*. *Infect. Immun.* **51**, 314–19.

Patrone, F., Dallegri, F. and Sacchetti, C. (1980). Stimulation of granulocyte adhesiveness by the chemotatic peptide N-formyl-L-methionyl-L-phenylalanine. *Res. Exp. Med.* **177**, 19–22.

Pinckart, R.N., Ludwig, J.C. and McManus, L.M. (1988). Platelet-activating factors. In: *Inflammation: Basic Principles and Clinical Correlates* (eds J.I. Gallin, I.M. Goldstein and R. Snyderman), pp. 139–67. Raven Press, New York.

Plaut, M., Pierce, J.H., Watson, C.J., Hanlay-Hyde, J., Nordan, R.P. and Paul, W.E. (1989). Mast cell lines produce lymphokines in response to cross-linkage of $Fc_\epsilon RI$ or to calcium ionophores. *Nature* **339**, 64–7.

Raulf, M., König, W., Köller, M. and Stüning, M. (1987). Release and functional characterization of the leukotriene D_4-metabolizing enzyme (dipeptidase) from human polymorphonuclear leukocytes. *Scand. J. Immunol.* **25**, 305–13.

Raulf, M., Alouf, J.E. and König, W. (1990). Effect of staphylococcal delta-toxin and bee venom peptide melittin on leukotriene induction and metabolism

of human polymorphonuclear granulocytes. *Infect. Immun.* **58**, 2678–82.

Robinson, C. and Holgate, S. (1985). Mast-cell dependent inflammatory mediators and their putative role in bronchial asthma. *Clin. Sci.* **68**, 103–12.

Rosoff, P.M., Walker, R. and Winberry, L. (1987). Pertussis toxin triggers rapid second messenger production in human T lymphocytes. *J. Immunol.* **139**, 2419–23.

Rossi, F., deTogni, P., Bellavite, P., Della Bianca, V. and Grzeskowiak, M. (1983). Relationship between the binding of N-formyl-methionyl-leucyl-phenylalanine and the respiratory response in human neutrophils. *Biochim. Biophys. Acta* **758**, 168–75.

Rossi, F., Della Bianca, V., Grzeskowiak, M. and Bazzoni, F. (1989). Studies on molecular regulation of phagocytosis in neutrophils Con A-mediated ingestion and associated respiratory burst independent of phosphoinositide turnover, rise in $[Ca^{2+}]$ and arachidonic acid release. *J. Immunol.* **142**, 1652–60.

Rouzer, C.A., Scott, W.A., Hamill, A.L., Liu, F., Katz, D.H. and Cohn, Z.A. (1982). Secretion of leukotriene C and other arachidonic acid metabolites by macrophages challenged with immunoglobulin E immune complexes. *J. Exp. Med.* **156**, 1077–86.

Salmon, J.A. and Higgs, G.A. (1987). Prostaglandins and leukotrienes as inflammatory mediators. In: *British Medical Bulletin: Inflammation – Mediators and Mechanisms* (ed. D.A. Willoughby), pp. 285–96. Churchill Livingstone, Edinburgh, London, Melbourne and New York.

Scheffer, J., König, W., Hacker, J. and Goebel, W. (1985). Bacterial adherence and hemolysin production from *Escherichia coli* induces histamine and leukotriene release from various cells. *Infect. Immun.* **50**, 271–8.

Scheffer, J., Vosbeck, K. and König, W. (1986). Induction of inflammatory mediators from human polymorphonuclear granulocytes and rat mast cells by haemolysin-positive and -negative *E. coli* strains with different adhesins. *Immunology* **59**, 541–8.

Scheffer, J., König, W., Braun, V. and Goebel, W. (1988). Comparison of four hemolysin-producing organisms (*Escherichia coli*, *Serratia marcescens*, *Aeromonas hydrophila*, *Listeria monocytogenes*) for release of inflammatory mediators from various cells. *J. Clin. Microbiol.* **26**, 544–51.

Schönfeld, W., Saak, A., Steinkamp, G., v.d.Hardt, H. and König, W. (1989). Histamine release from basophils in cystic fibrosis. *Clin. Exp. Immunol.* **76**, 434–9.

Serhan, C.N., Hamberg, M. and Samuelsson, B. (1984). Lipoxins: Novel series of biologically active compounds formed from arachidonic acid in human leukocyte. *Proc. Natl. Acad. Sci. USA* **81**, 5335–9.

Sha'áfi, R.I. and Molski, T.F.P. (1988). Activation of the neutrophil. In: *Progress in Allergy* (eds K. Ishizaka, P. Kallós, P.J. Lachmann and B.H. Waksman), pp. 1–64. S. Karger, Basel.

Siegel, M.I., McConnell, R.T., Porter, N.A. and

Cuatrecas, P. (1980). Arachidonate metabolism via lipoxygenase and 12L-hydroperoxy-5,8,10,14-icosatetraenoic acid peroxidase sensitive to antiinflammatory drugs. *Proc. Natl. Acad. Sci. USA* **77**, 308–12.

Sigal, E., Craik, C.S., Highland, E., Grunberger, D., Costello L.L., Dixon, A.F. and Nadel, J.A. (1988a). Molecular cloning and primary structure of human 15-lipoxygenase. *Biochem. Biophys. Res. Commun.* **157**, 457–64.

Sigal, E., Grunberger, D., Craik, C.S., Caughey, G.H. and Nadel, J.A. (1988b). Arachidonate 15-lipoxygenase (omega-6 lipoxygenase) from human leukocytes. *J. Biol. Chem.* **263**, 5328–32.

Snyderman, R. and Uhing, R.J. (1988). Phagocytic cells: Stimulus response coupling mechanisms. In: *Inflammation: Basic Principles and Clinical Correlates* (eds J.I. Gallin, I.M. Goldstein and R. Snyderman), pp. 309–23. Raven Press, New York.

Sordelli, D.O., Cerquetti, M.C., Hooke, A.M. and Bellanti, J.A. (1985). Effect of chemotactins released by *Staphylococcus aureus* and *Pseudomonas aeruginosa* on the murine respiratory tract. *Infect. Immun.* **49**, 265–9.

Spector, A.A., Gordon, J.A. and Moore, S.A. (1988). Hydroxyeicosatetraenoic acids (HETEs). *Prog. Lipid Res.* **27**, 271–323.

Sterzik, B., Jürgens, D. and Fehrenbach, F.J. (1986). Structure and function of CAMP factor of *Streptococcus agalactiae*. In: *Bacterial Protein Toxins* (eds P. Falmagne, J.E. Alouf, F.J. Fehrenbach, J. Jaljaszewicz and M. Thelestam), pp. 101–8. Gustav Fischer, Stuttgart.

Strnad, C.F. and Wong, K. (1985). Calcium mobilization in fluoride activated neutrophils. *Biochem. Biophys. Res. Commun.* **133**, 161.

Subramanian, N. and Bray, M.A. (1987). Interleukin 1 releases histamine from human basophils and mast cells *in vitro. J. Immunol.* **138**, 271–5.

Thomas, L.L., Arra, S.L. and Karns, B.K. (1987). Activation of human basophil histamine release by *E. coli* isolates. *J. Cell Biochem.* **11B**, 119.

Tomita, T., Mosmoi, K. and Kanegasaki, S. (1984). Staphylococcal delta toxin-induced generation of chemiluminescence by human polymorphonuclear leukocytes. *Toxicon* **22**, 957–65.

Tonnel, A.B., Gosset, P., Capron, M., Tomassini, M., Joseph, M. and Capron, A. (1989). Participation of Fc-εRII positive cells in asthma. In: *Progress in Allergy and Clinical Immunology* (eds W.J. Pichler, B.M. Stadler, C.A. Dahinden, A.R. Pecoud, P. Frei, C.H. Schneider and A.L. de Weck), pp. 172–8. Hogrefe & Huber. Stuttgart.

Ventur, Y., Scheffer, J., Hacker, J., Goebel, W. and König, W. (1990). Effects of adhesins from mannose-resistant Escherichia coli on mediator release from human lymphocytes, monocytes and basophils and from polymorphonuclear granulocytes. *Infect. Immun.* **58**, 1500–8.

Volpi, M., Nacchache, P.H., Molski, T.F.P., Shefcik, J., Huang, C.K., Marsh, M.L., Munoz, J., Becker, E.L. and Sha'áfi, R.I. (1985). Pertussis toxin inhibits fMet-Leu-Phe- but not phorbol ester-stimulated changes in rabbit neutrophils: role of G proteins in excitation response coupling. *Proc. Natl. Acad. Sci. USA* **82**, 2708–12.

Ward, P.A., Till, G.O., Kunkel, R. and Beauchamp, C. (1983). Evidence for the role of hydroxyl radical in complement and neutrophil dependent tissue injury. *J. Clin. Invest.* **72**, 789–801.

Wardlaw, A.J., Moqbel, R., Cromwell, O. and Kay, A.B. (1986). Platelet-activating factor – a potent chemotactic and chemokinetic factor for human eosinophils. *J. Clin. Invest.* **78**, 1701–6.

Warner, J.A., Yancey, K.B. and MacGlashan, D.W. (1987). The effect of pertussis toxin on mediator release from human basophils. *J. Immunol.* **139**, 161–5.

Wawryk, S.O., Novotny, J.R., Wicks, I.P., Wilkinson, D., Maher, D., Salvaris, E., Welch, K., Fecondo, J. and Boyd, A.W. (1989). The role of the LFA-1/ICAM-1 interaction in human leukocyte homing and adhesion. *Immunol. Rev.* **108**, 135–61.

Welch, D.F., Sword, C.P., Brehm, S. and Dusanic, D. (1979). Relationship between superoxide dismutase and pathogenic mechanisms of *Listeria monocytogenes*. *Infect. Immun.* **23**, 863–72.

Winkler, M., Erbs, G., Müller, F.E. and König, W. (1987). Epidemiologische Studien zur mikrobiellen Kolonisation von schwerverbrannten Patienten. *Zbl. Bakt. Hyg. B* **184**, 304–20.

Wodnar-Filipowicz, A., Heusser, C.H. and Maroni, C. (1989). Production of the haemopoietic growth factors GM-CSF and interleukin-3 by mast cells in response to IgE receptor mediated activation. *Nature* **139**, 150–2.

22

Protein Engineering of Microbial Toxins: Design and Properties of Novel Fusion Proteins with Therapeutic and Immunogenic Potential

John R. Murphy and Diane P. Williams

Evans Department of Clinical Research and the Department of Medicine, The University Hospital, Boston University Medical Center, Boston, MA 02118, USA

Introduction

Protein engineering is a relatively new and quickly developing area within the field of molecular biology that brings together recombinant DNA methodology with protein chemistry to provide new approaches to the study of protein structure and function. Traditional approaches to the study of protein structure involve physical, biochemical and genetic analysis. The application of protein engineering methodologies (e.g. domain substitution, or site-directed, cassette and in-frame deletion mutagenesis, etc.) at the level of the gene to produce selectively altered variant proteins brings a new and powerful edge to the study of protein structure and function. In addition, the techniques used in protein engineering allow for the substitution of native functional domains with heterologous sequences, thus giving rise to new hybrid or chimaeric proteins. Perhaps the ultimate goal of protein engineering is the elucidation of the rules which govern the formation of both secondary and tertiary conformation of proteins.

The ability to design 'new' proteins for novel applications is a natural outcome of protein engineering methodologies. While the field is still young, it is clear that this new-found ability has enormous potential for practical application in science and medicine. For example, Gutte and colleagues (Moser et al., 1983) have designed and synthesized a 24 amino acid polypeptide which is capable of binding 1,1-bis(p-chlorophenyl)-2,2,2,-trichloro-ethane (DDT). Since DDT is lipophilic and relatively insoluble in aqueous buffers, it is not readily degraded by the cytochrome P450 detoxification system (Lemberg and Barrett, 1973), and is therefore a long-term environmental hazard. The synthetic DDT-binding protein has been shown to bind DDT and facilitate its degradation. Recently, a synthetic gene encoding the DDT-binding protein has been assembled and the peptide has been successfully expressed in recombinant *Escherichia coli* (Moser et al., 1987). Thus, protein engineering methodologies have been used to create a new protein with a highly selective function. This is one of many examples where rational design of a protein and subsequent assembly of a synthetic gene has accelerated a process that would have occurred slowly, if at all, through evolution.

In addition to the design and synthesis of entirely new polypeptides and proteins, protein engineering methodologies have been applied to modify existing proteins with respect to their solubility, thermal and proteolytic stability, substrate specificity and affinity, and enzyme kinetics (reviewed in Offord, 1987; Leatherbarrow and Ferhst, 1986). These studies have confirmed and extended the concept that proteins, in general, are composed of distinct structural domains (Wetlaufer, 1973). Indeed, as

SOURCEBOOK OF BACTERIAL PROTEIN TOXINS
ISBN 0-12-053078-3

X-ray diffraction analysis of crystalline proteins accumulates, it is becoming apparent that structural domains often correlate with functional activities. Moreover, structural motifs may be more conserved during evolution than are specific amino acid sequences (reviewed by Creighton, 1983; Richardson, 1981).

In the field of microbial toxinology, protein engineering methods have been applied in two general areas: (i) the design and genetic construction of chimaeric toxins in which the native receptor-binding domain of either diphtheria or *Pseudomonas* exotoxin A have been replaced with polypeptide hormones or cell-specific growth factors, and (ii) the design and genetic construction of a variety of new immunogens by combining immunodominant epitopes from different proteins into a single polypeptide antigen. The clinical potential of these approaches toward the development of new cell receptor-specific cytotoxins and combined vaccines is enormous. While human clinical trials have only recently begun, the preliminary clinical findings are most promising.

Eukaryotic cell receptor-specific fusion toxins

Within the last several years there has been a confluence of research in the areas of eukaryotic cell growth factor biology and microbial toxinology. The joining of these fields has resulted in the development of several new highly potent and selective cytotoxins. Many of these new fusion toxins may play important roles as new biologicals for the treatment of specific malignancies or other disorders in which cells with particular surface receptors play a major role in pathogenesis (e.g. autoimmune disease, acute rejection of transplanted organs, etc.). It is now recognized that microbial toxins and cell-specific growth factors share many common attributes in their interaction with eukaryotic cells. For example, subsequent to binding to their respective surface receptors, these agents share common routes of entry into the cell: clustering of charged receptors in coated pits, internalization by receptor-mediated endocytosis and passage through an acidic compartment. Indeed, acidification of the endocytic vesicle is necessary for many biologic functions (e.g. release

and uptake of iron from transferrin) and is also required for the delivery of fragment A of diphtheria toxin across the endocytic vesicle membrane and into the cytosol. It is important to note, however, that in contrast to the wide tissue distribution of receptors for many of the microbial toxins, the distribution of growth factor receptors tends to be restricted to specific cell types and/or stages of the cell cycle. It is this tissue-restricted distribution of many growth factor receptors that has allowed for the development of receptor-targeted cytotoxins.

Biochemical and genetic analysis of both diphtheria toxin and *Pseudomonas* exotoxin A has revealed that these toxins contain at least three functional domains: (i) receptor binding (Middlebrook et al., 1978), (ii) membrane translocation, and (iii) enzymatically active (Uchida et al., 1971; Boquet et al., 1976; Hwang et al., 1987). Moreover, in the case of exotoxin A, crystallographic analysis has clearly shown three distinct structural regions (Allured et al., 1986) which correspond to functional domains (Hwang et al., 1987). Since fragments of both diphtheria toxin and exotoxin A that are devoid of their respective receptor-binding domain are non-toxic for intact cells, several investigators have attempted to target the enzymatic activity of the residual domains to selected cells. In general, two approaches have been used in the development of targeted cytotoxins: (i) chemical cross-linking of non-toxic fragments of toxins to ligands that bind specific cell surface receptors, or to monoclonal antibodies directed toward cell surface antigens, and (ii) genetic substitution of the diphtheria toxin or *Pseudomonas* exotoxin A native toxin receptor-binding domains with genes encoding either polypeptide hormones or growth factors to form toxin–growth factor fusion proteins, or fusion toxins.

Within the last three years a variety of diphtheria toxin and *Pseudomonas* exotoxin A-based fusion proteins have been described. In all instances, these hybrid toxins have been shown to selectively intoxicate target cells *in vitro*, and many of these fusion toxins have also been shown to have remarkably selective and highly potent activity *in vivo*. Indeed, the genetic substitution of the receptor-binding domain of either diphtheria toxin or exotoxin A with ligands that bind to receptors that are limited in distribution has allowed for the development of a new class of biological response modifier whose action is based on the selective elimination of target cells.

Diphtheria toxin structure and mode of action

The diphtheria toxin is a single polypeptide chain of 535 amino acids in length and contains four cysteine residues which form two disulphide bridges. The first disulphide bridge is formed between Cys^{186} and Cys^{201} and subtends an arginine-rich 14 amino acid loop; whereas, the second disulphide bridge is formed between Cys^{461} and Cys^{471}. Upon mild digestion with trypsin, intact diphtheria toxin is cleaved into an N-terminal 21.1-kDa fragment A and a 37.1-kDa fragment B. Fragment A of toxin is enzymatically active and catalyses the nicotinamide adenine dinucleotide (NAD^+)-dependent adenosine diphosphoribosylatin (ADPR) of eukaryotic elongation factor 2. Fragment B of diphtheria toxin carries the hydrophobic membrane-associating regions that facilitate the translocation of fragment A across the eukaryotic cell membrane and into the cytosol of sensitive cells. Fragment B also contains the native receptor-binding domain (Pappenheimer, 1977).

The diphtherial intoxication of a eukaryotic cell has been shown to require at least the following steps: (i) the binding of toxin to its cell surface receptor (Middlebrook et al., 1978), (ii) internalization by receptor-mediated endocytosis (Moya et al., 1985), (iii) upon acidification of the endosome, a partial denaturation of the hydrophobic membrane-associating domains (Dumont and Richards, 1988), which lead to (iv) membrane insertion (Kagan et al., 1981; Donovan et al., 1981), and (v) facilitated delivery of fragment A to the cytosol (Sandvig and Olsnes, 1982). Once delivered to the cytosol, fragment A catalyses a mono-specific ADP-ribosylation of elongation factor 2 which results in the inhibition of cellular protein synthesis and leads to the death of the cell. Under physiologic conditions, the ADP-ribosylation of elongation factor 2 proceeds to completion, and the delivery of a single molecule of fragment A to the cytosol of a cell has been shown to be lethal for that cell (Yamaizumi et al., 1978).

The structural gene for diphtheria toxin, tox,

has been shown to be carried by a family of closely related corynebacteriophages (Buck et al., 1985). The molecular cloning of fragments of the tox gene has allowed for the nucleotide sequence and deduced amino acid sequence to be determined (Kaczorek et al., 1983; Greenfield et al., 1983; Ratti et al., 1983). The structural gene for diphtheria toxin is carried on a HindIII to EcoRI restriction fragment of the bacteriophage genome. Leong et al. (1983a,b) have shown that the cloning of gene fragments that encode the tox promoter, signal sequence and fragment A of diphtheria toxin result in the constitutive expression and secretion of tox-related polypeptides into the periplasmic compartment of recombinant Escherichia coli. Recently, Bishai et al. (1987) have optimized the expression and final yield of tox-related products in E. coli. In these constructs the tox promoter and signal sequence were deleted, and expression and accumulation of toxin-related proteins in the cytoplasmic compartment was driven by the trc promoter. Following induction of expression with isopropyl-β-D-thiogalactopyranoside (IPTG), the final yield of tox-related proteins accounted for up to 8% of the total cellular protein.

Genetic construction of diphtheria toxin-related polypeptide hormone fusion proteins

The strategy that we have employed for the genetic construction of diphtheria toxin-related polypeptide hormone fusion genes involves the replacement of that portion of the tox structural gene which encodes the native toxin receptor-binding domain with DNA sequences encoding a particular peptide hormone or growth factor. For example, Murphy et al. (1986) and Williams et al. (1987) have recently described the assembly of fusion toxins in which synthetic genes encoding α-melanocyte stimulating hormone (α-MSH) and human interleukin 2 (IL2) were fused to a truncated form of the diphtheria toxin structural gene. In each instance, the resulting fusion protein encoded by the hybrid gene was found to be selectively cytotoxic toward eukaryotic cells that carried the appropriate receptors.

α-MSH–toxin (DAB₄₈₆–α-MSH)

α-Melanocyte stimulating hormone was chosen as the ligand for the genetic construction of the first recombinant fusion toxin. α-MSH is a 13 amino acid polypeptide that has been shown by Eberle and Schwyzer (1976) to interact with specific receptors on melanocytes through its C-terminal Glu–His–Phe–Arg–Trp and Gly–Lys–Pro–Val sequences. Thus, the structure/function orientation of α-MSH was analogous to native diphtheria toxin and allowed for the genetic fusion of the C-terminal end of a toxin-related protein with the N-terminal end of the peptide hormone. In addition, α-MSH, like native diphtheria toxin, has been shown to be internalized by receptor-mediated endocytosis (Varga et al., 1976). Finally, since α-MSH was only 13 amino acids in length, the oligodeoxyribonucleotides that were required to encode its primary sequence were readily synthesized in vitro.

The scheme adopted for the genetic construction of the chimaeric toxin genes takes advantage of unique restriction endonuclease sites within the structural gene for diphtheria toxin and involves the synthesis of modified cDNAs encoding for the peptide hormone. In the case of α-MSH–toxin, we have cloned a synthetic gene encoding α-MSH into a vector in which diphtheria toxin fragment B encoding sequences were truncated at a unique SphI site. The 5'-end of the α-MSH gene included an SphI site that was designed to maintain correct translational reading frame when fused to the fragment B SphI site. The fragment B/α-MSH fusion gene was then recloned into a vector encoding fragment A of diphtheria toxin to form the intact toxin-related peptide hormone fusion toxin (Murphy et al., 1986).

Recombinant E. coli that carried the diphtheria toxin-related α-MSH fusion gene were found to express low yields of a unique 56 000-kDa polypeptide that was selectively cytotoxic for MSH receptor-bearing human NEL-M1 malignant melanoma cells in vitro. Cell lines that do not bear the MSH receptor were universally resistant to the fusion toxin. Moreover, the cytotoxic action of α-MSH–toxin, like native diphtheria toxin, was found to require passage through an acidic compartment (Murphy et al., 1986). We (Shaw and Murphy, unpublished) have recently found that the action of α-MSH–toxin on both murine and human melanoma cell lines is specifically blocked by the presence of free α-MSH, thereby demonstrating the α-MSH receptor-mediated cytotoxic action of this fusion protein.

IL2–toxin (DAB₄₈₆–IL2)

Williams et al. (1987) have recently described the genetic construction and properties of IL2–toxin (hereafter designated DAB₄₈₆–IL2). The strategy that was employed in the construction of this fusion protein was analogous to that of α-MSH–toxin. In this instance, the diphtheria toxin-related portion of the chimaeric gene was carried on plasmid pABC508 (Bishai et al., 1987). This plasmid encodes the diphtheria tox promoter, signal sequence and the tox structural gene to the unique SphI site. A synthetic gene encoding the mature form of human interleukin 2 (IL2) was cloned in the pUC18 vector (Williams et al., 1988). By design the IL2 gene was modified at its 5'-end by the introduction of an SphI site that was positioned to maintain translational reading frame through the fusion junction such that Ala⁴⁸⁶ of diphtheria toxin was joined to Pro² of human IL2. In addition, the synthetic IL2-encoding gene employed E. coli codon usage bias and contains a number of unique restriction endonuclease sites (Williams et al., 1988). The genetic construction of DAB₄₈₆–IL2 involved the recloning of a 428-bp SphI – HindIII fragment from pDW15 into the SphI/HindIII sites of pABC508. In this instance, expression of DAB₄₈₆–IL2 is directed from the diphtheria tox promoter and the recombinant protein is secreted into the periplasmic compartment. Immunoblot analysis of recombinant E. coli using both anti-diphtheria toxin serum and monoclonal antibody to IL2 revealed a unique 68 000-kDa protein in the periplasm of recombinant E. coli (Williams et al., 1987). This value is in excellent agreement with the 68 086-kDa molecular weight that is deduced from the nucleotide sequence of the chimaeric toxin gene.

Interleukin 2 was chosen for the receptor-binding component for the second chimaeric toxin for a variety of reasons. IL2, like native diphtheria toxin and α-MSH, is internalized by receptor-mediated endocytosis (Weissman et al., 1986; Fujii et al., 1986; Duprez and Dautry-Versat, 1986). In addition, in normal tissue the distribution of the IL2 receptor appears to be limited to activated proliferating T lymphocytes, recently activated B lymphocytes, and macrophages; in these circumstances, an IL2 receptor-targeted cytotoxin might have therapeutic application where disease is mediated by IL2 receptor-bearing lymphocytes, such as autoimmune diseases and acute allograft rejection. Moreover, it is now clear that there are several forms of T and B cell leukaemia and lymphoma which express at least the p55 subunit (Tac antigen) of the IL2 receptor (Korsmeyer et al., 1983; Waldmann et al., 1984; Uchiyama et al., 1985; Sheibani et al., 1987; Ralfkiaer et al., 1987; Barnett et al., 1988). In the case of adult T cell leukaemia there is good evidence that the HTLV-I tax gene product activates a cytosolic factor which induces the expression of elevated numbers of IL2 receptors on the surface of leukaemic cells (Hattori et al., 1981; Ruben et al., 1988).

It is now known that the IL2 receptor exists in at least three forms: high-, intermediate- and low-affinity classes (review by Smith, 1988, 1989). The high-affinity IL2 receptor has been shown to be composed of at least two subunits: the Tac antigen which is a 55 000-Da glycoprotein (Uchiyama et al., 1981), and the Tic antigen which is a 75 000-Da glycoprotein (Sharon et al., 1986; Teshigawara et al., 1987; Smith personal communication). Interleukin 2 bound to either the high- or intermediate-affinity receptor is known to be rapidly internalized into the cell (Duprez and Dautry-Versat, 1986; Robb and Greene, 1987).

Williams et al. (1987) demonstrated that cell lines which expressed the high-affinity form of the IL2 receptor were sensitive to the action of DAB_{486}–IL2, whereas cell lines that were devoid of the IL2 receptor were resistant to the chimaeric toxin. These observations were confirmed and extended by Bacha et al. (1988) who demonstrated that the cytotoxic action of DAB_{486}–IL2 was mediated through the IL2 receptor and could be blocked by both free IL2 and monoclonal antibody to the 55-kDa subunit of the IL2 receptor. Moreover, Bacha et al. (1988) demonstrated that, like native diphtheria toxin, DAB_{486}–IL2 required passage through an acidic compartment in order to facilitate the delivery of the fragment A-associated ADP-ribosyltransferase to the cytosol of target cells, and that the inhibition of protein synthesis in intoxicated cells resulted from the ADP-ribosylation of elongation factor 2.

Recently, Waters et al. (1990) have demonstrated that T lymphocytes that bear the high-affinity form of the IL2 receptor are at least 1000-fold more sensitive to the action of DAB_{486}–IL2 than cell lines which bear either the intermediate (75-kDa) or low (55-kDa) subunit of the receptor. Typically, DAB_{486}–IL2 dose-response analysis of cell lines which bear the high-affinity form of the IL2 receptor show an IC_{50} of $0.5–1.0 \times 10^{-10}$ M (3–6ng/ml). In striking contrast, cell lines which bear either the low- or intermediate-affinity form of the IL2 receptor require exposure to DAB_{486}–IL2 concentrations of approximately 1×10^{-7} M (6.8 µg/ml) in order to achieve an IC_{50}. It is of particular interest to note that Waters et al. (1990) have demonstrated that human peripheral blood monocytes with natural killer (NK) activity, which are known to bear the 75-kDa intermediate-affinity form of the IL2 receptor, are as resistant to DAB_{486}–IL2 ($IC_{50} = 1 \times 10^{-7}$ M) as continuous cell lines which only express the p75 subunit. These results are especially significant in view of the report by Biron et al. (1989) who described a patient lacking NK cell function. This patient presented with an inability to control severe and life-threatening viral infections in spite of normal T and B cell function. To the extent that NK cells play a role in both viral and tumour surveillance, it is significant that they are at least 1 000-fold more resistant to the cytocidal action of DAB_{486}–IL2 than high-affinity IL2 receptor-bearing cells.

Since it is known that the p75 subunit of the IL2 receptor is rapidly internalized into the cell (Robb and Greene, 1987), the failure of DAB_{486}–IL2 to efficiently intoxicate cells bearing this subunit is most likely due to the conformation of the fusion toxin. Collins et al. (1988) have reported that Asp^{20} of native IL2 is essential for binding to the p75 intermediate-affinity receptor. Since the relative position of Asp^{20} in IL2 is Asp^{505} in DAB_{486}–IL2, it is likely that the fusion junction between the truncated form of diphtheria toxin and the growth factor places this residue in an internal and less accessible position for efficient receptor binding. Indeed, Waters et al. (1990) have shown by $[^{125}I]IL2$ competitive displacement experiments that 140-fold higher concentrations of DAB_{486}–IL2 are required to displace 50% of the radiolabelled ligand from the intermediate-affinity form of the receptor than are required for native recombinant IL2. This hypothesis is further supported by the observation that PE40–IL2, in which IL2 sequences are positioned at the N-terminal end of the fusion protein and may have a greater degree of freedom to interact with the IL2 receptor, is highly toxic for p75-bearing cells. In the case of PE40–IL2 there is only an 8- to 20-fold difference in the IC_{50} that is observed between high- and intermediate-affinity IL2 receptor-bearing cells (Lorberboum-Galski *et al.*, 1988b).

It is widely recognized that adult T cell leukaemia (ATL) is aetiologically associated with the retrovirus HTLV-I (Poiesz *et al.*, 1980; Hinuma *et al.*, 1981; Yoshida *et al.*, 1982). Clinically, ATL has been classified into four different types: acute, lymphoma, chronic and smoldering. Patients who present with acute-type ATL disease have increased numbers of leukaemic cells in peripheral blood, skin lesions, lymphadenopathy and visceral involvement. In addition, serum LDH and hypercalcaemia are often present. In the case of lymphoma-type ATL, peripheral blood leukaemic cells are absent; however, other clinical features are similar to that of acute-type disease. Patients with either chronic- or smoldering-type ATL present with mild clinical symptoms which include leukaemic cells in peripheral circulation, skin lesions, and sometimes a slight lymphadenopathy. The clinical course of patients with acute- and lymphoma-type ATL is progressive, and the 50% mortality rate generally falls within 5 months of diagnosis. At present, there is no effective chemotherapeutic regimen for the treatment of ATL.

Since the high-affinity form of the IL2 receptor is expressed on the surface of HTLV-I-infected transformed continuous cell lines and these cells are markedly sensitive to the cytotoxic action of DAB_{486}–IL2, Kiyokawa *et al.* (1989) have examined the sensitivity of T cells freshly withdrawn from ATL patients to the action of this fusion toxin. Both leukaemic T cells purified from lymph node aspirates and peripheral blood were extremely sensitive to DAB_{486}–IL2. In the case of leukaemic cells purified from aspirates of enlarged lymph nodes, dose-response analysis has demonstrated that the IC_{50} for DAB_{486}–IL2 was approximately 1×10^{-10} M. These results compare favourably with those obtained for high-affinity IL2 receptor-positive continuous cell lines (Waters *et al.*, 1990).

Leukaemic T cells purified from peripheral blood were also found to be sensitive to the cytotoxic action of DAB_{486}–IL2. In this circumstance, the IC_{50} for DAB_{486}–IL2 was reduced by approximately 10-fold. Since monoclonal antibody to the 55-kDa IL2 receptor was able to block the action of DAB_{486}–IL2 on both lymph node and peripheral blood leukaemic cells, Kiyokawa *et al.* (1989) concluded that the cytocidal action of the fusion toxin was mediated through the IL2 receptor.

While the reason(s) for the differences in DAB_{486}–IL2 sensitivity between lymph node and peripheral blood ATL cells is not known, it is likely that a combination of factors are important. For example, it is known that the sensitivity of a given cell line toward native diphtheria toxin is related to both its basal rate of protein synthesis, as well as the relative number of toxin receptors on the cell surface (Pappenheimer, 1977). Since only elongation factor 2 that is released from the ribosome is sensitive to ADP-ribosylation, the higher the basal rate of protein synthesis, the higher the apparent sensitivity of that cell line to the action of the toxin. In the case of ATL cells freshly withdrawn from patients, leukaemic T cells from lymph node aspirates were found to have a higher basal rate of protein synthesis than ATL cells purified from peripheral blood (Kiyokawa *et al.*, 1989). Uchiyama *et al.* (1985) have found that only 250 high-affinity IL2 binding sites per cell are present on the surface of peripheral blood ATL cells. At present there are no reports which describe the numbers of high-affinity IL2-binding sites on the surface of ATL cells purified from lymph nodes, or from ATL cells from patients with chronic- or smoldering-type disease. Nonetheless, since DAB_{486}–IL2 is highly potent and selectively active against high-affinity IL2 receptor-bearing cells, its administration may be a rational approach to the treatment of acute-, chronic- and lymphoma-type ATL.

Acute toxicology of DAB_{486}–IL2

While DAB_{486}–IL2 has been shown to be a potent cytocidal agent that is directed toward T lymphocytes that bear the high-affinity form of the IL2 receptor, cells which are devoid of the receptor are resistant to its action. Since the distribution of the high-affinity form of the IL2 receptor is limited, DAB_{486}–IL2 was not anticipated to have widespread systemic toxicity. Bacha and co-workers (unpublished) have compared the acute toxicity of diphtheria toxin with that of CRM45 and DAB$_{486}$–IL2. The lethal dose$_{50}$ (LD$_{50}$) of diphtheria toxin in sensitive species is between 100 and 150 ng/kg. In marked contrast, the LD$_{50}$ of DAB$_{486}$–IL2 in these species is greater than 3 mg/kg! Thus, DAB$_{486}$–IL2 is not a systemic toxin in the classic sense, even though it is a remarkably potent cytocidal agent for high-affinity IL2 receptor-bearing cells. It is of interest to note that the LD$_{50}$ for CRM45, a mutant form of diphtheria toxin which lacks the native receptor-binding domain, is 350 µg/kg which is intermediate between diphtheria toxin and DAB$_{486}$–IL2. In the case of CRM45 administration, the primary toxic lesion that is observed is that of acute tubular necrosis in the kidney. In a similar fashion, the primary toxic lesion observed in animals given high-dose bolus administration of DAB$_{486}$–IL2 is acute tubular necrosis. These results suggest that any non-specific toxicity of DAB$_{486}$–IL2 may be due to the partial degradation of the full-length fusion toxin and the release of a CRM45-like product.

In vivo action of DAB$_{486}$–IL2

Since the appearance of the high-affinity form of the IL2 receptor represents a pivotal point in the maturation of the immune response, DAB$_{486}$–IL2 has been used as an experimental therapeutic to suppress a variety of T cell-mediated reactions. For example, Kelley *et al.* (1988) have shown that daily administration of DAB$_{486}$–IL2 in low doses is able to abolish cell-mediated immunity in a murine model of delayed-type hypersensitivity. Interestingly, preimmunization of animals with diphtheria toxoid had little, if any, neutralizing effect on the action of DAB$_{486}$–IL2. These results are consistent with observations made by Zucker and Murphy (1984) in a study of monoclonal antibodies against diphtheria toxoid. This study clearly demonstrated that only those monoclonal antibodies that blocked the binding of native diphtheria toxin to its cell surface receptor were neutralizing antibodies both *in vitro* and *in vivo*. All other monoclonal antibodies to diphtheria toxoid failed to neutralize the toxin even though they bound to the protein. In the case of DAB$_{486}$–IL2, the diphtheria toxin receptor-binding

domain has been deleted and replaced with IL2 sequences, and as a result, one would anticipate that neutralizing antibodies to diphtheria toxin would not block the binding or activity of the fusion toxin.

Kirkman *et al.* (1989) have shown that a single 10-day course of DAB$_{486}$–IL2 therapy (1 µg/day) indefinitely prolongs murine heterotopic cardiac transplants in approximately 80% of the recipients. In these experiments, cardiac allografts from B10.BR donor mice were engrafted in C57B1/10 recipients. Allografts in the untreated controls, as well as mice treated with either CRM45 (a non-toxic 45-kDa fragment of diphtheria toxin) or DA197B$_{486}$–IL2 (a non-toxic ADP-ribosyltransferase minus mutant form of DAB$_{486}$–IL2) were uniformly rejected between days 10 and 22 post-transplantation. In a similar fashion, Pankewycz *et al.* (1989) have examined the effect of DAB$_{486}$–IL2 treatment on the survival of murine islet cell allografts in diabetic mice. In this instance, B6AF$_1$ mice were rendered diabetic by streptozotocin treatment and, when blood glucose levels reached 350 mg/dl, islet cells from DBA/2 donors were transplanted under the capsule of the kidney. Prolonged survival times were observed following DAB$_{486}$–IL2 administration (2 µg/day, 20 days). Histologic examination revealed that islets were totally replaced by infiltrating mononuclear cells in the untreated group, whereas in the DAB$_{486}$–IL2-treated group prominent islets were observed without sign of necrosis.

Pseudomonas exotoxin structure and mode of action

Pseudomonas exotoxin A is a 613 amino acid polypeptide (M_r 66 000) that is secreted by toxigenic strains of *Pseudomonas aeruginosa*. Crystallographic analysis to the 3.0 Å level of resolution has shown that the toxin can be divided into three distinct structural domains (Allured *et al.*, 1986). By expressing different regions of the exotoxin A structural gene in recombinant *E. coli*, Hwang *et al.* (1987) have shown that Domain I contains the eukaryotic cell receptor-binding region, and that Domain II contains sequences that facilitate the delivery of Domain III across the membrane and into the cytosol. Domain III of *Pseudomonas* exotoxin A is an ADP-ribosyltransferase and has been shown to catalyse an NAD-dependent ADP-ribosylation of eukaryotic elongation factor 2. This reaction is identical to that catalysed by diphtheria toxin fragment A (Iglewski and Kabat, 1975).

It is of particular interest to note that the structure/function relationships of *Pseudomonas* exotoxin A are the mirror image of native diphtheria toxin. While the native receptor-binding domain of diphtheria toxin is positioned in the C-terminal portion of the toxin, the receptor-binding domain of exotoxin A is positioned in the N-terminal region. In a similar fashion, the ADP-ribosyl-transferase of native diphtheria toxin is positioned in the N-terminal region, whereas the transferase activity of exotoxin A is located in the C-terminal region. The opposing configuration of these two extremely potent toxins provides the opportunity to design analogous fusion toxins that may have subtle but important differences. The selection of one or the other of these native toxins for receptor-binding domain substitution should be based upon the relative position of the binding domain of the new ligand. For example, if the binding domain is positioned in the N-terminal region of the ligand, *Pseudomonas* exotoxin A should be the toxin of choice. In a similar fashion, if the binding domain of the ligand is positioned on the C-terminal end, diphtheria toxin should be the toxin of choice. In this way, the relative structure/function relationships of the fusion toxin will reflect those of the component proteins.

Genetic construction and properties of *Pseudomonas* exotoxin A-based fusion toxins

PE40–TGFα

Chaudhary *et al.* (1987) have reported the fusion of transforming growth factor type α (TGFα) with a truncated form of *Pseudomonas* exotoxin A, PE40. The structural gene for the PE40 form of exotoxin A encodes a 40 000-Da polypeptide that is devoid of the native receptor-binding Domain I. TGFα has been shown to be closely related to epidermal growth factor (EGF) and to bind and stimulate growth of EGF receptor-bearing cells. In this genetic construct, TGFα was fused to the C-terminal end of PE40 through a hexapeptide linker. The fusion toxin was found to be highly toxic for EGF receptor-bearing cells *in vitro*. Recently, Edwards *et al.* (1989) have shown that PE40–TGFα binds approximately 100-fold less well to the EGF receptor than native TGFα. It is interesting to note that N-terminal deletions of 59 or 130 amino acids from the PE40 component of the fusion toxin, resulted in molecules that had higher receptor-binding affinities, but decreased potency.

IL2–PE40

The cDNA for human IL2 has been fused to the 5′-end of the structural gene for PE40 by Lorberboum-Galski *et al.* (1988a). The resulting fusion toxin, IL2–PE40, was shown to be highly potent (ID_{50} = 20 ng/ml) against high-affinity IL2 receptor-bearing HUT 102 cells *in vitro*. Moreover, Ogata *et al.* (1988) have shown that this fusion toxin is active against mitogen-stimulated murine spleen cells. These investigators also demonstrated that IL2–PE40 is immunosuppressive in allogeneic cytotoxic T cell response to antigen. Spleen cells from C57BL/6 (H-2^b) mice were stimulated with mitomycin C-treated BALB/c (H^d) spleen cells *in vitro*. Following incubation, the cytotoxic T cell response was measured using ^{51}Cr-labelled BALB/c cells as the target.

The ID_{50} of IL2–PE40 toxin action in this experimental system was found to be 1.8×10^{-10} M. Surprisingly, Lorberboum-Galski *et al.* (1988) have reported that IL2–PE40 is cytotoxic to cells which bear either the p55 or the p75 subunit of the IL2 receptor. As described above, Weissman *et al.* (1986) have shown that [^{125}I]-anti-Tac, which is directed to the p55 subunit of the IL2 receptor is not internalized by the cell. Paradoxically, IL2–PE40 is cytotoxic toward the p55-only bearing MT-1 cell line. The differential cytotoxicity for p55- or p75-only bearing cells versus high-affinity (p55 + p75) IL2 receptor-bearing cells is only 8- to 20-fold. In contrast, the differential cytotoxicity for the diphtheria toxin-based IL2 fusion toxin, DAB$_{486}$–IL2, on p55- or p75-only versus high-affinity IL2 receptor-bearing cells is greater than 1 000-fold.

In vivo IL2–PE40 has been shown to be a potent immunosuppressive agent. Case *et al.* (1989) has shown that this fusion toxin delays the onset and limits the severity of adjuvant-induced arthritis in the rat. In this model, rats are injected with a mycobacterial adjuvant which leads to T cell activation and proliferation and results in a mononuclear cell infiltration into the joints. Animals were treated two times each day by intraperitoneal injection. As compared to the controls, the arthritis index for the treated group was approximately 1/3 as great. In addition, IL2–PE40, like DAB_{468}–IL2 (described above), has been shown to prolong the survival of murine cardiac allografts (Lorberboum-Galski et al., 1989).

IL4–PE40

Interleukin 4 (IL4) is a 20 000-Da protein that stimulates the growth of B cells, stimulates the expression of class II major histocompatibility complex molecules, and enhances the production of IgG_1 and IgE antibodies. Ogata et al. (1989) have described the fusion of murine IL4 with PE40. The fusion toxin was shown to have cytotoxic activity against IL4 receptor-bearing murine cell lines in vitro. Although this fusion was specific for the murine IL4 receptor, the determination of cytotoxicity required a 48-h incubation period followed by a 4-h pulse with [³H]leucine. Since it is known that murine IL4 does not bind to the human IL4 receptor, the potential of a human IL4 fusion toxin as an experimental therapeutic agent remains unknown.

IL6–PE40

Siegall et al. (1988) have reported the fusion of B cell stimulatory factor 2, IL6, with PE40. The fusion toxin, IL6–PE40, was found to be selectively cytotoxic for the human myeloma U266 cell line ($ID_{50} = 8$ ng/ml). In contrast, IL6–[Asp^{553}]PE40, an ADP-ribosyltransferase minus mutant form of the fusion toxin, was found to have no cytotoxic activity. Since the action of IL6–PE40 on U266 cells could be blocked with either recombinant IL6, IL6–[Asp^{553}]PE40, or antibody to IL6, cytotoxicity appears to be mediated through the IL6 receptor.

The IL6 receptor has been reported to be present in increased numbers on human myeloma cells, cervical and bladder cell carcinomas, and cardiac myxomas (Hirano et al., 1987). IL6–PE40, or other IL6 receptor-targeted fusion toxins may have broad clinical potential; however, since Siegall et al. (1988) only reported results with the U266 cell line, the degree to which IL6–PE40 will be useful in eliminating IL6 receptor-bearing cells remains unclear.

CD4–PE40

Since it was well-known that gp120 of the human immunodeficiency virus (HIV) binds to the CD4 antigen on human T cells and that HIV-infected cells express gp120 on the cell surface, a cytotoxic agent directed towards cell surface bound gp120 would be anticipated to selectively eliminate HIV-infected cells. Chaudhary et al. (1988) have reported the genetic construction of a CD4–PE40 fusion toxin in which the gene encoding the N-terminal 178 amino acids of human CD4 was fused to the 5′-end of the PE40 structural gene. The resulting fusion toxin, CD4–PE40, was found to have an ID_{50} of 1.6×10^{-9} M against the 8E5 HIV-infected cell lline. This fusion toxin represents a novel therapeutic agent for the treatment of HIV infection.

Anti-Tac–PE40

Chaudhary et al. (1989) have described the genetic construction and properties of a unique fusion toxin in which the variable regions (Fv) of anti-Tac, a monoclonal antibody to the p55 subunit of the human IL2 receptor (see above), are linked to PE40. This fusion involved the genetic linkage of the light- and heavy-chain variable regions through a 4Gly–Ser–4Gly–Ser–4Gly–Ser encoding synthetic oligonucleotide, and then the modified Fv fragment was joined to the 5′-end of the structural gene for PE40. In this instance, the immunofusion toxin was highly potent against HUT 102 cells with an ID_{50} of 2.3×10^{-12} M. The biologic activity of this fusion protein was blocked by the addition of anti-Tac, but not by the monoclonal antibody OVB3, suggesting that toxicity is mediated through the p55 subunit of the IL2 receptor.

Since ^{125}I-labelled anti-Tac has been shown to remain on the cell surface and not be internalized by receptor-mediated endocytosis (Weissman et al., 1986), the high potency of this immunofusion toxin must be ascribed to the 4% of the p55 subunit that is involved in the formation of the high-affinity receptor. This report establishes a precedent for the formation of other immunofusion toxins with a wide array of monoclonal antibodies.

Genetic construction and properties of epitope fusion proteins

Recently, several investigators have used protein engineering techniques to combine independent immunodominant epitopes from different antigens into novel fusion proteins. The resulting fusion proteins have the promise of delivering a wide array of diverse antigenic determinants in a single molecular species. Kaczorek and associates have pioneered the use of a non-toxic diphtheria toxin-related protein, CRM 228, as the carrier of antigenic determinants from β-galactosidase, hepatitis virus type B and poliovirus (Zettlmeissl *et al.*, 1986; Phalipon and Kaczorek, 1987; Phalipon *et al.*, 1989). More recently, Sanchez and Hirst have described the genetic fusion of the heat-stable (ST) enterotoxin from *E. coli* to a variety of carrier proteins (Sanchez *et al.*, 1986, 1988a,b; Saarilahti *et al.*, 1989). In these studies, protein fusions containing the non-immunogenic ST toxin were assembled with the expectation that the novel structures would ellicit an immune response for protection against diarrhoea caused by enterotoxigenic *E. coli*. In all cases, the resulting fusion proteins were found to stimulate an immune response against antigenic determinants intrinsic to their component parts.

The construction of immunodominant fusion proteins represents a new and very exciting area in vaccine development. The ability to design new fusion proteins that carry a variety of immunogenic determinants will allow for the generation of new vaccines that can be tailor-made for both human and veterinary use.

Hepatitis B virus–diphtheria toxin-related fusion protein

Kaczorek and co-workers have evaluated the capacity of a diphtheria toxin-related protein, CRM 228, to present heterologous viral peptide antigens in a form that is recognized by the immune system. This work has involved the in-frame insertion of a hepatitis B DNA fragment that encodes a surface protein epitope into the *tox*-228 gene. Interestingly, when administered with complete Freund's adjuvant, this fusion protein stimulated the appearance of antibodies that were reactive with 22-nm HBsAg particles, as well as antibodies that were reactive to wild-type diphtheria toxin (Phalipon and Kaczorek, 1987). Although this genetic construct was found to disrupt a major neutralizing determinant of native diphtheria toxin, this work provides the precedent for the application of protein engineering methodologies to the development of epitope fusion proteins for vaccine development.

Poliovirus type I–diphtheria toxin-related fusion protein

Phalipon *et al.* (1989) have described the construction of an epitope fusion protein in which the 11 amino acids of the capsid protein VP1 were cloned into the C-terminal end of the *tox*-228 gene. The resulting fusion protein was found to elicit both neutralizing anti-diphtheria toxin and anti-polio virus antibodies in low titre.

Heat-stable enterotoxin-related fusion proteins

It is well-known that strains of *E. coli* which produce enterotoxins are able to cause diarrhoeal disease in both humans and domestic animals. These pathogenic strains produce a heat-labile enterotoxin (LT) and/or a heat-stable (ST) toxin. The latter toxin is of low molecular weight and is non-immunogenic. Sanchez and co-workers have produced a series of novel ST-containing fusion proteins. The gene for the STa form of heat-stable toxin was fused in correct translational reading frame to the C-terminal end of the gene encoding the binding (B) subunit of the heat-labile enterotoxin (Sanchez *et al.*, 1986). When combined with the native B subunit, the epitope fusion protein was found to co-assemble into the multimeric form of the LT enterotoxin. This complex form of LT was found to react with monoclonal antibodies to the A and B subunits of LT, as well as with antibodies to STa.

In a similar study, Sanchez *et al.* (1988b) reported the fusion of ST sequences to the N-terminal end of the cholera toxin B subunit. The epitope fusion protein was found to elicit antibodies to both cholera toxin and STa. It is of interest to note that expression of the chimaeric gene in *Vibrio cholera* resulted in the secretion of the fusion protein into the culture medium. More recently, Sanchez *et al.* (1988a) have reported the construction of STa fusion proteins with the A subunit of LT enterotoxin.

Saarilahti *et al.* (1989) have continued to refine the novel STa vaccines by the fusion of the STa gene to the 3'-end of the gene for *ompC* in *E. coli*. OmpC is a major outer membrane protein, and the fusion of STa sequences to OmpC should result in a bacterial cell-associated form of the enterotoxin. These investigators have shown that following immunization of rabbits with whole bacteria both anti-OmpC and anti-STa antibodies could be detected. This approach is particularly important, since it points the way for the development of live oral genetically engineered vaccine strains which carry epitope fusion proteins.

Conclusions

Within the last three years it has been possible, through advances that have been made in recombinant DNA methodologies, to design and genetically construct a variety of fusion toxins that are targeted toward specific cell surface receptors. Both Murphy and co-workers (Murphy *et al.*, 1986; Williams *et al.*, 1987) and Pastan and co-workers (Chaudhary *et al.*, 1987, 1988; Lorberboum-Galski *et al.*, 1988a; Siegall *et al.*, 1988) have demonstrated that it is possible to replace the native receptor-binding domain of either diphtheria toxin or *Pseudomonas* exotoxin A with eukaryotic polypeptide hormones or growth factors. Remarkably, in all instances that have been reported, the resulting fusion toxin has been shown to be extraordinarily potent and selectively toxic for only those cells which bear the appropriate targeted receptor. At pharmacologic levels, the action of many of the fusion toxins has been found to be restricted to receptor-bearing cells and, as a result, they are not toxic for the whole animal. In view of this restricted activity, many of the new fusion toxins may find clinical application in a variety of disease states for which there is, as yet, no effective therapy. Indeed, toward that end, the first of the genetically designed hybrid toxins, DAB_{468}–IL2, is currently in phase I human clinical trials for the treatment of IL2 receptor-positive leukaemias/lymphomas in patients who are refractory to currently available therapeutic regimens.

The application of protein engineering and recombinant DNA methods to the redesign of microbial toxins is only in its early stages. There is much to learn about the efficient delivery of the toxophore of these hybrid molecules to the cytosol of target cells. For example, it has been shown that the removal of 97 amino acids from the diphtheria fragment B portion of DAB_{486}–IL2 results in a fusion toxin that is 5- to 10-times more potent toward high-affinity IL2 receptor-bearing T lymphocytes ($IC_{50} = 0.6 - 1.0 \times 10^{-11}$ M) *in vitro* (Williams *et al.*, 1990; Kiyokawa *et al.*, 1991). Furthermore, genetic alterations in the region of the fusion junction between the toxophore and binding domains of the fusion toxins may markedly affect receptor-binding kinetics and affinity, and the genetic removal of protease-sensitive sites within these hybrid toxins may markedly affect pharmacodynamics *in vivo*.

The fusion toxins that have been developed over the past three years represent a new and novel class of biological response modifier. One can certainly envision the continued development of additional highly potent cytocidal agents for the treatment of specific malignancies. Equally exciting, however, is the potential for the development of new fusion proteins that will deliver a wide variety of enzymatic activities to the cytosol of selected cells. These fusion proteins would be expected to modulate the physiology of their target cells rather than bring about their elimination.

In a similar fashion, the application of protein engineering methodologies to construct unique fusion proteins that combine immunodominant epitopes from a variety of antigens represents a novel approach to the development of new vaccines. Since it has been demonstrated that these epitope fusion proteins stimulate an immune response against their component parts, it is clear that this area of research will continue to flourish. Indeed, the power of protein engineering for the development of a wide variety of new biologicals is enormous. Since all of the methods for this new trade are at hand, it is reasonable to say that the complexity and subtlety of these developments will be limited only by the imagination of the investigator.

Acknowledgements

Preparation of this manuscript was supported in part by Public Health Service grants AI-21628 and

AI-22882 from the National Institute of Allergy and Infectious Diseases, and CA-41746 and U01 CA-48626 from the National Cancer Institute.

References

Allured, V.S., Collier, R.J., Carroll, S.F. and McKay, D.B. (1986). Structure of exotoxin A of *Pseudomonas aeruginosa* at 3.0 Angstrom resolution. *Proc. Natl. Acad. Sci. USA* **83**, 1320–4.

Bacha, P., Williams, D.P., Waters, C., Williams, J.M., Murphy, J.R. and Strom, T.B. (1988). Interleukin-2 receptor targeted cytotoxicity: Interleukin-2 receptor-mediated action of a diphtheria toxin-related interleukin-2 fusion protein. *J. Exp. Med.* **167**, 612–22.

Barnettt, D., Wilson, G.A., Lawrence, A.C.K. and Buckley, G.A. (1988). The interleukin-2 receptor and its expression in the acute leukemias and lymphoproliferative disorders. *Dis. Marker* **6**, 133–9.

Biron, C., Byron, K.S. and Sullivan, J.L. (1989). Severe herpesvirus infections in an adolescent without natural killer cells. *N. Engl. J. Med.* **320**, 1731–5.

Bishai, W.R., Rappuoli, R. and Murphy, J.R. (1987). High level expression of a proteolytically sensitive diphtheria toxin fragment in *Escherichia coli*. *J. Bacteriol.* **169**, 5140–51.

Boquet, P., Silverman, M.S., Pappenheimer, Jr., A.M. and Vernon, B.W. (1976). Binding of Triton X-100 to diphtheria toxin, cross-reacting material 45, and their fragments. *Proc. Natl. Acad. Sci. USA* **73**, 449–53.

Buck, G.A., Gross, R.E., Wong, T.P., Lorea, T. and Groman, N. (1985). DNA relationships among some *tox*-bearing corynebacteriophage that carry the gene for diphtheria toxin. *Infect. Immun.* **49**, 679–84.

Case, J.P., Lorberboum-Galski, H., Lafyatis, R., Fitz-Gerald, D., Wilder, R.L. and Pastan, I. (1989). Chimeric cytotoxin IL2-PE40 delays and mitigates adjuvant-induced arthritis in rats. *Proc. Natl. Acad. Sci. USA* **86**, 287–91.

Chaudhary, V.K., FitzGerald, D.J., Adhya, S. and Pastan, I. (1987). Activity of a recombinant fusion protein between transforming growth factor α and *Pseudomonas* toxin. *Proc. Natl. Acad. Sci. USA* **84**, 4538–42.

Chaudhary, V.K., Mizukami, T., Fuerst, T.R., FitzGerald, D.J., Moss, B., Pastan, I. and Berger, E.A. (1988). Selective killing of HIV-infected cells by recombinant human CD4-*Pseudomonas* exotoxin. *Nature* **339**, 369–72.

Chaudhary, V.K., Queen, C., Junghaus, R.P., Waldmann, T.A., FitzGerald, D.J. and Pastan, I. (1989). A recombinant immunotoxin consisting of two antibody variable domains fused to *Psuedomonas* exotoxin. *Nature* **339**, 394–97.

Collins, L., Tsien, W.-H., Seals, C., Hakimi, J., Weber, D., Bailon, P., Hoskings, J., Greene, W.C., Toome,

V. and Ju, G. (1988). Identification of specific residues of human interleukin-2 that affect binding to the 70-kDa subunit (p70) of the interleukin-2 receptor. *Proc. Natl. Acad. Sci. USA* **85**, 7709–13.

Creighton, T.E. (1983). In: *Proteins: Structure and Molecular Properties*, W.H. Freeman and Co., New York, New York.

Donovan, J.J., Simon, M.I., Draper, R.K. and Montal, M. (1981). Diphtheria toxin forms transmembrane channels in planar lipid bilayers. *Proc. Natl. Acad. Sci. USA* **78**, 172–6.

Dumont, M.E. and Richards, F.M. (1988). The pH-dependent conformational change of diphtheria toxin. *J. Biol. Chem.* **263**, 5450–4.

Duprez, V. and Dautry-Versat, A. (1986). Receptor-mediated endocytosis of interleukin 2 in a human tumor T cell line. *J. Biol. Chem.* **261**, 15450–4.

Eberle, A.N. and Schwyzer, R. (1976). Hormone-receptor interactions. The message sequence of α-melanotropin: demonstration of two active sites. *Clin. Endocrin.* **5**, 41S–48S.

Edwards, G.M., DeFeo-Jones, D., Tai, J.Y., Vuocolo, G.A., Patrick, D.R., Heimbrook, D.C. and Oliff, A. (1989). Epidermal growth factor receptor binding is affected by structural determinants in the toxin domain of transforming growth factor-alpha-*Pseudomonas* exotoxin fusion proteins. *Molec. Cell Biol.* **9**, 2860–7.

Fujii, M., Sugamura, K., Sano, K., Naki, M., Sugita, K. and Hinuma, Y. (1986). High affinity receptor-mediated internalization and degradation of interleukin-2 in human T-cells. *J. Exp. Med.* **163**, 550–62.

Greenfield, L., Bjorn, M.J., Horn, G., Fong, D., Buck, G.A., Collier, R.J. and Kaplan, D.A. (1983). Nucleotide sequence of the structural gene for diphtheria toxin carried by *Corynebacterium diphtheriae*. *Proc. Natl. Acad. Sci. USA* **80**, 6853–7.

Hattori, T., Uchiyama, T., Toibana, T., Takatsuki, K. and Uchino, H. (1981). Surface phenotype of Japanese adult T-cell leukemia cells characterized by monoclonal antibodies. *Blood* **58**, 645–7.

Hinuma, Y., Nagata, K., Hanaoka, M., Naki, M., Matsumoto, T., Kinoshita, K.I., Shirakawa, S. and Miyoshi, I. (1981). Adult T-cell leukemia; antigen in an ATL cell line and detection of a patient with cutaneous T cell lymphoma. *Proc. Natl. Acad. Sci. USA* **78**, 6476–80.

Hirano, T., Taga, T., Yasukawa, K., Nikajima, K., Nakano, N., Takatsuki, F., Shimizu, M., Murashima, A., Tsunasawa, S., Sakiyama, F. and Kishimoto, T. (1987). Human B-cell differentiation factor defined by an anti-peptide antibody and its possible role in autoantibody production. *Proc. Natl. Acad. Sci. USA* **84**, 228–31.

Hwang, J., FitzGerald, D.J., Adhya, S. and Pastan, I. (1987). Functional domains of *Pseudomonas* exotoxin identified by deletion analysis of the gene expressed in *E. coli*. *Cell* **48**, 129–36.

Iglewski, B.H. and Kabat, D. (1975). NAD-dependent inhibition of protein synthesis by *Pseudomonas aerugi-*

nosa toxin. *Proc. Natl. Acad. Sci. USA* **72**, 2284–88.

Kagan, B.L., Finkelstein, A. and Colombini, M. (1981). Diphtheria toxin fragment forms large pores in phospholipid bilayer membranes. *Proc. Natl. Acad. Sci. USA* **78**, 4950–4.

Kelley, V.E., Bacha, P., Pankewycz, O., Nichols, J.C., Murphy, J.R. and Strom, T.B. (1988). Interleukin 2-diptheria toxin fusion protein can abolish cell-mediated immunity *in vivo*. *Proc. Natl. Acad. Sci. USA* **85**, 3980–4.

Kirkman, R.L. Bacha, P., Barrett, L.V., Forte, S., Murphy, J.R. and Strom, T.B. (1989). Prolongation of cardiac allograft survival in murine recipients treated with a diphtheria toxin-related interleukin-2 fusion protein. *Transplantation* **47**, 327–30.

Kiyokawa, T., Shirono, K., Hattori, T., Nishimura, H., Yamaguchi, K., Nichols, J.C., Strom, T.B., Murphy, J.R. and Takatsuki, K. (1989). Cytotoxicity of interleukin-2-toxin towards lymphocytes from patients with adult T-cell leukemia. *Cancer Res.* **49**, 4042–6.

Kiyokawa, T., Williams, D.P., Snider, C.E., Strom, T.B. and Murphy, J.R. (1991). Protein engineering of diphtheria toxin-related interleukin-2 fusion toxins to increase cytotoxic potency for high-affinity IL-2 receptor-bearing target cells. *Prot. Engng.* **4**, 463–8.

Korsmeyer, S.J., Greene, W.C., Cossman, J., Hsu, S.-M., Jensen, J.P., Neckers, L.M., Marshall, S.L., Bakhshi, A., Depper, J.M., Leonard, W.L., Jaffe, E.S. and Waldmann, T.A. (1983). Rearrangement and expression of immunoglobulin genes and expression of Tac antigen in hairy cell leukemia. *Proc. Natl. Acad. Sci. USA* **80**, 4522–6.

Leatherbarrow, R.J. and Ferhst, A. (1986). Protein engineering. *Protein Engng.* **1**:7–16.

Lemberg, R., and Barrett. (1973). In *Cytochromes*, pp. 73–96, Academic Press, London.

Leong, D., Coleman, K. and Murphy, J.R. (1983a). Cloned fragment A of diphtheria toxin is expressed and secreted into the periplasmic space of *Escherichia coli* K12. *Science* **220**, 515–7.

Leong, D., Coleman, K. and Murphy, J.R. (1983b). Cloned diphtheria toxin fragment A is expressed from the *tox* promoter and exported to the periplasmic space by the SecA apparatus of *Escherichia coli*. *J. Biol. Chem.* **258**, 15016–20.

Lorberboum-Galski, H., FitzGerald, D.J., Chaudhary, V., Adhya, S. and Pastan, I. (1988a). Cytotoxic activity of an interleukin 2–*Pseudomonas* exotoxin chimeric protein produced in *Escherichia coli*. *Proc. Natl. Acad. Sci. USA* **85**, 1922–6.

Lorberboum-Galski, H., Kozak, R.W., Waldmann, T.A., Bailon, P., FitzGerald, D.J.P. and Pastan, I. (1988b). Interleukin 2 (IL-2) PE40 is cytotoxic to cells displaying either the p55 or p70 subunit of the IL-2 receptor. *J. Biol. Chem.* **263**, 18650–6.

Lorberboum-Galski, H., Barrett, L.V., Kirkman, R.L., Ogata, M., Willingham, M.C., FitzGerald, D.J. and Pastan, I. (1989). Cardiac allograft survival in mice

treated with IL-2-PE40. *Proc. Natl. Acad. Sci. USA* **86**, 1008–12.

Middlebrook, J.L., Dorland, R.B. and Leppla, S.H. (1978). Association of diphtheria toxin with Vero cells: demonstration of a receptor. *J. Biol. Chem.* **253**, 7325–30.

Moser, R., Thomas, R.M. and Gutte, B. (1983). An artificial crystalline DDT-binding polypeptide. *FEBS Lett.* **157**, 247–51.

Moser, R., Frey, S., Munger, K., Hehlgans, T., Klausner, S., Langen, H., Winnaker, E.L., Mertz, R. and Gutte, B. (1987). Expression of a synthetic gene of an artificial DDT-binding polypeptide in *Escherichia coli*. *Protein Engng.* **1**, 339–43.

Moya, M., Dautry-Versat, A., Goud, B., Louvard, D. and Boquet, P. (1985). Inhibition of coated pit formation in Hep2 cells blocks the cytotoxicity of diphtheria toxin but not that of ricin toxin. *J. Cell Biol.* **101**, 548–59.

Murphy, J.R., Bishai, W., Borowski, M., Miyanohara, A., Boyd, J. and Nagle, S. (1986). Genetic construction, expression and melanoma-selective cytotoxicity of a diphtheria toxin-related α-melanocyte stimulating hormone fusion protein. *Proc. Natl. Acad. Sci. USA* **83**, 8258–62.

Offord, R.E. (1987). Protein engineering by chemical means? *Protein Engng.* **1**:151–157.

Ogata, M., Lorberboum-Galski, H., FitzGerald, D.J. and Pastan, I. (1988). IL-2-PE40 is cytotoxic for activated T lymphocytes expressing IL-2 receptors. *J. Immunol.* **141**, 4224–8.

Ogata, M., Chaudhary, V.K., FitzGerald, D.J. and Pastan, I. (1989). Cytotoxic activity of a recombinant fusion protein between interleukin 4 and *Pseudomonas* exotoxin. *Proc. Natl. Acad. Sci. USA* **86**, 4215–9.

Pankewycz, O., Mackie, J., Hassarjian, R., Murphy, J.R., Strom, T.B. and Kelley, V.E. (1989). Interleukin-2-diptheria toxin fusion protein prolongs murine islet cell engraftment. *Transplantation* **47**, 318–22.

Pappenheimer, A.M., Jr. (1977). Diphtheria toxin. *Ann. Rev. Biochem.* **46**, 69–94.

Phalipon, A. and Kaczorek, M. (1987). Genetically engineered diphtheria toxin fusion proteins carrying the hepatitis B surface antigen. *Gene* **55**, 255–63.

Phalipon, A. and Kaczorek, M. (1989). In *Vaccines 89*, Cold Spring Harbor Laboratory, Cold Spring Harbor, New York, pp.463–6.

Phalipon, A., Crainic, R. and Kaczorek, M. (1989). Expression of a poliovirus type i neutralization epitope on a diphtheria toxin fusion protein. *Vaccine* **7**, 132–6.

Poiesz, B., Ruscetti, F.W., Gazdar, A.F., Bunn, P.A., Minna, J.D. and Gallo, R.C. (1980). Detection and isolation of type c retrovirus particles from fresh and cultured lymphocytes of a patient with cutaneous T-cell lymphoma. *Proc. Natl. Acad. Sci. USA* **77**, 7415–9.

Ralfkiaer, E., Wantzin, G.L., Stein, H., Thomsen, K. and Mason, D.Y. (1986). T-cell growth factor receptor (Tac-antigen) expression in cutaneous lymphoid infiltrates. *J. Amer. Acad. Derm.* **15**, 628–37.

Ratti, G., Rappuoli, R. and Giannini, G. (1983). The complete nucleotide sequence of the gene coding for diphtheria toxin in the corynephage omega (*tox+*) genome. *Nucleic Acid Res.* **11**, 6589–95.

Richardson, J.S. (1981). In: *Advances in Protein Chemistry*, Vol. 34, pp. 167–339. Academic Press, New York.

Robb, R.J. and Greene, W.C. (1987). Internalization of interleukin-2 is mediated by the β chain of the high affinity interleukin-2 binding molecules in T cells. *J. Immunol.* **140**, 470.

Ruben, S., Poteat, H., Tan, T.-H., Kawakami, K., Roeder, R., Haseltine, W. and Rosen, C.A. (1988). Cellular transcription factors and regulation of IL-2 receptor gene expression by HTLV-I *tax* gene product. *Science* **241**, 89–92.

Saarilahti, H.T., Tapio-Palva, E., Holmgren, J. and Sanchez, J. (1989). Fusion genes encoding *Escherichia coli* heat-stable enterotoxin and outer membrane protein OmpC. *Infect. Immun.* **57**, 3663–5.

Sanchez, J., Uhlin, B.E., Grundstrom, T., Holmgren, J. and Hirst, T.R. (1986). Immunoactive chimeric ST-LT enterotoxins of *Escherichia coli* generated by in vitro gene fusion. *FEBS Lett.* **208**, 194–8.

Sanchez, J., Hirst, T.R. and Uhlin, B.E. (1988a). Hybrid enterotoxin LTA:STa proteins and their protection from degradation by in vivo association with B-subunits of *Escherichia coli* heat-labile enterotoxin. *Gene* **64**, 265–75.

Sanchez, J., Svennerholm, A.-M. and Holmgren, J. (1988b). Genetic fusion of a non-toxic heat-stable enterotoxin-related decapeptide antigen to cholera toxin B-subunit. *FEBS Lett.* **241**, 110–4.

Sandvig, K. and Olsnes, S. (1982). Entry of toxic proteins abrin, modeccin, ricin and diphtheria toxin into cells. II. Effect of pH, metabolic inhibitors, and ionophores and evidence for toxin penetration from endocytic vesicles. *J. Biol. Chem.* **257**, 7504–13.

Siegall, C.B., Chaudhary, V.K., FitzGerald, D.J. and Pastan, I. (1988). Cytotoxic activity of an interleukin 6–*Pseudomonas* exotoxin fusion protein on human myeloma cells. *Proc. Natl. Acad. Sci. USA* **85**, 9738–42.

Sharon, M., Klausner, R.D., Cullen, B.R., Chizzonite, R. and Leonard, W.L. (1986). Novel interleukin-2 receptor subunit detected by cross-linking under high-affinity conditions. *Science* **234**, 859–62.

Sheibani, K., Winberg, C.D., Van der Weld, S., Blayney, D.W. and Rappaport, H. (1987). Distribution of lymphocytes with interleukin-2 receptors (Tac antigens) in reactive lymphoproliferative processes, Hodgkin's disease, and non-Hodgkin's lymphomas. An immunohistologic study of 300 cases. *Am. J. Pathol.* **127**, 27–37.

Smith, K.A. (1988). The interleukin 2 receptor. *Adv. Immunol.* **42**, 165–79.

Smith, K.A. (1989). Interleukin-2: inception, impact and implications. *Science* **240**, 1169–76.

Teshigawara, K., Wang, H.-M, Kato, K. and Smith, K.A. (1987). Interleukin 2 high-affinity receptor expression requires two distinct binding proteins. *J. Exp. Med.* **165**, 223–38.

Uchida, T., Gill, D.M. and Pappenheimer, A.M. Jr. (1971). Mutation in the structural gene for diphtheria toxin carried by temperate phage β. *Nature (New Biol.)* **233**, 8–11.

Uchiyama, T., Broder, S. and Waldmann, T.A. (1981). A monoclonal antibody (anti-Tac) reactive with activated and functionally mature human T cells. I. Production of anti-Tac monoclonal antibody and distribution of Tac(+) cells. *J. Immunol.* **126**, 1393–7.

Uchiyama, T., Hori, T. and Tsudo, M. (1985). Interleukin-2 receptor (Tac antigen) expressed on adult T cell leukemia cells. *J. Clin. Invest.* **76**, 446–53.

Varga, J.M., Moellmann, K.A., Fritsch, P., Godawska, E. and Lerner, A.B. (1976). Association of cell surface receptors for melanotropin with the Golgi region in mouse melanoma cells. *Proc. Natl. Acad. Sci. USA* **73**, 559–62.

Waters, C.A., Schimke, P.A., Snider, C.E., Itoh, K., Smith, K.A., Nichols, J.C., Strom, T.B. and Murphy, J.R. (1990). Interleukin 2 receptor-targeted cytotoxicity. Receptor binding requirements for entry of a diphtheria toxin-related interleukin 2 fusion protein. *Eur. J. Immunol.* **20**, 785–91.

Waldmann, T.A., Greene, W.C., Sarin, P.S., Saxinger, C., Blayney, D.W., Blattner, W.A., Goldman, C.K., Bongiovanni, K., Sharrow, S., Depper, J.M., Leonard, W., Uchiyama, T. and Gallo, R.C. (1984). Functional and phenotypic comparison of human T cell leukemia/lymphoma virus positive adult T cell leukemia with human T cell leukemia/lymphoma virus negative Sezary leukemia, and their distribution using anti-Tac *J. Clin. Invest.* **73**, 1711–8.

Weissman, A.M., Harford, J.B., Svetlik, P.B., Leonard, W.L., Depper, J.M., Waldmann, T.A., Greene, W.C. and Klausner, R.D. (1986). Only high affinity receptor for interleukin-2 mediates internalization of ligand. *Proc. Natl. Acad. Sci. USA* **83**, 1463–6.

Wetlaufer, D.B. (1973). Nucleation, rapid folding, and globular intrachain regions in proteins. *Proc. Natl. Acad. Sci. USA* **70**, 697–701.

Williams, D.P., Parker, K., Bacha, P., Bishai, W., Borowski, M., Genbauffe, F., Strom, T.B. and Murphy, J.R. (1987). Diphtheria toxin receptor binding domain substitution with interleukin-2: Genetic construction and properties of a diphtheria toxin-related interleukin-2 fusion protein. *Protein Engng.* **1**, 493–8.

Williams, D.P., Snider, C.E., Strom, T.B. and Murphy, J.R. (1990). Structure/function analysis of interleukin-2-toxin (DAB$_{486}$-IL-2): Fragment B sequences required for the delivery of fragment A to the cytosol of target cells. *J. Biol. Chem.* **265**, 11885–9.

Yamaizumi, J., Mekada, E., Uchida, T. and Okada, Y. (1978). One molecule of diphtheria toxin fragment

A introduced into a cell can kill the cell. *Cell* **15**, 245–50.

Yoshida, M., Miyoshi, I. and Hinuma, Y. (1982). Isolation and characterization of retrovirus from cell lines of human adult T-cell leukemia and its implication in the disease. *Proc. Natl. Acad. Sci. USA* **79**, 2031–5.

Zettlmeissl, G., Kaczorek, M., Moya, M. and Streeck,

R.E. (1986). Expression of immunogenically reactive diphtheria toxin fusion proteins under the control of the P_r promoter of bacteriophage lambda. *Gene* **41**, 103–11.

Zucker, D. and Murphy, J.R. (1984). Monoclonal antibody analysis of diphtheria toxin. I. Localization of epitopes and neutralization of cytotoxicity. *Molec. Immunol.* **21**, 785–93.

Index